CARD & POCKET ON NEXT PAGE

Lecture Notes in Mathematics

Edited by A. Dold and B. Eckmann

794

Measure Theory Oberwolfach 1979

Proceedings of the Conference Held at Oberwolfach, Germany, July 1–7, 1979

Edited by D. Kölzow

Springer-Verlag
Berlin Heidelberg New York 1980

Editor
Dietrich Kölzow
Mathematisches
Universität Erlan
Bismarckstr. 1 1
8520 Erlangen
Federal Republi

AMS Subject Classifications (1980): 28-02, 28A40, 46Gxx, 60-02, 60Gxx, 60Hxx

ISBN 3-540-09979-4 Springer-Verlag Berlin Heidelberg New York
ISBN 0-387-09979-4 Springer-Verlag New York Heidelberg Berlin

Library of Congress Cataloging in Publication Data. Conference on Measure Theory, Oberwolfach, Ger., 1979. Measure theory, 1979. (Lecture notes in mathematics; 794) Bibliography: p. Includes index. 1. Measure theory--Congresses. I. Kölzow, Dietrich. II. Title. III. Series: Lecture notes in mathematics (Berlin); 794. QA3.L28. no. 794. [QA312]. 510s. [515.4'2] 80-13789

Printed in Germany

Printing and binding: Beltz Offsetdruck, Hemsbach/Bergstr.
2141/3140-543210

M. P. Ershov
Institut für Mathematik, Universität Linz, Altenberger Str. 69, A-4045 Linz, Austria

N. F. Falkner
Laboratoire de Calcul des Probabilités, Université de Paris VI, 9, Quai St. Bernard, Tour 46, F-75230 Paris, France

P. Georgiou
Department of Mathematics, University of Athens, 57, Solonos Street, Athens 143, Greece

C. Godet-Thobie
Département de Mathématiques, Université de Bretagne Occidentale, 6, Ave. Victor le Gorgeu, F-29283 Brest, France

V. Goodman
Department of Mathematics, University of Wisconsin, Madison, Wisconsin 53706, U.S.A.

S. Graf
Mathematisches Institut, Universität Erlangen-Nürnberg, Bismarckstr. 1 1/2, D-8520 Erlangen, Fed. Rep. of Germany

E. Grzegorek
Institute of Mathematics of the Polish Academy of Sciences, ul. Kopernika 18, PL-51-617 Wrocław, Poland

M. de Guzmán
Departamento de Ecuaciones Diferenciales, Universidad de Madrid, Madrid 3, Spain

W. Hackenbroch
Fachbereich Mathematik, Universität Regensburg, Universitätsstr. 31, D-8400 Regensburg, Fed. Rep. of Germany

W. Herer
Institute of Mathematics of the Polish Academy of Sciences, Sniadeckich 8, Warsawa, Poland

A. Hertle
Fachbereich Mathematik, Universität Mainz, Saarstr. 21, D-6500 Mainz, Fed. Rep. of Germany

A. D. Kappos
Lykabetton 29, Athens 135, Greece

A. Katavolos
Department of Mathematics, University of Crete, Iraklion, Greece

H.-G. Kellerer
Mathematisches Institut, Universität München, Theresienstr. 39, D-8000 München, Fed. Rep. of Germany

D. Kölzow
Mathematisches Institut, Universität Erlangen-Nürnberg, Bismarckstr. 1 1/2, D-8520 Erlangen, Fed. Rep. of Germany

J. Lembcke
Mathematisches Institut, Universität Erlangen-Nürnberg, Bismarckstr. 1 1/2, D-8520 Erlangen, Fed. Rep. of Germany

Losert, V.
Mathematisches Institut, Universität Wien, Strudlhofgasse 4, A-1090 Wien, Austria

W. A. J. Luxemburg
Alfred P. Sloan Laboratory of Mathematics and Physics, California Institute of Technology, Pasadena, California 91125, U.S.A.

F. Y. Maeda
Department of Mathematics, Hiroshima University, Hiroshima, Japan

G. Mägerl
Mathematisches Institut, Universität Erlangen-Nürnberg, Bismarckstr. 1 1/2, D-8520 Erlangen, Fed. Rep. of Germany

D. Maharam-Stone
Department of Mathematics, University of Rochester, Rochester, New York 14627, U.S.A.

P. R. Masani
Department of Mathematics, University of Pittsburgh, Pittsburgh, Pennsylvania 15260, U.S.A.

P. Mattila
Department of Mathematics, University of Helsinki, Hallituskatu 15, SF-00100 Helsinki 10, Finland

R. D. Mauldin
Department of Mathematics, North Texas State University, Denton, Texas 76203, U.S.A.

P. McGill
Mathematics Department, The New University of Ulster, Coleraine Co. Londonderry, Northern Ireland BT52 1SA, United Kingdom

P. Morales
Département de Mathématiques, Université de Sherbrook, Sherbrook, Quebec, Canada

K. Musiał
Institute of Mathematics, Wrocław University, Pl. Grunwaldski 2/4, PL-50-384 Wrocław, Poland

E. Pap
Šekspirova 26, YU-2100 Novi Sad, Yugoslavia

W. F. Pfeffer
Department of Mathematics, University of California at Davis, Davis, California 95616, U.S.A.

P. Prinz
Mathematisches Institut, Universität München, Theresienstr. 39, D-8000 München, Fed. Rep. of Germany

M. M. Rao
Department of Mathematics, University of California at Riverside, Riverside, California 92502, U.S.A.

L. Rogge
Fachbereich Wirtschaftswissenschaft und Statistik, Universität Konstanz, Postfach 7733, D-7750 Konstanz, Fed. Rep. of Germany

W. Schachermayer
Institut für Mathematik, Universität Linz, Altenberger Str. 69, A-4045 Linz-Auhof, Austria

D. Sentilles
Department of Mathematics, University of Missouri at Columbia, Columbia, Missouri 65211, U.S.A.

W. Słowikowski
Matematisk Institut, Aarhus Universitet, Universitetsparken, Ny Munkegade, DK-8000 Aarhus, Denmark

Ch. P. Stegall
Institut für Mathematik, Universität Linz, Altenberger Str. 69, A-4045, Linz-Auhof, Austria

A. H. Stone
Department of Mathematics, University of Rochester, Rochester, New York 14627, U.S.A.

W. Strauß
Mathematisches Institut A, Universität Stuttgart, Pfaffenwaldring 57, D-7000 Stuttgart, Fed. Rep. of Germany

L. Sucheston
Department of Mathematics, Ohio State University, Columbus, Ohio 43210, U.S.A.

M. Talagrand
Equipe d'Analyse, Université de Paris VI, 4, Place Jussieu, F-75230 Paris, France

E. G. F. Thomas
Mathematisch Instituut, Rijksuniversiteit Groningen, Postbus 800, Groningen, The Netherlands

S. Tomášek
Karl-Zörgiebel-Str. 48, D-6500 Mainz-Bretzenheim, Fed. Rep. of Germany

F. Topsøe
Matematisk Institut, København Universitet, Universitetsparken 5, DK-2100 København, Denmark

A. Volčič
Istituto di Matematica Applicata, Università di Trieste, Piazzale Europa 1, I-34100 Trieste, Italy

D. H. Wagner
Station Square 1, Paoli, Pennsylvania 19301, U.S.A.

H. von Weizsäcker
Fachbereich Mathematik, Universität Trier-Kaiserslautern, Pfaffenbergstr. 95, D-6750 Kaiserslautern, Fed. Rep. of Germany

R. F. Wheeler
Department of Mathematical Sciences, Northern Illinois University, DeKalb, Illinois 60115, U.S.A.

W. A. Woyczyński
Department of Mathematics, Cleveland State University, Cleveland, Ohio 44115, U.S.A.

NON PARTICIPATING CONTRIBUTORS

P. Erdös

Hungarian Academy of Sciences, Budapest, Hungary

A. D. Ioffe

Profsojuznaja 97-1-203, Moscow 117 279, U.S.S.R.

A. Millet

Department of Mathematics, The Ohio State University, 231 West 18 Ave., Columbus, Ohio 43210, U.S.A.

F. Terpe

Sektion Mathematik, Ernst-Moritz-Arndt-Universität, Jahnstr. 15a, DDR-22 Greifswald, G.D.R.

CONTRIBUTIONS

MEASURE EXTENSIONS AND MEASURABLE NEIGHBOURS OF A FUNCTION.

Dieter Bierlein

Introduction and summary.

In this paper we deal with the problem of making a given real function $f|M$ measurable by extending a given probability [1] measure $p|\mathfrak{A}$ on M. Our problem is a special case of the general measure extension problem, namely the case characterized by the target σ-algebra

$$\mathfrak{A}_f := {}^B(\mathfrak{A} \cup \mathfrak{K}_f), \quad \text{where } \mathfrak{K}_f := f^{-1}(\mathfrak{L}).$$

Without loss of generality we may assume that

$$f: M \to E := [0,1]$$

and accordingly $\mathfrak{L} := \mathfrak{L}(E)$.

We will investigate the set

$$\mathfrak{F}(p|\mathfrak{A}, f) := \{p_1|\mathfrak{A}_1 : \text{a probability with } p_1|\mathfrak{A} = p|\mathfrak{A}\}$$

of all measure extensions of $p|\mathfrak{A}$ which makes f measurable. In several cases a *measurable selection* technique can be used to obtain measure extensions of a special type, namely those which correspond in a specified way to a suitable $\mathfrak{A}$-measurable function $g|M$ (see [B3], section 5). Suitable functions in this context are just those which are *measurable neighbours* of f, where we call g a neighbour of f if

$$p^*(f = g) = 1.$$ [2]

In this paper we are concerned with the problem of the existence of measurable neighbours, with the problem of the existence of measure extensions onto $\mathfrak{A}_f$ for a function f *without* measurable neighbours, and with the problem of the existence of measure extensions *in addition* to those which correspond to measurable neighbours. In detail, after a summary of previous results regarding our problem (section 1) we give a criterion for the existence of an extension of a 0-1-measure onto $\mathfrak{A}_f$ (section 2), and investigate conditions for the existence

---

1) For the sake of simplicity: Any result for probability measures can here easily be extended to an analogous statement for σ-finite measures.

2) $p^*(f = g)$ means the outer measure of the set

$$\{f = g\} := \{x \in M: f(x) = g(x)\}.$$

of measurable neighbours (section 3). In section 4 we study measure extensions which do not correspond to measurable neighbours.

1. Preliminaries.

In this section several general statements illustrate the situation we are faced with when searching for measure extensions of a probability measure $p|\mathfrak{A}$ onto $\mathfrak{A}_f$.

1.1. The trivial case of a discrete measure $p|\mathfrak{A}$.

If $p|\mathfrak{A}$ is *discrete*, i.e. $\{x\} \in \mathfrak{A}$ for each $x \in M$ is true and there exists a countable subset T with $p(T) = 1$, then the completion $\tilde{p}$ of p is defined on the power set $\mathfrak{P}(M)$. Thus $\tilde{p}|\mathfrak{A}_f$ is the only measure on $\mathfrak{A}_f$ which is compatible with $p|\mathfrak{A}$. Therefore we have in this case

$$\mathfrak{F}(p|\mathfrak{A}, f) = \{\tilde{p}|\mathfrak{A}_f\}.$$

If $p|\mathfrak{A}$ is the *restriction of a discrete measure* $p_o|\mathfrak{A}_o$ then at least $\tilde{p}_o|\mathfrak{A}_f$ is an element of $\mathfrak{F}(p|\mathfrak{A}, f)$. If here $p|\mathfrak{A}$ itself is not discrete then, except for trivial cases, $p|\mathfrak{A}$ has more than one discrete extension and, in addition, extensions with a non-discrete component. At any rate here we have

$$\mathfrak{F}(p|\mathfrak{A}, f) \neq \emptyset.$$

Therefore the existence of an extension of $p|\mathfrak{A}$ onto $\mathfrak{A}_f$ depends on the properties of that component of $p|\mathfrak{A}$ which is *not* the restriction of a discrete measure.

1.2. The non-discrete case.

If $p|\mathfrak{A}$ is *not* the restriction of a discrete measure then, under an extremely weak set theoretical assumption (M), for each probability space $(M, \mathfrak{A}, p)$ there exists a function $f_o: M \to E$ such that

$$\mathfrak{F}(p|\mathfrak{A}, f_o) = \emptyset. \text{ [1)]}$$

This assumption can be characterized by the following condition:

<u>(M)</u> the set M is not "exorbitantly" large. [2)]

---

1) See [B1], Satz 1 C and footnote *.

2) See the precise definition of this condition in [U1] and a somewhat stronger condition in [BK].

It is not known yet if there exist sets of this type. Apart from these exceptions, for *any* set M and *any* probability measure $p|\mathfrak{A}$ which is not of the trivial type considered in section 1.1, there exists a real function $f_o|M$ which cannot be made measurable by measure extension.

1.3. A criterion for the existence of extensions of $p|\mathfrak{A}$ onto $\mathfrak{A}_f$.

Our measure extension problem can be transformed into the following one: Given a probability space $(M, \mathfrak{A}, p)$ and a function $f: M \to E$, find a measure μ defined on the product-σ-algebra $\mathfrak{A} * \mathfrak{L}$ such that the condition

$$(*) \quad \begin{cases} \mu(A \times E) = p(A) & \text{for all } A \in \mathfrak{A} \\ \mu^*(\mathrm{Gr}\ f) = 1 \end{cases}$$

is satisfied, where $\mathrm{Gr}\ f := \{(x, f(x)) : x \in M\}$ denotes the graph of f. This transformation of our problem is established by

Criterion 1. *There is a one-to-one-correspondence between the measures* $p_1|\mathfrak{A}_f$ *of* $\mathfrak{F}(p|\mathfrak{A}, f)$ *and the measures* $\mu|\mathfrak{A} * \mathfrak{L}$ *satisfying condition* (*).

Its *proof* is contained in the proof of Satz 3 A in [B1].

The transformed problem consists in looking for a measure $\mu|\mathfrak{A} * \mathfrak{L}$ which concentrates mass 1 "close" to Gr f such that the marginal condition resulting from $p|\mathfrak{A}$ is satisfied. Means for solving this problem can be found in [B1], § 3 and in [B2].

1.4. Measure extensions corresponding to measurable neighbours.

In several cases the task of concentrating mass 1 of measure μ *"close"* to Gr f can be mastered by distributing mass 1 *on* the graph of a suitable *measurable* function $g|M$, taking into account the marginal condition of (*). This is possible especially in those cases, where a measurable selection technique can be applied. [1]If measure $\mu|\mathfrak{A} * \mathfrak{L}$ is obtained in such a way, it satisfies the condition

$$(**) \quad \begin{cases} (*) \\ \text{There exists a function } g|M \text{ with } \mu(\mathrm{Gr}\ g) = 1. \end{cases}$$

1) See [L1], [L2], [Ku], and [B3], further in connection with some other measure extension problems [LR] and [Ru].

According to [B3], Lemma 2, any function $g|M$, for which there exists a measure $\mu|\mathfrak{A} * \mathfrak{L}$ satisfying (*) and $\mu(\mathrm{Gr}\ g) = 1$, is a neighbour of $f|M$. The converse is also true:

Lemma 1.1. *Let* $(M, \mathfrak{A}, p)$ *be a probability space and* $f: M \to E$. *Then a* $\mathfrak{A}$*-measurable function* $g|M$ *is a neighbour of* $f|M$, *if and only if there exists a measure* $\mu|\mathfrak{A} * \mathfrak{L}$ *satisfying condition* (*) *and* $\mu(\mathrm{Gr}\ g) = 1$.

Proof: It remains to prove that $p^*(f = g) = 1$ implies the existence of a measure $\mu|\mathfrak{A} * \mathfrak{L}$ with the properties mentioned above. Since $g|M$ is $\mathfrak{A}$-measurable, Gr g is a set of the type for which the existence of a measure $\mu|\mathfrak{A} * \mathfrak{L}$ with the following properties has been proved in [B2]:

(i) $\mu(A \times E) = p(A)$ for all $A \in \mathfrak{A}$

(ii) $\mu(\mathrm{Gr}\ g) = 1$.

Because of (ii) resp. (i) $p^*(f = g) = 1$ implies

(iii) $\mu^*(\mathrm{Gr}\ f) = \mu^*(\mathrm{Gr}\ f \cdot \mathrm{Gr}\ g) = p^*(f = g) = 1.$ ¤¤¤

According to the equivalence established by Criterion 1 and Lemma 1.1 the measurable neighbours of $f|M$ correspond to the measure extensions $p_1|\mathfrak{A}_f$ of a special type. This correspondence is one-to-one except for p-almost equal neighbours, since

$$\mu(\mathrm{Gr}\ g_1) = \mu(\mathrm{Gr}\ g_2) = 1,$$

in connection with marginal condition (i), implies

$$p(g_1 = g_2) = 1.$$

The set of all measure extensions $p_1|\mathfrak{A}_f$ corresponding (in the sense of Criterion 1) to a measure $\mu|\mathfrak{A} * \mathfrak{L}$ with property (**) will be denoted by

$$\mathfrak{F}^{\mathrm{m.n.}}(p|\mathfrak{A}, f),$$

where "m.n." can be interpreted, according to Lemma 1.1, as an abbreviation of "measurable neighbour".

2. The measure extension criterion in case of 0-1-measures.

Examples of cases, in which $\mathfrak{F}^{\mathrm{m.n.}} = \emptyset$ and $\mathfrak{F} \neq \emptyset$ are valid, can be found rather easily, if one chooses $p|\mathfrak{A}$ as a 0-1-measure without a discrete component; [1] for in case of a 0-1-measure only the p-almost constant functions are $\mathfrak{A}$-measurable, and consequently the functions having no measurable neighbours can be characterized in a convenient form. Now we will adapt Criterion 1 to the case of a (complete) 0-1-measure.

$p|\mathfrak{A}$ is a 0-1-measure, if

$$\mathfrak{A} = \mathfrak{N} + \mathfrak{E} \quad \text{with} \quad p|\mathfrak{N} = 0 \quad \text{and} \quad p|\mathfrak{E} = 1.$$

Using this notation we obtain

Criterion 2. *If* $p|\mathfrak{A}$ *is a complete 0-1-measure then there exists a one-to-one-correspondence between the measures* $p_1|\mathfrak{A}_f$ *of* $\mathfrak{F}(p|\mathfrak{A}, f)$ *and the measures* $p \times q|\mathfrak{A} * \mathfrak{L}$ *with a probability* $q|\mathfrak{L}$ *vanishing on*

$$\{B \in \mathfrak{L}\colon \bar{f}^{1}(B) \in \mathfrak{N}\}.$$

Proof: According to Criterion 1 it is sufficient to show that here a measure $\mu|\mathfrak{A} * \mathfrak{L}$ satisfies condition (*) iff there exists a measure $q|\mathfrak{L}$ with the properties

(P) $\quad \mu = p \times q$

(N) $\quad q|\{B \in \mathfrak{L}\colon \bar{f}^{1}(B) \in \mathfrak{N}\} = 0.$

1) Suppose condition (*) is valid for $\mu|\mathfrak{A} * \mathfrak{L}$.
Then the marginal property of μ relating to $p|\mathfrak{A}$ implies

$$\mu(A \times B) = \begin{cases} 0 & \text{for } A \in \mathfrak{N} \\ \mu(M \times B) =: q(B) & A \in \mathfrak{E}. \end{cases}$$

Consequently we have

$$\mu(A \times B) = p(A) \cdot q(B) \quad \text{for all } A \in \mathfrak{A} \text{ and } B \in \mathfrak{L},$$

and thus property (P). The probability $q|\mathfrak{L}$ defined above satisfies property (N), too; for, as a result of (P) and (*),

$$q(B) = \mu\left(\bar{f}^{1}(B) \times B\right) \le \mu_*(\overline{\mathrm{Gr}\ f}) = 0$$

holds for all

$$B \in \mathfrak{L} \text{ with } \bar{f}^{1}(B) \in \mathfrak{N}.$$

---

[1] See the example in [B3], section 5.

2) Suppose conditions (P) and (N) are valid for $\mu|\mathfrak{A} * \mathfrak{L}$. In order to prove (*) it is now sufficient to show that for each disjoint cover $S := \sum_{\nu} K_{\nu}$ of $X := \mathrm{Gr}\, f$ with

$$K_{\nu} = A_{\nu} \times B_{\nu}, \qquad A_{\nu} \in \mathfrak{A}, \qquad B_{\nu} \in \mathfrak{L}$$

$\mu(S) = 1$ is true.

To prove this we set

$$I_o := \{\nu : A_{\nu} \in \mathfrak{N}\}, \qquad I_1 := \{\nu : A_{\nu} \in \mathfrak{P}\},$$

$$N := \bigcup_{I_o} A_{\nu}, \qquad B := \bigcup_{I_1} B_{\nu}.$$

Since $A_{\nu} \in \mathfrak{P}$ for $\nu \in I_1$, the sets B_{ν} with $\nu \in I_1$ are disjoint just as the sets K_{ν}. From this and (P) we obtain

$$\mu(S) = \sum_{I_1} q(B_{\nu}) = q(B).$$

Since

$$\bar{f}^1(\bar{B}) \subset N \in \mathfrak{N}$$

and $p|\mathfrak{A}$ is complete, condition (N) yields

$$q(\bar{B}) = 0$$

and thereby

$$\mu(S) = q(B) = 1. \qquad ППП$$

The assumption of completeness of $p|\mathfrak{A}$ is used only in the proof of the second direction. Since in completing a measure its 0-1-property does not get lost, this assumption represents no essential restriction of the range of application of Criterion 2.

Condition (N) is illustrated by the statements of the following lemma.

Lemma 2.1. a) *A necessary condition for* (N) *is condition*

(N_1) $\quad q^*(f(M)) = 1.$

b) *Together with condition*

(N_2) $q^*(f(N)) = 0$ for all $N \in \mathfrak{N}$

condition (N_1) *is sufficient for* (N).

Proof: a) For any Borel set $B \subset \overline{f(M)}$ the set $\bar{f}^1(B)$ is empty and by condition (N) consequently $q(B) = 0$ holds.

b) For any $B \in \mathfrak{L}$ with $\bar{f}^1(B) \in \mathfrak{N}$ condition (N_2) yields

$$0 = q^*(f(\bar{f}^1(B))) = q^*(f(M)\cdot B)$$

and hence together with (N_1)

$$q(B) = q^*(f(M)\cdot B) + q_*(\overline{f(M)}\cdot B) = 0. \qquad ¤¤¤$$

Condition (N_2) is *not* necessary for (N), as one can see by the following *example*:

$$M = [-1,1], \qquad \mathfrak{A} = {}^B(\mathfrak{P}([-1,0)) + \mathfrak{N}(\lambda)),^{1)}$$

$$p(A) = \lambda(A\cdot E) \quad \text{for} \quad A \in \mathfrak{A},$$

$$f(x) = |x|, \qquad q|\mathfrak{L} = \lambda|\mathfrak{L},$$

where λ denotes the Lebesquemeasure on $\mathfrak{L}(E)$.

Here $p|\mathfrak{A}$ is a complete 0-1-measure and $q|\mathfrak{L}$ satisfies condition (N); but for $N_o := [-1,0]$ we have

$$N_o \in \mathfrak{N} \quad \text{and} \quad q(f(N_o)) = 1$$

in contradiction to condition (N_2). ¤¤¤

3. On the existence of measurable neighbours.

Let us now consider the set $\mathfrak{F}^{m.n.}(p|\mathfrak{A}, f)$ of the measure extensions corresponding to measurable neighbours and look for conditions for the existence of a measurable neighbour. We distinguish between several types of such conditions depending on the components of the quadruplet

[1] Generally, $\mathfrak{N}(p) := \{A \in \mathfrak{A}: p(A) = 0\}$ denotes the system of p-null sets for any measure $p|\mathfrak{A}$.

$(M,\mathfrak{A},p,f)$ the premise of the condition is concerned with. At first we establish some conditions concerned with $p|\mathfrak{A}$ under which *any* function $f|M$ has measurable neighbours. For this in section 3.1 we consider our question under the assumption that M can be decomposed into countably many p-atoms, while in section 3.2 we study 0-1-measures (i.e. the standardized case of a p-atom) and state a condition concerned with the system $\mathfrak{N}$ of p-null sets which is equivalent to the assertion that any function $f|M$ has a measurable neighbour relative to $(M,\mathfrak{A},p)$. In section 3.3 we establish a condition concerned with $\mathfrak{A}$ and f which is sufficient for f to have a measurable neighbour and, thus, for $\mathfrak{F}^{m.n.}(p|\mathfrak{A}, f)$ to be not empty for *any* probability measure p defined on $\mathfrak{A}$. Finally, in section 3.4 we state a general condition concerned with $(M,\mathfrak{A},p)$ and $f|M$ which is equivalent to f having *no* measurable neighbour.

3.1. Conditions concerned with the kernels of an atomic measure.

We refer to the probability space $(M,\mathfrak{A},p)$. A set A' is called p-*atom*, if

$$p(A') > 0$$

$$p(A) \in \{0,p(A')\} \quad \text{for all } A \in \mathfrak{A} \text{ with } A \subset A'.$$

The *kernel* of A' is defined as the set

$$D(A') := \bigcap \{A \subset A' : p(A) = p(A')\}.$$

Of course, D(A') need not belong to $\mathfrak{A}$. We speak of a p-*atom without* or *with kernel* resp. depending on whether the kernel is empty or not. For example, if $(M,\mathfrak{A},p) = (E, {}^{B}\mathfrak{N}(\lambda),\lambda)$, the set E is a p-atom without kernel. We list the following more general statements resulting from the definition of a p-atom and its kernel.

Remark 3.1. *Let* A_o *be a* p-*atom; then the following holds:*

a) $D(A_o) = A_o - \bigcup \{N \cdot A_o : N \in \mathfrak{N}(p)\}$.

b) *The kernel of* A_o *is empty, if and only if* $p^*(T) = 0$ *for any countable subset* T *of* A_o.

c) *If* T *is a non-empty subset of* $D(A_o)$ *then* $p^*(T) = p(A_o)$ *is true.*

d) *Any* p-*measurable function is* p-*almost constant on* A_o.

The following theorem gives an answer to the question of the existence of measurable neighbours in case of an atomic measure.

Theorem 3.1. *If* M *can be decomposed into countably many* p*-atoms with non-empty kernels, then any function* $f: M \to E$ *has measurable neighbours.*

Proof: Suppose $M = \sum_\nu A_\nu + N$, where A_ν is a p-atom with $D(A_\nu) \neq \emptyset$ and $N \in \mathfrak{N}(p)$. Now we define an $\mathfrak{A}$-measurable function $g: M \to E$ by

$$g|A_\nu = c_\nu \text{ with an abitrary number } c_\nu \in f(D(A_\nu)).$$

Then $g|M$ satisfies $p^*(f = g) = 1$ according to Remark 3.1 c) because of

$$\{f = g\} \cdot D(A_\nu) \neq \emptyset \quad \text{for any } \nu. \qquad ◻◻◻$$

Lemma 3.1. a) *If* $p|\mathfrak{A}$ *is the restriction of a discrete measure, then it satisfies the assumption of* Theorem 3.1.

b) *If* A_o *is a* p*-atom and the trace* $A_o \cdot \mathfrak{A}$ *is countably generated, then* A_o *is a* p*-atom with kernel.*

Proof: a) Suppose, $p_o|\mathfrak{A}_o$ is discrete with $p_o(\{x_n\}) > 0$ and $p_o|\mathfrak{A} = p|\mathfrak{A}$. Then we have $p^*(\{x_n\}) \geq p_o(\{x_n\}) > 0$. Therefore x_n is an element of a p-atom whose kernel is not-empty according to Remark 3.1 b).

b) Suppose, $\mathfrak{F}$ is a countable system of sets such that $^B\mathfrak{F} = A_o \cdot \mathfrak{A}$. Because

$$p(A) \in \{0, p(A_o)\} \quad \text{for any } A \in \mathfrak{F}$$

we can assume without loss of generality that $p|\mathfrak{F} = p(A_o)$. Then

$$D(A_o) = \bigcap_{A \in \mathfrak{F}} A = A_o - N, \quad \text{where } N := A_o \cdot \bigcup_{A \in \mathfrak{F}} \bar{A} \in \mathfrak{N}(p)$$

and consequently $p(D(A_o)) = p(A_o) > 0$ and $D(A_o) \neq \emptyset$. ◻◻◻

Whereas a p-atom *with kernel* "favours locally" (in a certain sense) the existence of measurable neighbours, the p-atoms *without kernel* show the opposite tendency:

Theorem 3.2. *If* M *contains a not exorbitantly large* p*-atom without kernel, then there exists a function* $f_o: M \to E$ *having no measurable neighbour.*

Proof: Let A_o be a p-atom without kernel satisfying the set theoretical condition (M) of section 1.2. According to Lemma 3.1 a) $p|A_o \cdot \mathfrak{A}$ is not the restriction of any discrete measure. Consequently there exists a function $f_o|A_o$ such that no measure $p_1|A_o \cdot \mathfrak{A}_f$ satisfies

$$p_1|A_o \cdot \mathfrak{A} = p|A_o \cdot \mathfrak{A},$$

as one can see by analogy to [B1], Satz 1 C and footnote * with respect to $p(A_o) > 0$. From this $\mathfrak{F}(p|\mathfrak{A}, f_o) = \emptyset$ and consequently

$$\mathfrak{F}^{m.n.}(p|\mathfrak{A}, f_o) = \emptyset.$$

Therefore f_o has no measurable neighbour. ¤¤¤

Moreover, the proof of Theorem 3.2 shows that, on the assumption of the theorem, there exist functions which *cannot* be made measurable by measure extension. Therefore the 0-1-measure without kernel (more precisely: the 0-1-measure whose basic set has an empty kernel) is here of specific interest as a normalized prototype of a p-atom without kernel.

3.2. An equivalence in case of 0-1-measures.

Let $p|\mathfrak{A}$ be a 0-1-measure on M with $\mathfrak{A} = \mathfrak{N} + \mathfrak{C}$ in the notation of section 2. Then M is a p-atom which, according to Remark 3.1 a), has a kernel iff M *cannot* be covered by sets of $\mathfrak{N}$. Herewith we obtain as a result of Theorem 3.1:

Remark 3.2. *If the system* $\{\mathfrak{N}_o \subset \mathfrak{N}: \bigcup \mathfrak{N}_o = M\}$ *is empty, then any function* $f: M \to E$ *has measurable neighbours.*

If M can be *covered* by sets of $\mathfrak{N}$, then the existence of a function f *without* measurable neighbours depends on how many null-sets are required to cover M:

Theorem 3.3. *Let* $p|\mathfrak{N} + \mathfrak{C}$ *be a complete 0-1-measure on* M. *Then any function* $f: M \to E$ *has measurable neighbours if and only if*

$$|\mathfrak{N}_o| > \aleph \quad \textit{for any} \quad \mathfrak{N}_o \subset \mathfrak{N} \quad \textit{with} \quad \bigcup \mathfrak{N}_o = M.$$

Considering Remark 3.2 and the trivial fact that there is no countable subsystem $\mathfrak{N}_o$ of $\mathfrak{N}$ satisfying $\bigcup \mathfrak{N}_o = M$, we can state Theorem 3.3 in the following equivalent form.

Theorem 3.3'. *Let* $p|\mathfrak{N}+\mathcal{C}$ *be a complete 0-1-measure on* M. *Then there exists a function* $f: M \to E$ *without a measurable neighbour if and only if there exists a subsystem* $\mathfrak{N}_o$ *of* $\mathfrak{N}$ *such that*

$$|\mathfrak{N}_o| = \mathfrak{X} \quad \textit{and} \quad \bigcup \mathfrak{N}_o = M.$$

Proof:

1) Suppose $|\mathfrak{N}_o| > \mathfrak{X}$ for any $\mathfrak{N}_o \subset \mathfrak{N}$ with $\bigcup \mathfrak{N}_o = M$. Then the system $\mathcal{L} := \left\{\{f = y\}: y \in E\right\}$ has the properties

$$|\mathcal{L}| \leq \mathfrak{X} \quad \text{and} \quad \bigcup \mathcal{L} = M$$

for any function $f: M \to E$. Therefore $\mathcal{L}$ is not a subsystem of $\mathfrak{N}$, i.e. there exists a $y_o \in E$ with $\{f = y_o\} \notin \mathfrak{N}$. Since $p|\mathfrak{N}+\mathcal{C}$ is a complete 0-1-measure, it follows that $p^*(f = y_o) = 1$. Thus $g|M := y_o$ is a measurable neighbour of f.

2) Suppose, $\mathfrak{N}_o := \{N_\tau: \tau \in E\}$ satisfies $\bigcup \mathfrak{N}_o = M$. Then we have

$$T(x) := \{\tau \in E: x \in N_\tau\} \neq \emptyset \quad \text{for any } x \in M.$$

We define a function $f: M \to E$ by choosing

$$f(x) \in T(x) \quad \text{for any } x \in M.$$

Then $f|M$ satisfies

$$\{f = \tau\} \subset N_\tau \quad \text{for any } \tau \in E.$$

According to Remark 3.1 d) any p-measurable function $g|M$ is p-almost constant on M; i.e. there exists a $y_o \in E$ such that

$$\{g = y_o\} =: \overline{N} \in \mathcal{C}$$

and thus

$$\{f = g\} \subset N \cup \{f = y_o\} \subset N \cup N_{y_o} \in \mathfrak{N}.$$

Therefore g is no neighbour of f. ∎∎∎

The assertion of Theorem 3.3 can be illustrated as follows: "Any function $f|M$ has a measurable neighbour iff M is *large* in comparison with the null-sets of the given 0-1-measure."

3.3. A sufficient condition concerned with $\mathfrak{A}$ and f.

In Game Theory and in other fields one takes an interest in sufficient conditions for *all* probability measures, defined on a given σ-algebra $\mathfrak{A}$, to have an extension onto $\mathfrak{A}_f$; i.e. conditions concerned only with M, $\mathfrak{A}$, and f. The following theorem provides a set of three conditions concerned with $\mathfrak{A}$ and f|M which is sufficient for the existence of a measurable neighbour, whereas any two of these conditions are not sufficient.

Theorem 3.4. a) *If* $\mathfrak{A}$ *and* f|M *satisfy the conditions*

(α) $\mathfrak{A}$ *is countably generated*

(β) $f(M) \in {}^S\mathfrak{L}$ [1]

(γ) $\mathfrak{A} \subset \mathfrak{E}_f$

then f|M *has measurable neighbours for any probability measure* p|$\mathfrak{A}$.

b) *Given two of the conditions* (α), (β), *and* (γ) *there exists a quadruplet* (M, $\mathfrak{A}$, p, f) *such that these two conditions are satisfied and* f|M *has no measurable neighbour.*

Proof of a): In the proof of Theorem 1 in [B3] it has been shown that (α), (β), and (γ) jointly imply the existence of an $\tilde{\mathfrak{A}}$-measurable function $f_1|M$ and of a measure $\mu|\tilde{\mathfrak{A}} * \mathfrak{L}$ with the properties

$$\mu(A \times E) = \tilde{p}(A) \quad \text{for any } A \in \tilde{\mathfrak{A}}$$

$$\mu^*(\mathrm{Gr}\, f) = 1$$

$$\mu(\mathrm{Gr}\, f_1) = 1$$

where $\tilde{p}|\tilde{\mathfrak{A}}$ denotes the completion of p|$\mathfrak{A}$. According to Lemma 1.1, $f_1|M$ is a neighbour of f|M relative to $\tilde{p}|\tilde{\mathfrak{A}}$, i.e. $\tilde{p}^*(f = f_1) = 1$ is valid. Since $f_1|M$ is $\tilde{\mathfrak{A}}$-measurable, there exists an $\mathfrak{A}$-measurable function g|M with $\tilde{p}(f_1 = g) = 1$. Because of

$$p^*(f = g) = \tilde{p}^*(f = g) = \tilde{p}^*(f = f_1) = 1$$

g|M is a neighbour of f|M.

---

1) Given a system $\mathfrak{G}$ of sets, ${}^S\mathfrak{G}$ denotes the system of the kernels of all Suslin-schemes builded with elements of $\mathfrak{G}$.

Proof of b): Examples of a quadruplet $(M, \mathfrak{A}, p, f)$, for which there exists no measure extension onto $\mathfrak{A}_f$ and both (β) and (γ) resp. (α) and (γ) are satisfied, are given in [B3], section 1.3 resp. section 4. Because of $\mathfrak{F} \supset \mathfrak{F}^{m.n.}$ it is enough to show that there exists a quadruplet such that (α), (β) and $\mathfrak{F} = \emptyset$ are valid.

Consider a measure space $(M_o, \mathfrak{A}_o, p_o)$ and a function $f_o|M_o$ where M_o, $p_o|\mathfrak{A}_o$ and $f_o|M_o$ have the properties mentioned in section 1.2 and where moreover $\mathfrak{A}_o$ is countably generated. Then there exists no measure on $\mathfrak{K}_{f_o}$ which is compatible with $p_o|\mathfrak{A}_o$. Now we set

$$M := M_o + N_o \quad \text{where } N_o := [0,1] - f_o(M_o)$$

$$\mathfrak{A} := \left\{A + T \colon A \in \mathfrak{A}_o,\ T \in \{\emptyset, N_o\}\right\}$$

$$p|\mathfrak{A}\colon \quad p(A + T) = p_o(A) \quad \text{for } A \in \mathfrak{A}_o,\ T \in \{\emptyset, N_o\}$$

$$f|M\colon \quad f(x) = \begin{cases} f_o(x) & \text{for } x \in M_o \\ x & x \in N_o. \end{cases}$$

Then, $\mathfrak{A}$ is a countably generated σ-algebra on M and

$$f(M) = f_o(M_o) + N_o = E,$$

i.e. the conditions (α) and (β) are fulfilled. Since $p|\mathfrak{A}_o = p_o|\mathfrak{A}_o$ and

$$\mathfrak{A}_f = {}^B(\mathfrak{A} \cup \mathfrak{K}_f) \supset M_o \cdot \mathfrak{K}_f = \mathfrak{K}_{f_o}$$

are valid, no measure on $\mathfrak{A}_f$ is compatible with $p|\mathfrak{A}$, i.e. $\mathfrak{F}(p|\mathfrak{A}, f)$ is empty. ◻◻◻

3.4. A condition equivalent to non-existence of a measurable neighbour.

Since we are interested especially in the measure extension problem for a quadruplet $(M, \mathfrak{A}, p, f)$ for which f has *no* measurable neighbour, we state a condition which is equivalent to f having no measurable neighbour:

Theorem 3.5. *Given a probability space* $(M, \mathfrak{A}, p)$, $f|M$ *has no measurable neighbour if and only if there exists a set* $A_o \in \mathfrak{A}$ *satisfying* $p(A_o) > 0$ *and the condition*

(O) $\quad p^*(\{f = g\} \cdot A_o) = 0$ *for any* $\mathfrak{A}$*-measurable function* $g|M$.

Proof: 1) Suppose that A_o is a set with the property mentioned above. Then

$$p^*(f = g) = p^*(\{f = g\} \cdot \overline{A_o}) \leq p^*(\overline{A_o}) < 1$$

is valid for any $\mathfrak{A}$-measurable function $g|M$.

2) Suppose $p^*(f = g) < 1$ for any $\mathfrak{A}$-measurable function $g|M$ and

$$s := \sup \{p^*(f = g) : g \ \mathfrak{A}\text{-measurable}\} = \lim_{n\to\infty} p^*(f = g_n).$$

Then we choose sets $A_n \in \mathfrak{A}$ such that

$$A_n \supset \{f = g_n\} \quad \text{with} \quad p(A_n) = p^*(f = g_n)$$

and set

$$S_n := A_n \cdot \bigcap_{i<n} \overline{A_i} \quad \text{and} \quad S := \bigcup_n A_n = \sum_n S_n .$$

The function $g_o|M$ defined by

$$g_o := \begin{cases} g_n & \text{on } S_n \quad \text{for all } n \in \mathbb{N} \\ 0 & \text{on } \overline{S} \end{cases}$$

is $\mathfrak{A}$-measurable and satisfies

$$s \geq p^*(f = g_o) \geq p^*(\{f = g_o\} \cdot S) = p(S) \geq \lim_{n\to\infty} p(A_n) = s,$$

since $p^*(\{f = g_n\} \cdot S_n) = p(S_n)$ is true for any $n \in \mathbb{N}$. Consequently

$$s = p^*(f = g_o) < 1$$

is valid and $A_o := \overline{S}$ satisfies $p(A_o) = 1 - s > 0$ and condition (0), since

$$g^* := \begin{cases} g_o & \text{on } S \\ g & \text{on } A_o \end{cases}$$

with any $\mathfrak{A}$-measurable function $g|M$ yields

$$p^*(\{f = g\} \cdot A_o) = p^*(f = g^*) - p^*(\{f = g_o\} \cdot S) \leq s - s = 0.$$

⌑⌑⌑

4. Measure extensions not corresponding to measurable neighbours.

In this section we deal with two problems:

1) The problem of the existence of measure extensions onto $\mathfrak{A}_f$ for a function, which has no measurable neighbour.

2) The problem of measure extensions onto $\mathfrak{A}_f$ existing besides those which correspond to measurable neighbours.

In other words, we are concerned with

$$\mathfrak{F} \text{ in case of } \mathfrak{F}^{m.n.} = \emptyset$$

and with

$$\mathfrak{F} - \mathfrak{F}^{m.n.} \quad \text{in case of } \mathfrak{F}^{m.n.} \neq \emptyset.$$

It is appropriate to consider the atomic components of $p|\mathfrak{A}$ and the component without p-atoms as two separated cases.

4.1. The case of 0-1-measures.

Using the notation introduced for 0-1-measures in section 2 and section 3.2 we state

Theorem 4.1. *Given a probability space* $(M, \mathfrak{A}, p)$ *where* $\mathfrak{A} = \mathfrak{N} + \mathfrak{E}$ *and a function* $f\colon M \to E$. *If there exists a probability measure* $q|\mathfrak{L}$ *such that*

(1) $\quad f(N) \in \mathfrak{N}(q)$ *for any* $N \in \mathfrak{N}$

(2) $\quad q^*(f(M)) = 1$,

and if

(3) $\quad \{f = x\} \in \mathfrak{N}$ *for any* $x \in E$

is valid, then

$$\mathfrak{F}^{m.n.}(p|\mathfrak{A}, f) = \emptyset \quad \text{and} \quad \mathfrak{F}(p|\mathfrak{A}, f) \neq \emptyset$$

are true and, in particular, $\mu = p \times q$ *satisfies condition* (*) *stated in Criterion 1.*

Proof: Let $g|M$ be any $\mathfrak{A}$-measurable function. Since $p|\mathfrak{A}$ is a 0-1-measure, there exists a number c such that $p(g = c) = 1$. Because of (3) we have

$$\{f = g\} \subset \{f = c\} \cup \{g \neq c\} \in \mathfrak{N}$$

and so $p^*(f = g) = 0$, i.e. f has no measurable neighbour.

The measure $\mu = p \times q$ satisfies $\mu^*(\mathrm{Gr}\ f) = 1$, since

$$A \in \mathcal{L},\quad B \in \mathcal{L} \text{ with } A \times B \subset \overline{\mathrm{Gr}\ f}$$

implies

$$B \subset \overline{f(A)} \subset f(\bar{A}) + \overline{f(M)}$$

and because of (1) and (2)

$$q(B) = 0. \qquad ¤¤¤$$

For example, the assumptions (2) and (3) are always satisfied if $f: M \to E$ is bijective and $p|\mathfrak{A}$ is the completion of a 0-1-measure without kernel. Theorem 4.1 shows that, in case of 0-1-measures, the combination $\mathfrak{F}^{m.n.} = \emptyset \neq \mathfrak{F}$ cannot be considered as an exception. This holds true for the combination $\mathfrak{F}^{m.n.} \neq \emptyset \neq \mathfrak{F} - \mathfrak{F}^{m.n.}$ as well, according to the following theorem.

Theorem 4.2. *For any probability space* $(M, \mathfrak{A}, p)$ *where* $\mathfrak{A} = \mathfrak{N} + \mathcal{L}$ *and* $p|\mathfrak{A}$ *is complete and for any function* $f: M \to E$, *the relations*

$$\mathfrak{F}^{m.n.}(p|\mathfrak{A}, f) \neq \emptyset \quad \textit{and}\ \mathfrak{F} - \mathfrak{F}^{m.n.}(p|\mathfrak{A}, f) \neq \emptyset$$

hold if and only if there exist two probability measures $q_i|\mathcal{L}$ *such that*

$$q_i|\{B \in \mathcal{L}: f^{-1}(B) \in \mathfrak{N}\} = 0$$

where q_1 *is a Dirac-measure and* q_2 *is not.*

Proof: According to Criterion 2 each extension p_1 of $\mathfrak{F}$ corresponds to a product measure $\mu = p \times q$ where q vanishes on $\{B \in \mathcal{L}: f^{-1}(B) \in \mathfrak{N}\}$. And $p_1 \in \mathfrak{F}^{m.n.}$ is valid iff there exists an $\mathfrak{A}$-measurable function $g|M$ with $\mu(\mathrm{Gr}\ g) = 1$. Therefore $p_1 \in \mathfrak{F}^{m.n.}$ holds iff there exists a number $c \in E$ with $q(\{c\}) = 1$, i.e. iff q is a Dirac-measure on $\mathcal{L}$. ¤¤¤

4.2. The case of measures without atoms.

If $p|\mathfrak{A}$ is a measure without atoms, we have $p^*(A) = 0$ and $\tilde{p}(A) = 0$ for any countable set A, as it is true in case of 0-1-measures without kernel (see Remark 3.1 b)). In this section, we will weaken the assumption of the absence of p-atoms even more by assuming that any countable subset of M has the *inner* measure 0. This implies that we also admit the case that $p|\mathfrak{A}$ has *both* extensions *with* and *without* discrete components. On the other hand, we assume - as in section 3.3.- that $\mathfrak{A}$ is countably generated. Under this assumption we can take advantage of an idea of H.G.Kellerer for the construction of a graph $G \subset E^2$ with $\lambda_2^*(G) = 1$ where λ_2 denotes the 2-dimensional Lebesgue-measure. According to this idea, the transfinite method of construction used in Oxtoby's proof of Satz 14.1 in [Ox] is modified such that G meets any $\mathcal{G}_\delta$- set A of E^2 with $\lambda_2(A) > 0$.

The *continuum hypothesis* required by that construction is assumed to be valid in this section, too.

Given any system $\mathcal{G}$ of sets, we set

$$\mathcal{G}_\delta := \left\{ \bigcap_{n=1}^{\infty} G_n : G_n \in \mathcal{G} \right\}$$

and establish

Remark 4.1. *Let* $(L, \mathcal{L}, q)$ *be a probability space,* $\mathcal{G}$ *an algebra with* ${}^B\mathcal{G} = \mathcal{L}$, *and* Y *a subset of* L. *Then* $q^*(Y) = 1$ *is equivalent to*

$$Y \cdot D \neq \emptyset \quad \textit{for any} \quad D \in \mathcal{G}_\delta \quad \textit{with} \quad q(D) > 0.$$

The proof follows from

$$q_*(\overline{Y}) = \sup \{q(D): \overline{Y} \supset D \in \mathcal{G}_\delta\}.$$

The following two theorems show that, under weak assumptions, there exist functions having measure extensions which do *not* correspond to measurable neighbours, also in the general case of measures, especially in case of those without atoms.

Theorem 4.3. *Assume that* $(M, \mathfrak{A}, p)$ *satisfies*

(1) $p_*(T) = 0$ *for any countable subsets* T *of* M

(2) $\mathfrak{A}$ *is countably generated.*

Then the following is true:

a) *Given any function* $g: M \to E$ *there exists a function* $f: M \to E$ *such that*

(i) $(p \times \lambda)^*(\mathrm{Gr}\, f) = 1$
where λ *denotes the Lebesgue-measure on* $\mathcal{L}$,

(ii) g *is a neighbour of* f.

b) *There exists a function* $f_1: M \to E$ *such that*

$$\mathfrak{F}^{m.n.}(p|\mathfrak{A}, f_1) \neq \emptyset \quad \textit{and} \quad \mathfrak{F} - \mathfrak{F}^{m.n.}(p|\mathfrak{A}, f_1) \neq \emptyset$$

hold.

Proof of a): Because of (2) there exists a countable system $\mathfrak{T}$ with ${}^{B}\mathfrak{T} = \mathfrak{A}$ which we can assume without loss of generality to be an algebra. Then we define the countable system $\mathfrak{R}$ of all "rectangles" $S \times I$ with S of $\mathfrak{T}$ and I of the semi-algebra $\mathfrak{J}_Q$ of the semi-open intervalls of E with rational numbers as extreme points:

$$\mathfrak{R} := \{S \times I: S \in \mathfrak{T}, I \in \mathfrak{J}_Q\}.$$

Then the algebra extension $\mathfrak{G} := {}^{A}\mathfrak{R}$ is also countable and fulfills

$${}^{B}\mathfrak{G} = \mathfrak{A} * \mathfrak{L}.$$

We set

$$p \times \lambda =: \mu, \qquad Gr\, f =: X,$$

$$\vartheta := \{D \in \mathfrak{G}_\delta: \mu(D) > 0\},$$

$$\vartheta_1 := \{A \times E: A \in \mathfrak{T}_\delta, \ p(A) > 0\}.$$

Because of

$$|\vartheta| = |\mathfrak{G}_\delta| = \aleph$$

we can represent ϑ as a transfinite series $\{D_\tau: \tau < \Omega\}$ where $\{\sigma: \sigma < \tau\}$ is countable for each $\tau < \Omega$ (here we utilize the continuum hypothesis). Proceeding inductively we choose now points $(x_\tau, y_\tau): \tau < \Omega$ from $M \times E$:

$$(x_1, y_1) \in D_1$$

$$(x_\tau, y_\tau) \in D_\tau \cdot \left(\overline{\{x_\sigma: \sigma < \tau\}} \times E\right) =: L_\tau \quad \text{for } 1 < \tau < \Omega,$$

such that the additional condition

$$y_\tau = g(x_\tau) \quad \text{for all } \tau \text{ with } D_\tau \in \vartheta_1$$

is satisfied. We can proceed in this way since assumption (1) implies

$$\mu^*(L_\tau) \geq \mu(D_\tau) - p_*(x_\sigma: \sigma < \tau) = \mu(D_\tau) > 0$$

and hence $L_\tau \neq \emptyset$ for any $\tau < \Omega$.

Then $f|M$ defined by

$$f(x) := \begin{cases} y_\tau & \text{for } x = x_\tau,\ \tau < \Omega \\ 0 & \text{otherwise} \end{cases}$$

is a map $M \to E$ with the properties

$$X \cdot D \neq \emptyset \qquad \text{for all } D \in \vartheta$$

$$\{f = g\} \cdot A \neq \emptyset \text{ for all } A \in \mathcal{T}_\delta \text{ with } p(A) > 0.$$

According to Remark 4.1 it follows

$$\mu^*(X) = 1 \quad \text{and} \quad p^*(f = g) = 1.$$

Proof of b): Let g be any $\mathfrak{A}$-measurable mapping $M \to E$. Then, according to part a), there exists a function $f_1|M$ with the properties

(i) $(p \times \lambda)^*(\mathrm{Gr}\, f_1) = 1$

(ii) g is a measurable neighbour of f_1.

Property (i) yields that $\mu = p \times \lambda$ corresponds to a p_1 of $\mathfrak{F}$ in the sense of Criterion 1; for any $\mathfrak{A}$-measurable function h we have, due to the Theorem of Fubini,

$$\mu(\mathrm{Gr}\, h) = \int_M \int_E \chi_{\mathrm{Gr}\, h}(x,y)\, d\lambda\, dp = 0;$$

hence the extension p_1 which corresponds to $\mu = p \times \lambda$ is an element of $\mathfrak{F} - \mathfrak{F}^{m.n.}$.

Property (ii) yields according to section 1.4 that $\mathfrak{F}^{m.n.}$ is not empty. ¤¤¤

Now it remains to answer the question of the existence of a function $f|M$ with $\mathfrak{F}^{m.n.} = \emptyset$ and $\mathfrak{F} \neq \emptyset$, under the assumptions (1) and (2) of Theorem 4.3. An useful contribution to this problem is the following Lemma of H. von Weizsäcker together with the method of construction used in its proof:

Lemma (H. von Weizsäcker).[1] *There exists a mapping* $z\colon E \to E$ *with the properties*

$$\lambda_2^*(\mathrm{Gr}\, z) = 1$$

1) According to a private communication. The assertion quoted here as a Lemma is a result contained in the proof of a theorem stating that $z|E$ has no *"extremal"* extension.

and

$$\lambda_g^*(\text{Gr } z) = 0 \quad \textit{for all measurable functions } g\colon E \to E,$$

where λ_g *means the measure which* λ *induces on* Gr g.

In the proof of this Lemma, the inductive definition of $f|M$ used in the proof of Theorem 4.3 a) is modified, in the case $(M,\mathfrak{A},p) = (E,\mathcal{L},\lambda)$, such that $f|M$ equals to any λ-measurable function g only on a countable subset.

We will show that the example of von Weizsäcker is no isolated case. On the contrary, there exist functions with $\mathcal{F}^{m.n.} = \emptyset$ and $\mathcal{F} \neq \emptyset$ for each probability space $(M,\mathfrak{A},p)$ which satisfies the following assumption (3) - as a sufficient condition for $\mathcal{F}^{m.n.} = \emptyset$ - besides the assumptions (1) and (2) of Theorem 4.3 - sufficient for the existence of a function f with

$$(p \times \lambda)^*(\text{Gr } f) = 1.$$

(3) M has a subset M' with power $\aleph$ such that

$$p^*(T) < p(M') \quad \text{for all countable subsets T of M'.}$$

Obviously $(E,\mathcal{L},\lambda)$ fulfills the three assumptions (1), (2), and (3). More generally, if $p|\mathfrak{A}$ has no p-atoms then assumption (3) is reduced to the extremely weak condition that there exists a subset M' of M with

$$|M'| = \aleph \quad \text{and} \quad p(M') > 0.$$

To prove the following theorem we use the assertion of

Remark 4.2. *If* $\mathfrak{A}$ *is countably generated then the set* $\mathfrak{M}$ *of all* $\mathfrak{A}$*-measurable functions on* M *has the power of the continuum.*

The *proof* results from the following statements:

1) $|\mathfrak{M}| \leq |\{G \in \mathfrak{A} * \mathcal{L}: G \text{ the graph of a function}\}| \leq |\mathfrak{A} * \mathcal{L}|$.

2) If $\mathfrak{G}$ is a (semi-) algebra then ${}^B\mathfrak{G} \subset {}^S\mathfrak{G}$ and, consequently, $|{}^B\mathfrak{G}| \leq |\mathfrak{G}|^{\aleph_0}$ holds.

3) If $\mathfrak{A}$ is countably generated then we may assume without loss of generality that $\mathfrak{A} * \mathcal{L}$ is generated by a countable semi-algebra $\mathfrak{G}$; from this together with 1) and 2)

$$|E| \leq |\mathfrak{M}| \leq \aleph = |E|$$

follows. ▫▫▫

Theorem 4.4. *Let* $(M, \mathfrak{A}, p)$ *satisfy the assumptions*

(1) $p_*(T) = 0$ *for any countable subsets* T *of* M

(2) $\mathfrak{A}$ *is countably generated*

(3) M *has a subset* M' *with the power of the continuum such that* $p^*(T) < p(M')$ *for any countable subsets* T *of* M'.

Then there exists a function $f_2 \colon M \to E$ *such that*

$$\mathfrak{F}^{m.n.}(p|\mathfrak{A}, f_2) = \emptyset \quad \text{and} \quad \mathfrak{F}(p|\mathfrak{A}, f_2) \neq \emptyset$$

Proof: We use the notations

$$\mu, \vartheta, D_\tau, L_\tau$$

defined in the proof of Theorem 4.3. According to Remark 4.2 the assumption(2) yields that the set $\mathfrak{M}$ of all $\mathfrak{A}$-measurable functions $f\colon M \to E$ has the power $\aleph$. Therefore, assuming the continuum hypothesis, we can represent $\mathfrak{M}$ in the form of a transfinite series $\{g_\tau : \tau < \Omega\}$ where $\{\sigma : \sigma < \tau\}$ is countable for each $\tau < \Omega$. We set $G_\tau := \mathrm{Gr}\ g_\tau$.

By analogy with the procedure in the proof of Theorem 4.3 we choose

$$(x_1, y_1) \in D_1 \cdot \overline{G_1} =: L_1'$$

$$(x_\tau, y_\tau) \in L_\tau \cdot \bigcap_{\sigma \leq \tau} \overline{G_\sigma} =: L_\tau' \quad \text{for } 1 < \tau < \Omega.$$

We can proceed in this way, since $L_\tau' \neq \emptyset$ for any $\tau < \Omega$; for we have

$$\mu^*(L_\tau') = \mu^*(L_\tau) = \mu(D_\tau) > 0,$$

since

$$L_\tau - L_\tau' \subset \bigcup_{\sigma \leq \tau} G_\sigma \in \mathfrak{N}(\mu)$$

holds as a consequence of the Theorem of Fubini.

Taking assumption (3) into account we set

$$M_1 := \{x_\tau : \tau < \Omega\}$$

$$M_2 := M' \cdot \overline{M_1} = \{x_\tau' : \tau < \Omega\}$$

$$M_3 := M - (M_1 + M_2).$$

In order to construct the wanted function $f_2|M$ on M_2 we choose

$$y'_\tau \in E - \{g_\sigma(x'_\tau): \sigma \leq \tau\} \quad \text{for all } \tau < \Omega.$$

Now we are ready to define $f_2|M$:

$$f_2(x) := \begin{cases} y_\tau & \text{for } x = x_\tau \in M_1 \\ y'_\tau & x = x'_\tau \in M_2 \\ 0 & x \in M_3. \end{cases}$$

Using the abbreviation $X_2 := \text{Gr } f_2$ we obtain:

1) $X_2 \cdot D \neq \emptyset$ for all $D \in \vartheta$

and therefore, according to Remark 4.1, $\mu^*(X_2) = 1$.

2) $\{f_2 = g_\sigma\} \subset M_3 + \{x_\tau: \tau < \sigma\} + \{x'_\tau: \tau < \sigma\} \subset \overline{M'} + T$ for $\sigma < \Omega$

where T is a countable subset of M'. Because of (3) this yields

$$p^*(f_2 = g_\sigma) \leq p(\overline{M'}) + p^*(T) < 1,$$

i.e. g_σ is no neighbour of f_2.

It follows from 1) that $p \times \lambda$ corresponds to a $p_1 \in \mathfrak{F}(p|\mathfrak{A}, f_2)$, according to Criterion 1, and from 2) that f_2 has no measurable neighbour. ППП

R e f e r e n c e s:

[BK] Banach, S. et C. Kuratowski: Sur une généralisation du problème de la mesure. Fund. Math. 14, 127 - 131 (1929).

[B1] Bierlein, D.: Über die Fortsetzung von Wahrscheinlichkeitsfeldern. Z. Wahrscheinlichkeitstheorie verw. Gebiete 1, 28 - 46 (1962).

[B2] Bierlein, D.: Die Konstruktion eines Maßes .... Z. Wahrscheinlichkeitstheorie verw. Gebiete 1, 126 - 140 (1962).

[B3] Bierlein, D.: Measure extensions according to a given function. Lect. Notes in Econ. and Math. Systems 157: Optimization and Operations Research, Proceedings, Bonn 1977, S. 25 - 35.

[Ku] Kurz, A.: Uniformisierung analytischer Mengen und eine Anwendung bei der Maßfortsetzung. Archiv Math. 29, 204 - 207 (1977).

[LR] Landers, D. and L. Rogge: On the extension problem for measures. Z. Wahrscheinlichkeitstheorie verw. Gebiete 30, 167 - 169 (1974).

[L1] Lehn, J.: Maßfortsetzungen und Aumann's Selektionstheorem. Z. Wahrscheinlichkeitstheorie verw. Gebiete 35, 265 - 268 (1976).

[L2] Lehn, J.: Prämeßbare Funktionen. Manuscripta math. 20, 141 - 152 (1977).

[Ox] Oxtoby, J.C.: Maß und Kategorie. Berlin, Heidelberg, New York: Springer 1971.

[Ru] Rupp, W.: Mengenwertige Maße und Maßfortsetzungen. Manuscripta math. 22, 137 - 150 (1977).

[Ul] Ulam, S.: Zur Maßtheorie in der allgemeinen Mengenlehre. Fund. Math. 16, 140 - 150 (1930).

Dieter Bierlein

Fakultät für Mathematik
Universität Regensburg
8400 Regensburg

A SURVEY OF SMALL BALL THEOREMS AND PROBLEMS

Jens Peter Reus Christensen

Abstract: We study the problem to what extent a measure is determined by its values on a family of balls and how it might possibly be computed from those values.

In the following we shall, if not specifically mentioned otherwise, work in a complete separable metric space (M,d). All the measures considered will be positive (unless called "signed") and bounded countably additive Borel measures unless otherwise specified. Let $\mathcal{S}$ be some family of balls and let u be a bounded positive Borel measure on M. We shall consider conditions on M for the following two closely related questions to have an affirmative solution:

I) If v is another bounded positive Borel measure with $v(B) = u(B)$ for all $B \in \mathcal{S}$ is then $u = v$? Equivalently if t is a bounded signed Borel measure with $t(B) = 0$ for all $B \in \mathcal{S}$ is then $t = 0$?

II) In the case where I) has an affirmative solution how can u be computed? We shall consider only one formula for u (or rather we shall consider conditions for this formula to hold). If $A \subseteq M$ is any Borel set do we have

$$u(A) = \inf\left\{\sum_i u(B_i) \mid B_i \in \mathcal{S},\ u(A \setminus \bigcup_i B_i) = 0\right\} \quad ?$$

Note that the last formula assumes we know the zero sets of u. We suppose that it is an effectively stronger condition to replace $u(A \setminus \bigcup_i B_i) = 0$ with $A \subseteq \bigcup_i B_i$ (see discussion in the problems at the end).

We shall in the present survey consider only two families of balls, the family of all balls and the family $\mathcal{S}_e$ of all balls with radies strictly less than e. Usually it will be of no consequence whether our balls are assumed to be open or closed since closed balls may be approximated with open balls and conversely. Whenever forced to make a decision we shall consider open balls.

Theorem 1. Let $\mathcal{S}$ be any family of balls. Then I) has an affirmative solution for all bounded measures if and only if for any probability measure u on M the linear span of $\{\chi_B \mid B \in \mathcal{S}\}$ is dense in $L^2(u)$.

The problem II) has an affirmative solution for any positive measure if and only if the convex cone generated by $\{\chi_B \mid B \in \mathcal{S}\}$ is dense in $L^2_+(u)$ for any probability measure u .

The problem II) has an affirmative solution for any positive measure u if and only if the "positivity principle" is valid: For any signed bounded measure v the condition $v(B) \geq 0$ for all $B \in \mathcal{S}$ implies that v is a positive measure.

Proof. The first part of the theorem can be proved by a really completely trivial Hahn-Banach argument. The density condition in the second part of the theorem can be seen to be equivalent with the positivity principle using an easy Hahn-Banach argument. The second part is less trivial.

Suppose now that the convex cone generated by $\{\chi_B \mid B \in \mathcal{S}\}$ is dense in $L^2_+(u)$ for some probability measure u . Let $A \subset M$ be a measurable set. For any $0 < b < 1$ there exists $B \in \mathcal{S}$ with $u(A \cap B) > bu(B) > 0$. We shall see this indirectly. Suppose therefore that this is not true for some $0 < b < 1$. Let

$$h_n = \sum_i \lambda_i^n \chi_{B_i^n}/u(B_i^n)$$

(finite sums with $\lambda_i^n > 0$) be choosen with $\|h_n - \chi_A\|_2 \to 0$. We see

$$\|h_n - \chi_A\|_2^2 = \int h_n^2 + \chi_A^2 - 2\sum_i \lambda_i^n \chi_A \cdot \chi_{B_i^n} /u(B_i^n) =$$

$$\int h_n^2 + \chi_A^2 - 2\sum_i \lambda_i^n \chi_{B_i^n}/u(B_i^n) + \int h_n(1-\chi_A) .$$

Now the last integral tends to zero and since all sides of the equations tends to zero a contradiction is readily obtained.

Let us consider all families of balls $B_i \in \mathcal{S}$ $(i \in I)$ with

1) $u(B_i) > 0$

2) $\sum_i u(B_i) \leq u(A \cap (\bigcup_i B_i))(1+\varepsilon)$

($\varepsilon > 0$ is presribed by the enemy).

When $B_i \in \mathcal{S}$ $(i \in I)$ is maximal (by inclusion) among such families we see $u(A \setminus (\bigcup_i B_i)) = 0$ because otherwise we could add some $B \in \mathcal{S}$ with $u(B \cap (A \setminus (\bigcup_i B_i)))/u(B)$ close to 1 to the family. Clearly this shows that formula II) is true.

Suppose now that formula II) holds for any measurable set A and any probability measure u. For each particular probability u it is easily seen that the formula gives an approximation of χ_A with elements of the convex cone generated by $\{\chi_B \mid B \in \mathcal{S}\}$ the approximation is however only in L^1 norm. Suppose to get a contradiction that the desired approximation does not hold in L^2 norm. Using the Hahn-Banach theorem we find $f \in L^2(u)$ strictly separating the cone from χ_A. For the measure v with density $|f|$ with respect to u we use the L^1 density to get a contradiction. This finishes the proof of Theorem 1.

We note that the proof of Theorem 1 is equally valid for any family of measurable set.

A locally finite 6-finite positive measure m on (M,d) is called uniformly distributed if and only if (by definition):

$$\forall_{x,y \in M} \forall_{r > 0}: m(B(x,r)) = m(B(y,r)) \quad .$$

The existence of a uniformly distributed measure (which is unique if it exists) places strong enough geometric restrictions on (M,d) for II) to hold for the family of small balls

$$\mathcal{S}_e = \{B(x,r) \mid r < e \ , \ x \in M\} \quad .$$

This is more or less clear from the results in [3] and [4] though not explicitly pointed out. We shall give a complete proof of II) for any bounded positive measure and the family $\mathcal{S}_e = \{B(x,r) \mid r < e, \ x \in M\}$ assuming the existence of a uniformly distributed measure m. This will be done by showing the "positivity principle".

Let v be a bounded signed measure on M with $v(B) > 0$ for any $B \in \mathcal{S}_e$. For $e > \varepsilon > 0$ we define

$$K_\varepsilon(x,y) = \begin{cases} (u(B(x,\varepsilon)))^{-1} & \text{if } d(x,y) < \varepsilon \\ 0 & \text{else} \end{cases}$$

We note that the mere existence of a uniformly distributed measure forces M to be locally compact. If φ is continuous with compact support we define

$$K_\varepsilon \ \varphi(x) = \int K_\varepsilon(x,y)\varphi(y)dm(y) \quad .$$

It is easily seen that

$$\| K_\varepsilon \ \varphi - \varphi \|_\infty \le w_\varepsilon(\varphi)$$

and since for small ε the support of $K_\varepsilon \varphi$ is uniformly compact we see by an easy Fubini argument that $\int \varphi(x)d\nu(x) \geq 0$ provided $\varphi \geq 0$. This proves the positivity principle (for a very closely related argument see [3]).

In the case of non locally compact metric spaces few really satisfying results are known (see discussion of the litterature at the end) although some sporadic results are known in special cases. The point is that we have at present no idea whether they are best possible. Much stronger results are likely to hold.

The only non locally compact complete separable metric space (M,d) for which question I) has been solved affirmatively for the family $\mathcal{S}_e$ of small balls and arbitrary measures seems to be Hilbert spaces (time of writing august 1979). The second seemingly much harder problem remain open even for Hilbert spaces.

<u>Theorem 2.</u> <u>Let</u> U <u>be the open unit ball in a real separable Hilbert space</u> H. <u>Let</u> ν <u>be a bounded signed measure on</u> U <u>which vanishes on balls contained in</u> U. <u>Then</u> ν <u>vanishes identically</u>.

<u>Proof</u>. Let us consider the kernel $G(x,y) = \exp(-\frac{1}{2}\|x-y\|_2^2)$. It is not difficult to show that this kernel is strictly positive definite in the sense that

$$\iint G(x,y)d\nu(x)d\nu(y) \geq 0$$

and the double integral vanishes if and only if the bounded signed measure ν vanishes identically. The kernel K on U^2 defined by the equation

$$K(x,y) = (1-\|x\|^2)(1-\|y\|^2)(1+\|x-y\|^2)^{-1} =$$

$$(1-\|x\|^2)(1-\|y\|^2)\int_0^\infty \exp(-r(1+\|x-y\|^2))dr$$

has the same property on U. The fact that the niveau sets

$$\{y \mid G(x,y) > a > 0\}$$

$$\{y \mid K(x,y) > a > 0\}$$

are balls centered in x in the first case and balls contained in U in the second case ($a > 0$ is arbitrary and $x \in U$ in the second case) makes it possible to show Theorem 2 (for details see [5]).

That question I) has an affirmative solution for the family $\mathcal{S}_e$ of small balls in a Hilbert space is an immediate consequence of Theorem 2. The question II) remains entirely open in the case of an infinite dimensional Hilbert space. It is of course slightly weaker than those statements to which counterexamples has been constructed by D. Preiss (see below).

For finite dimensional spaces affirmative solutions to question I) and II) has been known for a long time (see t.ex. [1]). It seems that the first example showing that I) does not always have an affirmative solution for the family of all balls was published by Roy O. Davies (see [7]). This very ingenious but also technically complicated example has later been substantially simplified (private communication). It is probably fair to say that even with this very easy example at hand it is not easy to get an idea about what geometric properties of the metric space are really involved in the question I) and II). For general Banach spaces but only for Gaussian measures C. Borell has obtained very strong results (see [2]). For general Banach spaces and general measures but only for the family of all balls Hoffmann-Jørgensen has obtained positive results on question I) under some smoothness assumptions on the norm.

Probably the most interesting recent negative results has been obtained by D. Preiss (see [9]). He shows that the classical Vitali covering theorems does not hold true for an infinite dimensional Hilbert space (see [9]). According to oral information available at the time of writing (7/9-1979) he has also shown that the classical differentiation theorems does not hold for an infinite dimensional Hilbert space, we have no written documentation available. However there are none of his results forbidding question II) to have an affirmative answer for an infinite dimensional Hilbert space. It would imply only some differentiation theorems weaker then the classical ones (the balls through which we differentiated could not be obtained to be centered).

We shall conclude the present survey by putting up a list of various problems with some relation to question I) and II).

Problem 1. The real problem is of course to find a reasonable simple geometric condition on the space (M,d) for question I) and II) to have affirmative solutions. This seems at present very ambitious. Probably a candidate should be some extremely weak form of finite dimensionality.

Problem 2. What stability properties does the class of metric spaces have for which question I) or II) has an affirmative solution for the class of all balls or the family $\mathcal{S}_e$ of small balls? Are t.ex. those properties preserved under Lipschitz equivalences?

Problem 3. What happens to question II) if we replace the condition $u(A\setminus(\bigcup_i B_i)) = 0$ with $A \subseteq \bigcup_i B_i$? It is highly likely that the question is totally different and much more difficult. Let G be any compact metrizable abelian group and $\mathcal{S}$ a translation invariant base for the topology on G, then question II) has an affirmative solution (the proof runs as a preceding one with the Haar measure instead of the uniformly distributed measure m) . The group G may be abelian and for the normalized Haar measure we may have

$$\inf\left\{\sum_i u(B_i) \mid B_i \in \mathcal{S},\ G \subseteq \bigcup_i B_i\right\} \geq a > 1 .$$

The group G is constructed as a countable infinite product of the finite abelian groups occuring in the proof of Theorem 1 in [6] (the really not very exciting details is left for the reader). Although the open sets are not balls but rather balls with many holes we feel that this example indicate that the stronger version of question II) are really very much stronger and very probably is equivalent with a weak form of finite dimensionality.

Problem 4. Although we dare to conjecture both question I) and II) to have affirmative solutions in the case where we assume that the complete separable metric space (M,d) admits a transitive group of isometries we see only a hopeful way of attack in the case of an abelian Polish group G with invariant metric d . The reason for our hope is that because G is abelian a convolution inequality

$$\chi_B * u \geq \chi_B * v$$

(u and v bounded positive measures) may be iterated to

$$\chi_B * u^{*n} \geq \chi_B * v^{*n}$$

for all $n > 1$. This suggest that potential theoretic methods should be applied! An important principle in the potential theory of locally compact abelian groups is, the principle of positivity of mass: If v is a bounded signed measure and f is a measurable function with $f(x) \geq a > 0$ on some non empty open set and $f \geq 0$ then the convolution inequality $f*v \geq 0$ implies $v(G) \geq 0$! We have been able to show that this principle is not valied in an infinite dimensional Hilbert space. However in the counterexamples we have been able to construct f has unbounded support! One should try to find out what kind of positivity principles do hold and apply them to our questions.

Problem 5. We leave as an exercise to the reader the following curious fact which we have unfortunately not been able to apply: In a complete separable metric space (M,d) question II) has an affirmative answer for any probability measure and any family of small balls $\mathcal{S}_e$ if and only if there does not exist some probability u , some a > 1 and some $\varepsilon > 0$ such that:

$$\int f(x)\chi_B(x)du(x) \geq a \cdot \int \chi_B(x)du(x)$$

for all $B \in \mathcal{S}_\varepsilon$ and some $f \in L^2_+(u)$ with $\int f(x)du(x) = 1$.

We feel satisfied, if we have been able to convince the reader that there are some interesting problems in this corner of geometric measure theory.

REFERENCES:

1) A. S. Besicovitch, A general form of the covering principle and relative differentiation of additive functions, Proc. Camb. Phil. Soc. 41 (1945), 103-110.

2) C. Borell, A note on Gaussian measures which agree on small balls, Ann. Inst. Henri. Poincare 13, (1977), 231-238.

3) Jens Peter Reus Christensen, On some measures analogous to Haar measure, Math. Scand. 26 (1970), 103-106.

4) Jens Peter Reus Christensen, Uniform measures and spherical harmonics, Math. Scand. 26 (1970), 293-302.

5) Jens Peter Reus Christensen, The small ball theorem for Hilbert spaces, Math. Ann. 237 (1978), 273-276.

6) Jens Peter Reus Christensen and Wojchiech Herer, On the existence of pathological submeasures and the construction of exotic topological groups, Math. Ann. 213 (1975), 203-210.

7) R. O. Davis, Measures not approximable or specificable by means of balls, Mathematica 18 (1971), 157-160.

8) J. Hoffmann-Jørgensen, Measures which agree on balls, Math. Scand. 37 (1975), 319-326.

9) D. Preiss, Gaussian measures and covering theorems, Commentationes Matematicae Universitatis Carolinae 20, 1 (1979).

A LONG JAMES SPACE

G. A. Edgar

This paper investigates some of the measurability properties of the James-type Banach space $J(\omega_1)$ obtained with an uncountable ordinal for index set. This space $J(\omega_1)$ is a second dual space with the Radon-Nikodym Property but is not weakly compactly generated. This answers a question of P. Morris reported in [1, p. 87]. (This question has also been answered by W. J. Davis, unpublished.) The space $J(\omega_1)$ is a dual RNP space, but it admits no equivalent weakly locally uniformly convex dual norm. This answers a question in Diestel-Uhl [1, p. 212]. The space $J(\omega_1)$ is a dual RNP space, but there is a bounded, scalarly measurable function on some probability space with values in $J(\omega_1)$ that is not Pettis integrable. The previously known "examples" of this phenomenon depend on the existence of a measurable cardinal [3, Example (1)]. The space is a dual RNP space, but the weak and weak* Borel sets are not the same. This answers a question asked in [10] and [4].

Other properties of this space can be found in the literature. For example, Hagler and Odell [6] have shown that every infinite-dimensional subspace of $J(\omega_1)$ contains an isomorphic copy of ℓ^2 .

We will use the following definitions for transfinite series and bases in a Banach space X . Let η be an ordinal, and let $x_\alpha \in X$ be given for each $\alpha < \eta$. The value (when it exists) of the series

$$\sum_{\alpha<\gamma} x_\alpha$$

is defined recursively as follows. If $\gamma = 0$, then

$$\sum_{\alpha<0} x_\alpha = 0 .$$

If $\gamma = \beta+1$ is a successor, then

$$\sum_{\alpha<\gamma} x_\alpha = \sum_{\alpha<\beta} x_\alpha + x_\beta ,$$

provided the series on the right-hand side converges. If γ is a limit, then

$$\sum_{\alpha<\gamma} x_\alpha = \lim_{\beta<\gamma} \left(\sum_{\alpha<\beta} x_\alpha\right) ,$$

where the limit is taken in the norm topology of X .

Supported in part by National Science Foundation grant MCS 77-04049 A01.

A transfinite sequence $(x_\alpha)_{\alpha<\eta}$ of vectors is called a basis for X iff for each $y \in X$, there is a unique sequence $(c_\alpha)_{\alpha<\eta}$ of scalars such that

$$y = \sum_{\alpha<\eta} c_\alpha x_\alpha .$$

Let η be an ordinal, and let $f: [0,\eta] \to \mathbb{R}$ be a function. The square variation of f is

$$(*) \qquad \sup\Big(\sum_{i=1}^{n} |f(\alpha_i) - f(\alpha_{i-1})|^2\Big)^{1/2}$$

where the sup is taken over all finite sequences $\alpha_0 < \alpha_1 < \alpha_2 < \ldots < \alpha_n$ in $[0,\eta]$. Let $J(\eta)$ be the set of all continuous functions f on $[0,\eta]$ with finite square variation and $f(0) = 0$. Then $J(\eta)$ is a Banach space with norm $(*)$. Alternately, let $\tilde{J}(\eta)$ be the set of all functions f on $[0,\eta[$ with finite square variation and $f(0) = 0$. For infinite η, the unique order preserving map of $[0,\eta[$ onto the non-limits in $[0,\eta]$ induces an isometry of $J(\eta)$ onto $\tilde{J}(\eta)$.

We begin by computing a basis for $J(\eta)$. If $\alpha \in [0,\eta]$, define $h_\alpha \in J(\eta)$ by

$$h_\alpha = \chi_{]\alpha,\eta]} .$$

Clearly, $\|h_\alpha\| = 1$. Define a projection P_α on $J(\eta)$ by

$$P_\alpha f = f\chi_{[0,\alpha]} + f(\alpha)\chi_{]\alpha,\eta]} .$$

PROPOSITION 1. The transfinite sequence $(h_\alpha)_{\alpha<\eta}$ is a basis for the Banach space $J(\eta)$.

Proof: Let $f \in J(\eta)$. I claim first that if γ is a limit ordinal, then $\lim_{\beta<\gamma} \|P_\beta f - P_\gamma f\| = 0$. Let $\epsilon > 0$. There exists a finite sequence $\alpha_0 < \alpha_1 < \ldots < \alpha_n$ in $[0,\eta]$ with

$$\|P_\gamma f\|^2 < \sum_{i=1}^{n} |P_\gamma f(\alpha_i) - P_\gamma f(\alpha_{i-1})|^2 + \epsilon .$$

Since $P_\gamma f$ is constant on $[\gamma,\eta]$, we may assume $\alpha_n \le \gamma$. Since f is continuous at γ, we may assume $\alpha_n < \gamma$. Consider $\beta \in]\alpha_n,\gamma[$. Then the same sequence $\alpha_0 < \alpha_1 < \ldots < \alpha_n$ shows that $\|P_\beta f\|^2 > \|P_\gamma f\|^2 - \epsilon$. Now $P_\beta f$ is constant on $[\beta,\gamma]$ and $(P_\gamma - P_\beta)f$ is constant on $[0,\beta]$, so

$$\|P_\gamma f\|^2 \ge \|P_\beta f\|^2 + \|(P_\gamma - P_\beta)f\|^2 .$$

Therefore $\|(P_\gamma - P_\beta)f\|^2 \leq \|P_\gamma f\| - \|P_\beta f\| < \epsilon$.

Now given $f \in J(\eta)$, define $c_\alpha = f(\alpha+1) - f(\alpha)$ for $\alpha \in [0,\eta[$. Then

$$f = \sum_{\alpha<\eta} c_\alpha h_\alpha .$$

This is proved by the equation $\sum_{\alpha<\gamma} c_\alpha h_\alpha = P_\gamma f$, which follows by induction on γ. □

COROLLARY. The space $J(\eta)$ is separable if and only if the ordinal η is countable.

Next we consider duals and preduals for $J(\eta)$. For $\alpha \in]0,\eta]$, define $e_\alpha \in J(\eta)^*$ by $e_\alpha(f) = f(\alpha)$. Then $\|e_\alpha\| = 1$.

PROPOSITION 2. The closed linear span Y of $\{e_\alpha : \alpha \in]0,\eta]$, α not a limit ordinal$\}$ is an isometric predual of $J(\eta)$ in the sense that Y^* is isometric to $J(\eta)$.

Proof: The space Y is a norming space of functionals for $J(\eta)$. Indeed, functionals of the form

$$\sum_{i=1}^{n} t_i(e_{\alpha_i} - e_{\alpha_{i-1}})$$

(where $\alpha_0 < \alpha_1 < \ldots < \alpha_n$ are non-limits and $\Sigma|t_i|^2 \leq 1$) have norm 1 and norm $J(\eta)$ isometrically. The unit ball $B = \{f \in J(\eta) : \|f\| \leq 1\}$ is compact in the topology of pointwise convergence on $\{e_\alpha : \alpha$ not a limit$\}$. To see this, consider a net f_θ in B . By taking a subnet, we may assume $f_\theta(\alpha)$ converges for all non-limits α , call that limit $f(\alpha)$. If $\alpha_0 < \alpha_1 < \ldots < \alpha_n$ are non-limits, then

$$\sum_{i=1}^{n} |f(\alpha_i) - f(\alpha_{i-1})|^2 \leq 1 .$$

From this it follows that the limits $\lim_{\alpha<\beta} f(\alpha)$ exist for limit ordinals β , call these limits $f(\beta)$. Then $f \in B$ and $f_\theta \to f$ pointwise on the non-limits.

Finally, since B is bounded, it is compact in the topology $\sigma(J(\eta),Y)$. Therefore $J(\eta) = Y^*$ isometrically (cf. [2, V.5.7]). □

We will see below that $J(\eta)^*$ has the Radon-Nikodym property. It follows from this that the isometric predual is unique [5].

PROPOSITION 3. The sequence $(e_\alpha)_{\alpha\in]0,\eta]}$ is a basis for $J(\eta)^*$.

Proof: Let $\ell \in J(\eta)^*$. We first claim that if γ is a limit, then $\lim_{\beta<\gamma}\ell(h_\beta)$ exists. Suppose it does not exist. Then there are real numbers $a < b$ and ordinals $\beta_0 < \beta_1 < \beta_2 < \ldots < \gamma$ with $\ell(h_{\beta_{2i}}) < a$, $\ell(h_{\beta_{2i+1}}) > b$. But for each n, we have $\|\sum_{i=1}^{n}(h_{\beta_{2i-1}} - h_{\beta_{2i}})\| = (2n)^{1/2}$, so

$$n(b-a) < \ell\left(\sum_{i=1}^{n}(h_{\beta_{2i-1}} - h_{\beta_{2i}})\right) \leq \|\ell\|(2n)^{1/2} ,$$

so $\|\ell\| = \infty$, a contradiction.

Define, for $\gamma \in]0,\eta]$,

$$u_\gamma = \begin{cases} \ell(h_{\gamma-1}) , & \gamma \text{ non-limit} \\ \lim_{\beta<\gamma} \ell(h_\beta) , & \gamma \text{ limit} . \end{cases}$$

Then $\lim_{\beta<\gamma} u_\beta = u_\gamma$ for limit ordinals γ. I claim that $\ell = \sum_{\alpha\in]0,\eta]}(u_\alpha - u_{\alpha+1})e_\alpha$. The series converges weak* to ℓ, so it is only required to show that the partial sums converge in norm at any limit ordinal γ. This calculation is the same as the one which shows the basis for the original James space is shrinking. See, for example, [9, p. 274, (d)]. □

Propositions 2 and 3 can be used to describe the canonical embedding of $J(\eta)$ into $J(\eta)^{**}$. In fact (if η is infinite), $J(\eta)^{**}$ is isometric to $\tilde{J}(\eta+1)$, and the set-theoretic inclusion $J(\eta) \to \tilde{J}(\eta+1)$ is the canonical embedding. This shows that $J(\eta)^{**}$ is isometric to $J(\eta+1)$, and isomorphic to $J(\eta)$ itself.

COROLLARY. $J(\eta)^*$ is separable if and only if η is countable.

With the understanding of $J(\eta)$ provided above, many of its properties can be determined.

PROPOSITION 4. The space $J(\eta)$ has the Radon-Nikodym property.

Proof: Consider the predual Y given in Proposition 2. By a result of Uhl [1, p. 82, Cor. 6], it suffices to show that every separable subspace of Y has separable dual. Let Z be a separable subspace of Y. Each element of Z is in the closed span of a countable set of e_α, so there is a countable set $R \subseteq [0,\eta]$ of nonlimits such that $Y_1 = \mathrm{cl\ sp}\{e_\alpha : \alpha \in R\}$ contains Z. Then the closure $\overline{R}$ is also countable; let η_1 be its order type. Then Y_1^* is isometric to $J(\eta_1)$, which is separable. □

PROPOSITION 5. The dual $J(\eta)^*$ has the Radon-Nikodym property.

PROOF: Any separable subspace of $J(\eta)$ is in the closed span of countably many vectors h_α. Therefore, as above, the dual of such a closed span is isometric to $J(\eta_1)^*$ for some countable ordinal η_1. □

PROPOSITION 6. If $\ell \in J(\eta)^*$, then ℓ is a Borel function on $(J(\eta),\text{weak*})$.

Proof: By Proposition 3, it suffices to show that e_γ is weak*-Borel for all $\gamma \in]0,\eta]$. If γ is a non-limit, then e_γ is weak*-continuous. Assume γ is a limit. The restriction map of $J(\eta)$ onto $J(\gamma)$ is weak*-continuous, so we may assume $\gamma = \eta$. For $\lambda \in \mathbb{R}$, $k \in \mathbb{N}$, $r \in \mathbb{Q}$, define

$$P_1(r) = \{f \in J(\eta): \|f\| \le r\}$$

$$P_2(r,k,\lambda) = \bigcup \{f \in J(\eta): \sum_{i=1}^{n} |f(\alpha_i)-f(\alpha_{i-1})|^2 > r^2 - \frac{1}{k^2}, \ f(\alpha_n) < \lambda - \frac{1}{k}\},$$

where the union is over all finite sequences $\alpha_0 < \alpha_1 < \ldots < \alpha_n$ of non-limits, and

$$P(r,k,\lambda) = P_1(r) \cap P_2(r,k,\lambda) .$$

Then $P_1(r)$ is weak*-closed and $P_2(r,k,\lambda)$ is weak*-open. But

$$\{f \in J(\eta): f(\eta) < \lambda\} = \bigcap_{k=1}^{\infty} \bigcup_{r>0} P(r,k,\lambda),$$

so $\{f: f(\eta) < \lambda\}$ is weak*-Borel. □

PROPOSITION 7. Suppose $\eta \ge \omega_1$, the least uncountable ordinal. Then e_{ω_1} is not a weak*-Baire function.

Proof: The set $R = \{h_\alpha: \alpha \in [0,\omega_1]\}$ is weak*-homeomorphic to $[0,\omega_1]$. Any real-valued continuous function on $[0,\omega_1]$ is constant on some interval $[\gamma,\omega_1]$ with $\gamma < \omega_1$, so any Baire function shares this property. But $e_{\omega_1}(h_\alpha) = 1$ for $\alpha < \omega_1$ and $e_{\omega_1}(h_{\omega_1}) = 0$. Therefore, e_{ω_1} is not a Baire function on R, and a fortiori on $J(\eta)$. □

Since $J(\eta)$ has the Radon-Nikodym property, the weak and weak*-universally measurable sets coincide [8], [3, Theorem 1.5]. For this reason, the following is somewhat surprising.

PROPOSITION 8. There is a weak-Borel subset of $J(\omega_1)$ which is not weak*-Borel.

Proof: Let $R = \{h_\alpha : \alpha \in [0,\omega_1]\}$. Then (R,weak*) is homeomorphic to $[0,\omega_1]$. But

$$\{f \in R: (e_{\alpha+1}-e_\alpha)(f) > \frac{1}{2}\} = \{h_\alpha\} , \alpha < \omega_1$$

$$\{f \in R: e_{\omega_1}(f) < \frac{1}{2}\} = \{h_{\omega_1}\} ,$$

so (R,weak) is discrete. So every subset of R is a weak-open set. But there is a subset of R which is not weak*-Borel. (A subset $A \subseteq [0,\omega_1[$ such that neither A nor its complement contains a closed unbounded set is not Borel.) □

Although the statement of the following proposition does not involve measurability questions, they are helpful in the proof.

PROPOSITION 9. The space $J(\omega_1)$ admits no equivalent dual norm that is weakly locally uniformly convex.

Proof: In a dual space with weakly locally uniformly convex norm, the weak and weak* topologies coincide on the surface of the unit ball. But then by [4, Theorem 2.1] the weak and weak* Borel algebras coincide on the entire space. So by Proposition 8, the space $J(\omega_1)$ admits no such norm. □

The terms used in the following can be found in [4].

PROPOSITION 10. The space $J(\omega_1)$ is not realcompact, not measure-compact, not Lindelof, not weakly compactly generated, not isomorphic to a subspace of a weakly compactly generated space, and fails the Pettis integral property.

Proof: We show that $J(\omega_1)$ is not realcompact; the other assertions follow from this. Define a zero-one measure μ on Baire $([0,\omega_1[)$ by $\mu(B) = 0$ iff B is countable, $\mu(B) = 1$ iff $[0,\omega_1[\setminus B$ is countable. The map $\varphi: [0,\omega_1[\to J(\omega_1)$ defined by $\varphi(\alpha) = h_\alpha$ is scalarly measurable since $e_\beta \circ \varphi$ is constant a.e. for each β . The image $\lambda = \varphi(\mu)$ is a zero-one measure on Baire $(J(\omega_1),\text{weak})$. But λ is not τ-smooth: If $A \subset [0,\omega_1[$ is countable, then

$$Z_A = \{f \in J(\omega_1): f(\alpha) = 0 \text{ for all } \alpha \in A, f(\omega_1) = 1\}$$

is a weak-zero-set, and $\lambda(Z_A) = 1$. But the collection Z_A decreases to $\emptyset$. Thus $(J(\omega_1),\text{weak})$ is not realcompact. □

If the continuum hypothesis holds, then there is a bijection $\theta: [0,1] \to [0,\omega_1[$, and $\varphi \circ \theta$ provides an example of a bounded, scalarly measurable function on $[0,1]$ which is not Pettis integrable with respect to Lebesgue measure, that is, $J(\omega_1)$ fails the Lebesgue-PIP. Similarly, if Martin's Axiom holds, $J(\omega_c)$ fails the Lebesgue-PIP, where ω_c is the least ordinal of power c .

Using the same measure space $[0,\omega_1[$ and the map $\alpha \to e_{\alpha+1}$, it can be shown similarly that the predual Y of $J(\omega_1)$ is not realcompact.

References

1. J. Diestel and J.J. Uhl, Vector Measures. Mathematical Surveys 15, American Mathematical Society, Providence, RI, 1977.

2. N. Dunford and J.T. Schwartz, Linear Operators I. Interscience, New York, 1957.

3. G.A. Edgar, Measurability in a Banach space. Indiana Univ. Math. J. 26 (1977), 663-677.

4. G.A. Edgar, Measurability in a Banach space II. Indiana Univ. Math. J. 28 (1979), 559-579.

5. G. Godefroy, Espaces de Banach: Existence et unicité de certains préduaux. Ann. Inst. Fourier Grenoble. 28, 3(1977), 87-105.

6. J. Hagler and E. Odell, A Banach space not containing ℓ_1 whose dual ball is not weak* sequentially compact. Illinois J. Math. 22 (1978), 290-295.

7. R. Haydon, Some more characterizations of Banach spaces containing ℓ_1. Math. Proc. Camb. Phil. Soc. 80 (1976), 269-276.

8. L. Schwartz, Propriété de Radon-Nikodym. Séminaire Maurey-Schwartz (1974-1975), Exp. no. V-VI.

9. I. Singer, Bases in Banach Spaces I. Die Grundlehren der Mathematischen Wissenschaften 154, Springer-Verlag, 1970.

10. M. Talagrand, Sur la structure borélienne des espaces analytiques, Bull. Sci. Math. (2) 101 (1977), 415-422.

Department of Mathematics
The Ohio State University
Columbus, Ohio 43210

Baire sets in product spaces

Hans G. Kellerer

If X is an arbitrary topological space (separation properties don't play any role in the sequel), the following notations will be used: $\mathcal{G}(X)$ is the family of open sets $G \subset X$, $\mathcal{C}(X)$ the space of continuous functions $f: X \to \mathbb{R}$ and $\mathcal{L}(X)$ the σ-algebra of Baire sets generated by $\mathcal{C}(X)$.

If now X is the product of a (finite or infinite) family of topological spaces X_i, $i \in I$, conditions are important in topological measure theory under which $\mathcal{L}(X)$ is the product of the corresponding σ-algebras $\mathcal{L}(X_i)$, $i \in I$, i.e.

(C) $\mathcal{L}(\prod_{i\in I} X_i) = \bigotimes_{i\in I} \mathcal{L}(X_i)$.

Because of the continuity of the projections $p_i: X \to X_i$ the inclusion " $\supset$ " holds in any case, hence the equation (C) is equivalent to

(C') f $\bigotimes_{i\in I} \mathcal{L}(X_i)$ - measurable for all $f \in \mathcal{C}(\prod_{i\in I} X_i)$.

By means of the Stone-Weierstraß theorem this easily yields as a well known sufficient condition the compactness of all spaces X_i (see [1], § 5, th. 4). An open problem, however, is the question, to what extent countability properties can substitute compactness. The main result of this note consists in a theorem which simultaneously covers the two most important cases of compact resp. second countable spaces.

To begin with the simplest case, I = {1,2} is assumed first. The following fact then turns out to be essential:

Lemma Let X_1, X_2 be second countable spaces, $X = X_1 \times X_2$ and $f \in \mathcal{C}(X)$. Then for any sets $A_i \subset X_i$ there exist sets $C_i \in \mathcal{L}(X_i)$ such that

$A_i \subset C_i$ for i = 1,2 and $f[C_1 \times C_2] \subset \overline{f[A_1 \times A_2]}$*).

Proof: Since the subset A_2 of X_2 is again a second countable and hence a

*) $\overline{B}$ denotes the closure of the set B.

separable space, A_2 contains a countable dense subset D_2. With the abbreviation $F' := \overline{f[A_1 \times A_2]}$ let

$$C_1 := \bigcap_{x_2 \in D_2} f_{x_2}^{-1}[F']^{*)}.$$

Then the continuity of f_{x_2} and the countability of D_2 imply $C_1 \in \mathcal{L}(X_1)$, while $D_2 \subset A_2$ implies $A_1 \subset C_1$. For all $x_1 \in C_1$ the closed set $f_{x_1}^{-1}[F']$ by definition contains the set D_2 and hence also the set $A_2 \subset \overline{D_2}$, i.e.

$$f[C_1 \times A_2] \subset F'.$$

Replacing $A_1 \times A_2$ by $C_1 \times A_2$ and repeating the argument similarly provides a set $C_2 \in \mathcal{L}(X_2)$ with $A_2 \subset C_2$ and

$$f[C_1 \times C_2] \subset \overline{f[C_1 \times A_2]} \subset \overline{F'} = F',$$

i.e. the sets C_i satisfy all requirements.

By this fact a first result can be established:

<u>Proposition 1</u> If X_1 and X_2 are second countable spaces, equation (C) holds for their product $X = X_1 \times X_2$.

Proof: Let $f \in \mathcal{C}(X)$ and $G' \in \mathcal{G}(\mathbb{R})$ be arbitrary. Since $\mathbb{R}$ is metrizable, for G' there exist sets $G'_n \in \mathcal{G}(\mathbb{R})$ satisfying

(1) $\overline{G'_n} \subset G'$ and $\bigcup_{n \in \mathbb{N}} G'_n = G'$,

and by the assumption for G'_n there exist sets $G^m_{in} \in \mathcal{G}(X_i)$ satisfying

(2) $f^{-1}[G'_n] = \bigcup_{m \in \mathbb{N}} (G^m_{1n} \times G^m_{2n})$ for $n \in \mathbb{N}$.

If for G^m_{in} sets C^m_{in} are chosen according to the lemma, it follows by (2) that

$$f^{-1}[G'_n] \subset \bigcup_{m \in \mathbb{N}} (C^m_{1n} \times C^m_{2n})$$

and by (1)

$$f[C^m_{1n} \times C^m_{2n}] \subset \overline{G'_n} \subset G'.$$

Combined this implies

$$f^{-1}[G'] = \bigcup_{m,n \in \mathbb{N}} (C^m_{1n} \times C^m_{2n}) \in \mathcal{L}(X_1) \otimes \mathcal{L}(X_2)$$

and thus the required measurability of f.

This proposition can be improved immediately:

<u>Proposition 2</u> If X_1 or X_2 is a second countable space, equation (C) holds for

---

*) f_{x_2} denotes the function $x_1 \mapsto f(x_1,x_2)$ (f_{x_1} analogously).

their product $X = X_1 \times X_2$.

Proof: Let $\{G'_n : n \in \mathbb{N}\}$ be a countable base for $\mathbb{R}$ and for instance $\{G_{1m} : m \in \mathbb{N}\}$ a countable base for X_1. Then for arbitrary $f \in \mathcal{C}(X)$ by an appropriate union sets $G^n_{2m} \in \mathfrak{G}(X_2)$ are obtained such that

$$f^{-1}[G'_n] = \bigcup_{m \in \mathbb{N}} (G_{1m} \times G^n_{2m}) \quad \text{for } n \in \mathbb{N},$$

If $\mathfrak{G}^*(X_2)$ denotes the topology generated by the sets G^n_{2m}, $m,n \in \mathbb{N}$, $\mathfrak{G}^*(X_2)$ has a countable base and is coarser than the topology $\mathfrak{G}(X_2)$. Moreover, f is continuous even with respect to the product topology of $\mathfrak{G}(X_1)$ and $\mathfrak{G}^*(X_2)$, and thus proposition 1 applies.

The assumption of a countable base in proposition 2 can be weakened neither to separability nor to the Lindelöf property as is shown by the following stronger counterexample:

Example Let X_1 be a discrete space with the power of the continuum and the space $X_2 = \{0,1\}^{X_1}$ be endowed with the product topology, i.e. especially

(1) X_1 locally compact and metrizable,

(2) X_2 compact and separable.

If p_{x_1} denotes the projection from X_2 to the factor $\{0,1\}$ corresponding to $x_1 \in X_1$, the definition

$$f(x_1,x_2) := p_{x_1}(x_2)$$

yields a continuous function, since for arbitrary (x^o_1,x^o_2) the set

$$U = \{(x_1,x_2): x_1 = x^o_1 \text{ and } p_{x^o_1}(x_2) = p_{x^o_1}(x^o_2)\}$$

defines a neighborhood of (x^o_1,x^o_2) in which f is constant. Now according to the introductory remark (or to the subsequent theorem) the compactness of the factors of X_2 implies the equation

$$\mathfrak{B}(X_2) = \bigotimes_{x_1 \in X_1} \mathfrak{B}(\{0,1\}),$$

which prevents f from being $\mathfrak{B}(X_1) \otimes \mathfrak{B}(X_2)$ - measurable; for otherwise f would depend only on a countable number of the coordinates x_1 and $p_{x_1}(x_2)$, $x_1 \in X_1$, in contradiction to the fact that X_1 is uncountable. The equation (C), therefore, doesn't hold in this example.

If, however, X_1 is locally compact and σ-compact, i.e. countable at infinity, separability of X_2 is sufficient:

Proposition 3 If X_1 is countable at infinity and X_2 is separable, equation (C) holds for their product $X = X_1 \times X_2$.

Proof: 1. According to the assumption there is a countable subfamily $\mathfrak{U}_1$ of $\mathfrak{G}(X_1)$ such that

(1) $\overline{U_1}$ compact for $U_1 \in \mathfrak{U}_1$ and $\cup \mathfrak{U}_1 = X_1$

and a countable dense subset D_2 of X_2. In addition let $\mathfrak{U}'$ be a countable base for $\mathbb{R}$ and $f \in \mathfrak{C}(X)$ be chosen arbitrarily.

$$\mathfrak{V}_1 := \{U_1 \cap \bigcap_{x_2 \in E_2} f_{x_2}^{-1}[U'] : U_1 \in \mathfrak{U}_1, E_2 \subset D_2 \text{ finite}, U' \in \mathfrak{U}'\}$$

then defines a countable subfamily of $\mathfrak{G}(X_1)$ such that

(2) $\overline{V_1}$ compact for $V_1 \in \mathfrak{V}_1$.

If $\mathfrak{G}^*(X_1)$ denotes the topology generated by $\mathfrak{V}_1$, $\mathfrak{G}^*(X_1)$ has a countable base and is coarser than the topology $\mathfrak{G}(X_1)$. In view of proposition 2 it suffices, therefore, to show that f is continuous even with respect to the product topology of $\mathfrak{G}^*(X_1)$ and $\mathfrak{G}(X_2)$.

2. For that purpose let $G' \in \mathfrak{G}(\mathbb{R})$ and $(x_1^o, x_2^o) \in f^{-1}[G']$ be arbitrary. Then because of the metrizability of $\mathbb{R}$ there exists a set $U' \in \mathfrak{U}'$ satisfying

(3) $f(x_1^o, x_2^o) \in U'$ and $\overline{U'} \subset G'$

and because of the continuity of f there exist sets $G_i^o \in \mathfrak{G}(X_i)$ satisfying

(4) $(x_1^o, x_2^o) \in G_1^o \times G_2^o \subset f^{-1}[U']$,

where it may be assumed without loss of generality that

(5) $G_1^o \subset U_1 \in \mathfrak{U}_1$.

Since $D_2^o := D_2 \cap G_2^o$ in view of $G_2^o \in \mathfrak{G}(X_2)$ is a dense subset of G_2^o, for every x_1 in the intersection of the sets $f_{x_2}^{-1}[\overline{U'}]$, $x_2 \in D_2^o$, the closed set $f_{x_1}^{-1}[\overline{U'}]$ with D_2^o also contains $x_2^o \in G_2^o$, i.e.

(6) $\bigcap_{x_2 \in D_2^o} f_{x_2}^{-1}[\overline{U'}] \subset f_{x_2^o}^{-1}[\overline{U'}]$.

3. The sets

$$G_1^{x_2} := U_1 \cap f_{x_2}^{-1}[U'] \quad \text{for } x_2 \in D_2^o$$

are relatively compact according to (1) and satisfy

$$\bigcap_{x_2 \in D_2^o} \overline{G_1^{x_2}} \subset \bigcap_{x_2 \in D_2^o} \overline{f_{x_2}^{-1}[U']} \subset \bigcap_{x_2 \in D_2^o} f_{x_2}^{-1}[\overline{U'}]$$

because of the continuity of f, hence according to (6) and (3)

$$\bigcap_{x_2\in D_2^o} \overline{G_1^{x_2}} \subset f_{x_2^o}^{-1}[\overline{U'}] \subset f_{x_2^o}^{-1}[G'].$$

Since the set $f_{x_2^o}^{-1}[G']$ is open, there exists a finite set $E_2^o \subset D_2^o$ such that

$$(7) \qquad \bigcap_{x_2\in E_2^o} \overline{G_1^{x_2}} \subset f_{x_2^o}^{-1}[G'].$$

4. Now according to (4) and (5) all the sets $G_1^{x_2}$ contain the set G_1^o, which yields

$$(8) \quad x_1^o \in V_1 := \bigcap_{x_2\in E_2^o} G_1^{x_2} \in \mathfrak{V}_1 .$$

The set $\overline{V_1} \times \{x_2^o\}$ is compact by (2) and contained in the open set $f^{-1}[G']$ by (7), i.e. there are finitely many sets $G_i^j \in \mathfrak{G}(X_i)$, $j \in J$, such that

$$(9) \quad \overline{V_1} \times \{x_2^o\} \subset \bigcup_{j\in J} (G_1^j \times G_2^j) \subset f^{-1}[G'].$$

Here $x_2^o \in G_2^j$ for all $j \in J$ may be assumed without loss of generality, which implies

$$(10) \quad x_2^o \in V_2 := \bigcap_{j\in J} G_2^j \in \mathfrak{G}(X_2).$$

Combining the relations (8) - (10) and using $\mathfrak{V}_1 \subset \mathfrak{G}^*(X_1)$ it follows that

$$V_1 \in \mathfrak{G}^*(X_1) \quad \text{and} \quad V_2 \in \mathfrak{G}(X_2),$$

$$(x_1^o, x_2^o) \in V_1 \times V_2 \in f^{-1}[G'],$$

which concludes the proof of the stronger continuity of f.

Since the Lindelöf property is common to compact resp. second countable spaces, it is natural to ask for an extension of proposition 1 to Lindelöf spaces. The Lindelöf property, however, not being preserved even under finite products, it is necessary to impose this condition on the product space. Then a reduction to proposition 1 is possible:

<u>Proposition 4</u> If $X = X_1 \times X_2$ is a Lindelöf space, equation (C) holds.

Proof: Let $\{G_n' : n \in \mathbb{N}\}$ be a countable base for $\mathbb{R}$ and $f \in \mathfrak{C}(X)$ be chosen arbitrarily. Since $\mathbb{R}$ is metrizable, G_n' and hence $f^{-1}[G_n']$ is a set of type F_σ. Consequently the subspace $f^{-1}[G_n']$ of X has again the Lindelöf property, i.e. there exist sets $G_{in}^m \in \mathfrak{G}(X_i)$ such that

$$f^{-1}[G_n'] = \bigcup_{m\in\mathbb{N}} (G_{1n}^m \times G_{2n}^m) \quad \text{for } n \in \mathbb{N}.$$

As in the proof of proposition 2 an application of proposition 1 now yields the assertion.

The sufficient condition in proposition 4 is still not a neccessary one for the validity of equation (C), as is seen by proposition 2 or 3. It is this formulation, however, which is suited best for a generalization to arbitrary products:

<u>Theorem</u> If $X = \prod_{i\in I} X_i$ is the product of topological spaces such that

(L) $\prod_{i\in T} X_i$ a Lindelöf space for every finite $T \subset I$,

then equation (C) holds.

Proof: 1. In view of assumption (L) and the metrizability of $\mathbb{R}$ a result of Engelking (see [2]) can be applied*): for any $f \in \mathcal{C}(X)$ there exist a countable subset J of I and a function $g \in \mathcal{C}(\prod_{i\in J} X_i)$ such that $f = g \circ p$ where p denotes the projection from X to $\prod_{i\in J} X_i$. Thus $I = \mathbb{N}$ may be assumed without loss of generality.

2. If in this case $x_n^o \in X_n$ are chosen arbitrarily,

$$f_n: x \mapsto f(x_1,\ldots,x_n,x_{n+1}^o,\ldots) \quad \text{for } n \in \mathbb{N}$$

defines a sequence of continuous functions such that $f_n \to f$. Since the function f_n depends only on the first n coordinates, $I = \{1,\ldots,n\}$ may be assumed without loss of generality.

3. In this case, however, the assertion follows from proposition 4 by induction, and the proof of the theorem is finished.

The result applied in part 1 of the proof is true not only under the assumption (L) but also for the product of an arbitrary family of separable spaces (see [3]), while part 2 of the proof is independent of any topological property. The following open problem, therefore, is encountered at this point: does proposition 1 remain valid, if both spaces X_i are assumed only to be separable?

The implicit condition (L) can be replaced by an explicit condition on the spaces X_i, $i \in I$, in the following way:

<u>Corollary</u> If $X = \prod_{i\in I} X_i$ is the product of topological spaces such that for every $i \in I$

(L') X_i second countable or σ-compact or a Suslin space,

then equation (C) holds.

---

*) The assumption of all factors X_i being T_1-spaces made there is easily seen to be superfluous.

Proof: Since the topological properties appearing in condition (L') are all preserved under finite products, in establishing condition (L) X may be assumed to be the product of a second countable space X_1, a σ-compact space X_2 and a Suslin space X_3. Now X_3 being the continuous image of a second countable space, the same is true for the product $X_1 \times X_3$, which accordingly as a continuous image of a Lindelöf space is a Lindelöf space itself. Therefore, if $X_2 = \bigcup_{n\in\mathbb{N}} X_2^n$ with compact sets X_2^n, each product $X_1 \times X_2^n \times X_3$ and finally $X = \bigcup_{n\in\mathbb{N}} (X_1 \times X_2^n \times X_3)$ is again a Lindelöf space.

Concluding, it should be remarked that, as is easily verified by checking the proofs, all the results obtained above carry over to "generalized Baire sets", which are generated by all continuous mappings $f: X \to X'$, provided that the space X' replacing $\mathbb{R}$ is still metrizable and second countable.

References

[1] Bourbaki, N.: Topologie générale, Ch. X, Paris (1949).

[2] Engelking, R.: On functions defined on Cartesian products, Fund. Math. 59 (1966), 221-231.

[3] Kellerer, H.G.: Stetige Funktionen auf Produkträumen, Arch. Math. 19 (1968), 79-82

ON A MEASURE EXTENSION THEOREM OF BIERLEIN

Jörn Lembcke
Mathematisches Institut
Bismarckstraße 1 1/2
D-8520 Erlangen, West-Germany

We want to give a new proof of the following result:

THEOREM. Let $(\Omega, \mathcal{S}, \mu)$ be a probability space and $(A_i)_{i\in I}$ a family of disjoint subsets of Ω. Then there exists a σ-additive measure ν extending μ to the σ-algebra $\mathcal{T}$ generated by $\mathcal{S} \cup \{A_i: i\in I\}$.

For the case of a countable index set I the Theorem is due to Bierlein [3, Satz 2B], whereas the full theorem was proved by Aschenl and Lehn [2, Corollary 3 of Theorem 1] using Bierlein's result. Recently Lipecki [6] gave a group-valued version of the Theorem for the case of a countable index set.

Our proof is based on an idea which turned out to be successful for the extension of inner regular measures (cf. [4] and [5]): We introduce an appropriate preorder relation on the set of all finitely additive extensions and show that this set has maximal elements each of which is σ-additive. Our first effort to use this concept for the proof of Bierlein's Theorem (cf. [4, Satz 3.3] unfortunately contained an error. In the present proof we will use a more direct approach with a slightly modified preorder relation.

Proof. Denote by F the set of all (finitely additive, positive) measures extending μ to $\mathcal{T}$. We preorder F by

$$\nu \prec \nu' \quad \text{iff} \quad \nu(A_i) \leqq \nu'(A_i) \quad \text{for every } i\in I.$$

The proof will be completed by showing the following two lemmas:

LEMMA 1. If $\nu \in F$ is $\prec$-maximal (i.e. $\nu' \in F$, $\nu \prec \nu'$ implies $\nu' \prec \nu$), then ν is σ-additive.

LEMMA 2. F has a $\prec$-maximal element.

REMARKS. 1. In general, $\prec$ is not a (partial) order relation, even if I has cardinality 1 (cf. [7, sect. 7, (i)]).

2. If I is finite and $\bigcup_{i\in I} A_i = \Omega$, then it is obvious that each $\nu \in F$ is $\prec$-maximal and hence, by Lemma 1, σ-additive. For the conclusion of

σ-additivity one may even drop the condition $\bigcup_{i \in I} A_i = \Omega$ and the disjointness of the sets A_i.

Although this remark ought to be well known, we could not find it in the literature.

3. If I is at most countable and $\bigcup_{i \in I} A_i = \Omega$, then it is easy to see that every σ-additive extension of μ to $\mathcal{T}$ is $\prec$-maximal. However, this need not be the case, if I is uncountable:
Let $\Omega = I$ be uncountable. Choose $\mathcal{S} = \{\Phi, \Omega\}$ and $A_i = \{i\}$ for $i \in \Omega$. Then $\mathcal{T} = \{T \subset \Omega\colon T$ or $\complement T$ is at most countable$\}$. Denote by ν the measure on $\mathcal{T}$ defined by $\nu(T) = 0$, if T is at most countable, and by $\nu(T) = 1$, else. Then ν is a σ-additive extension of the only probability measure μ on $\mathcal{S}$, but of course it is not $\prec$-maximal.

4. It is obvious that the Theorem holds true if μ is only a σ-finite σ-additive measure on the σ-algebra $\mathcal{S}$.

5. Our proof will show that Lemma 2 is valid, even if the sets A_i are arbitrary (not necessarily disjoint).

Proof of Lemma 1. Let $\nu \in F$ and $i \in I$. We claim that the measure ν_i on $\mathcal{T}$ defined by

$$\nu_i(T) = \nu(T \cap A_i) \qquad (T \in \mathcal{T})$$

is σ-additive. In fact, let (T_n) be a decreasing sequence in $\mathcal{T}$. The sets A_j $(j \in I)$ being pairwise disjoint, there exists a decreasing sequence (S_n^i) in $\mathcal{S}$ such that $T_n \cap A_i = S_n^i \cap A_i$ $(n \in \mathbb{N})$. Hence

$$\inf_{n \in \mathbb{N}} \nu_i(T_n) = \lim_{n\to\infty} \nu(T_n \cap A_i) = \lim_{n\to\infty} \nu(S_n^i \cap A_i) =$$

$$= \lim \mu(S_n^i) - \lim \nu(S_n^i \setminus A_i) = \mu(\bigcap S_n^i) - \inf \nu(S_n^i \setminus A_i) \leqq$$

$$\leqq \nu(\bigcap S_n^i) - \nu(\bigcap S_n^i \setminus A_i) = \nu(\bigcap S_n^i \cap A_i) = \nu(\bigcap T_n \cap A_i) =$$

$$= \nu_i(\bigcap T_n) \leqq \inf_{n \in \mathbb{N}} \nu_i(T_n),$$

which proves the σ-additivity of ν_i.

Then the measure $\lambda = \sum_{i \in I} \nu_i$ is also σ-additive with $\lambda \leqq \nu$.

Now suppose ν to be $\prec$-maximal. Consider the measure $\tau = \nu - \lambda$ on $\mathcal{T}$ and its (σ-additive) restriction ρ to $\mathcal{S}$. If ρ admitted an extension to a measure τ' on $\mathcal{T}$ with $\tau'(A_j) > 0$ for some $j \in I$, then, for $\nu' = \lambda + \tau'$, we would have $\nu' \in F$, $\nu' \succ \nu$ and $\nu'(A_j) > \nu(A_j)$, since $\nu(A_i) = \nu_i(A_i) = \lambda(A_i)$ $(i \in I)$. This contradicts the $\prec$-maximality of ν. Consequently, $\tau'(A_i) = 0$ for every $i \in I$ and every measure τ' extending ρ to $\mathcal{T}$. By [7, Theorem 1], we get $\rho^*(A_i) = 0$ for every $i \in I$, ρ^* denoting the outer measure derived from ρ. Hence $\mathcal{T}$ is contained in the ρ-completion of $\mathcal{S}$, and τ is therefore σ-additive. This implies the σ-additivity of $\nu = \lambda + \tau$.□

Although Lemma 2 is a special case of [5, Satz 3.3 and footnote 1, p.65], we want to give a direct proof using a combination of Łoś-Marczewski's Theorem 1 in [7] and the basic idea of the proof of Andenaes' Theorem 1 in [1].

<u>Proof of Lemma 2</u>. Let I be well-ordered by some relation $\sqsubseteq$ and denote by s the smallest element of I. For some $\ell \notin I$ we define $I_\ell = I \cup \{\ell\}$ and $i \sqsubset \ell$ for every $i \in I$. Let $\mathcal{A}_i$ be the algebra in Ω generated by $\mathcal{S} \cup \{A_j: j \sqsubset i\}$ $(i \in I_\ell)$. Then $\mathcal{A}_s = \mathcal{S}$ and $\mathcal{A}_\ell$ is the algebra generated by $\mathcal{S} \cup \{A_i: i \in I\}$.

We will define a transfinite sequence $(\nu_i)_{i \in I_\ell}$ of measures ν_i on $\mathcal{A}_i$ each of which extends those with smaller index:

(1) Let $\nu_s = \mu$.

(2) Suppose, that the measures ν_j have been defined for $j \sqsubset i \in I_\ell$.

(a) If there exists a greatest element $k \in I_\ell$ such that $k \sqsubset i$, then

$$\mathcal{A}_i = \{(B \cap A_k) \cup (C \cap \complement A_k): B, C \in \mathcal{A}_k\}.$$

In this case denote by $(\nu_k)^*$ and $(\nu_k)_*$, respectively, the outer and inner measure derived from ν_k (cf. [7, sect. 2]). Then, by [7, Theorem 1],

$$\nu_i: (B \cap A_k) \cup (C \cap \complement A_k) \mapsto (\nu_k)^*(B \cap A_k) + (\nu_k)_*(C \cap \complement A_k) \quad (B, C \in \mathcal{A}_k)$$

is a measure extending ν_k to $\mathcal{A}_i$.

(b) If there is no greatest strict minorant of i in I_ℓ, then

$$\mathcal{A}_i = \bigcup_{j \sqsubset i} \mathcal{A}_j.$$

In this case define ν_i to be the simultaneous extension of the measures ν_j with $j \sqsubset i$.

Transfinite recursion yields a measure ν_ℓ on $\mathcal{A}_\ell$ extending μ. Let ν be some measure on $\mathcal{T} = \sigma(\mathcal{A}_\ell)$ extending ν_ℓ (cf. [7, sect. 4, (i)]). Obviously, $\nu \in F$.

Let us prove that ν is $\prec$-maximal:

For $\lambda \in F$ with $\lambda \succ \nu$ let λ_i denote the restriction of λ to $\mathcal{A}_i$.

Then it is sufficient to prove $\lambda_i = \nu_i$ for every $i \in I_\ell$.

(1) $\lambda_s = \mu = \nu_s$, as $\lambda, \nu \in F$.

(2) Suppose, $\lambda_j = \nu_j$ for every $j \sqsubset i \in I_\ell$.

(a) If i has a greatest strict minorant $k \in I_\ell$, then, by [7, sect. 3, (i)], $\lambda_k = \nu_k$ implies

$$\lambda_i(A_k) \leqq (\nu_k)^*(A_k) = \nu_i(A_k).$$

Hence $\lambda_i(A_k) = \nu_i(A_k)$, since $\lambda \succ \nu$. Now it follows immediately from [7, sect. 7, (iii)] that $\lambda_i = \nu_i$.

(b) The case, where i has no greatest strict minorant, is obvious.□

REMARK 6. Evidently, the measure ν_ℓ defined in the proof of Lemma 2 is a maximal element of the set F_ℓ of all measures extending μ to the algebra $\mathcal{A}_\ell$ generated by $\mathcal{S} \cup \{A_i: i \in I\}$ (with respect to the preorder corresponding to $\prec$). One can show by transfinite induction that ν_ℓ is also an extreme point of F_ℓ (cf. the corresponding Observation 2 in Andenaes [1]). Then every extreme point of the set F' of all measures extending ν_ℓ to $\mathcal{T}$ is an extreme point and a maximal element of F. (The existence of an extreme point of F' follows from the preceding and Remark 4 applied to $(\mathcal{A}_\ell, \nu_\ell)$ and $\mathcal{T} \setminus \mathcal{A}_\ell$ instead of $(\mathcal{S}, \mu)$ and $\{A_i: i \in I\}$, respectively.)

I would like to thank Z. Lipecki for a number of valuable suggestions. In particular, Remarks 1 and 2 are essentially due to him.

REFERENCES.

[1] Andenaes, P. R.: Hahn-Banach extensions which are maximal on a given cone. Math. Ann. 188, 90-96 (1970).

[2] Ascherl, A. and Lehn, J.: Two principles for extending probability measures. Manuscripta Math. 21, 43-50 (1977).

[3] Bierlein, D.: Über die Fortsetzung von Wahrscheinlichkeitsfeldern. Z. Wahrscheinlichkeitstheorie verw. Geb. 1, 28-46 (1962)

[4] Lembcke, J.: Konservative Abbildungen und Urbilder regulärer Maße. Z. Wahrscheinlichkeitstheorie verw. Geb. 15, 57-96 (1970).

[5] Lembcke, J.: Reguläre Maße mit einer gegebenen Familie von Bildmaßen. Sitz.-Ber. Math.-Naturw. Kl. Bayer. Akad. Wiss. 1976, 61-115 (1977).

[6] Lipecki, Z.: A generalization of an extension theorem of Bierlein to group-valued measures. Preprint, 1979.

[7] Łoś, J. and Marczewski, E.: Extensions of measure. Fund. Math. 36, 267-276 (1949).

ONE-TO-ONE FUNCTIONS AND A PROBLEM ON SUBFIELDS

Dorothy Maharam and A.H. Stone
University of Rochester, Rochester NY 14627/USA

Let $(X, \mathcal{B})$ be a measurable space such that $\mathcal{B}$ has a countable separating system. In what follows, all subsets of X referred to are understood to be measurable, and "function" means "real-valued measurable function on X ". The real line is denoted by R , and the set of rational numbers by Q .

Theorem 1 Suppose functions f, g, h, c, d, ε, and a G_δ set $G^* \subset R$ are given, such that

$$f = g+h \ , \quad c \text{ and } d \text{ are never } 0 \ , \quad \varepsilon(x) > 0 \text{ for all } x \in X, \text{ and } G^* \supset Q \ .$$

Then (A) if c , d are everywhere positive, there exist one-to-one functions u, w, having disjoint ranges $u(X)$, $w(X)$, contained in some (Lebesgue) null σ-compact subset of $G^* \setminus Q$, such that, for all $x \in X$,

$$f(x) = c(x)u(x) + d(x)w(x) \quad \text{and}$$
$$|g(x) - c(x)u(x)| = |h(x) - d(x)w(x)| < \varepsilon(x) \ .$$

If, instead of (A), we assume

(B) there is no $x \in X$ for which $f(x) = 0 = c(x) \pm d(x)$, then we may further require the ranges of $|u|$ and $|w|$ to be disjoint.

The motivation for this theorem comes from the following situation. Suppose X is a Polish space with a completed regular probability measure m , and let $\mathcal{C}, \mathcal{D}$ be σ-subfields of the field $\mathcal{B}(X)$ of Borel sets of X . Suppose $\mathcal{D}$ is "properly contained" in $\mathcal{C}$, in the following sense: $\mathcal{D} \subset \mathcal{C}$ and, for each $D_0 \in \mathcal{D}$ of positive measure, there exists $C_0 \in \mathcal{C}$ such that $C_0 \subset D_0$ and, for all $D \in \mathcal{D}$, $m(C_0 \Delta D) \neq 0$. Define the "spectral field" of a real-valued function ϕ to consist of the sets $\phi^{-1}(B)$, $B \in \mathcal{B}(R)$. Then we have (subject to the checking of some details):

Theorem 2 In the situation above, given a $\mathcal{D}$-measurable function ψ, there exists a $\mathcal{C}$-measurable function ϕ such that (a) its spectral field is $\mathcal{C}$, (b) the conditional expectation $E(\phi | \mathcal{D}) = \psi$.

To prove Theorem 2, we twice apply the decomposition theorem of [1, Th. 4, p. 153]. First, decomposing $\mathcal{B}(X)$ with respect to $\mathcal{C}$ and replacing X by a suitable cross-section, we arrange without loss of generality that $\mathcal{C} = \mathcal{B}(X)$. The requirement on the spectral field of ϕ now becomes that ϕ is to be one-to-one. Next decompose $\mathcal{C}$ (= $\mathcal{B}(X)$) with respect to its sub-field $\mathcal{D}$; in effect X is replaced by a subset of the plane, consisting of a countable disjoint union of: atomic singletons, horizontal and vertical line segments (with measures equivalent to linear Lebesgue measure) , and the ordinate set of some positive function on a linear interval L (with plane Lebesgue measure). The field $\mathcal{B}$ consists of all vertical Borel cylinder sets, intersected with X . We partition X into a countable number of suitable $\mathcal{B}$ -measurable pieces, and deal with them separately; we use a trick, discussed later in connection with Theorem 1 , to combine the resulting partial functions ϕ into a single function that is still one-to-one. The atomic points and vertical intervals present no difficulty. On $(L \times R) \cap X$ we use the approximation aspects of Theorem 1. The other pieces are of the form $M_n \times A_n$, where M_n is a linear Borel set and A_n consists of exactly n points $(2 \leq n \leq \aleph_0)$; the possibility $n = 1$ is excluded by the assumption that $\mathcal{B}$ was properly contained in $\mathcal{E}$. The n - point case with $n > 2$ is easily reduced to the case $n = 2$, so it is enough to deal with $M_2 \times A_2$, or - with a change of notation - with $X \times \{0, 1\}$, where X is now a subset of R . Put $g(x) = \psi(x, 0)$, $h(x) = \psi(x, 1)$. The measure m induces measures m_i on $X \times \{i\}$ $(i = 0, 1)$; let their Radon-Nikodym derivatives with respect to Lebesgue linear measure be c , d respectively. (Thus $c(x) + d(x) = 1$ a.e.) Now apply Theorem 1, and define ϕ by: $\phi(x, 0) = u(x)$, $\phi(x, 1) = w(x)$. The requirement $E(\phi | \mathcal{B}) = \psi$ follows from the fact that $c(x)\,u(x) + d(x)w(x) = g(x) + h(x)$, and the one-to-one property of ϕ follows from the fact that u , w are one-to-one and have disjoint ranges.

As this outline shows, Theorem 1 states rather more than is actually needed for Theorem 2; however, the features involving G^* and Q are used in the proof of Theorem 2 (in ensuring global one-to-oneness), as well as in the proof of Theorem 1 itself. In outline this goes as follows.

First suppose (as a very special case) that f is non-negative, c and d are everywhere positive, f, c, d, 1/c, 1/d, $1/\varepsilon$ are all bounded, and $h = 0$ (identically). Using the countable separating system, we encode the points x of X , in a measurable way, as infinite "binary decimals". Taking a large enough positive integer b , one can define suitable functions u and w , with values expressed as "decimals" in scale b , in such a way that (for all $x \in X$) $f(x) = c(x)u(x) + d(x)w(x)$, $u(x) > 0 > w(x)$, and both u(x) and w(x) have rapidly increasing stretches of zeros, the nth stretch terminating with the nth binary digit of x . This ensures that u and w are one-to-one and have disjoint ranges. It also ensures that $u(x)$, $w(x)$ are irrational and contained in null Cantor sets. By controlling the stretches of zeros suitably, we get $u(x)$ and $w(x)$ in G^*.

To drop the requirement $f \geq 0$, we write $X_1 = \{x: f(x) \geq 0\}$, $X_2 = X \setminus X_1$. The preceding case, applied to $(-f)|X_2$, gives functions u_2, w_2 on X_2; we put $u'' = -u_2$, $w'' = -w_2$. Similarly we can find suitable functions u', w' on X_1 for $f|X_1$. We intend to combine u', u'' into u, and w', w'' into w. To ensure that u and w are one-to-one and have disjoint ranges, we replace G^*, in finding u' and w', by a smaller G_δ set (containing the rationals) disjoint from the ranges of u'' and w''.

The same trick permits us to find suitable functions u, w on the union of $\aleph_0$ sets, provided we can do so on each of those sets. Thus we can drop the requirement that the functions f, c, d, $1/c$, $1/d$, $1/\varepsilon$ are bounded.

To remove the restriction $h = 0$, first observe that (since c, d are still assumed to be everywhere positive) we may without loss of generality assume $c + d = 1$ (by replacing f by $f/(c+d)$, ε by $\varepsilon/(c+d)$, etc.). Next, partition X into countably many pieces, on each of which there are integers m, n (with $n > 0$) such that $\frac{m}{n} < \frac{h(x)}{d(x)} < \frac{m+1}{n}$ and $\varepsilon(x) < \frac{2}{n}$. On this piece apply the theorem obtained thus far to $f - \frac{m}{n}$ instead of to f. To the resulting u, w we add back m/n. As before, we take the precaution of successively replacing G^* by smaller G_δ sets to ensure that the resulting combined functions u, w are one-to-one and have disjoint ranges. This proves the assertion of Theorem 1 in Case A.

To derive Case B, choose a function η such that, for all $x \in X$,

$$0 < \eta(x) < \varepsilon(x)/2 \quad \text{and} \quad \left|\frac{g(x) + \eta(x)}{c(x)}\right| \neq \left|\frac{h(x) - \eta(x)}{c(x)}\right|.$$

The assumption in B ensures that this is possible (if η is chosen small enough). Now decompose X into countably many pieces, on the nth of which either

$$\left|\frac{g(x) + \eta(x)}{c(x)}\right| > \rho_n > \left|\frac{h(x) - \quad (x)}{c(x)}\right|$$

for some constant ρ_n, or a similar statement holds with the inequalities reversed. On the nth piece apply Case A, with c, d replaced by $|c|$, $|d|$; g, h replaced by $g+\eta$, $h-\eta$; and ε replaced by a small enough ε_n. The resulting functions u_n, w_n -- obtained, as usual, by using a suitable decreasing sequence of G_δ sets G_n^* -- combine to the u, w desired.

<u>Corollary 1</u> Every measurable function f on X can be expressed as the sum of two one-to-one measurable functions.

<u>Corollary 2</u> Every measurable function can be approximated arbitrarily well (to within $\varepsilon(x)$, ε being a preassigned positive measurable function) by a one-to-one measurable function.

There are several ways in which we hope to sharpen Theorem 2.

(i) The theorem should extend from Polish spaces to general σ-finite measure spaces (perhaps at the cost of discarding a null set). This should be a matter of standard technique.

(ii) One should be able to require $|\phi(x) - \psi(x)| < \varepsilon(x)$, given a positive (measurable) function ε.

(iii) The eventual aim is to deal, if possible, with an infinite sequence of σ-fields, each properly containing the next, given a function ψ that is measurable with respect to the intersection of the σ-fields.

<u>Reference</u>

[1] Dorothy Maharam, <u>Decomposition of measure algebras and spaces</u>, Trans. Amer. Math. Soc. 69 (1950) 142-160.

ROTATIONS OF THE CIRCLE

R. Daniel Mauldin
Department of Mathematics
North Texas State University
Denton, Texas 76203/USA

and

Paul Erdős
Hungarian Academy of Sciences
Budapest, Hungary

The ideas presented here having arisen from the consideration of the following problem of Erdős.

Let T be the unit circle and suppose S_1 and S_2 are subsets of T such that for each i, $i = 1,2$, there is an infinite subset R_i of T so that the sets rS_i, where $r \in R_i$ are pairwise disjoint. Is it true that the inner measure of $S_1 \cup S_2$ is zero?

The answer to this question is yes and we shall present two solutions to this problem. Neither solution is difficult. But, each seems to lead to some interesting problems which will be formulated here.

Our first solution which is contained in the following theorem is based on the fact that the circle group is amenable.

THEOREM 1. Let G be a locally compact T_2 group with left invariant Haar measure λ and such that G admits an invariant mean. Let $S_1,\ldots,S_k$ be subsets of G such that for each i, $1 \le i \le k$, there is an infinite subset R_i of G so that the sets rS_i, where $r \in R_i$, are pairwise disjoint and $\overline{R}_i$ is compact. Then the inner Haar measure of $\bigcup_{i=1}^{k} S_i$ is zero.

PROOF. Since G admits an invariant mean, there is a nonnegative, finitely additive extension μ, of λ to all subsets of G which is also left invariant.

It is enough to prove the theorem under the assumption that the sets S_i are pairwise disjoint.

Assume $\bigcup S_i$ has positive inner measure. Let F be a compact set $F \subset \bigcup S_i$, with $\lambda(F) > 0$. Let $D_i = F \cap S_i$. Now, the sets rD_i, where $r \in R_i$ are pairwise disjoint and $\bigcup\{rD_i : r \in R_i\}$ is a subset of the compact set $\overline{R}F$. Therefore, for each i, $\mu(D_i) = 0$.

This means $\mu(\bigcup D_i) = 0 = \mu(F) = \lambda(F) > 0$. This contradiction establishes the theorem.

Next we give an example to show that the conclusion of this theorem may be false if the group is not amenable.

Example. Let G be the orthogonal group on E^3. Notice that G acts transitively on S, the unit sphere of E^3. Let N be the north pole of S and let H be the stability subgroup of G at N. $H = \{g \in G : g(N) = N\}$. Then H is a closed subgroup of the compact group G. Let θ be the one-to-one map of the left coset space G/H onto $S: \theta(gH) = g(N)$.

Now, according to Hausdorff

$$S = A \cup B \cup C \cup D,$$

where A,B,C, and D are disjoint, D is countable and $(*) A \cong B \cup C$, $A \cong B$, $A \cong C$.

Let $E = \bigcup \{\theta^{-1}(d) : d \in D\}$. Since $\lambda(H) = 0$ and D is countable, $\lambda(E) = 0$. Let $S_1 = \bigcup\{\theta^{-1}(a) : a \in A\} S_2 = \bigcup\{\theta^{-1}(t) : t \in B \cup C\}$. Then $S_1 \cup S_2$ is a G_δ set with $\lambda(S_1 \cup S_2) > 0$. Also, because of (*), there are infinitely many pairwise disjoint translates of S_1 and S_2.

Thus, the first method of proof leads to the following problem.

PROBLEM. Let G be a locally compact T_2 group so that the conclusion of Theorem 1 holds. Is there a finitely additive left invariant extension of Haar measure to all subsets of G?

Before giving the second method of proof, let us state the following lemma.

LEMMA. Let G be a locally compact group with left invariant Haar measure λ. Let F be a compact set such that $0 < \lambda(F) < \infty$. For each positive integer n, there is a neighborhood V of e so that if $h_1, \ldots, h_n$ are points of V, then $\lambda(\bigcap\{h_i F : i \le n\}) > 0$.

The second method of proof is formulated for abelian groups.

THEOREM 2. Let G be a locally compact abelian group with Haar measure λ and let k be a positive integer. For each i, $1 \le i \le k$, let S_i and R_i be subsets of G such that R_i is infinite, $\overline{R}_i$ is compact, and the sets $S_i + g, g \in R_i$ are pairwise disjoint. The $\bigcup_{i=1}^{k} S_i$ has inner measure zero.

PROOF. Again, notice that we can and do assume that the sets S_i are disjoint. Let us assume that F is a compact set lying in $\bigcup S_i$ and $\lambda(F) > 0$.

Let $n > k$. According to the lemma, there is some neighborhood V of e so that if $h_p \in V, p = 1,\dots,n^k$, then $\lambda(M) > 0$, where

$$M = F \cap (\cap\{F - h_p : p \le n^k\}))$$

For each i, $1 \le i \le k$, obtain n+1 distinct points $g_{i0},\dots,g_{in}$ of R_i so that for each k-tuple, $p = (p_1,\dots,p_k)$ of the first n positive integers,

$$h_p = d_{1p_1} + \dots + d_{kp_k}$$

is in V, where $d_{it} = g_{i0} - g_{it}$.

According to the lemma, there is some x in $F \cap (\cap\{F - h_p : p = (p_1,\dots,p_k) \in \{1,\dots,n\}^k)$.

For each i, let $M_i = \{p : x + h_p \in S_i\}$. The sets M_i are pairwise disjoint and each k-tuple of the first n-integers is in some M_i.

A contradiction will be reached by examining the cardinalities of the sets M_i. Notice that M_i has the following property. If $(p_1,\dots,p_i,\dots,p_k) \in M_i$ and $(r_1,\dots,r_i,\dots,r_k)$ is such that $r_j = p_j$, if $j \ne i$ and $r_i \ne p_i$, then $(r_1,\dots,r_k) \notin M_i$. The reason for this is that if they were both in M_i, then $(x + d_{1p} + \dots + d_{ip_i} + d_{kp_k}) + g_{ip_i} = x + d_{1p_1} + \dots + g_{i0} + \dots + d_{kp_k} = (x + d_{1r_1} + \dots + d_{ir_i} + \dots + d_{kr_p}) + g_{ir_k}$. Thus, the sets $S_i + g_{ip_i}$ and $S_i + g_{ir_i}$ would not be disjoint.

Now, because of this property of the sets M_i, we know that, $\mathrm{card}(M_i) \le n^{k-1}$. Therefore,

$$n^k = \mathrm{card}(\cup M_i) \le kn^{k-1}.$$

This contradiction establishes the theorem. Q.E.D.

Both of these proofs raise the question of estimating the size of a subset S of T which can be partitioned into k sets each of which has n pairwise disjoint rotations.

As Mycielski pointed out to us, one can use the extension of Haar measure to all subsets of T and argue along the lines of Theorem 1 to obtain the following estimate.

Theorem 3. If $S \subset T$ and $S \subset \cup\{A_i : i \le k\}$ where each set A_i has n pairwise disjoint rotations then the inner measure of S is $\le k/n$.

The proof of Theorem 2 was based on some simple combinational properties of finite sets. The problem we pose is that of estimating the size of a measurable subset M of T from the fact that M possesses two sets of rotations which avoid the contradiction obtained in Theorem 2. We formulate this as follows.

STATEMENT 1. Let $0 < \alpha$. There is a positive integer $n_0(\alpha)$ so that if M is a measurable subset of T with $\lambda(M) > \alpha$ and R_1 and R_2 are subsets of T with $|R_1|, |R_2| > n_0(\alpha)$, then there are points g_{1i} of R_1, g_{2i} of R_2, $i = 0,1,\ldots,4$ such that

$$M \cap (\cap\{M - h_p : p = (p_1,\ldots,p_3) \in \{1,\ldots,3\}^2\}) \neq \phi,$$

where

$$h_p = (g_{10} - g_{1p_1}) + (g_{20} - g_{2p_2}).$$

In fact, statement 1 leads to the consideration of the following statement.

STATEMENT 2. Suppose $\alpha > 0$. There is a positive integer $t_0(\alpha)$ and a $\beta > 0$ so that if $M \subset T$, $\lambda(M) > \alpha$ and $R \subset T$, $|R| > t_0(\alpha)$, then there are points g_0, g_1, g_2, g_3 of R so that

$$\lambda\left(\bigcap_{i=0}^{3} M + (g_0 - g_i)\right) > \beta.$$

Clearly, statement 2 implies statement 1.

We have been unable to determine whether statement 2 is true. However, we have been led to the following statement.

STATEMENT 3. For each $c > 0$, there is an integer $\ell_0(c)$ and an integer $N_0(c)$ so that if $\ell > \ell_0(c)$ and $N > N_0(c)$ and

$$1 \le a_1 < a_2 < \ldots < a_t \le N \text{ where } t > cN,$$

then for each ℓ integers

$$b_1 < \ldots < b_\ell < N,$$

there is an arithmetic progession of three terms among the a's of difference some $b_j - b_i$.

At this time, we do not know which if any, of the preceding three statements is true.

THIN TREES AND GEOMETRICAL CRITERIA FOR LEBESGUE NULLSETS*)

Flemming Topsøe
University of Copenhagen

Summary and Introduction. Four years ago I had the opportunity to lecture at the first Oberwolfach conference entirely devoted to measure theory. In the conference proceedings I conjectured that a certain condition was necessary and sufficient for the conclusion of the classical Vitali theorem for Lebesgue measure to be true ([5], p. 197). Recall that Vitali type theorems for Lebesgue measure are results of the type that a given set $A \subseteq \mathbb{R}^N$ can be packed with sets from a given family $\mathcal{B}$ of subsets in $\mathbb{R}^N$ (usually balls), in the sense that there exists a packing (i.e. a collection of pairwise disjoint sets) $\mathcal{B}_0 \subseteq \mathcal{B}$ such that $|A \setminus \cup\{B \mid B \in \mathcal{B}_0\}|^* = 0$ where $|\cdot|$ denotes Lebesgue measure.

In a joint paper with Mejlbro it was pointed out that the classical Vitali theorem cannot be improved (I should have known!) but, at the same time, it was shown that the condition of the conjecture gave the precise answer to another problem! This problem is connected with a new type of "uniform" Vitali systems (the screened systems in the terminology of [3]) where the condition you impose on the points in the subset of $\mathbb{R}^N$ to be packed is the same for all points.

The two results, the classical and the uniform Vitali theorems (Theorems 1 and 2 of [3]), are not comparable and it lies nearby to search for a unified Vitali theorem. Such a result is not known. The pointwise version of the uniform Vitali theorem would have worked but this possibility is ruled out by a counterexample due to Michel Talagrand. Talagrand's example, as well as certain weakenings of the pointwise version, are reported in a joint work with Jørsboe and Mejlbro.

The present paper has two purposes. Firstly, a new technique involving trees is introduced (Lemmas 2, 3 and 5). This technique leads to substantial simplifications of proofs given in [2] and in [3] and also, the technique can be applied to obtain the classical result, so that it has a certain unifying effect. Secondly, new results are obtained, both results in the usual geometrical setting and also results about trees which we hope are of independant interest.

---

*) Research supported by the Danish Natural Science Research Council.

A main theme of the paper is geometrical criteria for Lebesgue nullsets. Some special cases are singled out in the next section as illustrative examples.

Applications to differentiation and density theorems are pointed out in [2] and [3]. The reader may also wish to consult de Guzman's monograph [1].

Further possibilities, not yet developed, may be contributions to the theory of singular integrals (perhaps our techniques are even relevant to such a classical subject as the study of pointwise convergence of Fourier series). Also, some of the results, especially those for trees, may be easy to extend to situations where other measures than Lebesgue measure are involved. However, there seems to be no possibilities for extension of the geometrical results to situations dealing with measures (even "nice" ones such as Gaussian measures) on infinite-dimensional spaces, cf. the recent results of David Preiss, [4].

Geometrical criteria for Lebesgue nullsets - some illustrative examples.

Let Z be a subset of N-dimensional euclidean space $\mathbb{R}^N$. Below we give various criteria, ensuring that $|Z| = 0$ where $|\cdot|$ denotes Lebesgue measure. Let us start with the following:

A. If Z permits Θ-conesections, then $|Z| = 0$.

The meaning with this is, that if, for every $x \in Z$, there exists a Θ-conesection with toppoint at x which is disjoint with Z (to be precise, we only require the interior to be disjoint with Z), then $|Z| = 0$. The situation is illustrated in the figure, where the signature //// is used to indicate "disjoint with Z".

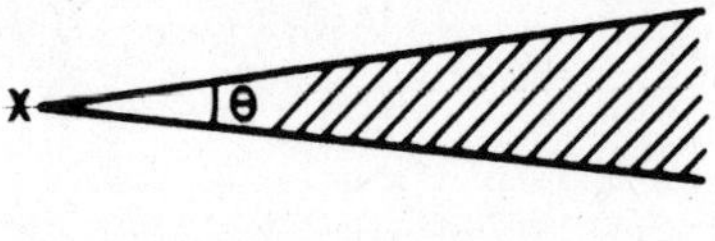

Here are some variants of the criterion A:

B. If Z permits conesections, then $|Z| = 0$.

This criterion differs from A only because the angle Θ is allowed to depend on x. Of course, as one may restrict attention to a sequence of Θ's converging to 0, B is a simple consequence of A.

A more interesting variant is the following:

C. <u>If</u> Z <u>permits convergent sequences of balls contained in a</u> Θ-<u>conesection, then</u> $|Z| = 0$.

The precise meaning with this should be clear from the figure. Of course, we may again permit Θ to depend on the point $x \in Z$ under consideration.

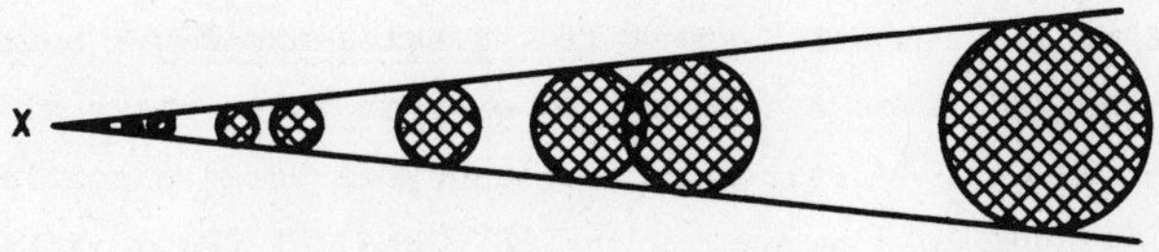

The restriction to conesections is not essential, for instance, we have:

D. <u>If</u> Z <u>permits cowhorns, then</u> $|Z| = 0$,

where a cowhorn is a figure something like this:

If we find that amusing, we may also note that if Z permits convergent sequences of balls contained in a cowhorn, then $|Z| = 0$.

All of the above results are immediate consequences of Lebesgue's density theorem or of the classical Vitali theorem (Theorem 1 of [3]).

Then, we shall discuss what happens if one replaces the conesections considered so far by other figures.

Let $\psi:]0,\infty[\to]0,\infty[$ be a given non-decreasing function. Then the figure obtained by rotating the function $t \to \psi(t)$ for $t \leq$ some t_0 around a ray is called a ψ-<u>trumpet</u>, cf. the figure (when ψ is chosen convex the illusion with trumpets is more convincing).

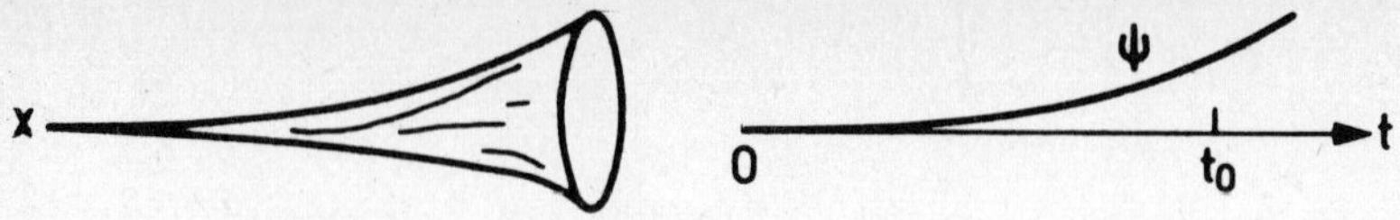

For $N \in \mathbb{N}$, we denote by Ψ_N the class of ψ-functions for which

$$\int_0^1 \left(\frac{\psi(t)}{t}\right)^N \frac{dt}{t} = \infty$$

or, equivalently, for which

$$\sum_{n=0}^{\infty} \left(\frac{\psi(2^{-n})}{2^{-n}}\right)^N = \infty$$

(cf. [3]).

E. <u>If</u> $\psi \in \Psi_N$ <u>and if</u> $Z \subseteq \mathbb{R}^N$ <u>permits</u> ψ-<u>trumpets</u>, <u>then</u> $|Z| = 0$.

This gives a generalization of criterion A, which corresponds to the case $\psi(t) = c \cdot t$ for some constant c . However, criterion C cannot be generalized; indeed, if ψ is such that every $Z \subseteq \mathbb{R}^N$ which permits convergent sequences of balls contained in a ψ-trumpet, is a Lebesgue nullset, then necessarily there are constants $c > 0$, $t_0 > 0$ such that $\psi(t) \geq c \cdot t$ for $t \subseteq t_0$ (cf. Theorem 1 of [3]). Nevertheless, there is an interesting strengthening of E:

F. <u>If</u> $\psi \in \Psi_N$ <u>and if</u> $Z \subseteq \mathbb{R}^N$ <u>permits convergent sequences of balls containing a</u> ψ-<u>trumpet in the shadowzone</u>, <u>then</u> $|Z| = 0$.

The precise meaning with this statement is explained in the figure.
Criterion F follows from Theorem 2 of [3] which also shows that the condition $\psi \in \Psi_N$ is necessary for the conclusion.

The pointwise version of F, where $\psi \in \Psi_N$ is allowed to depend on the point $x \in Z$ is not true (cf. Talagrand's example to be published in [2]). The reason why such a result would have been very satisfactory

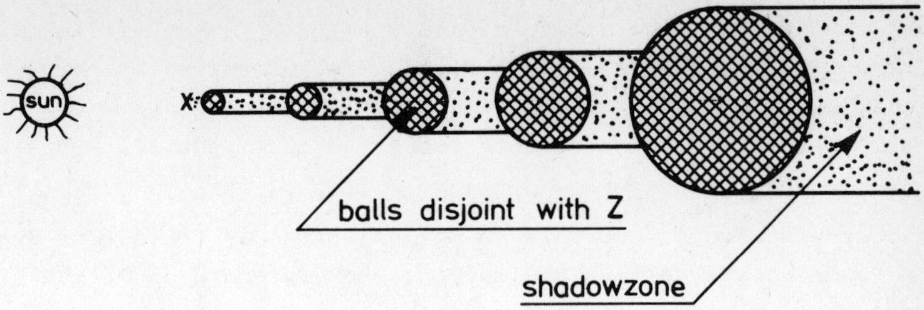

is that it would also contain criterion C since, as follows from Lemma 1 of [2], every convergent sequence of balls contained in a conesection contains a ψ-trumpet in the shadowzone for some $\psi \in \psi_N$.

Thin trees.

It turns out that the results we are aiming at are equally difficult to establish in all dimensions. For notational convenience, we shall therefore only develop the technique corresponding to the 1-dimensional case.

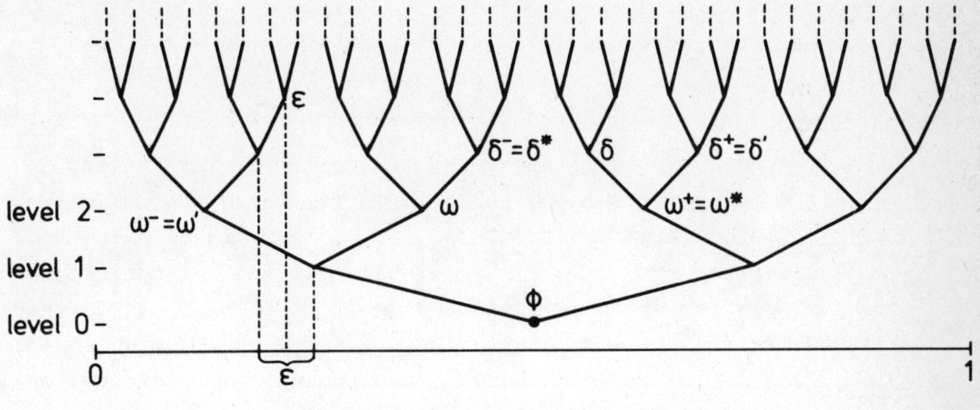

We shall work with the unit interval and the dyadic rational subintervals. The usual correspondence between the dyadic rational subintervals of [0,1] and the dyadic tree

$$2^{(\mathbb{N})} = \bigcup_{n=0}^{\infty} 2^n$$

is summarized in the figure. The elements of $2^{(\mathbb{N})}$ are finite multiindices of 0's and 1's. We shall not distinguish notationally between a multiindex and the corresponding dyadic rational interval. Thus, with reference to the figure, ε either means the multiindex $\varepsilon = 0011$ or the interval $\varepsilon = \left]\frac{3}{16}, \frac{4}{16}\right[$ (usually, we take the dyadic rational intervals as open intervals, but this is not essential as the set of dyadic rational points is a nullset and may thus be disregarded for the kind of investigations we are to carry out).

For $\delta \in 2^{(\mathbb{N})}$, we denote by δ^- the left and by δ^+ the right neighbour of δ. By δ' we denote the <u>dyadic</u> <u>neighbour</u> of δ, i.e. that interval which together with δ form a dyadic rational interval, and by δ^* we denote the other among the neighbours δ^- and δ^+, cf. the figure. These objects are not always defined, eg. if $\delta = \emptyset$, then neither δ', δ^-, δ^+ nor δ^* are defined. We denote the level of δ by $n(\delta)$, eg. $n(\emptyset) = 0$ and $n(0011) = 4$.

Let $\delta \in 2^{(\mathbb{N})}$ and $k \in \mathbb{N}$. Then $(2k+1)\delta$ denotes the (at most) $2k + 1$ multiindices obtained by taking δ together with the k (or less) neighbours at the same level lying to the left of δ (as many as the tree permits) and the k (or less) neighbours lying to the right of δ. Thus $3\delta = \{\delta^-, \delta, \delta^+\}$. We may also think of $(2k+1)\delta$ as the interval with length $|(2k+1)\delta| = (2k+1)|\delta|$ lying symmetrically around δ.

Let Z be a closed subset of $[0,1]$ such that the complement

$$\Gamma = [0,1] \setminus Z$$

is a union of dyadic rational intervals. We are mainly interested in criteria for Z to be a Lebesgue nullset. The criteria will be expressed in terms of a certain subtree T of $2^{(\mathbb{N})}$ associated with Z. First we consider the set $T_\gamma \subseteq 2^{(\mathbb{N})}$ of all multiindices corresponding to maximal dyadic intervals contained in Γ. Then T is defined as the set of $\varepsilon \in 2^{(\mathbb{N})}$ such that either $\varepsilon \in T_\gamma$ or else, no restriction of ε lies in T_γ. We also have, neglecting the dyadic rational points,

$$T = \{\emptyset\} \cup \{\varepsilon \mid (\varepsilon \cup \varepsilon') \cap Z \neq \emptyset\}$$

and, if $\mathrm{int}(Z) = \emptyset$, T is the only tree whose endpoints are T_γ.

The figure clarifies the definition of T. It also shows a decomposition

$$T = T_\alpha \cup T_\beta \cup T_\gamma$$

of T into α , β and γ-points. The γ-points are just the endpoints of T already considered. To define the α and β-points, we first associate with each $\varepsilon \in T$ the nearest endpoint in T extending ε with the convention that if there are several such points we take the one lying furthest to the right; the point thus defined is denoted $\gamma(\varepsilon)$. Thus, $\gamma(\emptyset) = 10$ and $\gamma(11) = 1110$ in the figure.

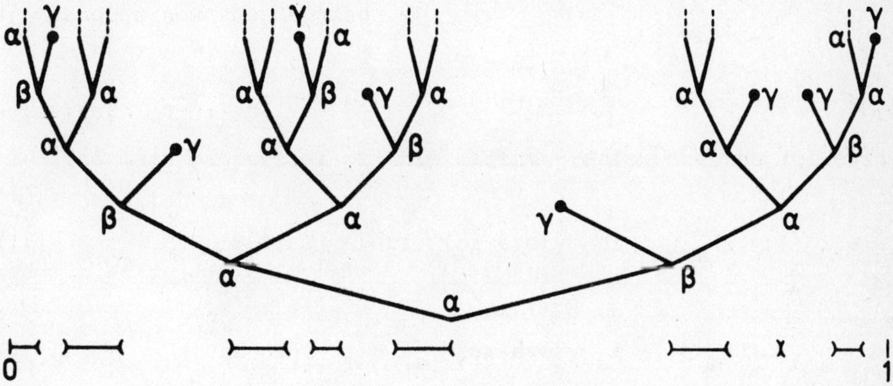

For $\varepsilon \in T$ we denote by $\Delta(\varepsilon)$ the number of levels $\gamma(\varepsilon)$ lies above ε . Thus $\Delta(\varepsilon) = 0$ if $\varepsilon \in T_\gamma$. By definition we put $\Delta(\varepsilon) = 0$ for any $\varepsilon \in 2^{(\mathbb{N})} \setminus T$. For $\varepsilon \in T$ we have

$$2^{-\Delta(\varepsilon)} = \frac{|\gamma(\varepsilon)|}{|\varepsilon|} .$$

Let $\varepsilon \in T \setminus T_\gamma$. If $\varepsilon = \emptyset$ then ε is an α-point by definition. Assume that $\varepsilon \neq \emptyset$. If $\Delta(\varepsilon) > \Delta(\varepsilon')$, then ε is an α-point, if $\Delta(\varepsilon) < \Delta(\varepsilon')$ then ε is a β-point and if $\Delta(\varepsilon) = \Delta(\varepsilon')$ then ε is an α-point if $\varepsilon' = \varepsilon^+$ and ε is a β-point if $\varepsilon' = \varepsilon^-$. With these conventions we have that if ε is an α-point, then all points between ε and $\gamma(\varepsilon)$ are β-points.

The infinite branches in T correspond to points in the set Z . We may then say that T is <u>thin</u> if there are so few infinite branches that $|Z| = 0$. Viewing T as the "lifetree" for a stopping time we see that another way of expressing the condition $|Z| = 0$ is that this stopping time is almost everywhere finite-valued. If, for some set $A \subseteq [0,1]$ we have $|Z \cap A| = 0$, then we say that T is <u>thin</u> at A .

By t we often denote an infinite multiindex, i.e. an element in $2^{\mathbb{N}}$. We may also view t as a point in $[0,1]$ in the usual way. For $t \in 2^{\mathbb{N}}$ and $n \in \mathbb{N} \cup \{0\}$, we denote by $t|n$ the restriction of t to a multiindex in 2^n .

By α_n we denote the set of α-points in the n'th level; α_n will also be identified with the corresponding subset of $[0,1]$, thus $|\alpha_n| = \#\alpha_n \cdot 2^{-n}$ ($\#$ = "number of elements in ..."). Similar notations are used for β_n and γ_n.

The following properties are easy to check:

(1) $\Gamma = \cup\{\gamma(\varepsilon) \mid \varepsilon \in T_\alpha\}$, a disjoint union,

(2) $Z = \{t \mid \text{for all } n, \; t|n \in T\}$,

(3) $\alpha_n \cup \beta_n \downarrow Z$,

(4) $\#\alpha_n = \#\alpha_{n-1} + \#\beta_{n-1}$; $n \geq 1$,

(5) $\#\beta_n + \#\gamma_n = \#\alpha_{n-1} + \#\gamma_{n-1}$; $n \geq 1$,

(6) $|\alpha_n| = \frac{1}{2}(|\alpha_{n-1}|+|\beta_{n-1}|)$; $n \geq 1$,

(7) $|\alpha_n| \downarrow \frac{1}{2}|Z|$.

<u>Lemma 1</u>. <u>A necessary and sufficient condition that the tree</u> T <u>be thin is that</u> $\sum_0^\infty \kappa_n = \infty$, <u>where</u> κ_n <u>is defined by</u>

(8) $$\kappa_n = \sum_{n=0}^{\infty} \frac{1}{\#\alpha_n} \sum_{\varepsilon\in\alpha_n} 2^{-\Delta(\varepsilon)} .$$

<u>Proof</u>. Put

$$\Gamma_n = \cup\{\gamma(\varepsilon) \mid \varepsilon \in \alpha_0 \cup \alpha_1 \cup \dots \cup \alpha_n\} \quad , \; Z_n = [0,1]\setminus\Gamma_n .$$

Then $\Gamma_n \uparrow \Gamma$ and $Z_n \downarrow Z$. Observe that $\alpha_n \subseteq Z_{n-1} \subseteq \alpha_n \cup \beta_n$, hence

$$|\alpha_n| \leq |Z_{n-1}| \leq |\alpha_n \cup \beta_n| \leq 2|\alpha_n| .$$

Note also, that $Z_{n-1}\setminus Z_n$ is the union of all $\gamma(\varepsilon)$ with $\varepsilon \in \alpha_n$. We then have

$$\begin{aligned} |Z_n| &= |Z_{n-1}| - |Z_{n-1}\setminus Z_n| \\ &= |Z_{n-1}| - |\alpha_n| \cdot \kappa_n \end{aligned}$$

$$\begin{cases} \leq |Z_{n-1}|(1-\frac{1}{2}\kappa_n) \\ \geq |Z_{n-1}|(1-\kappa_n) , \end{cases}$$

hence

$$\prod_{n=0}^{\infty} (1-\kappa_n) \leq |Z| \leq \prod_{n=0}^{\infty} (1-\tfrac{1}{2}\kappa_n)$$

so that $|Z| = 0$ if and only if $\sum_0^\infty \kappa_n = \infty$. ▯

Remark. The sufficiency of (8), which is the most useful part of the result, is in fact trivial; indeed, if $|Z| > 0$ then

$$\sum_0^\infty \kappa_n = \sum_0^\infty \frac{1}{|\alpha_n|} \sum_{\varepsilon\in\alpha_n} |\gamma(\varepsilon)| \leq \frac{2}{|Z|} \sum_{\varepsilon\in T_\alpha} |\gamma(\varepsilon)| \leq \frac{2}{|Z|} < \infty .$$

Corollary 1. If there exist numbers $(\varphi_n)_{n\geqslant 0}$ such that $\sum_0^\infty \varphi_n = \infty$ and such that $2^{-\Delta(\varepsilon)} \geq \varphi_{n(\varepsilon)}$ for every $\varepsilon \in 2^{(\mathbb{N})}$, then T is thin.

As we shall see in the next section, this criterion may be used to give a simple proof of the main result of [3]. To obtain deeper results - as those of [2] - we need more refined criteria. Even though (8) contains all the information, we shall introduce a new, very simple method.

Lemma 2. Let there be given a map: $f: T \setminus T_\gamma \to [0,\infty]$ and assume that the following two conditions hold:

(9) $\sum_0^\infty f(t|n) = \infty \ \forall t \in Z$,

(10) $\sum_{\varepsilon\in T\setminus T_\gamma} f(\varepsilon)\cdot|\varepsilon| < \infty$.

Then T is a thin tree.

If (9) is only assumed to hold for $t \in A \cap Z$, where A is some subset of $[0,1]$, then it can only be concluded that T is thin at A .

Proof. With a $*$ designating outer measure or integral, we have, as $Z \subseteq \alpha_n \cup \beta_n$ for all n ,

$$\infty \cdot |A\cap Z|^* = \int^*_{A\cap Z} \left(\sum_0^\infty f(t|n)\right)dt$$

$$\leq \sum_0^\infty \int_{\alpha_n\cup\beta_n} f(t|n)dt = \sum_0^\infty \sum_{\varepsilon\in\alpha_n\cup\beta_n} f(\varepsilon)|\varepsilon|$$

$$= \sum_{\varepsilon\in T\setminus T_\gamma} f(\varepsilon)|\varepsilon|$$

and this quantity is finite by (10). We conclude that $|A\cap Z|^* = 0$, in particular, $A \cap Z$ is Lebesgue measurable, and we may write $|A\cap Z|$ in place of $|A\cap Z|^*$. ▯

<u>Remark</u>. As we see from the proof, we have the formally stronger result that when (10) holds then

$$\int_Z \left(\sum_0^\infty f(t|n)\right)dt < \infty .$$

But this is less interesting as the result in this form looses its geometrical flavor.

One may also notice that for the last part, (10) may be weakened only considering ε's in $T\setminus T_\gamma$ which intersect A .

Sometimes the following extension of Lemma 2 is useful:

<u>Lemma</u> 3. <u>Let there be given a sequence</u> $(f_k)_{k\geq 1}$ <u>of mappings</u> f_k: $T\setminus T_\gamma \to [0,\infty]$. <u>Assume that the following two conditions are satisfied:</u>

$$\forall t \in Z \quad \exists k \geq 1: \sum_{n=0}^\infty f_k(t|n) = \infty ,$$

$$\forall k \geq 1: \quad \sum_{\varepsilon\in T\setminus T_\gamma} f_k(\varepsilon)\cdot|\varepsilon| < \infty .$$

<u>Then</u> T <u>is thin</u>.

<u>Proof</u>. For $k \geq 1$ put

$$Z_k = \{t \in Z \mid \sum_0^\infty f_k(t|n) = \infty\} .$$

Then as in the proof of Lemma 2, it is seen that $|Z_k| = 0$ for all k . As $Z = \bigcup_1^\infty Z_k$, it follows that $|Z| = 0$. ▯

In the next two theorems we single out what we consider to be the most interesting consequences in the dyadic setting of the lemmas just proved. A few definitions are needed. The <u>entropy</u> of the tree T is the number

$$H(T) = \sum_{\varepsilon\in T_\gamma} |\varepsilon| \log \frac{1}{|\varepsilon|} .$$

We also need a variant Δ' of the Δ-function defined by

$$\Delta'(\varepsilon) = \max\{\Delta(\varepsilon),\Delta(\varepsilon')\} .$$

<u>Theorem</u> 1. <u>In each of the cases described below, it can be concluded that</u> T <u>is thin</u>:

<u>Case</u> I <u>(the entropy criterion)</u>. The two conditions below hold:

(I') $\quad \sum_0^\infty 2^{-\Delta(t|n)} = \infty \quad \forall t \in Z$,

(I") $\quad H(T) < \infty$.

<u>Case</u> II. <u>There exist numbers</u> ξ_n ; $n \geq 0$ <u>such that</u>:

(II') $\quad \sum_0^\infty \xi_n \cdot 2^{-\Delta(t|n)} = \infty \quad \forall t \in Z$,

(II") $\quad \sum_0^\infty \xi_n < \infty$.

<u>Case</u> III (<u>the logarithmic criterion</u>)

(III') $\quad \sum_0^\infty \left|\log|\gamma(t|n)|\right|^{-1} \cdot 2^{-\Delta(t|n)} = \infty \quad \forall t \in Z$.

<u>Case</u> IV (<u>the neighbour criterion</u>)

(IV') $\quad \sum_0^\infty 2^{-\Delta'(t|n)} = \infty \quad \forall t \in Z$.

<u>If condition</u> (I'), <u>respectively</u> (II'), (III') <u>or</u> (IV') <u>are only assumed to hold for</u> $t \in Z \cap A$ <u>then we can only conclude that</u> T <u>is thin at</u> A .

<u>Proof</u>. First observe that (10) may be written as

$$\sum_{\varepsilon \in T \setminus T_\gamma} f(\varepsilon) \cdot |\varepsilon| = \sum_{\varepsilon \in T_\alpha} \sum_{\delta \in [\varepsilon, \gamma(\varepsilon)[} f(\delta) \cdot |\delta| ,$$

where $[\varepsilon, \gamma(\varepsilon)[$ is the set of multiindices δ between ε and $\gamma(\varepsilon)$ with ε included and $\gamma(\varepsilon)$ excluded. Also note that for all these δ we have $\gamma(\delta) = \gamma(\varepsilon)$.

<u>Case</u> I: Put $f(\varepsilon) = 2^{-\Delta(\varepsilon)}$. Then (9) follows directly from (I'). And (10) follows from (I") as

$$\sum_{\varepsilon \in T \setminus T_\gamma} f(\varepsilon) \cdot |\varepsilon| = \sum_{\varepsilon \in T_\alpha} \sum_{\delta \in [\varepsilon, \gamma(\varepsilon)[} |\gamma(\delta)|$$

$$= \sum_{\varepsilon \in T_\alpha} \Delta(\varepsilon) \cdot |\gamma(\varepsilon)| \leq \sum_{\varepsilon \in T_\alpha} n(\gamma(\varepsilon)) \cdot |\gamma(\varepsilon)|$$

$$= \sum_{\varepsilon \in T_\gamma} n(\varepsilon) \cdot |\varepsilon| = (\log 2)^{-1} \cdot H(T) .$$

<u>Case</u> II: Put $f(\varepsilon) = \xi_{n(\varepsilon)} \cdot 2^{-\Delta(\varepsilon)}$ and $s = \sum_0^\infty \xi_n$. Condition (9) follows from (II') and (10) follows from (II") as

$$\sum_{\varepsilon \in T \setminus T_\gamma} f(\varepsilon) \cdot |\varepsilon| = \sum_{\varepsilon \in T_\alpha} \sum_{\delta \in [\varepsilon, \gamma(\varepsilon)[} \xi_{n(\delta)} \cdot |\gamma(\delta)|$$

$$\leq s \cdot \sum_{\varepsilon \in T_\alpha} |\gamma(\varepsilon)| \leq s .$$

<u>Case</u> III: Put

$$f(\varepsilon) = \left|\log|\gamma(\varepsilon)|\right|^{-1} \cdot 2^{-\Delta(\varepsilon)} .$$

Then (9) follows from (III'). And (10) is automatically satisfied since

$$\sum_{\varepsilon\in T\setminus T_\gamma} f(\varepsilon)|\varepsilon| = \sum_{\varepsilon\in T_\alpha} \Delta(\varepsilon)\cdot\left|\log|\gamma(\varepsilon)|\right|^{-1}\cdot|\gamma(\varepsilon)|$$

$$\leq \sum_{\varepsilon\in T_\alpha} (\Delta(\varepsilon)+n(\varepsilon))\cdot\left|\log|\gamma(\varepsilon)|\right|^{-1}\cdot|\gamma(\varepsilon)|$$

$$= (\log 2)^{-1} \sum_{\varepsilon\in T_\alpha} |\gamma(\varepsilon)| \leq (\log 2)^{-1} .$$

Case IV: Here we take $f(\varepsilon) = 2^{-\Delta'(\varepsilon)}$. Then (9) follows from IV' and (10) is automatically fulfilled since, noting that $f(\varepsilon) = f(\varepsilon')$, we have

$$\sum_{\varepsilon\in T\setminus T_\gamma} f(\varepsilon)|\varepsilon| \leq \sum_{\varepsilon\in T_\alpha} f(\varepsilon)\cdot|\varepsilon| + \sum_{\varepsilon\in T_\alpha\setminus\{\emptyset\}} f(\varepsilon')\cdot|\varepsilon'|$$

$$\leq 2 \sum_{\varepsilon\in T_\alpha} f(\varepsilon)\cdot|\varepsilon| = 2 \sum_{\varepsilon\in T_\alpha} |\gamma(\varepsilon)| \leq 2 . \ \blacksquare$$

Remarks. Condition (I') - (IV') are of pointwise and local character. We consider (I') as the "natural" condition to impose. Note that all cases are strengthenings of this condition. Therefore, the results are really only interesting when we recall that by Talagrand's example (cf. [2]) it does not follows from (I') alone that T is thin.

Talagrands example also involves some manipulations with logarithms and it is natural to ask if, under the assumption of (I'), a condition like (I") is also necessary for T to be thin. In this connection it may be noted that the above proof shows that for criterion I, condition (I") may be replaced by the condition

$$\sum_{\varepsilon\in T_\alpha} \Delta(\varepsilon)\cdot|\gamma(\varepsilon)| < \infty$$

which is at least formally weaker than (I").

We also remark that both criteria II and IV imply Corollary I.

Finally, in accordance with the first remark to Lemma 2, the criteria of Theorem 1 hold in integrated form, for instance we have

$$(11) \quad \int_Z \left(\sum_0^\infty 2^{-\Delta'(t|n)}\right)dt < \infty$$

without any extra condition (compare with case IV), and if $H(T) < \infty$, we have

(12) $\int_Z (\sum_0^\infty 2^{-\Delta(t|n)})dt < \infty$

(compare with case I). It can be shown by examples much simpler than those needed for Talagrand's example that (12) may fail when $H(T) = \infty$. Indeed, one may consider trees of the form $T = T_{n_1p_1} \otimes T_{n_2p_2} \otimes \ldots$ where the definition of T_{np} and the meaning with the "tensor product" is indicated in the figure; for this to work, the T_{np}'s should be

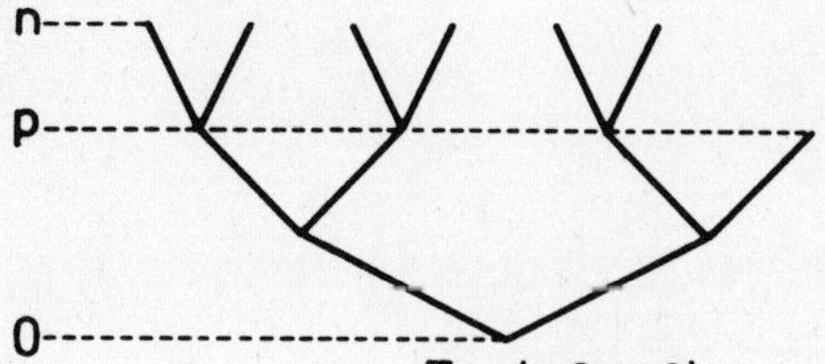

T_{np} (n=3, p=2)

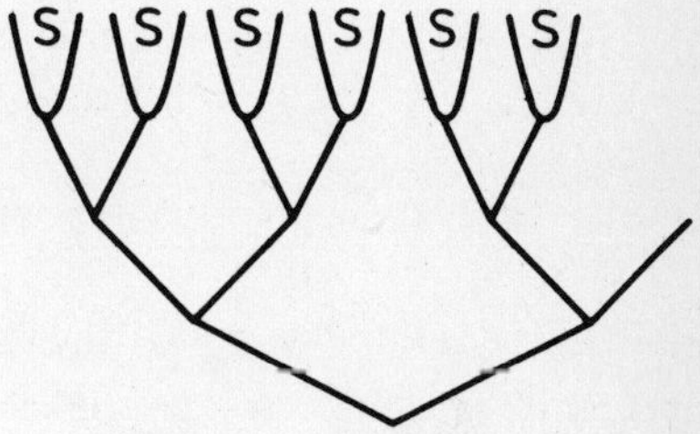

$T_{np} \otimes S$

chosen such that $\sum p_\nu 2^{-p_\nu} = \infty$ and $\sum 2^{-p_\nu} < \infty$ hold (one may take $n_\nu = p_\nu$ for all ν).

Theorem 2. Assume that, for every $t \in Z$, there exists a $k \in \mathbb{N}$ such that the set of multiindices $(2k+1)\cdot t|n$ contains a multiindex in $T_\gamma \cup (2^{(\mathbb{N})} \setminus T)$ for infinitely many values of n. Then T is thin.

Proof. For $k \in \mathbb{N}$ define $f_k: 2^{(\mathbb{N})} \to \{0,1\}$ by

$$f_k(\varepsilon) = \begin{cases} 1 & \text{if } (2k+1)\cdot\varepsilon \text{ intersects } T_\gamma \cup (2^{(\mathbb{N})} \setminus T) \\ 0 & \text{otherwise.} \end{cases}$$

The first condition of Lemma 3 holds by assumption. To verify the second condition fix $k \geq 1$, and associate with each ε for which $f_k(\varepsilon) = 1$ a point $\kappa(\varepsilon)$ in T_γ which is the restriction of a point in $(2k+1)\varepsilon \cap [T_\gamma \cup (2^{(\mathbb{N})} \setminus T)]$. Then

$$\sum_{\varepsilon \in T \setminus T_\gamma} f_k(\varepsilon)\cdot|\varepsilon| = \sum_{\kappa \in T_\gamma} \sum \left\{ |\varepsilon| \,\middle|\, \varepsilon \in T \setminus T_\gamma \,,\ f_k(\varepsilon) = 1 \,,\ \kappa(\varepsilon) = \kappa \right\}$$

$$\leq \sum_{\kappa \in T_\gamma} \sum_{\nu = n(\kappa)}^{\infty} (2k)\cdot 2^{-\nu} = 4k \sum_{\kappa \in T_\gamma} |\kappa| \leq 4k \,. \quad \blacksquare$$

Remarks. Theorem 2 is really the classical nullset criterion translated into the dyadic setting. As such the result is not really new. The main interest lies in the proof which shows that our technique (Lemmas 2 and 3) is flexible enough to lead to new types of criteria as those of Theorem 1 as well as to the classical criterion.

Theorems 1 and 2 are especially adapted to the dyadic setting. For the purpose of deriving purely geometrical result from dyadic results, the main purpose of the next section, some parts of Theorem 1 are more adequate when stated in a slightly different and, in fact, stronger form. The main argument is contained in the following variant of Lemma 2 where $\Delta_3 : 2^{(\mathbb{N})} \to \mathbb{N} \cup \{0\}$ denotes the function

$$\Delta_3(\varepsilon) = \min_{\delta \in 3\varepsilon} \Delta(\delta)$$

Lemma 4. Let $g: T \setminus T_\gamma \to [0,1]$ be a map which satisfies the condition that, for ε and δ in $T \setminus T_\gamma$, the implication

(13) $\delta \in 3\varepsilon \, , \, |\gamma(\varepsilon)| \leq |\gamma(\delta)| \Rightarrow g(\varepsilon) \leq g(\delta)$

holds. If

(14) $\sum_0^\infty g(t|n) 2^{-\Delta_3(t|n)} = \infty \quad \forall t \in Z$

and

(15) $\sum_{\varepsilon \in T \setminus T_\gamma} g(\varepsilon) \cdot |\gamma(\varepsilon)| < \infty$,

then T is thin.

Proof. For ε not in $T \setminus T_\gamma$ we put $g(\varepsilon) = 1$. Define $f: 2^{(\mathbb{N})} \to [0,1]$ by

$$f(\varepsilon) = g(\varepsilon) 2^{-\Delta_3(\varepsilon)} .$$

Then (9) holds by (14). To verify (10), we first define, for every $\varepsilon \in T \setminus T_\gamma$, the multiindex ε° such that $\varepsilon^\circ \in 3\varepsilon$ and $\Delta(\varepsilon^\circ) = \Delta_3(\varepsilon)$. Also, for ε not in $T \setminus T_\gamma$, we put $\gamma(\varepsilon) = \varepsilon$. We then have

$$\sum_{\varepsilon \in T \setminus T_\gamma} f(\varepsilon)|\varepsilon| = \sum_{\varepsilon \in T \setminus T_\gamma} g(\varepsilon) \cdot |\gamma(\varepsilon^\circ)|$$

$$\leq \sum_{\varepsilon \in T \setminus T_\gamma} g(\varepsilon^\circ) \cdot |\gamma(\varepsilon^\circ)| \; ;$$

the contribution to this sum coming from ε's with $\varepsilon^\circ \in T \setminus T_\gamma$ is bounded by

$$3 \cdot \sum_{\varepsilon \in T \setminus T_\gamma} g(\varepsilon)\,|\gamma(\varepsilon)|$$

which is finite by (15) and for all the remaining terms, $g(\varepsilon^\circ) = 1$ and $\gamma(\varepsilon^\circ) = \varepsilon^\circ$ so that the remaining contribution is

(16) $\sum\{|\varepsilon^\circ| \mid \varepsilon \in T\setminus T_\gamma \, , \, \varepsilon^\circ \in T_\gamma \cup CT\}$.

Observing that the only ε°'s which can occur in (16) are multiindices which are either obtained by adding 0's (if $\varepsilon^\circ = \varepsilon^+$) or by adding 1's (if $\varepsilon^\circ = \varepsilon^-$) to a multiindex in T_γ , we find that the sum (16) is bounded by 4, hence finite. ‖

It is then straightforward to establish the following result.

<u>Proposition</u> 1. <u>The criteria I and II of Theorem 1 are still valid when Δ is replaced by Δ_3</u> .

Before we state the variant of Theorem 1, case III we need, we introduce some notation. If $\varepsilon \notin T$ we put $\gamma(\varepsilon) = \varepsilon$. By $\gamma_3(\varepsilon)$ we denote the multiindex among $\gamma(\varepsilon^-)$, $\gamma(\varepsilon)$ and $\gamma(\varepsilon^+)$ which is of lowest level (with an arbitrary convention if this is not uniquely determined). Note that

$$2^{-\Delta_3(\varepsilon)} = \frac{|\gamma_3(\varepsilon)|}{|\varepsilon|}$$

<u>Proposition</u> 2. <u>If</u>

$$\sum_0^\infty \,|\log|\gamma_3(t|n)||^{-1}\, 2^{-\Delta_3(t|n)} = \infty \quad \forall t \in Z \cap A \ ,$$

<u>then</u> T <u>is thin at</u> A .

<u>Proof</u>. We apply Lemma 2 to f defined by

$$f(\varepsilon) = \frac{1}{n(\gamma_3(\varepsilon))}\, 2^{-\Delta_3(\varepsilon)}$$

and realise that the problem is to show that

$$s = \sum_{\varepsilon \in T \setminus T_\gamma} \frac{1}{n(\gamma_3(\varepsilon))}\, |\gamma_3(\varepsilon)|$$

is finite. The contribution to s coming from ε's for which either ε^- or ε^+ does not lie in T is easily seen to be finite (it is at most 2). The contribution from the other ε's is also finite since for each such ε, $\gamma_3(\varepsilon) \in T_\gamma$ and since, for each $\delta \in T_\gamma$, there are at most $3 \cdot n(\delta)$ such ε's for which $\gamma_3(\varepsilon) = \delta$. Therefore, the contribution in question is at most

$$\sum_{\delta \in T_\gamma} 3 \cdot n(\delta) \cdot \frac{1}{n(\delta)} \cdot |\delta| \leq 3 \quad . \blacksquare$$

The next result is a not entirely satisfactory variant of case IV of Theorem 1. The notation $\operatorname{cen}(\delta)$ where $\delta \in 2^{(\mathbb{N})}$ means the center of δ when δ is viewed as a dyadic rational interval.

Proposition 3. For every $c > 0$ define $f_c \colon T \setminus T_\gamma \to [0,1]$ by

$$(17) \quad f_c(\varepsilon) = \frac{1}{|\varepsilon|} \cdot \sup\{|\delta'| \wedge |\delta''| \;\big|\; \delta' \cup \delta'' \subseteq (\varepsilon \cup \varepsilon') \cap \Gamma ,$$

$$|\operatorname{cen}(\delta') - \operatorname{cen}(\delta'')| \geq c \cdot |\varepsilon|\} \; .$$

Assume that,

$$(18) \quad \forall t \in Z \quad \exists c > 0 \colon \sum_1^\infty f_c(t|n) = \infty \quad .$$

Then T is thin.

We conjecture that this still holds when $\varepsilon \cup \varepsilon'$ is replaced by 3ε in the definition of f_c, i.e. when instead of f_c we consider the function g_c defined by

$$(19) \quad g_c(\varepsilon) = \frac{1}{|\varepsilon|} \sup\{|\delta'| \wedge |\delta''| \;\big|\; \delta' \cup \delta'' \subseteq 3\varepsilon \cap \Gamma ,$$

$$|\operatorname{cen}(\delta') - \operatorname{cen}(\delta'')| \geq c \cdot |\varepsilon|\} \; .$$

Proof. By Lemma 3, it is enough to consider a fixed $c > 0$. In an attempt to verify the conjecture, assume that

$$\sum_1^\infty g_c(t|n) = \infty \quad \text{for all} \quad t \in Z \; .$$

Determine ν such that $c > 2^{-\nu}$. Determine, for each $\varepsilon \in T$, two rational dyadic intervals $\delta'(\varepsilon)$ and $\delta''(\varepsilon)$ such that

$$\delta'(\varepsilon) \cup \delta''(\varepsilon) \subseteq 3\varepsilon \cap \Gamma \; , \; |\operatorname{cen}(\delta'(\varepsilon)) - \operatorname{cen}(\delta''(\varepsilon))| \geq c \cdot |\varepsilon| \; ,$$

$$|\delta'(\varepsilon)| \leq |\delta''(\varepsilon)| \quad \text{and} \quad |\delta'(\varepsilon)| = g_c(\delta) \cdot |\varepsilon| \; .$$

If we can verify that

(20) $\sum_{\varepsilon \in T \setminus T_\gamma} |\delta'(\varepsilon)| < \infty$,

then T is thin by Lemma 2.

We consider various contributions to the sum (20).

Case 1: Contributions from ε's with $\varepsilon^* \notin T$. Each such ε is associated with the point $\omega(\varepsilon)$ defined by $\omega(\varepsilon) \in T_\gamma$, $\omega(\varepsilon) = \varepsilon^*|k$ for some (unique) k . Consider some $\omega \in T_\gamma$. Then even though there may be infinitely many ε's with $\omega(\varepsilon) = \omega$, their total contribution to the sum (20) is at most $2 \cdot |\omega|$ as is easily seen by the inequality $|\delta'(\varepsilon)| \leq |\varepsilon^*|$. Thus the total contribution to (20) coming from ε's under this case is at most $2\sum\{|\omega| \mid \omega \in T_\gamma\} = 2|\Gamma| \leq 2$.

Case 2: Contributions from ε's for which there exists $\kappa \in T_\gamma$ such that $\kappa \subseteq 3\varepsilon$ and $n(\kappa) \leq n(\varepsilon) + \nu$. For each such ε , choose $\kappa = \kappa(\varepsilon)$ with the stated properties. Employing the inequality $|\delta'(\varepsilon)| \leq 2^\nu |\kappa(\varepsilon)|$ it tollows from an argument similar to the onc used under case 1, that the total contribution to (20) in case 2 is at most $2^\nu \cdot 3\nu |\Gamma| \leq 3\nu \cdot 2^\nu$.

Case 3: Contributions from ε's not covered by case 1 or 2 and for which $\delta'(\varepsilon)|n(\varepsilon) = \delta''(\varepsilon)|n(\varepsilon)$ (= either ε , ε' or ε^*) . Let $m(\varepsilon)$ be the smallest integer m for which $\delta'(\varepsilon)|m \neq \delta''(\varepsilon)|m$. Since the distance between $\mathrm{cen}(\delta'(\varepsilon))$ and $\mathrm{cen}(\delta''(\varepsilon))$ exceeds $2^{-\nu} \cdot |\varepsilon|$, we have $m(\varepsilon) \leq n(\varepsilon) + \nu$. As we are not in case 2, both $\delta'(\varepsilon)|m(\varepsilon)$ and $\delta''(\varepsilon)|m(\varepsilon)$ are points in T , hence one of these points, say $\kappa(\varepsilon)$, is an α-point. Clearly, $|\delta'(\varepsilon)| \leq |\gamma(\kappa(\varepsilon))|$. As each $\kappa \in T_\gamma$ is of the form $\kappa(\varepsilon)$ for at most 3ν ε's, we find that the contribution to (20) in this case is at most $3\nu \cdot \sum\{|\gamma(\kappa)| \mid \kappa \in T_\alpha\} = 3\nu \cdot |\Gamma| \leq 3\nu$.

Case 4: Contributions from ε's for which there exists a k with $n(\varepsilon) \leq k \leq n(\varepsilon) + \nu$ such that either $\delta'(\varepsilon)|k$ or $\delta''(\varepsilon)|k$ is an α-point. Just as in case 3 we again find a finite contribution to (20) (at most $6 \cdot 2^{\nu+1}$) .

Of course, if it is known that $\delta'(\varepsilon) \cup \delta''(\varepsilon) \subseteq \varepsilon \cup \varepsilon'$ (as is the case if we work with f_c rather that with g_c) , then we are always in case 4. Proposition 3 is therefore fully proved. ‖

The conjecture could be established if it holds that when $\sum_1^\infty 2^{-\Delta^*(t|n)} = \infty$ for $t \in Z$, where $\Delta^*(\varepsilon) = \max(\Delta(\varepsilon), \Delta(\varepsilon^*))$, then T is thin. But this is hardly true since (10) may be infinite for f given by $f(\varepsilon) = 2^{-\Delta^*(\varepsilon)}$ (shown in collaboration with E. Sparre Andersen); compare with Theorem 1, case IV.

Some packing theorems of Vitali type.

The results of the previous section will be employed, not only to establish geometrical criteria for Lebesgue nullsets in the usual (non-dyadic) setting, but also, more generally, to prove packing theorems for Vitali systems in $\mathbb{R}^N$.

Recall, that a Vitali system $\mathcal{V}$ in $\mathbb{R}^N$ is a set of pairs $(A, \mathcal{B})$ where $A \subseteq \mathbb{R}^N$ and where $\mathcal{B}$ is a set of closed subset of $\mathbb{R}^N$ such that two conditions are satisfied: Firstly, if $(A, \mathcal{B}) \in \mathcal{V}$ and $B \subseteq A$, then $(B, \mathcal{B}) \in \mathcal{V}$. Secondly, if $(A, \mathcal{B}) \in \mathcal{V}$ and $F \subseteq \mathbb{R}^N$ is closed, then $(A \setminus F, \mathcal{B}_F) \in \mathcal{V}$ where $\mathcal{B}_F$ is the set of $B \in \mathcal{B}$ which are disjoint with F . We say that the packing theorem holds for $\mathcal{V}$ if, for every $(A, \mathcal{B}) \in \mathcal{V}$ the set A can be packed by sets from $\mathcal{B}$, i.e. there exists a sequence (perhaps finite) of pairwise disjoint sets (B_n) from $\mathcal{B}$ such that $|A \setminus \cup B_n|^* = 0$. For these notions we refer to [5].

We shall assume that $\mathbb{R}^N$ is provided with a norm $\|\cdot\|$. By $B[x,r]$ we denote the closed ball with center x and radius r . If B is a ball, $\operatorname{cen}(B)$ denotes its center and $\operatorname{rad}(B)$ its radius.

The Vitali system $\mathcal{V}$ is translation invariant if $(A, \mathcal{B}) \in \mathcal{V}$ implies $(x+A, x+\mathcal{B}) \in \mathcal{V}$ for all $x \in \mathbb{R}^N$ where $x + A$ and $x + \mathcal{B}$ are defined in the obvious manner.

Let $\mathcal{V}$ be a Vitali system in $\mathbb{R}^1$ and let $(A, \mathcal{B}) \in \mathcal{V}$. With $(A, \mathcal{B})$ we associate a tree $T = T_{A,\mathcal{B}}$ as follows. Put

$$(21) \quad \mathcal{I} = \{\text{dyadic rational intervals } I \mid I \cap A = \emptyset \text{ or } I \subseteq B \subseteq 5I \text{ for some } B \in \mathcal{B}\} ,$$

$$(22) \quad \mathcal{I}_{max} = \{\text{maximal intervals in } \mathcal{I}\}$$

and

$$(23) \quad \Gamma = \cup\{I \mid I \in \mathcal{I}\} = \cup\{I \mid I \in \mathcal{I}_{max}\} .$$

The tree T is then the tree having the intervals in $\mathcal{I}_{max}$ as γ-points.

Lemma 5. Let $\mathcal{V}$ be a translation invariant Vitali system in $\mathbb{R}^1$ and assume that the tree $T = T_{A,\mathcal{B}}$ is thin at A for every $(A, \mathcal{B}) \in \mathcal{V}$ with $A \subseteq [0,1]$. Then the packing theorem holds for $\mathcal{V}$.

Proof. Let $(A, \mathcal{B}) \in \mathcal{V}$ and assume that $A \subseteq [0,1]$. We consider the tree $T = T_{A,\mathcal{B}}$ and the related objects $\mathcal{I}_{max}$ and Γ as defined above. Let (I_n) be the family of all intervals in $\mathcal{J}_{max}$ which intersect A and choose, for each n, $B_n \in \mathcal{B}$ such that $I_n \subseteq B_n \subseteq 5I_n$. The I_n's are pairwise disjoint. It is well known (cf.[1], [5]) that for some absolute constant κ (eg. $\kappa = 27$ will do) it is possible to distribute $(5I_n)$ into κ disjoint sequences. For one of these, say for $(5I_{n_k})$ we have

$$\sum_k |5I_{n_k}| \geq \frac{1}{\kappa} \sum_n |5I_n| .$$

Now, as T is thin at A, we have

$$|A|^* = |A\cap\Gamma|^* \leq \sum_n |I_n| = \frac{1}{5} \sum_n |5I_n|$$

$$\leq \frac{\kappa}{5} \sum_k |5I_{n_k}| \leq \kappa \sum_k |B_{n_k}| .$$

Here, the B_{n_k}'s are pairwise disjoint. Then, by a standard technique (cf. (vi) p. 189 of [5]), also recalling that $\mathcal{V}$ is translation invariant, the packing theorem readily follows. ▮

Remarks. The proof shows that we may define the tree slightly differently not depending on A, by only considering I's with $I \subseteq B \subseteq 5I$ for some $B \in \mathcal{B}$. This tree we denote by $T_{\mathcal{B}}$.

It is easy, I think, to show that the condition of the lemma is also necessary for the packing theorem to hold.

We may now prove the main result of [2]:

Theorem 3. *Let* $A \subseteq \mathbb{R}^N$, *let* $\mathcal{B}$ *be a class of closed balls in* $\mathbb{R}^N$ *and let* $\psi \in \Psi_N$. *Assume that:*

(24) $\forall x \in A \quad \exists \delta_0 > 0 \quad \forall \delta \leq \delta_0 \quad \exists B \in \mathcal{B} : B \subseteq B[x,\delta] \ , \ \text{rad}(B) \geq \psi(\delta)$.

Then A *can be packed by sets from* $\mathcal{B}$.

Proof. By standard techniques, it is seen that the validity of the assertion is independant of the choice of norm. One may then work with the maximum norm. This has the advantage that the proof is essentially the same independant of the dimension. Therefore, we assume that $N = 1$.

Condition (24) specifies a certain translation invariant Vitali system in $\mathbb{R}^1$. To verify the condition of Lemma 5, let $(A, \mathcal{B})$ satisfy (24) and assume that $A \subseteq [0,1]$. We may assume that the same δ_0 works in (24) for all $x \in A$. To simplify, assume that $\delta_0 = 1$ works for all $x \in A$.

We shall apply Corollary 1 to show that the tree $T = T_{A,\mathcal{B}}$ is thin. Let $\varepsilon \in 2^{(\mathbb{N})}$ be given. Put $n = n(\varepsilon)$. Divide ε into 4 equal dyadic intervals and let ε_0 denote one of the 2 middle ones (having the center of ε as one of its endpoints). If $\varepsilon_0 \cap A = \emptyset$, then

$$2^{-\Delta(\varepsilon)} \geq \frac{1}{4} .$$

If $\varepsilon_0 \cap A \neq \emptyset$, we choose $x \in \varepsilon_0 \cap A$ and $B \in \mathcal{B}$ such that

$$B \subseteq [x-\frac{1}{4}2^{-n} , x+\frac{1}{4}2^{-n}] , \quad |B| \geq 2\cdot\psi(\frac{1}{4}2^{-n}) .$$

And then we choose I , a dyadic rational interval such that $I \subseteq B \subseteq 5I$. As $I \in \mathcal{J}$ and $I \subseteq \varepsilon$, we have

$$2^{-\Delta(\varepsilon)} = \frac{|\gamma(\varepsilon)|}{|\varepsilon|} \geq \frac{|I|}{|\varepsilon|} \geq \frac{1}{5}\frac{|B|}{|\varepsilon|} \geq \frac{2}{5}2^{n}\cdot\psi(\frac{1}{4}2^{-n}) .$$

It is now easy to check that Corollary 1 applies with $\varphi_n = \frac{2}{5}2^n\cdot\psi(\frac{1}{4}2^{-n})$. ∎

Remarks. The illustrative examples with "ψ-trumpets" mentioned earlier arise when we take as $\mathcal{B}$ the family of all balls disjoint with the set under investigation. For this application, it is convenient to replace the condition $B \subseteq B[x,\delta]$ appearing in (24) by the slightly weaker condition that $\mathrm{cen}(B) \in B[x,\delta]$ and $\mathrm{rad}(B) \leq \delta$ hold. That the corresponding variant of Theorem 3 is still valid either follows as a corollary (since the new conditions imply $B \subseteq B[x,2\delta]$) or by changing the proof slightly.

It may also be remarked that the proof could equally well have been based on Theorem 1, case IV.

With each family $\mathcal{B}$ of closed balls in $\mathbb{R}^N$ we associate a concentration function $\psi(x,t)$ defined for $x \in \mathbb{R}^N$ and $t > 0$ by

(25) $\psi(x,t) = \sup\{\mathrm{rad}(B) \mid B \in \mathcal{B} , B \subseteq B[x,t]\}$.

We say that $\mathcal{B}$ satisfies the entropy condition if, for every family (B_n) of pairwise disjoint sets from $\mathcal{B}$ such that $\cup B_n$ is bouded, we have

$$\sum_n |B_n| \log\frac{1}{|B_n|} < \infty .$$

Theorem 4. Let $\mathcal{B}$ be a family of closed balls in $\mathbb{R}^N$ which satisfies the entropy condition. Denote by A the set of $x \in \mathbb{R}^N$ for which $\psi(x,\cdot) \in \Psi_N$. Then A can be packed by sets from $\mathcal{B}$.

Proof. As in the proof of Theorem 3 we shall work in $\mathbb{R}^1$. Assume also that $A \subseteq [0,1]$. Consider the tree $T = T_{\mathcal{B}}$, cf. the remark following the proof of Lemma 5.

Let $t \in A$ and $n \in \mathbb{N}$ be given. Put $\varepsilon = t|n$. Determine $B \in \mathcal{B}$ such that

$$B \subseteq [t-2^{-n}, t+2^{-n}] \quad \text{and} \quad \mathrm{rad}(B) \geq \frac{1}{2}\psi(t,2^{-n}) .$$

Choose I, a dyadic rational interval, such that $I \subseteq B \subseteq 5I$. Then $I \subseteq 3\varepsilon$, hence

$$2^{-\Delta_3(t|n)} \geq \frac{|I|}{|\varepsilon|} \geq \frac{1}{5}\, 2^n\, \psi(t,2^{-n}) .$$

As $t \in A$, we have

$$\sum_0^\infty 2^{-\Delta_3(t|n)} = \infty .$$

It remains to be shown that $H(T) < \infty$ in order to apply Proposition 1.

Let (I_n) be the maximal dyadic intervals for which there exist B_n's in $\mathcal{B}$ with $I_n \subseteq B_n \subseteq 5B_n$. From the proof of Lemma 5, we can split $(5I_n)$ into κ sequences, say $(5I_n^\nu)_n$; $\nu = 1,\dots,\kappa$, such that each sequence $(5I_n^\nu)_n$ consists of pairwise disjoint sets. The contribution to $H(T)$ from $(I_n^\nu)_n$ is

$$\sum_n |I_n^\nu| \log(1/|I_n^\nu|) \leq \sum_n |B_n^\nu| \log(5/|B_n^\nu|)$$

which is finite by assumption since the B_n^ν's are disjoint as n varies. Thus $H(T) < \infty$, Proposition 1 (case I) applies and we conclude that T is thin at A. ▮

We can avoid any global conditions, such as the entropy condition, if we apply Theorem 2, part II of Proposition 1 or Proposition 2. The packing theorem you can deduce from Theorem 2 is just the classical result for which it is demanded that

$$\limsup_{t\to 0} \frac{\psi(x,t)}{t} > 0 \quad \forall x \in A .$$

The results you get from Proposition 1, case II and from Proposition 2 are the following:

Theorem 5. Let $f\colon]0,1] \to]0,\infty[$ be a non-decreasing function such *that*

$$\int_0^1 \frac{f(t)}{t}\, dt < \infty \quad .$$

Let $\mathcal{B}$ *be a family of closed balls in* $\mathbb{R}^N$ *and denote by* A *the set of* $x \in \mathbb{R}^N$ *for which*

$$\int_0^1 f(t) \left(\frac{\psi(x,t)}{t}\right)^N \frac{dt}{t} = \infty \; .$$

Then A *can be packed by sets from* $\mathcal{B}$.

Theorem 6. *Let* $\mathcal{B}$ *be a family of closed balls in* $\mathbb{R}^N$ *and denote by* A *the set of* $x \in \mathbb{R}^N$ *for which*

$$\int_0^1 |\log \psi(x,t)|^{-1} \left(\frac{\psi(x,t)}{t}\right)^N \frac{dt}{t} = \infty \quad .$$

Then there exist pairwise disjoint sets (B_n) *from* $\mathcal{B}$ *such that* $|A \setminus \cup B_n|^* = 0$.

The proofs follow closely the reasoning given in the proof of Theorem 4.

Theorem 6 is one half of a main result of [3] (Theorem 2). If the conjecture mentioned in connection with Proposition 3 is true, it easily implies the other half of that result.

Note. Part of the preparations for this paper were done while the author occupied a guestposition at the University of Münster.

REFERENCES

[1] De Guzman, M. (1975). Differentiation of integrals in R^N . Lecture Notes in Mathematics 481. Berlin: Springer.

[2] Jørsboe, O., Mejlbro, L. and Topsøe, F. (to appear). Some Vitali type theorems for Lebesgue measure.

[3] Mejlbro, L. and Topsøe, F. (1977). A precise Vitali theorem for Lebesgue measure. Math. Ann. 230, 183-193.

[4] Preiss, D. (1979). Gaussian measures and covering theorems. Commentationes Math. Univ. Carolinae. 20, 95-99.

[5] Topsøe, F. (1976). Packings and coverings with balls in finite dimensional normed spaces. Lecture Notes in Mathematics 541, 187-199. Berling: Springer.

REMARK ON EXTREMAL MEASURE EXTENSIONS

Heinrich v. Weizsäcker

Let $(X,\mathcal{B},\lambda)$ be a finite measure space. Let $\mathcal{A}$ be another σ-algebra over X such that $\mathcal{B} \subset \mathcal{A}$. The set H of all extensions of λ to a measure on $\mathcal{A}$ is convex. If there is an extension one might ask whether there is even an extremal extension, i.e. when does $H \neq \emptyset$ imply ex $H \neq \emptyset$? Even if $X = [0,1]$ and $\mathcal{A}$ is the Borel σ-algebra the answer may be negative as is shown by the following simple and well known example: Take $\mathcal{B} = \{B \subset [0,1] : B$ or $[0,1]\setminus B$ is countable$\}$ and as λ the restriction of Lebesgue measure to $\mathcal{B}$. If, on the other hand, $(X,\mathcal{A})$ is analytic and $\mathcal{B}$ is countably generated then always $H \neq \emptyset$ and there are enough extreme points of H for a Choquet type theorem (see e.g. [1]; a fairly general Choquet theorem for non compact sets of measures may be found in [2]). This suggests the

QUESTION: Is ex $H \neq \emptyset$ whenever $H \neq \emptyset$ and $\mathcal{A}$ and $\mathcal{B}$ are countably generated?

The purpose of this note is to give a negative answer under the following set theoretic assumption which is implied either by the continuum hypothesis or by Martin's axiom:

(*) Every set $A \subset [0,1]$ of cardinality $< 2^{\aleph_0}$ is a Lebesgue nullset.

THEOREM: Let $([0,1],\mathcal{B},\lambda)$ be the Lebesgue measure space with the Borel-σ-algebra $\mathcal{B} = \mathcal{B}([0,1])$. If (*) holds, then there is a countably generated σ-algebra $\mathcal{A}$ containing $\mathcal{B}$ such that λ can be extended to $\mathcal{A}$ but there is no extremal extension to $\mathcal{A}$.

PROOF: 1. Using (*) we construct a map $z:[0,1] \longrightarrow [0,1]$ such that

a) $\lambda_2^*(\text{Graph } z) = 1$ where λ_2 denotes two dimensional Lebesgue measure,

b) $\lambda^*\{s : z(s) = g(s)\} = 0$ for every Borel map $g:[0,1] \longrightarrow [0,1]$.

To do this let $(x_\alpha)_{\alpha<2^{\aleph_0}}$, $(F_\alpha)_{\alpha<2^{\aleph_0}}$ and $(g_\alpha)_{\alpha<2^{\aleph_0}}$ be well-orderings of $[0,1]$, of all sets $F \in \mathcal{B}([0,1]^2)$ such that $\lambda_2(F) > 0$ and of all Borel maps $g:[0,1] \longrightarrow [0,1]$ respectively. Now define ordinals $\beta(\alpha)$ and real numbers $y(\alpha)$ by induction as follows: Choose

$$\beta(\alpha) = \min\{\beta < 2^{\aleph_0} : \beta \notin \{\beta(\gamma) : \gamma < \alpha\} \text{ and } \lambda\{y : (x_\beta,y) \in F_\alpha\} > 0\}.$$

Since $\lambda_2(F_\alpha) > 0$ by Fubini's theorem and using (*) it is easy to see that the set in the brackets is non empty. Choose y_α such that $(x_{\beta(\alpha)}, y_\alpha) \in F_\alpha \setminus \bigcup_{\gamma<\alpha} \text{Graph } g_\gamma$. Define z on the set $\{x_{\beta(\alpha)} : \alpha < 2^{\aleph_0}\}$ by $z(x_{\beta(\alpha)}) = y_\alpha$. Extend z to [0,1] in such a way that $z(x_\beta) \neq g_\gamma(x_\beta)$ for all β,γ satisfying $\gamma < \beta$. This property implies b) by (*). Also z satisfies a) since $Z \cap F_\alpha \neq \phi$ for every α where Z = Graph z.

2. Let $\mathcal{A}$ be the σ-algebra generated by $\mathcal{B}$ and by the sets $z^{-1}([0,a])$, a rational. Then $\mathcal{A}$ is countably generated. $([0,1],\mathcal{A})$ is isomorphic to $(Z, \mathcal{B}([0,1]^2) \cap Z)$ via the bijection $s \longmapsto (s,z(s))$. Now let G denote the set of all probability measures on $\mathcal{B}([0,1]^2)$ whose marginal measure on the first coordinate is λ. Then the above isomorphism establishes an <u>affine</u> bijection between H and $\{\mu \in G : \mu^*(Z) = 1\}$. So by a) the set H is not empty. On the other hand
$\text{ex } H \simeq \text{ex } \{\mu \in G : \mu^*(Z) = 1\} = \{\mu \in \text{ex } G : \mu^*(Z) = 1\}$
since the relation $\mu^*(Z) = 1$ defines a face in G. But it is well known that every $\mu \in \text{ex } G$ is the "Lebesgue measure on Graph g" for some Borel function $g : [0,1] \longrightarrow [0,1]$. By b) this is incompatible with $\mu^*(Z)=1$, so ex H = ϕ which proves the theorem.

It would be interesting to know whether the above question can be decided in ZF + AC only.

<u>References:</u>

[1] YERSHOV, M.P.: The Choquet theorem and stochastic equations. Analysis Math. 1 (1975), 259 - 271.

[2] v.WEIZSÄCKER, H. and WINKLER, G.: Integral Representation in the set of Solutions of a Generalized Moment Problem. To appear in Math. Ann. (1979).

EXTENSIONS OF A σ-ADDITIVE MEASURE TO THE PROJECTIVE COVER

Robert F. Wheeler
Department of Mathematical Sciences
Northern Illinois University
DeKalb, Illinois 60115 U.S.A.

If X is a completely regular Hausdorff space, then there is an (essentially unique) extremally disconnected space $E(X)$, called the projective cover or absolute of X, and a perfect irreducible map κ of $E(X)$ onto X. The study of Baire measures on X and $E(X)$ was undertaken in [28]. The emphasis there is on the following problem: if μ and ν are finitely-additive Baire measures on X and $E(X)$, respectively, say that ν is a functional (resp., measure) extension of μ if $\nu(f\circ\kappa) = \mu(f)$ for every bounded continuous real-valued function f on X (resp., $\nu(\kappa^{-1}(B)) = \mu(B)$ for every Baire set B of X). Then, if μ is known to be σ-additive, what conditions on X will insure that some, or all, extensions ν of μ are again σ-additive?

In [28] it is shown that, for the rather large class of spaces which have the weak cb property [13], all extensions will indeed be σ-additive. In this paper the topological aspects of the situation are refined and systematized. In particular, the weak cb property is factored into two others: one guarantees that every functional extension is actually a measure extension (the converse always being true), while the other insures that every measure extension of a σ-additive measure is again σ-additive. In considering these new properties, a natural connection emerges between the extension results mentioned above, and the classical problem of extending a σ-additive Baire measure to a closed-regular Borel measure.

The paper is organized as follows: In Section 1 the basic topological measure theory of X and $E(X)$ is sketched, and all definitions are given. The known relationships among various properties are summarized in Table I, some of the proofs being deferred until later sections. In Section 2 the relation between functional and measure extensions is investigated; in Section 3 the weak and strong lifting properties introduced in [28] are studied, and some permanence properties are given. Section 4 contains a discussion of the Borel extension problem.

Section 5 is centered about nine key counter-examples, which are referred to by their abbreviations throughout the paper. They are the Dieudonné plank (D), the Mack-Johnson space (MJ), Michael's product space (MPS), the pseudocompact extension of D (PD), M. E. Rudin's Dowker space (RDS), the Sorgenfrey plane (SP), the Tychonoff plank (T), the join of T and D (T#D), and the deleted join of T with itself ($T\#T^-$). Almost all measure-theoretic pathology known to the author in the study of X and E(X) is revealed by one or more of these spaces. Their behavior with respect to eleven topological properties is summarized in Table II.

The paper concludes in Section 6 with a list of open problems.

Some of this research was carried out under the auspices of the Montague Research Institute in Stockton, California. The author extends his thanks to the MRI for their continuing hospitality.

1. Topological preliminaries. X always denotes a completely regular Hausdorff space. A subset F of X is regular closed if $F = cl_X int_X F$. Then $RC(X)$, the collection of regular closed subsets of X, is a complete Boolean algebra under the operations $\bigvee F_\alpha = cl_X(\bigcup F_\alpha)$, $\bigwedge F_\alpha = cl_X int_X(\bigcap F_\alpha)$, and $F' = cl_X(X - F)$. The projective cover, or absolute, of X is defined to be $E(X) = \{p\colon\ p$ is an ultrafilter in $RC(X)$ which converges to a point of $X\}$. Note that the filter requirement is: $F, G \in p \Rightarrow F \wedge G \in p$, so that $int_X F \cap int_X G$ must be non-empty. If $F \in RC(X)$, let $\lambda(F) = \{p \in E(X)\colon\ F \in p\}$, and take the collection of all $\lambda(F)$ as a base for the topology of $E(X)$. The topology is extremally disconnected (i.e., the closure of every open set is again open), and the sets $\lambda(F)$ are precisely the clopen sets.

There is a natural map $\kappa\colon E(X) \to X$ which sends each convergent ultrafilter to its limit. The map κ is perfect (a continuous closed surjection such that $\kappa^{-1}(x)$ is compact for each $x \in X$) and irreducible (the image of a closed proper subset of $E(X)$ is a proper subset of X. If $F \in RC(X)$, then $\kappa(\lambda(F)) = F$. If (F_n) is a decreasing sequence in $RC(X)$, then $F_n \downarrow \emptyset$ if and only if $\lambda(F_n) \downarrow \emptyset$.

Now $F \to \lambda(F)$ is a Boolean isomorphism of $RC(X)$ onto $clop(E(X))$, and $F \to cl_{\beta X}F$ is a Boolean isomorphism of $RC(X)$ onto $RC(\beta X)$, where βX is the Stone-Cech compactification of X. On forming $E(\beta X)$ and $\bar{\kappa}\colon E(\beta X) \to \beta X$, the following diagrams result:

$$\begin{array}{ccc} E(X) & \longrightarrow \beta E(X) = & E(\beta X) \\ \kappa \downarrow & & \downarrow \bar{\kappa} \\ X & \longrightarrow & \beta X \end{array} \qquad\qquad \begin{array}{ccc} clop(E(X)) & \longleftrightarrow & clop(E(\beta X)) \\ \lambda \updownarrow & & \updownarrow \bar{\lambda} \\ RC(X) & \longleftrightarrow & RC(\beta X) \end{array}$$

The diagram on the left commutes; it should be noted that $E(\beta X)$ is the Stone space of $RC(X)$, $E(X) = \bar{\kappa}^{-1}(X)$, and $\kappa = \bar{\kappa}|X$. The diagram on the right illustrates that the four Boolean algebras shown are naturally isomorphic.

The basic facts of topological measure theory are given by Varadarajan [27]; other useful references include Knowles [11] and Sentilles [23]. Let

$C^*(X) = \{f\colon\ f$ is a bounded continuous real-valued function on $X\}$

$M^+(X) = \{\Phi\colon\ \Phi$ is a positive linear functional on $C^*(X)\}$.

A subset A of X is called a zero-set (resp., cozero-set) if it has the form $\{x\colon f(x) = 0\}$ (resp., $\{x\colon f(x) > 0\}$) for some $f \in C^*(X)$. Let $Ba^*(X)$ (resp., $Ba(X)$) denote the algebra (resp., σ-algebra) of Baire sets generated by the zero-sets of X, and let $Bo(X)$ denote the σ-algebra of Borel sets generated by the closed sets.

Any $\Phi \in M^+(X)$ is represented by a finitely-additive, zero-set regular measure μ on $Ba^*(X)$, and by a compact-regular Borel measure μ^β on $Bo(\beta X)$: $\Phi(f) = \int_X f\,d\mu = \int_{\beta X} f^\beta\,d\mu^\beta$, where f^β is the unique continuous extension of f to βX.

Functionals and measures may be classified according to their additivity properties; for σ- and τ-additivity the following characterizations are standard (Z and Q stand for zero-sets and compact sets, respectively):

	Φ	μ	μ^β
$M^+_\sigma(X)$	$f_n \downarrow 0 \Rightarrow \Phi(f_n) \to 0$	$Z_n \downarrow \emptyset \Rightarrow \mu(Z_n) \to 0$	$Z \subset \beta X - X \Rightarrow \mu^\beta(Z) = 0$
$M^+_\tau(X)$	$f_\alpha \downarrow 0 \Rightarrow \Phi(f_\alpha) \to 0$	$Z_\alpha \downarrow \emptyset \Rightarrow \mu(Z_\alpha) \to 0$	$Q \subset \beta X - X \Rightarrow \mu^\beta(Q) = 0$

If μ is σ-additive, then μ has a unique countably-additive zero-set regular extension to $Ba(X)$, and it will always be assumed that this extension has been made.

<u>Example 1.1.</u> Let $p \in \beta X$

$\Phi_p(f) = f^\beta(p)$ $\qquad$ $\mu_p(Z) = 1,\ p \in cl_{\beta X} Z$; $= 0$, otherwise $\qquad$ $\mu_p^\beta(A) = 1,\ p \in A$; $= 0$, otherwise

Φ_p is τ- (resp., σ-) additive iff $p \in X$ (resp., $p \in \nu X$, the Hewitt realcompactification of X [4]).

A space X is called measure-compact if $M_\sigma^+(X) = M_\tau^+(X)$. Measure-compactness implies realcompactness, as 1.1 shows, but the converse fails [17].

If ϕ: $Y \to X$ is any continuous surjection, then the map $f \to f \circ \phi$ is an isometric embedding of $C^*(X)$ into $C^*(Y)$. The adjoint ϕ^* defined by $\phi^*\nu(f) = \nu(f \circ \phi)$ maps $M^+(Y)$ onto $M^+(X)$, by the Hahn-Banach Theorem; moreover, if ν is σ- or τ-additive, so is $\phi^*\nu$. Thus for $Y = E(X)$ and $\phi = \kappa$, $\kappa^*(M_\sigma^+(E(X))) \subset M_\sigma^+(X)$; it is shown in [28] that $\kappa^{*-1}(M_\tau^+(X)) = M_\tau^+(E(X))$.

Definition 1.2. Let $\mu \in M^+(X)$, $\nu \in M^+(E(X))$. Then

(a) ν is a functional extension of μ if $\kappa^*\nu = \mu$;

(b) ν is a measure extension of μ if $\nu(\kappa^{-1}(B)) = \mu(B)$ for every $B \in Ba^*(X)$.

Definition 1.3. [28] (a) X has the weak lifting property (WLP) if for each $\mu \in M_\sigma^+(X)$, $\exists\, \nu \in M_\sigma^+(E(X))$ with $\kappa^*\nu = \mu$; (b) X has the strong lifting property (SLP) if $\kappa^{*-1}(M_\sigma^+(X)) = M_\sigma^+(E(X))$.

Definition 1.4. [12, 13] X is a [weak] cb-space if for each sequence (F_n) of [regular] closed sets with $F_n \downarrow \emptyset$, there is a sequence (Z_n) of zero-sets with $F_n \subset Z_n \downarrow \emptyset$.

Any countably compact space is a cb-space; any pseudocompact space is weak cb.

Definition 1.5. [14] X is [weakly] δ-normally separated if each [regular] closed set and zero-set disjoint from it can be completely separated by a continuous function on X. (Call these properties WδNS and δNS).

Definition 1.6. X has property (*) [(**)] if for each sequence (F_n) of [regular] closed sets with $F_n \downarrow \emptyset$, there is a sequence (U_n) of cozero-sets with $F_n \subset U_n \downarrow \emptyset$.

Any space with (*) is countably metacompact, but the converse fails, as D shows.

Definition 1.7. X is a Marik space if every $\mu \in M_\sigma^+(X)$ admits at least one closed-regular countably-additive Borel extension.

Table I summarizes relationships among spaces with these properties and some better-known classes of topological spaces. The following "factorization theorem" is useful.

Proposition 1.8. (a) cb $\leftrightarrow$ (weak cb + countably paracompact);
(b) cb $\leftrightarrow$ (δNS + (*)); (c) weak cb $\leftrightarrow$ (WδNS + (**)).

Proof. (a) is easy, using Ishikawa's characterization of countable paracompactness [8]: for each sequence (F_n) of closed sets with $F_n \downarrow \emptyset$, there is a sequence (H_n) of regular closed sets with $F_n \subset \text{int}_X H_n \subset H_n \downarrow \emptyset$. (b) and (c): Using lower semicontinuous functions, Mack [14] proved that cb $\Rightarrow$ δNS and weak cb $\Rightarrow$ WδNS. We give a direct proof of these facts here. Let F be [regular] closed, Z a zero-set, $F \cap Z = \emptyset$. Let $Z = f^{-1}(0)$, where $f \geq 0$, and let $U_n = \{x: f(x) < 1/n\}$. If $F \cap U_n = \emptyset$ for any n, then certainly F and Z can be completely separated. Otherwise, $(\text{cl}_X(F \cap U_n))$ is a sequence of [regular] closed sets decreasing to the empty set. Choose a sequence (Z_n) of zero-sets with $\text{cl}_X(F \cap U_n) \subset Z_n \downarrow \emptyset$. Then $W = \bigcap_{n=1}^{\infty}(Z_n \cup (X - U_n))$ is a zero-set containing F and disjoint from Z, so F and Z can be completely separated.

The remainder of the proof of (b) and (c) is routine.

Remark 1.9. Any pseudocompact, non-countably compact space (e.g., T̲) is weak cb but not countably paracompact; MJ̲ is countably paracompact, but not weak cb. RDS̲ is normal, hence δNS, but fails (*); T̲ has (*), but fails to be δNS. MJ̲ is WδNS, but fails (**); T̲#T̲⁻ has (**) but fails WδNS.

Remark 1.10. If X is normal, then countable paracompactness, (*), and the cb-property are equivalent. RDS̲ is normal and weak cb, but not countably paracompact. In comparing (*), countable paracompactness, and weak cb, I do not know if there is a space with the first two properties, but not the third. There cannot be a space which has the last two properties, but fails the first; the other six combinations are all possible, as Table II shows.

TABLE I.

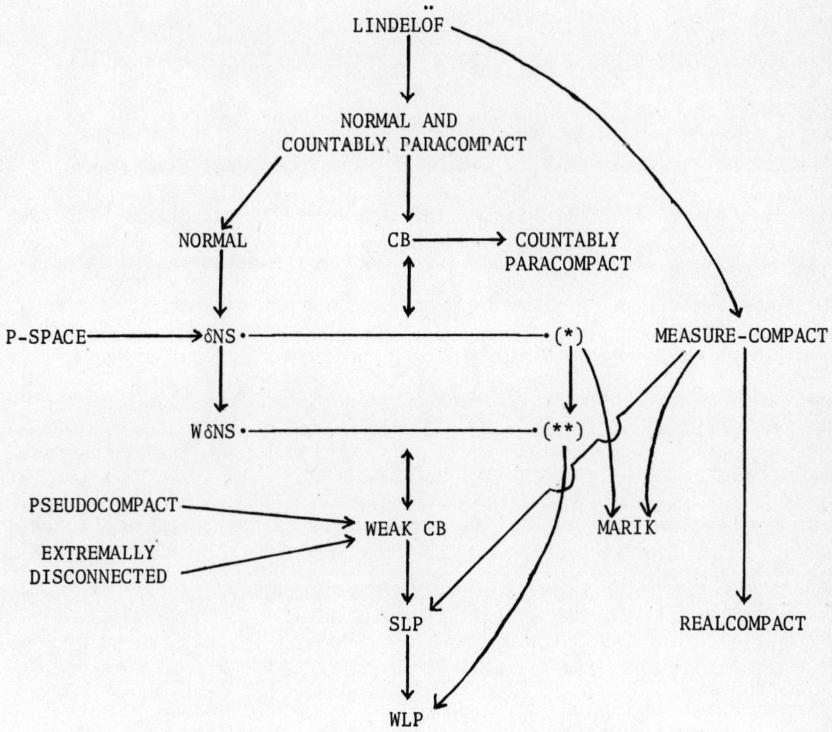
LINDELÖF
NORMAL AND
COUNTABLY PARACOMPACT
NORMAL
CB
COUNTABLY
PARACOMPACT
P-SPACE
δNS
(*)
MEASURE-COMPACT
WδNS
(**)
PSEUDOCOMPACT
WEAK CB
MARIK
EXTREMALLY
DISCONNECTED
SLP
REALCOMPACT
WLP

2. Functional and measure extensions. Let $\mu \in M^+(X)$, $\nu \in M^+(E(X))$, both considered as (finitely-additive) Baire measures.

Proposition 2.1. ν is a measure extension of μ if (and only if) $\nu(\kappa^{-1}(Z)) = \mu(Z)$ for every zero-set Z of X.

Proof. Every member of $Ba^*(X)$ can be written as a disjoint union of sets $Z_1 - Z_2 = Z_1 - (Z_1 \cap Z_2)$, and a short calculation yields the result. If ν is σ-additive, then so is μ, and we can use zero-set regularity to obtain $\nu(\kappa^{-1}(B)) = \mu(B)$ for every B in the σ-algebra $Ba(X)$.

Proposition 2.2. ν is a functional extension of μ if and only if $\nu^\beta(\overline{\kappa}^{-1}(B)) = \mu^\beta(B)$ for every Borel set B of βX.

Proof. This is a standard fact about compact spaces.

Proposition 2.3. Every measure extension is a functional extension.

Proof. If $f \in C^*(X)$, then sets of the form $\{x \in X: \ a \le f(x) < b\}$ are members of $Ba^*(X)$. Since $\int_X f\, d\mu$ and $\int_{E(X)} (f \circ \kappa) d\nu$ can be approximated using finite Baire partitions of the respective spaces, the result follows in a straightforward way.

The converse fails; it is shown in [28] that every functional extension is a measure extension if and only if X is WδNS. In fact, if X is not WδNS, then there is a $\{0,1\}$-valued μ with a $\{0,1\}$-valued functional extension ν which is not a measure extension.

Lemma 2.4. If ν is a functional extension of μ, then

(a) $\mu(U) \le \nu(\kappa^{-1}(U))$ for every cozero-set $U \subset X$

(b) $\mu(Z) \ge \nu(\kappa^{-1}(Z))$ for every zero-set $Z \subset X$.

Proof. By the Alexandroff Representation Theorem [27, p. 165], $\mu(U) = \sup\{\mu(f): \ 0 \le f \le \chi_U, \ f \in C^*(X)\} = \sup\{\nu(f \circ \kappa): \ 0 \le f \le \chi_U, \ f \in C^*(X)\} \le \sup\{\nu(g): \ 0 \le g \le \chi_{\kappa^{-1}(U)}, \ g \in C^*(E(X))\} = \nu(\kappa^{-1}(U))$. The proof of (b) is similar, using the fact that $\mu(Z) = \inf\{\mu(f): \ \chi_Z \le f \le 1, \ f \in C^*(X)\}$.

Proposition 2.5. If ν is a σ-additive functional extension of μ, then ν is a measure extension.

Proof. Fix a zero-set Z, and choose a sequence (U_n) of cozero-sets with $U_n \downarrow Z$. Then $\kappa^{-1}(U_n) \downarrow \kappa^{-1}(Z)$, and so $\nu(\kappa^{-1}(Z)) = \lim \nu(\kappa^{-1}(U_n)) \geq \lim \mu(U_n) = \mu(Z)$, using 2.4 (a). The result follows from 2.1 and 2.4 (b).

Existence of functional extensions is guaranteed by the Hahn-Banach Theorem; existence of measure extensions for every $\mu \in M^+(X)$ follows from a recent result of Bachman and Sultan [1]. If T is a set, and L is a lattice of subsets, let $MR(L)$ denote the space of bounded non-negative L-regular measures on $A(L)$, the algebra generated by L. If $L_1 \subset L_2$ are two lattices of subsets of T, then L_2 is said to be L_1-countably paracompact if $A_n \downarrow \emptyset$ in L_2 implies existence of a sequence (B_n) in L_1 with $A_n \subset \tilde{B}_n \downarrow \emptyset$.

Theorem 2.6. (Bachman and Sultan). If L_1 and L_2 are lattices of subsets of T, then every $\mu \in MR(L_1)$ has a measure extension to a $\nu \in MR(L_2)$. If L_2 is L_1-countably paracompact, and μ is σ-additive, then every measure extension ν is also σ-additive.

Corollary 2.7. Every $\mu \in M^+(X)$ has at least one measure extension.

Proof. Let $T = E(X)$, $L_1 = \{\kappa^{-1}(Z)\colon\ Z$ a zero-set of $X\}$, $L_2 = \{W\colon\ W$ a zero-set of $E(X)\}$, and apply the theorem.

Proposition 2.8. If X has (**), then all measure extensions of a σ-additive measure are σ-additive.

Proof. It is easily seen that if X has (**), then L_2 is L_1-countably paracompact (argue as in Theorem 10 of [28]). Now the result follows from 2.6.

Remark 2.9. (**) does not imply that all functional extensions of a σ-additive measure are σ-additive. $T\#T^-$ has (*), hence (**), but if p is the "corner point," and μ the corresponding σ-additive measure (as in 1.1), then some functional extensions (coming from points in $\beta E(T) - E(T)$) are σ-additive, and some (coming from points in $\beta E(T^-) - E(T^-)$) are not. This happens because $E(T)$ is pseudocompact and $E(T^-)$ is σ-compact, hence realcompact.

3. The weak and strong lifting properties. Results in 2.3, 2.7, and 2.8 combine to yield:

Proposition 3.1. Let X be a space with (**). Then (a) X has the WLP; and (b) X is measure-compact if and only if $E(X)$ is measure-compact.

Proof. It is observed in [28] that (a) always implies (b).

As 2.9 shows, a space with (**), or even (*), need not satisfy the SLP (but cf. 4.6). However, it is a central result of [28] that the combination of (**) and WδNS (i.e., the weak cb property) is enough to force the SLP. There are at least three ways to see this. One is to combine the remark after 2.3 with 2.8. An "external" proof involving βX and $\beta E(X)$, given in [28], is a second alternative. We offer here a third description of the weak cb property, which makes it immediately clear that functional extensions of σ-additive measures must be σ-additive.

Theorem 3.2. The following are equivalent: (a) If $f_n \downarrow 0$ in $C^*(E(X))$, then there is a sequence (g_n) in $C^*(X)$ with $f_n \le g_n \circ \kappa \downarrow 0$; (b) X is a weak cb space.

Proof. (a) $\Rightarrow$ (b): Let (F_n) be a sequence of regular closed sets with $F_n \downarrow \emptyset$. Let $f_n = \chi_{\lambda(F_n)} \in C^*(E(X))$. Then $f_n \downarrow 0$, and we choose g_n as above. Letting $Z_n = \{x\colon\ g_n(x) \ge 1\}$, we have $F_n \subset Z_n \downarrow \emptyset$.

(b) $\Rightarrow$ (a): Assume first that X is Lindelöf, and let $f_n \downarrow 0$ in $C^*(E(X))$. Let $G = \{g \in C^*(X)\colon\ g \circ \kappa \ge f_n$ for some $n\}$. We show that $G \downarrow 0$. Given $x_0 \in X$ and $\epsilon > 0$, $\exists\, n_0$ such that $\kappa^{-1}(x_0) \subset U = \{p \in E(X)\colon\ f_{n_0}(p) < \epsilon\}$, because $\kappa^{-1}(x_0)$ is compact. Since $\kappa(\tilde{U})$ is closed in X and misses x_0, $\exists$ an open set V with $x_0 \in V$ and $\kappa^{-1}(V) \subset U$. Choose $g \in C^*(X)$ with $g(x) \ge \epsilon\ \forall x$, $g(x_0) = \epsilon$, and $g(x) \ge \|f_{n_0}\|_\infty$ on $X - V$. Then $g \circ \kappa \ge f_{n_0}$, so $g \in G$, and $G \downarrow \emptyset$. For fixed m, $X = \bigcup_{g \in G} \{x\colon\ g(x) < 1/m\}$, so there is a countable subcover given by $\{g_\alpha\colon\ \alpha \in A_m\}$. Enumerate $\bigcup A_m$ as $\{\alpha_\kappa\}_{\kappa=1}^\infty$, and let $g_n = \inf\{g_{\alpha_1}, \ldots, g_{\alpha_n}\}$. Then $g_n \downarrow 0$, so $g_n \circ \kappa \downarrow 0$, and each $g_n \circ \kappa \ge f_j$ for some j. The conclusion follows.

Now let X be any weak cb space, and let $f_n \downarrow 0$ in $C^*(E(X))$. Fix m, and let $Z_m = \bigcap_{n=1}^{\infty} \{p \in \beta E(X):\ f_n^\beta(p) \geq 1/m\}$. Then Z_m is a zero-set of $\beta E(X)$, and $Z_m \subset \beta E(X) - E(X)$. In [28, Th. 5] it is proved that there is a zero-set W_m of βX with $\overline{\kappa}(Z_m) \subset W_m \subset \beta X - X$. Let $Y_m = \overline{\kappa}^{-1}(\beta X - W_m)$. Then $Y_m = E(\beta X - W_m)$, where $\beta X - W_m$ is σ-compact, hence Lindelöf, and $Y_m \subset \beta E(X) - Z_m$. Now $((f_n^\beta - 1/m) \vee 0) \downarrow 0$ on Y_m as $n \to \infty$. Thus by the previous argument for a Lindelöf space, there is a sequence $(h_{mn})_{n=1}^{\infty}$ in $C^*(X)$ with $(f_n - 1/m) \vee 0 \leq h_{mn} \circ \kappa \downarrow 0$ as $n \to \infty$. Thus $f_n \leq (h_{mn} + 1/m) \circ \kappa$ for all m and n.

Let $g_n = \min_{i,j \leq n} (h_{ij} + 1/i)$, $n = 1, 2, \ldots$. Then $f_n \leq g_n \circ \kappa\ \forall n$, and (g_n) is a decreasing sequence. Finally, let $x_0 \in X$ and $\epsilon > 0$ be given. Choose m such that $1/m < \epsilon/2$, and n such that $h_{mn}(x_0) < \epsilon/2$. Then if $r = \max(m,n)$, $g_r(x_0) \leq h_{mn}(x_0) + 1/m < \epsilon$. Thus $g_n \downarrow 0$, to complete the proof.

Measure-compactness is also a sufficient condition for the SLP [28]. Necessary conditions for either lifting property seem difficult to obtain. In particular, <u>MPS</u> has the SLP, but is not WδNS; <u>T#D</u> has the WLP, but fails (**).

Now we turn to permanence properties of the WLP and SLP.

<u>Proposition 3.3</u>. The SLP is preserved by regular closed subspaces.

<u>Proof</u>. Let $F \in RC(X)$, and let $G = cl_X(X - F)$ be the Boolean complement of F. Then $E(X)$ is the disjoint union of $\lambda(F)$ and $\lambda(G)$, and $\lambda(F)$ may be identified with $E(F)$. Then (with i and j the natural embeddings)

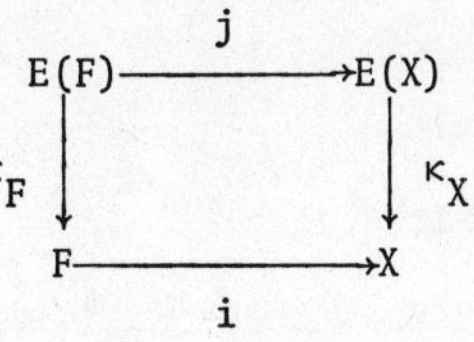

$\kappa_X \circ j = i \circ \kappa_F$. If $\mu \in M_\sigma^+(F)$, $\nu \in M^+(E(F))$, and $\kappa_F^*(\nu) = \mu$, then $\kappa_X^*(j^*(\nu)) = i^*(\mu) \in M_\sigma^+(X)$, and so $j^*(\nu) \in M_\sigma^+(E(X))$, since X has the SLP. It follows easily that $\nu \in M_\sigma^+(E(F))$, and so F has the SLP.

This result is not true if "regular closed" is replaced by "closed" or if "SLP" is replaced by "WLP." <u>PD</u> has the SLP, but the closed subspace <u>D</u> fails even the WLP. <u>T#D</u> has the WLP, but its regular closed subspace <u>D</u> does not.

<u>Proposition 3.4</u>. If ϕ: $Y \to X$ is perfect and irreducible, and X has the

WLP or SLP, then so does Y.

Proof. $\phi\circ\kappa_Y$: $E(Y) \to X$ is a perfect irreducible map of $E(Y)$ onto X, so $E(Y) = E(X)$, and $\kappa_X = \phi\circ\kappa_Y$. Both results now follow by straight-forward calculations.

The converse of 3.4 fails: let $X = \underline{D}$, Y the projective cover, $\phi = \kappa$. Then Y trivially has the SLP, but X fails the WLP.

Proposition 3.5. Let $X = \bigcup_{\beta\in B} A_\beta$, where $\{A_\beta\}$ is a locally-finite family of regular closed sets, and $\text{int}_X A_\beta \cap \text{int}_X A_\gamma = \emptyset$ for $\beta \neq \gamma$. Then $E(X) = \sum_{\beta\in B} E(A_\beta)$, the topological sum of the $E(A_\beta)$.

Proof. Under the stated conditions, the natural map ϕ: $\sum A_\beta \to \bigcup A_\beta = X$ is perfect and irreducible. Thus, as in 3.4, $E(X) = E(\sum A_\beta)$. It is easily verified that E and $\sum$ commute, to complete the argument.

This raises the question: if each A_β is known to have the WLP or SLP, what can be said about X? In this regard, note that $\underline{MJ}$ can be expressed as a countable union of A_β's in the prescribed way; each A_β is cb (since it is countably compact), and therefore has the SLP, but $\underline{MJ}$ has only the WLP.

If X is the topological sum of the A_β's, then something can be said.

Proposition 3.6. Let $X = \sum_{\beta\in B} A_\beta$, where card B is non-measurable. If each A_β has the WLP, then so does X.

Proof. Let $\mu \in M_\sigma^+(X)$; then $C = \{\beta \in B\colon \mu(A_\beta) > 0\}$ is countable. If $T \in Ba(X)$, then $\mu(T) = \sum_{\beta\in C} \mu(T \cap A_\beta)$. Indeed if this fails for some T_0, let $\mu(T_0) - \sum_{\beta\in C} \mu(T_0 \cap A_\beta) = \delta > 0$. For each $H \subset B$, let $Y_H = \bigcup\{A_\beta\colon \beta \in H - C\}$, and define η: $P(B) \to R$ by $\eta(H) = \mu(T_0 \cap Y_H)$. It may be verified that η is countably additive, $\eta(\{\beta\}) = 0 \ \forall \beta \in B$, and $\eta(B) = \delta > 0$. This contradicts the assumption that card B is non-measurable.

Now let $\mu_\beta = \mu|Ba(A_\beta)$ for each β. Then $\mu_\beta \in M_\sigma^+(A_\beta)$, so $\exists\, \nu_\beta \in M_\sigma^+(E(A_\beta))$ with $\kappa_\beta^*(\nu_\beta) = \mu_\beta$. Define ν: $Ba(E(X)) \to R$ by $\nu(D) = \sum_{\beta\in B} \nu_\beta(D \cap E(A_\beta)) = \sum_{\beta\in C} \nu_\beta(D \cap E(A_\beta))$. Then it is easy to see that ν is countably additive and $\kappa^*(\nu) = \mu$.

We conclude this section by proving that the SLP is preserved by products with compact spaces. It is convenient to begin with some topological considerations. Let X be an arbitrary completely regular space, Y compact. We define a map $h\colon E(X \times Y) \to E(X)$ in such a way that the diagram

$$\begin{array}{ccc} E(X\times Y) & \xrightarrow{\ h\ } & E(X) \\ {\scriptstyle \kappa_{X\times Y}}\downarrow & & \downarrow{\scriptstyle \kappa_X} \\ X\times Y & \xrightarrow[\ \pi\ ]{} & X \end{array}$$

commutes. Such a map h is called an absolute of the projection π; the properties of such maps have been studied by Sapiro [22].

Suppose $p \in E(X \times Y)$. If $F \in p$, then $\pi(F)$ is closed, because a projection along a compact factor is a perfect map, and has dense interior, because π is continuous and open. Thus $\pi(F) \in RC(X)$.

Now we show that $q = \{\pi(F)\colon F \in p\}$ is an ultrafilter in $RC(X)$. If $F_1, F_2 \in p$, then $\mathrm{int}_{X\times Y}(F_1) \cap \mathrm{int}_{X\times Y}(F_2) \neq \emptyset$, and so $\mathrm{int}_X\pi(F_1) \cap \mathrm{int}_X\pi(F_2) \neq \emptyset$. If $F \in p$, $G \in RC(X)$, and $\pi(F) \subset G$, then $F \subset G \times Y$, so $G \times Y \in p$, and $\pi(G \times Y) = G \in q$. Thus q is a filter in $RC(X)$. Now let $H \in RC(X)$, and suppose $H \wedge \pi(F) \neq \emptyset \ \forall F \in p$. Then $\mathrm{int}_X H \cap \pi(F) \neq \emptyset$, and so $\mathrm{int}_{X\times Y}(H\times Y) \cap F \neq \emptyset$. Since F is regular closed, $\mathrm{int}_{X\times Y}(H \times Y) \cap \mathrm{int}_{X\times Y}F \neq \emptyset$. Thus $H \times Y \in p$, so $\pi(H \times Y) = H \in q$.

Since $p \in E(X \times Y)$, $\lim p = (x_0,y_0)$ exists. If F is a regular closed neighborhood of x_0, then $F \times Y$ is a regular-closed neighborhood of (x_0,y_0), so $F \times Y \in p$, and therefore $F \in q$. We have $\lim q = x_0$; since q is a convergent ultrafilter, $q \in E(X)$.

<u>Theorem 3.7.</u> If X has the SLP, and Y is compact, then $X \times Y$ has the SLP.

<u>Proof.</u> Define $h\colon E(X \times Y) \to E(X)$ by $h(p) = q$. The preceding paragraph shows that $\kappa_X \circ h = \pi \circ \kappa_{X\times Y}$. Moreover, if $F \in RC(X \times Y)$, then $h(\lambda_{X\times Y}(F)) \subset \lambda_X(\pi(F))$, and this implies that h is continuous on $E(X \times Y)$.

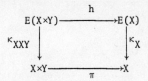

Let $\mu \in M_\sigma^+(X \times Y)$, $\nu \in M^+(E(X \times Y))$, $\kappa_{X\times Y}^*(\nu) = \mu$. Then $\kappa_X^*(h^*(\nu)) = \pi^*(\mu) \in M_\sigma^+(X)$, so $h^*(\nu) \in M_\sigma^+(E(X))$, since X has the SLP. To prove that ν is σ-additive, it suffices to show that if (C_n) is a sequence of clopen sets in $E(X \times Y)$ with $C_n \downarrow \emptyset$, then $\nu(C_n) \to 0$. Now $C_n = \lambda_{X\times Y}(F_n)$, where each $F_n \in RC(X \times Y)$ and $F_n \downarrow \emptyset$. Since Y is compact, $\pi(F_n) \downarrow \emptyset$, and so $D_n = \lambda_X(\pi(F_n)) \downarrow \emptyset$ in $E(X)$. Note that $h(C_n) \subset D_n \ \forall n$, from the first paragraph of the proof.

Now if $f_n = \chi_{D_n} \in C^*(E(X))$, then $f_n \downarrow 0$, and $\chi_{C_n} \le f_n \circ h \ \forall n$. Thus $\nu(C_n) \le \nu(f_n \circ h) = h^*(\nu)(f_n) \downarrow 0$, since $h^*(\nu)$ is σ-additive. Thus ν is also σ-additive, to complete the proof.

4. The Borel extension problem. A well-known result of Marik [15] asserts that if X is normal and countably paracompact, then every $\mu \in M_\sigma^+(X)$ admits a unique extension to a closed-regular countably-additive Borel measure. Without normality the uniqueness of the extension is, in general, lost. The general result of Bachman and Sultan [1], quoted above as Theorem 2.6, insures that any $\mu \in M^+(X)$ admits a finitely-additive closed regular Borel extension. However, even if μ is countably-additive, it may not be possible to form a countably-additive Borel extension.

Example 4.1. Let p be the "corner point" of $\underline{D}$, μ the corresponding σ-additive measure (as in 1.1). The top edge Z_0 is a discrete zero-set of cardinal $\aleph_1$; $\mu(\{x\}) = 0$ for each $x \in Z_0$, but $\mu(Z_0) = 1$. Since every subset of Z_0 is closed and therefore a Borel set, μ cannot have a countably-additive Borel extension, by a well-known result of Ulam [26].

Proposition 4.2. Measure-compact spaces and spaces with property (*) are Marik spaces.

Proof. It is known (e.g., [10]) that any τ-additive measure admits a (net-additive), closed-regular Borel extension, so the result holds for measure-compact spaces. The other half follows at once from Theorem 2.6 of Bachman and Sultan: Let $T = X$, L_1 = all zero-sets, L_2 = all closed sets, and note that the condition "L_2 is L_1-countably paracompact" is precisely property (*). This extends the existence portion of Marik's result, since every normal and countably paracompact space has (*).

Remark 4.3. If X has either property mentioned in 4.2, then so does E(X); hence E(X) will also be a Marik space.

It is apparently an open question whether every countably paracompact space is a Marik space. We do have:

Proposition 4.4. (a) If X is countably paracompact, then E(X) is a Marik space; (b) if also X has (**), then X is a Marik space.

Proof. (a) Countable paracompactness is preserved (both ways) by perfect maps [7]; but E(X), being extremally disconnected, is also weak cb, hence cb, and therefore has (*). (b) countable paracompactness + (**) implies (*), using the characterization given in the proof of 1.8 (a).

Remark 4.5. RDS is normal and weak cb, but not Marik; SP is realcompact, but not Marik. In the other direction, T is Marik, but not countably paracompact; MJ is Marik, but is not measure-compact, and does not have (*), or even (**).

Note also that T#D is Marik, but its regular closed, C-embedded subset D is not; PD is pseudocompact and Marik, but its closed, C^*-embedded subset D is not.

Remark 4.6. Moran [19] and Haydon [6] have shown that if X is metacompact, and every closed discrete subset has non-measurable cardinal, then $\mu \in M_\sigma^+(X)$ has a closed-regular countably-additive Borel extension if and only if it is τ-additive. Thus in this case, X is Marik if and only if it is measure-compact, and property (*) implies the SLP.

5. The counter-examples. They are listed here in alphabetical order, and some proofs that properties do or do not hold are indicated. Table II summarizes the information about the spaces; the notation (CH) indicates that the continuum hypothesis was assumed in verifying the presence or absence of a given property. There are two gaps: I do not know if MPS satisfies (*) or (**).

Of the nine spaces, only RDS is normal; D and MPS are metacompact;

TABLE II.

	D	MJ	MPS	PD	RDS	SP	T	T#D	$T\#T^{-}$
COUNTABLY PARACOMPACT	NO	YES	NO	NO	NO	NO	NO	NO	NO
WEAK CB	NO	NO	NO	YES	YES	YES	YES	NO	NO
(*)	NO	NO		NO	NO	NO	YES	NO	YES
(**)	NO	NO		YES	YES	YES	YES	NO	YES
δNS	NO	YES	NO	NO	YES	NO	NO	NO	NO
WδNS	NO	YES	NO	YES	YES	YES	YES	NO	NO
WLP	NO	YES	YES(CH)	YES	YES	YES	YES	YES	YES
SLP	NO	NO	YES(CH)	YES	YES	YES	YES	NO	NO
MARIK	NO	YES	YES(CH)	YES	NO	NO(CH)	YES	YES	YES
REALCOMPACT	NO	NO	YES	NO	NO	YES	NO	NO	NO
MEASURE-COMPACT	NO	NO	YES(CH)	NO	NO	NO	NO	NO	NO

PD and T are pseudocompact. None is cb; indeed any cb-space would automatically satisfy the first nine properties in Table II.

D, the Dieudonné plank [25, Ex. 89].

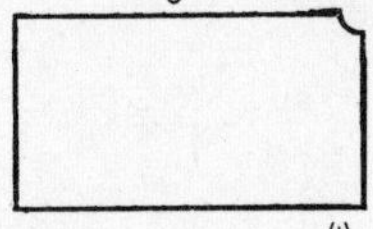

D fails the WLP [28], and is not Marik, by 4.1. To see that D is not WδNS, partition $[0,\omega_1)$ into a sequence (A_n) of uncountable disjoint subsets. Then $F = \bigcup_{n=1}^{\infty}\{(\alpha,n)\colon\ \alpha \in A_n \text{ or } \alpha = \omega_1\}$ is regular closed, and Z_0 is a disjoint zero-set, but F and Z_0 cannot be completely separated.

MJ, the Mack-Johnson space [13, pp. 240-241].

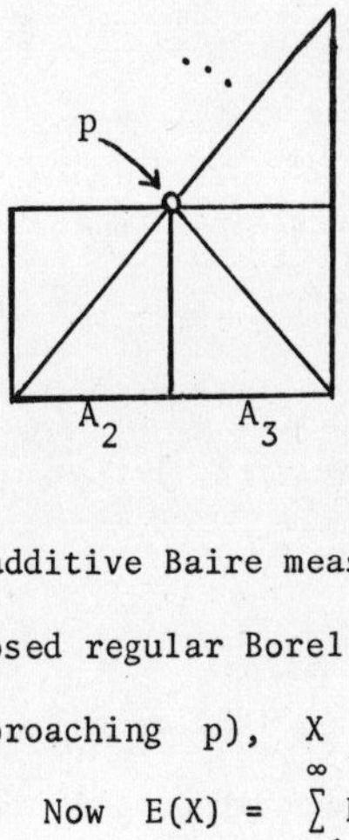

Let $A = \{(\alpha,\beta)\colon\ 0 \le \alpha < \omega_1,\ 0 \le \beta \le \omega_1,\ \alpha \le \beta\}$. To form MJ, attach countably many copies of A alternately along the diagonal and the edge $\{(\alpha,\omega_1)\colon\ 0 \le \alpha < \omega_1\}$. The Hewitt realcompactification is obtained by adding $p = (\omega_1,\omega_1)$, and is σ-compact. If $p \in cl_{\upsilon X}Z$, then Z contains a deleted neighborhood of p; this leads to a proof that X = MJ is δNS. The only purely σ-additive Baire measures are multiples of μ_p (as in 1.1). Since μ_p has many closed regular Borel extensions (form a "Dieudonné measure" on any ordinal edge approaching p), X is Marik.

Now $E(X) = \sum_{n=1}^{\infty} E(A_n)$, by 3.5. Since each A_n is countably compact, so is $E(A_n)$ [7]; thus $\upsilon E(X) = \sum_{n=1}^{\infty} \upsilon E(A_n) = \sum_{n=1}^{\infty} \beta E(A_n)$. On the other hand, $E(\upsilon X) = E(X \cup \{p\}) = [\sum_{n=1}^{\infty} E(A_n)] \cup \kappa^{-1}(p)$. Then $\sum_{n=1}^{\infty} (\beta E(A_n) - E(A_n)) \subset \kappa^{-1}(p)$, and the containment is proper, since the RHS is compact and the LHS is not. Thus μ_p has some functional extensions which are σ-additive and some which are not; hence MJ has the WLP but not the SLP.

MPS, Michael's product space [16; 25, Ex. 85].

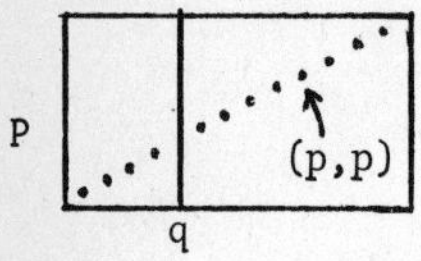

P = irrationals, Q = rationals, $\hat{R} = R$ with the irrationals made discrete. Then MPS $= \hat{R} \times P$. This space is not countably paracompact [14]: if $F_n = \{(q,p) \in Q \times P: |q - p| \leq 1/n\}$, then (F_n) is a sequence of closed sets with $F_n \downarrow \emptyset$, but there is no sequence (H_n) of regular closed sets with $F_n \subset \text{int}_X H_n \subset H_n \downarrow \emptyset$. Also, MPS is not WδNS: express P as a countable disjoint union of dense subsets A_n, and let $F = (Q \times P) \cup \bigcup_{n=1}^{\infty} H_n$, where $H_n = \{(p,p'): p \in A_n, |p - p'| \geq 1/n\}$. Then F is regular closed, but cannot be completely separated from the diagonal, which is a zero-set. Thus MPS is not weak cb [13, p. 238].

Assuming the continuum hypothesis, MPS is measure-compact [18, p. 508] and therefore is Marik and has the SLP.

PD, the pseudocompact extension of D.

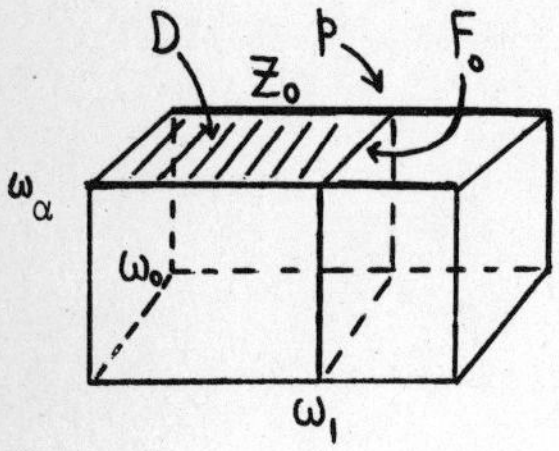

Noble [20, Construction 2.3] gives a method for embedding any completely regular T_2 space X as a closed C^*-embedded subspace of a pseudocompact space PX. Following his construction, let $\aleph_\alpha$ be a cardinal greater than card βD such that α is a non-limit ordinal, let ω_α be the smallest ordinal of cardinal $\aleph_\alpha$, and let PD $= (\beta D \times [0,\omega_\alpha)) \cup (D \times \{\omega_\alpha\})$, topologized as a subspace of $\beta D \times [0,\omega_\alpha]$. If $T = \beta D \times [0,\omega_\alpha)$, then T is countably compact and locally compact, and $\beta T = \beta(\text{PD}) = \beta D \times [0,\omega_\alpha]$ (see [4, 9K.5]). In what follows we identify βD with $\beta D \times \{\omega_\alpha\}$. Note that PD contains the vertical "line segment" $[0,\omega_\alpha)$ under p, but not p itself.

PD is not δNS: Let W be a zero-set of βD with $W \cap D = Z_0$, the top edge of D. Then $(W \times [0,\omega_\alpha]) \cap$ PD is a zero-set of PD which cannot be completely separated from F_0, the right edge of D.

PD is a Marik space: since PD is pseudocompact, every $\Phi \in M^{+}(X)$ is σ-additive [27, p. 172]. It will be enough to show that there is a closed-regular countably-additive Borel measure λ on PD such that

$\Phi(f) = \int_{PD} f d\lambda \ \forall f \in C^{*}(PD)$. Indeed in this case the unique Baire measure μ corresponding to Φ must be $\lambda | Ba(PD)$, so λ is a Borel extension of μ.

Now Φ does correspond to a compact-regular Borel measure ν on $\beta(PD) = \beta T$. Since T is locally compact, it is open in βT. Similarly, $D - F_0$ is open in βD, and F_0 is σ-compact; thus PD is a Borel set in $\beta(PD)$, and $\nu_1 = \nu | Bo(PD)$ is compact-regular on $Bo(PD)$.

If $\nu(\{p\}) = c > 0$, let V be the vertical segment $[0,\omega_\alpha)$ under p, and define

$$\nu_2(B) = \begin{cases} c, & \text{if } B \cap V \text{ contains an unbounded closed subset of } V \\ 0, & \text{otherwise} \end{cases}$$

This "Dieudonne measure" is closed-regular on $Bo(PD)$; cf. Example 4.1.

Finally, let (K_n) be an increasing sequence of compact subsets of $S = \beta D - (D \cup \{p\})$ such that $\nu(\bigcup_1^\infty K_n) = \nu(S)$, and let $\gamma_j(B) = \nu(B \cap [K_j - \bigcup_{i<j} K_i])$. Then γ_j is a compact-regular Borel measure on $K_j \times \{\omega_\alpha\}$. Now $T_j = K_j \times [0,\omega_\alpha)$ is closed in PD, countably compact, and satisfies $\beta T_j = K_j \times [0,\omega_\alpha]$. Since any countably compact space is cb, and therefore Marik, there is a closed-regular countably-additive Borel measure λ_j on T_j which represents the same functional on $C^{*}(T_j)$ as γ_j. Then $\lambda = \nu_1 + \nu_2 + \sum_1^\infty \lambda_j$ is a closed-regular Borel measure on PD corresponding to Φ.

RDS, Mary Ellen Rudin's Dowker space [21].

This famous example of a normal non-countably paracompact space is not realcompact [21]. Hardy and Juhasz [5] have shown that RDS is weak cb, although its Alexandroff duplicate is not. RDS is a P-space [21] (every G_δ-set is open, every zero-set is clopen), so it surely is δNS.

If X is any normal P-space, $\mu \in M_\sigma^{+}(X)$, and ν is a countably-additive closed-regular Borel extension of μ, then μ and ν have the same range. Indeed if $B \in Bo(X)$, we can use the P-space property to find a closed set F and an open set O with $F \subset B \subset O$ and $\nu F = \nu O$. Then normality yields a zero-set Z with $F \subset Z \subset O$.

This observation leads to a proof that RDS is not Marik. Simon [24] proved that RDS is closed-complete (i.e., every ultrafilter of closed sets with the countable intersection property is fixed). Gardner [3, Th. 3.5] showed that every {0,1}-valued regular Borel measure on a closed-complete space is τ-additive. Hence if $p \in \nu(\text{RDS}) - \text{RDS}$, then the {0,1}-valued Baire measure μ_p (as in 1.1) is σ-additive, but has no regular Borel extension.

SP, the Sorgenfrey plane [25, Ex. 84]

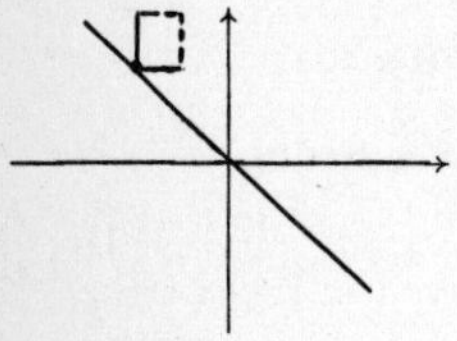

SP is realcompact but not measure-compact [17]. However, Gardner has observed [3, p. 101] that SP is Borel measure-compact, assuming the continuum hypothesis. By Theorem 3.3 of the same paper, SP is not Marik.

Badé [2] has shown that every regular closed subset of SP is a zero-set; hence SP is weak cb. The irrational and rational points of the negative diagonal are disjoint closed sets, the former being a zero-set; but they cannot be completely separated, so SP is not δNS.

To see that SP fails (*), identify [0,1] with an interval of length one on the negative diagonal. Let $[0,1] = \sum_1^\infty A_n$, where each A_n has Lebesgue outer measure 1, and let $F_n = \bigcup_{k=n}^\infty A_k$. Then each F_n is closed in SP, but there cannot be any sequence (B_n) of Baire sets in SP with $F_n \subset B_n \downarrow \emptyset$. Indeed a result of Badé [2, Th. 2.1] implies that $Ba(\text{SP}) = Bo(R^2)$, so that each $B_n \cap [0,1]$ would be a Euclidean Borel subset of [0,1] with Lebesgue measure 1. Such a sequence could not then have empty intersection.

T, the Tychonoff plank [25, Ex. 87]

T is pseudocompact, but not countably compact. However, the open subspace

$[0,\omega_1) \times [0,\omega_0]$ is countably compact; this leads easily to a proof that T has (*).

T#D, the join of the Tychonoff plank and the Dieudonné plank.

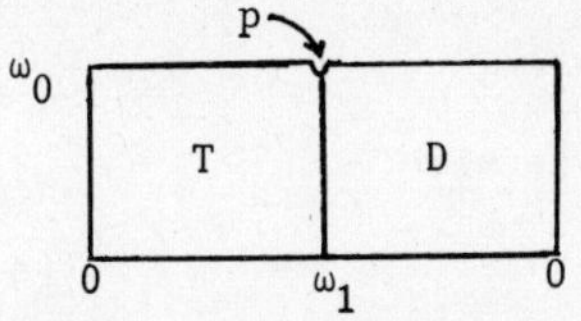

This space is obtained by joining T and D along the edge $\{(\omega_1,n): n \in N\}$. T#D has the WLP but not the SLP [28]. Multiples of μ_p (as in 1.1) are the only purely σ-additive measures, and the Dieudonné measure on the top edge of T yields regular Borel extensions. Hence T#D is Marik, although its regular closed, C-embedded subset D is not. The proof that T#D is not WδNS is similar to the proof for D.

$T\#T^-$, the deleted join of the Tychonoff plank with itself.

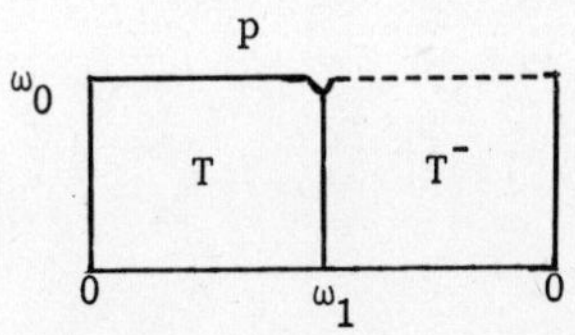

Join two copies of T along the edge $\{(\omega_1,n): n \in N\}$, but delete the top edge of one. The space is not WδNS (T^- and the top edge of T). The Hewitt realcompactification is obtained by adding $p = (\omega_1,\omega_0)$; it follows (see Remark 2.9) that $T\#T^-$ is Marik, and has the WLP, but not the SLP. If (F_n) is a sequence of closed sets with $F_n \downarrow \emptyset$, then F_n is a subset of T^- (including the common edge) for all large n. It follows that the space has property (*).

6. Some open questions.

Q1: What topological condition on X is equivalent to the statement: If $\mu \in M_\sigma^+(X)$, then every functional extension of μ is a measure extension? Note that MPS has this property, although it is not WδNS.

Q2: Must a realcompact space have the WLP? the SLP?

Q3: Does every almost realcompact, non-realcompact space fail the WLP? (See [28]).

Q4: Propositions 3.4, 3.5, 3.6, and 3.7 are surely not the best possible results in the area of permanence properties. What improvements can be made?

Q5: Is every countably paracompact space a Marik space? Is every pseudocompact space a Marik space?

Q6: Must a measure-compact space satisfy (*)? (**)? MPS is a possible counter-example.

Q7: Is it true that X is Marik if and only if E(X) is Marik?

REFERENCES

1. G. Bachman and A. Sultan, On lattice tight measures and various types of measure extensions and measure repleteness, to appear in Pacific J. Math.

2. W. Badé, Two properties of the Sorgenfrey plane, Pacific J. Math. 51 (1974), 349-354.

3. R. Gardner, The regularity of Borel measures and Borel measure-compactness, Proc. London Math. Soc. 30 (1975), 95-113.

4. L. Gillman and M. Jerison, Rings of continuous functions, Springer-Verlag, New York, 1976.

5. K. Hardy and I. Juhasz, Normality and the weak cb property, Pacific J. Math. 64 (1976), 167-172.

6. R. Haydon, On compactness in spaces of measures and measurecompact spaces, Proc. Lond. Math. Soc. 29 (1974), 1-16.

7. M. Henrikson and J. Isbell, Some properties of compactifications, Duke Math. J. 25 (1958), 83-105.

8. F. Ishikawa, On countably paracompact spaces, Proc. Japan Acad. 31 (1955), 686-687.

9. A. Kato, Union of realcompact spaces and Lindelöf spaces, to appear.

10. R. Kirk, Locally compact, B-compact spaces, Kon. Ned. Akad. van Wetensch. Proc. Ser. A 72 (1969), 333-344.

11. J. Knowles, Measures on topological spaces, Proc. London Math. Soc. 17 (1967), 139-156.

12. J. Mack, On a class of countably paracompact spaces, Proc. Amer. Math. Soc. 16 (1965), 467-472.

13. ______ and D. Johnson, The Dedekind completion of C(X), Pacific J. Math. 20 (1967), 231-243.

14. ______, Countable paracompactness and weak normality properties, Trans. Amer. Math. Soc. 148 (1970), 265-272.

15. J. Marik, The Baire and Borel measure, Czech. Math. J. 7 (82) (1957), 248-253.

16. E. Michael, The product of a normal space and a metric space need not be normal, Bull. Amer. Math. Soc. 69 (1963), 375-376.

17. W. Moran, The additivity of measures on completely regular spaces, J. London Math. Soc. 43 (1968), 633-639.

18. ______, Measures and mappings on topological spaces, Proc. London Math. Soc. 19 (1969), 493-508.

19. ______, Measures on metacompact spaces, Proc. London Math. Soc. 20 (1970), 507-524.

20. N. Noble, Countably compact and pseudocompact products, Czech. Math. J. 19 (94) (1969), 390-397.

21. M. Rudin, A normal space X for which $X \times I$ is not normal, Fund. Math. 73 (1971), 179-186.

22. L. Sapiro, On absolutes of topological spaces and continuous mappings, Soviet Math. Doklady 17 (1976), 147-151.

23. D. Sentilles, Bounded continuous functions on a completely regular space, Trans. Amer. Math. Soc. 168 (1972), 311-336.

24. P. Simon, A note on Rudin's example of Dowker space, Comm. Math. Univ. Carolinae 12 (1971), 825-834.

25. L. Steen and J. Seebach, Counterexamples in topology, second edition, Springer-Verlag, New York, 1978.

26. S. Ulam, Zur Masstheorie in der allgemeinen Mengenlehre, Fund. Math. 16 (1930), 140-150.

27. V. Varadarajan, Measures on topological spaces, Amer. Math. Soc. Transl. 48 (1965), 161-228.

28. R. Wheeler, Topological measure theory for completely regular spaces and their projective covers, to appear in Pacific J. Math.

SOME SELECTION THEOREMS FOR PARTITIONS OF SETS WITHOUT TOPOLOGY

M.P.Ershov
(Univ.Linz, Austria)

We prove three theorems on measurable selections in non-topological cases which are improvements of the result of [1].

We will not consider here topological applications. One might find them in the papers which motivated our results. The technique here is a modification of that of M.Sion [6] , and Theorem 2 below is closely related with Theorem 4.1 of [6]. There is also a relationship between Theorem 1 and Theorem II.6.1 in J.Hoffmann-Jørgensen[4]. Results on continuous selections motivated Theorem 3 (first of all, it is M.Hasumi's paper [3]; see also S.Graf [2]).

A different and very interesting approach to non-topological forms of measurable and continuous selections was recently proposed by G.Mägerl [5]. A very detailed survey, where one can find other non-topological selection theorems, was given by D.H.Wagner [7] ,[8] to whom the author is indebted for useful remarks and suggestions.

Let X be a non-empty set and $\mathscr{X}$ a class of its subsets. Points x and x' from X are called $\mathscr{X}$-equivalent if there is no set in $\mathscr{X}$ containing only one of them. Classes of $\mathscr{X}$-equivalence are called atoms of $\mathscr{X}$. The atom of $\mathscr{X}$ containing a given point $x \in X$ will be denoted by $a_{\mathscr{X}}(x)$. If $\mathscr{X}$ coincides with the set of its atoms, it is called a partition of X.

Let $\mathscr{A}$ be a partition of X. A set $X' \subset X$ is called a selection of $\mathscr{A}$ if each atom of $\mathscr{A}$ contains one and only one point of X'. Let X' be a selection of $\mathscr{A}$. The function $x \longmapsto a_{\mathscr{A}}(x)X'$ is called a selector of $\mathscr{A}$. There is an obvious one-to-one correspondence between selections and selectors.

Given an $\mathscr{X} \subset 2^X$ and a set $H \subset X$, the saturation of H with respect to $\mathscr{X}$ is the set

$$\langle H \rangle_{\mathscr{X}} := \{ x \in X : a_{\mathscr{X}}(x)H \neq \emptyset \}.$$

If $\mathscr{H} \subset 2^X$,

$$\langle \mathscr{H} \rangle_{\mathscr{X}} := \{ \langle H \rangle_{\mathscr{X}} : H \in \mathscr{H} \}.$$

For set-theoretic operations we use the following notation:

$^-$ complement;

$\cap$, $\cap_{\mathfrak{m}-}$, $\cap_{\mathfrak{m}}$, $\cap_a$ intersections respectively: finite, of cardinality $<\mathfrak{m}$, $\leq \mathfrak{m}$, arbitrary; analogously for unions;

$\mathcal{S}$ Souslin operation.

Let $\mathcal{O}$ be one of the above operations, then $(\mathcal{O})\mathcal{H}$ stands for the class obtained from sets of $\mathcal{H}$ with operation $\mathcal{O}$; e.g. $(^-)\mathcal{H} = \{\bar{H} : H \in \mathcal{H}\}$. If $\mathcal{O}$ is one or several of the above operations, $\mathcal{H}(\mathcal{O})$ stands for the minimal class containing $\mathcal{H}$ and closed with respect to $\mathcal{O}$. If $\mathcal{H} = \mathcal{H}(\mathcal{O})$, we say that $\mathcal{H}$ is an $(\mathcal{O})$-*class*. We write $\alpha(\mathcal{H})$ and $\sigma(\mathcal{H})$ in place of $\mathcal{H}(^-, \cap)$ and $\mathcal{H}(^-, \cap_{\aleph_0})$ respectively.

In simple formulas, the operations $^-$ and $\cap$ have respectively the first and second priority, the latter symbol being omitted.

A class $\mathcal{H} \subset 2^X$ is called *compact* if any its subclass with empty intersection contains a finite subclass also with empty intersection.

In the following theorems, fixed are: a set X, its partition $\mathcal{A}$ and a class $\mathcal{H} \subset 2^X$ of cardinality $\mathfrak{m}$, and the two conditions are assumed:

(I) *for each* $a \in \mathcal{A}$, *the class* $\{aH : H \in \mathcal{H}\}$ *is compact*;

(II) *atoms of* $\mathcal{A} \cup \mathcal{H}$ *are singletons*.

Theorem 1. Let $\mathcal{X} \subset 2^X$ *be an* $(\cap_{\mathfrak{m}-}, \cup_{\mathfrak{m}-})$*-class containing* $\mathcal{H}(^-)$ *and such that*

$$\langle (\cap)\mathcal{H} \rangle_{\mathcal{A}} \subset \mathcal{X}.$$

There exists a selection X' *of* $\mathcal{A}$ *such that*

$$\overline{X'} \in (\cup_{\mathfrak{m}})\mathcal{X}.$$

Theorem 2. Let $\mathfrak{m} \leq \aleph_i$ $(i = 0, 1)$.

Then there exists a selector f *of* $\mathcal{A}$ *such that*

$$f^{-1}(\mathcal{H}) \subset \alpha(\langle (\cap)\mathcal{H} \rangle_{\mathcal{A}}) \quad (i=0),$$

$$f^{-1}(\mathcal{H}) \subset \sigma(\langle (\cap)\mathcal{H} \rangle_{\mathcal{A}}) \quad (i=1).$$

Let $\mathfrak{m}$ be a cardinal number. By an $\mathfrak{m}$-*operation* we will call any function $\tau : 2^{2^X} \longrightarrow 2^{2^X}$ with the properties:

(a) for any function $f : X \longrightarrow X$ and any class $\mathcal{C} \subset 2^X$,

$$f^{-1}(\tau(\mathcal{C})) \subset \tau(f^{-1}(\mathcal{C}));$$

(b) the relations $\tau(\mathcal{C}) \subset \tau(\mathcal{C}')$ and $C \in \tau(\mathcal{C}')$ imply

$$\tau(\mathcal{C} \cup \{C\}) \subset \tau(\mathcal{C}');$$

(c) for any initial ordinal ξ with $|\xi| \leq \mathfrak{n}$ and any increasing transfinite ξ-sequence $\{\mathcal{C}_\alpha\}_{\alpha<\xi} \subset 2^{2^X}$, the inclusions $\tau(\mathcal{C}_\alpha) \subset \tau(\mathcal{C})$ $(\alpha<\xi)$ imply

$$\tau\Big(\bigcup_{\alpha<\xi} \mathcal{C}_\alpha\Big) \subset \tau(\mathcal{C}).$$

Examples of $\mathfrak{n}$-operations (for infinite $\mathfrak{n}$):

(i) $\tau : \mathcal{C} \longmapsto (\mathcal{O})\mathcal{C}$ where $\mathcal{O} = \cap$ or $\cup$, $= \cap_{\mathfrak{m}-}$ or $\cup_{\mathfrak{m}-}$ (if $\mathfrak{m}$ is weakly inaccessible, or $\mathfrak{m} < \mathfrak{n}$ and $\mathfrak{n}$ is weakly inaccessible), $= \cap_{\mathfrak{m}}$ or $\cup_{\mathfrak{m}}$ for infinite $\mathfrak{m}$, $= \cap_a$ or $\cup_a$, $= \mathcal{S}$;

(ii) $\tau : \mathcal{C} \longmapsto (\cup_{\mathfrak{m}})(\cap_{\mathfrak{m}})\mathcal{C}$ if $\mathfrak{n} > \mathfrak{m} \geq \aleph_0$ is weakly inaccessible;

(iii) $\tau : \mathcal{C} \longmapsto \mathcal{C}(\mathcal{O})$ where $\mathcal{O}$ is any subset of the set $\{^-, \cap, \cup, \cap_{\mathfrak{m}-}, \cup_{\mathfrak{m}-}, \cap_{\mathfrak{m}'}, \cup_{\mathfrak{m}'}, \cap_a, \cup_a, \mathcal{S}\}$; in particular, $\mathcal{C} \longmapsto \mathcal{C}(\cap, \cup_a)$ (the topology generated by $\mathcal{C}$) or $\mathcal{C} \longmapsto \sigma(\mathcal{C})$;

(iv) $\tau : \mathcal{C} \longmapsto \sigma(\mathcal{C})(\mu)$ or $\mathcal{C} \longmapsto \sigma(\mathcal{C})(\wedge)$ (the completion with respect to a measure μ or the universal completion).

Theorem 3. Let a class $\mathcal{X} \subset 2^X$ be such that $\mathcal{X} = \langle \mathcal{X} \rangle_{\mathcal{A}}$ and, for any $\mathcal{H}' \subset \mathcal{H}$ with card $\mathcal{H}' < \mathfrak{n}$,

$$\langle \mathcal{H} \rangle_{\mathcal{A} \cup \mathcal{H}'} \subset \tau(\mathcal{X} \cup \mathcal{H}')$$

where τ is an $\mathfrak{n}$-operation.

Then there exists a selector f of $\mathcal{A}$ such that

$$\tau(f^{-1}(\mathcal{X} \cup \mathcal{H})) \subset \tau(\mathcal{X}).$$

Proof. In all the three theorems a selector is constructed in the same way.

Let us somehow well-order $\mathcal{H}$:

$$\mathcal{H} = \{H_\alpha : \alpha < \xi\}$$

where ξ is the initial ordinal number of cardinality $\mathfrak{n}$.

For each $a \in \mathcal{A}$, define inductively:

$$f_0(a) = a,$$

$$f_{\alpha+1}(a) = \begin{cases} aH_\alpha & (aH_0 \neq \emptyset) \\ a & (aH_0 = \emptyset) \end{cases},$$

$$f_\beta(a) = \bigcap_{\alpha<\beta} f_\alpha(a) \text{ for limit } \beta .$$

From conditions (I) and (II) it follows that

$$f : x \longmapsto f_{\xi}(a_{\mathcal{A}}(x))$$

is a selector of $\mathcal{A}$.

To check the assertions of the theorems we use different induction procedures.

1. Put

$$X_{\alpha} = \bigcup_{a \in \mathcal{A}} f_{\alpha}(a) \quad (\alpha \leq \xi).$$

It is clear that

$$X' := X_{\xi} = f(X)$$

is a selection of $\mathcal{A}$.

By construction

$$X_{\alpha+1} = X_{\alpha}(H_{\alpha} \cup \overline{\{H_{\alpha}\}_{\alpha}})$$

where, for each $H \subset X$,

$$\{H\}_{\alpha} := \bigcup_{a \in \mathcal{A}: f_{\alpha}(a) H \neq \emptyset} f_{\alpha}(a).$$

It is easy to see that, for each $H \subset X$,

$$\{H\}_{\alpha+1} = \{HH_{\alpha}\}_{\alpha} H_{\alpha} \cup \{H\}_{\alpha} \overline{\{H_{\alpha}\}_{\alpha}}$$

and by condition (I), for limit β and each $H \in \mathcal{H}(\cap_{a}, \cup)$,

$$\{H\}_{\beta} = \bigcap_{\alpha < \beta} \{H\}_{\alpha}.$$

The induction assumption: for each $\alpha < \beta$ and $H \in (\cap)\mathcal{H}$,

$$X_{\alpha} \in (^{-})\mathcal{X},$$

$$X_{\alpha} \overline{\{H\}_{\alpha}} \in (^{-})\mathcal{X}.$$

Verification:

$$X_{0} = X = H\overline{\overline{H}} \in (^{-})\mathcal{X},$$

$$X_{0}\overline{\{H\}_{0}} = \overline{\langle H \rangle_{\mathcal{A}}} \in (^{-})\mathcal{X}.$$

It is easy to check that $(^{-})\mathcal{X}$ is an $(\cap_{w-}, \cup_{w-})$-class. Therefore

$$X_{\alpha+1} = X_{\alpha} H_{\alpha} \cup X_{\alpha} \overline{\{H_{\alpha}\}_{\alpha}} \in (^{-})\mathcal{X},$$

$$X_{\alpha+1} \overline{\{H\}_{\alpha+1}} = X_{\alpha}(H_{\alpha} \cup \overline{\{H_{\alpha}\}_{\alpha}})(\overline{\{HH_{\alpha}\}_{\alpha}} \cup \overline{H_{\alpha}})(\overline{\{H\}_{\alpha}} \cup \{H_{\alpha}\}_{\alpha})$$

$$(X_{\alpha} H_{\alpha} \subset \{H_{\alpha}\}_{\alpha})$$

$$= X_{\alpha}(\cdots\cdots)(\cdots\cdots\cdots)(\cdots\; H_{\alpha}) \in (^{-})\mathcal{X}.$$

For limit β, by construction

$$X_\beta = \bigcap_{\alpha<\beta} X_\alpha \in (^-)\mathcal{X}$$

and

$$X_\beta\overline{\{H\}_\beta} = (\bigcap_{\alpha<\beta} X_\alpha)(\bigcup_{\alpha<\beta}\overline{\{H\}_\alpha})$$

$$= \bigcap_{\alpha<\beta} \bigcup_{\alpha<\alpha'<\beta} X_{\alpha'}\overline{\{H\}_{\alpha'}} \in (^-)\mathcal{X} .$$

[Here we have used the simple proposition: If $\{A_\alpha\}_{\alpha<\beta}$ and $\{A'_\alpha\}_{\alpha<\beta}$ are decreasing and increasing transfinite sequences of sets respectively, then

$$\bigcup_{\alpha<\beta} \bigcap_{\alpha<\alpha'<\beta} A_{\alpha'}A'_{\alpha'} = (\bigcap_{\alpha<\beta} A_\alpha)(\bigcup_{\alpha<\beta} A'_\alpha) = \bigcap_{\alpha<\beta} \bigcup_{\alpha<\alpha'<\beta} A_{\alpha'}A'_{\alpha'} .]$$

Thus $X_\alpha \in (^-)\mathcal{X}$ for each $\alpha<\xi$, and

$$\overline{X'} = \bigcup_{\alpha<\xi} \overline{X_\alpha} \in (\cup_{\mathfrak{w}})\mathcal{X} .$$

2. For each $\alpha<\xi$ and $H\subset X$, put

$$[H]_\alpha = \{x: f_\alpha(a_{\mathcal{A}}(x))H\neq\emptyset\}.$$

It is easy to see that

$$f^{-1}(H_\alpha) = [H_\alpha]_\alpha .$$

The induction assumption: for each $\alpha<\beta$ and $H\in(\cap)\mathcal{H}$,

$$[H]_\alpha \in \begin{cases} \alpha(\langle(\cap)\mathcal{H}\rangle_{\mathcal{A}}) & (i=0) \\ \sigma(\langle(\cap)\mathcal{H}\rangle_{\mathcal{A}}) & (i=1) \end{cases}.$$

Verification:

$$[H]_0 = \langle H\rangle_{\mathcal{A}} .$$

Direct calculation gives

$$[H]_{\alpha+1} = [HH_\alpha]_\alpha \cup [H]_\alpha \overline{[H_\alpha]_\alpha}$$

$$\in \begin{cases} \alpha(\langle(\cap)\mathcal{H}\rangle_{\mathcal{A}}) & (i=0) \\ \sigma(\langle(\cap)\mathcal{H}\rangle_{\mathcal{A}}) & (i=1) \end{cases}.$$

For $i = 0$, this completes the proof. Let $i = 1$; for limit β, by condition (I)

$$[H]_\beta = \bigcap_{\alpha<\beta} [H]_\alpha \in \sigma(\langle(\cap)\mathcal{H}\rangle_{\mathcal{A}}).$$

Thus, for each $\alpha<\xi$,

$$f^{-1}(H_\alpha) = [H_\alpha]_\alpha \in \begin{cases} \alpha(\langle(\cap)\mathcal{H}\rangle_{\mathcal{A}}) & (i=0) \\ \sigma(\langle(\cap)\mathcal{H}\rangle_{\mathcal{A}}) & (i=1) \end{cases}.$$

3. Put

$$\mathcal{H}_\beta = \{H_\alpha : \alpha < \beta\} \qquad (\mathcal{H}_0 = \emptyset).$$

The induction assumption: for each $\alpha < \beta$,

$$\tau(f^{-1}(\mathcal{X} \cup \mathcal{H}_\alpha)) \subset \tau(\mathcal{X}).$$

Verification: By the assumptions, $f^{-1}(\mathcal{X}) = \mathcal{X}$; hence

$$\tau(f^{-1}(\mathcal{X})) = \tau(\mathcal{X}).$$

It is easy to see that

$$f^{-1}(H_\alpha) = f^{-1}(\langle H_\alpha \rangle_{\mathcal{A} \cup \mathcal{H}_\alpha}).$$

Therefore by the assumptions, for each $\alpha < \xi$ (ξ is an initial ordinal number!),

$$f^{-1}(H_\alpha) \in f^{-1}(\tau(\mathcal{X} \cup \mathcal{H}_\alpha)).$$

By property (a) of $\mathcal{N}$-operations, for each $\alpha < \beta$,

$$f^{-1}(H_\alpha) \in \tau(f^{-1}(\mathcal{X} \cup \mathcal{H}_\alpha)) \subset \tau(\mathcal{X}).$$

If β is non-limit, then by property (b) of $\mathcal{N}$-operations,

$$\tau(f^{-1}(\mathcal{X} \cup \mathcal{H}_\beta)) = \tau(f^{-1}(\mathcal{X} \cup \mathcal{H}_{\beta-1}) \cup \{f^{-1}(H_{\beta-1})\}) \subset \tau(\mathcal{X}).$$

If β is limit, then

$$\tau(f^{-1}(\mathcal{X} \cup \mathcal{H}_\beta)) = \tau(\bigcup_{\alpha<\beta} f^{-1}(\mathcal{X} \cup \mathcal{H}_\alpha)) \subset \tau(\mathcal{X})$$

by property (c) of $\mathcal{N}$-operations.

References

1 Ershov M.P., On selectors in abstract spaces, Uspekhi Matematičeskih Nauk 33 5(1978).

2 Graf S., A measurable selection theorem for compact-valued maps, Manuscripta Math. 27(1979), 341 - 352.

3 Hasumi M., A continuous selection theorem for extremally disconnected spaces, Math.Ann 179(1969), 83 - 89.

4 Hoffmann-Jørgensen J., The theory of analytic spaces, Various Publication Series, No.10, Mat.Inst.Aarhus Univ., Aarhus 1970.

5 Mägerl G., A unified approach to measurable and continuous selections, Trans.Amer.Math.Soc. 245(1978), 443 - 452.

6 Sion M., On uniformization of sets in topological spaces, Trans.Amer.Math.Soc. 96(1960), 237 - 245.

7 Wagner D.H., Survey of measurable selection theorems, Siam J. Control and Optimization 15 5(1977), 859 - 903.

8 Wagner D.H., Survey of measurable selection theorems: an update Preprint 1979.

SOME RESULTS ABOUT MULTIMEASURES AND THEIR SELECTORS

C. Godet-Thobie
Département de Mathématiques
Faculté des Sciences de Brest (France)

Introduction and Notations

Let $(\Omega, \mathcal{a})$ a measurable space and X a locally convex vector space. A multimeasure is a map M from $\mathcal{a}$ to the family $P_o(x)$ of non empty subsets of X which satisfies a property of the following type:

If $A_n \in \mathcal{a}$ and $A = \bigcup_{n \in N} A_n$, $M(A) = \sum_{n \in N} M(A_n)$.

Several definitions are possible for $\sum$ and give different multimeasures. In a first part, we compare these definitions and give some examples. In a second part, we give some results about the existence and properties of selectors of a multimeasure.

I. If M is a map from $\mathcal{a}$ to $P_o(x)$, we have

Definition 1: M is a "strong multimeasure" iff

1) M is punctually additive, i.e.

$A, B \in \mathcal{a}$ $A \cap B = \phi$ $M(A \cup B) = M(A) + M(B)$

2) for every disjoint séquence $(A_n)_{n \in N}$ of $\mathcal{a}$, for every x_n of $M(A_n)$, the series x_n is commutatively convergent and M(A) is equal to $\{x \in X \mid \exists\ x_n \in M(A_n),\ x = \sum x_n\}$.

If the values of M are closed, we define

Definition 2: M is a "normal multimeasure" iff

1) M is additive in the following sense:

$\forall A \in \mathcal{a}, \forall B \in \mathcal{a}$ $A \cap B = \phi$, $M(A \cup B) = \overline{M(A) + M(B)}$ denoted $M(A) \dot{+} M(B)$.

2) M is σ-additive with respect to the uniformity of Hausdorff [In particular, if X is metrizable and D the distance of Hausdorff $\lim_{n \to \infty} D[M(A), \dot{\sum}_{i \leq n} M(A_i)] = 0$].

It is possible to show (Cf. prop.6 p.57 of [3]), when M has bounded closed values and X sequentially complete 2) is equivalent to 2').

2') $\forall x_n \in M(A_n)$, the series x_n is commutatively convergent and M(A) is equal to to $\overline{\{\sum_{n \in N} x_n,\ x_n \in M(A_n)\}}$.

The more simple and pleasant definition is the following

Definition 3: If the values of M are closed, M is a "weak multimeasure" iff, for every $x' \in X'$, $m_{x'}(\cdot) = \varphi(x', M(\cdot))$ is a real measure in $]-\infty, +\infty]$ $(\varphi(x', M(A)) = \sup\{x'(x),\ x \in M(A)\})$.

It is easy to prove the propositions 1 and 2.

Proposition 1

If M is a strong multimeasure with values in $\mathcal{P}_o(x)$, the map $\overline{M}$ defined from $\mathfrak{a}$ by $\overline{M}(A) = \overline{M(A)}$ for every A of $\mathfrak{a}$ is a normal multimeasure.

Proposition 2

If M is a normal (resp. weak) multimeasure whose values are weakly compact sets of X, if N is an additive function from $\mathfrak{a}$ to the family of weakly compact sets of X and if, for every A of $\mathfrak{a}$, $N(A) \subset M(A)$, N is a normal (resp. weak) multimeasure.

Proposition 3

If X is equipped by the $\sigma(X,X')$-topology and M has convex weakly compact values, the following conditions are equivalent:

1) *M is a normal multimeasure*
2) *M is a weak multimeasure.*

Summing up of the demonstration

The difficult point is 2 ⇒ 1. Since the additivity of M is evident, if $(A_m)_{n\in N}$ is a disjoint family of $\mathfrak{a}$, $B_n = \bigcup_{i\leq n} A_n$ and $A = \bigcup_{n\in N} A_n$, we must see that $M(B_n)$ converges to $M(A)$ with respect to the Hausdorff uniformity. If $\Omega = N \cup \{\omega\}$ is the Alexandroff compactification of N, we define the multifunction Γ from Ω by $\Gamma(n) = M(B_n)$, for every n of N, and $\Gamma(\omega) = M(A)$. In virtue of Corollary of theorem 2, Ch. 0 of [3] Γ is continued at ω and following [5] th. 3.3, since the Vietoris topology with respect to the $\sigma(X,X')$-topology and the topology associated to the Hausdorff uniformity coincide on the family of convex weakly compact sets of X, $M(B_n)$ converges to $M(A)$, that is, M is a normal multimeasure.

We give now some examples (Cf [2] and [3]).

An example of a strong multimeasure

$(\Omega,\mathfrak{a},\lambda)$, λ a positive measure, X a Banach space and Q a bounded set of X. If ξ is a family of vector measures from $\mathfrak{a}$ to X satisfying, for every m of ξ, for every A of $\mathfrak{a}$, $m(A) \in \lambda(A)\cdot Q$, M defined by

$M(A) = \{ \sum_{i\leq n} m_i(A_i),\ m_i \in \xi,\ (A_i)_{i\leq n}$ finite partition of A$\}$

is a strong multimeasure.

An example of weak multimeasure which is not normal

$X = \mathbb{R}^2$ and C the graph of the parabol $y = (x-1)^2$, $\Omega = \mathbb{N}$ and $\mathfrak{a} = \mathcal{P}(\Omega)$, M is defined by

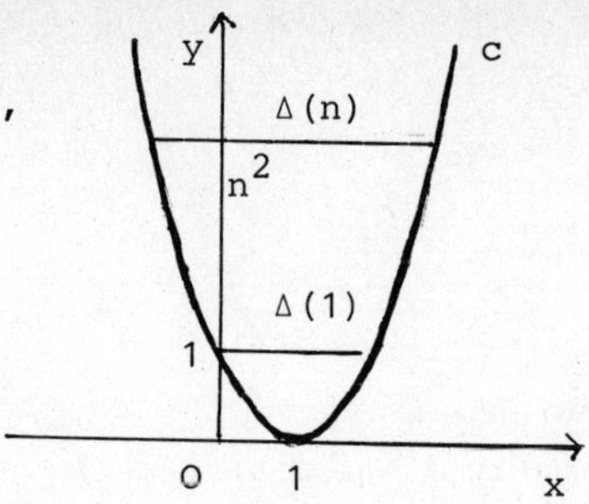

if A is finite, $M(A) = \sum_{n \in A} \Delta(n)$

if A is finite, $M(A) = \mathbb{R}^2$. For every z of $\mathbb{R}^2$, $\lim_{n \to \pm\infty} \varphi(z, \Delta(n)) = +\infty$. Then, for every z of $\mathbb{R}^2$, for every $A = \cup A_n$ ($\cup$ disjoint), $\varphi(z, M(A)) = \sum_{n \in N} \varphi(z, M(A_n))$. But, for every x_n of $M(A_n)$, x_n is not summable.

<u>A classical example</u>

$(\Omega, \mathcal{A}, \lambda)$ a measure space. If Γ is a measurable multifunction from Ω to the family of closed convex bounded sets of X, such that $\varphi(x', \Gamma(\cdot))$ is integrable for every x' of X', if every measurable section of Γ is Pettis integrable, we note, for every A of $\mathcal{A}$,

$$M(A) = \int_A \Gamma d\lambda = \overline{\{\int f d\lambda, f \in S_\Gamma\}}$$

it is true that

$$\varphi(x', \int_A M(t) d\lambda(t)) = \int_A \varphi(x', \Gamma(t)) d\lambda(t)$$

and M is a weak multimeasure.

II. Now, we are interested in weak multimeasures (called briefly multimeasures) and we are concerned with selectors of M.

<u>Definition 4</u>: "<u>Selectors</u>" of M are measures $m: \mathcal{A} \to X$ such that, for every A of $\mathcal{A}$, m(A) belongs to M(A). The set of all selectors of M is denoted by S_M. According to Propositions 2 and 3, if values of M are convex and weakly compact sets of X, a map m from $\mathcal{A}$ to X is a selector of M if and only if m is additive.

If M has compact values, we have

<u>Theorem 1</u>

1) <u>If</u> M <u>has compact values</u>, S_M <u>is not empty and</u>, <u>for every</u> A <u>of</u> $\mathcal{A}$, $M(A) = \{m(A) \mid m \text{ additive selector}\}$.

2) <u>If</u> M <u>is compact convex valued</u>, S_M <u>is convex compact for the pointwise convergence topology and if</u> $\ddot{S}_M$ <u>is the set of extremal points of</u> S_M, <u>for every</u> A <u>of</u> $\mathcal{A}$,

$$M(A) = \{m(A), m \in S_M\} = \overline{co}\{m(A), m \in \ddot{S}_M\}$$

<u>Proof</u>

Given a lexicographic order on X, for every A of $\mathcal{A}$, there exists a lexicographic maximum m(A) for M(A). It is easy to prove m is additive and belongs to S_M ([3] th. 3 p. 63). Then, $M(A) = \{m(A), m \in S_M\}$ is shown by an argument of projective limit of measures ([2] p.III 8).

The last equality of 2 proceeds from Krein-Milman theorem ([3] th.6, p. 67).

If M has no compact values but if there are properties of separability in $(\Omega,\mathfrak{A})$ or X, A. Costê established the three following theorems ([2]). Other results are given in [1] for multimeasures whose values are subsets of $\mathbb{R}^n$.

Theorem 2

If $(\Omega,\mathfrak{A},\lambda)$ is a finite positive measure space, if $\mathfrak{A}_\lambda$ quotient algebra is separable, if M is closed bounded valued such that $M \ll \lambda$ (i.e. $\lambda(A) = 0 \Rightarrow M(A) = \{0\}$) for every A of $\mathfrak{A}$, $M(A) = \overline{\{m(A), m \in S_M\}}$.

Theorem 3

If X is separable and if M has closed convex bounded values, for every A of $\mathfrak{A}$, $M(A) = \overline{\{m(A), m \in S_M\}}$.

Theorem 4

If M has convex closed bounded values and X has Radon-Nikodym property, for every A of $\mathfrak{A}$, $M(A) = \overline{\{m(A), m \in S_M\}}$.

The existence of a selector m of M is proved with same technics of 1 of th. 1, the following lemma and theorem is proved with this result of Phelps: if X has R. N. property, every convex closed bounded set of X is equal to the closed convex hull of its strongly exposed points [6].

Lemma: If A,B,C are closed convex bounded sets of X, such that $C = A \dotplus B$, for every z strongly exposed point of C, there exists a unique $(x,y) \in A \times B$, such that $z = x + y$ where x(resp. y) is a strongly exposed point of A (resp. B).

I think it is unknown if a closed bounded valued multimeasure has always a selector and if the answer is positive, is M(A) equal to $\overline{\{m(A), m \in S_M\}}$?

We finish with a result concerning measurable families of selectors of a multimeasure.

Theorem 5

Let X be a separable Fréchet space and $\mathcal{B}(X)$ its Borel σ-algebra, $(\Omega,\mathcal{C})$ a complete space, $(T,\mathfrak{A},\lambda)$ a separable measurable space and λ a finite positive measure on $\mathfrak{A}$. If M is a multimeasure from $\mathfrak{A}$ to the convex $\sigma(X,X')$ compact subsets of X such that $M \ll \lambda$, if x is a measurable map from Ω to X (measurable for $\mathcal{C}$ and $\mathcal{B}(X)$) such that, for every ω of Ω, $x(\omega)$ belongs to M(T), then there exists a measurable family

$(m_\omega)_{\omega\in\Omega}$ of selectors of M such that

1) for every ω of Ω, for every A of $\mathfrak{A}$, $m_\omega(A) \in M(A)$.
2) for every A of $\mathfrak{A}$, the map $\omega \to m_\omega(A)$ is measurable with respect to $\mathcal{C}$ and $\mathcal{B}(X)$.
3) For every ω of Ω, $m_\omega(T) = x(\omega)$.

Summing up of demonstration

We consider the metric quotient space (Θ,φ) of the measure space $(T,\mathfrak{A},\lambda)$, D a dense sequence in Θ and for every λ-continuous measure m from $\mathfrak{A}$ to X, $\dot{m}$ is the uniformly continuous function associated with m. According to a result of Hoffmann-Jørgensen ([4] p. 239) the set

$$H = \{ \dot{m} \mid m \ \lambda\text{-continuous}\}$$

equipped with the topology of pointwise convergence on D is shown to be a standard space. Now we verify that the multifunction Γ from Ω into H defined by

$$\Gamma(\omega) = \{\dot{m} \in H \mid m(A) \in M(A), A \in \mathfrak{A} \text{ and } m(T) = x(\omega)\}$$

has non-empty values and its graph belongs to $\mathcal{C} \otimes \mathcal{B}(H)$ (where $\mathcal{B}(H)$ is the Borel σ-algebra of H). From the von Neumann selection theorem, it follows that Γ admits a measurable selection. Since the map $\dot{m} \to m(A)$ is a Borel mapping for every A of $\mathfrak{A}$, the theorem is proved.

By entirely similar methods, we prove the following variant

Theorem 6

Let $(\Omega,\mathcal{C})$ be a complete measurable space, $(T,\mathfrak{A},\lambda)$ a separable measurable space with λ a finite positive measure on $\mathfrak{A}$. If M is a strong multimeasure whose values are closed convex subsets of $\mathbb{R}^n$ and $M \ll \lambda$, if x is a measurable map from Ω to $\mathbb{R}^n$ (for $\mathcal{C}$ and $\mathcal{B}(\mathbb{R}^n)$) such that, for every ω of Ω, $x(\omega) \in M(T)$, then there exists a measurable family $(m_\omega)_{\omega\in\Omega}$ of selectors of M such that, for every ω of Ω, $m_\omega(T) = x(\omega)$.

[1] ARTSTEIN, Z., Set valued measures. Trans Amer.Math.Soc. 165 (1972), 63-125

[2] COSTE, A., Contribution à la théorie de l'intégration multivoque. Thèse, Paris 1977

[3] GODET-THOBIE, C., Multimesures et multimesures de transition. Thèse, Montpellier 1975

[4] HOFFMANN-JØRGENSEN, J., The theory of analytic spaces. Aarhus 1970

[5] MICHAEL, E., Topologies on spaces of subsets. Trans Amer.Math.Soc. 71 (1951), 152-162

[6] PHELPS, R., Dentability and extreme points in Banach spaces.

MEASURABLE WEAK SELECTIONS

Siegfried Graf

0. Summary

An existence theorem for measurable selections of a multivalued map is proved, without "countability" assumptions on the range. This theorem is used to show the existence of generalized measurable selections, so-called weak selections. The last result yields a generalization of Edgar's existence theorem for weak sections (cf. [2]) and, in additon, the existence of preimage measures in some so far unresolved cases. Moreover, the question of uniqueness for weak selections is settled. From this conditions for the uniqueness of preimage measures - similar to those of Yershov [27], Eisele [3], and Lohn-Mägerl [15] - are derived.

1. Introduction

If Φ is a multivalued map from a set X to a set Y, then a selection for Φ is a (singlevalued) map φ: $X \to Y$ such that $\varphi(x) \in \Phi(x)$ for all $x \in X$. According to the axiom of choice every multivalued map has a selection. But if $(X,\mathfrak{A})$ and $(Y,\mathfrak{B})$ are measurable spaces the existence of a measurable selection for Φ becomes a serious problem. In the case that $\Phi = p^{-1}$, for a measurable map p: $Y \to X$, von Neumann's celebrated measurable choice theorem (cf. [20]) solves the problem for some natural spaces X, Y and continuous maps p. After the publication of this result the problem has attracted the attention of many mathematicians and more general existence theorems for measurable selections have been proved (see for instance [28],[13], [8], and, for further literature, [2],[21] , and [26]). In most of these theorems $(Y,\mathfrak{B})$ is assumed to satisfy some kind of countability condition (separability, metrizability etc.). For <u>continuous</u> selections Hasumi [7] succeeded in deriving an existence theorem without countability conditions. Using this result the following existence theorem for <u>measurable</u> selections will be proved, which does not contain a countability assumption on the range, either:

> Let Φ be an upper semi-continuous compact-valued multivalued map from a topological space X to a regular topological space Y. If $\mathfrak{A}$ is a σ-field on X containing all open sets and μ a finite measure on $\mathfrak{A}$, such that $(X,\mathfrak{A},\mu)$ is complete and admits a strong lifting, then there is a measurable selection for Φ.

A special case of this theorem has been conjectured by J.P.R. Christensen, and was communicated to the author by H. von Weizsäcker. This special case of the theorem has also independently been proved by M. Talagrand [24].

Another existence theorem without countability assumptions on the range of the multivalued map Φ has been proved by Edgar [2] for generalized selections (so-called weak sections) in the special case $\Phi = p^{-1}$, for a measurable map p from Y to X. Edgar calls a measurable map φ from an arbitrary measure space $(X,\mathfrak{A},\mu)$ to $(Y,\mathfrak{B})$ a weak section for p if $\varphi^{-1}p^{-1}(A)$ is equivalent to A, modulo μ, for every $A \in \mathfrak{A}$. Inspired by Edgar's definition and the development in the theory of measurable selections succeeding von Neumann [20], we shall generalize this definition to arbitrary multivalued maps. There seems to be no obvious way in doing so but as it turns out the following definition includes the above one in some interesting cases and also proves its usefulness in applications: Given a finite measure space $(X,\mathfrak{A},\mu)$, a Hausdorff space Y, and a multivalued map Φ from X to Y, then a measurable map $\varphi: X \to Y$ is called a <u>weak selection</u> for Φ iff, for every open subset U of Y, $\varphi^{-1}(U)$ is μ-almost-everywhere contained in any measurable cover of $\Phi^{-1}(U)$. - It will be shown that Φ admits a measurable weak selection φ with $\varphi(\mu)$ a Radon measure if

$$(1) \qquad \mu(X) = \sup\{\mu^*(\Phi^{-1}(K)) : K \subset Y,\ K \text{ compact}\}.$$

If $\Phi = p^{-1}$ for a measurable map $p: Y \to X$ and

$$(2) \qquad \mu^*(p(K)) = \inf\{\mu^*(p(U)) : K \subset U,\ U \text{ open}\}$$

for all compact $K \subset Y$ then a weak selection for Φ is a weak section for p provided that $\varphi(\mu)$ is a Radon measure on Y. A combination of the two last results yields, on the one hand, a generalization of the main theorem of Edgar [2], and, on the other hand, the existence of a preimage measure for μ with respect to p, provided conditions (1) and (2) are satisfied.

Finally the question of uniqueness for weak selections is investigated. Several necessary and sufficient conditions for the uniqueness of weak selections (up to weak μ-eqivalence) are given, which, in return, imply conditions for the uniqueness of preimage measures. The latter ones are similar to those of Yershov [27], Eisele [3], and Lehn-Mägerl [15]. While these authors consider continuous or measurable maps between analytic spaces, the results, presented here, hold for certain measurable maps from compact spaces to arbitrary finite measure spaces.

<u>Acknowledgment.</u> The author is indebted to Prof. D. Kölzow for the introduction to this field and his helpful criticism, to Dr. G. Mägerl for many useful discussions.

2. Preliminaries

a) Basic notations and definitions

By $\mathbb{N}$ (resp. $\mathbb{R}$) we denote the natural (resp. real) numbers. Let Z be any set, and D, E subsets of Z. By D^c we mean the complement of D in Z, by $E \setminus D$ the relative complement $E \cap D^c$, and by $E \triangle D$ the symmetric difference $(E \setminus D) \cup (D \setminus E)$. We write card Z for the cardinality of Z.

Let $(X,\mathfrak{A},\mu)$ be a finite measure space, i.e. X is a set, $\mathfrak{A}$ a σ-field of subsets of X and μ a nonnegative real valued (countably additive) measure on $\mathfrak{A}$. Throughout the paper we assume that $0 < \mu(X) < +\infty$. We write $\mathfrak{A}_\mu$ for the completion of $\mathfrak{A}$ with respect to μ. The canonical extension of μ to $\mathfrak{A}_\mu$ will again be denoted by μ. Two sets $A,A' \in \mathfrak{A}_\mu$ are called μ-equivalent, iff $\mu(A \triangle A') = 0$. In this case we write $A \sim A'$ $[\mu]$ or, if no confusion can arise, $A \sim A'$. A set $A \in \mathfrak{A}_\mu$ is called μ-negligible iff $A \sim \emptyset$. Let μ^* denote the outer measure induced by μ on the collection of all subsets of X. If E is an arbitrary subset of X then there exists a set $A \in \mathfrak{A}$ with $E \subset A$ and $\mu^*(E) = \mu(A)$. Such a set A is called a measurable cover of E. If A' is another measurable cover of E then $A \sim A'$. By E^* we denote an arbitrarily chosen measurable cover of E. If $(E_n)_{n\in\mathbb{N}}$ is a sequence of subsets of X then $\bigcup\{E_n^*: n\in\mathbb{N}\}$ is a measurable cover of $\bigcup\{E_n: n\in\mathbb{N}\}$. If $E \subset X$ and $A \in \mathfrak{A}_\mu$ then $A \cap E^*$ is a measurable cover of $A \cap E$. Given two sets $E_1,E_2 \subset X$ let $E_1 \lesssim E_2$ stand for $\mu(E_1^* \setminus E_2^*) = 0$. For $E_1,E_2 \in \mathfrak{A}_\mu$ we have $E_1 \lesssim E_2$ if and only if $\mu(E_1 \setminus E_2) = 0$, hence $E_1 \sim E_2$ if and only if $E_1 \lesssim E_2$ and $E_2 \lesssim E_1$. Let $(Y,\mathfrak{B})$ be a measurable space, i.e. Y a set and $\mathfrak{B}$ a σ-field of subsets of Y. A map $f: X \to Y$ is called $\mathfrak{A}$-$\mathfrak{B}$-measurable iff $f^{-1}(B) \in \mathfrak{A}$ for all $B \in \mathfrak{B}$. The equality $\nu(B) = \mu(f^{-1}(B))$ $(B \in \mathfrak{B})$ defines a measure on $\mathfrak{A}$, it is called the image measure of μ w.r.t. f and is denotedby $f(\mu)$. Conversely, if $p: Y \to X$ is a $\mathfrak{B}$-$\mathfrak{A}$-measurable map, then each measure ν on $\mathfrak{B}$ with $\nu(p^{-1}(A)) = \mu(A)$ for all $A \in \mathfrak{A}$ is called a preimage measure of μ w.r.t. p.

b) Liftings and lifting topologies

A <u>lifting</u> for the finite measure space $(X,\mathfrak{A},\mu)$ is a map $\theta: \mathfrak{A} \to \mathfrak{A}$ with

(i) $A \sim A'$ implies $\theta(A) = \theta(A')$

(ii) $\theta(A) \sim A$

(iii) $\theta(A) \cap \theta(A') = \theta(A \cap A')$

(iv) $\theta(X) = X$, $\theta(\emptyset) = \emptyset$ and

(v) $\theta(A^c) = \theta(A)^c$ for all $A, A' \in \mathfrak{A}$.

From these properties one can deduce

(vi) $\theta(A) \cup \theta(A') = \theta(A \cup A')$ and

(vii) $A \lesssim A'$ implies $\theta(A) \subset \theta(A')$ for all $A,A' \in \mathfrak{A}$.

D. Maharam [19] has shown that every complete finite measure space has a lifting.

The following proposition is a combination of results due to Maharam [19] and A. Ionescu Tulcea [9].

2.1. Proposition

Let θ be a lifting for the complete finite measure space $(X,\mathfrak{A},\mu)$ and $\mathcal{T}_\theta = \{A \in \mathfrak{A}: A \subset \theta(A)\}$. Then $\mathcal{T}_\theta$ is a topology on X with the following properties

(i) $(X,\mathcal{T}_\theta)$ is extremally disconnected, i.e. the closure $\overline{U}$ of each set $U \in \mathcal{T}_\theta$ is again in $\mathcal{T}_\theta$.

(ii) The Borel field generated by $\mathcal{T}_\theta$ is equal to $\mathfrak{A}$.

(iii) For every $E \subset X$ the closure $\overline{E}$ of E w.r.t. $\mathcal{T}_\theta$ is a measurable cover of E.

(iv) For a filtering increasing family $(U_i)_{i\in I}$ in $\mathcal{T}_\theta$ the equality
$$\mu(\bigcup\{U_i: i\in I\}) = \sup\{\mu(U_i): i\in I\}$$
holds.

In what follows the topology $\mathcal{T}_\theta$ is called the lifting topology on X corresponding to θ.

For a topological space Z let $\mathfrak{B}(Z)$ denote the Borel field of Z, i.e. the σ-field generated by the open subsets of Z.
A finite measure ν on $\mathfrak{B}(Z)$ is called regular iff it satisfies one of the following equivalent conditions

(I) $\nu(B) = \sup\{\nu(F): F \subset B, F \text{ closed}\}$

(II) $\nu(B) = \inf\{\nu(U): B \subset U, U \text{ open}\}$

for all $B \in \mathfrak{B}(Z)$.

2.2. Proposition

Let $(X,\mathfrak{A},\mu)$ be a complete finite measure space and θ a lifting for $(X,\mathfrak{A},\mu)$. Moreover, let X be equipped with the corresponding lifting topology $\mathcal{T}_\theta$, Y a regular topological space and $f: X \to Y$ a continuous map. Then the image measure $\nu = f(\mu)$ on $\mathfrak{B}(Y)$ is regular.

Proof: Let $\mathcal{P}$ be the class of all $B \in \mathfrak{B}(Y)$ such that B satisfies (I) and (II). We shall show that $\mathcal{P}$ contains all open subsets of Y. To this end let $U \subset X$ be open. Because Y is regular we have $U = \bigcup \mathfrak{V}$, where $\mathfrak{V} = \{V \subset Y: \overline{V} \subset U, V \text{ open}\}$ and, therefore,
$$f^{-1}(U) = \bigcup\{f^{-1}(V): V \in \mathfrak{V}\} = \bigcup\{f^{-1}(\overline{V}): V \in \mathfrak{V}\} .$$
Since $(f^{-1}(V))_{V\in\mathfrak{V}}$ is a filtering increasing family in $\mathcal{T}_\theta$ Proposition 2.1(iv) yields
$$\mu(f^{-1}(U)) = \sup\{\mu(f^{-1}(V)): V \in \mathfrak{V}\} = \sup\{\mu(f^{-1}(\overline{V})): V \in \mathfrak{V}\},$$
hence $\nu(U) = \sup\{\nu(F): F \subset U, F \text{ closed}\}$.

Since U was an arbitrary open subset of Y we deduce that $\mathcal{P}$ contains all open subsets of Y. By standard arguments it can be shown that $\mathcal{P}$ is stable under relative complements and countable unions of pairwise disjoint sets. Due to a variant of the monotone class theorem (cf. Bauer [1], p. 18, Satz 2.3) this implies $\mathcal{P} = \mathfrak{B}(Y)$. Thus the proposition is proved.

Before we can formulate an easy consequence of the above proposition we must introduce the notion of strong lifting. Let X be a topological space, $\mathfrak{A}$ a σ-field of subsets of X containing all open sets, and μ a finite measure on $\mathfrak{A}$. A lifting θ for $(X,\mathfrak{A},\mu)$ is said to be strong iff $U \subset \theta(U)$ for all open sets $U \subset X$. A lifting is obviously strong if and only if the original topology on X is weaker than the lifting topology $\mathcal{T}_\theta$ corresponding to θ.

The existence of a strong lifting is, for instance, known in each of the following cases

(i) X is a second countable topological space, μ is strictly positive on nonempty open sets, and $(X,\mathfrak{A},\mu)$ is complete (see [4]),

(ii) X is a metrizable locally compact space, μ is a Radon measure on X whose support is X, and $\mathfrak{A}$ is the σ-field of μ-measurable subsets of X (see [11]),

(iii) X is a compact group, μ is a Haar measure on X, and $\mathfrak{A}$ is the σ-field of μ-measurable subsets of X (see [10]).

For a long time it was an open problem whether every Radon measure on a compact space whose support is the whole space had a strong lifting. But in 1978 Losert [16] constructed a counterexample.

We are now able to state the following corollary of Proposition 2.2.

2.3. Corollary

If X is a regular topological space and μ a finite Borel measure on X such that $(X,\mathfrak{B}_\mu(X),\mu)$ admits a strong lifting then μ is regular.

Proof: Let θ be a strong lifting for $(X,\mathfrak{B}_\mu(X),\mu)$ and $\mathcal{T}_\theta$ the corresponding lifting topology. Then the identity map id_X is continuous as a map from X equipped with the topology $\mathcal{T}_\theta$ to X with the original topology. Thus the corollary follows immediately from Proposition 2.2.

c) Relations and correspondences

Given two sets X and Y, a relation $\Phi \subset X \times Y$, and subsets A of X and B of Y let $\Phi(A) = \bigcup_{x\in A}\{y \in Y: (x,y) \in \Phi\}$ and $\Phi^{-1}(B) = \bigcup_{y\in B}\{x \in X: (x,y) \in \Phi\}$. As usual, if $x \in X$ and $y \in Y$, we write $\Phi(x)$ (resp. $\Phi^{-1}(y)$) for $\Phi(\{x\})$

(resp. $\Phi^{-1}(\{y\})$). A relation $\Phi \subset X \times Y$ is called a correspondence (multi-valued map) from X to Y iff $\Phi^{-1}(Y) = X$. A selection for a correspondence Φ from X to Y is a map $\varphi: X \to Y$ with $\varphi(x) \in \Phi(x)$ for all $x \in X$. If Y is a topological space then a correspondence Φ from X to Y is called closed-valued (resp. compact-valued) iff $\Phi(x)$ is closed (resp. compact) for all $x \in X$. If X is also a topological space, then a correspondence Φ is upper semi-continuous iff $\Phi^{-1}(F)$ is closed for every closed set $F \subset Y$. For example, if X and Y are compact spaces and $p: Y \to X$ is a continuous map onto, then p^{-1} is a compact-valued upper semi-continuous correspondence.

3. On the existence of measurable selections

Combining the results on lifting topologies with the following selection theorem of Hasumi [7] we shall get some interesting facts on the existence of measurable selections.

3.1. Theorem (Hasumi [7])
Let X be an extremally disconnected topological space, Y a regular Hausdorff space, and Φ a compact-valued upper semi-continuous correspondence from X to Y. Then there exists a contninuous selection for Φ.

Though Hasumi [7] assumed X to be Hausdorff his proof works unchanged in the situation of the above theorem and will, therefore, be omitted.

In what follows $(X,\mathfrak{A},\mu)$ will always be a finite measure space with $\mu(X) > 0$. The condition of finiteness is imposed to avoid technical difficulties. With suitable modifications most of our results also hold for nonfinite measure spaces which admit a lifting and which are finitely determined.

As an immediate consequence of Hasumi's theorem we get the main result of this section.

3.2. Theorem
Let X be a topological space, $\mathfrak{A}$ a σ-field of subsets of X containing all open subsets of X, and μ a finite measure on $\mathfrak{A}$, such that $(X,\mathfrak{A},\mu)$ is a complete measure space which has a strong lifting. If Φ is a compact-valued upper semi-continuous correspondence from X to a regular topological space Y then there exists an $\mathfrak{A}$-$\mathfrak{B}(Y)$-measurable selection φ for Φ such that $\varphi(\mu)$ is a regular Borel measure on Y.

Proof: Let θ be a strong lifting for $(X,\mathfrak{A},\mu)$ and $\mathcal{T}_\theta$ the corresponding lifting topology. Obviously Φ is also upper semi-continuous as a correspondence from $(X,\mathcal{T}_\theta)$ to Y. Thus it follows from Proposition 2.1 (i) and Theorem 3.1 that there is a selection φ for Φ which is continuous as a map from $(X,\mathcal{T}_\theta)$ to Y. For obvious reasons (cf. Proposition 2.1 (ii)) such a map is $\mathcal{A}$-$\mathcal{B}(Y)$-measurable and the regularity of $\varphi(\mu)$ is deduced from Proposition 2.2.

3.3. Remarks

a) It is worthwile to list some special cases of the above theorem which follow immediately from the existence theorems for strong liftings.

(i) If X is a second countable topological space, μ is strictly positive on nonempty open sets, and $(X,\mathcal{A},\mu)$ is complete then every compact-valued upper semi-continuous correspondence from X to a regular Hausdorff space Y has an $\mathcal{A}$-$\mathcal{B}(Y)$-measurable selection.

(ii) If X is a compact group, μ a Haar measure on X, and $\mathcal{A}$ the σ-field of μ-measurable sets in X then the conclusion of the theorem holds.

b) As H. v. Weizsäcker told the author the statement of the above theorem, in the special case that X and Y are compact and Φ is the inverse of a continuous map from Y onto X, is a conjecture of J.P.R. Christensen. In this special case the theorem has also independently been proved by M. Talagrand [24]. It is possible that Christensen or v. Weizsäcker has a different proof of this special case of the theorem.

c) Losert [17] gave a characterization of those (completely regular) topological measure spaces $(X,\mathcal{A},\mu)$ such that every compact-valued upper semi-continuous correspondence from X to a completely regular space admits a Baire measurable selection. He showed that there is a compact space X with a Radon measure μ without a strong lifting such that every compact-valued upper semi-continuous correspondence from X to a completely regular space admits a Baire measurable selection. He also constructed a compact space with a Radon measure for which this conclusion is not true.

Since every lifting is strong with respect to the corresponding lifting topology we get the following corollary of Theorem 3.2 which will be used in the next section.

3.4. Corollary

Let $(X,\mathcal{A},\mu)$ be complete, θ a lifting for $(X,\mathcal{A},\mu)$, and X equipped with the corresponding lifting topology $\mathcal{T}_\theta$. Moreover let Φ be a compact-valued upper semi-continuous correspondence from X to a regular topological space Y. Then there exists an $\mathcal{A}$-$\mathcal{B}(Y)$-measurable selection φ for Φ such that $\varphi(\mu)$ is a regular Borel measure on Y.

4. On the existence of measurable weak selections

The following topological lemma will provide the basic tool for the proof of our main result.

4.1. Lemma

Let X be a topological space, Y a compact space, and $\Phi \subset X \times Y$ a relation with $\overline{\Phi^{-1}(Y)} = X$. Then the closure $\overline{\Phi}$ of Φ in $X \times Y$ with the product topology is a compact-valued upper semi-continuous correspondence with

$$\overline{\Phi}^{-1}(B) \subset \bigcap\{\overline{\Phi^{-1}(U)}: B \subset U,\ U \text{ open}\}$$

for all $B \subset Y$.

Proof:

(1) First we shall prove the identity

$$\overline{\Phi}(x) = \bigcap\{F \subset Y: x \notin \overline{\Phi^{-1}(F^c)},\ F \text{ closed}\}$$

for all $x \in X$.

α) For $(x,y) \in X \times Y \setminus \overline{\Phi}$ there are open neighbourhoods U of x and V of y with $U \times V \cap \Phi = \emptyset$, or, what is the same, $\Phi^{-1}(V) \cap U = \emptyset$. Thus $x \notin \overline{\Phi^{-1}(V)}$ and $y \notin V^c$ which implies

$$y \notin \bigcap\{F \subset Y: x \notin \overline{\Phi^{-1}(F^c)},\ F \text{ closed}\}.$$

β) For $x \in X$ and $y \notin \bigcap\{F \subset Y: x \notin \overline{\Phi^{-1}(F^c)},\ F \text{ closed}\}$ there exists an open set V in Y with $y \notin V^c$ and $x \notin \overline{\Phi^{-1}(V)}$, hence $y \in V$ and we can find an open neighbourhood U of x with $U \cap \Phi^{-1}(V) = \emptyset$, which implies $(x,y) \in U \times V$ and $U \times V \cap \Phi = \emptyset$, hence $y \notin \overline{\Phi}(x)$.

(2) For $x \in X$ let $\mathcal{F}_x = \{F \subset Y: x \notin \overline{\Phi^{-1}(F^c)},\ F \text{ closed}\}$. It is easy to see that $\mathcal{F}_x$ is a nonempty collection of nonempty closed sets in Y, which is stable under finite intersections. Since Y is compact we deduce, therefore, that $\overline{\Phi}$ is a compact-valued correspondence.

(3) Given a closed set $F \subset Y$ we deduce from (1) that

$$\overline{\Phi}^{-1}(F) = \{x \in X: F \cap \bigcap \mathcal{F}_x \neq \emptyset\}.$$

Since Y is compact, this last identity together with (2) implies

$$\begin{aligned}\overline{\Phi}^{-1}(F) &= \{x \in X: F \cap E \neq \emptyset \text{ for all } E \in \mathcal{F}_x\}\\ &= \{x \in X: E \notin \mathcal{F}_x \text{ for closed } E \subset Y \text{ with } E \cap F = \emptyset\}\\ &= \bigcap\{\overline{\Phi^{-1}(E^c)}: E \subset F^c,\ E \text{ closed}\}\\ &= \bigcap\{\overline{\Phi^{-1}(U)}: F \subset U,\ U \text{ open}\}.\end{aligned}$$

Thus $\overline{\Phi}^{-1}(F)$ is closed and, since F was an arbitrary closed subset of Y, the correspondence $\overline{\Phi}$ is upper semi-continuous.

(4) For an arbitrary subset B of Y we deduce

$$\overline{\Phi}^{-1}(B) = \{x \in X\colon B \cap \bigcap \mathcal{T}_x \neq \emptyset\}$$
$$\subset \{x \in X\colon B \cap E \neq \emptyset \text{ for all } E \in \mathcal{T}_x\}$$
$$= \bigcap \{\overline{\Phi^{-1}(U)}\colon B \subset U,\ U \text{ open}\}.$$

The statement of the lemma results from combining (1) to (4).

In the rest of this section $(X,\mathfrak{A},\mu)$ is again a finite measure space with $\mu(X) > 0$, Y is a Hausdorff space, and $\Phi \subset X \times Y$ a relation. As before $\mathcal{B}(Y)$ denotes the Borel field of Y while $\mathcal{L}(Y)$ stands for the σ-field generated by all closed G_δ-sets. If Y is normal then $\mathcal{L}(Y)$ is just the Baire σ-field of Y (cf. Bauer [1], p. 167).

A map $\varphi\colon X \longrightarrow Y$ is called an $\mathfrak{A}$-$\mathcal{B}(Y)$- (resp. $\mathfrak{A}$-$\mathcal{L}(Y)$-) <u>measurable weak selection</u> for Φ iff φ is $\mathfrak{A}$-$\mathcal{B}(Y)$- (resp. $\mathfrak{A}$-$\mathcal{L}(Y)$-) measurable and $\varphi^{-1}(U) \subseteq \overline{\Phi}^{-1}(U)$ for every open set $U \subset Y$.

If Φ is a correspondence then every $\mathfrak{A}$-$\mathcal{B}(Y)$- (resp. $\mathfrak{A}$-$\mathcal{L}(Y)$-) measurable selection for Φ is obviously an $\mathfrak{A}$-$\mathcal{B}(Y)$- (resp. $\mathfrak{A}$-$\mathcal{L}(Y)$-) measurable weak selection for Φ.

The next lemma leads up to the problem of existence for measurable weak selections.

<u>4.2. Lemma</u>

Let Y be compact and $\mu^*(\Phi^{-1}(Y)) = \mu(X)$. Then there exists a correspondence Φ^* from X to Y with the following properties

(i) $\Phi \subset \Phi^*$

(ii) $(\Phi^*)^{-1}(F) \in \mathfrak{A}_\mu$ *for all closed* $F \subset Y$

(iii) *There is an $\mathfrak{A}_\mu$-$\mathcal{B}(Y)$-measurable selection φ for Φ^*, such that $\varphi(\mu)$ is a regular Borel measure on Y.*

(iv) *Every $\mathfrak{A}_\mu$-$\mathcal{B}(Y)$- (resp. $\mathfrak{A}_\mu$-$\mathcal{L}(Y)$-) measurable weak selection for Φ^* is an $\mathfrak{A}_\mu$-$\mathcal{B}(Y)$- (resp. $\mathfrak{A}_\mu$-$\mathcal{L}(Y)$-) measurable weak selection for Φ and vice versa.*

<u>Proof:</u> Let θ be a lifting for $(X,\mathfrak{A}_\mu,\mu)$ and $\mathcal{T}_\theta$ the corresponding lifting topology. Since we have $\mu^*(\Phi^{-1}(Y)) = \mu(X)$ Proposition 2.1(iii) together with properties (i) and (iv) in the definition of a lifting imply $\overline{\Phi^{-1}(Y)} = X$. From Lemma 4.1 we know that $\overline{\Phi}$ is a compact-valued, upper semi-continuous correspondence. Thus $\Phi^* = \overline{\Phi}$ fulfills conditions (i) and (ii) of Lemma 4.2. According to Corollary 3.4 there is an $\mathfrak{A}_\mu$-$\mathcal{B}(Y)$-measurable selection φ for $\overline{\Phi}$, such that $\varphi(\mu)$ is a regular Borel measure on Y, which proves $\overline{\Phi}$ to satisfy condition (iii). Now, let ψ be an arbitrary $\mathfrak{A}_\mu$-$\mathcal{B}(Y)$- (resp. $\mathfrak{A}_\mu$-$\mathcal{L}(Y)$-) measurable weak selection for $\overline{\Phi}$. For $U \subset Y$ open we get from Lemma 4.1 and Proposition 2.1(iii)

$$\psi^{-1}(U)^* \subseteq \bar{\Phi}^{-1}(U)^* \subseteq \overline{\Phi^{-1}(U)} \sim \Phi^{-1}(U)^*.$$

This shows that ψ is an $\mathfrak{A}_\mu$-$\mathfrak{B}(Y)$- (resp. $\mathfrak{A}_\mu$-$\mathfrak{L}(Y)$-) measurable weak selection for Φ.

Conversely, if ξ is an $\mathfrak{A}_\mu$-$\mathfrak{B}(Y)$- (resp. $\mathfrak{A}_\mu$-$\mathfrak{L}(Y)$-) measurable weak selection for Φ, then we deduce from Lemma 4.1 and Proposition 2.1(iii)

$$\xi^{-1}(U)^* \subseteq \Phi^{-1}(U)^* \sim \overline{\Phi^{-1}(U)} = \overline{\bar{\Phi}^{-1}(U)} \sim \bar{\Phi}^{-1}(U)^*$$

for $U \subset Y$ open. Hence ξ is an $\mathfrak{A}_\mu$-$\mathfrak{B}(Y)$- (resp. $\mathfrak{A}_\mu$-$\mathfrak{L}(Y)$-) measurable weak selection for $\bar{\Phi}$.

This completes the proof of the lemma.

Before we shall formulate the main theorem of this section let us recall the definition of a Radon measure on an arbitrary Hausdorff space. A finite Borel measure ν on a Hausdorff space Y is called a <u>Radon measure</u> iff $\nu(B) = \sup\{\nu(K): K \subset B, K \text{ compact}\}$ for all $B \in \mathfrak{B}(Y)$. For the basic facts about Radon measures on arbitrary topological spaces we refer the reader to [22].

<u>4.3. Theorem</u>

<u>Let</u> $(X,\mathfrak{A},\mu)$ <u>be a finite measure space</u>, Y <u>a Hausdorff space, and</u> $\Phi \subset X \times Y$. <u>If</u> $\mu(X) = \sup\{\mu^*(\Phi^{-1}(K)): K \subset Y, K \text{ compact}\}$, <u>then there exists an</u> $\mathfrak{A}_\mu$-$\mathfrak{B}(Y)$-<u>measurable weak selection</u> φ <u>for</u> Φ <u>such that</u> $\varphi(\mu)$ <u>is a Radon measure on</u> Y.

<u>Proof:</u> Let $(K_n)_{n\in\mathbb{N}}$ be a sequence of compact subsets of Y with

$$\mu(X) = \sup\{\mu^*(\Phi^{-1}(K_n)): n \in \mathbb{N}\}.$$

According to our assumptions such a sequence exists. For $n \in \mathbb{N}$ define

$$A_n = \Phi^{-1}(K_n)^* \setminus \bigcup_{i=1}^{n-1} \Phi^{-1}(K_i)^*.$$

If $L = \{n \in \mathbb{N}: \mu(A_n) > 0\}$ then $(A_n)_{n\in L}$ is a countable or finite family of mutually disjoint sets in $\mathfrak{A}$ with $\bigcup\{A_n: n \in L\} \sim X$.

For $n \in L$ we define $\mathfrak{A}_n = \{A \cap A_n: A \in \mathfrak{A}\}$ and $\mu_n = \mu|\mathfrak{A}_n$.

Then $(A_n,\mathfrak{A}_n,\mu_n)$ is a finite measure space. For $\Phi_n = \Phi \cap (A_n \times K_n)$ we have

$$\mu_n^*(\Phi_n^{-1}(K_n)) = \mu^*(A_n \cap \Phi^{-1}(K_n)) = \mu(A_n \cap \Phi^{-1}(K_n)^*) = \mu(A_n) = \mu_n(A_n) .$$

By Lemma 4.2(iii) and (iv) there is an $\mathfrak{A}_{n,\mu_n}$-$\mathfrak{B}(K_n)$-measurable weak selection $\varphi_n: A_n \longrightarrow K_n$ for Φ_n such that $\nu_n = \varphi_n(\mu_n)$ is a regular Borel measure on K_n.

Let a fixed $y \in Y$ be chosen and define $\varphi: X \longrightarrow Y$ by

$$\varphi(x) = \begin{cases} \varphi_n(x), & \text{for } x \in A_n \text{ and } n \in L \\ y \quad, & \text{for } x \notin \bigcup\{A_n: n \in L\} \end{cases} .$$

Then φ is $\mathfrak{A}_\mu$-$\mathfrak{B}(Y)$-measurable because we have

$$\varphi^{-1}(B) = (\varphi^{-1}(B) \setminus \bigsqcup_{n\in L} A_n) \cup \bigsqcup_{n\in L} \varphi_n^{-1}(B \cap K_n)$$

for all $B \in \mathcal{B}(Y)$, and $\varphi^{-1}(B) \setminus \bigsqcup_{n\in L} A_n$ is μ-negligible while $\varphi_n^{-1}(B \cap K_n) \in \mathfrak{A}_{n,\mu_n} \subset \mathfrak{A}_\mu$. Moreover

$$\mu(\varphi^{-1}(B)) = \sum_{n\in L} \mu_n(\varphi_n^{-1}(B \cap K_n)) = \sum_{n\in L} \nu_n(B \cap K_n).$$

From this last equation and the fact that each ν_n is a Radon measure we deduce by a standard argument that $\varphi(\mu)$ is also a Radon measure. It remains to show that φ is an $\mathfrak{A}_\mu$-$\mathcal{B}(Y)$-measurable weak selection for Φ. If $U \subset Y$ is open then $U \cap K_n$ is open in K_n, hence

$$\varphi_n^{-1}(U \cap K_n) \lesssim \Phi_n^{-1}(U \cap K_n) = A_n \cap \Phi^{-1}(U \cap K_n) \subset \Phi^{-1}(U).$$

This leads to

$$\varphi^{-1}(U) \sim \bigsqcup_{n\in L} \varphi_n^{-1}(U \cap K_n) \lesssim \Phi^{-1}(U)$$

and completes the proof.

Let us now introduce a notion of regularity for subsets of Y which is essential for our further considerations. We call a subset B of Y a <u>Φ-regular</u> set iff $\mu^*(\Phi^{-1}(B)) = \inf\{\mu^*(\Phi^{-1}(U)) : B \subset U, U \text{ open}\}$.

<u>4.4 Examples</u>

(i) Obviously every open subset of Y is Φ-regular.

(ii) Let Y be a normal space, $\Phi(x)$ compact for all $x \in X$, and assume that $\Phi^{-1}(U) \in \mathfrak{A}$ for all open sets $U \subset Y$. If $F \subset Y$ is a closed G_δ-set then F is Φ-regular and $\Phi^{-1}(F) \in \mathfrak{A}$.

(iii) Let Y again be normal and $\Phi(x)$ compact for all $x \in X$. If $\Phi^{-1}(F)$ for all closed sets $F \subset Y$ then every closed G_δ-set in Y is Φ-regular.

(iv) Let X be an arbitrary topological space, Y a Hausdorff space, μ a finite regular Borel measure on X, and $p: Y \longrightarrow X$ a continuous map. If $\Phi = p^{-1}$ then every $B \in \mathcal{B}(Y)$ is Φ-regular.

The proofs of the above statements are straightforward and will, therefore, be omitted.

Next we will examine how Φ-regular sets fit into the context of measurable weak selections.

<u>4.5. Lemma</u>

<u>If φ is an $\mathfrak{A}_\mu$-$\mathcal{B}(Y)$- (resp. $\mathfrak{A}_\mu$-$\mathcal{L}(Y)$-) measurable weak selection for Φ then $\varphi^{-1}(B) \lesssim \Phi^{-1}(B)$ holds for all Φ-regular sets $B \in \mathcal{B}(Y)$ (resp. $B \in \mathcal{L}(Y)$).</u>

Proof: Let $B \in \mathcal{B}(Y)$ (resp. $B \in \mathcal{L}(Y)$) be Φ-regular. Then there exists a decreasing sequence $(U_n)_{n\in\mathbb{N}}$ of open subsets of Y with $B \subset U_n$ and

$$\mu^*(\Phi^{-1}(B)) = \inf\{\mu^*(\Phi^{-1}(U_n)) : n \in \mathbb{N}\}.$$

This implies $\Phi^{-1}(B)^* \sim \bigcap_{n\in\mathbb{N}} \Phi^{-1}(U_n)^*$. According to our assumptions we have $\varphi^{-1}(B) \subset \varphi^{-1}(U_n)^* \lesssim \Phi^{-1}(U_n)^*$ and hence $\varphi^{-1}(B) \lesssim \Phi^{-1}(B)$.
Thus the lemma is proved.

For the statement of our next proposition we need one more definition. Let, for the moment, $(Y,\mathcal{L})$ be a measurable space and $p: Y \longrightarrow X$ a $\mathcal{L}$-$\mathfrak{A}$-measurable map. An $\mathfrak{A}_\mu$-$\mathcal{L}$-measurable map $f: X \longrightarrow Y$ is said to be a weak section for p iff $f^{-1}p^{-1}(A) \sim A$ for all $A \in \mathfrak{A}$. If f is a weak section for p then $f(\mu)$ is obviously a preimage measure of μ w.r.t. p.

4.6. Proposition

Let $(X,\mathfrak{A},\mu)$ be a finite measure space, Y a Hausdorff space, and $p: Y \longrightarrow X$ a $\mathcal{B}(Y)$-$\mathfrak{A}$-measurable map satisfying the conditions

(1) $\mu(X) = \sup\{\mu^*(p(K)) : K \subset Y,\ K \text{ compact}\}$

and (2) $\mu^*(p(K)) = \inf\{\mu^*(p(U)) : K \subset U,\ U \text{ open}\}$ *for every compact subset* K *of* Y.

Then there exists an $\mathfrak{A}_\mu$-$\mathcal{B}(Y)$-*measurable weak section* φ *for* p *with* $\varphi(\mu)$ *a Radon measure on* Y.

Proof: Since $\Phi = p^{-1}$ satisfies the assumptions of Theorem 4.3 there is an $\mathfrak{A}_\mu$-$\mathcal{B}(Y)$-measurable weak selection φ for Φ with $\nu = \varphi(\mu)$ a Radon measure on Y. Consider an arbitrary set $A \in \mathfrak{A}$. Then

$$\nu(p^{-1}(A)) = \sup\{\nu(K) : K \subset p^{-1}(A),\ K \text{ compact}\}.$$

Hence there exists an increasing sequence $(K_n)_{n\in\mathbb{N}}$ of compact subsets of Y with $K_n \subset p^{-1}(A)$ and $\bigcup_{n\in\mathbb{N}} K_n \sim p^{-1}(A)\ [\nu]$. This implies $\bigcup_{n\in\mathbb{N}} \varphi^{-1}(K_n) \sim \varphi^{-1}p^{-1}(A)$. Since every compact subset of Y is p^{-1}-regular (by condition (2)) we deduce by means of Lemma 4.5

$$\varphi^{-1}(K_n) \lesssim p(K_n)^* \lesssim A,$$

hence $\bigcup_{n\in\mathbb{N}} \varphi^{-1}(K_n) \lesssim A$ and, therefore, $\varphi^{-1}p^{-1}(A) \lesssim A$.

By the same argument we can show

$$(\varphi^{-1}p^{-1}(A))^c = \varphi^{-1}p^{-1}(A^c) \lesssim A^c,$$

hence

$$A \lesssim \varphi^{-1}p^{-1}(A).$$

Thus $\varphi^{-1}p^{-1}(A) \sim A$ and the proof is completed.

4.7. Remarks

a) Let $(Y,\mathfrak{L})$ be a measurable space and $p: Y \longrightarrow X$ a $\mathfrak{L}$-$\mathfrak{A}$-measurable map. A necessary condition for the existence of a preimage measure of μ w.r.t. p and, therefore, for the existence of a weak section for p is $\mu^*(p(Y)) = \mu(X)$. The above proposition is a partial converse of this statement. For similar considerations in the case of analytic spaces we refer the reader to Yershov [27].

b) If, in the above situation, $\varphi: X \longrightarrow Y$ is an $\mathfrak{A}_\mu$-$\mathfrak{L}$-measurable weak section for p, then $\nu = \varphi(\mu)$ is an extreme point in the convex set of all preimage measures of μ w.r.t. p.
Proof: Suppose ν_1, ν_2 are preimage measures of μ w.r.t. p such that there is an $r \in]0,1[$ with $\nu = r\nu_1 + (1-r)\nu_2$. Then the ν_i $(i = 1,2)$ are absolutely continuous w.r.t. ν, hence, by the Radon-Nikodym theorem, there are ν-integrable functions $h_i: Y \longrightarrow \mathbb{R}$ with $\int_B h_i \, d\nu = \nu_i(B)$ for all $B \in \mathfrak{L}$, especially for every $A \in \mathfrak{A}$

$$\mu(A) = \nu_i(p^{-1}(A)) = \int_{p^{-1}(A)} h_i \, d\nu = \int_{\varphi^{-1}p^{-1}(A)} h_i \circ \varphi \, d\mu = \int_A h_i \circ \varphi \, d\mu \qquad (i = 1,2).$$

This implies that $h_i \circ \varphi$ is equal to 1 μ-a.e., hence for arbitrary $B \in \mathfrak{L}$

$$\nu_i(B) = \int_B h_i \, d\nu = \int_{\varphi^{-1}(B)} h_i \circ \varphi \, d\mu = \mu(\varphi^{-1}(B)) = \nu(B) \qquad (i = 1,2).$$

Thus our assertion is proved.

In the special case that X and Y are compact spaces, $\mathfrak{A} = \mathfrak{L}(X)$, $\mathfrak{L} = \mathfrak{L}(Y)$, p a continuous map from Y onto X, and μ a finite measure on $\mathfrak{A}$ Edgar [2] proved the converse of the above statement, i.e. he showed that every extreme point in the set of all Baire preimage measures of μ w.r.t. p is the image measure of μ under some $\mathfrak{A}_\mu$-$\mathfrak{L}$-measurable weak section for p.
In [5] it is shown that a Radon measure ν on a Hausdorff space Y is an extreme point in the set of all preimages of a measure μ on X w.r.t. a measurable map $p: Y \longrightarrow X$ if and only if ν is the image of μ w.r.t. some measurable weak section for p.
For a measurable map p from a Polish space Y to another one, say X, Yershov [28] has shown that the set of preimage measures of a given finite Borel measure μ on X w.r.t. p contains enough extreme points (of the kind described above) to have a Choquet-type integral representation theorem.

c) It should also be mentioned that Proposition 4.6 imlies the following extension theorem for measures:
Let X be a Hausdorff space, $\mathfrak{A}$ a σ-subfield of $\mathfrak{B}(X)$, and μ a finite measure on $\mathfrak{A}$ with

$$(1) \quad \mu(X) = \sup\{\mu^*(K): K \subset X, K \text{ compact}\}$$

and (2) $\mu^*(K) = \inf\{\mu^*(U): K \subset U,\ U \text{ open}\}$

for all compact sets $K \subset X$.

Then there is a Radon measure ν on X with $\nu|_{\mathfrak{A}} = \mu$.

Proof: Take for p the identity map in X, which is certainly $\mathcal{B}(X)$-$\mathfrak{A}$-measurable and also satisfies the assumptions of Proposition 4.6. Thus there is an $\mathfrak{A}_\mu$-$\mathcal{B}(X)$-measurable weak section φ for p with $\nu = \varphi(\mu)$ a Radon measure on Y. Obviously $\nu|_{\mathfrak{A}} = \mu$.

Our next corollary indicates the connection of our results to several known theorems.

4.8. Corollary

Let X be an arbitrary topological space, μ a finite regular Borel measure on X, Y a Hausdorff space, and p: $Y \longrightarrow X$ a continuous map with

$$\mu(X) = \sup\{\mu^*(p(K)): K \subset Y,\ K \text{ compact}\}.$$

Then there exists a $\mathcal{B}_\mu(X)$-$\mathcal{B}(Y)$-measurable weak section φ for p such that $\varphi(\mu)$ is a Radon measure on X.

Proof: According to Example 4.4(iv) every compact subset of Y is p^{-1}-regular. Thus Proposition 4.6 is applicable and yields the statement of the corollary.

4.9. Remarks

a) If, in the situation of the above corollary, X is completely regular in addition, then the corollary remains true, if one replaces $\mathcal{B}(X)$ by the Baire σ-field in X and μ by any Baire measure ν on X with

$$\nu(X) = \sup\{\nu^*(p(K)): K \subset Y,\ K \text{ compact}\},$$

for one can easily see that every compact set $K \subset Y$ is p^{-1}-regular in this case (cf. Example 4.4(iv)).

Thus Theorem 1.3, p. 633 of Edgar [2] is also a consequence of our Proposition 4.6, because the existence of a Radon preimage measure for ν implies

$$\nu(X) = \sup\{\nu^*(p(K)): K \subset Y,\ K \text{ compact}\}.$$

b) Since the image measure of μ under a weak section for p is a preimage measure of μ w.r.t. p the above corollary also implies an existence theorem for Radon preimage measures stated in Schwartz [22] (Theorem 12, p. 39).

c) Given a $\mathcal{B}(Y)$-$\mathfrak{A}$-measurable map p from a Hausdorff space Y to a finite measure space $(X,\mathfrak{A},\mu)$ a necessary condition for the existence of a Radon preimage measure of μ w.r.t. p is the following one:

$$\mu(A) = \sup\{\mu^*(p(K)): K \subset p^{-1}(A),\ K \text{ compact}\} \qquad (A \in \mathfrak{A}).$$

Under certain additional assumptions (cf. Lembcke [14]) this condition is also sufficient for the existence of a preimage measure. But an easy example due to Lembcke shows that the existence of a Radon preimage measure does not imply the existence of an $\mathfrak{A}_\mu$-$\mathcal{B}(Y)$-measurable weak section for p.

5. On the uniqueness of measurable weak selections

Again $(X,\mathfrak{A},\mu)$ will always be a finite measure space with $\mu(X) > 0$ and Y a Hausdorff space.

We call two $\mathfrak{A}_\mu$-$\mathcal{B}(Y)$- (resp. $\mathfrak{A}_\mu$-$\mathcal{L}(Y)$-) measurable maps φ_1, φ_2: $X \longrightarrow Y$ <u>weakly Borel</u> (resp. <u>Baire</u>) <u>equivalent</u> iff $\varphi_1^{-1}(B) \sim \varphi_2^{-1}(B)$ for all $B \in \mathcal{B}(Y)$ (resp. $B \in \mathcal{L}(Y)$).

5.1. Proposition

<u>Let $\Phi \subset X \times Y$ be a relation with</u>

(1) $\mu(X) = \sup\{\mu^*(\Phi^{-1}(K)) : K \subset Y,\ K \text{ compact}\}$

<u>and</u> (2) <u>every compact subset of Y is Φ-regular.</u>

<u>Then the following statements are equivalent:</u>

(i) <u>There exists, up to weak Borel equivalence, exactly one $\mathfrak{A}_\mu$-$\mathcal{B}(Y)$-measurable weak selection φ for Φ with $\varphi(\mu)$ a Radon measure on Y.</u>

(ii) <u>For all disjoint compact sets $K_1, K_2 \subset Y$ the equality</u>

$$\mu^*(\Phi^{-1}(K_1 \cup K_2)) = \mu^*(\Phi^{-1}(K_1)) + \mu^*(\Phi^{-1}(K_2)) \qquad \underline{\text{holds}}.$$

<u>Proof:</u>

(i) $\Longrightarrow$ (ii). Suppose there are disjoint compact sets $K_1, K_2 \subset Y$ with $\mu^*(\Phi^{-1}(K_1 \cup K_2)) \neq \mu^*(\Phi^{-1}(K_1)) + \mu^*(\Phi^{-1}(K_2))$. Let φ: $X \longrightarrow Y$ be an $\mathfrak{A}_\mu$-$\mathcal{B}(Y)$-measurable weak selection for Φ with $\varphi(\mu)$ Radon. For $i \in \{1,2\}$, let $X_i = \Phi^{-1}(K_i)^*$, $\mathfrak{A}_i = \{A \in \mathfrak{A} : A \subset X_i\}$, and $\mu_i = \mu|_{\mathfrak{A}_i}$. Define $\Phi_i = \Phi \cap (X_i \times K_i)$. Since obviously $\mu_i^*(\Phi_i^{-1}(K_i)) = \mu_i(X_i)$ Theorem 4.3 implies the existence of an $\mathfrak{A}_{i,\mu}$-$\mathcal{B}(K_i)$-measurable weak selection $\tilde{\varphi}_i$: $X_i \longrightarrow K_i$ for Φ_i with $\tilde{\varphi}_i(\mu_i)$ a Radon measure on K_i. Define

$$\varphi_i(x) = \begin{cases} \varphi(x), & x \notin X_i \\ \tilde{\varphi}_i(x), & x \in X_i \end{cases} .$$

Then φ_i is an $\mathfrak{A}_\mu$-$\mathcal{B}(Y)$-measurable map and for every open set $U \subset Y$ we have

$$\begin{aligned} \varphi_i^{-1}(U) &= (\varphi^{-1}(U) \setminus X_i) \cup \tilde{\varphi}_i^{-1}(U \cap K_i) \\ &\subsetneq (\Phi^{-1}(U) \setminus X_i) \cup \Phi_i^{-1}(U \cap K_i) \subset \Phi^{-1}(U) \end{aligned} .$$

Thus φ_i is an $\mathfrak{A}_\mu$-$\mathcal{B}(Y)$-measurable weak selection for Φ.

We claim that $\varphi_i(\mu)$ is a Radon measure on Y. Given $\varepsilon > 0$ and $B \in \mathcal{B}(Y)$ let C_i be a compact set with $C_i \subset B$ and $\varphi(\mu)(B) < \varphi(\mu)(C_i) + \frac{\varepsilon}{2}$, and let D_i be a compact set with $D_i \subset B \cap K_i$ and $\tilde{\varphi}_i(\mu_i)(B \cap K_i) \leq \tilde{\varphi}_i(\mu_i)(D_i) + \frac{\varepsilon}{2}$. Then

$$\begin{aligned}\varphi_i(\mu)(B) &= \mu(\varphi_i^{-1}(B)) = \mu((\varphi^{-1}(B) \setminus X_i) \cup \tilde{\varphi}_i^{-1}(B \cap K_i)) \\ &= \mu(\varphi^{-1}(B) \setminus X_i) + \mu_i(\tilde{\varphi}_i^{-1}(B \cap K_i)) \\ &\leq \mu(\varphi^{-1}(C_i) \setminus X_i) + \mu_i(\tilde{\varphi}_i^{-1}(D_i)) + \varepsilon \\ &= \mu((\varphi^{-1}(C_i) \setminus X_i) \cup \tilde{\varphi}_i^{-1}(D_i)) + \varepsilon \\ &\leq \mu((\varphi^{-1}(C_i \cup D_i) \setminus X_i) \cup \tilde{\varphi}_i^{-1}((C_i \cup D_i) \cap K_i)) + \varepsilon \\ &= \mu(\varphi_i^{-1}(C_i \cup D_i)) + \varepsilon \\ &= \varphi_i(\mu)(C_i \cup D_i) + \varepsilon \qquad .\end{aligned}$$

This proves our claim.

It remains to show that φ_1 and φ_2 are not weakly Borel equivalent. According to Lemma 4.5 we have

$$\varphi_1^{-1}(K_1) = (\varphi^{-1}(K_1) \setminus X_1) \cup X_1 \sim X_1$$

since φ is an $\mathfrak{A}_\mu$-$\mathfrak{B}(Y)$-measurable weak selection for Φ and K_1 is Φ-regular. For the same reasons

$$\varphi_2^{-1}(K_1) = \varphi^{-1}(K_1) \setminus X_2 \subseteq X_1 \setminus X_2$$

which leads to

$$\mu(\varphi_2^{-1}(K_1)) \leq \mu(X_1) - \mu(X_1 \cap X_2),$$

but

$$\begin{aligned}\mu(X_1 \cap X_2) &= \mu(X_1) + \mu(X_2) - \mu(X_1 \cup X_2) \\ &= \mu^*(\Phi^{-1}(K_1)) + \mu^*(\Phi^{-1}(K_2)) - \mu^*(\Phi^{-1}(K_1 \cup K_2)) > 0,\end{aligned}$$

hence

$$\mu(\varphi_2^{-1}(K_1)) < \mu(\varphi_1^{-1}(K_1)).$$

Thus our last assertion is proved.

(ii) $\Longrightarrow$ (i). Let $\varphi_1, \varphi_2 : X \longrightarrow Y$ be two $\mathfrak{A}_\mu$-$\mathfrak{B}(Y)$-measurable weak selections for Φ such that $\varphi_1(\mu)$ and $\varphi_2(\mu)$ are Radon measures on Y. Let $K \subset Y$ be a compact set. Then there is an increasing sequence $(K_n)_{n\in\mathbb{N}}$ of compact sets in K^c with

$$\varphi_i(\mu)(K^c) = \sup\{\varphi_i(\mu)(K_n) : n \in \mathbb{N}\} \qquad (i = 1,2).$$

This implies

$$\begin{aligned}\mu(X) = \varphi_i(\mu)(Y) &= \varphi_i(\mu)(K) + \varphi_i(\mu)(K^c) \\ &= \varphi_i(\mu)(K) + \sup\{\varphi_i(\mu)(K_n) : n \in \mathbb{N}\} \\ &\leq \mu^*(\Phi^{-1}(K)) + \sup\{\mu^*(\Phi^{-1}(K_n)) : n \in \mathbb{N}\} \\ &= \sup\{\mu^*(\Phi^{-1}(K \cup K_n)) : n \in \mathbb{N}\} \leq \mu(X)\end{aligned}$$

and hence

$$\mu(\varphi_i^{-1}(K)) = \mu^*(\Phi^{-1}(K)) \qquad (i = 1,2).$$

From this last equality we deduce

$$\varphi_i^{-1}(K) \sim \Phi^{-1}(K)^* \qquad (i = 1,2)$$

and, therefore,

$$\varphi_1^{-1}(K) \sim \varphi_2^{-1}(K) \ .$$

If $B \in \mathcal{B}(Y)$ is arbitrary then there exists an increasing sequence $(C_n)_{n\in\mathbb{N}}$ of compact subsets of B with $\varphi_i(\mu)(B) = \varphi_i(\mu)(\bigcup_{n\in\mathbb{N}} C_n)$ (i = 1,2), i.e.

$$\varphi_1^{-1}(B) \sim \bigcup_{n\in\mathbb{N}} \varphi_1^{-1}(C_n) \sim \bigcup_{n\in\mathbb{N}} \varphi_2^{-1}(C_n) \sim \varphi_2^{-1}(B) \quad .$$

This shows φ_1 and φ_2 to be weakly Borel equivalent; and the proof of the proposition is completed.

Let us now state a similar proposition for Baire measurable weak selections. Since its proof consists in a slight modification of the proof of the preceding proposition it will be omitted.

5.2. Proposition

Let $(X,\mathfrak{A},\mu)$ be a finite measure space, Y a compact space, $\Phi \subset X \times Y$ with $\mu(X) = \mu^*(\Phi^{-1}(Y))$ and such that every compact G_δ-set in Y is Φ-regular. Then the following conditions are equivalent:

(i) *There is, up to weak Baire equivalence, exactly one $\mathfrak{A}_\mu$-$\mathcal{B}_0(Y)$-measurable weak selection for Φ.*

(ii) *For all compact G_δ-sets $K \subset Y$ the equality*

$$\mu^*(\Phi^{-1}(K)) + \mu^*(\Phi^{-1}(K^c)) = \mu(X)$$

holds.

5.3. Corollary

If, in the situation of the above proposition, Y is metrizable in addition then the following statements are equivalent:

(i) *There is, up to weak Borel equivalence, exactly one $\mathfrak{A}_\mu$-$\mathcal{B}(Y)$-measurable weak selection for Φ.*

(ii) $\mu^*(\{x \in X: \mathrm{card}(\Phi(x)) \geq 2\}) = 0$ *and* $\Phi^{-1}(F) \in \mathfrak{A}_\mu \cap \Phi^{-1}(Y)$ *for all closed sets* $F \subset Y$.

Proof:

(i)$\Longrightarrow$(ii). Since Y is metrizable and compact there exists a countable collection $\mathcal{K}$ of compact subsets of Y such that for different points $y_1, y_2 \in Y$ there is a $K \in \mathcal{K}$ with $y_1 \in K$ and $y_2 \notin K$. According to Lemma 4.2 there is a correspondence Φ^* from X to Y with $\Phi \subset \Phi^*$ such that $\Phi^{*-1}(F) \in \mathfrak{A}_\mu$ for all closed sets $F \subset Y$ and such that Φ and Φ^* have the same $\mathfrak{A}_\mu$-$\mathcal{B}(Y)$-measurable weak selections. Thus, an $\mathfrak{A}_\mu$-$\mathcal{B}(Y)$-measurable weak selection for Φ^* is uniquely determined up to weak Borel equivalence. For $N = \bigcup\{\Phi^{*-1}(K) \cap \Phi^{*-1}(K^c): K \in \mathcal{K}\}$ we, therefore, deduce by means of Proposition 5.2 that $\mu^*(N) = 0$ (In a metrizable compact space all compact sets are G_δ-sets and all Borel sets are Baire sets.). For $x \in X$ with $\mathrm{card}(\Phi^*(x)) \geq 2$ there is a set $K \in \mathcal{K}$ with

$$x \in \Phi^{*-1}(K) \cap \Phi^{*-1}(K^c) \subset N,$$

and hence $$\mu^*(\{x \in X: \mathrm{card}(\Phi^*(x)) \geq 2\}) = 0.$$

Since $\Phi \subset \Phi^*$ this implies

$$\mu^*(\{x \in X: \mathrm{card}(\Phi(x)) \geq 2\}) = 0$$

and

$$\Phi^* \cap ((\Phi^{-1}(Y) \setminus N) \times Y) = \Phi \cap ((\Phi^{-1}(Y) \setminus N) \times Y).$$

For a closed set $F \subset Y$ this leads to

$$\begin{aligned}\Phi^{-1}(F) &= (\Phi^{-1}(F) \cap N) \cup (\Phi^{-1}(F) \setminus N) \\ &= (\Phi^{-1}(F) \cap N) \cup (\Phi^{*-1}(F) \cap (\Phi^{-1}(Y) \setminus N)),\end{aligned}$$

hence $\Phi^{-1}(F) \in \mathfrak{A}_\mu \cap \Phi^{-1}(Y)$, since $A := (\Phi^{*-1}(F) \setminus N) \cup (\Phi^{-1}(F) \cap N) \in \mathfrak{A}_\mu$ and $A \cap \Phi^{-1}(Y) = \Phi^{-1}(F)$.

(ii) $\Longrightarrow$ (i). Let K be a compact set in Y. Then

$$\Phi^{-1}(K) \cap \Phi^{-1}(K^c) \subset \{x \in X: \mathrm{card}(\Phi(x)) \geq 2\}$$

and there are sets $A, A' \in \mathfrak{A}_\mu$ with $A \cap \Phi^{-1}(Y) = \Phi^{-1}(K)$ and $A' \cap \Phi^{-1}(Y) = \Phi^{-1}(K^c)$. Thus $\mu^*(A \cap A' \cap \Phi^{-1}(Y)) = 0$, which implies $\mu(A \cap A') = 0$, since $\mu^*(\Phi^{-1}(Y)) = \mu(X)$. Hence $\mu(\Phi^{-1}(K)^* \cap \Phi^{-1}(K^c)^*) = 0$ and, therefore, $\mu^*(\Phi^{-1}(K)) + \mu^*(\Phi^{-1}(K^c)) = \mu(X)$, from which condition (ii) follows by means of Proposition 5.2.

A counter-example below will show that implication (i) $\Longrightarrow$ (ii) of the corollary does not generalize to arbitrary compact spaces Y.

For the rest of the paper we will be concerned with conditions which ensure the uniqueness of preimage measures.

5.4. Lemma

Let $(Y,\mathfrak{B})$ be a measurable space, $p: Y \longrightarrow X$ a $\mathfrak{B}$-$\mathfrak{A}$-measurable map and $\varphi_1, \varphi_2: X \longrightarrow Y$ two $\mathfrak{A}_\mu$-$\mathfrak{B}$-measurable weak sections for p. Then the following statements are equivalent:

(i) $\varphi_1(\mu) = \varphi_2(\mu)$

(ii) $\varphi_1^{-1}(B) \sim \varphi_2^{-1}(B)$ for all $B \in \mathfrak{B}$.

Proof: Since the implication (ii) $\Longrightarrow$ (i) is obviously true it remains to prove its converse. To this end let $B \in \mathfrak{B}$ be arbitrary. Then we have

$$\begin{aligned}\mu(\varphi_1^{-1}(B) \triangle \varphi_2^{-1}(B)) &= \mu(\varphi_1^{-1}(B) \triangle \varphi_1^{-1}p^{-1}\varphi_2^{-1}(B)) \\ &= \mu(\varphi_1^{-1}(B \triangle p^{-1}\varphi_2^{-1}(B))) \\ &= \mu(\varphi_2^{-1}(B \triangle p^{-1}\varphi_2^{-1}(B)))\end{aligned}$$

$$= \mu(\varphi_2^{-1}(B) \triangle \varphi_2^{-1}p^{-1}\varphi_2^{-1}(B))$$
$$= \mu(\varphi_2^{-1}(B) \triangle \varphi_2^{-1}(B)) = 0,$$

hence $\varphi_1^{-1}(B) \sim \varphi_2^{-1}(B)$.

5.5. Theorem

Let $(X,\mathfrak{A},\mu)$ *be a finite measure space,* Y *a Hausdorff space, and* $p: Y \to X$ *a* $\mathfrak{B}(Y)$-$\mathfrak{A}$-*measurable map with*

$$(1)\ \mu(X) = \sup\{\mu^*(p(K)): K \subset Y \text{ compact}\}$$

and

$$(2)\ \mu^*(p(K)) = \inf\{\mu^*(p(U)): K \subset U,\ U \text{ open}\} \text{ for every compact subset } K \text{ of } Y.$$

Then the following statements are equivalent:

(i) *There exists an, up to weak Borel equivalence, unique* $\mathfrak{A}_\mu$-$\mathfrak{B}(Y)$-*measurable weak selection* φ *for* p^{-1} *with* $\varphi(\mu)$ *a Radon measure on* Y.

(ii) *There exists an, up to weak Borel equivalence, unique* $\mathfrak{A}_\mu$-$\mathfrak{B}(Y)$-*measurable weak section* φ *for* p *with* $\varphi(\mu)$ *a Radon measure on* Y.

(iii) *There exists a unique Radon measure* ν *on* Y *with* $p(\nu) = \mu$.

(iv) *For all disjoint compact sets* $K_1, K_2 \subset Y$ *the equality*
$$\mu^*(p(K_1 \cup K_2)) = \mu^*(p(K_1)) + \mu^*(p(K_2)) \quad \text{holds.}$$

Proof:

(i) $\Longrightarrow$ (iv). This is an immediate consequence of Proposition 5.1.

(iv) $\Longrightarrow$ (iii). Let ν be a Radon measure on Y with $p(\nu) = \mu$ and let K be a compact subset of Y. Then there exists an increasing sequence $(K_n)_{n\in\mathbb{N}}$ of compact subsets of K^c with $\nu(K^c) = \sup\{\nu(K_n): n \in \mathbb{N}\}$, hence $K \cup \bigcup_{n\in\mathbb{N}} K_n \sim Y$. Because of $p(\nu) = \mu$, $K \subset p^{-1}(p(K)^*)$, and $\bigcup_{n\in\mathbb{N}} K_n \subset p^{-1}(\bigcup_{n\in\mathbb{N}} p(K_n)^*)$ we get

$$\nu(Y) \leq \nu(p^{-1}(p(K)^*) + \nu(p^{-1}(\bigcup_{n\in\mathbb{N}} p(K_n)^*))$$
$$= \mu^*(p(K)\ ) + \sup\{\mu^*(p(K_n)): n \in \mathbb{N}\}$$
$$= \sup\{\mu^*(p(K \cup K_n)): n \in \mathbb{N}\} \leq \mu(X) = \nu(Y),$$

hence $\nu(p^{-1}(p(K)^*) \cap p^{-1}(\bigcup_{n\in\mathbb{N}} p(K_n)^*)) = 0$ and since $K \subset p^{-1}(p(K)^*)$ and $p^{-1}(\bigcup_{n\in\mathbb{N}} p(K_n)^*) \supset \bigcup_{n\in\mathbb{N}} K_n \sim K^c\ [\nu]$ this implies

$$\nu(K) = \nu(p^{-1}(p(K)^*)) = \mu^*(p(K)).$$

Thus ν is uniquely determined on compact sets in Y which implies (iii) since ν is a Radon measure.

(iii) $\Longrightarrow$ (ii). This implication follows immediately from Lemma 5.4 combined with Proposition 4.6.

(ii) $\Longrightarrow$ (i). Since, under the assumptions of the theorem, every $\mathfrak{A}_\mu$-$\mathfrak{B}(Y)$-

measurable weak selection φ for p^{-1} with $\varphi(\mu)$ a Radon measure on Y is an $\mathfrak{A}_\mu$-$\mathfrak{B}(Y)$-measurable weak section for p the imlication is obviously true.

Our next theorem is the analogue of Theorem 5.5 for Baire measures on compact spaces.

5.6. Theorem

Let $(X,\mathfrak{A},\mu)$ be a finite measure space, Y a compact space, $p: Y \to X$ a $\mathfrak{L}(Y)$-$\mathfrak{A}$-measurable map such that every compact G_δ-set is p^{-1}-regular and $\mu^*(p(Y))=\mu(X)$. Then the following statements are equivalent:

(i) There exists an, up to weak Baire equivalence, unique $\mathfrak{A}_\mu$-$\mathfrak{L}(Y)$-measurable weak selection for p^{-1}.

(ii) There exists an, up to weak Baire equivalence, unique $\mathfrak{A}_\mu$-$\mathfrak{L}(Y)$-measurable weak section for p.

(iii) There exists a unique Baire measure ν on Y with $p(\nu) = \mu$.

(iv) For all compact G_δ-sets $K \subset Y$ the equality $\mu^*(p(K)) + \mu^*(p(K^c)) = \mu(X)$ holds.

(v) If $K \subset Y$ is a compact G_δ-set then $p(K) \in \mathfrak{A}_\mu \cap p(Y)$ and $\mu^*(p(K) \cap p(K^c)) = 0$.

(vi) If $\mathfrak{L}_0 := \{p^{-1}(A): A \in \mathfrak{A}\}$ and $\mu_0: \mathfrak{L}_0 \to \mathbb{R}$ is the measure defined by $\mu_0(p^{-1}(A)) = \mu(A)$ for $A \in \mathfrak{A}$, then the completion of $(Y,\mathfrak{L}_0,\mu_0)$ contains $\mathfrak{L}(Y)$.

Proof:

(i) $\Longrightarrow$ (iv). This is an immediate consequence of Proposition 5.2.

(iv) $\Longrightarrow$ (v). This implication can be proved by an easy calculation similar to that in the second part of the proof of Corollary 5.3.

(v) $\Longrightarrow$ (vi). Let $K \subset Y$ be an arbitrary compact G_δ-set. Then we have

$$p^{-1}(p(K)^*) \setminus p^{-1}(p(K^c)^*) \subset p^{-1}(p(K)^*) \setminus K^c = K \subset p^{-1}(p(K)^*) \quad \text{and}$$

$$\mu_0(p^{-1}(p(K)^*) \setminus p^{-1}(p(K^c)^*)) = \mu^*(p(K)) - \mu(p(K)^* \cap p(K^c)^*)$$
$$= \mu^*(p(K)) = \mu_0(p^{-1}(p(K)^*))$$

(since $p(K) \cap p(K^c) \sim \emptyset$, $p(K) \in \mathfrak{A}_\mu \cap p(Y)$, $p(K^c) \in \mathfrak{A}_\mu \cap p(Y)$ and $\mu^*(p(Y)) = \mu(X)$ imply $\mu(p(K)^* \cap p(K^c)^*) = 0$).

Thus K is in the completion of $(Y,\mathfrak{L}_0,\mu_0)$ which, therefore, contains all Baire sets in Y.

(vi) $\Longrightarrow$ (iii) is obvious.

(iii) $\Longrightarrow$ (ii) is an immediate consequence of Lemma 5.4 combined with Proposition 4.6.

(ii) $\Longrightarrow$ (i). By arguments similar to those used in the proof of Proposition 4.6 it can be seen that, under the assumptions of the theorem, every $\mathfrak{A}_\mu$-$\mathfrak{L}(Y)$-measurable weak selection for p^{-1} is an $\mathfrak{A}_\mu$-$\mathfrak{L}(Y)$-measurable weak section for p.

In the case that $(X,\mathfrak{A},\mu)$ is countably generated modulo μ and Y is an analytic metric space the equivalence of (iii) and (vi) was proved by Yershov [27] for arbitrary $\mathfrak{B}(Y)$-$\mathfrak{A}$-measurable maps p: $Y \to X$ with $\mu^*(p(Y)) = \mu(X)$.

<u>5.7. Corollary</u>

<u>Let</u> $(X,\mathfrak{A},\mu)$ <u>be a finite measure space</u>, Y <u>a metrizable compact space</u>, p: $Y \to X$ <u>a</u> $\mathfrak{B}(Y)$-$\mathfrak{A}$-<u>measurable map with</u> $\mu^*(p(Y)) = \mu(X)$ <u>and such that every compact subset of</u> Y <u>is</u> p^{-1}-<u>regular</u>. <u>Then the following conditions are equivalent</u>:

(i) <u>There is a unique Borel measure</u> ν <u>on</u> Y <u>with</u> $p(\nu) = \mu$.

(ii) $\mu^*(\{x \in X: \text{card}(p^{-1}(x)) \geq 2\}) = 0$ <u>and</u> $p(K) \in \mathfrak{A}_\mu \cap p(Y)$ <u>for all compact subsets</u> K <u>of</u> Y.

<u>Proof:</u> Since Baire and Borel sets in Y are identical the corollary is an immediate consequence of the equivalence of (i) and (iii) in Theorem 5.6 combined with Corollary 5.3.

<u>5.8. Remarks</u>

a) The above corollary obviously implies the following result of Eisele [3]: Let X and Y be compact spaces, $\mathfrak{A} = \mathfrak{B}(X)$, μ a Radon measure on X, and p: $Y \to X$ a continuous map onto. If Y is metrizable then the existence of a unique Borel preimage measure for μ w.r.t. p is equivalent to $\mu^*(\{x \in X: \text{card}(p^{-1}(x)) \geq 2\}) = 0$.
This result was generalized by Lehn-Mägerl [15] to arbitrary surjective measurable maps between analytic spaces. Their gneralization does not (immediately) follow from the above corollary.

b) The implication (i) $\Rightarrow$ (ii) in Corollary 5.7 does not generalize to arbitrary nonmetrizable compact spaces Y. The following is an example where this implication fails to hold.

Let $X = [0,1]$, $\mathfrak{A}$ be the Borel field in X, and μ Lebesgue measure on $\mathfrak{A}$. Denote by $\mathfrak{F}$ the field generated by $\{[s,t[: 0 \leq s \leq t \leq 1\}$ and let Y be its Boolean representation space, i.e. the elements of Y are the ultrafilters in $\mathfrak{F}$, a set $U \subset Y$ is open-closed if and only if there is an $F \in \mathfrak{F}$ with $U = \{y \in Y: F \in y\}$, and the open-closed sets form a base for the topology of Y. It is easy to see that all elements of $\mathfrak{F}$ are finite unions of intervals (cf. Bauer [1], p. 23). Define p: $Y \to X$ in such a way that, for $y \in Y$, p(y) is the uniquely determined element of X to which the ultrafilter y converges. Then p is a continuous map onto. For $x \in]0,1[$ let y_1 be an ultrafilter in $\mathfrak{F}$ containing $\{[t,x[: t < x\}$ and y_2 an ultrafilter in $\mathfrak{F}$ containing $\{[x,t[: x < t\}$. Then $y_1 \neq y_2$, $p(y_1) = p(y_2) = x$ and hence

$$]0,1[\subset \{x \in X: \operatorname{card}(p^{-1}(x)) \geq 2\}$$

which implies $\mu^*(\{x \in X: \operatorname{card}(p^{-1}(x)) \geq 2\}) \neq 0$.
According to Example 4.4(iv) every set $B \in \mathfrak{B}(Y)$ is p^{-1}-regular.
Now let $K_1, K_2 \subset Y$ be disjoint compact sets. Then there are disjoint open-closed sets $U_1, U_2 \subset Y$ with $K_i \subset U_i$ $(i = 1,2)$. Let $F_1, F_2 \in \mathfrak{F}$ be such that $U_i = \{y \in Y: F_i \in y\}$. Then $F_1 \cap F_2 = \emptyset$ and $p(U_i) \subset \overline{F_i}$. But F_i is a finite union of intervals and, therefore, $\mu(\overline{F_1} \cap \overline{F_2}) = 0$, hence

$$\mu^*(p(K_1 \cup K_2)) = \mu^*(p(K_1)) + \mu^*(p(K_2)).$$

Thus, according to Theorem 5.5, there exists a unique Radon measure ν on Y with $p(\nu) = \mu$.

The example also shows that there exists a compact space Y such that the implication (i) $\Longrightarrow$ (ii) in Corollary 5.3 does not hold.

References:

[1] H. Bauer, Wahrscheinlichkeitstheorie und Grundzüge der Maßtheorie, de Gruyter, Berlin 1968

[2] G. A. Edgar, Measurable Weak Sections, Illinois J. Math. 20 (1976), 630-646

[3] K. T. Eisele, On the Uniqueness of Preimages of Measures, in: Measure Theory, Proc. of the Conference Held at Oberwolfach 1975 (A. Bellow, D. Kölzow eds.), Lecture Notes in Math. 541, p. 5-12, Springer, Berlin-Heidelberg-New York 1976

[4] S. Graf, On the Existence of Strong Liftings in Second Countable Topological Spaces, Pacific J. Math. 58 (1975), 419-426

[5] S. Graf, Induced σ-homomorphisms and a Parametrization of Measurable Sections via Extremal Preimage Measures, submitted for publication

[6] P. R. Halmos, Measure Theory, van Nostrand, Princeton 1950

[7] M. Hasumi, A Continuous Selection Theorem for Extremally Disconnected Spaces, Math. Ann. 179 (1969), 83-89

[8] C. J. Himmelberg and F. S. van Vleck, Some Selection Theorems for Measurable Functions, Canadian J. Math. 21 (1969), 394-399

[9] A. Ionescu Tulcea, Liftings Compatible with Topologies, Bull. Soc. Math. de Grèce 8 (1967), 116-126

[10] A. Ionescu Tulcea and C. Ionescu Tulcea, On the Existence of a Lifting Commuting with the Left Translations of an Arbitrary Locally Compact Group, Proc. Fifth Berkeley Symp. on Math. Stat. and Probability, p. 63-67, Univ. of California Press 1967

[11] A. Ionescu Tulcea and C. Ionescu Tulcea, Topics in the Theory of Lifting, Springer, Berlin-Heidelberg-New York 1969

[12] D. Kölzow, Differentiation von Maßen, Lecture Notes in Math. 65, Springer, Berlin-Heidelberg-New York 1968

[13] K. Kuratowski and C. Ryll-Nardzewski, A General Theorem on Selectors, Bull. Acad. Polon. 13 (1965), 397-403

[14] J. Lembcke, Konservative Abbildungen und Fortsetzung regulärer Maße, Z. Wahrscheinlichkeitstheorie verw. Geb. 15 (1970), 57-96

[15] J. Lehn and G. Mägerl, On the Uniqueness of Preimage Measures, Z. Wahrscheinlichkeitstheorie verw. Geb. 38 (1977), 333-337

[16] V. Losert, A Measure Space without the Strong Lifting Property, Math. Ann. 239 (1979), 119-128

[17] V. Losert, A Counterexample on Measurable Selections and Strong Lifting, in this volume

[18] G. Mägerl, Zu den Schnittsätzen von Michael und Kuratowski & Ryll-Nardzewski, Thesis, Erlangen 1977

[19] D. Maharam, On a Theorem of von Neumann, Proc. Amer. Math. Soc. 9 (1958), 987-994

[20] J. von Neumann, On Rings of Operators. Reduction Theory, Ann. Math. 50 (1949), 401-485

[21] T. Parthasarathy, Selection Theorems and Their Applications, Lecture Notes in Math. 263, Springer, Berlin-Heidelberg-New York 1972

[22] L. Schwartz, Radon Measures on Arbitrary Topological Spaces and Cylindrical Measures, Oxford University Press, London 1973

[23] M. Sion, On Uniformization of Sets in Topolocigal Spaces, Trans. Amer. Math. Soc. 96 (1960), 237-244

[24] M. Talagrand, Non-existence de certaines sections mesurables et application à la théorie du relèvement, C. R. Acad. Sc. Paris 286 (Sér. A) (1978), 1183-1185

[25] M. Talagrand, Non-existence de certaines sections mesurables et contre-exemples en théorie du relèvement, in this volume

[26] D. H. Wagner, Survey of Measurable Selection Theorems, SIAM J. Control Optimization 15 (1977), 859-903

[27] M. P. Yershov (Ershov), Extensions of Measures and Stochastic Equations, Theor. Prob. Appl. 19 (1974), 431-444

[28] M. P. Yershov (Ershov), The Choquet Theorem and Stochastic Equations, Analysis Mathematica 1 (1975), 259-271

Mathematisches Institut der
Universität Erlangen-Nürnberg
Bismarckstr. 1 1/2
D 8520 Erlangen

REPRESENTATION THEOREMS FOR MEASURABLE MULTIFUNCTIONS

A.D. Ioffe

If T, X are sets and F is a set-valued mapping from T into X (i.e. a mapping from T into 2^X), then by a representation of F we mean a pair (Y,h), where Y is a set and the mapping $h: T \times Y \to X$ is such that $h(t,Y) = F(t)$ for all t.

In connection with representations, the main interest lies so far in existence of a representation having certain properties, topological, measurable etc.. A fairly complete account on now existing representation theorems is contained in Wagner's survey [1]. (Note that a different terminology is used there.)

All of the representation theorems contained in [1] and, very likely, all of the representation theorems obtained by now deal with the case when the range space X is metrizable. Here we present two representation theorems free of this restriction. There is also a number of results stated as corollaries (which means only that they follow rather easily from the theorems).

1. PRELIMINARIES

Throughout the paper X is a regular Hausdorff topological space with weight $X = \tau$, T is a set and $\mathcal{M}$ is a field of subsets of T having the property that the union of any subfamily $\mathcal{M}' \subset \mathcal{M}$ with card $\mathcal{M}' < \tau$ belongs to $\mathcal{M}$. If λ is a cardinal, then $\omega(\lambda)$ is the minimal ordinal of cardinality λ and $\Omega(\lambda)$ is the collection of ordinals $\alpha < \omega(\lambda)$.

Let for any $\alpha < \omega(\lambda)$ a cardinal number μ_α be given, and $M = \{\mu_\alpha\}$. We denote by

$$\mathbb{N}(M,\lambda) = \prod_{\alpha < \omega(\lambda)} \Omega(\mu_\alpha)$$

the generalized "Baire" space which is, in other words, the collection of sequences $i = (i_1,\dots,i_\alpha,\dots)$ $(\alpha < \omega(\lambda))$ of ordinals such that $i_\alpha < \omega(\mu_\alpha)$ for any α. Given $i \in \mathbb{N}$, we denote

$$i|\alpha = (i_1, \ldots, i_\alpha)$$

and, more generally, if $Q \subset \Omega(\lambda)$,

$$i|Q = \{ i_\alpha \mid \alpha \in Q \}.$$

Further

$$\mathcal{N}_{i|Q} = \{ j \in \mathbb{N} \mid j|Q = i|Q \},$$
$$\mathcal{N}_{i|\alpha} = \{ j \in \mathbb{N} \mid j|\alpha = i|\alpha \},$$
$$\mathcal{N}^-_{i|\alpha} = \{ j \in \mathbb{N} \mid j|\xi = i|\xi \ , \ \forall\ \xi < \alpha \},$$
$$\mathcal{N}_{\alpha\beta} = \{ j \in \mathbb{N} \mid j_\alpha = \beta \}.$$

If $\mu_\alpha = \mu$ for all α, we write $\mathbb{N}(\mu,\lambda)$ instead of $\mathbb{N}(M,\lambda)$. From now on, Q will be a finite subset of $\Omega(\lambda)$.

The collections $\mathcal{N}^* = \{ \mathcal{N}_{i|Q} \}$ and $\mathcal{N} = \{ \mathcal{N}_{i|\alpha} \}$ are topology bases in $\mathbb{N}$. The first topology will be called weak and the second strong. For $\lambda = \aleph_0$ both topologies coincide. We shall mostly deal with the strong topology.

A family

$$\mathcal{U} = \{ U_{i|Q} \mid i \in \mathbb{N}(M,\lambda),\ \ Q \subset \Omega(\lambda) \ \text{ is finite} \}$$

of subsets of X will be called (M,λ)-sieve if

$$\bigcup_Q U_{i|Q} = T,$$

$$U_{i|Q} \subset U_{i|Q'}, \text{ if } Q' \subset Q,$$

$$\bigcup_{j \in \mathcal{N}_{i|Q}} U_{j|Q \cup Q'} = U_{i|Q}, \text{ if } Q \cap Q' = \emptyset.$$

The sieve is fine if the sets

$$U_{i|\alpha} = \bigcap_{\max Q \leq \alpha} U_{i|Q}$$

satisfy the condition

$$\bigcup_{j \in \mathcal{N}^-_{i|\alpha}} U_{j|\alpha} = \bigcap_{\xi < \alpha} U_{i|\xi}.$$

An (M,λ)-sieve $\mathcal{U} \subset X$ will be called compatible with the topology if the filter generated by the sets $\overline{U}_{i|Q}$ (i fixed) converges if $i \in \mathbb{N}$ is such that $\bigcap_Q \overline{U}_{i|Q} \neq \emptyset$. (The bar denotes the closure.)

A set $A \subset X$ is called $\mathcal{U}$ -complete if $\bigcap_Q A \cap \overline{U}_{i|Q} \neq \emptyset$ whenever $i \in \mathbb{N}$ is such that $A \cap \overline{U}_{i|Q} \neq \emptyset$ for any Q.

Let Y be another set,and let $\mathcal{V}$ be an (N,λ)-sieve in Y (N may differ from M but λ is the same). A mapping $h: Y \to X$ will be called uniform (w.r.t. $\mathcal{U}$ and $\mathcal{V}$) if for any $j \in \mathbb{N}(N,\lambda)$ there is an $i \in \mathbb{N}(M,\lambda)$ such that $h(V_{j|Q}) \subset \overline{U}_{i|Q}$ for any (finite) $Q \subset \Omega(\lambda)$. If the elements of $\mathcal{V}$ are open and $\mathcal{U}$ is compatible with the topology in X , then a uniform mapping $h: Y \to X$ is continuous.

Let F be a set-valued mapping from T into X. We write, as usual,

$$F^{-}(A) = \{ t \in T \mid F(t) \cap A \neq \emptyset \} .$$

If $\mathcal{Z}$ is a family of subsets of T, then we shall say, following [1], that F is $\mathcal{Z}$ -measurable (weakly $\mathcal{Z}$ -measurable) if $F^{-}(A)$ belongs to $\mathcal{Z}$ for any closed (open) $A \subset X$.

By $\mathcal{M}_\tau$ we denote the collection of unions of at most τ elements of $\mathcal{M}$.

2. UNIFORM REPRESENTATION THEOREM

THEOREM 1. Assume that there are (M,λ)-sieves $\mathcal{U} = \{U_{i|Q}\}$ in X and $\mathcal{R} = \{R_{i|Q}\}$ in T with $\text{card}\,\mathcal{U} = \text{card}\,\mathcal{R} = \tau$ such that $\mathcal{U}$ is compatible with the topology in X, $\mathcal{R}$ is fine and $\mathcal{R} \subset \mathcal{M}$. Let F be a set-valued mapping from T into X with nonempty and $\mathcal{U}$ -complete values and such that

$$R_{i|Q} \subset F^{-}(\overline{U}_{i|Q}) , \quad \forall \; i, Q .$$

Then there is a mapping $h: T \times \mathbb{N}(M,\lambda) \to X$ such that

(i) the mapping $t \to h(t,i)$ is $\mathcal{M}_\tau$ -measurable for any $i \in \mathbb{N}$;

(ii) the mapping $i \to h(t,i)$ is uniform from $(\mathbb{N},\mathcal{N})$ into $(X,\mathcal{U})$ for any $t \in T$;

(iii) for any $t \in T$,

$$F_R(t) \subset h(t,\mathbb{N}) \subset F(t) ,$$

where

$$F_R(t) = \{ x \in X \mid \exists \; i \in \mathbb{N} : \; x \in U_{i|Q} , \; t \in R_{i|Q} , \; \forall \; Q \} .$$

PROOF. For any $\alpha < \omega(\lambda)$, let $\sigma_\alpha : \Omega(\mu_\alpha) \to \Omega(\mu_\alpha)$ be a bijection, and let $\sigma : \mathbb{N} \to \mathbb{N}$ be defined by

$$(\sigma(i))_\alpha = \sigma_\alpha(i_\alpha) .$$

We set

$$S^{\sigma}_{i|\alpha} = R_{i|\alpha} \setminus \bigcup \{R_{j|\alpha} \mid j \in \mathcal{N}^{-}_{i|\alpha} \;,\; \sigma_\alpha(j_\alpha) < \sigma_\alpha(i_\alpha) \;,$$

$$T^{\sigma}_{i|\alpha} = \bigcap_{\xi \leq \alpha} S^{\sigma}_{i|\xi}$$

Then

$$T^{\sigma}_{i|\alpha} \subset T^{\sigma}_{i|\xi} \;, \quad \text{if} \quad \alpha > \xi \;,$$

$$T^{\sigma}_{i|\alpha} \cap T^{\sigma}_{j|\alpha} = \emptyset \;, \quad \text{if} \quad i|\alpha \neq j|\alpha \;,$$

$$\bigcup_{i \in \mathcal{N}_{\alpha\beta}} T^{\sigma}_{i|\alpha} \in \mathcal{M} \;, \quad \forall\, \alpha, \beta \;,$$

$$\bigcup_{i} T_{i|\alpha} = T \;, \quad \forall\, \alpha \quad .$$

The first two relations are obvious. The second two can be proved by induction using the fact that $\mathcal{R}$ is fine.

Therefore for any $t \in T$ there is exactly one $i = i(t,\sigma)$ such that

$$t \in T_{i|\alpha} \quad \text{for any} \quad \alpha \quad .$$

It follows that $F(t) \cap \overline{U}_{i|Q} \neq \emptyset$ for any finite $Q \subset \Omega(\lambda)$ and, since $F(t)$ is $\mathcal{U}$-complete and $\mathcal{U}$ is compatible with the topology, that there is a unique $x = x(t,\sigma)$ such that $x \in F(t)$ and $x \in \overline{U}_{i|Q}$ for any finite $Q \subset \Omega(\lambda)$.

We have

$$t \in T^{\sigma}_{i|Q} = \bigcup_{j \in \mathcal{N}_{i|Q}} T^{\sigma}_{j|\alpha} \Rightarrow x(t,\sigma) \in F(t) \cap \overline{U}_{i|Q}$$

whenever $\max Q \leq \alpha$, and if $U \subset X$ is open, then $x(t,\sigma) \in U$ iff the inclusions $\overline{U}_{j|Q} \subset U$, $t \in T_{j\,Q}$ are valid for certain $j|Q$ so that

$$\{t \mid x(t,\sigma) \in U\} = \bigcup \{T_{j|Q} \mid \overline{U}_{j|Q} \subset U\} \in \mathcal{M}_\tau .$$

Furthermore, for any $j \in \mathbb{N}$ let us define a bijection $\sigma(j)\colon \mathbb{N} \to \mathbb{N}$ by $(\sigma(j,i))_\alpha = \sigma_\alpha(j,i_\alpha)$, where

$$\sigma_\alpha(j,\beta) = \begin{cases} 1 \;, & \text{if} \quad \beta = j_\alpha \;, \\ j_\alpha, & \text{if} \quad \beta = 1 \;, \\ \beta \;, & \text{otherwise} \quad . \end{cases}$$

Then $h(t,j) = x(t,\sigma(j))$ is the desired mapping.

Indeed, (i) follows from what has been proved. If $k|\alpha = j|\alpha$ then $i(t,\sigma(k))|\alpha = i(t,\sigma(j))|\alpha$ so that $T_{i|\alpha}^{\sigma(k)} = T_{i|\alpha}^{\sigma(j)}$ and both $x(t,\sigma(k))$ and $x(t,\sigma(j))$ belong to the same $\overline{U}_{i|Q}$ whenever $\max Q \leq \alpha$. This proves (ii).

Finally, if $x \in F_R(t)$ and $j \in \mathbb{N}$ is such that $x \in U_{j|Q}$, $t \in R_{j|Q}$ for all Q, then $i(t,\sigma(j)) = j$ so that $x(t,\sigma(j)) = x$.

<u>COROLLARY</u> 1.1. In addition to the hypotheses of the theorem, let us assume that all $U_{i|Q}$ are closed and

$$U_{i|Q} \cap U_{j|Q} = \emptyset \quad \text{if} \quad i|Q \neq j|Q .$$

Then the conclusion of the theorem is valid with (i) replaced by a stronger property:

(i') $\{t \mid h(t,j) \in U_{i|Q}\} \in \mathcal{M}$ for any i, j, Q.

Let $(X,\mathcal{V})$ be a uniform space, $\mathcal{V} \subset 2^{X \times X}$ being the uniform structure in X. Then the space $B(T,X)$ of all mappings from T into X supplied with the structure of the uniform convergence is also a uniform space.

<u>COROLLARY</u> 1.2. Let $(X,\mathcal{V})$ be a Hausdorff uniform space with the cardinality of the uniform structure $\mathcal{V}$ equal to $\lambda \leq \tau$ *), and let F be an either $\mathcal{M}$-measurable or weakly $\mathcal{M}$-measurable set-valued mapping from T into X with nonempty and complete values $F(t)$ which are initially compact up to λ. Then there are a complete uniform space $(Y,\mathcal{W})$ with

$$\text{weight } Y \leq \sum_{\nu < \lambda} \tau^{\nu} \quad , \quad \text{card } \mathcal{W} = \lambda$$

and a mapping $h: T \times Y \to X$ such that

(i) $h(.,y)$ is $\mathcal{M}_{\tau}$-measurable for any $y \in Y$;

(ii) the mapping $y \to h(.,y)$ from Y into $B(T,X)$ is uniformly continuous;

(iii) (Y,h) represents F.

(Recall that a set A is initially compact up to λ if any centralized system of closed subsets of A containing less than λ elements has a nonempty intersection.)

<u>PROOF</u>. If $V \in \mathcal{V}$ and $x \in X$, we set $V(x) = \{z \in X \mid (x,z) \in V\}$. Now let $\mathcal{V} = \{V_{\alpha} \mid \alpha < \omega(\lambda)\}$, and let $\{x_{\gamma} \mid \gamma < \omega(\tau)\}$ be a dense subset of X. Then we define $U_{i|Q}$ and $R_{i|Q}$ by

---

*) See Editor's remark on page 12!

$$U_{i|Q} = \begin{cases} \bigcap_{\alpha \in Q} \operatorname{int} V(x_{i_\alpha}) , & \text{if } F \text{ is weakly measurable,} \\ \bigcap_{\alpha \in Q} \overline{V}(x_{i_\alpha}) , & \text{if } F \text{ is measurable} \quad ; \end{cases}$$

$$R_{i|Q} = F^{-}(U_{i|Q}).$$

Then $\mathcal{U} = \{U_{i|Q}\}$ is obviously compatible with the topology and $\mathcal{R} = \{R_{i|Q}\}$ is fine thanks to the initial compactness of $F(t)$. It remains to take $Y = \mathbb{N}(\tau,\lambda)$ and apply the theorem.

<u>COROLLARY</u> 1.3. (cf. Ioffe [2, 3], Srivastava [4])
Let (X,ρ) be a metric space and F a weakly $\mathcal{M}$-measurable set-valued mapping from T into X with nonempty and complete values. Then there are a complete metric space (Y,d) with weight Y equal to τ and a mapping $h: T\times Y \to X$ such that

(i) $h(.,y)$ is $\mathcal{M}_\tau$-measurable for any $y \in Y$;

(ii) for any $t \in T$, $y, v \in Y$,

$$\rho(h(t,y),h(t,v)) \le (1 + \rho(h(t,y),h(t,v)))d(y,v) ;$$

(iii) (Y,h) represents F.

<u>PROOF</u>. Define $U_{i|Q}$ and $R_{i|Q}$ as in the proof of Corollary 1.2 with $\mathcal{V} = \{V_n\}$ $(n = 1,2,\ldots)$, $V_n = \{(x,z) \mid \rho(x,z) < 2^{-n}\}$, and let the distance in $\mathbb{N}(\tau,\aleph_0)$ be defined by

$$d(i,j) = \sum_{n=1}^{\infty} 2^{-(n-1)}\delta(i_n,j_n) ,$$

where

$$\delta(\beta,\gamma) = \begin{cases} 1 , & \text{if } \beta \ne \gamma , \\ 0 , & \text{if } \beta = \gamma . \end{cases}$$

This result embraces earlier theorems of the author and Srivastava (see also Wagner [1]). In the first of them X was assumed separable and in the second only continuity of $h(t,.)$ was stated, not uniform continuity implied by (ii). This property appears to be useful in many instances as follows in particular from the corollaries below.

<u>COROLLARY</u> 1.4. Let (X,ρ) be a separable metric space, let T be a second countable zero-dimensional topological space, and let F be a set-valued mapping from T into X with nonempty and complete values. Then the following two properties are equivalent:

(a) F is lower semicontinuous,

(b) there are a complete metric space (Y,d) and a (jointly) continuous mapping $h: T\times Y \to X$ having the properties (ii) and (iii) of Corollary 1.3.

PROOF. Implication (b)$\Rightarrow$(a) is obvious. Let F be l.s.c., that is to say, $F^-(U)$ is open whenever U is open. For any $n = 1,2,\ldots$, let $\{U_{nm}\}$ $(m = 1,2,\ldots)$ be a family of open balls with diameter not exceeding 2^{-n} covering X, and for any $i\in \mathbb{N} = \mathbb{N}(\aleph_0,\aleph_0)$ let

$$U_{i|n} = \bigcap_{r=1}^{n} U_{ri_r}$$

Take a family $\{P^s_{i|n}\}$ $(i\in \mathbb{N},\ n, s = 1,2,\ldots)$ of clopen subsets of T such that

$$P^s_{i|n} \subset F^-(U_{i|n}) ;$$

$$P^s_{i|(n+1)} \subset P^s_{i|n} ;$$

$$\bigcup_{j\in \mathcal{N}_{i|n}} P^s_{j|(n+1)} = P^s_{i|n} ;$$

$$\bigcup_i P^s_{i|1} = T ;$$

$$\bigcup_s P^s_{i|n} = F^-(U_{i|n})$$

(such a family can be easily built by induction), then take a bijection $\phi: \mathbb{N}\times\mathbb{N} \to \mathbb{N}$ and set

$$R_{i|n} = \bigcap_{m\le n} P^{k_m}_{l|m} ,$$

where $\phi(i) = (l,k)\in \mathbb{N}\times\mathbb{N}$.

Finally taking $\mathfrak{M}$ to be the field of clopen subsets of T, we arrive at the desired result. (Joint continuity of h is due to the uniformity property (ii) of Corollary 1.3 and to the fact that the unions of elements of $\mathfrak{M}$ are open sets.)

COROLLARY 1.5 (cf. Hasumi [5], Graf [6]). Assume that X is second countable, T is an extremally disconnected topological space and F is an upper semicontinuous set-valued mapping from T into X with nonempty and compact values. Then there are a compact metric space Y and a (jointly) continuous mapping $h: T\times Y \to X$ such that

(i) $h(t,y) \subset F(t)$ for any t, y ;

(ii) whenever $x(t)$ is a continuous selection of F, there is a continuous mapping $y(.): T \to Y$ such that $x(t) = h(t,y(t))$.

PROOF. Assume that X is compact (otherwise take a second countable compactification of X). Then for certain family $M = \{\mu_n\}$ $(n = 1,2,\ldots)$ of integers, there is an $(M,\aleph_0)$-sieve $\mathcal{U}$ in X formed by open sets and compatible with the topology.

Let

$$R_{i|Q} = \text{int } F^{-}(\overline{U}_{i|Q}) ,$$

and let $\mathfrak{M}$ be the field of clopen subsets of T. It follows from [7] (Lemma 6.4) that $\mathcal{R} = \{R_{i|Q}\}$ is a sieve in T so that we can apply the theorem.

The mapping h will be then jointly continuous by the same reason as in the preceding corollary and $Y = \mathbb{N}(M,\aleph_0)$ will be compact because all μ_n are finite. Furthermore, if $x(t)$ is a continuous selection of F, then $x(t) \in F_R(t)$ for any t (which is immediate from the definition of F_R) so that $G(t) = \{y \mid h(t,y) = x(t)\} \neq \emptyset$ for any t. The set-valued mapping G is compact-valued and u.s.c. and it remains to take an arbitrary continuous selection $y(t)$ of G.

Observe that the continuous selection theorem of Hasumi does not require that X be second countable. There are some difficulties however in extending the above result to this general setting, mainly connected with the fact that, for higher cardinalities, the field of clopen sets does not possess the additivity property as stated at the beginning of the first section. As follows from Graf's selection theorem [6], this property is not necessary for a selection to exist. One can only assume that any subfamily $\mathfrak{M}' \subset \mathfrak{M}$ with card $\mathfrak{M}' < \tau$ has a supremum in $\mathfrak{M}$, the property which the field of clopen subsets does possess. However, the uncertainty in knowing what extra points (beyond those belonging to the union of elements of $\mathfrak{M}'$) the supremum of $\mathfrak{M}'$ contains makes difficult (if not impossible) the proof that a selection can be drawn through any $x \in F_R(t)$.

3. COMPACT-VALUED MAPPINGS. THE METHOD OF GRAF

Let $\mathcal{U}$ be a semifield of subsets of X (i.e. $\emptyset \in \mathcal{U}$, $X \in \mathcal{U}$ and $\mathcal{U}$ contains finite unions and intersections). A mapping $R: \mathcal{U} \to 2^T$ is called U-homomorphism if $R(\emptyset) = \emptyset$, $R(X) = T$, $R(A \cup B) = R(A) \cup R(B)$ for any $A,B \in \mathcal{U}$. If R is a monotone U-homomorphism (i.e. if

$A \subset B$, then $R(A) \subset R(B)$), then R can be extended to all of 2^X by setting

$$R(A) = \bigcap \{ R(U) \mid U \in \mathcal{U} , A \subset U \} .$$

If $U \subset X$ is open, then $k(U)$ will denote the smallest cardinal ν for which there is a family $\mathcal{U}' \subset \mathcal{U}$ with card $\mathcal{U}' = \nu$ and such that the unions of interiors of elements of $\mathcal{U}'$ as well as of their closures coincides with U.

<u>THEOREM</u> 2. Let $\mathcal{U} = \{ U_\xi \mid \xi < \omega(\tau) \}$ be a semifield of subsets of X such that $U_\xi \subset \overline{\text{int } U_\xi}$ and the interiors of U_ξ form a topology base in X. Let F be a set-valued mapping from T into X with nonempty and compact values, and let R be a U-homomorphism from the field $\mathcal{A}$ generated by $\mathcal{U}$ into 2^T such that $R(U) \in \mathcal{M}$ for all $U \in \mathcal{U}$ and $R(A) \subset F^-(\overline{A})$ for all $A \in \mathcal{A}$.

Then there are a set $Y \subset \mathbb{N}(\tau,\tau)$ and a mapping $h: T \times Y \to X$ such that

(i) $\{ t \mid h(t,y) \in U \} \in \mathcal{M}_{k(U)}$ for any open $U \subset X$;

(ii) for any $t \in T$, the mapping $y \to h(t,y)$ is continuous from the strong topology of $\mathbb{N}(\tau,\tau)$ into X ;

(iii) for any $t \in T$,

$$F^o_R(t) \subset h(t,Y) \subset F(t) ,$$

where

$$F^o_R(t) = \{ x \in X \mid t \in R(U) \text{ whenever } x \in \text{int } U , U \in \mathcal{U} \}.$$

<u>PROOF</u>. Let Y be the collection of all bijections σ from $\Omega(\tau)$ onto itself. Given a $\sigma \in Y$, we define the following transfinite sequence of U-homomorphisms:

$$R_o(A) = R(A) ,$$

$$\cdots\cdots\cdots\cdots\cdots\cdots$$

$$S^\sigma_\xi(A) = \bigcap_{\eta < \xi} R^\sigma_\eta(A) ,$$

$$R^\sigma_\xi(A) = S^\sigma_\xi(A \cap U_{\sigma(\xi)}) \cup (S^\sigma_\xi(A) \setminus S^\sigma_\xi(U_{\sigma(\xi)})).$$

Then (see [6])

$$R^\sigma_\xi(A) \subset R^\sigma_\eta(A) , \quad \text{if } \xi > \eta ;$$

$$R^\sigma_\xi(U_{\sigma(\xi)}) \cap R^\sigma_\xi(CU_{\sigma(\xi)}) = \emptyset ;$$

$$R^\sigma_\xi(U_{\sigma(\ )}) = R^\sigma(U_{\sigma(\eta)}), \quad \text{if } \xi > \eta ;$$

$$R^{\sigma}_{\xi}(U) \in \mathcal{M} \quad , \quad \text{if} \quad U \in \mathcal{U} \quad .$$

These four relations show that

$$R^{\sigma}(A) = \bigcap_{\xi} R^{\sigma}_{\xi}(A)$$

is a Boolean homomorphism (i.e. U-homomorphism with $R^{\sigma}(A \cap B) = R^{\sigma}(A) \cap R^{\sigma}(B)$) from $\mathcal{A}$ into $\mathcal{M}$.

For any $t \in T$, let

$$\mathcal{A}(t) = \{ A \in \mathcal{A} \mid t \in R^{\sigma}(A) \} .$$

Then $\mathcal{A}(t)$ is a filter base, hence, since $F(t)$ is compact and X is Hausdorff, there is exactly one $x = h(t,\sigma)$ such that for any $A \in \mathcal{A}(t)$, $x \in F(t) \cap \overline{A}$.

It follows that $t \in R^{\sigma}(A)$ whenever $x \in \operatorname{int} A$ $(A \in \mathcal{A})$ for otherwise we would have $t \in R^{\sigma}(CA)$ and $x = h(t,\sigma) \in \overline{CA}$. Therefore if $U \subset X$ is open and $\mathcal{U}' \subset \mathcal{U}$ is such that the union of interiors of elements of $\mathcal{U}'$ coincides with U as well as the union of the closures of the elements, then

$$\{ t \mid h(t,\sigma) \in U \} = \bigcup_{V \in \mathcal{U}'} R^{\sigma}(V) .$$

which implies (i).

The proof of (ii) is similar to that for Theorem 1. To prove (iii), it suffices, for given t and $x \in F^{o}_{R}(t)$, to find $\sigma \in Y$ such that

$$x \notin \overline{U}_{\sigma(\xi)} \Rightarrow t \notin R^{\sigma}(U_{\sigma(\xi)}).$$

A sufficient condition for this implication to be valid is that

$$x \notin \overline{U}_{\sigma(\xi)} \Rightarrow \exists\, \eta < \xi : x \in \operatorname{int} U_{\sigma(\eta)},\ U_{\sigma(\eta)} \cap U_{\sigma(\xi)} = \emptyset \quad ;$$

$$U \in \mathcal{U} , \quad x \in \operatorname{int} U \Rightarrow t \in R^{\sigma}_{\xi}(U) \qquad \text{for all } \xi \quad .$$

A bijection σ for which these two properties are satisfied can be obtained from any other bijection, say, identical by the formula

$$\sigma(\xi) = \begin{cases} \xi \ , & \text{if } \xi \notin D \ n(D), \\ n(\xi) \ , & \text{if } \xi \in D \quad , \\ n^{-1}(\xi), & \text{if } \xi \in n(D) \quad , \end{cases} \qquad (*)$$

where $D \subset \Omega(\tau)$ and injection $n\colon D \to \Omega(\tau)$ are such that

$$D \cap n(D) = \emptyset \ , \quad n(\xi) > \xi \qquad \text{for all } \xi \in D,$$

and D and n are defined by induction as follows: if we have already defined $D_{\gamma} = \{ \xi \in D \mid \xi < \gamma \}$, then $\gamma \in D$ if and only if either $x \notin \bar{U}_{\sigma_\gamma(\gamma)}$ and for no $\eta < \gamma$ the relations

$$x \in \operatorname{int} U_{\sigma_\gamma(\eta)}, \quad U_{\sigma_\gamma(\eta)} \cap U_{\sigma_\gamma(\gamma)} = \emptyset$$

hold in which case the relations have to be valid for certain $\eta > \gamma$;

or $t \in S_\gamma^{\sigma_\gamma}(U_{\sigma_\gamma(\gamma)})$ and there is $\eta > \gamma$ such that

$$x \in \operatorname{int} U_{\sigma_\gamma(\gamma)}, \quad t \notin S_\gamma^{\sigma_\gamma}(U_{\sigma_\gamma(\gamma)} \cap U_{\sigma_\gamma(\eta)}).$$

In both cases we set $n(\gamma) = \sigma_\gamma(\gamma)$.

Here the bijection σ_γ is defined by (*) with D_γ instead of D.

The "selection" part of this proof is a slight modification of the arguments of Graf [6]. The advantage of this modification is the possibility to weaken measurability assumption on F as in the representation counterpart of the selection theorem of Sion [8] (see Corollary 2.1 below). Compared with the method we used to prove Theorem 1 (which, in turn, is a modification of the one used in [3]), the method of Graf allows to obtain better measurability properties of a representation but is less advantegeous as far as topological properties are concerned. It is still a question if the conclusions of Theorems 1 and 2 can be united. Corollary 1.1 shows that the answer is yes at least if X is sufficiently disconnected.

<u>COROLLARY</u> 2.1. Let X be a compact Hausdorff space and F a set-valued mapping from T into X with nonempty and closed values which is either $\mathcal{M}$-measurable or weakly $\mathcal{M}$-measurable. Then F can be represented by a pair (Y,h) such that $Y \subset \mathbb{N}(\tau,\tau)$, h is continuous in y and for any cozero set U of a continuous function on X and any $y \in Y$

$$\{t \mid h(t,y) \in U\} \in \mathcal{M}.$$

<u>PROOF</u>. Assume that F is weakly measurable. Let $\mathcal{U}'$ be an open topology base with card $\mathcal{U}' = \tau$. Let $\mathcal{C}$ be a dense subset of $C(X)$ (the space of continuous functions with the topology of uniform convergence) with card $\mathcal{C} = \tau$, and let $\mathcal{U}''$ be formed by the sets $\{x \in X \ f(x) < a\}$ with $f \in \mathcal{C}$ and rational a. Take $\mathcal{U}$ to be the semifield generated by $\mathcal{U}' \cup \mathcal{U}''$ and apply the theorem. The reformulation of the proof for measurable F is obvious.

<u>COROLLARY</u> 2.2. Assume that the topology in X is defined by a family $\{\rho_\xi\}$ $(\xi < \omega(\tau))$ of pseudometrics. Let F be a ($\mathcal{M}$-measurable or weakly $\mathcal{M}$-measurable) set-valued mapping from T into

X with nonempty and compact values. Then there are $Y \in \mathbb{N}(\tau, \tau)$ and $h: T \times Y \to X$ such that (Y,h) represents F, h is continuous in y and for any ξ, x, y and $\varepsilon > 0$

$$\{t \mid \rho_\xi(h(t,y),x) < \varepsilon\} \in \mathfrak{m} .$$

PROOF. Let $\mathcal{U}$ be the semifield generated by all sets $U(\xi,\eta,\delta) = \{t \mid \rho_\xi(x_\eta, F(t)) < \delta\}$ if F is weakly measurable or by their closures if F is measurable, where $\{x_\eta \mid \eta < \omega(\tau)\}$ is a dense subset of X and δ is running through the collection of positive rationals, and let $R(A) = F^-(A)$. Then

$$\{t \mid \rho_\xi(h(t,\sigma),x) < \varepsilon\} = \bigcup_{n=1}^{\infty} R^\sigma(U(\xi,\eta_n,\delta_n)) ,$$

where $\rho_\xi(x_{\eta_n},x) \to 0$ and $\rho_\xi(x_{\eta_n},x) + \delta_n \uparrow \varepsilon$.

REFERENCES

1 . D.H. Wagner, Survey on measurable selection theorems; an update, manuscript (1979).

2 . A.D. Ioffe, Representation theorems for multifunctions and analytic sets, Bull. Amer. Math. Soc., 84 (1978), 142 - 144.

3 . A.D. Ioffe, Single-valued representation of set-valued mappings, Trans. Amer. Math. Soc., 252 (1979), 133 - 145.

4 . S.M. Srivastava, Studies in the theory of measurable multifunctions, Thesis, Indian Stat. Inst. (1978).

5 . M. Hasumi, A continuous selection theorem for extremally disconnected spaces, Math. Ann., 179 (1970), 83 - 89.

6 . S. Graf, A measurable selection theorem for compact-valued maps, manuscripta mathematica, 27(1979), 341-352.

7 . S. Graf, A selection theorem for Boolean correspondences, J. Reine Angew. Math., 295(1977), 169-186.

8. M. Sion, On uniformization of sets in topological spaces, Trans. Amer. Math. Soc., 96 (1960), 237 - 244.

Editor's remark: In Corollary 1.2. it is sufficient to require that there is a basis for the uniform structure with cardinality not greater than τ.

A COUNTEREXAMPLE ON MEASURABLE SELECTIONS AND STRONG LIFTING

V. Losert

Institut für Mathematik
Universität Wien

Strudlhofgasse 4, A-1090 Wien
Austria

In Theorem 5 of [2] S. Graf proved a measurable-selection-theorem for topological measure spaces which admit a strong lifting. Since it has been shown in [6] that there exist Radon measures without a strong lifting, one can ask, to what extent the assumption of the existence of a strong lifting is necessary. In this paper we give a condition which is slightly weaker than that of a strong lifting, and which is necessary and sufficient in order to get such a selection theorem. We give an example of a Radon measure on a compact space, which does not fulfill this condition. On the other hand, we show that the original counterexample of [6] fulfills this condition (at least under assumption of the continuum hypothesis). This proves that the new condition is really weaker than the existence of a strong lifting. The construction shows also the existence of strong Baire liftings on spaces of the type $\prod_{i\in I} X_i$ with X_i compact metrizable, card $I \leq \aleph_2$ and measures $\mu = \otimes\mu_i$ (compare with [1] and [8]).

Acknowledgement: I would like to thank S. Graf who led my interest to these questions and M. Talagrand with whom I had also discussions on some parts of this paper.

Notations: X,Y,Z will always be completely regular, Hausdorff topological spaces. $\mathfrak{B}_0(X)$ resp. $\mathfrak{B}(X)$ shall denote the σ-algebra of Baire-resp. Borelsets on X. If λ is a finite, σ-additive Borel measure on X, $\mathfrak{B}_\lambda(X)$ shall denote the σ-algebra of λ-measurable sets (i.e. the completion of $\mathfrak{B}(X)$ with respect to λ) and $L^\infty(X,\lambda)$ the space of equivalence classes of bounded, measurable, complex-valued functions on X. C(X) stands for the space of bounded, continuous, complex valued functions on X, $\mathfrak{K}^*(X)$ for the family of nonempty, compact subsets of X. A map F: $X \to \mathfrak{K}^*(Y)$ is called upper semicontinuous (u.s.c.) if $\{x \in X: F(x) \subseteq U\}$ is open in X for each open subset U of Y. A map f: $X \to Y$ is called a selection for F, if $f(x) \in F(x)$ for all $x \in X$.

If $A,B \in \mathfrak{B}_\lambda(X)$ we write $A \sim B$ if $\lambda((A\setminus B) \cup (B\setminus A)) = 0$. A Boolean algebra homomorphism $l: \mathfrak{B}_\lambda(X) \to \mathfrak{B}_\lambda(X)$ is called a <u>lifting</u>, if $A \sim l(A)$ for all $A \in \mathfrak{B}_\lambda(X)$, $A \sim B$ implies $l(A) = l(B)$. l is called <u>strong</u>, if $l(U) \supseteq U$ for all open subsets U of X, l is called a <u>Borel</u> (<u>Baire</u>) lifting if $l(A) \in \mathfrak{B}(X)(\mathfrak{B}_0(X))$ for all $A \in \mathfrak{B}_\lambda(X)$.

Theorem 1: Let X be a completely regular topological space. λ a Borel probability measure whose support is X (i.e. $\lambda(U)>0$ if U is open, nonempty). The following statements are equivalent:

(i) For each completely regular space Y and each u.s.c. map $F: X \to \mathfrak{K}^*(Y)$ there exists a $\mathfrak{B}_\lambda(X) - \mathfrak{B}_o(Y)$-measurable selection.

(ii) If Z_o denotes the spectrum of $L^\infty(X,\lambda)$, Z the preimage of X with respect to the canonical map $\pi: Z_o \to \beta X$ (βX denotes the Stone Čech compactification of X), $F(x) = \pi^{-1}(x)$ for $x \in X$, then there exists a $\mathfrak{B}_\lambda(X) - \mathfrak{B}_o(Z)$-measurable selection for F.

(iii) There exists a Boolean algebra homomorphism $\rho: \mathfrak{B}_\lambda(X) \to \mathfrak{B}_\lambda(X)$ such that $A \sim B$ implies $\rho(A) = \rho(B)$ and $\rho(U) \supseteq U$ for each open subset U of X.

Proof: (i) $\Rightarrow$ (ii) trivial

(ii) $\Rightarrow$ (i) We consider X (resp. Y) as a subspace of their Stone Čech compactifications βX (resp. βY). Put $G = \{(z,y) \in Z \times Y: y \in F(\pi(z))\}$ and let G_o be its closure in $Z_o \times \beta Y$. We have a canonical map $p_o: G_o \to Z_o$ which is surjective, since G_o is compact and $p_o(G_o)$ contains the dense subspace Z of Z_o. Z_o is extremally disconnected ([9] 25.5.2), consequently there exists a continuous section j_o for p_o, i.e. $p_o \cdot j_o = id_{Z_o}$ ([9] 24.7.1). Let $p_1: G_o \to \beta Y$ be the coordinate projection. We will show that $p_1 \cdot j_o(z) \in F(\pi(z))$ for $z \in Z$. Since by (ii) there exists a measurable section $j: X \to Z$ for π, it follows then that $p_1 \circ j_o \circ j$ is a measurable selection for F.

It is clearly sufficient to show that $p_o^{-1}(z) = \{z\} \times F(\pi(z))$ for $z \in Z$. Assume the contrary, i.e. there exists $(z,u) \in G_o$ with $u \notin F(\pi(z))$. Since $F(\pi(z))$ is compact in Y, it is also closed in βY. Consequently there exists $f \in C(\beta Y) = C(Y)$ such that $f(u) = 1$, $f = 0$ on $F(\pi(z))$. $f \circ F$ is clearly also an u.s.c. map $X \to \mathfrak{K}^*(\mathbb{C})$. Now if $(z,u) = \lim\ (z_\iota, u_\iota)$ with $(z_\iota, u_\iota) \in G_o$, then $(\pi(z), f(u)) = \lim\ (\pi(z_\iota), f(u_\iota))$. The upper semicontinuity of $f \circ F$ implies that $\{(a,\alpha): \alpha \in f \circ F(x),\ x \in X\}$ is closed in $X \times \mathbb{C}$. Consequently there exists $v \in F(\pi(z))$ with $(\pi(z), f(u)) = (\pi(z), f(v))$ - a contradiction.

(ii) $\Rightarrow$ (iii): Each $A \in \mathfrak{B}_\lambda(X)$ defines an idempotent in $L^\infty(X,\mu)$ and this corresponds to a clopen subset $A\hat{}$ of Z_o. If $j: X \to Z$ is a measurable section for π, we define $\rho(A) = j^{-1}(A\hat{})$. This belongs to $\mathfrak{B}_\lambda(X)$, defines a Boolean algebra homomorphism and since $A \sim B$ implies $A\hat{} = B\hat{}$ we have also $\rho(A) = \rho(B)$. If $U \subseteq X$ is open, then its characteristic function satisfies $c_U = \sup\ \{f \in C(X): f \le c_U\}$ (pointwise). This gives $c_{U\hat{}} \ge$ $\ge \sup\ \{f \circ \pi: f \in C(X),\ f \le c_U\}$ and it follows that $c_{\rho(U)} = c_{U\hat{}} \circ j \ge$ $\ge \sup\ \{f \circ \pi \circ j: f \in C(X),\ f \le c_U\} = c_U$ (since $\pi \circ j = id_X$).

(iii) $\Rightarrow$ (ii): Let M^∞ be the algebra of bounded, $\mathfrak{B}_\lambda(X)$-measurable, complex valued functions on X, ρ defines an algebra homomorphism $\bar{\rho}: L^\infty(X,\lambda) \to M^\infty$ in the same way as in the case of a lifting (see [4]

p. 36). The dual map carries evaluation functionals at points of X into multiplicative functionals on $L^\infty(X,\lambda)$, i.e. elements of Z_o. This defines a map $j: X \to Z_o$. Each clopen subset of Z_o has the form $A^\wedge$ for some $A \in \mathfrak{B}_\lambda(X)$ (same notation as above). Since $\bar{\rho}(c_{A^\wedge}) = c_{\rho(A)}$ we have $j^{-1}(A^\wedge) = \rho(A)$. This shows that j is $\mathfrak{B}_\lambda(X) - \mathfrak{B}_o(Z_o)$-measurable. If $p: M^\infty \to L^\infty(X,\lambda)$ is the natural projection, then $\bar{\rho} \circ p$ acts as the identity on C(X) (this is a consequence of the property $\rho(U) \supseteq U$ for U open). But this means exactly that $\pi \circ j = id_X$ and it follows in particular that $j(X) \subseteq Z$.

Remark: Similar statements hold, if we require other measurability properties for the selection of F. For example, the existence of a $\mathfrak{B}(X) - \mathfrak{B}_o(Y)$ (or $\mathfrak{B}_o(X) - \mathfrak{B}_o(Y)$)-measurable selection in (i) is equivalent to the additional property $\rho(A) \in \mathfrak{B}(X)$ (or $\mathfrak{B}_o(X)$) for all $A \in \mathfrak{B}_\lambda(X)$ in (iii). If ρ is a strong lifting in the ordinary sense one gets always $\mathfrak{B}_\lambda(X) - \mathfrak{B}(Y)$-measurability from [7] Theorem 4. In general this holds iff ρ satisfies $\bigcup_{i \in I} \rho(A_i) \in \mathfrak{B}_\lambda(X)$ for arbitrary $A_i \in \mathfrak{B}_\lambda(X)$ and an arbitrary index set I.

Theorem 2: There exists a compact space $\bar{X}$ and a Radon probability measure λ on $\bar{X}$ with supp $\lambda = \bar{X}$ such that the equivalent properties of Theorem 1 are not fulfilled.

Proof: The construction is similar to that of [6]. We use the same notations as in that paper and will indicate only what has to be changed.

Lemma 1: Set S be a compact, metrizable, zero-dimensional space, ν a continuous probability measure on S (i.e. $\nu(\{x\}) = 0$ for all $x \in S$) with supp $\nu = S$. Then there exist measurable subsets M and N of S such that $M \cup N = S$, $\nu(M \cap N) > 0$ and $\nu(F \setminus M) > 0$, $\nu(F \setminus N) > 0$ for all nonempty clopen subsets F of S.

Proof of Lemma 1: Let $\{F_n\}_{n=1}^\infty$ be an enumeration of all nonempty clopen subsets of S. By induction we construct open dense subsets M_n and N_n of S such that the following holds: $M_n \supseteq M_{n+1}$, $N_n \supseteq N_{n+1}$,

$$M_n \cup N_n = S,\ \nu(F_n \setminus M_n) > 0,\ \nu(F_n \setminus N_n) > 0,\ \nu(M_n \cap N_n) \geq \frac{1}{2} + \frac{1}{2^{n+1}} .$$

For $n = 0$ we start with $M_o = N_o = S$.

We assume now that M_n and N_n have been constructed. Since ν is continuous, there exist disjoint, closed, nowhere dense subsets M_n', N_n' of $F_{n+1} \cap M_n \cap N_n$ such that $\nu(M_n') > 0$, $\nu(N_n') > 0$ and $\nu(M_n' \cup N_n') < 2^{-n-2}$. Now we put $M_{n+1} = M_n \setminus M_n'$, $N_{n+1} = N_n \setminus N_n'$ and it is easily seen that these sets have the desired properties.

Finally we take $M = \bigcap_{n=1}^\infty M_n$, $N = \bigcap_{n=1}^\infty N_n$ to get our result.

We take up the construction of the space $(\bar{X}, \lambda)$:

Let S be a compact, metrizable, zero dimensional space, ν a continuous probability measure on S with supp $\nu = S$. Choose measurable subsets M and N as in the preceding lemma. We fix some nonempty clopen subset $F \neq S$. Put $\mathfrak{M} = \{M, N, F, S \setminus F\}$. Let J be an index set of cardinality $\geq \aleph_2$.

Put $T = S^J$ and write μ for the product of the measures ν.

We use the same notational conventions as in [6]. Put
$I = \{A_{j_1} \times \ldots \times A_{j_k} : j_1, \ldots, j_k \in J,\ A_{j_i} \in \mathfrak{M}$, at least one A_{j_i} is not clopen$\}$.

As in [6] we define $X = T \times T^I$ and
$\lambda(A \times B_{C_1} \times \ldots \times B_{C_n}) = \sum_{\Gamma \subseteq \mathbb{N}_n} \mu(A \cap \bigcap_{k \in \mathbb{N}_n \setminus \Gamma} (C_k \cap B_{C_k}) \cap \bigcap_{l \in \Gamma} \complement C_l) . \prod_{l \in \Gamma} \mu(B_{C_l})$.
(where $\mathbb{N}_n = \{1, \ldots, n\}$).

In the same way as in [6] we define $X_o = T_o \times T^I$ and λ_o on X_o. The Lemmas 1,2,4,5 of [6] remain valid, also the description of the support of λ_o in Lemma 3 of [6]. But it follows easily from the fact that $M \cup N = S$ that the support of λ is no longer X. Therefore we put $\bar{X} = \operatorname{supp} \lambda$.
We claim that on $(\bar{X}, \lambda)$ there does not exist a function ρ with the properties of (iii) of Theorem 1.
The proof goes by contradiction. We assume that there exists such a function ρ. Put $\rho_1(A) = p^{-1}(\rho(A))$, where $p: X_o \to X \supseteq \bar{X}$ denotes the canonical map. $\rho_1(A)$ is a measurable subset of X_o. Consequently there exists a Baire set $d_o(A)$ such that $d_o(A) \sim \rho_1(A)$. $d_o(A)$ depends on countably many coordinates from I and J. We apply Lemma 5 of [6] and its corollary to $N = d_o(A) \setminus \rho_1(A)$ and $B = d_o(A)$. In the same way as in [6] we get the following conclusion:

(*) There exist countable subsets $I_1(A) \subseteq I$, $J_1(A) \subseteq J$ and a measurable subset $d(A) \subseteq X_o$ which depends only on $I_1(A)$ and $J_1(A)$ such that $d(A) \cap \operatorname{supp} \lambda_o \subseteq \rho_1(A)$ and $d(A) \sim \rho_1(A)$.

The next step is to prove a result corresponding to Lemma 6 of [6].

<u>Lemma 2</u>: Assume that $A, B, A \cap B \in I$ and that the sets $d(A)$ and $d(B)$ of (*) have been constructed. Assume that $(t, t_C), (t', t_C) \in X_o$ are given such that $(t, t_C) \in d(B) \cap \operatorname{supp} \lambda_o$, $(t', t_C) \in \operatorname{supp} \lambda_o$, $p_T(t) = p_T(t')$, $A \cap B \notin t \cup t'$. Then $(t', t_C) \notin d(A)$.

Proof of Lemma 2: Assume that $(t', t_C) \in d(A)$.
We choose countable sets $I_1 \supseteq I_1(A) \cup I_1(B)$, $J_1 \supseteq J_1(A) \cup J_1(B)$ such that $A, B \in I_1$ and each element of I_1 depends only on J_1. Since J is uncountable, there exists $j \in J \setminus J_1$.

We define $C_1 = A \cap B \times F_j$, $C_2 = A \cap B \times \complement F_j$. Then $C_1, C_2 \in I \setminus I_1$ (see the definition of I).

Define $t_o = p_T(t) = p_T(t')$. Choose a clopen subset $E \subseteq T$ such that $t_o \in E$ and put $D = \complement E$. A similar computation as in [6] shows that $\lambda(A \cap B \cap E \times D_{C_1} \times D_{C_2}) = 0$. It follows then from the properties of ρ that $\emptyset = \rho(A \cap B \cap E \times D_{C_1} \times D_{C_2}) = \rho(A) \cap \rho(B) \cap E \times D_{C_1} \times D_{C_2}$. Since $A \cap B \notin t \cup t'$ we have also $C_1, C_2 \notin t \cup t'$. This makes it possible, as in [6], to construct an element $(t_o, t_C') \in \rho(A) \cap \rho(B) \cap E \times D_{C_1} \times D_{C_2}$ – a contradiction. This proves Lemma 2.

For the final step of the proof of Theorem 2 we consider the sets M_j and $N_j (j \in J)$. We choose sets $d(M_j)$ and $d(N_j)$ according to (*) and countable sets $I_1(j)$ and $J_1(j)$ on which these two sets depend. Now let J_o be a subset of J of cardinality $\aleph_1$. Define $J_o^- = \bigcup_{j \in J_o} J_1(j)$. Then card $J_o^- = \aleph_1$. Therefore we may choose $k \in J \setminus J_o^-$ and $j \in J_o \setminus J_1(k)$. In particular $k \notin J_1(j)$. We have $M \cup N = S$. Consequently $d(M_j) \cup d(N_j) \sim \rho_1(M_j \cup N_j) = p^{-1}(\overline{X}) \supseteq \text{supp}\ \lambda_o$. Put $A = M \cap N$. Then $\nu(A) > 0$ by the construction of M and N. It follows that we may assume that $\hat{A_j} \cap \hat{A_k} \cap d(M_j) \cap d(M_k) \neq \emptyset$ (otherwise replace M by N at one or both places; $A^\wedge$ stands for the clopen subset of T_o corresponding to A - see [6]).

We choose an element $(s, t_C)_{C \in I} \in \hat{A_j} \cap \hat{A_k} \cap d(M_k) \cap \text{supp}\ \lambda_o$. Write $s = (s_i)_{i \in J}$. By the properties of M there exist elements $t_j^!$, $t_k \in (S \setminus M)^\wedge$ such that $p_S(t_j^!) = p_S(s_j)$, $p_S(t_k) = p_S(s_k)$. Put $t_i^! = t_i = s_i$ for $i \neq j,k$ and $t_j = s_j$, $t_k^! = s_k$. Finally define $t = (t_i)$, $t' = (t_i^!)$. We have $p_T(t) = p_T(t') = p_T(s)$ and if we can show that (t, t_C), $(t', t_C) \in \text{supp}\ \lambda_o$, we get a contradiction to Lemma 2 (with $A=M_j$, $B=M_k$).

As in [6] it suffices to show that $I_1(j) \cap (t \cup t') \subseteq s$ (and similarly for k). The fact that $I_1(j) \cap t \subseteq s$ follows from the construction of t as in [6]. For the second part assume that $C \in I_1(j) \cap t' \setminus s$. Then $C = C^{(1)} \times C^{(2)}$ where $C^{(1)} \subseteq S^{J_1(j) \setminus \{j\}}$ and $C^{(2)} \subseteq S_j$. Since $t'|_{\mathfrak{B}(S^{J \setminus \{j\}})} = s|_{\mathfrak{B}(S^{J \setminus \{j\}})}$ we conclude that $C^{(2)} \in t' \setminus s$ or equivalently $C^{(2)} \in t_j \setminus s_j$. Since $p_S(t_j) = p_S(s_j)$, $C^{(2)}$ has to be either M_j or N_j. Since $A_j = M_j \cap N_j$ and by construction $A_j \in s_j$ we get a contradiction. This finishes the proof of Theorem 2.

<u>Remark</u>: If Z_o denotes the spectrum of $L^\infty(\overline{X}, \lambda)$, it follows in particular from (ii) of Theorem 1, that there exists no $\mathfrak{B}_\lambda(\overline{X})$-$\mathfrak{B}_o(Z_o)$ - measurable section for the canonical map $\pi : Z_o \to \overline{X}$. In [10] a similar result was proved by Talagrand. But instead of the σ-algebra $\mathfrak{B}_\lambda(\overline{X})$ of λ-measurable sets, he proved it for the σ-algebra $\mathfrak{D}\mathfrak{C}(\overline{X})$ of 'sets determined by countably many open sets'. This σ-algebra contains also the Borel sets but in general not their completion.
In the last section we want to give a positive result for a more special class of spaces (cp. with [1],[8]).

<u>Theorem 3</u>: Let X_i be compact, metrizable spaces, λ_i probability measures on X_i with supp $\lambda_i = X_i (i \in I)$, $X = \prod_{i \in I} X_i$, $\lambda = \bigotimes_{i \in I} \lambda_i$ and assume that card $I \leq \aleph_2$. Under assumption of the continuum hypothesis, there exists a strong Baire lifting for (X, λ).

<u>Proof</u>: The lifting l will be defined by transfinite induction. First we need an auxiliary notion:

Let Σ be a subalgebra of $\mathfrak{B}_\lambda(X)$. We consider maps d with the following properties:

$$
(P)\quad \begin{cases} d : \Sigma \to \mathfrak{B}_o(X) \\ A \sim B \text{ implies } d(A) = d(B) \\ \quad d(A \cap B) = d(A) \cap d(B) \\ \lambda(d(A) \setminus A) = 0 \\ A^o \subseteq d(A) \qquad\qquad (A^o \text{ denotes the interior of } A) \end{cases}
$$

(This is essentially the definition of a strong lower density (see [4] p.36 and 64) with the exception that we do not require $d(A) \sim A$).

Lemma 3: If card $\Sigma \leq \aleph_1$ and d is given with the properties (P), then there exists a strong lifting l of Σ such that $d(A) \subseteq l(A) \subseteq \complement d(\complement A)$ for all $A \in \Sigma$.

Proof: This is similar to that of [1] Lemma 1, or [5]. Write $\Sigma = \{A_\alpha : \alpha < \alpha_0\}$ with $\alpha_0 \leq \omega_1$ the first uncountable ordinal. Let Σ_α be the algebra generated by $\{A_\beta : \beta < \alpha\}$. l is defined on Σ_α by induction on α. Only the step from α to $\alpha+1$ is non trivial and we may assume that $A_\alpha \notin \Sigma_\alpha$. Then it follows that $\Sigma_{\alpha+1} = \{(B \cap A_\alpha) \cup (C \setminus A_\alpha) : B, C \in \Sigma_\alpha\}$. It is easily seen that we get an extension of l with the required properties if we choose a Baire set $l(A_\alpha)$ equivalent to A_α such that $\bigcap_{B \in \Sigma_\alpha} \complement d(B \cup \complement A_\alpha) \cup l(B) \supseteq$ $\supseteq l(A_\alpha) \supseteq \bigcup_{C \in \Sigma_\alpha} d(C \cup A_\alpha) \setminus l(C)$ (use the fact that $(B \cap A_\alpha) \cup (C \setminus A_\alpha) =$ $= (B \cup \complement A_\alpha) \cap (C \cup A_\alpha)$ and the intersection property of d). It follows from the properties of d and l that the set on the right is contained in the left one. Since Σ_α is countable for $\alpha < \alpha_0$, both sets are Baire-measurable, the right set is contained in A_α, the left one contains A_α up to a set of measure zero. Therefore such a choice is possible.

Lemma 4: Let $\{I_n\}$ be an increasing family of subsets of I, Σ_0 be subalgebra of $\mathfrak{B}_\lambda(X)$ consisting of those subsets of $X = \prod_{i \in I} X_i$ which depend only on coordinates from I_n for some $n \geq 1$. Assume that l is a strong Baire lifting of Σ_0. For $A \in \mathfrak{B}_\lambda(X)$ define $d(A) = \bigcup\{l(B) \cap U : B \in \Sigma_0$, U open in $X, \lambda(B \cap U \setminus A) = 0\}$. Then d has the properties (P) on $\mathfrak{B}_\lambda(X)$.

Proof: The main task is to show that $d(A) \in \mathfrak{B}_0(X)$. In the definition of d it suffices to take U from a basis for the topology of X. Since for $A \sim B$ evidently $d(A) = d(B)$, we may assume that A depends only on a countable subset of coordinates $I' \subseteq I$. Now assume that U is a product of open subsets of $X_i (i \in I)$. Then it may be decomposed in the following fashion: $U = U_1 \times U_2 \times U_3$, where U_1 depends only on I_n for some n, U_2 depends on $I' \setminus \bigcup_n I_n$ and U_3 is independent of $\bigcup_n I_n \cup I'$. If $\lambda(B \cap U \setminus A) = 0$, then it follows from the fact that $B \setminus A$ depends on $\bigcup_n I_n \cup I'$ that $\lambda(B \cap U_1 \times U_2 \setminus A) = 0$ holds too. Since $l(U_1) \supseteq U_1$ (l is strong), it suffices to consider open sets which depend only on I'. Thus there are countably many choices for U. If U and n are fixed, the family of sets B which depend on I_n and satisfy $\lambda(B \cap U \setminus A) = 0$ has a least upper bound in the measure algebra of (X, λ). Consequently d(A) is the union of countably many Baire sets, hence $d(A) \in \mathfrak{B}_0(X)$ and $\lambda(d(A) \setminus A) = 0$. All other properties of d are easily proved.

Proof of Theorem 3: Let $I = \{i_\alpha : \alpha < \alpha_0\}$ be a well-ordering of I such that card $\alpha \leq \aleph_1$ for $\alpha < \alpha_0$. For $\alpha < \alpha_0$ let Σ_α be the sub-σ-algebra of $\mathfrak{B}_\lambda(X)$ of those sets which depend only on coordinates i_β with $\beta < \alpha$. We define the lifting l_α on Σ_α by induction on α such that $l_\alpha|_{\Sigma_\beta} = l_\beta$ for $\beta < \alpha$. If α is a limit ordinal with uncountable cofinality then $\Sigma_\alpha = \bigcup_{\beta < \alpha} \Sigma_\beta$ and there is nothing to do.

In the other cases we use Lemma 4 to define a function d on $\mathfrak{B}_\lambda(X)$ with the properties (P). By lemma 3 there exists a strong Baire lifting l_α on Σ_α such that $d(A) \subseteq l_\alpha(A) \subseteq \complement d(\complement A)$ for $A \in \Sigma_\alpha$ (card $\Sigma_\alpha \leq \aleph_1$ by use of the continuum hypothesis). Since for $A \in \Sigma_\beta$ $(\beta < \alpha)$ $d(A) = \complement d(\complement A) = l_\beta(A)$, we get $l_\alpha(A) = l_\beta(A)$, i.e. l_α is an extension of l_β for $\beta < \alpha$.

Remark: If $\pi : Y \to \{0,1\}^I$ is continuous, surjective (Y compact), Theorem 3 combined with Theorem 1 shows the existence of a $\mathfrak{B}_0(\{0,1\}^I)$- $\mathfrak{B}_0(Y)$-measurable section for π. This follows also from [10] combined with [1] or [8]. In particular, Theorem 1 shows the existence of a map ρ as in (iii) for the measure λ' on $\{0,1\}^I$ defined in [6]. It follows also from the remark after Theorem 1 that the section for π is $\mathfrak{B}_\lambda(\{0,1\}^I)$ - - $\mathfrak{B}(Y)$-measurable too, where λ denotes the ordinary product measure on $\{0,1\}^I$.

References:

[1] Fremlin D.H., On two theorems of Mokobodzki, Preprint 1977.

[2] Graf S., A measurable selection theorem for compact valued maps, Manuscripta Math. 27, 341-352 (1979).

[3] Hasumi M., A continuous selection theorem for extremally disconnected spaces, Math.Ann. 179, 83-89 (1969).

[4] Ionescu Tulcea A. and C., Topics in the theory of lifting, Berlin-Heidelberg-New York, Springer 1969.

[5] Lloyd S.P., Two lifting theorems, Proc. AMS 42, 128-134 (1974).

[6] Losert V., A measure space without the strong lifting property, Math.Ann. 239, 119-128 (1979).

[7] Maharam D., On a theorem of von Neumann, Proc.AMS 9, 987-994 (1958).

[8] Mokobodzki G., Rélèvement borélien compatible avec une classe d'ensembles négligeables. Application à la désintégration des mesures. Séminaire de Probabilités IX pp. 539-543, Lecture Notes in Math. 465, Berlin-Heidelberg-New York, Springer 1975.

[9] Semadeni Z., Banach spaces of continuous functions, Monografie Matematiyczne, Warszawa, PWN-Polish Scientific Publishers 1971.

[10] Talagrand M., En général il n'existe pas de rélèvement linéaire borélien fort, C.R.Acad.Sci.Paris, Sér.A 287, 633-636 (1978).

SOME SELECTION THEOREMS AND PROBLEMS

R. Daniel Mauldin
Department of Mathematics
North Texas State University
Denton, Texas 76203/USA

Let I be the closed unit interval, [0,1]. Let B be a Borel subset of $I\times I$ such that for each $x, B_x=\{y:(x,y)\in B\}\neq \phi$. Using the axiom of choice, we find that the Borel set B contains a uniformization (= the graph of some function f mapping I onto I). The question was raised concerning how nice or describable the function f is in the famous letters exchanged among Baire, Borel, Hadamard and Lebesgue [1]. Novikov gave the first example of a Borel subset of $I\times I$ which does not possess a Borel uniformization [2]. Kondo proved that every such Borel set B possesses a uniformization which is coanalytic [3]. Yankov [4] and von Neumann [5] proved that B contains the graph of a function f which is measurable with respect to the σ-algebra generated by the analytic subsets of I. In fact, they proved this result assuming only that B is an analytic set. Whether every Borel set B possesses a uniformization which is the difference of two coanalytic sets seems to be an unsolved problem.

Various extensions of these uniformization problems have been considered not only for their intrinsic interest but for their applications. One such problem was discussed by A.H. Stone at the 1975 Oberwolfach conference on Measure Theory [6]. The problem is essentially that of filling up a set with pairwise disjoint uniformizations so that the uniformizations are indexed in some reasonable manner. For simplicity let me formulate the problem as follows.

PARAMETRIZATION PROBLEM.

Let B be a subset of $I\times I$ so that for each x, B_x is uncountable. A parametrization of B is a map g from $I\times I$ onto B so that for each x, $g(x,\cdot)$ maps I onto $\{x\}\times B_x$. Given a description of B how describable can a parametrization of B be?

Stone was particularly interested in the case where B is a Borel set and he inquired about the existence of a universally measurable parametrization. Wesley proved by forcing techniques that the answer is yes. Cenzer and I proved the following theorem [7].

THEOREM 1.

Let B be an analytic subset of $I\times I$ so that for each x, B_x is uncountable.

Then there is a parametrization g of B such that both g and g^{-1} are measurable with respect to S(I×I).

Here S(I×I) denotes the family of C-sets of Selivanovskii, the smallest family of subsets of I×I containing the open sets and closed under operation (A). S(I×I) is a very nice family of universally measurable sets.

In the course of proving this theorem, we showed that such an analytic set B possesses $2^{\aleph_0}$ pairwise disjoint uniformizations which are the graphs of functions which are measurable with respect to the σ-algebra generated by the analytic subsets of I. Whether such an analytic set can be filled up by pairwise disjoint uniformizations each of which is the graph of function measurable with respect to this σ-algebra is an unsolved problem. Whether the function g chosen so that both g and g^{-1} are measurable with respect to the σ-algebra generated by the analytic subsets of I×I is also unsolved. Necessary and sufficient conditions for a Borel parametrization are given in the following theorem [8].

THEOREM 2.

Let B be a Borel subset of I×I such that for each x, B_x is uncountable. The following are equivalent.

1. B has a Borel parametrization;
2. there is an atomless conditional probability distribution μ so that for each x, $\mu(x,B_x) > 0$;
3. B contains a Borel set M such that for each x, M_x is a nonempty perfect subset of I.

In connection with this I showed that there is a closed uncountable subset B of I×I such that for each x, B_x is uncountable and yet B does not have a Borel parametrization.

In view of this last result, let me pose the following problem which seems to be unsolved.

PROBLEM.

Let B be a Borel subset of I×I so that for each x, B_x is a closed uncountable set. Does B have $2^{\aleph_0}$ pairwise disjoint Borel uniformizations?

That such a Borel set B has a Borel uniformization was proven by Novikov [9]. That B possesses $\aleph_1$ Borel uniformizations was shown by Larman [10]. We can obtain Larman's result from the following selection theorem.

THEOREM 3.

There are Borel measurable maps, $f_\alpha, \alpha < \omega_1$ from 2^I into I such that for each closed set $K, f_\alpha(K) \in K$ and if K is uncountable, then $f_\alpha(K) \neq f_\beta(K)$, if $\alpha \neq \beta$.

Before proving this theorem, let us indicate how this gives $\aleph_1$ pairwise disjoint Borel measurable uniformizations. One simply takes the map ϕ from I into 2^I defined by $\phi(x) = B_x$. It is well known that this is a Borel measurable map. For $\alpha < \omega_1$, set $g_\alpha(x) = f_\alpha(\phi(x))$. The graphs of the functions g_α are pairwise disjoint Borel uniformizations of B.

PROOF OF THEOREM 3.

Let τ be a Borel isomorphism of I onto 2^I. For each $\alpha < \omega_1$, let D_α be the αth derived set map of 2^I onto 2^I. D_1 is a Borel measurable map of class 2. Thus, for each $\alpha < \omega_1$, D_α is a Borel measurable map. (The exact class of these maps is an unsolved problem [11]). Let $P = \{F \in 2^I : F \text{ is perfect}\}$. Let $Z_\alpha = D_\alpha^{-1}(\phi)$ and let $h: 2^I \to B$ by $h(E) = \{x \in E : x \text{ is an isolated point of } E\}$.

For each ordinal α, $0 \leq \alpha < \omega_1$, let B_α be the graph of $D_\alpha \circ \tau$ in $I \times I$, let

$$T_\alpha = (D_\alpha \circ \tau)^{-1}(P),$$

let

$$K_\alpha = T_\alpha - \bigcup\{T_\gamma : \gamma < \alpha\};$$

and let

$$M_\alpha = B_\alpha \cap (K_\alpha \times I).$$

Notice that for each α, M_α is a Borel subset of $I \times I$ and if $x \in \pi_1(M_\alpha) = K_\alpha$, then $(M_\alpha)_x$ is a perfect set. According to Theorem 2 each M_α has continuumly many pairwise disjoint Borel uniformizations. So, let $\{V^\alpha_\beta\}_{\beta < \omega_1}$ be a family of $\aleph_1$ pairwise disjoint Borel uniformizations of M_α.

For each $0 \leq \alpha < \omega_1$, let

$$E^\alpha = \bigcup_{\gamma \leq \alpha} V^\gamma_\alpha \cup (\tau^{-1}(Z_\alpha) \times I) \cap B \cup \mathscr{G}r((h \circ D_\alpha \circ \tau) | I - T_\alpha)$$

We note the following properties of these sets: (1) E^α is a Borel subset of $I \times I$, (2) $E^\alpha \subset B$, (3) for each x in I, E^α_x is a nonempty countable set and finally (4) if $0 \leq \alpha < \beta < \omega_1$ and Y is an uncountable compact subset of I, then

$$E^\alpha{}_{\tau^{-1}(Y)} \cap E^\beta{}_{\tau^{-1}(Y)} = \phi$$

For each α, let g_α be a Borel map of I into I which uniformizes E^α. Let $f^\alpha(X) = g_\alpha(\phi^{-1}(X))$ for each $X \in 2^I$. Then $\{f_\alpha : \alpha < \omega_1\}$ is a family of Borel measurable selectors for 2^I such that if $\alpha \neq \beta$ and X is an uncountable closed subset of I, $f_\alpha(X) \neq f_\beta(X)$. Q.E.D.

The preceding proof is a variation on an argument of Larman[10]. I was simply unable to obtain Theorem 3 as a corollary of the results in [10].

There is one closely related problem which seems to bear stating.

PROBLEM.

Let B be a Borel subset of $I \times I$ so that for each x, B_x is uncountable and is the union of countably many compact sets. Does B have $2^{\aleph_0}$ pairwise disjoint Borel uniformization?

Again it was shown by Larman that B has $\aleph_1$ Borel uniformizations provided each B_x is also a G_δ set [10]. It may appear that this problem can be reduced to the previous problem by employing the beautiful result of Saint-Raymond [12] which states that B is the union of countably many Borel sets B_n so that for each n and each x, B_{nx} is compact. If one could insure that for some n each B_{nx} is uncountable then this last problem would be reduced to the previous problem. However, we give the following example.

EXAMPLE.

There is an F_σ subset B of $I \times I$ so that for each x, B_x is uncountable and yet B does not contain a Borel set each section of which an is an uncountable compact set.

Let $\{A_n\}_{n=1}^{\infty}$ be an increasing sequence of analytic subsets of I so that $\bigcup A_n = I$ such that there does not exist a sequence of Borel sets D_n so that for each n, $D_n \subset A_n$ and $\bigcup D_n = I$. The existence of such sets was shown by Leo Harrington. I understand that this is a classical result although I do not know any reference.

For each n, let M_n be a closed subset of $I \times [1/2n+1, 1/2_n]$ so that $(M_n)_x$ is uncountable if and only if x is in A_n. Let $B = \bigcup M_n$. For each n, let $T_n = \bigcup \{M_p : p \leq m\}$. Suppose B contains a Borel set K so that each K_x is compact and uncountable. Notice that for each x there is some n so that $K_x \subset T_{nx}$. For each n, let $D_n = \{x : K_x \subset T_{nx}\}$. Then each set D_n is a Borel set, $D_n \subset A_n$ and $\bigcup D_n = I$. This contradiction establishes that the F_σ set B has the required properties. □

Let me mention that one can easily obtain the following improvement of Larman's theorem [10,p.244] by combining the results of Saint-Raymond with those of Larman.

THEOREM 4.

Let B be a Borel subset of $I \times I$ so that for each x, B_x is uncountable and is the union of countably many compact sets. Then B has $\aleph_1$ pairwise disjoint Borel uniformizations.

PROOF.

First, express $B = \bigcup\{B_n : n \geq 1\}$ where each B_n is a Borel set and for every x, B_{nx} is compact. For each n, let g_n map $I\times[1/2n+1\ 1/2n]$ onto $I\times I$ by $g_n(x,y) = (x, t_n(y))$, where $t_n(y)$ is a linear map of $[1/2n+1, 1/2n]$ onto $[0,1]$.

Set $B_0 = \phi$ and for each positive integer n, $M_n = B_n - B_{n-1}$. Let $K_n = g_n^{-1}(M_n)$. Let $K = \bigcup\{K_n : n \geq 1\}$. Clearly, K is a Borel subset of $I\times I$ and for each x, K_x is both a K_σ and a G_δ set. Also, if $(x,y) \in K$, there is a unique n so that $(x,y) \in K_n$. So, define $g(x,y) = g_n(x,y) \in K$. Then g is a Borel isomorphism of K onto B so that for each x, $g(x,\cdot)$ maps $\{x\}\times K_x$ onto $\{x\}\times B_x$. According to Larman's theorem, there exists $\aleph_1$ pairwise disjoint Borel uniformizations of K. The images of these uniformizations of K are pairwise disjoint Borel uniformizations of M.Q.E.D.

REFERENCES

1. Cinq lettres sur la theorie des ensembles, Bull. Soc. Math., France, 33(1905).

2. P. S. Novikov, Sur les fonctions implicites measurables B, Fund. Math. 17(1931), 8-25.

3. M. Kondo, L'uniformisation des complementaries analytiques, Proc. Imp. Acad., Japan 13(1937), 287-290.

4. W. Yankov, Sur l'uniformisation des ensembles A, Dokl. Akad. Nauk SSSR (N.S.) 30(1941), 597-598.

5. J. von Neumann, On rings of operators; reduction theory, Ann. of Math. 30(1949), 401-485.

6. A. H. Stone, Measure Theory, Lecture Notes in Math, Vol. 541, Springer Verlag, Berlin and New York, 1976, 43-48.

7. D. Cenzer and R.D. Mauldin, Measurable parametrizations and selections, Trans. Amer. Math. Soc. 245(1978), 399-408.

8. R. D. Mauldin, Borel parametrizations, Trans. Amer. Math. Soc., 250 (1979), 223-234.

9. P. S. Novikov, Doklady Akad. Nauk. SSSR(N.S.) 3(1934).

10. D. G. Larman, Projecting and uniformizing Borel sets with K_σ-sections II, Mathematika 20(1973), 233-246.

11. K. Kuratowski, Some problems concerning semi-continuous set-valued mappings, Set-Valued Mappings, Selections and Topological Properties of 2^X, Lecture Notes in Math, vol. 171, Springer-Verlag, Berlin and New York, 1970, 45-48.

12. J. Saint-Raymond, Boreliens a coupes K , Bull. Soc. Math., France 104(1976), 389-400.

NON-EXISTENCE DE CERTAINES SECTIONS MESURABLES ET CONTRE-EXEMPLES EN THEORIE DU RELEVEMENT

M. Talagrand
Equipe d'Analyse, Université Paris VI
4 Place Jussieu, F-75230 Paris Cedex 05, France

I) Notations. Pour un espace compact X, on désignera par $\mathcal{B}a(X)$ (resp. $\mathcal{B}(X)$) la tribu des boréliens de Baire (resp. boréliens) de X. La tribu $\mathcal{B}(X)$ est assez difficile à manipuler; on a avantage à considerer la tribu plus générale $\mathcal{D}\mathcal{O}(X)$ des ensembles qui "ne dépendent que d'un nombre dénombrable d'ouverts", c'est à dire que $A \in \mathcal{D}\mathcal{O}(X)$ ssi il existe une suite (U_n) d'ouverts de X telle que A soit réunion quelconque d'atomes de la tribu engendrée par la suite (U_n). Par exemple, si X est métrisable, tout sous-ensemble de X appartient à $\mathcal{D}\mathcal{O}(X)$, les éléments de $\mathcal{D}\mathcal{O}(X)$ peuvent donc être très irréguliers.

On désigne par $M_1^+(X)$ l'ensemble des mesures de probabilité sur X, muni de la topologie vague.

Supposons que X soit le support d'une mesure $\mu \in M_1^+(X)$. Un relèvement linéaire (resp. multiplicatif) ρ de $L^\infty(\mu)$ est une application linéaire positive (resp. et multiplicative) de $L^\infty(\mu)$ dans $\mathcal{L}^\infty(\mu)$ telle que $\rho(1_X) = 1_X$ et que $\rho(f)$ appartienne à la classe de f pour tout $f \in L^\infty(\mu)$. On désigne par $\mathcal{B}(\rho)$ la tribu engendrée par les fonctions $\rho(f)$ pour $f \in L^\infty(\mu)$. Un des problèmes de la théorie du relèvement est de rendre $\mathcal{B}(\rho)$ le plus petit possible. On dit que ρ est fort si $\rho(f)$ est continue lorsque f appartient à la classe d'une fonction continue.

Désignons par S le spectre de $L^\infty(\mu)$, par π la surjection canonique de S sur X, et par $\sim$ l'isomorphisme canonique de $L^\infty(\mu)$ sur $\mathcal{C}(S)$.

II) Relèvement et séléction (aucun des résultats de ce paragraphe n'est original).

Pour $x \in X$, on peut définir $\tau(x) \in M_1^+(S)$ par $\tau(x)(\hat{f}) = \rho(f)(x)$ pour $f \in L^\infty(\mu)$. Pour $g \in \mathcal{C}(S)$, l'application $x \to \tau(x)(g)$ est donc $\mathcal{B}(\rho)$ mesurable. Mais on voit sans peine que la tribu de Baire de $M_1^+(S)$ est la moins fine rendant mesurable les applications $\nu \to \nu(g)$ pour $g \in \mathcal{C}(S)$. Il en résulte que τ est $\mathcal{B}(\rho)$ à $\mathcal{B}a(M_1^+(S))$ mesurable. D'autre part, si $f \in L^\infty(\mu)$ est dans la classe d'une fonction continue, on a:

$$\rho(f)(x) = \tau(x)(\tilde{f}) = \tau(x)(f\circ\pi) = \pi(\tau(x))(f).$$

Il en résulte que si ρ est fort, $\pi(\tau(x))$ est la mesure de Dirac δ_x en x, et ainsi que $\tau(x)$ est portée par $\pi^{-1}(x)$.

Le lemme suivant montre que le couple (S,π) possède une propriété d'universalité.

LEMMA 1: Soit Y un espace compact, et θ une surjection continue de Y sur X. Il existe alors une application continue η de S dans Y telle que $\theta\circ\eta = \pi$.

PREUVE. Soit $\nu \in M_1^+(Y)$ telle que $\theta(\nu) = \mu$. Soit S' le spectre de $L^\infty(\nu)$ et π' l'application canonique de S' dans Y. Pour $f \in L^\infty(\mu)$, la classe de $f\circ\theta$ dans $L^\infty(\nu)$ ne depend que de celle de f. On a donc un homomorphisme naturel d'algèbres de $\mathcal{C}(S)$ dans $\mathcal{C}(S')$, et donc une application continue φ de S' dans S. Pour $f \in \mathcal{C}(X)$ l'élément $f\circ\pi\circ h$ de $\mathcal{C}(S')$ correspond à la classe de la fonction $f\circ\theta$ sur Y, c'est à dire $f\circ\pi\circ h = f\circ\theta\circ\pi'$ et donc $\pi\circ h = \theta\circ\pi'$. D'autre part, puisque S est Stonien, il est classique qu'il existe une application continue ψ de S dans S' telle que $h\circ\psi = id_S$. Si on pose $\eta = \pi'\circ\psi$, on a

$$\theta\circ\eta = \theta\circ\pi'\circ\psi = \pi\circ h\circ\psi = \pi \qquad \text{Q.E.D.}$$

THEOREME 2. Soit Y un espace compact, et θ une surjection continue de Y sur X. Soit μ une mesure sur X, de support X, et ρ un relèvement linéaire de $L^\infty(\mu)$. Il existe alors une application φ de X dans $M_1^+(Y)$, qui est $\mathcal{B}(\rho)$ à $\mathcal{B}a(M_1^+(Y))$ mesurable, et telle que $\theta(\varphi(x))(f) = \rho(f)(x)$ pour tout $f \in \mathcal{C}(X)$. En particulier si ρ est fort, $\varphi(x)$ est portée par $\theta^{-1}(x)$ pour $x \in X$.

PREUVE. Avec les notations précédentes on pose $\varphi(x) = \eta(\tau(x))$. Q.E.D.

III) Non existence de certaines sections.

On va montrer que dans certains cas, il n'existe pas d'application φ du type décrit dans le théorème précédent qui soit $\mathcal{D}\mathcal{O}(X)$ à $\mathcal{B}a(M_1^+(Y))$ mesurable; on en déduira la non-existence de certains types de relèvements.

Désignons par I un ensemble fixé de cardinal $\geq \aleph_2$. Le lemme combinatoire suivant sera précieux.

LEMMA 3. Soit I' une partie de I de cardinal $\geq \aleph_2$. A tout élément $i \in I'$ supposons associé un ensemble dénombrable D_i. Alors il existe $i_1, i_2, i_3 \in I'$ tels que $i_\ell \notin D_{i_s}$ pour $\ell, s \in \{1,2,3\}$, $\ell \neq s$.

PREUVE. Soit $L \subset I'$ avec card $L = \aleph_1$. Posons $A = \bigcup_{j \in L} D_j$. On a card $A = \aleph_1$. Soit $L' \subset I' \setminus A$ avec card $L' = \aleph_1$. Posons $A' = \bigcup_{j \in L'} D_j$. On a card $A' = \aleph_1$. Il suffit alors de choisir $j_1 \in I' \setminus (A \cup A')$, $j_2 \in L' \setminus D_{j_1}$, $j_3 \in L \setminus (D_{j_1} \cup D_{j_2})$.
Q.E.D.

Posons $Z = \{0,1\}^I$ et désignons par X l'ensemble des fermés non vides de Z, muni de la topologie de Hausdorff. Pour une partie finie J de I désignons par p_J la projection naturelle de Z sur $\{0,1\}^J$. Une base de la topologie de X est formée des ouvert-fermés $V(J,A) = \{F \in X;\ p_J(F) = A\}$ où A est une partie de $\{0,1\}^J$.

Le résultat suivant est crucial.

THEOREME 4. Il n'existe pas d'application $F \to \mu_F$ de X dans $M_1^+(Z)$ qui soit $\mathfrak{B}\mathfrak{O}(X)$ à $M_1^+(Z)$ mesurable et telle que pour tout $F \in X$ on ait $\mu_F(F) = 1$, i.e. μ_F est portée par F.

PREUVE. Pour tout $i \in I$, posons $B_i = \{\mathfrak{z} \in Z;\ (i) = 1\}$. L'application $F \to \mu_F(B_i)$ est $\mathfrak{B}\mathfrak{O}(X)$ mesurable. Il existe donc une suite d'ouverts $(U_n^i)_n$ de X telle que l'application $F \to \mu_F(B_i)$ soit constante sur les atomes de la tribu engendrée par les $(U_n^i)_n$.

Chaque U_n^i est réunion d'ouverts de la forme $V(J,A)$ pour $J \subset I$, J finie, $A \subset \{0,1\}^J$. On peut donc écrire $U_n^i = \bigcup_p U_{n,p}^i$, où chaque $U_{n,p}^i$ est réunion d'ouverts de la forme $V(J,A)$ avec card $J \le p$. En remplaçant la suite $(U_n^i)_n$ par la suite $(U_{n,p}^i)_{n,p}$ on peut donc supposer que pour tout n et tout i on a

$$U_n^i = \bigcup_\alpha V(J_\alpha, A_\alpha)$$

où la réunion est prise sur un ensemble $\Omega_{i,n}$ d'indices et où card $J_\alpha \le a_{i,n} < \infty$ pour tout α.

Pour toute partie dénombrable D de I posons
$G_D = \{\mathfrak{z} \in Z;\ j \in D \Rightarrow \mathfrak{z}(j) = 0\}$.

Pour $i \notin D$, posons
$H_D^i = \{F \in X;\ F \subset G_D,\ F \cap B_i = \emptyset,\ F \cap (Z \setminus B_i) \neq \emptyset\}$.

Fixons i. Désignons par M^i l'ensemble des entiers n verifiant la condition suivante:

"Il existe une partie dénombrable E_n^i de I, $i \notin E_n^i$, telle que pour toute partie dénombrable D de I, $i \notin D$, $D \supset E_n^i$, on ait $H_D^i \cap U_n^i = \emptyset$". Posons $E^i = \bigcup_{n \in M^i} E_n^i$. On a $i \notin E^i$, et pour toute partie dénombrable D de I, $i \notin D, D \supset E^i$, on a $H_D^i \cap U_n^i = \emptyset$ pour tout $n \in M^i$.

Posons $N^i = \mathbb{N} \setminus M^i$. Pour $n \in N^i$ et pour toute partie dénombrable D de $\mathbb{N}$, $i \notin D$, on a $H_D^i \cap U_n^i \neq \emptyset$. Pour $n \in N^i$, $i \notin D$, posons

$$K_{n,D}^i = \{J_\alpha ; \alpha \in \Omega_{n,i} ; H_D^i \cap V(J_\alpha, A_\alpha) \neq \emptyset\}.$$

Par hypothèse cet ensemble est non vide. L'ensemble des parties de I de cardinal $\leq a_{i,n}$ est compact pour la topologie canonique. Donc pour i et n fixés les ensembles $K_{n,D}^i$ possèdent une valeur d'adhérence J_n^i selon l'ensemble filtrant croissant des partiesdénombrables de I ne contenant pas i. On pose $L_i = E^i \cup \bigcup_{n \in N} J_n^i$. C'est une partie dénombrable de I.

Pour chaque i, soit

$$W_i = \left(\bigcup_{n \in N^i} U_n^i\right) \cap \left(\bigcup_{n \in M^i} X \setminus U_n^i\right) .$$

C'est un atome de la tribu engendrée par les $(U_n^i)_n$. Il existe donc $c_i \in \mathbb{R}^+$ avec $F \in W_i \Rightarrow \mu_F(B_i) = c_i$. Il existe $p \in \mathbb{R}^+$ tel que si on pose $I' = \{i \in I ; c_i \in [\beta, \beta + 1/3]\}$ on ait card $I' \geq \aleph_2$.

D'après le lemme 3, il existe $i_1, i_2, i_3 \in I'$ avex $i_\ell \notin L_{i_s}$ pour $\ell, s \in \{1,2,3\}$, $\ell \neq s$.

On va montrer qu'il existe $F, G \in X$, vérifiant les propriétés suivantes:

a) $F \in W_{i_1} \cap W_{i_2}$; $G \in W_{i_1} \cap W_{i_2} \cap W_{i_3}$.

b) La projection $p_{\{i_1,i_2\}}$ de F sur $\{0,1\}^{\{i_1,i_2\}}$ est l'ensemble $\{(1,0);(0,1)\}$. La projection de G sur $\{0,1\}^{\{i_1,i_2,i_3\}}$ est l'ensemble $\{(1,0,0),(0,1,0),(0,0,1)\}$.

Avant de procéder à la construction, on va montrer que l'existence de F et G est une contradiction. En effet, d'après b), F est la réunion disjointe de $F \cap B_{i_1}$ et $F \cap B_{i_2}$, et G est la réunion disjointe de $G \cap B_{i_1}$, $G \cap B_{i_2}$, $G \cap B_{i_3}$. D'après a) on a $\mu_F(B_{i_\ell}) = c_{i_\ell}$ pour $\ell = 1,2$, $\mu_G(B_{i_\ell}) = c_{i_\ell}$ pour $\ell = 1,2,3$. On en déduit donc, puisque $\mu_F(F) = \mu_G(G) = 1$, que $c_{i_1} + c_{i_2} = 1$, $c_{i_1} + c_{i_2} + c_{i_3} = 1$. On a donc $c_{i_3} = 0$. Ceci montre que $\beta = 0$. Mais alors $c_{i_1} + c_{i_2} \leq 2/3 < 1$, ce qui est la contradiction cherchée.

On va procéder maintenant à la construction de F. Celle de G est analogue, avec seulement des complications d'écriture.

On va supposer par exemple que N^{i_1} et N^{i_2} sont tous deux infinis (les autres cas sont analogues). Enumérons $N^{i_1} = (n_k)$ et $M^{i_1} = (m_k)$. On pose $L_1 = L_{i_1} \cup L_{i_2} \setminus \{i_1\}$, $L_2 = L_{i_1} \cup L_{i_2} \setminus \{i_2\}$.
Ainsi $J_n^{i_1} \subset L_1$ et $J_n^{i_2} \subset L_2$ pour tout n. Ceci permet de construire par induction sur k des suites J_k et J'_k de parties finies de I vérifiant les conditions suivantes:

i) $J_k \in K^{i_1}_{n_k,L_1}$; $J'_k \in K^{i_2}_{m_k,L_2}$

ii) $i_2 \notin J_k$, $i_1 \notin J'_k$

iii) $(J_k \setminus L_1) \cap \left(\bigcup_{\ell<k} (J_\ell \setminus L_1) \cup (J'_\ell \setminus L_1)\right) = \emptyset$

$(J'_k \setminus L_2) \cap \left(\bigcup_{\ell<k} (J_L \setminus L_2) \cup (J'_\ell \setminus L_2)\right) = \emptyset$

$(J'_k \setminus L_2) \cap (J_k \setminus L_2) = \emptyset$.

Pour tout k il existe $A_k \subset \{0,1\}^{J_k}$ et $A'_k \subset \{0,1\}^{J'_k}$ tels que $H_L^{i_1} \cap V(J_k,A_k) \neq \emptyset$, $H_L^{i_2} \cap V(J'_k,A'_k) \neq \emptyset$ et que

$$V(J_k,A_k) \subset U^{i_1}_{n_k},\ V(J'_k,A'_k) \subset U^{i_2}_{m_k}.$$

Posons $L = L_1 \cap L_2 = L_{i_1} \cup L_{i_2} \setminus \{i_1,i_2\}$. Posons $p_k = p_{J_k}$, $q_k = p_{J'_k}$.
Posons enfin
$F = G_L \cap p^{-1}_{\{i_1,i_2\}}\{(0,1),(1,0)\} \cap \bigcap_k p_k^{-1}(A_k) \cap q_k^{-1}(A'_k)$

Avant de montrer que F satisfait à a) et à b), une remarque s'impose. Pour tout k, on a $H^{i_1}_{L_1} \cap V(J_k,A_k) \neq \emptyset$. Cela implique que $A_k \subset \{t \in \{0,1\}^{J_k} ; i \in J_k \cap L_1 \Rightarrow t(i) = 0\}$ et on a une conclusion comparable pour A'_k.

Montrons que A satisfait à b). On a par définiton $p_{\{i_1,i_2\}}(F) \subset \{(0,1),(1,0)\}$. On va montrer qu'il existe $x \in F$ tel que $x(i_1) = 0$, $x(i_2) = 1$. Pour chaque k, soit $x_k \in A_k$, $y_k \in A_k$ choisis de sorte que si $i_1 \in J_k$, on ait $x_k(i_1) = 0$ et que si $i_2 \in J'_k$, on ait $y_k(i_2) = 1$. (Un tel élément existe par définition de H_D^i). Il suffit de prouver qu'il existe $x \in Z$ tel que $x(i_1) = 0$, $x(i_2) = 1$ et $p_k(x) = x_k$, $q_k(x) = y_k$. Pour le voir, on définit x de la façon suivante:

$x(i_1) = 0$, $x(i_2) = 1$
$x(i) = 0$ si $i \in L$ ou si i n'appartient à aucun J_k ou J'_k
$x(i) = x_k(i)$ si $i \in J_k$
$x(i) = y_k(i)$ si $i \in J'_k$.
La compatibilité de ces conditions résulte de ce que si $i \in J_k \cap L$, $x_k(i) = 0$, si $i \in J'_k \cap L$, $y_k(i) = 0$, et que par construction les ensembles $J_k \setminus L$, $J'_k \setminus L$ sont deux à deux disjoints.

Ceci montre que F vérifie b). Prouvons que $F \in W_{i_1}$.

On a $F \in H^{i_1}_{E_1}$ (puisque $E_1 \subset L$, $F \subset G_L$, et d'après b)). Ceci montre que pour $n \in M^{i_1}$ on a $F \in X \setminus U^{i_1}_n$. Il faut montrer que pour $n \in N^{i_1}$ on a $F \in U^{i_1}_n$. Il suffit de montrer que $F \in V(J_k, A_k)$ où k est tel que $n = n_k$. Or $p_k(F) \subset A_k$ par construction, et le fait que $p_k(F) = A_k$ se montre par des arguments très semblables à ceux utilisés pour montrer que F satisfait à b). On a de même $F \in W_{i_2}$, ce qui termine la démonstration.
Q.E.D.

THEOREME 5. a) X est support d'une mesure de Radon.
b) Si μ est une mesure de Radon telle que X soit le support de μ, il n'existe pas de relèvement linéaire fort ρ de $L^\infty(\mu)$ tel que $\mathcal{B}(\rho) \subset \mathfrak{B}\mathfrak{o}(X)$.

PREUVE. a) Pour tout entier k considérons l'application h_k de Z_k dans X qui envoie un k-uple de Z sur la réunion de ses points. Elle est continue. Si ν désigne la mesure de Haar de Z désignons par μ_k la mesure image de $\nu^{\otimes k}$ par h_k. Son support est l'ensemble X_k des $F \subset X$ de cardinal $\leq k$. Mais l'ensemble $\bigcup_k X_k$ est dense dans X, donc le support de $\sum 2^{-k}\mu_k$ est égal à X.

b) Soit $Y = \{(z,F); z \in Z, F \in X, x \in F\}$. C'est un fermé de $Z \times X$. Désignons par Θ la projection Y sur X. Soit φ une application de X dans $M^+_1(Y)$ telle que $\varphi(F)$ soit portée par $\Theta^{-1}(F)$ pour $F \in X$, Alors $\varphi(F)$ est une mesure de Radon μ_F sur F, donc sur Z. Le théorème 4 montre que φ ne peut pas être $\mathfrak{B}\mathfrak{o}(X)$ à $\mathcal{B}a(M^+_1(Y))$ mesurable, et on conclut avec le théorème 2. Q.E.D.

La pathologie qui empêche ici l'existence de bons relèvements est une pathologie du compact, contrairement à se qui se passe dans l'exemple de Losert [2] où c'est une pathologie de la mesure.

IV) Le cas de $\{0,1\}^I$.

Le théorème 2 montre que l'existence de relèvements forts implique celle de sections. On va montrer dans ce paragraphe que pour la mesure de Haar μ de $\{0,1\}^I$, (sous certaines conditions sur I) l'existence de relèvements non nécessairement forts implique celle de sections. Il n'est pas connu si μ admet un relèvement de Baire (i.e. $\mathcal{B}(\rho) \subset \mathcal{B}a$ pour I grand. Le meilleur résultat dans cette direction (dû à Mockobodski, impublié) affirme que sous l'hypothèse du continu c'est le cas si card $I \leq \aleph_2$; ce relèvement est également fort. Il nous semble que l'intérêt des résultats de cette section est qu'ils pourraient peut-être être utilisés pour montrer la non-existence d'un relèvement de Baire.

Le lemme suivant est un cas particulier du théorème de Erdös-Rado Nous en donnons une preuve pour la commodité du lecteur. Un cardinal sera dit régulier s'il est égal à sa cofinalité.

LEMMA 6. *Soit I un ensemble de cardinal régulier $\geq \aleph_2$. Supposons donnée une famille $(D_i)_{i\in I}$ de parties dénombrables de I. Il existe alors des parties E et J de I avec card E < card I; card J = card I et $i,j \in J,\ i \neq j \Rightarrow D_i \cap D_j \subset E$.*

PREUVE. Montrons que la condition suivante est vérifiée.
$\exists\ E \subset I$, card E < card I, $\forall\ E' \subset I$, $E' \cap E = \emptyset$, card E' < card I, $\exists\ i \notin E'$, $D_i \cap E' = \emptyset$. (c)
En effet, dans le cas contraire on peut construire par induction sur $\alpha < \aleph_1$ des parties E_α disjointes de I, de cardinal < card I, telles que $i \notin E_\alpha \Rightarrow D_i \cap E_\alpha \neq \emptyset$. On a $\bigcup_{\alpha<\aleph_1} E_\alpha \neq I$, car I n'est pas de cofinalité $\aleph_1$. Mais si $i \notin \bigcup_{\alpha<\aleph_1} E_\alpha$ on a $D_i \cap E_\alpha \neq \emptyset$ pour tout $\alpha < \aleph_1$, ce qui est impossible puisque D_i est dénombrable.

La condition (c) et la régularité de card I permettent alors de construire par induction pour $\alpha <$ card I des $i_\alpha \in I$ tels que $\alpha \neq \beta \Rightarrow D_{i_\alpha} \cap D_{i_\beta} \subset E$. Q.E.D.

Le résultat suivant est crucial.

PROPOSITION 6. *Soit I un ensemble tel que $\aleph$=card I soit régulier, > card $\mathbb{R}$, et tel que $\lambda < \aleph \Rightarrow \lambda^{\mathbb{N}} < \aleph$. Soit $X = \{0,1\}^I$ et φ une application de X dans $M_1^+(X)$ telle que pour toute $g \in \mathcal{C}(X)$, on ait $\psi(x)(g) = g$ μ p.p. Il existe alors un compact L de X, homéomorphe à X, et une*

<u>rétraction continue</u> h <u>de</u> X <u>sur</u> L <u>telle que</u> $h(\psi(x)) = \delta_x$ <u>pour tout</u> $x \in L$.

<u>PREUVE</u>. Pour chaque i, il existe deux ensembles A_i et B_i de $\mathcal{B}a(X)$, tels que $\mu(A_i) = \mu(B_i) = 1/2$ et que

$$A_i \subset \{x \in X; \psi_x(\{y \in x; y(i) = 1\}) = 1\}$$

$$B_i \subset \{x \in X; \psi_x(\{y \in X; y(i) = 0\}) = 1\}.$$

Soit D_i une partie dénombrable de I telle que A_i et B_i ne dépendent que des coordonnées de rang appartenant à D_i. Soient J et E les parties de I fournies par le lemme précédent. On peut supposer que $J \cap E = \emptyset$ et que pour $i \in J$, card $D_i \setminus E$ est constant, mettons infini pour fixer les idées. Pour chaque i, soit η_i une bijection de $D_i \setminus E$ sur $\mathbb{N}$, telle que $\xi_i(i) = 1$. Ceci met $D_i \cup E$ en bijection avec $\mathbb{N} \cup E$ donc $\{0,1\}^{D_i \cup E}$ avec $T = \{0,1\}^{\mathbb{N}} \times \{0,1\}^E$. Pour $i \in J$, soient A_i' et B_i' les ensembles de $\mathcal{B}a(T)$ transportés de A_i et B_i par cette bijection (on peut naturellement regarder A_i et B_i comme éléments de $\mathcal{B}a(\{0,1\}^{D_i \cup E})$), chaque élément de $\mathcal{B}a(T)$ ne dépend que d'un nombre dénombrable de coordonnées. On a donc card $\mathcal{B}a(T) \leq (\text{card } E)^{\mathbb{N}} \times \text{card } \mathbb{R} < \aleph$.

On peut donc supposer que tous les A_i' sont égaux à un certain A et tous les B_i' à un certain B. Si ν désigne la mesure de Haar de T, on a

$$\nu(A \Delta \{(u,v) \in \{0,1\}^{\mathbb{N}} \times \{0,1\}^E; u(1) = 1\}) = 0$$

$$\nu(B \Delta \{(u,v) \in \{0,1\}^{\mathbb{N}} \times \{0,1\}^E; u(1) = 0\}) = 0.$$

Il en résulte que l'on peut trouver $a,b \in \{0,1\}^{\mathbb{N}}$ et $c \in \{0,1\}^E$ tels que $a(1) = 1$, $b(1) = 0$, $(a,c) \in A$, $(b,c) \in B$.

Désignons par ξ l'application de $\{0,1\}^J$ dans X donnée par:

$$\xi(t)(i) = \begin{cases} c(i) & \text{si } i \in E \\ 0 & \text{si } i \notin \bigcup_{j \in J} D_j \cup E \\ a(\eta_j(i)) & \text{si } i \in D_j \setminus E \text{ et } t(j) = 1 \\ b(\eta_j(i)) & \text{si } i \in D_j \setminus E \text{ et } t(j) = 0. \end{cases}$$

Il est clair que ξ est continue. Si $t(j) \neq t'$, alors pour i tel que $\eta_j(i) = 1$ on a $\xi(t)(i) = 1$, $\xi(t')(i) = 0$, ce qui montre que ξ est injecture . Soit L son image, qui est donc homéomorphe à $\{0,1\}^J$, donc à X.

Désignons par p la projection canonique de X sur $\{0,1\}^J$. L'application $\xi \circ p$ de X dans L est continue. Pour prouver que c'est une rétraction de X, il suffit de prouver que pour $t \in \{0,1\}^J$ on a $p \circ \xi(t) = t$.

Ceci résulte du fait que $j \in D_j$ et que $\eta_j(j) = 1$, donc

$$\xi(t)(j) = \begin{cases} b(1) = 1 & \text{si } t(j) = 1 \\ c(1) = 0 & \text{si } t(j) = 0. \end{cases}$$

Prouvons que pour $x \in L$ on a $\xi \circ p(\psi(x)) = x$. Il suffit de montrer que pour $t \in \{0,1\}^J$ on a $p(\psi(\xi(t))) = \delta_t$. Mais pour $j \in J$, on a

$$t(j) = 1 \Rightarrow \xi(t) \in B_j \Rightarrow \psi(\xi(t))(\{y; y(j) = 1\}) = 1$$

$$t(j) = 0 \Rightarrow \xi(t) \in C_j \Rightarrow \psi(\xi(t))(\{y; y(j) = 1\}) = 0$$

ce qui suffit. Q.E.D.

On en déduit

<u>THEOREME 7</u>. <u>Soit</u> I <u>un ensemble tel que</u> card I <u>soit régulier</u>, > card $\mathbb{R}$, <u>et tel que</u> $\lambda < \aleph \Rightarrow \lambda^{\mathbb{N}} < \aleph$. <u>Supposons que la mesure de Haar</u> μ <u>de</u> $X = \{0,1\}^I$ <u>admette un relèvement linéaire</u> ρ <u>non nécessairement fort, tel que</u> $\mathcal{B}(\rho) \subset \mathcal{B}a(X)$ (<u>resp</u>. $\mathcal{B}(X)$, $\mathcal{D}\sigma(X)$). <u>Alors, pour tout espace compact</u> Y <u>et toute surjection</u> θ <u>de</u> Y <u>sur</u> X <u>il existe une application</u> φ <u>de</u> X <u>dans</u> $M_1^+(Y)$, <u>qui est</u> $\mathcal{B}a(X)$ (<u>resp</u>. $\mathcal{B}(X)$, $\mathcal{D}\sigma(X)$) <u>à</u> $\mathcal{B}a(M_1^+(Y))$ <u>mesurable et telle que</u> $\varphi(x)$ <u>soit portée par</u> $\theta^{-1}(x)$ <u>pour</u> $x \in X$. <u>Si</u> ρ <u>est multiplicatif, il existe une application</u> φ <u>de</u> X <u>dans</u> Y <u>qui est</u> $\mathcal{B}a(X)$ (<u>resp</u>. $\mathcal{B}(X)$, $\mathcal{D}\sigma(X)$) <u>à</u> $\mathcal{B}a(Y)$ <u>mesurable et telle que</u> $\theta \circ \varphi = id$.

<u>PREUVE</u>. Supposons d'abord ρ linéaire. L'application ψ de X dans $M_1^+(X)$ définie par $\psi(x)(g) = \rho(g)(x)$ pour $x \in X$ et $g \in \mathcal{C}(X)$ vérifie les conditions du lemme 6. Soient L et h comme dans ce lemme. Puisque L est homéomorphe à X, on peut supposer que θ est à valeurs dans L. Soit

$$z = \{(x,y) \in X \times Y; \theta(y) = h(x)\}.$$

C'est un compact. Désignons par θ' la projection de Z sur X, qui est surjective. D'après le théorème 2 il existe une application φ' de X dans $M_1^+(Z)$, qui est $\mathcal{B}a(X)$ (resp. $\mathcal{B}(X)$, $\mathcal{D}\sigma(X)$) à $\mathcal{B}a(M_1^+(Z))$ mesurable, et telle que $\theta'(\varphi'(x)) = \psi(x)$. Pour $x \in L$, on a donc $h \circ \theta'(\varphi'(x)) = h(\psi(x)) = \delta_x$, donc $\varphi'(x)$ est portée par l'ensemble $\{(u,y); \theta(y) = h(u) = x\}$.

Désignons par q la projection Z sur Y. Si pour $x \in L$ on pose $\varphi(x) = q(\varphi'(y))$, $\varphi(x)$ est donc portée par $\theta^{-1}(y)$, et φ est $\mathcal{B}a(L)$ (resp. $\mathcal{B}(L)$), $\mathcal{D}\sigma(L)$) à $\mathcal{B}a(M_1^+(Y))$ mesurable, car la trace sur L de $\mathcal{B}a(X)$, $\mathcal{B}(X)$, $\mathcal{D}\sigma(X)$ n'est autre que $\mathcal{B}a(L)$, $\mathcal{B}(L)$, $\mathcal{D}\sigma(L)$ respectivement.

La preuve du cas où ρ est multiplicatif est analogue. Q.E.D.

<u>Remarque</u>. a) Les cardinaux de la forme $\aleph_{\alpha+1}$, où $\aleph_\alpha$ est de la forme

$(K)^{\mathbb{N}}$, vérifient les hypothèses du théorème 7.

b) Nous savons prouver le théorème 7 sous des conditions plus générales portant sur $\aleph$, mais l'intérêt de ce théorème est d'être valable pour des cardinaux arbitrairement grands.

Bibliographie

[1] G.A. EDGAR, Measurable weak sections. Ill. J. Math. 20 (1976), 630-646

[2] V. LOSERT, A Measure Space Without the Strong Lifting Property. Math. Ann. 239 (1979), 119-128

[3] M. TALAGRAND, Non existence de certaines sections et applications à la théorie du relèvement. C.R.A.S. t.286 (1978), 1183-1185

[4] M. TALAGRAND, En général il n'existe pas de relèvement borélien fort. C.R.A.S. t.289 (1978), 633-634

[5] H.von WEIZSÄCKER, Some Negative Results in the Theory of Lifting, in Measure Theory, Proceedings of the Conference Held at Oberwolfach, June 1975 (A. Bellow, D. Kölzow, editors). Lecture Notes in Mathematics No.541. Springer-Verlag, Berlin-Heidelberg-New York, 1976, 159-172

SURVEY OF MEASURABLE SELECTION THEOREMS: AN UPDATE

Daniel H. Wagner

Daniel H. Wagner, Associates
Paoli, Pennsylvania, U.S.A. 19301

1. Introduction. In [WG3], we presented a survey of results on the existence of a measurable selection of a set-valued function, attempting to cover all work up to the early part of 1977. Coverage of Russian literature in [WG3] was rather incomplete, and a very useful supplement, [IF3], was supplied by Ioffe soon thereafter.

The present paper undertakes to update the survey [WG3], for results which have become known in the ensuing period through most of 1979. It also includes mention of some papers overlooked in [WG3], some cases of which were significant oversights, and in some other respects corrects [WG3]. Undoubtedly coverage of recent Russian work will again be deficient, and it is to be hoped that this deficiency will be filled elsewhere.

The basics of the problem area under review are as follows: Let $(T, \mathcal{M})$ be a measurable space, X be a topological space, and $\phi \neq F(t) \subset X$ for $t \in T$. Does there exist a measurable selection of F, i.e., a measurable map $f: T \to X$ such that $f(t) \in F(t)$ for $t \in T$? Beyond this question, one is interested in the existence of a countable family of measurable selections of F which is pointwise dense in the values of F (a Castaing representation), or stronger still, whether one can "fill" the graph of F with measurable selections which are nicely parameterized. The first result of the latter form with uncountable values of F was obtained by Wesley; see [WG3, § 10]. An especially rich advent of results in this vein has appeared in 1977-79 -- see § 4 below.

General definitions and notation are given in § 2. In § 3 we repeat some of the main results reviewed in [WG3], supplanted by [IF3]. The final section summarizes those corrections to [WG3] which are not noted in prior parts of the paper. The intervening sections review a variety of topics. Following is an index of sections, with the principal new contributors parenthetically noted:

6. G_δ -valued functions (Srivastava, Burgess, Debs, Srivatsa)
7. Compact-valued functions (Graf, Losert, Talagrand)
8. Measurable weak selections (Edgar, Graf)
9. Optimal measurable selections (Rieder, Dolecki, Bain)
10. Stopping time theory (Dellacherie, Meyer)
11. Convex-valued functions (Castaing, Valadier, Debs)
12. Borel selections (Levin, Saint-Raymond, Martin)
13. Generalizations of Fundamental Measurable Selection Theorem (Mägerl, Miller, Le Van Tu)
14. Reduction principle (Maitra, B. V. Rao, Sarbadhikari)
15. Extensions of measurable functions (Maitra)
16. Measurable implicit functions (Castaing, Valadier)
17. Special topics (Baggett, Ramsey, Doberkat)
18. Corrections to previous survey

Acknowledgment

There are substantial areas which could be properly considered part of measurable selection theory but which are not reviewed here, because they have so much specialized technology of their own. These include continuous selections, multi-valued differential equations, and set-valued measures. The latter topic was reviewed rather superficially as § 15 of [WG3]; a significant subsequent reference is [CS8]. General theorems on continuous selections were noted in [WG3, § 13], and in § 13 below we do note briefly a unified approach to continuous selection theory and measurable selection theory by Mägerl.

The bibliography is coded in sequel to those of [WG3] and [IF3]. An underlined reference citation refers to the bibliography of [WG3], possibly to that part added in proof. The present bibliography ends with updates of several citations that were given incompletely in [WG3].

2. Preliminaries. We largely follow usages in [WG3, § 2], much of which is repeated here. For every set S we define $\mathscr{P}(S) = \{A : \phi \neq A \subset S\}$. When $\mathscr{L}$ is a set of sets, $\mathscr{L}_\sigma$ $\{\mathscr{L}_\delta\}$ is the set of countable unions $\{$intersections$\}$ of members of $\mathscr{L}$. We denote the set of real numbers by R, Euclidean n-space by R^n, the set of natural numbers by ω, and the set of irrational numbers by ω^ω (topologized as usual). We also denote $A \triangle D = (A \setminus D) \cup (D \setminus A)$.

Suppose D is topologized. By $\mathscr{B}(D)$ we mean the σ-algebra of Borel sets of D. By $\mathscr{B}_0(D)$ we mean the Baire σ-algebra of D, i.e., the smallest σ-algebra w.r.t. which every bounded continuous $f: D \to R$ is measurable (measurability is defined below). Always $\mathscr{B}_0(D) \subset \mathscr{B}(D)$ and if D is metrizable, $\mathscr{B}_0(D) = \mathscr{B}(D)$. Also, for $A \subset D$ we denote the closure of A by cl A. The topological weight of A is denoted wt A.

We fix $(T, \mathscr{M}, \mu)$ as a measure space with μ nonnegative. It should be recognized that we include the case where no measure is present, i.e., when one is dealing with the measurable space $(T, \mathcal{M})$; one may let μ be the trivial measure given by $\mu(S) = 0$ for $S \in \mathcal{M}$ to bring this case into our framework. For theorem statements which do not mention properties of μ, in fact, one may just as well consider that μ is not present.

We fix X as a topological space, and we reserve F to mean $F: T \to \mathcal{P}(X)$, i.e., F is a set-valued function. We say F is (adjective)-valued if F(t) is (adjective) for $t \in T$. We define the graph of F, denoted Gr F, by

$$\text{Gr } F = (T \times X) \cap \{(t, x) : x \in F(t)\}.$$

For $A \subset X$, we define $F^{-}(A) = T \cap \{t : F(t) \cap A \neq \phi\}$. Definitions on F extend to any set-valued function. For $D \subset Y \times Z$, we let π_Y and π_Z be the projections on D to Y and Z respectively.

Suppose $f: T \to X$. We say f is a <u>measurable function</u> if $f^{-1}(B) \in \mathcal{M}$ for $B \in \mathcal{B}(X)$. We say F is <u>measurable</u> (as a set-valued function) if $F^{-}(C) \in \mathcal{M}$ for all closed $C \subset X$ and F is <u>weakly</u> measurable if $F^{-}(U) \in \mathcal{M}$ for all open $U \subset X$.

N. B. In some important newer works, what is here and usually historically called "weak measurability" of F is there called "measurability," and [SVS4] uses "strong measurability" for our "measurability" of F. The reason for this trend is that weak measurability of F has become a much more frequent hypothesis than measurability of F. Readers must check definitions in each paper, and authors must be explicit about same.

More generally, suppose $\mathcal{L}$ is a family of subsets of T. We define {weak} $\mathcal{L}$-measurability of F to mean $F^{-}(E) \in \mathcal{L}$ for closed {open} $E \subset X$. Still more generally, if $\mathcal{D}$ is a family of subsets of X, $\mathcal{L}$ - $\mathcal{D}$ -measurability of F means $F^{-}(E) \in \mathcal{L}$ for $E \in \mathcal{D}$; this is called $\mathcal{D}$-measurability if $\mathcal{L} = \mathcal{M}$. Measurability of $f: T \to X$ is generalized similarly.

A <u>selection</u> of F is a function $f: T \to X$ such that $f(t) \in F(t)$ for $t \in T$. We denote

$$\mathcal{S}(F) = \{f : f \text{ is a measurable selection of } F\}.$$

Following <u>[RC6]</u>, we say $\{f_1, f_2, \ldots\} \subset \mathcal{S}(F)$ is a <u>Castaing representation</u> of F if $\{f_1(t), f_2(t), \ldots\}$ is dense in F(t) for $t \in T$. The term "representation" is used in a different way in § 4.

If $\mathcal{N}$ is a σ-algebra over X and $f: T \to X$ is $\mathcal{N}$-measurable, we denote by $f(\mu)$ the measure ν on $\mathcal{N}$ given by $\nu(D) = \mu(f^{-1}(D))$ for $D \in \mathcal{N}$.

If Y and Z are topological spaces and $A \subset T \times Y$, we say $f: A \to Z$ is a <u>Carathéodory</u> map if $f(t, \cdot)$ is continuous for $t \in \pi_T(A)$ and $f(\cdot, y)$ is measurable for $y \in \pi_Y(A)$.

If T is topologized and F is closed-valued, we say F is <u>upper {lower} semi-continuous</u>, abbreviated usc {lsc}, as a set-valued function, if for each closed {open} $A \subset X$, $F^{-}(A)$ is closed {open}.

By a <u>Polish space</u> is meant a (not necessarily complete) homeomorph of a complete separable metric space. We say S is a Suslin {Lusin} space if S is topologized as a Hausdorff space and there exist a Polish space P and a continuous surjective {bijective} $\varphi: P \to S$. We define a weakly Suslin space in the same way without the Hausdorff requirement. A Suslin set in a topological space is a subset which is a Suslin space. We say A is co-Suslin in S if $S \setminus A$ is Suslin.

Suppose $\mathscr{L}$ is a family of sets. By $\sigma(\mathscr{L})$ we mean the σ-algebra generated by $\mathscr{L}$ and by $\Sigma(\mathscr{L})$ we mean the family of sets obtained from $\mathscr{L}$ by the Suslin operation (see, e.g., [WG3, § 2]). If $\Sigma(\mathscr{L}) = \mathscr{L}$, we say $\mathscr{L}$ is a Suslin family. When D is topologized, by $\mathscr{C}(D)$ we mean the smallest family of sets which is a Suslin family, is a σ-algebra, and contains $\mathscr{B}(D)$; these are Selivanovski's " C-sets, " and are universally measurable w.r.t. $\mathscr{B}(D)$.

If $\mathscr{A}$ and $\mathscr{D}$ are families of sets, we denote $\mathscr{A} \times \mathscr{D} = \{A \times D : A \in \mathscr{A} \text{ and } D \in \mathscr{D}\}$. If $\mathscr{A}$ and $\mathscr{D}$ are σ-algebras, we denote $\mathscr{A} \otimes \mathscr{D} = \sigma(\mathscr{A} \times \mathscr{D})$.

We close this section with some recent interesting examples by Himmelberg, Parthasarathy, and Van Vleck [HPV2] and Nishiura [NI].

EXAMPLES 2.1. In [HPV2] are given examples with the following properties:

(i) $T = [0,1]$, $\mathscr{M} = \mathscr{B}(T)$, $X = \omega^\omega$, Gr F is closed, $F^-(K) \in \mathscr{M}$ for compact $K \subset X$, and F is not weakly measurable;

(ii) $T = [0,1]$, $\mathscr{M} = \mathscr{B}(T)$, $X = T \times \omega^\omega$, $G: T \to \mathscr{P}(X)$ and F are closed-valued and weakly measurable, and $F \cap G$ is not weakly measurable (adaptation of Kaniewski's example of [WG3, § 2]);

(iii) $T = [0,1]$, $\mathscr{M} = \mathscr{B}(T)$, $X = \omega^\omega$, $f: T \times X \to R$ is continuous, $F(t) = \{x : f(t,x) = 0\}$ for $t \in T$, and F is not weakly measurable;

(iv) $T = [0,1]$, $\mathscr{M} = \mathscr{B}(T)$, $X = \omega^\omega$, Gr F is closed, and $\mathscr{S}(F) = \phi$ (compare Example 4.6 below and remark preceding it);

(v) $f: [0,1] \times \omega^\omega \to R$ is continuous, $\{x : f(t,x) = 0\} \neq \phi$ for $t \in [0,1]$, and there is no Borel $r: [0,1] \to \omega^\omega$ such that $f(t, r(t)) = 0$ for $t \in [0,1]$;

(vi) F is countable-valued (not closed-valued) and measurable and $\mathscr{S}(F) = \phi$.

For (iv), from Sierpinski's [SP2] (or see [KU1, p. 485]), take Suslin $A_1, A_2 \subset T$ such that $T \setminus A_1$ and $T \setminus A_2$ are disjoint and not separable by Borel sets. Take continuous $h_i : \omega^\omega \to A_i$ for $i = 1, 2$. Let $F = h_1^{-1} \cup h_2^{-1}$. This is a modification by Darst of an example with Gr F Borel, rather than closed, by Novikov [NO1]. To obtain (v), take f with Gr F as zero set. (In [HPV2], (iv) and (v) are given together.)

To obtain (vi), let $T = \{A : \phi \neq A \subset R \text{ and } A \text{ is countable}\}$, $\mathscr{M} = \sigma(\{\{A : A \cap B \neq \phi, A \in T\} : B \subset R \text{ is closed}\})$. (Then $\mathscr{M}$ is the Borel σ-algebra of the "upper semi-finite" topology on T.) Let $F(A) = A$ for $A \in T$. In [HPV2], [BD, Cor. 2] is applied to show the desired properties.

Prior to [HPV2], Nishiura [NI] gave an example related to (i) as follows:

(vi) $T = X = \omega^\omega$, μ is a complete probability measure, F is closed-valued, $F^-(K) \in \mathcal{M}$ for compact $K \subset X$, and F is not measurable: Let $\mathcal{M} = \{A : A \subset T$ is meager or comeager$\}$, $\mu(A) = 0$ if A is meager, $\mu(A) = 1$ if A is comeager, and $F(t) = \{t\}$ for $t \in T$.

3. Review of principal results in previous survey. In this section, we review some of the most important theorems previously surveyed in [WG3].

First is the Fundamental Measurable Selection Theorem.

THEOREM 3.1 [WG3, Th. 4.1]. *If F is weakly measurable and closed-valued and X is Polish, then* $\mathcal{S}(F) \neq \phi$.

Theorem 3.1 was first proved as given (and in more generality) by Kuratowski and Ryll-Nardzewski [KRN]. For reasons discussed in detail in [WG3, p. 867, p. 901], I feel that credit should be shared with Rokhlin and Castaing. A great deal of measurable selection theory has been deduced from Theorem 3.1. An especially useful direct generalization is the following.

THEOREM 3.2 [WG3, Th. 4.2]. *Suppose either*

(i) F *is complete-valued and* X *is separable metric, or*

(ii) F *is closed-valued,* $\mathcal{M}$ *is a Suslin family, and* X *is metric Suslin.*

Then F *is weakly measurable iff* F *has a Castaing representation.*

To correct the record in [WG3], we note that under (i), Theorem 3.2 was first proved by Valadier [VA2, Th. 0.3]. The equivalence expressed in Theorem 3.2 is especially useful for manipulation of set-valued functions by operations on the value sets, as demonstrated particularly well by Rockafellar, e.g., [RC6].

Also in connection with [WG3, Th. 4.2], Himmelberg and Van Vleck [HV10], and subsequently and independently Maitra and Rao [MR3], have discovered an error in and given counterexamples to [HV6, Ths. 1'(i), 2'(ii)], which is given also as [WG3, Th. 4.2e((xi) $\Rightarrow$ (ix))]. However, it is shown in [HV10] that if X is separable metric, F is complete-valued, and $F^-(K) \in \mathcal{M}$ for compact $K \subset X$, then there exist selections $\{f_1, f_2, \ldots\}$ of F such that $F(t) = \mathrm{cl}\{f_i(t) : i = 1, 2, \ldots\}$ and $f_i(K) \in \mathcal{M}$ for compact $K \subset X$.

Following is a version of the Yankov-von Neumann theorem, which is the basic graph-conditioned measurable selection theorem.

THEOREM 3.3 [WG3, Cor. 5.2]. *Suppose* T *and* X *are Polish,* Gr F *is Suslin, and* $\mathcal{M}$ *contains the Suslin sets of* T. *Then* $\mathcal{S}(F) \neq \phi$.

The attribution of Theorem 3.3 to Yankov and von Neumann is discussed in detail in [WG3, p. 900]. Additional clarification is given in [IF3] pertaining to a confusing point arising from Russian-to-French translation. There is no doubt that Yankov was the first to publish a statement and proof of this result, with $T \subset R = X$, and there seems to be no disagreement that von Neumann deserves independent credit. It is easily deduced in a little more generality from Theorem 3.1 -- see [WG3, Th. 5.3]. Successive generalizations by Aumann, Sainte-Beuve, and Leese have resulted in the following.

THEOREM 3.4 [WG3, Th. 5.10]. Suppose $\mathcal{M}$ is a Suslin family, X is a weakly Suslin space, and Gr F $\in \mathcal{M} \otimes \mathcal{B}(X)$. Then F has a Castaing representation.

When one wants a Borel function selection, it is natural to turn to uniformization theory, since under weak conditions a function which is a Borel set is also a Borel function, notably [WG3, Th. 12.4] due to Baker or [WG3, Th. 12.5] due to Hoffman-Jørgensen. The latter states: When $f: T \to X$ and T and X are Suslin, f is a Borel function iff f is a Borel set iff f is a Suslin set.

In fact in [WG3] we gave insufficient attention (in § 12) to the theorem of Stschegolkov (we thank Valadier for calling this to our attention), the following generalization of which from R to Polish spaces was given by Brown and Purves [BP].

THEOREM 3.5 [WG3, § 12]. Suppose T and X are Polish, F is σ-compact-valued, Gr F is Borel, and $\mathcal{M} = \mathcal{B}(T)$. Then $\mathcal{S}(F) \neq \phi$.

In [WG3, § 12], it is incorrectly stated that this follows from [SN]. A subsequent proof was given by Saint-Raymond [SRY1, Cor. 12], who in the process showed (Corollaire 10) that under the hypothesis of Theorem 3.5, Gr F is a countable union of Borel sets with compact vertical sections.

The Suslin type concept formulated by Leese has proved very useful for manipulation of measurable set-valued functions and for its unifying effect on a great deal of measurable selection theory. Following [LE2] and [WG3, § 6], we say F is of {weak} Suslin type if there exists a Polish space P, a continuous $\varphi: P \to X$, and a {weakly} measurable $G: T \to \mathcal{P}(X)$ such that $F(t) = \varphi(G(t))$ for $t \in T$. If F is of weak Suslin type, F has a Castaing representation.

If $\mathcal{M}$ is a Suslin family and X is Hausdorff, the set of set-valued functions of Suslin type is closed under the Suslin operation (on value sets). If also X is weakly Suslin and Gr F $\in$ $\mathcal{M} \otimes \mathcal{B}(X)$, F is of Suslin type. Other criteria are given in [WG3, § 6] and [LE2]. Applications to measurable selection theory are extensive.

Following is a strong theorem on measurable implicit functions due to Leese. Others, conditioning g to be instead a continuous, Borel, or Carathéodory map, are given in [WG3, § 7].

THEOREM 3.6 [WG3, Th. 7.1]. Suppose Y is a separable metric space, $g: \text{Gr } F \to Y$, $h: T \to Y$ is measurable, $h(t) \in g(\{t\} \times F(t))$ for $t \in T$, $\mathcal{M}$ is a Suslin family, X is Hausdorff, F is of Suslin type, and g is $\mathcal{M} \otimes \mathcal{B}(X)$-measurable. Then there exists $f \in \mathcal{S}(F)$ such that $h = g(\cdot, f(\cdot))$.

The strongest result in [WG3] on optimal measurable selections is the following due to Schäl.

THEOREM 3.7 [WG3, Th. 9.1(iii)]. Suppose $u: \mathrm{Gr}\, F \to R$ is $\mathcal{M} \otimes \mathcal{B}(X)$-measurable, $u(t, \cdot)$ is usc on $F(t)$ for $t \in T$, F is measurable and compact-valued, X is separable metric, and u is the limit of a decreasing sequence of Carathéodory maps. Then there exists $f \in \mathcal{S}(F)$ such that $u(t, f(t)) = \sup\{u(t,x) : x \in F(t)\}$.

We conclude this section with Wesley's parameterization result, which is given in [WG3] in a little more general form, simply by superimposing σ-algebra isomorphisms on Wesley's own version which is given here. (We thank C. Becker for pointing out that [WG3, Th. 10.3] does not follow from Wesley's result, Theorem 3.8 below, because, denoting by $\mathcal{L}_n$ the set of n-dimensional Lebesgue measurable sets, $\mathcal{L}_2 \not\subset \mathcal{L}_1 \otimes \mathcal{L}_1$. However, [WG3, Th. 10.3] does follow from Theorem 4.3 below.)

THEOREM 3.8. Suppose $T = X = [0,1]$, μ is Lebesgue measure, Gr F is Borel, and F is uncountable-valued. Then there exists $f: T \times [0,1] \to \mathrm{Gr}\, F$ such that:

(a) for $t \in T$, $f(t, \cdot)$ is a one-to-one Borel function on $[0,1]$ onto $F(t)$;

(b) for $y \in [0,1]$, $f(\cdot, y) \in \mathcal{S}(F)$;

(c) f is a measurable function w.r.t. 2-dimensional Lebesgue measure.

Theorem 3.8 was proved by Wesley in his 1973 thesis [WE1] using Cohen's forcing methods. It represents a good transition point between our review up to 1977 and our review of subsequent work, since this result inspired much of the latter, i.e., the considerable advances that have been made in "representation," i.e., "parameterization" results since early 1977.

We should also note precursors to Wesley-type parameterizations in the form of Lusin's result [WG3, Cor. 10.2] in similar form for countable-valued F, Larman's findings [LA1, 2] that under reasonable assumptions Gr F contains an uncountable disjoint family of Borel selections of F, and results of Ekeland and Valadier [EV] giving a related type of representation for linear X.

We next turn to new work on representation of set-valued functions.

4. Representation of measurable set-valued functions. In this section we review results subsequent to Wesley's on "filling" Gr F with measurable selections, parameterized in a nice fashion, sometimes one-to-one. Considerable advances have been made in this area in a fairly short time in papers of, in approximate chronological order, Bourgain, Cenzer and Mauldin, Ioffe, Mauldin, Srivastava, Sarbadhikari and Srivastava, Graf, and Srivatsa. Particularly striking are the cases where certain types of representations are given with direct converses. These are summarized at the end along with a list of cases where the parameterizations are one-to-one.

We say F is induced by (Z, f) if $f: T \times Z \to X$ and $F(t) = f(t, Z)$ for $t \in T$. We then denote by $\hat{f}$ the map on $T \times Z$ onto Gr F defined by $\hat{f}(t, z) = (t, f(t, z))$ for $t \in T$, $z \in Z$. In this section $I = [0, 1]$.

Ioffe [IF3] uses the terminology "(Z, f) is a representation of F" instead of "F is induced by (Z, f)," and in the case where Z is Polish and f is a Carathéodory map he refers to this as an "analytic" representation, generalizing the classic motion of analyticity. The "induced" usage we take from Srivastava [SVS4].

An important breakthrough was achieved by Bourgain [BRG1] (thesis dated March, 1977) and independently by Cenzer and Mauldin [CM2] (submitted April 10, 1977, sent to me March 24, 1977, and verbally announced two weeks earlier) in proving Wesley's result without forcing methods or other metamathematics. In the process they also generalized Wesley's result in separate significant ways.

The main results of [CM2] are summarized in the next two theorems.

THEOREM 4.1 [CM2, Ths. 6, 9]. Suppose $T = X = I$, Gr F is Borel {Suslin} and F is uncountable-valued. Then there exists f such that (I, f) induces F, $\hat{f}$ is one-to-one and $\mathscr{C}(T \times I)$-measurable, $\hat{f}^{-1}$ is $\mathscr{C}(T \times I)$-measurable, and for $t \in T$, $f(t, \cdot)$ is a Borel isomorphism {is a σ-algebra isomorphism of $(I, \sigma(\Sigma(\mathscr{B}(I))))$ onto $(F(t), \sigma(\Sigma(\mathscr{B}(F(t)))))$}.

THEOREM 4.2 [CM2, Th. 4]. Suppose $T = I$, $X = \omega^\omega$, F is uncountable-closed-valued, and Gr F is Suslin. Then there exists f such that (I, f) induces F, $\hat{f}$ is one-to-one and $\mathscr{C}(T \times I)$-measurable, $\hat{f}^{-1}$ is $\mathscr{C}(I \times \omega^\omega)$-measurable and for $t \in T$, $f(t, \cdot)$ is a Borel isomorphism.

Independently of [CM2] and [BRG1], Wesley in [WE4] (announced in [WE3]) obtained f as in Theorem 3.8 with $\hat{f}$ one-to-one, and $\hat{f}$ and $\hat{f}^{-1}$ absolutely measurable and preserving verticals. He again used forcing methods. The results of [CM2] are stronger.

Bourgain's generalization of Wesley's theorem is Theorem 4.3 given next. It conditions μ heavily, but has the merit (uniquely among the results reported here) of having F induced by a Carathéodory f with each $f(t, \cdot)$ one-to-one. The parameterization space $\omega^\omega \cup \omega$ in Theorem 4.3 has as topology the topological sum of the usual topology on ω^ω with the discrete topology on ω. Shortly after [BRG1], Ioffe ([IF5], submitted summer, 1977) independently obtained Theorem 4.3; his statement has D = Gr F and the conclusion that the parameterization space is Polish rather than specifically $\omega^\omega \cup \omega$, however Theorem 4.3 easily follows.

Srivastava has conjectured (verbal communication) that Bourgain's proof of Theorem 4.3 can be used to obtain ω^ω as the parameterization space, providing each F(t) is a condensed set. (For a related conjecture, see Remark 4.27 below.) In [BRG2, Prop. 3.39], Bourgain gives, after a very lengthy chain of definitions, a version of Theorem 4.3 in which no measure is used and which is possibly more general (not clear). The condition on $\pi_T(\text{Gr } F \setminus D)$ is replaced by $\pi_T(\text{Gr } F \,\Delta\, D) \in \mathscr{N}$, where $\mathscr{N}$ is a σ-ideal of subsets of T which is related to a family $\mathscr{L}$ of subsets of T closed under finite union and intersection with $\{T \setminus A : A \in \mathscr{L}\} \subset \Sigma(\mathscr{L})$, and which in turn is related to $\mathscr{M}$. A converse is [BRG2, Th. 3.43].

THEOREM 4.3 [BRG1, Th. 2.20; or IF5, Th.]. Suppose μ is σ-finite and complete, X is Polish, F is uncountable-Borel-valued, and there exists $D \subset \mathrm{Gr}\, F$ such that $D \in \mathscr{M} \otimes \mathscr{B}(X)$ and $\mu(\pi_T(\mathrm{Gr}\, F \setminus D)) = 0$. Then there exists a Carathéodory map f such that $(\omega^\omega \cup \omega, f)$ induces F and $f(t, \cdot)$ is one-to-one for $t \in T$.

In [MD2], Mauldin made further important advances by giving necessary and sufficient conditions for the inducing map to be a Borel isomorphism, i.e., an isomorphism between σ-algebras of Borel sets. Following Definition 4.4 (used, e.g., in [BRN]) these results are summarized as Theorem 4.5.

DEFINITION 4.4. We say $\nu : T \times \mathscr{B}(X) \to [0, \infty)$ is a conditional measure distribution if for $t \in T$, $\nu(t, \cdot)$ is a measure on $\mathscr{B}(X)$ and for $B \in \mathscr{B}(X)$, $\nu(\cdot, B)$ is a Borel function.

THEOREM 4.5 [MD2, Ths. 2.4, 1.2, 2.9]. Suppose T and X are uncountable Lusin spaces, Gr F is Borel, and F is uncountable-valued. Then the following are equivalent:

(a) for some f, F is induced by (X, f) and $\hat{f}$ is a Borel isomorphism, i.e., $\hat{f}$ is a "Borel parameterization" of F;

(b) there is a conditional measure distribution ν on $T \times \mathscr{B}(X)$ such that for $t \in T$, $\nu(t, F(t)) = 1$, and $\nu(t, \cdot)$ is nonatomic;

(c) there exists a non-empty compact perfect $G(t) \subset F(t)$, for $t \in T$, such that Gr G is Borel;

(d) there exist $\phi \neq H(t) \subset F(t)$ for $t \in T$ and h such that H is induced by (X, h) and $\hat{h}$ is a Borel isomorphism.

Also, (a) is implied by

(e) F is nonmeager-valued, and X has no isolated points.

Examples as follows show that Theorem 4.5 cannot be improved in certain ways. It is also recalled in [MD2, Ex. 3.1] that classic examples (e.g., take F as the H of Example 4.7) show that we can have $T = X = I$, Gr F is Borel, F is uncountable-G_δ-valued, $\mathscr{M} = \mathscr{B}(T)$, and $\mathscr{S}(F) = \phi$.

EXAMPLE 4.6 [MD2, Ex. 3.2]. We can have $T = X = I$, Gr F is compact, F is uncountable-valued, and (a) fails in Theorem 4.5. Take Suslin $A_1, A_2 \subset I$ so that $A_1 \cup A_2 = I$ and there exist no Borel K_1, K_2 with $K_1 \subset A_1$, $K_2 \subset A_2$ and $K_1 \cup K_2 = I$. Take closed $B_1 \subset I \times [0, 1/3]$ and closed $B_2 \subset I \times [2/3, 1]$ so that $A_i = \{t : \pi_T^{-1}(t) \cap B_i \text{ is uncountable}\}$ for $i = 1, 2$. Such sets are assured by [SP2]. Let $\mathrm{Gr}\, F = B_1 \cup B_2$. It is shown in [MD2] that F has the desired properties.

EXAMPLE 4.7 (Srivastava [MD2, Ex. 3.3]). We cannot replace "X has no isolated points" with "F is G_δ-valued" in (e) of Theorem 4.5 and have (e) imply (a): Let $T = I$, $X = I \cup \{2\}$, and $\operatorname{Gr} F = \operatorname{Gr} H \cup (T \times \{2\})$, where $H = B_1 \cup B_2$, B_1 and B_2 are Borel with uncountable G_δ vertical sections, $\pi_T(B_1 \cup B_2) = T$, and $T \setminus \pi_T(B_1)$ and $T \setminus \pi_T(B_2)$ cannot be separated by a Borel set. Another example is [SVS4, Ex. 4.4.4].

The following typographical corrections to [MD2] are noted: In Theorem 1.2, change "M ... in $X \times Y$" to "M ... in B," and in the proof when defining S_n, H, and G change respectively T^n to k^n, $\cap$ to $\cup$, and $\cap$ to $\cup$. In Example 3.1 define $B = B_1 \cup B_2$. In Example 3.3 change $[0, 1/2] \times \{1\}$ to $[0, 1/2] \cup \{1\}$.

In [MD3], Mauldin discusses Theorems 4.1 and 4.5, extends Larman's [LA1, 2] work on finding a large number of disjoint Borel uniformizations of a Borel subset of $I \times \omega^\omega$, and presents some interesting history and examples.

Parameterizations when Gr F is co-Suslin are given by Cenzer and Mauldin as follows.

THEOREM 4.8 [CM1, Ths. 6.1, 6.2]. Suppose T and X are uncountable Polish spaces, X has no isolated points, Gr F is co-Suslin, and either

(i) F is nonmeager-valued, or

(ii) condition (b) of Theorem 4.5 holds.

Then

(a) F is induced by (X, f) for some f such that $\hat{f}$ is one-to-one and $\hat{f}$ and $\hat{f}^{-1}$ are $\sigma(\Sigma(\mathscr{B}(T \times X)))$-measurable;

(b) if $\mathscr{M} = \mathscr{B}(T)$, card $\mathscr{S}(F) \geq 2^{\aleph_0}$.

Ioffe [IF4] (see also [IF1] cited in [IF3]) developed comprehensive results on inducing F by a Polish space and a Carathéodory map. Fundamentally he showed in Theorem 4.9 below that such a representation holds under the hypothesis of Theorem 3.1. In Theorem 4.14 he gives necessary as well as sufficient conditions and relates such set-valued functions to those of Suslin type. An application of Theorem 4.9 includes a generalized Castaing representation, Theorem 4.11. The main results of [IF4] follow.

THEOREM 4.9 [IF4, Cor. 1.1]. Suppose X is Polish with complete metric ρ and F is closed-valued and weakly measurable. Then there exist Z with complete metric r and a Carathéodory map f such that (Z, f) induces F and such that

$$\rho(f(t, z), f(t, w)) \leq [1 + \rho(f(t, z), f(t, w))]\, r(z, w) \quad \text{for } t \in T,\ z \in Z. \tag{4.1}$$

If moreover X is compact, we may take Z to be compact.

Note that in (4.1) the r factor does not depend on t.

THEOREM 4.10 [IF4, Cor. 1.2]. Suppose T and X are Polish, $\mathscr{M} = \mathscr{B}(T)$, X is metrized by the complete metric ρ and is σ-compact, F is closed-valued, and Gr F is Borel. Then there exist a Polish Z with complete metric r and a Borel function f such that (Z, f) induces F and (4.1) holds. If moreover X is compact, we may take Z to be compact.

THEOREM 4.11 [IF4, Cor. 1.3]. Suppose F and X are as in Theorem 4.9. Then there exists a countable $Q \subset \mathscr{S}(F)$ such that $F(t) = \{f(t) : f \in \text{cl } Q\}$ for $t \in T$, where cl Q is w.r.t. the supremum metric. If X is moreover compact, then Q may be chosen precompact.

THEOREM 4.12 [IF4, Cor. 1.4]. Suppose X is a separable Banach space, F is closed-convex-valued, and $\mathscr{M} = \sigma(F^-(\mathscr{G}))$, where $\mathscr{G} = \{U : U \subset X \text{ is open}\}$. Then there exist a closed convex subset Z of a separable Banach space Y and a Carathéodory $f : T \times Y \to X$ such that $f(t, \cdot)$ is linear and non-expansive, for $t \in T$, and $(Z, f|(T \times Z))$ induces F.

In [IF2, Th. 1], Theorem 4.12 is given with X a Fréchet space. In [IF4] it is stated that "Fréchet" is valid if there is a fixed bound on the values of F, but not as given here. In [IF4, Cor. 1.4] the assumption on $\mathscr{M}$ is omitted -- we have taken the one given here from a preprint.

THEOREM 4.13 [IF4, Th. 4]. Suppose (Z, f) induces F, X and Z are Polish, and f is a Carathéodory map. Then $\text{Gr } F \in \Sigma(\mathscr{M} \otimes \mathscr{B}(X))$.

THEOREM 4.14 (IF4, Cor. 5.4]. Suppose $\mathscr{L}$ is a family of subsets of T, $\phi \in \mathscr{L}$, and $\mathscr{M} = \sigma(\Sigma(\mathscr{L}))$. Consider the following:

(i) there exist a Polish Z and a Carathéodory f such that (Z, f) induces F;

(ii) F is of weak Suslin type;

(iii) $\text{Gr } F \in \Sigma(\mathscr{M} \times \mathscr{B}(X))$.

We have:

(a) if X is Suslin, then (iii) $\Rightarrow$ (ii) $\Rightarrow$ (i);

(b) if X is Polish, $\mathscr{M} = \mathscr{L}$, and $\mathscr{M}$ is a Suslin family, then (i) $\Leftrightarrow$ (ii) $\Leftrightarrow$ (iii).

In [IF4, Cors. 5.5, 5.6], Ioffe shows that, i.a., with respect to $\mathscr{M}$ a Suslin family and X Polish, the property of F being induced by (Z, f) with Z Polish and f a Carathéodory map is preserved under countable union, intersection and cartesian product (on set values). Leese has raised the question: Is this family closed under the Suslin operation?

Srivastava's thesis [SVS4] contributes importantly to representation theory by generalizing some previous results and adding new types pertaining to G_δ-valued functions and to nonseparable X.

Of the results of [SVS4], we first give three theorems on representations of closed-valued maps. Theorem 4.15 generalizes the main content of Theorem 4.9 to the framework of Kuratowski and Ryll-Nardzewski [KRN]. It was inspired by the announcement in [IF2] (see bibliography of [IF3]) and has a proof which is relatively easy (within representation theory) using the method of [KRN]. Theorem 4.16 generalizes Theorem 4.15 to nonseparable X. The proofs of these two theorems draw heavily on Maitra and Rao [MR3] -- see § 14 below. Theorem 4.17

affords a closed map in the parameterizing coordinate, but in other respects Theorem 4.15 is stronger.

THEOREM 4.15 [SVS4, Th. 1.2.3; or SVS1, Th. 4]. Suppose $\mathscr{L}$ is a field of subsets of T, X is Polish, and F is weakly $\mathscr{L}_\sigma$-measurable and closed-valued. Then there exists f such that (ω^ω, f) induces F, $f(t, \cdot)$ is continuous for $t \in T$ and $f(\cdot, z)$ is $\mathscr{L}_\sigma$-measurable for $z \in \omega^\omega$.

THEOREM 4.16 [SVS4, Th. 1.2.6; or SVS1, Th. 8]. Suppose λ is an infinite cardinal, $\mathscr{L}$ is a λ-field (see § 14 below) of subsets of T, X is complete metric, $\mathrm{wt}\, X \leq \lambda$, and F is closed-valued and weakly $\mathscr{L}_{\lambda+}$-measurable. Then there exists f such that (λ^ω, f) induces F, $f(t, \cdot)$ is continuous for $t \in T$ and $f(\cdot, z)$ is $\mathscr{L}_{\lambda+}$-measurable for $z \in \lambda^\omega$, where λ^ω has the product of discrete topologies on λ.

THEOREM 4.17 [SVS4, Th. 1.5.3]. Suppose $\mathscr{L}$ is a field over T, X is Polish, and F is closed-valued and $\mathscr{L}$-measurable. Then there exists f such that (ω^ω, f) induces F, $f(t, \cdot)$ is continuous and closed for $t \in T$, and $f(\cdot, z)$ is $\mathscr{L}_{\delta\sigma}$- measurable for $z \in \omega^\omega$.

Ioffe in turn has asserted Theorem 4.18 (communicated in correspondence and to appear in [IF6], not yet received by me), accompanied by the ensuing corollary, and a representation version, Theorem 4.20, of Graf's Theorem 7.1 below. Theorem 4.18 in part generalizes Theorem 4.15 but not strictly, since it assumes $\mathscr{L}$ rather than $\mathscr{L}_\sigma$-measurability. Define a uniform space to be of type (λ, θ) if the cardinal λ is the weight of the space and θ is the smallest cardinal of a base of the uniform structure.

THEOREM 4.18 [IF6]. Suppose X is a complete uniform space of type (λ, θ), $\theta < \lambda$, $\mathscr{L}$ is a λ-field, and F is closed-valued and weakly $\mathscr{L}$-measurable. Then some (Z, f) induces F, where Z is complete uniform of type not exceeding $(\Sigma_{\xi<\theta} \lambda^\xi, \theta)$, $f(\cdot, z)$ is $\mathscr{L}_{\lambda+}$-measurable for $z \in Z$, and $f(t, \cdot)$ is uniformly continuous for $t \in T$, uniformly over T.

COROLLARY 4.19 [IF6]. Suppose X is a metrizable space, $\mathrm{wt}\, X \leq \lambda$, $\mathscr{L}$ is a λ-field, and F is compact-valued and weakly $\mathscr{L}$-measurable. Then there exist Z and f such that (Z, f) induces F, Z is a compact metrizable space, $f(\cdot, z)$ is $\mathscr{L}_{\lambda+}$-measurable for $z \in Z$, and $f(t, \cdot)$ is continuous for $t \in T$. This continuity is uniform in the same sense as in Theorem 4.18.

THEOREM 4.20 [IF6]. Under the hypothesis of Graf's [GF2, Th. 1], i.e., Theorem 7.1 below, there exists (Z, f) inducing F (compact-valued), where Z is a completely regular space and f is Carathéodory w.r.t. the measurability properties of Theorem 7.1.

Srivastava gives the following representations of compact-valued and σ-compact-valued maps. Theorem 4.22 generalizes the main content of Ioffe's Theorem 4.10 by omitting "F is closed-valued."

THEOREM 4.21 [SVS4, Th. 1.3.1]. Suppose X is second countable metric and F is compact-valued and weakly measurable. Then there exists a Carathéodory map f such that (C, f) induces F, where C is the Cantor set.

THEOREM 4.22 [SVS4, Th. 1.4.4; or SVS1, Ths. 2, 4]. Suppose T is Suslin, X is Polish, $\mathcal{M} \subset \mathcal{B}(T)$, F is σ-compact-valued, and Gr F $\in \mathcal{M} \otimes \mathcal{B}(X)$. Then there exist a locally compact, second countable, metrizable space Z and a Carathéodory map f such that (Z, f) induces F.

The most important results of Srivastava's [SVS4] are on representations of G_δ-valued functions. He characterizes "analytic" representations of such in Theorem 4.23. Further results, with Sarbadhikari, on finding a Borel isomorphism representation, are given in Theorem 4.24.

THEOREM 4.23 [SVS4, Th. 5.1.1; or SVS1, Th. 1.1]. Suppose T is Suslin, X is Polish, and $\mathcal{M} \subset \mathcal{B}(T)$. Then the following are equivalent:

(a) F is weakly measurable and G_δ-valued and Gr F $\in \mathcal{M} \otimes \mathcal{B}(X)$;

(b) there exists a Carathéodory map f such that (ω^ω, f) induces F and f(t, ·) is closed for t $\in$ T.

THEOREM 4.24 [SS, Ths. 3.2, 4.1, 5.1; or SVS4, Ths. 4.3.2, 4.4.1, 4.5.3]. Under the hypothesis and (a) of Theorem 4.23, we have

(a) there exists a Carathéodory map f such that (ω^ω, f) induces F and f(t, ·) is open for t $\in$ T;

(b) if F(t) has no isolated points for t $\in$ T, then there exists an $\mathcal{M} \otimes \mathcal{B}(X)$-measurable map f such that F is induced by (X, f) and f(t, ·) is a Borel isomorphism for t $\in$ T;

(c) F is of weak Suslin type, in fact there exist a weakly measurable closed-valued $G: T \to \mathcal{P}(\omega^\omega)$ and a continuous open surjective $\varphi: \omega^\omega \to X$ such that $F(t) = \varphi(G(t))$ for t $\in$ T.

Theorems 4.22, 4.23, and 4.24 are given in the cited references with the stronger assumption that $\mathcal{M}$ is countably generated and T is Polish rather than Suslin. Srivastava advises that they hold as stated here and that specifically this improvement holds in [SVS4, 1.4.4, 2.3.1, 2.3.3, 4.3.2, 4.4.1, 4.5.3, 5.1.1]. Theorem 4.24(b) relates to Mauldin's Theorem 4.5 [(e) $\Rightarrow$ (a)] above and has a corollary [SVS4, Cor. 4.4.5] which extends a result of Larman [LA1, 2; or WG3, § 10, § 12].

Very recently Srivatsa [SVT] has given a version of Theorem 4.23 [(a) $\Rightarrow$ (b)] using a hypothesis close to that of Debs' [DB2, Th.] (see Theorem 6.6 below) in place of (a) -- [SVT] has $\mathcal{L}$ a field while [DB2] has $\mathcal{L}$ a clan. Srivatsa's ensuing interesting example shows that the converse of Theorem 4.24(a) fails, answering a question in [SVS4]; if the converse held we would have a parallel to Theorem 4.23 in terms of each $f(t,\cdot)$ being an open map.

THEOREM 4.25 [SVT, Th. 4.1]. Suppose X is Polish, $\mathscr{L}$ is a field of subsets of T, F is weakly $\mathscr{L}$-measurable, and Gr F $\in (\mathscr{L} \times \mathscr{G})_{\sigma\delta}$, where $\mathscr{G} = \{U : U \subset X \text{ is open}\}$. Then there exists f such that (ω^ω, f) induces F, $f(t,\cdot)$ is continuous and open relative to F(t) for $t \in T$, and $f(\cdot, z)$ is $\mathscr{L}_\sigma$-measurable for $z \in \omega^\omega$.

EXAMPLE 4.26 [SVT, Ex. 1]. We can have T and X Polish, a G_δ-valued F induced by (ω^ω, f) with f Carathéodory, $f(t,\cdot)$ open for $t \in T$, and Gr F $\notin \mathscr{M} \otimes \mathscr{B}(X)$: Let T and X be uncountable and Polish, $\mathscr{M} = \mathscr{B}(T)$, $x_0 \in X$, $A \subset T$ be non-Borel Suslin, and $C: A \to \mathscr{P}(\omega^\omega)$ with Gr C closed. Let F(t) = X for $t \in A$ and $F(t) = X \setminus \{x_0\}$ for $t \in X \setminus A$. Also let

$$K(t) = \begin{cases} [(X\setminus\{x_0\}) \times \omega^\omega] \cup [\{x_0\} \times C(t)] & \text{for } t \in A \\ (X\setminus\{x_0\}) \times \omega^\omega & \text{for } t \in T\setminus A. \end{cases}$$

In [SVT] it is shown that Theorem 4.23 [(a) $\Rightarrow$ (b)] above applies to K, and that taking f as the inducing map thus obtained followed by projection to X, f has the desired properties.

REMARK 4.27. Srivastava has conjectured (verbal communication) that in Theorem 4.24(b) we may obtain each $f(t,\cdot)$ additionally to be continuous, which would be a Bourgain-Ioffe type of representation with no measure present. He poses this as a very interesting and difficult problem.

We complete this section with Graf's parameterization of measurable weak sections (see § 8) via the set of extremal preimage measures. Edgar [ED] had previously used the same identification, under continuous p, without asserting measurability of the identification. An earlier use of E was by Ershov (Yershov) [ER2] for a different purpose, in measure extension problems. Contemporaneous work of Ershov (which we have not seen) is relevant [GF3].

THEOREM 4.28 [GF3, Cor. 1 to Th. 4]. Suppose $0 < \mu(T) < \infty$, $\mathscr{M}$ is countably generated, $\{t\} \in \mathscr{M}$ for $t \in T$, X is Suslin, $p: X \to T$ is $\mathscr{B}(X)$-$\mathscr{M}$-measurable and surjective, E is the set of extreme points in $\{\nu : \nu$ is a non-negative measure on $\mathscr{B}(X)$ and $\mu = p(\nu)\}$, $E \neq \phi$, $\mathscr{E} = \sigma(\{\{\nu : \nu \in E$ and $\nu(B) \in A\} : A \in \mathscr{B}(R),\ B \in \mathscr{B}(X)\})$, $\mathscr{M}_\mu$ is the μ-completion of $\mathscr{M}$, $\mathscr{M}_0$ is a σ-algebra, $\mathscr{M} \subset \mathscr{M}_0 \subset \mathscr{M}_\mu$, and there exists an $\mathscr{M}_0$-measurable selection of p^{-1}. Then there exists an $(\mathscr{M}_0 \otimes \mathscr{E})$-measurable $g: T \times E \to X$ such that

(a) for $\nu \in E$, $g(\cdot,\nu)$ is an $\mathscr{M}_0$-measurable selection of p^{-1} and $g(\cdot,\nu)(\mu) = \nu$;

(b) whenever $f: T \to X$ is an $\mathscr{M}_\mu$-measurable weak section of p (see § 8), there exists $\nu \in E$ with $f = g(\cdot,\nu)$, μ-a.e.

Theorem 4.28 is given in more generality as [GF3, Th. 4]. Examples where the condition on

$\mathcal{M}_0$ is satisfied are noted after [GF3, Cor. to Th. 3]. It should be noted that in Theorem 4.28 the parametrized sections need not fill Gr p^{-1}, i.e., we do not have $g(t, E) = p^{-1}(t)$ for $t \in T$. However, this result has the remarkable property that it parameterizes not only all, modulo μ, of $\mathcal{S}(p^{-1})$ but also all, modulo μ, of the larger family of $\mathcal{M}_\mu$-measurable weak sections of p. This raises questions of relating properties of subsets of $\mathcal{S}(p^{-1})$, and more generally properties of subsets of the set of $\mathcal{M}_\mu$-measurable weak sections, to properties of subsets of extremal preimage measures. Some results in this direction have been given by Graf.

We summarize the most important "iff" results on representations as follows:

(1) When T and X are uncountable Polish, Gr F is Borel, and F is uncountable-valued, we have that F has a Borel parameterization precisely when F has a compact-perfect-valued subfunction with Borel graph, i.e., precisely when $T \times \mathcal{B}(X)$ has a conditional measure distribution which is pointwise a nonatomic probability measure (Mauldin, Theorem 4.5).

(2) When X is Polish and $\mathcal{M}$ is a Suslin family, we have that F has an analytic representation (Ioffe's term) precisely when F is of weak Suslin type (Ioffe, Theorem 4.14), i.e., precisely when Gr $F \in \Sigma(\mathcal{M} \times \mathcal{B}(X))$ (Leese [LE2]).

(3) When T and X are Polish and $\mathcal{M} \subset \mathcal{B}(T)$, we have that F has an analytic representation by some f with each $f(t, \cdot)$ a closed map precisely when F is G_δ-valued and weakly measurable and Gr $F \in \mathcal{M} \otimes \mathcal{B}(X)$ (Srivastava, Theorem 4.23).

Finally we note that among the foregoing results, the following have representations which are one-to-one in the second (i.e., parameterization) coordinate and for that coordinate we note mapping properties: Theorems 4.1 (Cenzer and Mauldin -- Borel or analytic (i.e., $\sigma(\Sigma(\mathcal{B}(X)))$ measurability), 4.2 (Cenzer and Mauldin -- $\mathcal{C}(I)$-measurability), 4.3 (Bourgain, Ioffe -- continuity), 4.5 (Mauldin -- Borel measurability), 4.8 (Cenzer and Mauldin -- analytic measurability), and 4.24(b) (Sarbadhikari and Srivastava -- Borel measurability).

5. Partitions. In this section we suppose $\mathcal{Q}$ is a partition of T. We let X = T and let F be given by $t \in F(t) \in \mathcal{Q}$ for $t \in T$. In this context, $F^-(A)$ is often called the $\mathcal{Q}$-saturation of A for $A \subset T$. We define $\hat{\mathcal{S}}(F)$ to be the set of selections of F which are constant on each member of $\mathcal{Q}$ (without requiring measurability). Thus $\mathcal{S}(F) \cap \hat{\mathcal{S}}(F)$ is the set of selections of $\mathcal{Q}$ which are measurable maps. The newer partition results reviewed here are largely concerned with whether $\mathcal{S}(F) \cap \hat{\mathcal{S}}(F) \neq \phi$. In [WG3, § 11], results are generally in terms of finding $f \in \mathcal{S}(F)$ whose range (there called the selection and often called a transversal) is a measurable set. (Our description of the Kallman-Mauldin result in addendum (ix) in proof to [WG3] should have more explicitly asserted $\mathcal{S}(F) \cap \hat{\mathcal{S}}(F) \neq \phi$, although that is loosely implied by the beginning of § 11.)

The principal recent results on measurable selections of partitions have been by Miller, Burgess, Srivastava, and Ershov. Miller's main results are the following two theorems. For α an ordinal, a function g is α-Borel {Borel of ambiguous class α} if for open U, $g^{-1}(U)$ is Borel of additive class α {Borel of ambiguous class α}.

THEOREM 5.1 [MI3, Th. A]. Suppose G is a Polish topological group acting continuously on a Polish space Y, $T \subset Y$ is an invariant Borel set of ambiguous class $\alpha \geq 1$, $\mathcal{Q} = \{T \cap Gy : y \in Y\}$, and $\mathcal{Q}$ is countably separated by projections from T to $\mathcal{Q}$ which are Borel of ambiguous class α. Then there exists an α-Borel function $f \in \hat{\mathcal{S}}(F)$.

THEOREM 5.2 [ML3, Th. B]. Suppose T is Polish, F is G_δ-valued, $F^-(U)$ is Borel of ambiguous class $\gamma \geq 0$ for each basic open $U \subset T$, and $\alpha = \sup\{\gamma + \beta : \beta < \gamma\}$. Then there exists an α-Borel function $f \in \hat{\mathcal{S}}(F)$.

Theorems 5.1 and 5.2 are milestones in two chains of results. Both theorems rely on Miller's Theorem 13.2 below. An additional common link in these two chains to be described is Burgess' Theorem 5.3 below.

Without Borel classifications, Theorem 5.1 was obtained by Burgess [BS3]. By replacing measure-theoretic arguments by category-theoretic arguments, he extended to action by a Polish group (without compactness assumptions) a result of Effros [EF] on action by a locally compact group, which in turn enlarged on Mackey [MC2], who obtained an a.e. Borel selection. (The paper [EF] was referenced on page 884 of [WG3], but was omitted from the bibliography.) A complicated generalization of Theorem 5.1, with Borel action rather than continuous action, is given as [MI3, Th. 3.2]. Topological group applications of selection results in [DI] and [RN] are given in [ML1].

Theorem 5.2 pertains to a chain of results on selections of partitions into G_δ sets. This topic of course interrelates with that of G_δ-valued functions in general, reviewed in § 6. As noted there, G_δ-valued hypotheses seldom appear in [WG3] and often appear herein, although much prior work was with closed-valued F, including selections of partitions.

Probably the first result on measurable selection of a partition into G_δ's as such was in Miller's [ML2], which gave Theorem 5.2 under $\gamma = 1$. Earlier special cases of Theorem 5.2 were given by Kallman and Mauldin [KLM] and Kuratowski and Maitra [KMT]. Srivastava [SVS4, Th. 3.2.3] gave Theorem 5.2 without Borel classifications. Shortly thereafter, the latter was generalized by Miller to Theorem 5.2. His methods were used by Burgess to apply the latter's important result on countably generated partitions, Theorem 5.3 below, to obtain Theorem 5.4 on G_δ-valued F. Miller has also generalized the G_δ-valued F in Theorem 5.2 to a Baire-space-valued F [ML3, Th. 3.4].

A great deal of the work of Burgess and Miller on selections draws heavily on Baire category theory of Vaught. Burgess' principal partition results are from [BS5] given next. Burgess observes [BS2, § 1.2] that in Theorem 5.3 one cannot change "$\mathcal{M} = \mathcal{C}(T)$" (see § 2 for definition) to "$\mathcal{M} = \mathcal{B}(T)$": Let T be Polish, $f: T \to T$ be continuous with non-Borel range and $\mathcal{Q} = \{f^{-1}(t) : t \in T\}$ (example from [MR2]). However, one can use instead of $\mathcal{C}(T)$, the successively larger families, Blackwell's Borel programmable sets [BS2, Th. I], Kolmogorov's R-sets [BS2, § 1.5], Solovay's strongly Δ_2^1 sets [BS2, Th. III], or the Baire property sets [BS2, § 1.6]. One can also use $\mathcal{M}$ whenever μ is σ-finite, complete, and regular Borel [BS2, § 1.7]. The strongly Δ_2^1 (also called absolutely Δ_2^1) sets have been metamathematically defined by Solovay, who showed that they are universally measurable.

THEOREM 5.3 [BS5, Prop. 3]. *Suppose* T *is Polish,* $\mathcal{M} = \mathcal{C}(T)$, *and* $\mathcal{Q}$ *is generated by a countable subset of* $\mathcal{M}$. *Then* $\mathcal{S}(F) \cap \hat{\mathcal{S}}(F) \neq \phi$.

THEOREM 5.4 [BS5, Prop. 5]. *Suppose* T *is Polish,* $\mathcal{M} = \mathcal{C}(T)$, F *is* G_δ *-valued, and* F *is weakly measurable. Then* $\mathcal{S}(F) \cap \hat{\mathcal{S}}(F) \neq \phi$.

The following lemma of Burgess, en route to proving Theorem 5.3, follows Kaniewski's [KA1] generalization of Kondō's theorem.

THEOREM 5.5 [BS5]. *Suppose the equivalence relation given by* $\mathcal{Q}$ *is the restriction to* $T \times T$ *of a Suslin subset of* $Y \times Y$, *with* Y *Polish and* T *a co-Suslin subspace of* Y. *Suppose* F *is closed-valued and* $\mathcal{M} = \sigma(\Sigma(\mathcal{B}(T)))$. *Then there exists* $f \in \mathcal{S}(F) \cap \hat{\mathcal{S}}(F)$ *such that range* f *is co-Suslin.*

Kaniewski [KA1] obtains $f \in \hat{\mathcal{S}}(F)$ with range f co-Suslin. The lemma statement in [BS5] has f in $\hat{\mathcal{S}}(F)$ and a $\mathcal{C}(T)$-measurable map, that being what Burgess applies. However, the proof in [BS5], which uses the construction of [KA1], yields the conclusion stated in Theorem 5.5. Theorem 5.5 is probably the first measurable selection result derived from Kondō's uniformization of a co-Suslin set. The paper [BS4] is superseded by [BS3, 5].

While it probably does not come under the heading of *measurable* selection theory, we note the following 1975 result of Burgess and Miller, referring to [BM] for the definition of Σ_k^1 strong well-ordering. Related results are given in [BS1]. Miller advises that Kuratowski treated the Σ_2^1 case in a 1972 Berkeley colloquium.

THEOREM 5.6 [BM, Th. 1.6(a)]. *Assume there exists a* Σ_k^1 *strong well-ordering of* ω^ω. *Suppose* T *is a Polish space, and* $\mathcal{E}$ *is an equivalence relation over* T *inducing* $\mathcal{Q}$. *Then there exists* $f \in \hat{\mathcal{S}}(F)$ *such that* f *is a* Σ_k^1 *subset of* $\mathcal{E}$.

Following are two partition results of Srivastava's, with the members of $\mathcal{Q}$ respectively σ-compact and G_δ in a larger space.

THEOREM 5.7 [SVS4, Th. 3.3.1 and Proof]. *Suppose* T *is Polish,* F *is* σ*-compact-valued, and* $\mathcal{M} = \mathcal{B}(T) \cap \{F^-(A) : A \subset T\}$ (*whence* $\mathcal{S}(F) \subset \hat{\mathcal{S}}(F)$). *Then the following are equivalent:*

(a) $\mathcal{M}$ is countably generated;

(b) the equivalence relation associated with $\mathcal{Q}$ is in $\mathcal{M} \otimes \mathcal{B}(T)$;

(c) the equivalence relation associated with $\mathcal{Q}$ is Suslin in $T \times T$ and there exists $f \in \hat{\mathcal{S}}(F)$ such that range f is Borel;

(d) $\mathcal{S}(F) \neq \phi$.

THEOREM 5.8 [SVS4, Cor. 3.2.5]. Suppose T is a Borel subset of a Polish space Y, F is weakly measurable and each $F(t)$ is a G_δ in Y. Then $\mathcal{S}(F) \cap \hat{\mathcal{S}}(F) \neq \phi$.

Srivastava [SVS4, p. 44] adapts an example of [KLM] to show that in Theorem 5.8 we cannot replace "each $F(t)$ is a G_δ in Y" by "each $F(t)$ is closed in T and F is lsc."

Ershov (Yershov) [ER4] has given a very abstract and rather general result, Theorem 5.13 below, on selections of partitions. His [ER4] supersedes [ER3]. He describes his technique as a modification of that of Sion [SN]. In presentation at Oberwolfach, he advised that the main result of [SN] can be obtained from Theorem 5.13. Unfortunately, this and other applications to topologized cases are not given in [ER4]; they should make an interesting future publication. It is noted in [ER4] that Theorem 5.13 was motivated by work on continuous selections, notably [HS] with reference also to [GF2], and that there is a relationship between Theorem 5.10 and Hoffman-Jørgensen's [HJ, Th. II.6.1; or WG3, Th. 11], which is an earlier non-topological partition selection theorem.

CONVENTION 5.9. Through Theorem 5.13 we adopt the following. Topology on T (= X) is irrelevant. We denote $2^A = \{D : D \subset A\}$. For $\mathcal{L} \subset 2^A$, we denote $\cup \mathcal{L} = \cup_{D \in \mathcal{L}} D$, similarly for $\cap$, and we say $\mathcal{L}$ is compact if whenever $\mathcal{L}' \subset \mathcal{L}$ and $\cap \mathcal{L}' = \phi$, there exists a finite $\mathcal{L}'' \subset \mathcal{L}'$ such that $\cap \mathcal{L}'' = \phi$. We fix $\mathcal{H} \subset 2^T$ such that $\{A \cap H : H \in \mathcal{H}\}$ is compact for $A \in \mathcal{Q}$ and the partition generated by $\mathcal{Q} \cup \mathcal{H}$ consists of singletons. We denote $\eta = \text{card}\, \mathcal{H}$.

THEOREM 5.10 [ER4, Th. 1]. Suppose $\mathcal{R} \subset 2^T$, $\cap \mathcal{L} \in \mathcal{R}$ and $\cup \mathcal{L} \in \mathcal{R}$ whenever $\mathcal{L} \subset \mathcal{R}$ and card $\mathcal{L} < \eta$, $\{T \setminus H : H \in \mathcal{H}\} \subset \mathcal{R}$, and $F^-(\cap \mathcal{K}) \in \mathcal{R}$ for finite $\mathcal{K} \subset \mathcal{H}$. Then there exists $f \in \hat{\mathcal{S}}(F)$ and $\mathcal{K}_0 \subset \mathcal{R}$ such that card $\mathcal{K}_0 \leq \eta$ and $T \setminus$ range $f = \cup \mathcal{K}_0$.

THEOREM 5.11 [ER4, Th. 2]. Suppose $\eta = \aleph_0 \{\eta = \aleph_1\}$ and let $\mathcal{L} = \{F^-(\cap \mathcal{K}) : \mathcal{K} \subset \mathcal{H}$ is finite$\}$. Then there exists $f \in \hat{\mathcal{S}}(F)$ such that for $H \in \mathcal{H}$, $f^{-1}(H)$ is in the Boolean algebra (i.e., field) generated by $\mathcal{L}$ $\{$is in $\sigma(\mathcal{L})\}$.

DEFINITION 5.12. Let ζ be a cardinal. We say $\tau : 2^{2^T} \to 2^{2^T}$ is a ζ-operation if

(i) $f^{-1}(\tau(\mathcal{L})) \subset \tau(f^{-1}(\mathcal{L}))$ whenever $f : T \to T$ and $\mathcal{L} \subset 2^T$;

(ii) if $\mathcal{L}, \mathcal{L}' \subset 2^T$, $\tau(\mathcal{L}) \subset \tau(\mathcal{L}')$ and $L \in \tau(\mathcal{L}')$, then $\tau(\mathcal{L} \cup \{L\}) \subset \tau(\mathcal{L}')$;

(iii) whenever ξ is an initial ordinal, card $\xi \leq \zeta$, $\mathcal{L} \subset 2^T$, $\mathcal{L}_\alpha \subset \mathcal{L}_\beta \subset 2^T$ for $\alpha \leq \beta < \xi$, and $\tau(\mathcal{L}_\alpha) \subset \tau(\mathcal{L})$ for $\alpha < \xi$, we have $\tau(\cup_{\alpha<\xi} \mathcal{L}_\alpha) \subset \tau(\mathcal{L})$.

As examples of ξ-operations for infinite ξ, [ER4] mentions closure with respect to such usual set-theoretic operations as $\setminus, \subset, \cup, \Sigma, \sigma$, completion of a σ-algebra, and generalizations of these.

THEOREM 5.13 [ER4, Th. 3]. Suppose $\mathscr{R} \subset 2^T$, $\mathscr{R} = F^-(\mathscr{R})$, τ is an η-operation, and for $H \in \mathscr{H}$ and $\mathscr{H}' \subset \mathscr{H}$ with card $\mathscr{H}' < \eta$ we have that the saturation of H w.r.t. the partition generated by $\mathscr{H}' \cup \mathscr{Q}$ belongs to $\tau(\mathscr{R} \cup \mathscr{H}')$. Then there exists $f \in \hat{\mathscr{S}}(F)$ such that

$$\tau(f^{-1}(\mathscr{R} \cup \mathscr{H})) \subset \tau(\mathscr{R}).$$

We now drop Convention 5.9.

Maitra and B. V. Rao [MR2] have shown that if T is Polish, F is closed-valued, $\alpha > 0$, and $F^-(U)$ is Borel of additive class α for open $U \subset T$, then there is $f \in \hat{\mathscr{S}}(F)$ so that range f is of multiplicative class α. Sarbadhikari [SR3] has weakened "Polish" to "complete metric."

Dellacherie gives the following in the course of a general treatise on the theory of sets which are analytic in the sense of Sion [SN, or WG3, p. 872] -- he bases his treatment on an equivalent definition in terms of capacity theory.

THEOREM 5.14 [DC4, Th. II.29]. Suppose T is an analytic space in the sense of Sion, $\mathscr{M} = \{F^-(B) : B \subset T \text{ is Borel}\}$, and $\mathscr{M}$ is separable. Then there exists $f \in \hat{\mathscr{S}}(F)$ which is a universally measurable function.

We close this section on an historical note. The following partition selection theorem was given by Bourbaki [BO3] in 1958. As noted in [WG3, § 4], Dixmier [DI] gave the same result in 1962 under weak measurability of F.

THEOREM 5.15 [BO3]. Suppose T is Polish, $\mathscr{M} = \mathscr{B}(T)$, and F is closed-valued and measurable. Then there exists $f \in \hat{\mathscr{S}}(F)$ such that range f is Borel (i.e., $\mathscr{S}(F) \cap \hat{\mathscr{S}}(F) \neq \phi$).

6. G_δ-valued functions. The main results on measurable selections of G_δ-valued functions come from Srivastava [SVS4, Ch. 2; SVS3], Burgess [BS5, 6, 7], Miller [MI2, 3], Debs [DB2], and Srivatsa [SVT]. The papers [MI2, 3; BS5] pertain to partitions and are reviewed in § 5. They also supersede [SVS3, Th. 5.1; or SVS4, Th. 3.2.3] on partitions into G_δ's. In [WG3] the sole results cited on G_δ-valued F as such were by Aronszayn (see [SV]), who in 1964 found a measurable implicit function under constant G_δ-valued F, and by Larman [LA1, 2], who found an uncountable disjoint family of Borel selections of a σ-compact-G_δ-valued F.

Following is a corollary of Theorem 4.23.

THEOREM 6.1 [SVS4, Th. 2.3.2, Remark and Cor. 2.3.3; or SVS3, Th. 4.2]. Suppose T is a Suslin space, X is Polish, $\mathscr{M} \subset \mathscr{B}(T)$, F is G_δ-valued and weakly measurable and $\text{Gr } F \in \mathscr{M} \otimes \mathscr{B}(X)$. Then F has a Castaing representation.

Srivastava has shown in [SVS3, 4] that "$\text{Gr } F \in \mathscr{M} \otimes \mathscr{B}(X)$" cannot be dropped from

Theorem 6.1. Recently, Srivatsa went farther in showing by Example 6.2 given next that this condition cannot be replaced by "Gr F is Suslin." However, very recently Burgess has shown, Theorem 6.3 (a) below, that a replacement by "Gr F is co-Suslin" is valid provided one assumes $\mathcal{M} = \mathcal{B}(T)$ and is content to conclude $\mathcal{S}(F) \neq \phi$. Also, "F is G_δ-valued" is relaxed to "each F(t) is nonmeager in cl F(t)," thereby generalizing Theorem 12.5(iii) below of Sarbadhikari [SR]. In the wake of Theorem 6.3(a), Burgess found Theorem 6.3(b) which changes "nonmeager" to "comeager" and obtains a Castaing representation of F. Srivastava has at the same time shown by Example 6.4 that in Theorem 6.3(a) we cannot reach the latter conclusion and by Example 6.5 that in Theorem 6.3 we cannot relax "$\mathcal{M} = \mathcal{B}(T)$" to "$\mathcal{M} \subset \mathcal{B}(T)$ and is countably generated" as is done in Theorem 6.1. Theorem 6.3(a) also considerably strengthens Debs' [DB2, Cor. 6] (a consequence of Theorem 6.6 below), where it is assumed that F is $(F_\sigma \cap G_\delta)$-valued. It appears that in 6.1 through 6.5, Srivastava, Burgess, and Srivatsa have achieved a very well rounded set of results on selections of a G_δ-valued function.

EXAMPLE 6.2 [SVT, Ex. 2]. We can have $T = X = \omega^\omega$, $\mathcal{M} = \mathcal{B}(T)$, F is G_δ-valued, Gr F is Suslin, and $\mathcal{S}(F) = \phi$: Take $P \subset \omega^\omega$ and $C : P \to \mathcal{P}(\omega^\omega \times \omega^\omega)$ such that Gr C is co-Suslin and $\{C(t) : t \in P\} = \{Q : Q \subset \omega^\omega \times \omega^\omega$ is co-Suslin$\}$. Let $D(t) = \{z : (t, z) \in C(t)\}$ for t such that this is non-empty. Then Gr D is co-Suslin and Kondō's theorem [KO or WG3, § 12] gives us a selection d of D such that d is a co-Suslin subset of $\omega^\omega \times \omega^\omega$. Let $F(t) = \omega^\omega \times \omega^\omega$ for $t \notin$ domain D and $F(t) = (\omega^\omega \times \omega^\omega) \setminus \{(t, d(t))\}$ for $t \in$ domain D. Then F has the desired properties as is shown in [SVT].

THEOREM 6.3 [BS7]. Suppose T is a Suslin space, $\mathcal{M} = \mathcal{B}(T)$, X is Polish, F is weakly measurable, and Gr F is co-Suslin. Then

(a) if F(t) is nonmeager in cl F(t) for $t \in T$, then $\mathcal{S}(F) \neq \phi$;

(b) if F(t) is comeager in cl F(t) for $t \in T$, then F has a Castaing representation.

EXAMPLE 6.4 (Srivastava [BS7]). We cannot conclude in Theorem 6.3(a) that F has a Castaing representation: Take $T = X = [0, 1]$, $\mathcal{M} = \mathcal{B}(T)$, and $H : T \to \mathcal{P}(X)$ such that Gr H is a Borel subset of $T \times (\omega^\omega \cap [0, \frac{1}{2}])$, H(t) is dense and an F_σ in $\omega^\omega \cap [0, \frac{1}{2})$ for $t \in T$, and $\mathcal{S}(H) = \phi$. This follows Kallman and Mauldin [KLM, Ex. 8]. Let $F(t) = H(t) \cup \{3/4\}$ for $t \in T$. Suppose $\{f_1, f_2, \dots\}$ is a Castaing representation of F. Let

$$T_1 = f_1^{-1}([0, \tfrac{1}{2})),$$

$$T_n = f_n^{-1}([0, \tfrac{1}{2})) \setminus \bigcup_{i<n} T_i \qquad \text{for } n = 2, 3, \dots,$$

and $f(t) = f_n(t)$ whenever $t \in T_n$, $0 < n \in \omega$. Then $f \in \mathcal{S}(H)$ in contradiction.

EXAMPLE 6.5 (Srivastava [BS7]). We can have $T = X = [0, 1]$, $\mathcal{M}$ is a countably generated, $\mathcal{M} \subset \mathcal{B}(T)$, F is weakly measurable and G_δ-valued, Gr F is open, and $\mathcal{S}(F) = \phi$: Let

$\mathcal{M} = \{\phi, T\}$ and $F(t) = X \setminus \{t\}$ for $t \in T$.

Debs' main theorem of [DB2] is given next, followed by two corollaries. A clan is a family of sets which is closed under differences and finite union.

THEOREM 6.6 [DB2]. Suppose $\mathcal{L}$ is a clan of subsets of T, X is Polish, $\mathcal{G} = \{U : U \subset X$ is open$\}$, Gr F $\in (\mathcal{M} \times \mathcal{G})_{\sigma\delta}$, and F is weakly $\mathcal{L}_\sigma$-measurable. Then F has an $\mathcal{L}_\sigma$-measurable selection.

COROLLARY 6.7 [DB2, Cor. 1]. Suppose $\mathcal{L}$ is a clan of subsets of T, X is Polish, and F is closed-valued and weakly $\mathcal{L}_\sigma$-measurable. Then F has an $\mathcal{L}_\sigma$-measurable selection.

COROLLARY 6.8 [DB2, Cor. 3]. Suppose T is a metric space, X is Polish, Gr F is a G_δ, $\alpha > 0$ is an ordinal, and $F^-(U)$ is Borel of class α for open $U \subset X$. Then F has a selection which is a Borel function of class α.

In [DB, Cor. 3], metrizability of T is omitted and $\alpha = 0$ is permitted. Srivastava asserts that the changes here are needed. He further advises that in [DB2, Cor. 4] (a partition selection statement), the selection obtained in the proof is not constant on equivalence classes.

Srivatsa has recently given the following with hypothesis somewhat similar to that of Debs' Theorem 6.6 above.

THEOREM 6.9 [SVT, Th. 3.1]. Suppose X is Polish, $\mathcal{L}$ is a family of subsets of T closed under finite intersection and countable union and satisfying the weak reduction principle [see § 14], $\phi \in \mathcal{L}$, $T \in \mathcal{L}$, and Gr F $\in (\mathcal{L} \times \mathcal{G})_{\sigma\delta}$, where $\mathcal{G} = \{U : U \subset X$ is open$\}$. Let $\mathcal{H} = \{A : A \in \mathcal{L}$ and $T \setminus A \in \mathcal{L}\}$. Then F has an $\mathcal{H}_\sigma$-measurable selection.

Also recently, Burgess has obtained the following two theorems, with G_δ-valued F, as part of a heirarchal study. Theorem 6.11 is a $\mathcal{C}(T)$ parallel to much of Srivastava's Theorem 6.1 above on $\mathcal{B}(T)$. A key concept is a "uniform" family of sets. Uniformity affords preservation of measurability of functions under composition. Intuitively, I find this concept appealing, having encountered so much in measurable selection theory where such preservation is important. Burgess observes that if T is Polish, $\mathcal{C}(T)$ is the smallest σ-algebra $\mathcal{N}$ such that $\mathcal{N} \supset \Sigma(\mathcal{B}(T))$ and $g \circ h$ is $\mathcal{N}$-measurable whenever $g : T \to T$ and $h : T \to T$ are $\mathcal{N}$-measurable. In proving Theorem 6.12, Burgess uses considerable ad hoc machinery, drawing heavily on category-theoretic work of Vaught, Moschovakis, Kechris, i.a.

DEFINITION 6.10 [BS6, § 2]. Suppose T is Polish, $\mathcal{M} \supset \mathcal{B}(T)$, and $g^{-1}(A) \in \mathcal{M}$ for $A \in \mathcal{M}$ whenever $g : T \to T$ is measurable (i.e., $g^{-1}(\mathcal{M}) \subset \mathcal{M}$ whenever $g^{-1}(\mathcal{B}(T)) \subset \mathcal{M}$). We then say that $\mathcal{M}$ is uniform.

THEOREM 6.11 [BS6, § 8(d)]. Suppose T is an uncountable Polish space, T = X, $\mathcal{M}$ is uniform, F is $(G_\delta \cap F_\sigma)$-valued and weakly measurable, and Gr F $\in \mathcal{M} \otimes \mathcal{M}$. Then there exists $f \in \mathcal{S}(F)$ such that $f(t) = f(t')$ whenever $t, t' \in T$ and $F(t) = F(t')$.

THEOREM 6.12 [BS6, § 8(f)]. Suppose T and X are uncountable Polish spaces, $\mathcal{M} = \mathcal{C}(T)$, F is G_δ-valued and weakly measurable, and $\operatorname{Gr} F \in \mathcal{C}(T \times X)$. Then $\mathcal{S}(F) \neq \phi$.

In [BS6, § 8] it is asserted that counterexamples show that none of the properties of F may be omitted from Theorem 6.12.

7. Compact-valued functions. The principal new work on selections of compact-valued functions has been by Graf [GF1, 2], Losert [LO1, 2], and Talagrand [TA3, 4, 5].

Graf's main result in [GF2], Theorem 7.1 next, has complicated hypotheses (mainly avoiding metrisability of X and restrictions on wt X) and conclusions. However, he deduces quite a few consequences, notably the subsequent results taken from [GF2]. Included are two consequences on closed-valued F, Theorems 7.3 and 7.6.

The function Φ in the next theorem is an abstraction of the map F^-. In Theorem 13.2 we see an abstraction of F^- in a very similar way, by Miller.

THEOREM 7.1 [GF2, Th. 1]. Suppose X is regular Hausdorff, F is compact-valued, $\mathcal{L}$ is a field of subsets of T, and there exists a map Φ on the set of closed subsets of X to $\mathcal{L}$ such that $\Phi(\phi) = \phi$, $\Phi(X) = T$, and $\Phi(A \cup B) = \Phi(A) \cup \Phi(B)$ and $\Phi(A) \subset F^-(A)$ for closed $A, B \subset X$. Suppose the topology of X has a base $\mathcal{D}$ such that $\mathcal{H}$ has a supremum in $\mathcal{L}$ whenever $\mathcal{H} \subset \mathcal{L}$ and $\operatorname{card} \mathcal{H} < \operatorname{card} \mathcal{D}$. Then there exists a selection f of F such that for open $U \subset X$, $f^{-1}(U) = \bigcup \mathcal{K}$ (see 5.9) where $\operatorname{card} \mathcal{K} \leq \operatorname{card} \mathcal{J}$ whenever $\mathcal{J} \subset \mathcal{D}$ and $\operatorname{cl} J \subset \bigcup \mathcal{J} = U$ for $J \in \mathcal{J}$.

One consequence of Theorem 7.1 noted in [GF2] is Sion's [SN, Th. 4.1]. Others follow.

THEOREM 7.2 [GF2, Th. 3]. Suppose X is compact Hausdorff, $\operatorname{wt} X \leq \aleph_1$, and F is compact-valued and measurable. Then F has a $\mathcal{B}_0(X)$-measurable selection.

It is stated in [GF2] that Theorem 7.2 was proved by Talagrand (unpublished) for X metric instead of compact. Upon reviewing a preliminary draft of this paper at Oberwolfach, Talagrand denied this (Graf's information was second hand). However, to keep us honest he proceeded to prove it and improved $\aleph_1$ to $2^{\aleph_0}$ (still unpublished). Leese [LE5, Th. 4.2; or WG3, Th. 4.10(ii)] achieves $\mathcal{S}(F) \neq \phi$ with the same condition on F and the sole condition on X that $\mathcal{B}(X)$ be generated by a family of cardinality at most $\aleph_1$ consisting of closed sets.

THEOREM 7.3 [GF2, Cor. 1 to Th. 3]. Suppose X is completely regular Hausdorff, F is closed-valued and measurable, and $T = \bigcup_{i \in \omega} F^-(K_i)$ with $K_i \subset X$ compact and $\operatorname{wt} K_i \leq \aleph_1$ for $i \in \omega$. Then F has a $\mathcal{B}_0(X)$-measurable selection.

DEFINITION 7.4. Suppose T is topologized and $\mathcal{M} \supset \mathcal{B}(T)$. Let $n = \mathcal{M} \cap \{A : [B \in \mathcal{M}$ and $\mu(B) < \infty]$ implies $\mu(A \cap B) = 0\}$.

We say l is a lifting of $(T, \mathcal{M}, \mu)$ if $l : \mathcal{M} \to \mathcal{M}$ is a Boolean homomorphism such that

(i) [A, B $\in \mathcal{M}$ and A $\triangle$ B $\in$ n] implies l(A) = l(B);

(ii) l(A) $\triangle$ A $\in$ n for A $\in \mathcal{M}$.

If also the following holds, l is a strong lifting:

(iii) l(A) $\subset$ A for closed A $\subset$ T, equivalently U $\subset$ l(U) for open U $\subset$ T.

The following theorem is deduced in [GF1] from a continuous selection theorem of Hasumi [HS]. Replacing the conclusion by $\mathcal{S}(F) \neq \phi$, it is given as [GF2, Th. 5], which follows from Theorem 7.1 above. Graf notes in [GF2] that Theorem 7.5 (from [GF1], written earlier) had been independently proved by J. P. R. Christensen and also by M. Talagrand.

THEOREM 7.5 [GF1, Th. 3.2]. Suppose T is topologized, μ is complete, $0 < \mu(T) < \infty$, $\mathcal{M} \supset \mathcal{B}(T)$, (T, $\mathcal{M}$, μ) admits a strong lifting, F is compact-valued and usc, and X is regular Hausdorff. Then there exists f $\in \mathcal{S}(F)$ such that f(μ) is regular Borel on $\mathcal{B}(X)$.

It is noted in [GF2] that if $\mathcal{M} \supset \mathcal{B}(T)$ and μ is complete, then (T, $\mathcal{M}$, μ) admits a strong lifting in each of the following cases:

(a) T is second countable and μ is σ-finite (Graf);

(b) T has topological weight at most $\aleph_1$ and μ is Radon (Fremlin, under continuum hypothesis);

(c) T is locally compact metric and μ is Radon (A. and C. Ionescu-Tulcea);

(d) T is a locally compact group and μ is a Haar measure (A. and C. Ionescu-Tulcea).

THEOREM 7.6 [GF2, Cor. 2 to Th. 5]. Suppose (T, $\mathcal{M}$, μ) is as in Theorem 7.5, X is Hausdorff, F is closed-valued and usc, and there exist compact $K_1, K_2, \ldots \subset X$ with $T \setminus \bigcup_{i=1}^{\infty} F^-(K_i)$ a μ-null set. Then $\mathcal{S}(F) \neq \phi$.

THEOREM 7.7 [GF2, Th. 6]. Suppose T is a Baire topological space, X is completely regular Hausdorff, F is compact-valued and usc, and $\mathcal{M}$ is the σ-algebra of Baire property subsets of T. Then F has a $\mathcal{B}_0(X)$-measurable selection. If also T is second countable, $\mathcal{S}(F) \neq \phi$.

THEOREM 7.8 [GF2, Th. 7]. Suppose T is a separable Baire topological space, X is completely regular Hausdorff, $\mathcal{M} = \mathcal{B}(T)$, and F is compact-valued and usc. Then under the continuum hypothesis, F has a $\mathcal{B}_0(X)$-measurable selection.

It is noted in [GF2] that with T separable and compact, Theorem 7.8 was proved independently by Evstigneev [ES2].

Losert [LO1] has very importantly produced a compact T with Radon μ such that (T, $\mathcal{M}_\mu$, μ) does not admit a strong lifting, where $\mathcal{M}_\mu$ is the μ-completion of $\mathcal{B}(T)$. This result, with Graf's Theorem 7.5 above, led Losert [LO2, Th. 2] to produce, by a variation on the deep construction of [LO1], a compact T and a Radon μ such that there is a usc compact-valued F with $\mathcal{S}(F) = \phi$, specifically such that the equivalent conditions of the following theorem are not satisfied. To say that F has no $\mathcal{M}_\mu - \mathcal{B}_o(X)$ measurable selection is much stronger than saying that $\mathcal{S}(F) = \phi$. Also, (c) of Theorem 7.9 is weaker than the existence of a strong lifting.

THEOREM 7.9 [LO2, Th. 1]. Suppose T is a completely regular Hausdorff space, μ is a Borel probability measure, and $\mu(U) > 0$ whenever $\phi \neq U \subset T$ and U is open. Then the following are equivalent:

(a) Whenever X is completely regular Hausdorff and F is usc and compact-valued, F has an $\mathcal{M}_\mu$-$\mathcal{B}_o(X)$-measurable selection.

(b) Letting Z_o be the spectrum of $L^\infty(T, \mu)$, π be the canonical map on Z_o to the Stone-Cech compactification of T, and $F = \pi^{-1}$, F has an $\mathcal{M}_\mu(T) - \mathcal{B}_o(\pi^{-1}(T))$-measurable selection.

(c) There exists a Boolean algebra homomorphism $\rho : \mathcal{M}_\mu \to \mathcal{M}_\mu$ such that [A, B $\in \mathcal{M}$ and $\mu(A \triangle B) = 0$] implies $\rho(A) = \rho(B)$, and $\rho(U) \supset U$ for open $U \subset T$.

Losert's [LO2] had significant antecedents, in Talagrand's [TA3, 4], in addition to [GF2]. In fact, the following example of a continuous map on a compact space without a measurable selection of its inverse was previously given in [TA3]. Talagrand generalizes on this in [TA4] and enlarges further in [TA5].

EXAMPLE 7.10 [TA3, Th. 4]. Let I be a set such that card I $\geq \aleph_2$, $Z = \{0,1\}^I$, $X = Z \times Z$, and $T = \{\{a,b\} : a, b \in Z\}$. Let Z have the product of discrete topologies and let T inherit its topology from the space of closed subsets of Z. Let $\mathcal{M}$ be the σ-algebra of subsets of T of form $\eta^{-1}(A)$, where $A \subset \{0,1\}^\omega$ and $\eta : T \to \{0,1\}^\omega$ is given by $\eta(t)_i = \chi_{U_i}(t)$ for $t \in T$, $i \in \omega$, with $U_i \subset T$ open for $i \in \omega$. Let θ be the canonical map on X onto T. Then $\mathcal{M} \supset \mathcal{B}(T)$ and $\mathcal{M}$ is a Suslin family. Also T and X are compact and θ is continuous. It is shown in [TA3] that θ^{-1} has no $\mathcal{B}_o(X)$-measurable selection.

It is noted in [TA3, Th.] that it is easy to show that if T, X are compact, $f : X \to T$ is continuous and surjective, $\text{wt}\, X \leq \aleph_1$, then f^{-1} has a selection which is $\mathcal{B}_o(T) - \mathcal{B}_o(X)$ measurable, and that von Weizsäcker [WEI] has shown that this fails without the condition on X.

We note two papers which should have been mentioned in [WG3]. In 1971, Hermes [HE4] found an $L_1([0,1])$ conditionally compact sequence of Lebesgue measurable selections of an equicontinuous (Hausdorff metric) sequence of uniformly bounded compact-valued maps into $\mathcal{P}(R^n)$ -- variation conditions are imposed. Ahmed and Teo [AT] showed that in Cole's [CL1] "closed-bounded-convex" may be weakened to "weakly compact."

8. Measurable weak selections. The concept of measurable weak section was introduced by Edgar in [ED]. It pertains to a near-selection f of the inverse of $p : X \to T$ in the following sense: p o f is so close to the identity map on T that $\mu([p \circ f]^{-1}(A) \triangle A) = 0$ for $A \in \mathcal{M}$. Edgar

showed existence of measurable weak sections in [ED, Th. 1.3]. His concept and this result have been generalized by Graf [GF1], whose results are summarized below. Throughout this section $0 < \mu(T) < \infty$.

DEFINITION 8.1. We denote by μ^* the outer measure induced by μ. As before we denote by $\mathcal{M}_\mu$ the μ-completion of $\mathcal{M}$. Following [GF1] we say A^* is a measurable cover of $A \subset T$ if $A \subset A^* \in \mathcal{M}$ and $\mu^*(A) = \mu(A^*)$. We say $f: T \to X$ is an $\mathcal{M}_\mu$-measurable weak selection of F if f is $\mathcal{M}_\mu$-measurable and $\mu(f^{-1}(U)^* \setminus F^-(U)^*) = 0$ for open $U \subset X$, where * denotes measurable cover. Following Edgar [ED], if $p: X \to T$ is $\mathcal{B}(X)$-$\mathcal{M}$-measurable and $f: T \to X$ is $\mathcal{M}_\mu$-measurable we say f is an $\mathcal{M}_\mu$-measurable weak section of p iff $\mu(f^{-1}(p^{-1}(A)) \triangle A) = 0$ for $A \in \mathcal{M}$. These definitions are given in [ED] and [GF1] for more general σ-algebras.

Note that in Theorem 8.2 the assumption on measurability of F is very weak.

THEOREM 8.2 [GF1, Th. 4.3]. Suppose X is Hausdorff, and $\mu(T) = \sup\{\mu^*(F^-(K)) : K \subset X$ and K is compact$\}$. Then there exists an $\mathcal{M}_\mu$-measurable weak selection f of F such that $f(\mu)$ is Radon.

THEOREM 8.3 [GF1, Prop. 4.6]. Suppose X is Hausdorff, $p: X \to T$, is $\mathcal{B}(X)$-$\mathcal{M}$ measurable, $\mu(T) = \sup\{\mu^*(p(K)) : K \subset X$ is compact$\}$, and $\mu^*(p(K)) = \inf\{\mu^*(p(U)) : K \subset U, U \subset X$ is open$\}$ for compact $K \subset X$. Then there exists an $\mathcal{M}_\mu$-measurable weak section f of p such that $f(\mu)$ is Radon.

Edgar [ED, Th. 3.1] proved Theorem 8.3 under the principal additional assumptions that T and X are completely regular and p is continuous.

COROLLARY 8.4 [GF1, Cor. 4.8]. Suppose T is topologized, $\mathcal{M} = \mathcal{B}(T)$, μ is Borel regular, X is Hausdorff, $p: X \to T$ is continuous, and $\mu(T) = \sup\{\mu^*(p(K)) : K \subset X$ is compact$\}$. Then the conclusion of Theorem 8.3 holds.

In § 5 of [GF1], Graf gives results on uniqueness of measurable weak selections and measurable weak sections up to sets of μ measure 0. Applications are given in [GF1] and [ED] to existence of pre-image measures, extensions of measures, and related topics. In particular, one may note that existence of an extension of μ to a measure on a σ-algebra $\mathcal{N} \supset \mathcal{M}$ is implied by existence of an $\mathcal{M}$-$\mathcal{N}$-measurable weak section of the identity map of T. Under some conditions the converse holds. Among papers on the measure extension problem are [LN1], [LM], [LB], [ALE], [ER1], and [MRR].

The following theorem of Graf generalizes Edgar's [ED, Th. 3.8] which has T and X compact, p continuous, and $\mathcal{M} = \mathcal{B}(T)$.

THEOREM 8.5 [GF3, Th. 3 and Cor.]. Suppose $p: X \to T$ is $\mathcal{B}(X)$-$\mathcal{M}$-measurable, X is Hausdorff, $0 < \mu(T) < \infty$, and ν is a Radon measure on $\mathcal{B}(X)$. Then the following are equivalent:

(a) ν is an extreme point in the set of preimage measures on $\mathscr{B}(X)$ under p and μ;

(b) there exists an $\mathscr{M}_\mu$-measurable weak section f of p with $\nu = f(\mu)$.

If also $\mathscr{M}$ is countably separated and p is surjective, then these are equivalent to the following:

(c) there exists an $\mathscr{M}_\mu$-measurable selection f of p^{-1} such that $\nu = f(\mu)$.

9. Optimal measurable selections. Rieder [RI1] has taken an interesting approach to finding maximal or ϵ-maximal measurable selections -- see [WG3, § 9]. Paraphrasing his definition, supposing for $G \in \mathscr{G}$ we have $T_G \in \mathscr{M}$ and $G: T_G \to \mathscr{P}(X)$, we say $\mathscr{G}$ is a selection class if $\mathscr{S}(G) \neq \phi$ for $G \in \mathscr{G}$. Assume $u: \text{Gr } F \to R \cup \{-\infty, \infty\}$ and let $v(t) = \sup \{u(t, x) : x \in F(t)\}$ for $t \in T$. Rieder's main results follow.

THEOREM 9.1 [RI1, Cors. 3.2 and 3.7]. Suppose $\mathscr{G}$ is a selection class, $F \in \mathscr{G}$, and for $c \in R$, $G_c \in \mathscr{G}$, where G_c is determined by $\text{Gr } G_c = \{(t, x) : u(t, x) \geq c\}$. Then v is a measurable function and for $\epsilon > 0$ there exists $f \in \mathscr{S}(F)$ such that for $t \in T$, $u(t, f(t)) \geq v(t) - \epsilon$ when $v(t) < \infty$ and $u(t, f(t)) \geq 1/\epsilon$ when $v(t) = \infty$. If also X is separable metric and G_c is compact-valued for $c \in R$, then there exists $g \in \mathscr{S}(F)$ such that $u(t, g(t)) = v(t)$ for $t \in T$.

The "selection class" approach seemed to have the merit of separating the focus on basic measure selection theory from the focus on those features of problems which are peculiar to optimal measurable selections. A similar approach to measurable implicit functions, for example, suggests itself.

Unfortunately Example 2.4 of [RI1] is incorrect, being a restatement of the erroneous 4.2e ((xi) $\Rightarrow$ (ix)) of [WG3], attributed to [HV6], noted in § 3 above. This does not affect the main results of [RI1], summarized as Theorem 9.1 above. Of the statements of [RI1] which are affected, Rieder advises that there is no problem in clearing up 4.1, 4.2, 4.3, or 4.4, but 2.7 and 4.8(b)(c) remain open questions (personal correspondence).

Rieder has been more general, e.g., by treating measurability w.r.t. an arbitrary σ-algebra on X. He applies Theorem 9.1 to obtain some known results and new generalizations of such.

Dolecki has investigated selections which are optimal with respect to convex cones. In his two theorems given next, think of a - b belonging to a convex cone as meaning that b is preferred to a.

THEOREM 9.2 [DK, Th. I.4]. Suppose X is a separable Banach space, $C: T \to \mathscr{P}(X)$ and F are closed-valued and measurable, and for $t \in T$, C(t) is a convex cone, and there exist $\epsilon_t > 0$ and a linear continuous $\varphi_t : X \to R$ such that $C(t) \subset \{x: \varphi_t(x) \geq \epsilon_t \|x\|\}$. Suppose also that the space spanned by $\{\varphi_t : t \in T\}$ is separable and there exists a measurable $\alpha : T \to X$ such that $F(t) \supset \alpha(t) + C(t)$ for $t \in T$. Then there exists $f \in \mathscr{S}(F)$ such that for $t \in T$, when $x \in F(t)$ and $f(t) - x \in C(t)$ we have $x = f(t)$.

In [DK, Th. I.4] measurability of F is omitted. We infer from [DK, Rem. I.6] that this is an oversight, and note that "F is closed-valued and measurable" may be replaced by "F + C is closed-valued and measurable." In the next theorem "Borel measurable" is not defined in [DK] (or here).

THEOREM 9.3 [DK, Th. II.5]. Suppose X and Y are separable Banach spaces, F is closed-valued and Borel measurable, $C: T \to \mathscr{P}(Y)$ satisfies the condition on $C: T \to \mathscr{P}(X)$ in Theorem 9.2, $g: T\times X \to Y$ is a Carathéodory map, $g(\cdot, F(\cdot))$ is closed-valued, and for $t \in T$ there exists $y_0(t) \in Y$ such that $g(t, F(t)) - y_0(t) \subset C(t)$. Then there exists $f \in \mathscr{S}(F)$ such that $g(t, F(t)) - g(t, f(t)) \subset C(t)$ for $t \in T$.

Bain [BAI] has investigated optimal measurable selections of conditional expectation of "integrands." In the following we assume $\tilde{u}: T\times R^n \to R \cup \{\infty\}$ is a normal integrand bounded above, meaning that (we adapt minimization terminology to maximization), defining $G(t) = \{(x, y): x \in R^n \text{ and } y \le \tilde{u}(t,x)\}$ for $t \in T$, we have G is measurable and closed-valued, and there exists a finitely integrable $b: T \to R$ such that $b(t) \ge \tilde{u}(t, x)$ for $t \in T$, $x \in R^n$.

THEOREM 9.4 [BAI, Th. 1]. Suppose $X = R^n$, μ is a probability measure, $\mathscr{N}$ is a sub-σ-algebra of $\mathscr{M}$, and F is compact-valued and $\mathscr{N}$-measurable. Then there exists an $\mathscr{N}$-measurable selection v^* of F such that whenever v is another such,

$$E[\tilde{u}(\cdot, v^*(\cdot)) | \mathscr{N}] \ge E[\tilde{u}(\cdot, v(\cdot)) | \mathscr{N}], \qquad \mu\text{-a.e.},$$

where $E[\cdot | \mathscr{N}]$ denotes conditional expectation conditioned on $\mathscr{N}$.

Bain proves this by reducing it to a problem in deterministic optimal measurable selection. Most of the proof is in finding a normal integrand g w.r.t. $\mathscr{N}$ such that $g(t, v(t)) = E[\tilde{u}(\cdot, v(\cdot)) | \mathscr{N}](t)$, μ-a.s., whenever $v: T \to R^n$ is $\mathscr{N}$-measurable. He notes antecedents by Bismut [BI2] and Castaing and Valadier [CV3] using convexity (concavity for maximizers) assumptions.

Theorem 9.2 of [WG3] is a result of Brown and Purves [BP, Cor. 1] giving ϵ-optimal Borel function selections. Theorem 3(i) of [BP] (not cited in [WG3]) is a related result which avoids assuming F is σ-compact-valued but has the selection universally measurable rather than a Borel function. Bertsekas and Shreve [BS] have improved this, principally by allowing u (as above) to have the property that $\{(t,x): u(t,x) > a\}$ is Suslin for $a \in R$ rather than a Borel function. This clarifies (vii) of the addenda in proof to [WG3].

Nowak [NW, Th. 2.1] has shown existence of ϵ-optimal Borel selections under Borel assumptions similar to those of [HPV] -- see [WG3, Th. 9.1(ii)]. He has u bounded above and F closed-valued rather than compact-valued.

Mühlbauer's thesis [MB] finds optimal and ϵ-optimal measurable selections when X is a linear space (usually separable Banach), X' its algebraic dual, and for a fixed $u': T \to X'$, $u(t,x) = u'(t)x$ so that v relates closely to the support function of F, i.e., $v(t) = \sup\{u'(t)x : x \in F(t)\}$ for $t \in T$. In the main part of [MB], these results are applied to statistical hypothesis testing.

10. Stopping time theory. In [WG3], I was greatly remiss in failure to mention contributions to measurable selection theory by C. Dellacherie and P. A. Meyer, motivated by their deep work in probability theory. (Some remarks on their work were made by Ioffe in [IF3].) Tracing the history of this work was not easy, so I am most grateful for a very informative account given to me by Dellacherie at Oberwolfach, which has provided the main content of this section.

Throughout this section, μ is a complete probability measure. We fix for each $s \geq 0$ a σ-algebra $\mathcal{M}_s \subset \mathcal{M}$ such that

(i) $\mathcal{M}_s \subset \mathcal{M}_{s'}$ for $0 \leq s \leq s'$,

(ii) $\mathcal{M}_s = \bigcap_{s'>s} \mathcal{M}_{s'}$ for $s \geq 0$,

(iii) $A \in \mathcal{M}_0$ whenever $A \in \mathcal{M}$ and $\mu(A) = 0$.

One calls $\{\mathcal{M}_s : s \geq 0\}$ a filtration and thinks of $\mathcal{M}_s$ as the set of events known at time s. We define the σ-algebra $\mathcal{O}$ of optional sets ("bien mesurable" sets in earlier work) {resp. the σ-algebra $\mathcal{P}$ of predictable sets ("prévisible" in French) -- not to be confused with $\mathcal{P}$ fixed in § 2} as the σ-algebra on $T \times [0,\infty)$ generated by the set of all $g^{-1}([a,\infty))$ for which $a \in R$ and $g: T \times [0,\infty) \to [0,\infty)$ is such that

(i) $g(\cdot, s)$ is $\mathcal{M}_s$-measurable for $s \geq 0$,

(ii) $g(t, \cdot)$ is right {resp. left} continuous for $t \in T$.

It can be shown that $\mathcal{P} \subset \mathcal{O} \subset \mathcal{M} \otimes \mathcal{B}([0,\infty))$ with equality when $\mathcal{M}_0 = \mathcal{M}$. We further say that $f: T \to [0,\infty]$ is an optional {resp. a predictable} time if (note that "$<\infty$" may make this less than the full graph)

$$[\![f]\!] \equiv \{(t,s) : t \in T,\ 0 \leq s < \infty,\ f(t) = s\} \in \mathcal{O} \qquad \{\text{resp. } \in \mathcal{P}\}.$$

The main result in this section is the following.

THEOREM 10.1. Suppose $X = [0,\infty)$, $\text{Gr } F \in \mathcal{O}$ {resp. $\text{Gr } F \in \mathcal{P}$} and $\epsilon > 0$. Then there exists an optional {resp. predictable} time f such that $[\![f]\!] \subset \text{Gr } F$, i.e., $[\![f]\!]$ is a selection of F, and $\mu(\pi_T(\text{Gr } F)) \leq \mu(\{t : f(t) < \infty\}) + \epsilon$.

Theorem 10.1 affords proof that an optional {resp. predictable} time is precisely a "stopping time" {resp. an "announceable" time (in French, annonçable)}. Dellacherie has further remarked that this theorem "is one of the very fundamental results in the so-called 'general theory of stochastic processes.'" It fails if $\epsilon = 0$.

The history of this result began with Choquet's 1954 fundamental paper on capacities [CQ] which has the following.

THEOREM 10.2 [CQ, Ex. 37.1]. Suppose $T \subset R = X$, $\mathcal{M} = \mathcal{B}(T)$, Gr F is Suslin, and $\epsilon > 0$. Then there exists a compact $K_\epsilon \subset$ Gr F such that (μ^* is the outer extension of μ) $\mu^*(\pi_T(\text{Gr } F)) \leq \mu(\pi_T(K_\epsilon)) + \epsilon$.

This follows from the capacitability theorem, the map taking $H \subset R^2$ to $\mu^*(\pi_T(H))$ being the simplest example of a Choquet capacity not a measure. As observed by Choquet, we may obtain K_ϵ to be a uniformization, e.g., the second coordinate infima of K_ϵ form a lsc "selection within ϵ" of F. This is readily generalized to an abstract probability space and Suslin X, via K_ϵ with compact vertical sections, using the facts that for X Suslin, $(X, \mathcal{B}(X))$ is σ-algebra isomorphic to $(A, \mathcal{B}(A))$ for some Suslin $A \subset R$ and that in this situation one may assume $T \subset R$ and $\mathcal{M}$ is the μ-completion of $\mathcal{B}(T)$.

The first version of Theorem 10.1 in the optionality-predictability context was given by Meyer in [ME1]. His proof was greatly simplified and in some ways corrected by Cornea and Licea [CLI]. Both proofs follow Choquet's idea but in a more complicated situation, dealing with abstract capacities and semi-compact pavings. Dellacherie ([DC2, Ch. IV, § 2] and in earlier work) gave a more general statement and proof using Choquet's selection within ϵ for $\mathcal{M} \otimes \mathcal{B}([0, \infty))$ with abstract T. All of this work (discussed also in [ME2] and [DM]) was without knowledge of Yankov-von Neumann.

Again the substance of this discussion is due to Dellacherie. Finally, we also note that Federer's version [WG3, Th. 5.12; or WS, Th. 4.1] of Yankov-von Neumann gives a selection within ϵ.

11. Convex-valued functions. We begin our discussion of convex-valued functions with a review of the Castaing-Valadier text [CV3], which is undoubtedly the best general reference on the subject and is one of the very few texts dealing significantly with measurable set-valued functions. The latter topic as such, including selection theory, is treated primarily in Chapter III of [CV3]. Later chapters treat primarily applications in convex analysis, motivated by applications to theory of control and optimization. Chapters I and II address basics of convex analysis and use of the Hausdorff metric.

In Chapter III, fundamentals are systematically given on various concepts of measurability of set-valued functions, on their interrelationships, and on existence of measurable selections and Castaing representations. Usually, X is separable metric (not necessarily linear) and the values of F are closed or complete. Alternatively, T and X are Suslin and F has a measurable graph in some sense. With X linear, the key to measurability of convex-valued functions is that of their support functions. Compact-valued maps are often treated as point-valued maps via the Hausdorff metric. Most of the facts in Chapter III are in [WG3], without proofs. A notable exception is an abstract implicit function theorem, 16.1 below. Historical comment is meager and attribution is largely by giving a literature list, for the most part uncited. Leese's then recent theory of Suslin type could be used to good advantage (this comment is neutral regarding later chapters).

After Chapter III, always X is linear and F is convex-valued and usually F is compact-valued. Chapter IV gives measurable parametric Carathéodory representations (see [WG3, § 8] for finite-dimensional versions), Choquet representations, and results on $\ddot{\mathscr{S}}(F) = \mathscr{S}(\ddot{F})$ (the umlaut refers to the set of extreme points), compactness and convexity of $\int F$ (generalizing Lyapunov), and $\int F = \int \ddot{F}$ (bang-bang principle). Chapter V gives results on compactness of and integral representations of $\mathscr{S}(F)$, suitably topologized. Chapters VI and VII respectively treat multivalued differential equations and, in the spirit of Rockafellar, convex integral functionals and their duals. Conditional expectations arise in Chapter VIII.

For other reviews, see Ekeland [EK] and Levin [LV4].

We now review new results with X linear and F convex-valued. These are by Debs on $\mathscr{S}(\ddot{F}) \neq \phi$, by Castaing on Carathéodory selections, by Rieder on fixed point selections, by Saint-Pierre on selections which yield well-behaved bases, by Levin on Borel measurability of $\ddot{F}$, and by Phan Van Chuong on density of $\mathscr{S}(\ddot{F})$ in $\mathscr{S}(F)$.

Debs' principal tool is [DB1, Th. 7], which finds a measurable selection of the "frontier" points of F with respect to the order on measures over X induced by a cone of balayage of X. By definition, point masses at frontier points are maximal w. r. t. this order.

THEOREM 11. 1 [DB1, Th. 9]. Suppose $\mathscr{L}$ is a clan (see § 6) over T, X is a locally convex linear separable metric space, and F is $\mathscr{L}_\sigma$-measurable, $\mathscr{L}_\sigma$-weakly measurable, and convex-compact-valued. Then F has an $\mathscr{L}_\sigma$-measurable selection.

COROLLARY 11. 2 [DB1, Cor. 10]. Suppose $\mathscr{L}$ and X are as in Theorem 11. 1, X' is the dual of X topologized by $\sigma(X', X)$, $G: T \to \mathscr{P}(X')$ is weakly $\mathscr{L}$-measurable and convex-compact-valued. Then $\ddot{G}$ has an $\mathscr{L}_\sigma$-measurable selection.

Debs [DB1, Th. 12] applies Theorem 11. 1 to generalize a result of Castaing of the form $\ddot{\mathscr{S}}(F) = \mathscr{S}(\ddot{F})$ modulo μ-negligible sets. He further applies Théorème 7 to obtain maximal simplicial measures, generalizing results of Castaing [CV3, Ths. IV-11, IV-13] and Talagrand [TA2].

We cite what appears to be the main result, by Castaing, on selections which are Carathéodory maps.

THEOREM 11. 3 [CA23]. Suppose Y is a Hausdorff space, ν is a bounded complete Radon measure on a σ-algebra $\mathscr{N}$ over Y, Z is a Polish space, $T = Y \times Z$, $\mathscr{M} = \mathscr{N} \otimes \mathscr{B}(Z)$, X is a separable Banach space, F is convex-closed-valued and measurable, and $F(y, \cdot)$ is lsc for $y \in Y$. Then F has a Carathéodory selection.

Castaing [CA23] observes that Cellina [CE] independently treated the same problem under different hypotheses. He further notes that when Y is Suslin, Z is compact metric, and F(·, Z) is usc for $z \in Z$, Theorem 11.3 results directly from Yankov-von Neumann and Michael's theorem on continuous selections. Earlier results by Castaing in this vein were in [CA17, 18, 22].

Rieder [RI2] has obtained the following fixed point selection theorem using Theorem 3.1 and the Glicksberg-Fan extension of the Kakutani fixed point theorem.

THEOREM 11.4 [RI2, Th. 1]. Suppose X is a subset of a locally convex linear Hausdorff space, T and X are Borel subsets of Polish spaces, $\mathcal{M} = \mathcal{B}(T)$, Gr F $\in \mathcal{B}(T \times X)$, $G: T \times X \to \mathcal{P}(X)$ is measurable, F and G are compact-convex-valued, Gr G is closed, and $G(t, x) \subset F(t)$ for $t \in T$, $x \in X$. Then there exists $f \in \mathcal{S}(F)$ such that $f(t) \in G(t, f(t))$ for $t \in T$.

Saint-Pierre in [SA5] deals primarily with Borel graph properties of convex-valued F and associated profile and exposed point subfunctions, with $X = R^n$. Among the results is the following.

THEOREM 11.5 [SA5, Prop. 5 and Cor. to Prop. 4]. Suppose $X = R^n$ and T is topologized and has the property that whenever $A \subset T$, $g: A \to R^n$, and g is a Borel subset of $T \times R^n$, we have A is Borel (e.g., T is Lusin). Suppose also F is convex-valued and Gr F is Borel. Then

(a) Gr ri F, Gr Aff F, and Gr $F^{\perp}$ are Borel;

(b) if $\mathcal{M} = \mathcal{B}(T)$, $\mathcal{S}(F) \neq \phi$;

(c) there exist Borel functions $f_1, \ldots, f_n : T \to R^n$ such that for $t \in T$, $\{f_1(t), \ldots, f_n(t)\}$ is an orthonormal basis of R^n, and whenever F(t) has dimension k, $\{f_1(t), \ldots, f_k(t)\}$ is a basis of the space spanned by Aff F(t).

Here Ri D, Aff D, and $D^{\perp}$ are respectively the relative interior of D, affine variety associated with D, and orthogonal complement of D.

Levin [LV3] has given conditions for the graph of the profile $\ddot{F}$ of F to be Borel. We thank Ershov for translation of this result -- see also § 12.

THEOREM 11.6 [LV, Th.]. Suppose X is a subset of a locally convex space, with the induced topology, such that $X = \bigcup_{i=1}^{\infty} K_i$ with each K_i compact metric and any convex compact metric subset of X contained in some K_i, and such that there exists a Borel $\lambda: X \to R$ such that $\lambda \mid K$ is continuous and strictly convex whenever $K \subset X$ is compact metric. Suppose T is a Borel subset of a Polish space, Gr F $\in \mathcal{B}(T \times X)$, and F is convex-compact-metric-valued. Then Gr $\ddot{F} \in \mathcal{B}(T \times X)$.

Phan Van Chuong has recently given results on the density of $\mathcal{S}(\ddot{F})$ in $\mathcal{S}(F)$, with T = [0, 1], Lebesgue μ, and locally convex Suslin X[PV, Th. 2] or Banach X [PV, Th. 3]. Analogous results were given in [CV3] for the weak topology. In [PV] finer topologies are found for the density results to hold. Preliminarily a measurably parametrized Krein-Milman result is obtained.

12. Borel selections. Here we give results on selections which are Borel functions, by Levin, Saint-Raymond, Martin, Srivastava, Sarbadhikari, and Dellacherie. Results in this topic, more than most, are also scattered elsewhere in the paper.

Levin [LV3] has given several conditions (in the vein of [WG3, Th. 4.2]) equivalent to existence of a Castaing representation by Borel functions. The proof uses Novikov's [NO2] (1939), the first work affording such a representation. For translation and interpretation of the results of [LV3], we are indebted to Ershov.

THEOREM 12.1 [LV3, Prop. 2.3]. Suppose T is a Borel subset of a Polish space, X is metrized by ρ and is σ-compact, $\mathcal{M} = \mathcal{B}(T)$, and F is closed-valued. Then the following are equivalent:

(a) $\mathrm{Gr}\, F \in \mathcal{B}(T \times X)$;

(b) F is measurable;

(c) F is weakly measurable;

(d) $F^{-}(K) \in \mathcal{M}$ for compact $K \subset X$;

(e) $\rho(x, F(\cdot))$ is a Borel function for $x \in X$;

(f) $\rho(\cdot, F(\cdot))$ is a Borel function;

(g) F has a Castaing representation.

Srivastava has observed (verbal communication) that in Theorem 12.1 one may weaken the condition on T to T being Suslin, and, as a consequence of his [SVS4, Th. 1.4.4 and proof] (see Theorem 4.22 above) one may add the following to the equivalence:

(h) F is induced by (Z, f) with Z Polish and f Carathéodory;

(i) F is induced by (Z, f) with Z σ-compact metric and f Carathéodory;

(j) F is induced by $(\omega \times C, f)$, where C is the Cantor set, with f Carathéodory.

Srivastava has called our attention to the following result of Saint-Raymond [SRY2], and to the significance of the composition being of class α instead of being increased to $\alpha + 1$.

THEOREM 12.2 [SRY2, Lems. 2, 3]. Suppose T and X are compact metric spaces, Y is a separable metric space, $p: X \to T$ is continuous, $\alpha \geq 1$ is a countable ordinal and $\varphi: X \to Y$ is Borel of class α (see § 5). Then p^{-1} has a Borel selection f of class 1 such that $\varphi \circ f$ is of class α.

Dellacherie has called our attention to, and Burgess has helped to interpret, the fact that [MN], an important result of Martin, is a novel selection-theoretic statement. Theorem 12.3 follows from Martin's finding that one of the players of a Gale-Stewart game on a Borel set has a winning strategy; players I and II alternatively choose an integer, and I wins if the sequence is in a prechosen Borel subset of ω^ω.

THEOREM 12.3 [MN]. Suppose T and X are Polish and $B \in \mathscr{B}(T \times X)$. Then there exists either a Borel $f: T \to X$ such that $f \subset B$ or a Borel $g: X \to T$ such that $g^{-1} \subset ((T \times X)/B)$. If $T = X = \omega^\omega$, then there exists a continuous such f or g.

Srivastava gives the following as a corollary to his $\gamma = 1$ case of Theorem 5.2 above.

THEOREM 12.4 [SVS4, Cor. 3.2.7; or SVS3, Cor. 5.4]. Suppose X and Y are Polish, $p: X \to Y$ is Borel of class 1, $T = p(X)$, and $p(U)$ is Borel in T for open $U \subset X$, $F = p^{-1}$, and $\mathscr{M} = \mathscr{B}(T)$. Then T is Borel in Y and $\mathscr{S}(F) \neq \phi$.

In [WG3, § 11], my mention of Sarbadhikari's [SR] was too brief. Her main result follows as Theorem 12.5; under (iii) it is generalized by Theorem 6.3.

THEOREM 12.5 [SR, Ths. 1,2]. Suppose T and X are Polish, $\mathscr{M} = \mathscr{B}(T)$, and Gr $F \in \mathscr{B}(T \times X)$. Then $\mathscr{S}(F) \neq \phi$ if any of the following holds:

(i) F(t) has a point isolated in F(t) for $t \in T$;

(ii) F(t) has a point which is not a point of condensation in F(t) for $t \in T$;

(iii) F is nonmeager-valued or comeager-valued.

Dellacherie [DC3, Th. 19] has given several conditions, under compact metric T and X and Borel Gr F, which are equivalent to F having a Borel selection, as a generalization of Stschegolkov's theorem (see Theorem 3.5 above). In part, these pertain to existence of a compact-valued subfunction of F with Suslin or co-Suslin graph. This paper has interesting notes on belated western awareness of Yankov [JN].

13. Generalizations of Fundamental Measurable Selection Theorem. We note generalizations of Theorem 3.1 above in four quite different directions.

First we note briefly Mägerl's [MG2, 3] approach to unification of the theories of continuous selections and measurable selections -- as mentioned in § 1, beyond this point we do not endeavor to review continuous selection theory as such. Mägerl's main result is given as Theorem 13.1, but we refer to [MG2 or 3] for much of the terminology. The approach generalizes on methods and the main results of the classical selection papers of Kuratowski and Ryll-Nardzewski [KRN] (itself more general than Theorem 3.1) and Michael [MI1].

THEOREM 13.1 [MG2, Satz 1.11; or MG3, Th.]. Suppose $(T, \mathscr{L})$ is a (k, α)-paracompact paved space, X is Polish and k-bounded, H is an α-convex, compatible hull-operator on X, F is closed-valued and weakly $\mathscr{L}$-measurable, and $F(t) = H(F(t))$ for $t \in T$. Then F has an $\mathscr{L}$-measurable selection.

Miller, in generalizing the main theorem and proof of [KRN], obtains Theorem 13.2, used in proving Theorems 5.1 and 5.2 above. Note that the condition on $F^{\#}$ is satisfied by F^{-}; a similar generalization of the F^{-} operator is used by Graf (see § 7). Dropping "F is closed-valued" and changing the conclusion to "$\mathscr{S}(\mathrm{cl}\ F) \neq \phi$" could also be done in, e.g., [KRN].

THEOREM 13.2 [ML3, Th. 1.1]. Suppose $F^{\#}(U) \subset T$ for open $U \subset X$, $\bigcup_{i \in \omega} F^{\#}(U_i) = F^{\#}(\bigcup_{i \in \omega} U_i)$ whenever $U_i \subset X$ is open for $i \in \omega$, $F^{\#}(X) = T$, and $F^{\#}(U) \subset F^{-}(U)$ for open $U \subset X$. Suppose $\mathscr{L}$ is a field of subsets of T, X is Polish, and $F^{\#}(U) \in \mathscr{L}_\sigma$ for open $U \subset X$. Then cl F has an $\mathscr{L}_\sigma$-measurable selection.

Le Van Tu [LT] extends Theorem 3.1 (not strictly, since 3.1 has weak measurability of F) to a generalization of Polish spaces. He assumes the continuum hypothesis and denotes the first uncountable ordinal by Ω. He calls X as conditioned in the next theorem an Ω-Polish space and F thin if it satisfies the new condition there. He applies Theorem 13.3 to extend Robertson's [RB, Lem. 1] and to obtain a McShane-Warfield type lifting result [MW].

THEOREM 13.3 [LT, Th. 3]. Suppose $\mathscr{D}$ is a family of closed Polish subspaces of X whose union is X, card $\mathscr{D} \leq \Omega$, F is closed-valued and measurable, and $\{D \cap A : D \in \mathscr{D}$ and $F^{-}(D \cap A) \neq \phi\}$ is countable for closed proper $A \subset X$. Then $\mathscr{S}(F) \neq \phi$.

Christensen [CH, Th. 4.2] deduced the following result (also given as [DC4, Th. II.14] for X Polish) from Hoffman-Jørgensen's abstract selection theorem [HJ, Th. II.6.1; CH, Th. 4.1; or WG3, Th. 11.1] for partitions. In [DC4] it is observed that the Fundamental Measurable Selection Theorem, Theorem 3.1 above, is a corollary. Burgess [BS6, § 8(b)] has observed that Theorem 13.4 applies to strengthen Theorem 3.1 to Corollary 13.5. Here we denote $\hat{D} = \{A : A \subset D$ and A is closed$\}$ for D a closed subset of a topological space.

THEOREM 13.4 [CH, Th. 4.2]. Suppose X is metric Suslin, $T = \hat{X}$, $\mathscr{N} = \sigma(\{\hat{D} : D \subset X$ and D is closed$\})$, i.e., $\mathscr{N}$ is the Effros σ-algebra, $\mathscr{M} = \sigma(\Sigma(\mathscr{N}))$, and $F(A) = A$ for $A \in T$. Then $\mathscr{S}(F) \neq \phi$. If moreover X is Polish, F has an $\mathscr{N}$-measurable selection.

COROLLARY 13.5 (Dellacherie [BS6, § 8(b)]). Suppose X is Polish and F is closed-valued and weakly measurable. Then there exists $f \in \mathscr{S}(F)$ such that $f(t) = f(t')$ whenever $F(t) = F(t')$ for $t, t' \in T$.

According to [DC4], Saint-Raymond has obtained the following.

THEOREM 13.6 [SRY3]. Suppose the hypothesis of Theorem 13.4 holds with X metric Lusin. Then F has an $\mathscr{N}$-measurable selection iff X is the union of a Polish space with a σ-compact separable metric space.

14. Reduction principle. Maitra and B. V. Rao [MR3], Sarbadhikari [SR2], and Srivastava [SVS4] have given new results with reduction principle conditions, in the spirit of [MR1]. Results of [SVS4] on this topic are given in § 4. The others deal with closed-valued F and

compact-valued F respectively. These are summarized here.

Here α, β are ordinals. Suppose $\mathscr{L}$ is a set of subsets of T and λ is an infinite cardinal. Let λ^+ be the cardinal successor to λ. We say $\mathscr{L}$ is λ-additive {λ-multiplicative} if whenever $\beta < \lambda$ and $A_\alpha \in \mathscr{L}$ for $\alpha < \beta$, we have $\bigcup_{\alpha<\beta} A_\alpha \in \mathscr{L}\{\bigcap_{\alpha<\beta} A_\alpha \in \mathscr{L}\}$. Let $\mathscr{L}_\lambda$ be the smallest λ-additive set containing $\mathscr{L}$ and let $\mathscr{L}^c = \{T \setminus A : A \in \mathscr{L}\}$. We say $\mathscr{L}$ is a λ-field over T if $\phi \in \mathscr{L}$, $\mathscr{L}$ is λ-additive and $\mathscr{L} = \mathscr{L}^c$. We say $\mathscr{L}$ satisfies the λ-reduction principle if whenever $\beta < \lambda$ and $\{A_\alpha : \alpha < \beta\} \subset \mathscr{L}$, there exists a pairwise disjoint $\{B_\alpha : \alpha < \beta\} \subset \mathscr{L}$ such that $B_\alpha \subset A_\alpha$ for $\alpha < \beta$, and $\bigcup_{\alpha<\beta} B_\alpha = \bigcup_{\alpha<\beta} A_\alpha$. The λ-weak reduction principle adds: $\bigcup_{\alpha<\beta} A_\alpha = T$. Although Theorems 14.1 and 14.2 do not state a reduction principle condition as such, [MR3] notes and uses the fact, due to Kuratowski, that if $\mathscr{L}$ is a λ-field over T, $\mathscr{L}_{\lambda^+}$ satisfies the λ^+-reduction principle.

THEOREM 14.1 [MR3, Th. 2]. Suppose $\mathscr{H}$ is a λ^+-additive and λ-multiplicative set of subsets of T and ϕ, $T \in \mathscr{H}$. Then the following are equivalent:

(a) $\mathscr{H} = \mathscr{L}_{\lambda^+}$ for some λ-field $\mathscr{L}$ over T;

(b) if X is complete metric, wt $X \le \lambda$, and F is closed-valued and weakly $\mathscr{H}$-measurable, then there exist $\mathscr{H}$-measurable $f_\alpha : T \to X$ for $\alpha < \lambda$ such that $F(t) = \mathrm{cl}\{f_\alpha(t) : \alpha < \lambda\}$ for $t \in T$;

(c) statement (b) holds with $X = \{0, 1\}$, topologized discretely.

The implication (c) $\Longrightarrow$ (b) seems to be remarkable and especially interesting. Following is a direct generalization of the very useful Theorem 3.2(i) above.

THEOREM 14.2 [MR3, Th. 3]. Suppose $\mathscr{L}$ is a λ-field over T, X is complete metric, wt $X \le \lambda$, and F is closed-valued. Then the following are equivalent:

(a) F is weakly $\mathscr{L}_{\lambda^+}$-measurable;

(b) there exist $\mathscr{L}_{\lambda^+}$-measurable $f_\alpha : T \to X$ for $\alpha < \lambda$ such that $F(t) = \mathrm{cl}\{f_\alpha(t) : \alpha < \lambda\}$, $t \in T$.

In [MR3, Th. 4], Theorem 14.2 is specialized to Borel functions of class α.

THEOREM 14.3 [SR2]. Suppose $\lambda > \aleph_0$, $\mathscr{L}$ is λ-additive and λ-multiplicative and satisfies the λ-weak reduction principle, X is regular Hausdorff, wt $X \le \lambda$, for any family of open subsets of X there is a subfamily of cardinality less than λ with the same union, and F is compact-valued and $\mathscr{L}$-measurable. Then F has an $(\mathscr{L} \cap \mathscr{L}^c)_\lambda$-measurable selection.

15. Extensions of measurable functions. Maitra [MT3] has given some new results on extensions of measurable functions, summarized in this section. We note that in giving Theorem 16.1 of [WG3] on this topic our citation of Maitra and Rao was incorrect as to choice of paper and of corollary number and, as pointed out in correspondence from Valadier, we rendered the statement into something trivial.

If Φ is a family of subsets of S and $A \subset S$, we denote $\Phi \cap A = \{B \cap A : B \in \Phi\}$.

THEOREM 15.1 [MT3, Th. 1]. Suppose Φ is a countably additive, finitely multiplicative family of subsets of S, $\phi \in \Phi$, $S \in \Phi$, Φ satisfies the countable reduction principle (see § 14), X is Polish, $T \subset S$, and F is closed-valued and weakly $\Phi \cap T$-measurable. Then there exist $T \subset T^* \in \Phi_\delta$ and a $\Phi \cap T^*$-measurable $g: T^* \to X$ such that $g(t) \in F(t)$ for $t \in T$.

THEOREM 15.2 [MT3, Th. 2]. Suppose S, Φ, X, and F are as in Theorem 15.1 relaxed to the weak countable reduction principle, $T = S$, $T \setminus A \in \Phi$, and f is a $\Phi \cap A$-measurable selection of $F|A$. Then there exists a Φ-measurable selection g of F such that $g|A = f$.

COROLLARY 15.3 [MT3, Cor. 1]. Suppose $\mathscr{L}$ is a field of subsets of T, X is Polish, $A \subset T$, and $f: A \to X$ is $\mathscr{L}_\sigma \cap A$-measurable. Then there are $A \subset A^* \in \mathscr{L}_{\sigma\delta}$ and an $\mathscr{L}_\sigma \cap A^*$-measurable $g: A^* \to X$, such that $g|A = f$.

COROLLARY 15.4 [MT3, Cor. 2]. Suppose $\mathscr{L}$ is a field of subsets of T, X is Polish, F is $\mathscr{L}_\sigma$-weakly measurable and closed-valued, $A \subset T$, and f is an $\mathscr{L}_\sigma \cap A$-measurable selection of $F|A$. Then there is an $\mathscr{L}_{\delta\sigma}$-measurable selection g of F such that $g|A = f$.

16. Measurable implicit functions. In contrast to pre-1977 work, recent results on measurable implicit functions as such have been sparse. However, we note some special results achieved in control theory and the following result of Castaing and Valadier [CV3]. The novelty of the latter is in the role of G, which in other work has been played by a point-valued map.

THEOREM 16.1 [CV3, Th. III.38]. Suppose $(Y, \mathscr{N})$ is a measurable space, X is Suslin, $\mathrm{Gr}\, F \in \mathscr{M} \otimes \mathscr{B}(X)$, $G: T \to \mathscr{P}(Y)$, $\mathrm{Gr}\, G \in \mathscr{M} \otimes \mathscr{N}$, $g: \mathrm{Gr}\, F \to Y$ is $\mathscr{M} \otimes \mathscr{B}(X)$-$\mathscr{N}$-measurable, and $g(t, F(t)) \cap G(t) \neq \phi$ for $t \in T$. Then there exists a selection f of F such that f is universally measurable w.r.t. $\mathscr{M}$ and $g(t, f(t)) \in G(t)$ for $t \in T$.

In [WR], Warga applies Theorem 3.1 to solve a measurable implicit function problem arising in control theory, using in part Lipschitz continuity without differentiability. In [HK, Lemma 5.4], Hajek obtains measurable selections of constant set-valued functions as approximate solutions to an implicit function problem in discontinuous differential equations. Nababan [NA, Lem 2.2] finds a measurable map u into R^n such that $f(t, u(t), u(\alpha_1(t)), \ldots, u(\alpha_m(t))) = z(t)$ for given measurable z, continuous f and α_i, and $0 < \alpha_m(t) \leq \alpha_{m-1}(t) \leq \ldots \leq \alpha_1(t) \leq t$, among other conditions. Various typographical errors make interpretation difficult. He applies this to show existence of optimal controls under time-delayed arguments of the controls.

17. Special topics. Several additional results are noted in this section.

Baggett and Ramsay [BRM] use functional analytic methods to obtain first a selection-theoretic version of Hahn-Banach, Theorem 17.1(a) below, and from this various classic measurable selection results.

THEOREM 17.1 [BRM, Th. 1]. Suppose G is a subspace of a separable normed linear space H over R, T and X are closed unit balls in the conjugate spaces of G and H respectively, $r: X \to T$ is the restriction map, $F = r^{-1}$, and $\mathcal{M} = \mathcal{B}(T)$. Then r is surjective and there exists f such that

(a) $f \in \mathcal{S}(F)$, and

(b) e is an extreme point of T implies f(e) is an extreme point of X.

This is applied to obtain [FM, Lem. 5] of Federer and Morse, somewhat overly attributed to [SK] on pg. 866 of [WG3], by embedding a compact metric space Z as the unit ball of the conjugate space of the space of continuous functions on Z to R. This is easily generalized to σ-compact X, which generalization is applied in turn to obtain Yankov-von Neumann and Dixmier's [DI, Lem. 2].

Doberkat [DO1,2] has made interesting applications of measurable selection theory to the theory of stochastic automata. Suppose X is σ-compact and Polish, F is closed-valued, T^* and X^* are free semi-groups generated by T and X respectively, word length in X^* is denoted $| \ |$, $\mathcal{M}^*$ is the σ-algebra on T^* induced by $\mathcal{M}$, X^* is topologized by topological sum from X, and P and P^* are respectively the sets of probability measures on X and X^*, topologized by weak convergence (i.e., by the Prohorov metric). A stochastic transformation over T and X is a measurable $L: T^* \to P^*$ such that $L(v)(X^{|v|}) = 1$ for $v \in T^*$ and L(vt) induces L(v) under the map which takes a point in $X^{|v|+1}$ into the first $|v|$ coordinates for $v \in T^*$, $t \in T$. Say F^* is a measurable automaton transformation over T and X if

(i) $F^*: T^* \to \mathcal{P}(X^*)$ is measurable and closed-valued;

(ii) $F^*(v) \subset X^{|v|}$ for $v \in T^*$;

(iii) $F^*(v)$ is $F^*(vx)$ with last coordinate removed, for $v \in T^*$, $t \in T$.

Doberkat's main result [DO1, Th. 4.4] is that F^* is a measurable automaton transformation iff there exists a stochastic transformation L over T and X such that $F^*(v)$ is the support of L(v) for $v \in T^*$. His Corollary 4.7 says that F is measurable iff there is a measurable $K: T \to P$ such that F(t) is the support of K(t) for $t \in T$.

Also treating X as a set of measures is M. Rao's [RA]. In [WG3, addendum (xi)], [RA] is incorrectly (pointed out by Valadier) identified with set-valued measures. The result of [RA] is as follows ([RA] is not explicit as to topology on X, presumably by the Prohorov metric).

THEOREM 17.2 [RA, Th. 2]. Suppose T is a compact metric space, B is a separating linear subspace of the space C(T) of complex continuous functions on T, B contains constants, X is the (Polish) space of probability measures concentrated on the Choquet boundary of B, $\mathcal{M} = \mathcal{B}(T)$, and for $t \in T$, $F(t) = X \cap \{\lambda : \lambda$ represents t w.r.t. $B\}$. Then $\mathcal{S}(F) \neq \phi$.

Mauldin [MD1, Th. C] and later Maitra [MT2] have given new proofs of the following theorem due to Lusin -- (ii) follows from (i) and [WG3, Cor. 10.2], the latter coming from

[LS, pg. 244].

THEOREM 17.3. Suppose T and X are Polish, F is countable-valued, and Gr F is Suslin. Then

(a) [LS, pg. 247] there exists $G: T \to \mathscr{P}(X)$ such that G is countable-valued, $F(t) \subset G(t)$ for $t \in T$, and Gr G is Borel;

(b) if $\mathscr{M} = \sigma(\Sigma(\mathscr{B}(T)))$, then $\mathscr{S}(F) \neq \phi$.

Salinetti and Wets [SWE] have investigated convergence of a sequence of measurable closed-valued functions and have related this to existence of a convergent sequence of measurable selections and of a convergent sequence of Castaing representations.

Saint-Pierre [SA1, 3] principally proves a characterization of $\Sigma(\mathscr{F}(T \times X))$ (where $\mathscr{F}$ means "family of closed sets of") when T is Hausdorff and X is analytic, i.e., the image of ω^ω under a usc set-valued map: It coincides with the images of $T \times \omega^\omega$ under a compact-valued usc $G: T \times \omega^\omega \to \mathscr{P}(T \times X)$ such that for $t \in T$, $z \in \omega^\omega$, $G(t, z) \subset \{t\} \times X$. Applications include selection theorems; the principal of these, [SA3, Th. 4.3.1], was achieved by Leese in more general form as [LE5, Th. 5.5] (submitted May 1975).

In [SA2, 4], Saint-Pierre gives a natural extension of the Novikov separation theorem, by separating an analytic set of sequences of $\mathscr{F}$-Suslin sets by complements of such sets, and equivalently separating Suslin schemes.

Kozek and Suchanecki [KS2] apply a version of Theorem 3.4 above to show the following. Suppose $\mu(T) = 1$, $\mathscr{N}$ is a μ-complete sub-σ-algebra of $\mathscr{M}$, X is a Banach space, Gr F $\in \mathscr{N} \otimes \mathscr{B}(X)$, $f \in \mathscr{S}(F)$ is integrable, and g is a conditional expectation of f w.r.t. $\mathscr{N}$. Then $g \in \mathscr{S}(F)$ and f(t) is in the same face as g(t) for a.e. $t \in T$. Their results are applied to extend Jensen's inequality and to characterize equality cases. (In at least the preprint version of [KS2], one should delete (viii) from the "Survey Theorem" (the error mentioned after Theorem 3.2 above) and references to (viii), with no effect on the main results.) A precursor is [KS1].

In [VA9], Valadier corrects the proof of an embedding theorem in [DAP].

On an historical note, in 1953 the following measurable "lifting" result (see remarks after [WG3, Lem. 7.5]) was given by Doob.

THEOREM 17.4 [DOO, p. 603, Th. 1.5]. Suppose $f_1, \ldots, f_n : T \to R$, $\mathscr{N}$ is the smallest σ-algebra over T w.r.t. which $f_1, \ldots, f_n$ are measurable, $\mathscr{N} \subset \mathscr{M}$, and $g: T \to R$ is $\mathscr{N}$-measurable. Then there exists a Borel $h: R^n \to R$ such that $g(t) = h(f_1(t), \ldots, f_n(t))$ for $t \in T$.

18. Corrections to previous survey. We note corrections to [WG3] in addition to those given above:

p. 870 Valadier ([VA2, prop. I.14; or VA3, prop. 14]), gave a result similar to

Theorem 4.9, in some ways more general.

p. 872 In Corollary 5.4 "continuous" may be weakened to "Borel," since Gr F is still Suslin ([WG3, Th. 12.5]), as observed in [SW, Ch. II, end of proof of Th. 13].

p. 877 Add "$A \in \mathcal{M} \otimes \mathcal{B}(X)$ and $g: A \to Y$" to the hypothesis of Lemma 7.5.

p. 879 Theorem 8.4(ii) follows from Valadier's [VA3, Prop. 6].

p. 891 Delete the 10th and 11th lines of § 17 and replace by "... set-valued functions in abstract terms with rather good generality and attention to minimizing assumptions." Also delete the next to last sentence of the fourth paragraph of § 17 (refers to [CV2]).

p. 893 The date of the Copenhagen Colloquium cited under [CA1] was 1965, not 1975 (of priority significance).

p. 894 Delete [CV2] (confused by Wagner with [CV3]).

p. 896 [HV8] has been replaced by [HV11].

p. 901 To the result stated in addendum (ii), add the hypothesis "$\mathcal{M} = \mathcal{B}(T)$."

The following names were misspelled: Godet-Thobie on p. 889, Sarbadhikari on p. 899, and Wets on p. 902.

ACKNOWLEDGEMENT

It is a pleasure to acknowledge the extensive help I have received from others in preparation of this survey. As always, responsibility rests with the author.

Foremost I must record my profound debt to Shashi Srivastava, who patiently imparted abundant knowledge and numerous corrections, particularly in regard to § 4 and § 6. Extremely important help of similar character was received from John Burgess, particularly on § 5 and § 6. Further major sources of critique and other help have been Claude Dellacherie (§ 10), Siegfried Graf (§ 7 and § 4), Alexander Ioffe (§ 4), Dietrich Kölzow, Daniel Mauldin (§ 4), and Douglas Miller (§ 5). Thanks for helpful communications are also due to Cornelius Becker, Gerald Edgar, M. P. Ershov, Charles Himmelberg, Stephen Leese, Gerhard Mägerl, Ulrich Rieder, Michel Talagrand, Michel Valadier, and Fred Van Vleck.

It is interesting to compare the group acknowledged above with the corresponding group in [WG3]. The names have largely turned over. More significantly, the same is seen in comparing the pre-1977 and post-1976 bibliographies.

BIBLIOGRAPHY

The first part of the following bibliography consists of new references. It is coded in sequel to three previous bibliographies: the main bibliography of the paper cited below as [WG3], the bibliography of the addenda in proof to [WG3], and the bibliography in the paper cited below as [IF3]. For example, [CA23] by Castaing cited below is in sequel to 22 papers of his cited in [WG3]. We have not carried over the categorization by primes used in [WG3]. The second part of the bibliography gives updates to some incomplete citations in [WG3].

To find a reference whose coding is underlined, one goes to the bibliography in [WG3], possibly to the addenda in proof. Only a few of these are in the update section below.

New Bibliography

[AT] N. U. Ahmed and K. L. Teo, Comments on a selector theorem in Banach spaces, J. Opt. Th. Appl., 19 (1976), 117-118.

[ALE] A. Ascherl and J. Lehn, Two principles for extending probability measures, Manuscripta Math. 21 (1977), 43-50.

[BRM] L. W. Baggett and A. Ramsey, A functional analytic proof of a selection lemma, preprint.

[BA1] J. A. Bain, Optimal measurable selections for conditional expectations, preprint.

[BI2] J-M. Bismut, Intégrales convexes et probabilités, J. Math. Anal. Appl., 42 (1973), 639-673.

[BO3] N. Bourbaki, Topologie Général, Hermann, 1958.

[BRG1] J. Bourgain, Meetbare beschrijring van Baire Verzamelingen (Measurable description of Baire sets), Thesis, Vrije Universiteit Brussel, 1977.

[BRG2] ______, A stabilization property and its applications in the theory of sections, Séminaire Choquet (Initiation à l'Analyse), 1977/78, no. 5.

[BS1] J. P. Burgess, A selector principle for Σ_1^1 equivalence relations, Michigan Math. J., 24 (1977), 65-76.

[BS2] ______, A measurable selection theorem, preprint.

[BS3] ______, A selection theorem for group actions, Pacific J. Math., 80 (1979), 333-336.

[BS4] ______, Two selection theorems, Bull. Math. Soc. Grèce, 18 (1977), 121-136.

[BS5] ______, Sélections mesurables pour relations d'équivalences analytiques à classes G_δ, preprint.

[BS6] ______, Classical hierarchies from a modern standpoint, Part I. C-sets, preprint.

[BS7] ______, Careful choices, preprint.

[BM] J. P. Burgess and D. E. Miller, Remarks on invariant descriptive set theory, Fund. Math., 90 (1975), 53-75.

[CA23] C. Castaing, À propos de l'existence des sections séparément mesurables et séparément continues d'une multiapplication séparément mesurable et séparément semi-continue inférieurement, Séminaire d'Analyse Convexe, Montpellier (1976), Exposé no. 6.

[CE] A. Cellina, A selection theorem, Rend. Sem. Mat. Univ. Padova, 55 (1976), 143-149.

[CQ] G. Choquet, Theory of capacities, Annales de l'Institut Fourier 5 (1953/54), 131-295.

[CLI] A. Cornea and G. Licea, Une démonstration unifiée des théorèmes de sections de P. A. Meyer, Z. Wahrscheinlichkeitstheorie und Verw. Gebiete, 10 (1968), 198-202.

[CS8] A. Costé, Contribution à la théorie de l'intégration multivoque, Thèse, l'Université Pierre et Marie Curie, Paris 6, 1977.

[DB1] G. Debs, Sélections maximales d'une multi-application, preprint.

[DB2] ______, Sélection d'une multi-application à valeurs G_δ, preprint.

[DC2] C. Dellacherie, Capacités et Processus Stochastiques, Springer, 1972.

[DC3] ______, Ensembles analytiques: théorèmes de séparation et applications, Sém. Probabilité IX, 1973/74, 336-372, Lecture Notes in Math., Vol. 465, Springer, 1975.

[DC4] ______, Un cours sur les ensembles analytiques, London Symposium on Analytic Sets, 1979.

[DM] C. Dellacherie and P. A. Meyer, Un nouveau théorème de projection et de section, Sém. Probabilité IX, 1973/74, 239-245, Lecture Notes in Math., Vol. 465, Springer, 1975.

[DO1] E. E. Doberkat, On a representation of measurable automaton-transformations by stochastic automata, J. Math. Anal. Appl., 69 (1979), 455-468.

[DO2] ______, Deterministic representations of stochastic and nondeterministic automata, preprint.

[DK] S. Dolecki, Extremal measurable selections, Bull. Acad. Polon. Sci. Sér. Sci. Math. Astronom. Phys., 25 (1977), 355-360.

[DOO] J. L. Doob, Stochastic Processes, Wiley, 1953.

[ED] G. A. Edgar, Measurable weak sections, Illinois J. Math., 20 (1976), 630-646.

[EF] E. G. Effros, Transformation groups and C*-algebras, Ann. of Math., 81 (1965), 38-55.

[EK] I. Ekeland, Review of [CV3], Bull. Amer. Math. Soc., 84 (1978), 950-957.

[ER1] M. P. Ershov (Yershov), Extensions of measures and stochastic equations, Theory Probability Appl., 19 (1974), 431-444.

[ER2] ______, The Choquet theorem and stochastic equations, Analysis Mathematica, 1 (1975), 259-271.

[ER3] ______, On selectors in abstract spaces, Uspekhi, Mat. Nauk, 33 (1978), 205-206 (Russian).

[ER4] ______, Some selection theorems for partitions of sets without topology, these Proceedings.

[ES2] I. V. Evstigneev, Measurable choice and the continuum axiom, Soviet Math. Dokl., 19 (1978), 1-5.

[GF1] S. Graf, Measurable weak selections, these Proceedings.

[GF2] ______, A measurable selection theorem for compact-valued maps, Manuscripta Math. 27, (1979), 341-352.

[GF3] ______, Induced σ-homomorphisms and a parametrization of measurable sections via extremal preimage measures, preprint.

[HK] O. Hajek, Discontinuous differential equations I, preprint.

[HS] M. Hasumi, A continuous selection theorem for extremally disconnected spaces, Math. Ann., 179 (1969), 83-89.

[HE4] H. Hermes, On continuous and measurable selections and the existence of solutions of generalized differential equations, Proc. Amer. Math. Soc., 29 (1971), 535-542.

[HPV2] C. J. Himmelberg, T. Parthasarathy, and F. S. Van Vleck, On measurable relations, Fund. Math., to appear.

[HV10] C. J. Himmelberg and F. S. Van Vleck, Corrigendum and addendum to "Multifunctions on abstract measurable spaces and application to stochastic decision theory," Ann. Mat. Pura. Appl., to appear.

[HV11] ______, Existence of solutions for generalized differential equations with unbounded right-hand side, preprint.

[IF3] A. D. Ioffe, Survey of measurable selection theorems: Russian literature supplement, SIAM J. Control and Opt., 16 (1978), 728-732.

[IF4] ______, Single-valued representations of set-valued mappings, Trans. Amer. Math. Soc., 252 (1979), 133-145.

[IF5] ______, One-to-one Carathéodory representation theorem for multifunctions with uncountable values, Fund. Math., to appear.

[IF6] ______, Representation theorems for measurable multifunctions, these Proceedings.

[KS1] A. Kozek and Z. Suchanecki, Conditional expectations of selectors and Jensen's inequality, Proc. Conference on Mathematical Statistics, Wisła, 1978.

[KS2] ______, Multifunctions of faces for conditional expectations of selectors and Jensen's inequality, Institute of Mathematics, Polish Academy of Sciences, Preprint 154, 1978.

[LT] Le Van Tu, Omega-Polish spaces and measurable selections, J. Austral. Math. Soc., Ser. A 23 (1977), 257-265.

[LV3] V. L. Levin, On Borel selections of many-valued mappings, Siberian Math. J., 19 (1978), 617-623 (in Russian).

[LV4] ______, Review of [CV3], Math. Rev. 7169, 57 (1978), 941.

[LO1] V. Losert, A measure space without the strong lifting property, Math. Ann., 239 (1979), 119-128.

[LO2] ______, A counterexample on measurable selections and strong lifting, preprint.

[MG2] G. Mägerl, Zu den Schnittsätzen von Michael und Kuratowski & Ryll-Nardzewski, Dissertation, Universität Erlangen-Nürnberg, 1977.

[MG3] ______, A unified approach to measurable and continuous selections, Trans. Amer. Math. Soc., 245 (1978), 443-452.

[MT2] A. Maitra, Analytic sets with countable sections, Fund. Math., to appear.

[MT3] ______, Extensions of measurable functions, preprint.

[MR3] A. Maitra and B. V. Rao, Generalizations of Castaing's theorems on selectors, Colloq. Math., to appear.

[MRR] A. Maitra, B. V. Rao, and K. P. S. Bhaskara Rao, A problem in extension of measures, Illinois J. Math., 23 (1979), 211-216.

[MN] D. A. Martin, Borel determinancy, Ann. of Math., 102 (1975), 363-371.

[MD1] R. D. Mauldin, The unboundedness of the Cantor-Bendixson order of some analytic sets, Pacific J. Math., 74 (1978), 167-177.

[MD2] ______, Borel parametrizations, Trans. Amer. Math. Soc., 250 (1979), 223-234.

[MD3] ______, Some selection theorems and problems, these Proceedings.

[ME1] P. A. Meyer, Une présentation de la théorie des ensembles sousliniens. Application aux processes stochastiques, Séminaire de Théorie de Potential. Institut Henri Poincairé, 7e année, 201-218.

[ME2] ______, Une nouvelle démonstration des théorèmes de section, Sém. Probabilité III, 1967/68, pp. 155-159. Lecture Notes in Math., Vol. 88, Springer, 1969.

[ML1] D. E. Miller, On the measurability of orbits in Borel actions, Proc. Amer. Math. Soc., 63 (1977), 165-170.

[ML2] ______, A selector for equivalence relations with G_δ orbits, Proc. Amer. Math. Soc., 72 (1978), 365-369.

[ML3] ______, Borel selectors for separated quotients, preprint.

[MB] J. A. Mühlbauer, Bayes-Losungen für Sequenztests mit kontinuierlicher Zeit und nicht notwendig endlich vielen Hypothesen, Inaugural-Dissertation, Marburg/Lahn (1977).

[NA] S. Nababan, A Fillipov-type lemma for functions involving delays and its application to time-delayed optimal control problems, J. Opt. Th. and Appl., 27 (1979), 357-376.

[NI] T. Nishiura, Two counterexamples for measurable relations, Rocky Mountain J. Math., 9 (1979), 499-501.

[NW] A. Nowak, Discounted dynamic programming on Euclidean spaces, Applicationes Mathematicae 16 (1979), 465-473.

[PV] Phan Van Chuong, Densité des sections extremales d'une multiapplication mesurable, Séminaire d'Analyse Convexe, Montpellier, 1979, Exposé No. 5.

[RI1] U. Rieder, Measurable selection theorems for optimization problems, Manuscripta Math., 24 (1978), 115-131.

[RI2] ______, Measurable selection of fixed points, preprint.

[SA1] J. Saint-Pierre, Caractérisations des $\mathscr{F}$-sousliniens d'un produit, C. R. Acad. Sci. Paris Sér. A, 283 (1976), 583-585.

[SA2] ______, Séparation simultanée d'un ensemble analytique de suites d'ensembles $\mathscr{F}$-Sousliniens, Ibid. Sér. A, 285 (1977), 933-936.

[SA3] ______, Caractérisation des $\mathscr{F}$-sousliniens d'un produit et théorèmes de sections, Séminaire D'Analyse Convexe, Montpellier (1977), Exposé no. 8.

[SA4] ______, Séparation d'un schema de Souslin, Ibid, Exposé no. 16.

[SA5] ______, Points exposés d'une multi-application à valeurs convexes de R^n, Ibid, 1978, Exposé no. 7.

[SRY1] J. Saint-Raymond, Boréliens à coupes K_σ, Bull. Soc. Math. France, 104 (1976), 389-400.

[SRY2] ______, Fonctions boréliennes sur un quotient, Bull. Sci. Math. 2e Sér., 100 (1976), 1-7.

[SRY3] ______, Sélections mesurables pour la structure borélienne d'Effros, Bull. Sci. Math., 2e Sér., 101 (1977), 353-355.

[SWE] G. Salinetti and R. J-B. Wets, On the convergence of closed-valued measurable multifunctions, preprint.

[SR2] H. Sarbadhikari, A selection theorem for multifunctions, Proc. Amer. Math. Soc., 71 (1978), 285-288.

[SR3] ______, Selection theorems for partitions of complete metric spaces, Bull. Acad. Polon. Sci., Sér. Sci. Math. Astronom. Phys., to appear.

[SS] H. Sarbadhikari and S. M. Srivastava, Parametrization of G_δ-valued multifunctions, Trans. Amer. Math. Soc., to appear.

[SW] L. Schwartz, Radon measures on arbitrary topological spaces and cylindrical measures, Oxford University Press, 1973.

[SBE1] S. Shreve and D. Bertsekas, Equivalent stochastic and deterministic optimal control problems, Proc. IEEE Conf. on Decision and Control, 1976, 705-709.

[SBE2] ______, Universally measurable policies in dynamic programming, Mathematics of Operations Research, 4 (1979), 15-30.

[SVS1] S. M. Srivastava, A representation theorem for closed valued multifunctions, Bull. Pol. Acad. Sci. Sér. Sci. Math. Astronom. Phys., to appear.

[SVS2] ______, A representation theorem for G_δ-valued multifunctions, Amer. J. Math., to appear.

[SVS3] ______, Selection theorems for G_δ-valued multifunctions, Trans. Amer. Math. Soc., 254 (1979), 283-293.

[SVS4] ______, Studies in the theory of measurable multifunctions, Thesis, Indian Stat. Inst. (1978).

[SVT] V. V. Srivatsa, Existence of measurable selectors and parametrizations for G_δ-valued multifunctions, preprint.

[TA2] M. Talagrand, Sélection mesurable de mesures maximales simpliciales, Bull. Sci. Math., 2e Sér. 102 (1978), 49-56.

[TA3] ______, Non-existence de certaines sections mesurables et application à la théorie du relèvement, C. R. Acad. Sci. Paris Sér. A, 286 (1978), 1183-1185.

[TA4] ______, En général il n'existe pas de relèvement linéaire borélien fort, Ibid, 287 (1978), 633-634.

[TA5] ______, Non-existence de certaines sections mesurables et contre-exemples en théorie du relèvement, these Proceedings.

[VA9] M. Valadier, Sur le plongement d'un champ mesurable d'espaces metriques dans un champ trivial, Ann. Inst. Henri Poincaré, Sect. B. to appear; also in Séminaire d'Analyse Convexe, Montpellier (1977), Exposé 14.

--- M. P. Yershov -- see Ershov.

[WG3] D. H. Wagner, Survey of measurable selection theorems, SIAM J. Control and Opt., 15 (1977), 859-903.

[WR] J. Warga, An implicit function theorem without differentiability, Proc. Amer. Math. Soc., 69 (1978), 65-69.

[WEI] H. von Weizsäcker, Some negative results in the theory of lifting, Measure Theory (Proc. Conf. Oberwolfach, 1975), pp. 159-172. Lecture Notes in Math., Vol. 541, Springer, 1976.

[WE4] E. Wesley, On the existence of absolutely measurable selection functions, preprint.

Update of Previous Bibliography

[CV3] C. Castaing and M. Valadier, Convex analysis and measurable multifunctions, Lecture Notes in Mathematics, Vol. 580, Springer, 1977.

[CM1] D. Cenzer and R. D. Mauldin, Inductive definitions: Measure and category, Advances in Math., to appear.

[CM2] ______, Measurable parametrizations and selections, Trans. Amer. Math. Soc., 245 (1978), 399-408.

[HPV] C. J. Himmelberg, T. Parthasarathy, and F. S. Van Vleck, Optimal plans for dynamic programming problems, Mathematics of Operations Research, 1 (1976), 390-394.

[KLM] R. R. Kallman and R. D. Mauldin, A cross section theorem and application to C*-algebras, Proc. Amer. Math. Soc., 69 (1978), 57-61.

[KA2] J. Kaniewski, A selection theorem for partitions of Borel sets into compact subsets, Bull. Acad. Polon. Sci. Sér. Sci. Math. Astron. Phys., 25 (1977), 1073-1081.

[LE5] S. J. Leese, Measurable selections and the uniformisation of Souslin sets, Amer. J. Math., 100 (1978), 19-41.

[LN2] J. Lehn, Remark on measurable graph theorems, Proc. Amer. Math. Soc., 63 (1977), 46-48.

[LM] J. Lehn and G. Mägerl, On the uniqueness of preimage measures, Z. Wahrscheinlichkeitstheorie und Verw. Gebiete, 33 (1977), 333-337.

[SR] H. Sarbadhikari, Some uniformization results, Fund. Math., 97 (1977), 209-214.

[VA4] M. Valadier, Convex integrands on Souslin locally convex spaces, Pac. J. Math., 59 (1975), 267-276.

Addendum in proof

To the new reference on set-valued measures mentioned in section 1 should be added Godet-Thobie, these Proceedings.

Bain has withdrawn [BAI] (section 9), having learned via Dynkin that a more general treatment was given by Dynkin and Evstigneev (referenced in [ES]).

ALMOST STRONG LIFTINGS AND τ-ADDITIVITY

by

A.G.A.G.Babiker and W.Strauss

1. Introduction

Let X be a completely regular space and μ a finite topological measure (i.e. a Baire or a Borel measure on X). In this note we discuss the existence of strong liftings for the associated topological measure space. It is shown that when X is locally metrizable, the τ-additivity of μ relative to the given topology on X, which is always necessary for the existance of almost strong liftings, is sufficient to ensure that all liftings are almost strong. A stronger τ-additivity criterion for a given lifting to be almost strong is given and this is used to show that a τ-additive Lebesgue measure space may admit liftings which are not almost strong.

2. Notation and definitions

Throughout X denotes a topological space with a completely regular topology $\mathcal{T}$, $\mathcal{A}$ the σ-algebra of all Baire subsets of X and $\mathcal{B}$ the σ-algebra of all Borel subsets of X. By a Baire (Borel) measure on X we mean a finite, non-negative countably additive set function defined on $\mathcal{A}$ ($\mathcal{B}$). Furthermore we assume that Borel measures are regular in the sense of inner approximation by closed sets. [Note that the regularity of Baire measures is a consequence of finiteness and countable additivity]. A Baire (Borel) measure is said to be τ-additive if $\lim \mu(Q_\alpha) = 0$ for any decreasing net $\{Q_\alpha\}$ of zero (closed) sets such that $Q_\alpha \downarrow \emptyset$, [16]. The measure μ is said to be completion regular [5], if for any Borel set B there exist Baire sets A_1 and A_2 such that $A_1 \subset B \subset A_2$ and $\mu(A_1) = \mu(A_2)$. Denote by $\mathcal{A}(\mu)$ ($\mathcal{B}(\mu)$) the completion of $\mathcal{A}$ ($\mathcal{B}$) w.r.t. μ. A topological measure space is the quadriple $(X, \mathcal{T}, \mathcal{F}, \mu)$ where $\mathcal{F}$ is either $\mathcal{A}(\mu)$ or $\mathcal{B}(\mu)$.

A map $\rho: \mathcal{F} \to \mathcal{F}$ is called a lifting [7], if for all $A, B \in \mathcal{F}$

(i) $\mu[\rho(A) \Delta A] = 0$

(ii) $\mu(A \Delta B) = 0 \Rightarrow \rho(A) = \rho(B)$

(iii) $\rho(A \cap B) = \rho(A) \cap \rho(B)$

(iv) $\rho(A \cup B) = \rho(A) \cup \rho(B)$

(v) $\rho(X) = X$, $\rho(\emptyset) = \emptyset$.

ρ is a monotone lifting [9] if ρ satisfies (i), (ii), (v) and

(vi) $A \subset B \Rightarrow \rho(A) \subset \rho(B)$.

Clearly every lifting is a monotone lifting.

A lifting (monotone lifting) ρ is said to be strong if $\rho(G) \supset G$ for all open $G \in \mathfrak{A}$.

ρ is almost strong if there is a set $N \in \mathcal{F}$ with $\mu(N)=0$ and $\rho(G) \supset G \setminus N$ for all open $G \in \mathfrak{A}$.

3. Proposition

If $(X, \mathcal{T}, \mathcal{F}, \mu)$ has an almost strong monotone lifting then μ is τ-additive. Furthermore if μ is a Baire measure then $\mathcal{B} \subset \mathfrak{A}(\mu)$ i.e. μ is completion regular.

Proof:

Let ρ be an almost strong lifting and (Z_α) a decreasing net of zero sets such that $Z_\alpha \searrow \emptyset$. Then, with $G_\alpha = X \setminus Z_\alpha$ we have G_α is open Baire and $G_\alpha \nearrow X$. Write $c = \sup_\alpha \mu(G_\alpha)$. Since ρ is almost strong we can find a set $N \in \mathcal{F}$ with $\mu(N) = 0$ such that $\rho(G_\alpha) \supset G_\alpha \setminus N$ for all α. Find an increasing sequence (α_n) such that $c = \lim_n \mu(G_{\alpha_n})$ and let $H = \bigcup_n G_{\alpha_n}$. Clearly, for each α we have $\mu(G_\alpha \setminus H) = 0$. Hence $\rho(H) \supset \rho(G_\alpha) \supset G_\alpha \setminus N$ for all α, i.e. $\rho(H) \supset X \setminus N$ and so $c = \mu(H) = \mu(\rho(H)) = \mu(X)$. Thus $\lim_\alpha \mu(Z_\alpha) = 0$ and the τ-additivity of μ is established. If $\mathcal{F} = \mathfrak{A}(\mu)$, then the τ-additivity of μ gives a unique extension υ of μ to the Borel sets $\mathcal{B}$. υ satisfies

$$\upsilon(G) = \sup\{\mu(Z) : Z \text{ zero set, } Z \subset G\}$$

for all open G.

Also since every open set G is a union of open Baire sets, $\rho(G) \cup N \supset G$. But $\rho(G) \cup N \in \mathfrak{A}(\mu)$, i.e. every open, and hence every Borel set is μ-measurable.

We discuss the following question: Under what conditions is τ-additivity sufficient for the existence of an almost strong monotone lifting? It follows from proposition (3) that any τ-additive non-completion regular Baire measure (and hence any non-completion regular Baire measure on a compact space) provides an example where almost strong monotone lifting do not exist. (See [3] for an example of a compact space supporting a non-atomic non-completion regular measure.) For Borel measures, Losert [11] has given an example of a Borel measure μ on compact space X such that $(X, \mathcal{T}, \mathcal{B}(\mu), \mu)$ does not admit a strong lifting. On the other hand, Fremlin [6] has shown that under the Continuum Hypothesis τ-additivity is sufficient for the existence of strong liftings when X has a base of cardinal c for its topology. In the following theorem, we show that for locally metrizable spaces τ-additivity is not only sufficient for the existence of strong liftings but it insures that every lifting is almost strong.

4. Theorem

Let X be locally metrizable und μ a Borel measure on X. Then every monotone lifting (and hence every lifting) for $(X,\mathcal{T},\mathfrak{F},\mu)$ is almost strong if and only if μ is τ-additive.

Proof:

Since μ is finite and complete there exists at least one lifting for $(X,\mathcal{T},\mathfrak{F},\mu)$ [12]. The necessity of τ-additivity then follows from proposition (3).

Conversely, suppose that μ is τ-additive and ρ a monotone lifting for $(X,\mathcal{T},\mathfrak{F},\mu)$. We may assume that $\operatorname{supp}(\mu)=X$. Write, $E=\bigcup\{(G\setminus\rho(G)) : G$ is an open Baire subset of $X\}$ and let F be any μ-measurabe envelope of E and Q a closed set such that $Q\subset F$. We prove that $\mu(Q)=0$. Suppose that $\mu(Q)=\delta>0$. Local metrizability of X and τ-additivity of μ give an open metrizable set $U\subset X$ such that $\mu(U\cap Q)>0$. Since every open subset of X, and hence every open subset of U, has positive measure, U satisfies the countable Chain Condition. U, being metrizable, must therefore be separable and hence second countable [8]. p.120. Let $\{V_n\}$ be a countable base for the topology of U and write $D=\bigcup_n[V_n\setminus\rho(V_n)]$. It follows that $\mu(D)=0$ and for every open set $V\subset U$ $\rho(V)\cup D\supset V$. Thus for any open set $G\subset X$ we have, $[G\setminus\rho(G)]\cap U\subset(G\cap U)\setminus\rho(G\cap U)\subset D$.

Hence $E\cap U\subset D$ and so $E\cap U$ is measurable and $\mu(E\cap U)=0$. Therefore $\mu(F\cap U)=0=\mu(Q\cap U)$-a contradiction. We have proved that $\mu^*(E)=0$ i.e. ρ is almost strong.

The following two corollaries are immediate from theorem (4) and proposition (3).

5. Corollary.

For a Baire measure μ on a locally metrizable space X, the following are equivalent:

(i) Every monotone lifting for $(X,\mathcal{T},\mathfrak{F},\mu)$ is almost strong.

(ii) μ is τ-additive and completion regular.

6. Corollary.

Every τ-additive Borel measure on a locally metrizable space is completion regular outside a null set.

Theorem (4) implies that under a certain mild set theoretic assumption every lifting of a metric measure space is almost strong. Let us recall that a cardinal number p is said to be non-measurable if the only continuous measure on the discrete space of cardinal p is the zero measure. Otherwise p is called measurable [14] and [15]. It is consistent with the usual axioms of set theory to assume that all cardinal numbers are non-measurable. Every measure on a paracompact space X is τ-additive, provided that X contains no closed discrete subspace of measurable cardinal [16] . Thus,

7. Corrollary.
Suppose that every closed discrete subspace of a metric space X has non-measurable cardinal. Then every monotone lifting (and hence every lifting) for $(X,\mathcal{T},\mathcal{F},\mu)$ is almost strong.

8. Remarks.
(a) The cardinality restriction of corollary (7) does not imply the τ-additivity of μ in locally metrizable spaces. In [1] an example of a non-τ-additive measure is given on a locally metrizable, locally compact and realcompact space X.

(b) Defining liftings as appropriate multiplicative linear operators on $\mathcal{L}^\infty(X,\mu)$ the algebra of all bounded μ-measurable functions, a strong lifting ρ is characterized by the condition: $\rho(f)=f$, for all $f\in C^*(X)$ the space of all bounded continuous realvalued functions on X. Suppose that $X=\mathrm{supp}(\mu)$, the support of μ. Then for each $x\in X$, the evaluation $f\to f(x)$; $f\in C^*(x)$ can be extended to a character χ_x of $\mathcal{L}^\infty(X,\mu)$ in such a way that $\chi_x(g) = \chi_x(h)$ if and only if $g\equiv h$ ω.r.t.μ. [10], p.132. For any almost strong lifting ρ let N be the universal null sut such that $\rho(f)|N = f|N$ for all $f\in C^*(X)$. We can define ρ' by

$$\begin{aligned}\rho'(g)(x) &= \rho(g)(x),\ x\in X\setminus N\\ &= \chi_x(g),\quad x\in N\ .\end{aligned}$$

Clearly ρ' is a strong lifting. Thus from theorem (4) we have if X is locally metrizable and $\mathrm{supp}(\mu)=X$ then $(X,\mathcal{T},\mathcal{F},\mu)$ has a strong lifting.

The following Lemma shows that, in general, a stronger form of τ-additivity is required for the existence of strong liftings. For a given lifting ρ of $(X,\mathcal{T},\mathcal{F},\mu)$ let $\mathcal{T}_\rho = \{E\in\mathcal{F} : E\subset\rho(E)\}$ be the topology associated with ρ[7], p.54. Clearly ρ is strong if and only if $\mathcal{T}\subset\mathcal{T}_\rho$.

9. Lemma.
A lifting ρ for $(X,\mathcal{T},\mathcal{F},\mu)$ is almost strong if and only if μ is τ-additive relative to $\mathcal{T}_\rho\vee\mathcal{T}$, the coarsest topology finer than $\mathcal{T}_\rho$ and $\mathcal{T}$.

Proof.
Since in either case μ is τ-additive relative to $\mathcal{T}$ and hence in the Baire case μ admits a unique Borel extension, we may assume that μ is

a Borel measure.

Suppose that ρ is almost strong. Note that μ is always τ-additive relative to $\mathcal{T}_\rho$ [7], p.4o. Let N be a null set such that $G \subset \rho(G) \cup N$ for all $G \in \mathcal{T}$. Let $\{H_\alpha\}$ be an increasing net of elements of $\mathcal{T}_\rho \vee \mathcal{T}$ such that $H_\alpha \uparrow X$. Since ρ is strong on $X \setminus N$ and so $\mathcal{T}|(X \setminus N) \subset \mathcal{T}_\rho|(X \setminus N)$, for each α there is $U_\alpha \in \mathcal{T}_\rho$ such that $H_\alpha \setminus N \subset U_\alpha \setminus N$, $\mu(U_\alpha) = \mu(H_\alpha)$ and $\{U_\alpha\}$ is increasing. Thus

$$\mu(H_\alpha) = \mu(U_\alpha) \to \mu(\bigcup_\alpha U_\alpha) = \mu(X \setminus N) = \mu(X)$$

i.e. μ is τ-additive relative to $\mathcal{T}_\rho \vee \mathcal{T}$.

Conversely suppose that μ is τ-additive relative to $\mathcal{T}_\rho \vee \mathcal{T}$. For each $G \in \mathcal{T}$, $G \setminus \rho(G) = G \cap \rho(G^c) \in \mathcal{T}_\rho \vee \mathcal{T}_\mu$ and $\mu[G \cap \rho(G^c)] = 0$. By τ-additivity $\mu[\bigcup_{G \in \mathcal{T}} (G \cap \rho(G^c))] = 0$ i.e. ρ is almost strong.

We now apply Lemma (8) to show that a τ-additive Lebesgue measure may admit liftings which are not almost strong. Let us recall that a non-atomic measure μ on X is said to be a Lebesgue measure if there exist a μ-null set $E \in \mathcal{F}$, a Lebesgue set $D \subset I$ the unit interval with standard Lebesgue measure $m(D)=0$ and a bimeasurable bijection $\varphi : x \setminus E \to I \setminus D$ [13] and [2]. It can be easily shown [4] that if μ is a tight Lebesgue measure then every lifting of $(X, \mathcal{T}, \mathcal{F}, \mu)$ is almost strong. The following example shows that the assumption of tightness of μ cannot be omitted.

10. Example.

Let X be the unit interval I with the topology generated by all left-closed right-open intervals $[a,b) = \{x \in X : a \leq x < b\}$ $a,b \in I$. This topology is clearly finer than the unsual topology on I, and since X is hereditary Lindelöf [8], p.59, every open subset of X is a Borel subset of I. Thus the identity map $i: X \to I$ is Borel bimeasurable. So the standard Lebesgue measure m on I induces a Borel measure μ on X which is Lebesgue.

$(X, \mathcal{T}, \mathcal{F}, \mu)$ has a lifting ρ such that $\rho((a,b]) = (a,b]$ [7], p.123. Thus for any $x \in X$, $\{x\} = (0,x] \cap [x,1) \in \mathcal{T}_\rho \vee \mathcal{T}$. i.e. $\mathcal{T}_\rho \vee \mathcal{T}$ is the discrete topology on X, and so μ is not τ-additive relative to $\mathcal{T}_\rho \vee \mathcal{T}$ by Lemma (7) ρ is not almost strong.

11. Example

Define $D = \{(1/n,k/n) : 1 \leq k \leq n \text{ and } n\in N\}$, $E = \{(0,y) : 0 \leq y \leq 1\}$, $C = D\cup E$. Define a topology on C such that each $d\in D$ is isolated, and for each $e = (0,y)\in E$ the sets $T_n = \{(x,z) : 0 \leq x \leq 1/n, |z-y| \leq x\}$ form a neighborhood base of e for $n\in N$. Then C is locally compact, locally metrizable, and realcompact. Let Y be the standard non-measurable set on the unit interval I and ψ the measure on Y defined by $\psi(B) = m^*(B)$ for Borel $B\subset Y$ when m is ordinary Lebesgue measure on I. Since Y is m-thick, ψ is a measure. Moreover Y is separable metric and therefore ψ is τ-additive and $\psi(K) = 0$ for compact K.

Let $X = C\times Y$ be the product space and enumerate $D = (d_n)$. If μ_n is the measure on X induced by the ψ-measure on $\{d_n\}\times Y$, put

$$\mu = \sum_{n=1}^{\infty} \frac{1}{2^n} \mu_n .$$

Then $\mu(X) = 1$, supp $(\mu) = X$, μ is purely τ-additive, i.e. τ-additive and $\mu(K) = 0$ for all compact K, and X is locally metrizable but not metrizable.

References

[1] A.G.A.G.Babiker, Some measure theoretic properties of completely regular spaces; Rend. Fisci. Mat., Academia Nazionale dei Lincei Rome 59 (1975), 677-681.

[2] A.G.A.G.Babiker, Lebesgue measures on topological spaces; Mathematika 24(1977),52-59.

[3] A.G.A.G.Babiker and J.D.Knowles, An example concerning completion regular measures, images of measurable sets and measurable selections; Mathematika 25(1978),120-124.

[4] A.G.A.G.Babiker,Lifting properties of Lebesque measures on topological spaces, (to appear).

[5] S.K.Berberian, Measure and Integration; New York 1965.

[6] D. Fremlin, On two theorems of Mokobodzki, (preprint).

[7] A. and C.Ionescu Tulcea, Topics in the Theory of Lifting; Berlin-Heidelberg-New York 1969.

[8] J.L.Kelly, General Topology, New York 1955.

[9] D.Kölzow, Differentiation von Maßen; Lecture Notes in Mathematics No. 65, Berlin-Heidelberg-New York 1968

[10] V.L.Levin, Convex integral functionals and the theory of lifting; Russian Math.Surveys 30(1975),119-184, from Uspekhi Mat.Nauk 30 (1975).

[11] V.Losert, A measure space without the strong lifting property, Math.Ann. 239(1979),119-128.

[12] D.Maharam, On a theorem of von Neumann; Proc.Amer.Math.Soc.9, (1958),987-994.

[13] V.A.Rochlin, On the fundamental ideas of measure theory; Math. Sb.(N.S.)25(67)1949, 107-150(Russian). AMS translation 71(1952) (English).

[14] A.Tarski, Über unerreichbare Kardinalzahlen; Fund.Math.30(1938), 68-89.

[15] S.Ulam, Zur Maßtheorie der allgemeinen Mengenlehre; Fund.Math. 16(1931),140-150.

[16] V.S.Varadarajan, Measures on topological spaces; Math.Sb.(N.S.) 55(97)1961,33-100. (Russian),AMS Translations 48,14-228 (English).

A.G.A.G. Babiker
Mathematisches Institut A
Universität Stuttgart
Pfaffenwaldring 57
D-7000 Stuttgart 80
Federal Republic of Germany

and

School of Mathematical Sciences
University of Khartoum
P.O.Box 321
Khartoum, Sudan.

W. Strauß
Mathematisches Institut A
Universität Stuttgart
Pfaffenwaldring 57
D-7000 Stuttgart 80
Federal Republic of Germany

MEASURE SPACES IN WHICH EVERY LIFTING IS AN ALMOST $\mathcal{H}$-LIFTING

by

A.G.A.G.Babiker and W.Strauß

$\mathcal{H}$-liftings were introduced by V.L.Levin in [5] where he showed that the existence of "good" integral representations for the subdifferentials of convex integral functionals is equivalent to the existence of certain $\mathcal{H}$-liftings.

For a measure space $(X,\mathcal{F},\mu)$ and a fixed subset $\mathcal{H}$ of $\mathcal{L}^\infty = \mathcal{L}^\infty(X,\mu)$ a lifting ρ of $\mathcal{L}^\infty$(compare [4],Def.2) is said to be an <u>$\mathcal{H}$-lifting</u> if $\rho(f)=f$ for all $f\in\mathcal{H}$, and it is called an <u>almost $\mathcal{H}$-lifting</u> if there exists a null set N such that $\rho(f)(x) = f(x)$ for all $x\in X\setminus N$ and all $f\in\mathcal{H}$. We call $(X,\mathcal{F},\mu)$ an $\mathcal{H}$-space if every lifting of $\mathcal{L}^\infty(X,\mu)$ is an almost $\mathcal{H}$-lifting. In this note we develop a criterion which characterizes $\mathcal{H}$-spaces, among the complete decomposable measure spaces, in terms of the net-additivity of μ over certain subalgebras of $\mathcal{L}^\infty(X,\mu)$. We show that this criteria is localizable and we use it to obtain and extend some known results concerning strong liftings.

When X is a Hausdorff topological space, μ a Borel measure on X and $\mathcal{H} = C^*(X)$ the space of bounded continuous functions on X, $C^*(X)$-liftings (almost $C^*(X)$-liftings) are known as strong liftings (almost strong liftings)[4], Chapt. VIII. The problem of the existence of such liftings received the attention of a number of authors [1],[2],[4],[6],[7] and [8]. For a Radon measure on a locally compact metrizable space it is known, not only that strong liftings exist, but every lifting is almost strong [4], Chapt. VII, Th. 8 (see also [1] for an extension of this). Under the continuum hypothesis, strong liftings are shown to exist when X has a base, of cardinal at most c, for its topology. On the other hand, Losert [6] has provided an example of a compact measure space admitting no strong lifting.

1. <u>Permissible algebras</u>.

We assume throughout that $(X,\mathcal{F},\mu)$ is a complete decomposable measure space. Under the pointwise operations $\mathcal{L}^\infty = \mathcal{L}^\infty(X,\mu)$ is both an algebra and a lattice. By a subalgebra of $\mathcal{L}^\infty$ we shall always mean a subalgebra of $\mathcal{L}^\infty$ containing 1 and closed under the lattice operations. We denote by $\mathcal{T}$ the topology on L^∞ inherited from the topology of local mean convergence on L^1_{loc}.

We say that a subalgebra $\mathcal{A}$ of $\mathcal{L}^\infty$ is <u>permissible</u> if the following conditions are satisfied:

(i) There exists a null set N such that, for all $f, g \in \mathcal{A}$, $f \equiv g \Rightarrow f(x) = g(x)$, for all $x \in X \setminus N$.

(ii) $\pi(\mathcal{A})$ is $\mathcal{C}$-closed.

(iii) $\pi(\mathcal{A}) \cap L^1$ is cofinal in $L^\infty \cap L^1$, i.e. for each $f \in \mathcal{L}^\infty \cap \mathcal{L}^1$, there exists $g \in \mathcal{A} \cap \mathcal{L}^1$ such that $g \geq f$ a.e. (μ).

With this definition the following lemma is immediate.

2. Lemma

Suppose that $\mathcal{A}$ is a permissible subalgebra of $\mathcal{L}^\infty$, $f_n \in \mathcal{A}$ and $f = \limsup_n f_n \in \mathcal{L}^\infty$. Then there exists $f' \in \mathcal{A}$ such that $f' \equiv f$.

3. Subspaces We now show that permissibility is inherited by certain subspaces. For $G \in \mathcal{F}$, let

$$\mathcal{F}_G = \{E \in \mathcal{F}: E \subset G\},$$
$$\mu_G = \mu|\mathcal{F}_G,$$
$$\mathcal{L}^\infty_G = \mathcal{L}^\infty|G = \mathcal{L}^\infty(G, \mu_G);$$
$$\mathcal{A}_G = \mathcal{A}|G \quad .$$

Proposition

Suppose that $\mathcal{A}$ is a permissible subalgebra of $\mathcal{L}^\infty$, $G \in \mathcal{F}$, and $\mu(G) < \infty$. Then $\mathcal{A}_G$ is a permissible subalgebra of $\mathcal{L}^\infty_G$.

It is plain that $\mathcal{A}_G$ is a subalgebra satisfying (iii), and (ii) follows by use of lemma 2. To prove (i) it is sufficient to show that there is a null set $N \subset G$ s.t. for all $f \in \mathcal{A}$ with $fI_G \equiv 0$ follows $f(x) = 0$ for all $x \in G \setminus N$. Cindition (iii) for $\mathcal{A}$ ensures

$$c = \inf\{\langle g, \mu\rangle \mid g \in \mathcal{A},\ 0 \leq g \leq 1,\ g \geq I_G \text{ a.e. } (\mu)\} < \infty \quad ,$$

so choose a decreasing sequence in $\mathcal{A}$ with $0 \leq g_n \leq 1$, $g_n \geq I_G$ a.e.(μ), and $\lim_n \langle g_n, \mu\rangle = c$. By lemma 2 find $h \in \mathcal{A}$ s.t. $h \equiv g = \lim_n g_n$ and $0 \leq h \leq 1$. Then $N'' = \{x \in G \mid h(x) = 0\}$ is a null set. If N' is the null set s.t. for all $f \in \mathcal{A}$ with $f \equiv 0$ follows $f(x) = 0$ for all $x \in X \setminus N'$, then $N = N'' \cup (N' \cap G)$ is the desired null set.

4. The criteria

For $\mathcal{D} \subset \mathcal{L}^\infty$ we denote by $\text{alg}(\mathcal{D})$ the uniformly closed subalgebra of $\mathcal{L}^\infty$ generated by the elements of $\mathcal{D}$.

We say that μ is net additive on $\mathcal{D}$ if for any (pointwise) increasing net $\{f_\alpha\}_{\alpha \in a}$ of elements of $\mathcal{D}$, the pointwise limit $f = \lim_\alpha f_\alpha \in \mathcal{L}^0$ and, for all $E \in \mathcal{F}_0$

$$\lim_\alpha \langle f_\alpha, \mu_E\rangle = \langle f, \mu_E\rangle \ .$$

The last equality is meant to be in its fullest sense, i.e. $<f,\mu_E>$ may take the value ∞, in which case $\lim <f_\alpha, \mu_E> = \infty$.

In the case when $\mathcal{D}$ is a uniformly closed subalgebra of $\mathcal{L}^\infty$, it is easily seen that μ is net additive if and only if for any subfamily $\mathcal{E} \subset \mathcal{D}^+$ of nonnegative elements of $\mathcal{D}$, the pointwise supremum $f = \sup \{e: e \in \mathcal{F}\}$ is $\mathcal{F}$-measurable and for all $F \in \mathcal{F}_o$

$$<f,\mu_F> = \sup \{<e,\mu_F> : e \in \mathcal{E}\}.$$

Theorem 1

The measure space $(X,\mathcal{F},\mu)$ is an $\mathcal{H}$-space if and only if for any permissible subalgebra $\mathcal{A}$ of $\mathcal{L}^\infty$, μ is net additive on $\text{alg}(\mathcal{A} \cup \mathcal{H})$.

First assume $(X,\mathcal{F},\mu)$ to be an $\mathcal{H}$-space. Then choose a decomposition. If $\mathcal{A}$ is admissible then by proportion 3 on each $\mathcal{A}_C$ for $C\in \mathcal{C}$ can be constructed a lifting ρ_C of $\mathcal{L}^\infty{}_C$ s.t. outside a null set $\mathcal{A}_C \subseteq \rho(\mathcal{L}^\infty{}_C)$. Picking the ρ_C together we get a lifting ρ of $\mathcal{L}^\infty$ with $\mathcal{A}|X\setminus N' \subset \rho(\mathcal{L}^\infty)|X\setminus N'$ for some null set N'. By hypothesis ρ is an almost $\mathcal{H}$-lifting and this infers $\text{alg}(\mathcal{A} \cup \mathcal{H})|X\setminus N \subset \rho(\mathcal{L}^\infty)|X\setminus N$ for a null set N. Now the τ-additivity on $\text{alg}(\mathcal{A} \cup \mathcal{H})$ follows by an appeal to Theorem 3 in Chapt. III of [4], since μ is net additive on $\rho(\mathcal{L}^\infty)$.

Conversely, suppose that μ is net additive on $\text{alg}(\mathcal{A} \cup \mathcal{H})$ for all permissible $\mathcal{A}$. Let ρ be any lifting of $\mathcal{L}^\infty$. Then clearly $\mathcal{A} = \rho(\mathcal{L}^\infty)$ is permissible; so μ is net additive on $\text{alg}(\mathcal{A} \cup \mathcal{H})$.

Let A be the set of finite subsets of $\mathcal{H}$, partially ordered by inclusion, and for each $\alpha \in A$ let

$$f_\alpha = \sum_{h\in\alpha} |h - \rho(h)|.$$

Now apply τ-additivity of the measure μ to the family (f_α) giving $\sum_{h\in\mathcal{H}} |h - \rho(h)| = 0$ outside a null set, i.e. ρ is an almost strong $\mathcal{H}$-lifting. The above net additivity criterion for $\mathcal{H}$-spaces is not valid if we relax the assumption of decomposability. Complete, contracted locally determined and non-decomposable measure spaces exist [3], sec. 31, ex. 9. Any such measure space cannot admit any lifting [9] and so is trivially an $\mathcal{H}$-space for any $\mathcal{H} \subset \mathcal{L}^\infty$. However, it is not possible for μ to be net additive on $\mathcal{L}^\infty$ in this case, as this would imply that points of X constitute a decomposition of $(X,\mathcal{F},\mu)$.

5. Applications

We now state applications of the criteria in theorem 1. The first one says that the property of being an $\mathcal{H}$-space is a local one.

Theorem 2

Let $(X,\mathcal{F},\mu)$ be a complete decomposable measure space and $\mathcal{H}\subset\mathcal{L}^\infty(X,\mu)$. Suppose that there is a family $\mathcal{G}$ of elements of $\mathcal{F}_0$ such that:

(a) $\forall\, G_1, G_2 \in \mathcal{G},\ G_1 \cup G_2 \in \mathcal{G}$,

(b) $Y = \bigcup\{G: G\in \mathcal{G}\}\in \mathcal{F}$ and $\mu(X\setminus Y) = 0$,

(c) μ is net additive on $\mathcal{G}$,

(d) $\forall\, G \in \mathcal{G}$. $(G, \mathcal{F}_G, \mu_G)$ is an $\mathcal{H}_G$-space.

Then $(X,\mathcal{F},\mu)$ is an $\mathcal{H}$-space.

Theorem 3

Let $(X,\mathcal{F},\mu)$ be a complete decomposable measure space and $\mathcal{H}\subset\mathcal{L}^\infty(X,\mu)$. Then $(X,\mathcal{F},\mu)$ is an $\mathcal{H}$-space if and only if there exists a decomposition $\mathcal{C}$ such that $(X_C, \mathcal{F}_C, \mu_C)$ is an $\mathcal{H}_C$-space for all $C \in \mathcal{C}$.

We now give some applications of the above results to strong liftings. Through the rest of this note, we assume that $(X,\mathcal{F},\mu)$ is a complete decomposable measure space and that X is equipped with a completely regular Hausdorff topology in such a way that every Baire subset of X is measurable. Let $\mathcal{A}(\mathcal{B})$ denote the σ-algebra of all Baire (Borel) subsets of X, and C^* the space of all bounded continuous realvalued functions on X. Call $(X,\mathcal{F},\mu)$ a Baire (Borel) measure space, and μ a Baire (Borel) measure if $\mathcal{F}$ is identical with the completion of $\mathcal{A}(\mathcal{B})$ w.r.t.μ. We assume that Baire and Borel measures are regular in the sense of inner approximation by zero sets (sets of the form $f^{-1}(0), f\in C^*$) in the Baire case and closed sets in the Borel case. By a topological measure space we mean either a Baire or a Borel measure space. We now give an extension of a result proved in [7] for Radon measures.

Theorem 4

Let $(X,\mathcal{F},\mu)$ be a topological measure space which admits a decomposition $\mathcal{C}$ consisting of metrizable subsets of X. If X has a non-measurable cardinal then every lifting of $\mathcal{L}^\infty$ is almost strong.

Another application of theorem 2 is theorem 3 from [1] which was proved there directly.

REFERENCES

[1] A.G.A.G. Babiker and W.Strauß, Almost strong liftings and τ-additivity; this proceedings.

[2] D. Fremlin, On two theorems of Mokobodzki; (preprint).

[3] P.R. Halmos, Measure theory; Van Nostrand (1955).

[4] A. and C. Ionescu Tulcea, Topics in the theory of lifting; Springer-Verlag (1969).

[5] V.L. Levin, Convex integral functionals and the theory of lifting; Russian Math. Surveys 30,2 (1975), 119-184 from Uspekhi Math. Nauk 30, 2 (1975).

[6] V. Losert, A measure space without the strong lifting property; Math. Ann. 239 (1979), 119-128.

[7] R.J. Maher, Strong liftings on topological measured spaces; Studies in Probability and Ergodic Theory. Advances in Mathematics Supplementary Studies, 2 (1978), 155-166.

[8] G. Mokobodzki, Relèvement borélien compatible avec une classe d'ensembles négligables. Application à la désintegrations des mesures; Sém. de Prababilités IX (1974-5), Springer Lecture Notes No. 465.

[9] W. Strauß, Retraction number, liftings and the decomposability of measure spaces; Bull. Acad. Pol. Sci. 23 (1975), 27-33.

A.G.A.G. Babiker
Mathematisches Institut A
der Universität Stuttgart
Pfaffenwaldring 57
D 7000 Stuttgart 80
The Federal Republic of Germany

and,

School of Mathematical Sciences
University of Khartoum
P.O.Box 321
Khartoum
Sudan

W. Strauß
Mathematisches Institut A
der Universität Stuttgart
Pfaffenwaldring 57
D 7000 Stuttgart 80
The Federal Republic of Germany

LIFTING COMPACT SPACES

A. Bellow[1]
Northwestern University
Evanston, IL 60201/USA

This paper is to a large extent self-contained, its main purpose being to formalize and study the notions of <<lifting compact function>> and <<lifting compact space>>. Applications to Banach spaces and locally convex spaces will be given elsewhere. The paper is divided as follows:

§1. Notation and terminology.

§2. Abstract lifting induced by a real lifting and construction of a Radon measure.

§3. Extension of measures.

§4. Functions that are <<lifting compact in T relative to ρ>>.

§5. Examples.

§6. Lifting compact spaces.

[1]This research was supported by the National Science Foundation. The main results of this paper were announced in [1].

§1. Notation and terminology

Throughout this paper $(\Omega,\mathcal{F},\mu)$ will always denote a complete probability space, $M_R^\infty = M_R^\infty(\Omega,\mathcal{F},\mu)$ the space of all bounded $f: \Omega \to R$ which are $\mathcal{F}$-measurable, $L_R^\infty = L_R^\infty(\Omega,\mathcal{F},\mu)$ the corresponding Banach space of all equivalence classes (mod μ) and $f \to \tilde{f}$ the canonical mapping of M_R^∞ onto L_R^∞. Finally we shall denote by $\rho: M_R^\infty \to M_R^\infty$ a lifting of M_R^∞ (see [4]); alternatively one can think of ρ as an isomorphic injection of the algebra L_R^∞ into M_R^∞, which is a cross-section of the canonical mapping $f \to \tilde{f}$.

All the topological spaces considered in this article will be assumed to be Hausdorff.

Let X be a completely regular space and let $C^b(X)$ be the space of all real-valued bounded continuous functions on X. By the Baire σ-field in X we mean the smallest σ-field with respect to which every $h \in C^b(X)$ is measurable and by the Borel σ-field in X we mean the σ-field spanned by the open sets in X. A Borel measure on X (respectively a Baire measure on X) is a finite, positive, σ-additive measure defined on the Borel σ-field in X (respectively on the Baire σ-field in X). If $T \subset X$, we denote by $i_{T,X}: T \to X$ the "canonical injection" of T into X: $i_{T,X}(t) = t$ for all $t \in T$.

Let $(\Omega_1,\mathcal{F}_1)$ be a measurable space and γ a probability on $(\Omega_1,\mathcal{F}_1)$. We denote by γ^* the outer measure (in the sense of Carathéodory) generated by γ. If Y is an arbitrary subset of Ω_1, the trace measurable space is $(Y,\mathcal{F}_1 \cap Y)$ and the measure γ_Y induced by γ is defined as follows: Let $\tilde{Y}$ be a measurable cover of Y, i.e., $\tilde{Y} \supset Y$, $\tilde{Y} \in \mathcal{F}_1$ such that $\gamma(\tilde{Y}) = \gamma^*(Y)$. If $B \in \mathcal{F}_1 \cap Y$, let $\tilde{B} \subset \tilde{Y}$, $\tilde{B} \in \mathcal{F}_1$ be any set such that $\tilde{B} \cap Y = B$; then

$$\gamma_Y(B) = \gamma^*(B) = \gamma(\tilde{B}).$$

If $(\Omega_2,\mathcal{F}_2)$ is another measurable space and $u: \Omega_1 \to \Omega_2$ a measurable mapping, we recall that the image measure $u(\gamma)$ is the measure on $(\Omega_2,\mathcal{F}_2)$ defined by

$$u(\gamma)(A) = \gamma(u^{-1}(A)) \quad \text{for } A \in \mathcal{F}_2$$

and that

$$(u(\gamma))^*(Y) \geq \gamma^*(u^{-1}(Y)) \quad \text{for all } Y \subset \Omega_2.$$

We now recall the definition of two important classes of Borel measures, namely the τ-additive measures (these are called normal measures in [11] and τ-regular in [13]) and the Radon measures. For a systematic treatment of measures on topological spaces see [13], Appendix II. See also [11], p. xxiii.5 - xxiii.7 for a succinct account of the basic facts that are collected below.

A Borel measure ν on X is said to be outer regular if

$$\nu(B) = \inf\{\nu(U) \mid U \text{ open}, U \supset B\}$$

for every Borel set B in X.

A Borel measure ν on X is called τ-additive if

$$\nu(\bigcup_\alpha U_\alpha) = \sup_\alpha \nu(U_\alpha)$$

for any increasing net (U_α) of open sets. We have:

1°) Every τ-additive measure is outer regular; hence such a Borel measure is completely determined by its values on the open sets.

2°) For a τ-additive measure on X, the integral passes to the limit for increasing nets of lower semi-continuous functions on X that are ≥ 0.

3°) The Borel measure induced by a τ-additive measure on an arbitrary part of X is τ-additive.

4°) The image of a τ-additive measure under a continuous mapping is τ-additive.

A Borel measure ν on X is called a Radon measure if:

$$\nu(B) = \sup\{\nu(K) \mid K \text{ compact, } K \subset B\}$$

for every Borel set B in X.

5°) Every Radon measure is τ-additive.

6°) The Borel measure induced by a Radon measure ν on a ν^*-measurable subset of X is a Radon measure.

7°) The image of a Radon measure under a continuous mapping is a Radon measure.

We recall also the notion of universal measurability: The completely regular space X is said to be universally (Radon) measurable if whatever be the completely regular space Z, containing X as a subspace, and whatever be the Radon measure μ on Z, X is μ^*-measurable.

8°)(Sunyach [12]). For the completely regular space X the following are equivalent assertions: a) X is universally measurable; b) there is a compact space Y, containing X as a subspace, such that X as a set in Y is universally measurable (i.e., for every Radon measure μ on Y, X is μ^*-measurable); c) every τ-additive measure on X is a Radon measure. In particular every compact space, every K_σ space (i.e., a space which is a countable union of compact sets), more generally every space X which is a $K_{\sigma\delta}$ in some compactification, is universally measurable.

Finally we recall the following classical facts (see for instance [13] pp. 264-265 and p. 277):

9°)(Alexandrov). If α and β are Baire measures on X and if

$$\int h d\alpha = \int h d\beta \quad \text{for every } h \in C^b(X),$$

then $\alpha = \beta$.

10°)(Riesz). Let K be a compact space. The formula

$$T(h) = \int h d\mu \quad \text{for } h \in C(K)$$

establishes a one-to-one correspondence between positive linear functionals T on C(K) and Radon measures μ on K.

We conclude this section by reviewing some basic facts of topology.

Remark. Let T be a completely regular space and let $H \subset C^b(T)$. Consider the product space

$$R^H = \prod_{h \in H} R_h, \quad R_h = R \quad \text{for each } h \in H$$

and the mapping $\Phi: t \to (h(t))_{h \in H}$ of T into R^H. Then the following are equivalent assertions:

a) H determines the topology of T, in the sense that the weakest topology on T making every $h \in H$ continuous coincides with the topology of T; equivalently, the set $\{h^{-1}(U) | h \in H, U \subset R \text{ open}\}$ is a subbase for the topology of T.

b) For a net (t_α) in T and an element $t \in T$ we have $t_\alpha \to t$ in T if and only if $h(t_\alpha) \to h(t)$ for each $h \in H$; equivalently Φ is a homeomorphic embedding of T onto $\Phi(T) \subset R^H$.

Let $pr_h: R^H \to R_h$ be the coordinate projection mapping and note that for every $h \in H$ we have

$$pr_h(\Phi(t)) = h(t) \quad \text{for } t \in T.$$

With this notation we may now state:

Proposition 1.1. - Let T be a completely regular space and let $H \subset C^b(T)$ be a uniformly closed algebra containing the constants and determining the topology of T. The homeomorphic embedding $\Phi: t \to (h(t))_{h \in H}$ of T onto $\Phi(T) \subset R^H$ carries every $h \in H$ into $pr_h|\Phi(T)$. Let $K = \overline{\Phi(T)}$, closure in R^H. Then K is compact, $\Phi(T)$ is dense in K and for each $h \in H$, $pr_h|\Phi(T)$ extends uniquely to $pr_h|K$, an element of C(K). Furthermore the extension mapping $pr_h|\Phi(T) \to pr_h|K$ is an isomorphism of the algebra $\{pr_h|\Phi(T) \mid h \in H\}$ onto C(K).

Proof: The set $\{pr_h|\Phi(T) \mid h \in H\}$ is clearly a uniformly closed algebra of $C^b(\Phi(T))$ containing the constants and the extension mapping $pr_h|\Phi(T) \to pr_h|K$ is easily checked to be an algebra isomorphism which preserves the supremum norm. Hence $\{pr_h|K \mid h \in H\}$ is a uniformly closed algebra of C(K) containing the constants; since it obviously separates the points of K, it must coincide with C(K) (Stone-Weierstrass theorem). This finishes the proof.

Remark. With the notation of Proposition 1.1, if we identify T with its homeomorphic image $\Phi(T)$, then $H = C(K)|T$, i.e. H coincides with the restrictions to T of the continuous functions on K.

§2. Abstract lifting induced by a real lifting and construction of a Radon measure

We begin by recalling the definition of the abstract lifting induced by a lifting ρ of M_R^∞ (see [4]. pp. 51-53). Let K be a compact space, g: $\Omega \to K$ a Baire measurable mapping. For $\omega \in \Omega$ fixed, the map $h \to \rho(h \circ g)(\omega)$ of C(K) into R is a character of C(K) (= a multiplicative linear functional) and hence there is a unique element of K which we will denote by $\rho_K(g)(\omega)$ satisfying

(1) $$\rho(h \circ g)(\omega) = h \circ \rho_K(g)(\omega) \quad \text{for each } h \in C(K).$$

This defines $\rho_K(g)$ as a mapping of Ω into K; clearly $\rho_K(g): \Omega \to K$ is Baire measurable.

Note also that the "Baire image measures on K" given by $\rho_K(g)(\mu)$ and $g(\mu)$ coincide, since for every $h \in C(K)$

$$\int hd\rho_K(g)(\mu) = \int h \circ \rho_K(g)d\mu = \int h \circ gd\mu = \int hdg(\mu).$$

We now have:

Theorem 2.1. - Let K be a compact space, g: $\Omega \to K$ a Baire measurable mapping and ρ a lifting of M_R^∞. Then

1) $\rho_K(g): \Omega \to K$ is Borel measurable.

2) The Borel image measure on K, $\nu^\# = \rho_K(g)(\mu)$, is a Radon measure.

Proof: Statement 1) was proved in [4] (see p. 52). For completeness, however, we include the proof. To simplify notation we write $\phi = \rho_K(g)$; then $\rho_K(\phi) = \phi$. We show that for any open set $U \subset K$, $1_U \circ \phi$ is $\mathcal{F}$-measurable and

(2) $$\rho(1_U \circ \phi) \geq 1_U \circ \phi.$$

For this purpose, let $\mathcal{H} = \{h \in C(K) \mid h \leq 1_U\}$. Then $1_U \circ \phi = \sup_{h \in \mathcal{H}} h \circ \phi$ (pointwise supremum), the set $\mathcal{H}$ is directed upward for "$\leq$", and $\rho(h \circ \phi) = h \circ \phi$ for every $h \in \mathcal{H}$. By the theorem on the «measurability of the upper envelope» (see [4], p. 40, Theorem 3), we deduce that $1_U \circ \phi$ is $\mathcal{F}$-measurable and satisfies (2).

2) It obviously suffices to show that the Borel measure $\nu^\# = \phi(\mu)$ is τ-additive. Hence let (U_α) be an increasing net of open sets in K and let $U = \bigcup_\alpha U_\alpha$. We have: $1_U \circ \phi = \sup 1_{U_\alpha} \circ \phi$ (pointwise supremum), the net $(1_{U_\alpha} \circ \phi)_\alpha$ is increasing and $\rho(1_{U_\alpha} \circ \phi) \geq 1_{U_\alpha} \circ \phi$ for each α; also $\nu^\#(U) = \int 1_U \circ \phi d\mu$, and $\nu^\#(U_\alpha) = \int 1_{U_\alpha} \circ \phi d\mu$. Applying once more the theorem on the «measurability of the upper envelope», we can pass to the limit:

$$\nu^\#(U) = \sup_\alpha \nu^\#(U_\alpha).$$

This completes the proof of the theorem.

Remark. - The Borel measure $\nu^\#$ is the (unique) Radon extension of the Baire image measure $g(\mu)$ on K; hence $\nu^\#$ is independent of the particular lifting ρ of M_R^∞.

§3. Extension of measures

This section contains essentially a review of the extension of Baire measures to Borel measures.

From now on (unless explicitly stated otherwise), T will be a completely regular space and $K \supset T$ a compactification of T, i.e., a compact space K such that T is a (topological) subspace of K; we will use the notation

$$H = C(K)|T = \{h'|T \mid h' \in C(K)\}$$

(typically such a situation arises as in Proposition 1.1; see also the Remark following it).

The σ-field on T generated by the elements of H is precisely the trace on T of the Baire σ-field in K

$$\sigma(H) = (\text{Baire } \sigma\text{-field in K}) \cap T;$$

$\sigma(H)$ is obviously a sub-σ-field of the Baire σ-field in T.

We begin with some elementary but useful observations which we collect in the form of lemmas.

Lemma 3.1. - For each $u \in C_+^b(T)$ we have

$$u = \sup\{h \in H \mid 0 \leq h \leq u\},$$

and for each $U \subset T$ open in T we have

$$1_U = \sup\{h \in H \mid 0 \leq h \leq 1_U\}.$$

Proof: Standard.

Let now $f: \Omega \to T$ be measurable as a mapping of $(\Omega, \mathcal{F})$ into $(T, \sigma(H))$; this is equivalent with saying that $i_{T,K} \circ f: \Omega \to K$ is Baire measurable (i.e. f regarded "as a mapping of Ω into K" is Baire measurable). Let $\lambda = f(\mu)$ be the image probability on $(T, \sigma(H))$.

The formula

$$S(h') = \int h d\lambda \quad \text{for } h' \in C(K) \text{ and } h = h'|T$$

defines a positive linear functional on C(K). We then have:

Lemma 3.2. - There is a unique Radon measure $\lambda^{\#}$ on K satisfying

(I) If $h' \in C(K)$ and $h = h'|T$ then

$$\int h' d\lambda^{\#} = \int h d\lambda.$$

Furthermore,

(II) $\lambda^{\#}$ is an extension of the Baire image measure $(i_{T,K} \circ f)(\mu)$ on K;

(III) In particular if $A \in \sigma(H)$ and $A = A' \cap T$, where A' is a Baire set in K, then

$$\lambda^{\#}(A') = \lambda(A).$$

Proof: The existence and uniqueness of the Radon measure $\lambda^{\#}$ on K satisfying

(I) follows from 10°) of §1.

(II) By 9°) of §1 it suffices to note that $\lambda^{\#}$ and $(i_{T,K} \circ f)(\mu)$ agree on every $h' \in C(K)$:

$$\int h' d(i_{T,K} \circ f)(\mu) = \int h' \circ (i_{T,K} \circ f) d\mu = \int h \circ f d\mu$$

$$= \int h d\lambda = \int h' d\lambda^{\#}.$$

(III) We have:

$$\lambda^{\#}(A') = (i_{T,K} \circ f)(\mu)(A') = \mu((i_{T,K} \circ f)^{-1}(A'))$$

$$= \mu(f^{-1}(A' \cap T)) = \mu(f^{-1}(A)) = \lambda(A).$$

<u>Remark.</u> $\lambda^{\#}$ is the (unique) Radon extension of the Baire measure $(i_{T,K} \circ f)(\mu)$ on K. Hence (see Theorem 2.1 and the Remark following it):

$$\lambda^{\#} = \rho_K(i_{T,K} \circ f)(\mu).$$

This fact will not be needed however until Section 4.

With the above notation we have:

<u>Proposition 3.3. - Let</u> $f: \Omega \to T$ <u>be measurable as a mapping of</u> $(\Omega, \mathfrak{F})$ <u>into</u> $(T, \sigma(H))$, <u>let</u> $\lambda = f(\mu)$ <u>the image probability on</u> $(T, \sigma(H))$ <u>and</u> $\lambda^{\#}$ <u>the Radon measure on</u> K <u>satisfying</u> (I) <u>of Lemma</u> 3.2. <u>Then the assertions below are equivalent</u>:

(i) λ <u>is "tight on</u> $\sigma(H)$," i.e. <u>for each</u> $\varepsilon > 0$ <u>there is</u> $C \subset T$ <u>compact such that</u>

$$B \in \sigma(H),\ B \supset C \Rightarrow \lambda(B) \geq 1 - \varepsilon.$$

(ii) <u>For each</u> $\varepsilon > 0$ <u>there is</u> $C \subset T$ <u>compact such that</u> $h \in H$, $0 \leq h \leq 1$, $h|C = 1 \Rightarrow \int h d\lambda \geq 1 - \varepsilon$.

(iii) T <u>is</u> $(\lambda^{\#})^*$<u>-measurable and of measure</u> 1.

(iv) λ <u>admits a Borel extension which is a Radon measure on</u> T. <u>This Radon extension is given precisely by the measure induced by</u> $\lambda^{\#}$ <u>on</u> T, <u>namely</u> $(\lambda^{\#})_T$.

<u>Proof</u>: It is obvious that (iv) $\Rightarrow$ (i).

(i) $\Rightarrow$ (ii). Let $\varepsilon > 0$ and C be as in (i). Let $h \in H$, $0 \leq h \leq 1$, $h|C = 1$; then $B = \{h \geq 1\} \supset C$ and $B \in \sigma(H)$ so that

$$\int h d\lambda \geq \lambda(B) \geq 1 - \varepsilon.$$

(ii) $\Rightarrow$ (iii). By (ii) there is an increasing sequence (C_n) of compact sets in T such that

$$\inf\{\int h d\lambda \mid h \in H,\ 0 \leq h \leq 1,\ h|C_n = 1\} \geq 1 - \frac{1}{n}.$$

For $h' \in C(K)$, $0 \leq h' \leq 1$, $h' = 1$ on C_n and $h = h'|T$ we have

$$\int h' d\lambda^{\#} = \int h d\lambda \geq 1 - \frac{1}{n},$$

and hence

$$\lambda^{\#}(C_n) \geq 1 - \frac{1}{n}.$$

Thus $\bigcup_{n=1}^{\infty} C_n$ is Borel and of full $\lambda^{\#}$-measure. As $\bigcup_{n=1}^{\infty} C_n \subset T \subset K$, assertion (iii) is proved.

(iii) ⇒ (iv). It is clear that $(\lambda^{\#})_T$ is a Radon measure on T (see statement 6°) in §1). We show that $(\lambda^{\#})_T$ is an extension of λ. For $A \in \sigma(H)$ we have

$$A = A' \cap T, \ A' \text{ a Baire set in } K$$

and hence by (III) of Lemma 3.2 above:

$$(\lambda^{\#})_T(A) = (\lambda^{\#})^*(A' \cap T) = \lambda^{\#}(A') = \lambda(A).$$

This completes the proof.

Note. In the case when $H = C^b(T)$ ($\sigma(H)$ coincides with the σ-field of all Baire sets in T) and λ satisfies (i), we say simply that λ is tight.

With the notation of Proposition 3.3 we have:

Corollary 1. - *Suppose that $T \subset K_\sigma \subset K$. Then the Radon measure $\lambda^{\#}$ is supported by K_σ.*

Proof: Let (K_n) be a sequence of compact sets in K such that $K_\sigma = \bigcup_{n=1}^{\infty} K_n$. Suppose that $\lambda^{\#}(K_\sigma) < 1$. Let $D \subset (K_\sigma)^c$ a compact set such that $\lambda^{\#}(D) > 0$. For each n let $h'_n \in C(K)$ such that

$$0 \leq h'_n \leq 1, \ h'_n | D = 1, \ h'_n | K_n = 0.$$

Let $u'_n = \inf(h'_1,\dots,h'_n)$ and $u_n = u'_n | T$. Then $u'_n | D = 1$, $u'_n | K_1 \cup \cdots \cup K_n = 0$, $u'_n \searrow$ and $\lim_n u'_n(x) = 0$ for $x \in \bigcup_{n=1}^{\infty} K_n$. By (I) of Lemma 3.2 we deduce:

$$\int u'_n d\lambda^{\#} = \int u_n d\lambda \searrow 0;$$

but $\int u'_n d\lambda^{\#} \geq \lambda^{\#}(D) > 0$. This contradiction proves our assertion.

Corollary 2. - *If T is a $K_{\sigma\delta}$ in K then $\lambda^{\#}$ is supported by T and thus λ extends to a Radon measure on T.*

Proposition 3.4. - *Let $f: \Omega \to T$ be measurable as a mapping of $(\Omega, \mathfrak{F})$ into $(T, \sigma(H))$, let $\lambda = f(\mu)$ the image probability on $(T, \sigma(H))$ and let $\lambda^{\#}$ the Radon measure on K satisfying (I) of Lemma 3.2. Then the assertions below are equivalent:*

(j) λ *is "τ-smooth on $\sigma(H)$," i.e. if (h_α) is a decreasing net in H, converging pointwise to 0, then $\lim_\alpha \int h_\alpha d\lambda = 0$.*

(jj) $(\lambda^{\#})^*(T) = 1$.

(jjj) λ *admits a Borel extension which is τ-additive. This τ-additive extension is given precisely by the measure induced by $\lambda^{\#}$ on T, namely $(\lambda^{\#})_T$.*

Proof: It is obvious that (jjj) ⇒ (j).

(j) ⇒ (jj). Since $\lambda^{\#}$ is outer regular, it suffices to show that if $U \supset T$ and U is open in K, then $\lambda^{\#}(U) = 1$. To see this let (h'_α) be an increasing net in C(K) such that

$$1_U = \sup_\alpha h'_\alpha.$$

Let $h_\alpha = h'_\alpha \mid T$; then $h_\alpha \in H$, (h_α) is an increasing net in H converging pointwise to the constant function 1 on T and hence $\lim_\alpha \int h_\alpha d\lambda = 1$. We deduce

$$\lambda^\#(U) = \sup_\alpha \int h'_\alpha d\lambda^\# = \sup_\alpha \int h_\alpha d\lambda = 1.$$

(jj) $\Rightarrow$ (jjj). It is clear that $(\lambda^\#)_T$ is a τ-additive measure on T (see statement 3°) in §1). To show that $(\lambda^\#)_T$ is an extension of λ the argument is the same as that in the last part of the proof of Proposition 3.3.

Note. - In the case when $H = C^b(T)$ ($\sigma(H)$ coincides with the σ-field of all Baire sets in T) and λ satisfies (j), we say simply that λ is τ-smooth.

Remarks. - 1) If T is Lindelöf (i.e. every open cover of T contains a countable subcover), then every probability λ on $(T,\sigma(H))$ is "τ-smooth on $\sigma(H)$" (see for instance [9]).

This is easily checked directly. In fact let (h_α) be an increasing net in H, converging pointwise to 1. Let $a < 1$. Let $U_\alpha = \{h_\alpha > a\}$. The U'_αs form an increasing net and an open cover of T by sets belonging to $\sigma(H)$. Since

$$\int h_\alpha d\lambda \geq a\lambda(U_\alpha) \quad \text{for each } \alpha,$$

we deduce, using the Lindelöf property of T, that $\lim_\alpha \int h_\alpha d\lambda \geq a$. This proves the assertion.

2) Propositions 3.3 and 3.4 are the analogs of Theorems 2.4 and 2.5 of [7] (see also [13]).

§4. Functions that are <<lifting compact in T relative to ρ>>

The study of the abstract lifting induced by a lifting ρ of M^{∞}_{R} was initiated in [4]; it was subsequently developed in [5],[16] and finally in [3] from which the following notion emerges (see [1]):

Definition 1. - Let T be a completely regular space, K a compactification of T and ρ a lifting of M^{∞}_{R}. We say that a mapping f: $\Omega \to T$ is <<lifting compact in T relative to K and ρ>> if:

a) The mapping $i_{T,K} \circ f: \Omega \to K$ is Baire measurable (i.e. f regarded "as a mapping of Ω into K" is Baire measurable).

b) There is $\Omega_0 \in \mathfrak{F}$ with $\mu(\Omega_0) = 1$ such that

$$\rho_K(i_{T,K} \circ f)(\omega) \in T \quad \text{for each } \omega \in \Omega_0.$$

Remark. - Let f: $\Omega \to T$ be such that $i_{T,K} \circ f: \Omega \to K$ is Baire measurable and suppose there are $\Omega_0 \in \mathfrak{F}$ with $\mu(\Omega_0) = 1$ and $g: \Omega_0 \to T$ such that

$$(1) \qquad \rho(h \circ (i_{T,K} \circ f))(\omega) = h \circ g(\omega) \quad \text{for } \omega \in \Omega_0$$

and for each $h \in \mathcal{H}$, where $\mathcal{H} \subset C(K)$ is a set separating the points of K. Then we have

$$(2) \qquad g(\omega) = \rho_K(i_{T,K} \circ f)(\omega) \quad \text{for } \omega \in \Omega_0$$

and hence f is <<lifting compact in T relative to K and ρ>>.

It is easily seen that Definition 1 above does not depend on the particular compactification $K \supset T$:

Proposition 4.1. - Let f: $\Omega \to T$, let $\Omega_0 \in \mathfrak{F}$ with $\mu(\Omega_0) = 1$, let $g: \Omega_0 \to T$ and let $H_0 \subset C^b(T)$ be a set of functions determining the topology of T. Suppose that for each $h \in H_0$, the mapping $h \circ f$ is measurable and

$$(3) \qquad \rho(h \circ f)(\omega) = h \circ g(\omega) \quad \text{for all } \omega \in \Omega_0.$$

Then f: $\Omega \to T$ is Baire measurable and for each $v \in C^b(T)$ we have

$$(4) \qquad \rho(v \circ f)(\omega) = v \circ g(\omega) \quad \text{for all } \omega \in \Omega_0.$$

Proof: This is an easy consequence of the theorem on the <<measurability of the upper envelope>> (see [4], p. 40, Theorem 3). Consider the probability space $(\Omega_0, \mathfrak{F}_0, \mu_0)$, where $\mathfrak{F}_0 = \mathfrak{F} \cap \Omega_0$, $\mu_0 = \mu|\mathfrak{F}_0$ and denote by ρ_0 the "restriction" of ρ to $M^{\infty}_{R}(\Omega_0, \mathfrak{F}_0, \mu_0)$.

By the linearity, the multiplicativity of a lifting and its continuity for the supremum norm topology, relation (3) remains valid for every h belonging to the uniformly closed algebra $A(H_0)$ spanned by H_0 and the constants.

We divide the proof into two steps:

Step I. We show that for each $u \in C^b_+(T)$ we have:

α) $u \circ g \in M^{\infty}_{R}(\Omega_0, \mathfrak{F}_0, \mu_0)$.

β) $\rho_0(u \circ g) \geq u \circ g$.

γ) There is a null-set A_u depending on u, $A_u \subset \Omega_0$ such that

$$\omega \in \Omega_0,\ \omega \notin A_u \Rightarrow u \circ g(\omega) \leq u \circ f(\omega).$$

It suffices to let (h_α) be an increasing net of elements of $A(H_0)$ with $0 \leq h_\alpha \leq u$ and $u = \sup_\alpha h_\alpha$ (see Lemma 3.1), to note that we have the pointwise relations:

$$(5)\qquad \begin{cases} u \circ g = \sup\limits_\alpha h_\alpha \circ g \\ u \circ f = \sup\limits_\alpha h_\alpha \circ f \end{cases} \quad \text{and } \rho_0(h_\alpha \circ f) = \rho_0(h_\alpha \circ g) = h_\alpha \circ g \quad \text{for each } \alpha,$$

and to use the theorem on the <<measurability of the upper envelope>> (see [4], p. 40, Theorem 3; also p. 5, Theorem 4).

Step II. Let $v \in C^b(T)$. There is a constant $c > 0$ such that $-c \leq v \leq c$. By Step I applied to $u = c - v$ and $u = v + c$, we deduce:

α') $v \circ g \in M^\infty_R(\Omega_0, \mathcal{F}_0, \mu_0)$.

β') $\rho_0(v \circ g) = v \circ g$.

γ') There is a null-set N_v depending on v, $N_v \subset \Omega_0$ such that

$$\omega \in \Omega_0,\ \omega \notin N_v \Rightarrow v \circ g(\omega) = v \circ f(\omega).$$

This completes the proof.

Note. - Let f be lifting compact in T relative to K and ρ and let $\Omega_0 \in \mathcal{F}$ be as in b) of Definition 1. By Proposition 4.1, given any other compactification $L \supset T$, f is lifting compact in T relative to L and ρ and $\rho_L(i_{T,L} \circ f)(\omega) = \rho_K(i_{T,K} \circ f)(\omega)$ for all $\omega \in \Omega_0$. *If we set*

$$(6)\qquad \rho(f)(\omega) = \rho_K(i_{T,K} \circ f)(\omega) \quad \textit{for } \omega \in \Omega_0,$$

then the mapping $\rho(f): \Omega_0 \to T$ *is unambiguously defined, is measurable Borel* (see Theorem 2.1) *and satisfies the identity*:

$$(7)\qquad \rho(v \circ f)(\omega) = v \circ \rho(f)(\omega) \quad \textit{for each } \omega \in \Omega_0,\ v \in C^b(T).$$

We say from now on simply that f is <<*lifting compact in* T *relative to* ρ>>.

We may now derive the following corollary from Propositions 3.3 and 3.4:

Theorem 4.2. - *Let* $f: \Omega \to T$ *be measurable as a mapping of* $(\Omega, \mathcal{F})$ *into* $(T, \sigma(H))$ *and let* $\lambda = f(\mu)$ *be the image probability measure on* $(T, \sigma(H))$. *Consider the following assertions*:

(1) λ *is tight on* σ(H) (*equivalently* λ *admits a Borel extension which is a Radon measure on* T).

(2) *For any lifting* ρ *of* M^∞_R, $f: \Omega \to T$ *is* <<*lifting compact in* T *relative to* ρ>>.

(3) *There exists a lifting* ρ *of* M^∞_R *such that* $f: \Omega \to T$ *is* <<*lifting compact in* T *relative to* ρ>>.

(4) λ *is* τ-*smooth on* σ(H) (*equivalently* λ *admits a Borel extension which is* τ-*additive*).

Then (1) ⇒ (2) ⇒ (3) ⇒ (4). *If* T *is universally measurable then* (1) ≡ (2) ≡

(3) ≡ (4).

Proof: Let ρ be a lifting of M_R^∞ and let $\lambda^{\#} = \rho_K(i_{T,K} \circ f)(\mu)$ be the Radon measure on K (see the Remark following Lemma 3.2). We note that

(α) $(\lambda^{\#})^*(A) \geq \mu^*(\rho_K(i_{T,K} \circ f)^{-1}(A))$ for each set $A \subset K$;

(β) In particular if A is $(\lambda^{\#})^*$-measurable, then $\rho_K(i_{T,K} \circ f)^{-1}(A)$ is μ^*-measurable and of the same measure.

The implications (1) ⇒ (2) and (3) ⇒ (4) follow from Proposition 3.3 and (β) above, respectively from Proposition 3.4 and (α) above. The implication (2) ⇒ (3) is obvious.

Finally if T is universally measurable (see statement 8°) in §1), then obviously (1) ≡ (2) ≡ (3) ≡ (4), completing the proof.

Remarks. - 1) The implication (1) ⇒ (2) is due to von Weizsäcker [16]. The implication (3) ⇒ (4) is due to Edgar and Talagrand [3]. It is also observed in [3] that the implications (1) ⇒ (2) and (3) ⇒ (4) are not in general reversible; for this see also Example I and Example III in §5 of the present paper.

2) The implication (2) ⇒ (3) is not in general reversible; this is shown by Example II in §5 of the present paper.

3) Let $f: \Omega \to T$ be <<lifting compact in T relative to ρ>>; then $\rho(f): \Omega_0 \to T$, where $\Omega_0 \in \mathcal{F}$ and $\mu(\Omega_0) = 1$ and $\rho(f)$ is Borel measurable. Let $\nu = \rho(f)(\mu)$ be the Borel image measure on T. Then with the notation used in Theorem 4.2 and in its proof (see in particular (α)), ν is an extension of λ and in fact $\nu = (\lambda^{\#})_T$.

We now turn to stability properties.

Proposition 4.3. - *Let T be an arbitrary subspace of a compact metrizable space K. Let $f: \Omega \to T$ be such that $i_{T,K} \circ f: \Omega \to K$ is Baire measurable. Then f is <<lifting compact in T relative to ρ>> and in fact, with the exception of a μ-negligible set in Ω,*

$$\rho(f)(\omega) = f(\omega).$$

Proof: Let $\mathcal{H} \subset C(K)$ be a countable set separating the points of K. For each $h \in \mathcal{H}$ we have

$$h \circ \rho_K(i_{T,K} \circ f)(\omega) = h \circ f(\omega) \quad \mu\text{-a.e.}$$

There is then a μ-negligible set $A \subset \Omega$ such that

$$\omega \notin A \Rightarrow h \circ \rho_K(i_{T,K} \circ f)(\omega) = h \circ f(\omega) \quad \text{for each } h \in \mathcal{H}.$$

We deduce

$$\omega \notin A \Rightarrow \rho_K(i_{T,K} \circ f)(\omega) = f(\omega) \in T.$$

This shows that f is lifting compact in T relative to K and ρ and hence finishes the proof.

Proposition 4.4. - *Let S and T be completely regular spaces. Let $f: \Omega \to S$ be lifting compact in S relative to ρ and let $h: S \to T$ be continuous. Then $h \circ f: \Omega \to T$*

is lifting compact in T *relative to* ρ *and in fact, with the exception of a* μ*-negligible set in* Ω*, we have*

$$\rho(h \circ f)(\omega) = h \circ \rho(f)(\omega).$$

Proof: Let $\Omega_0 \in \mathcal{F}$ with $\mu(\Omega_0) = 1$ such that

$$\rho(v \circ f)(\omega) = v \circ \rho(f)(\omega) \quad \text{for each } \omega \in \Omega_0,\ v \in C^b(S). \tag{8}$$

Let $u \in C^b(T)$; then $u \circ h \in C^b(S)$ and by (8) we have:

$$\rho((u \circ h) \circ f)(\omega) = (u \circ h) \circ \rho(f)(\omega) \quad \text{for each } \omega \in \Omega_0.$$

Thus

$$\rho(u \circ (h \circ f))(\omega) = u \circ (h \circ \rho(f))(\omega)$$

for each $\omega \in \Omega_0$ and each $u \in C^b(T)$. By the uniqueness of the abstract lifting we deduce

$$\rho(h \circ f)(\omega) = h \circ \rho(f)(\omega) \quad \text{for each } \omega \in \Omega_0,$$

proving the desired assertion.

Proposition 4.5. - *Let* $(T_k)_{k \in N}$ *be a sequence of completely regular spaces. For each* $k \in N$ *let* $f_k: \Omega \to T_k$ *be lifting compact in* T_k *relative to* ρ*. Then the product map* $(f_k)_{k \in N}: \Omega \to \prod_{k \in N} T_k$ *is lifting compact in* $\prod_{k \in N} T_k$ *relative to* ρ *and in fact, with the exception of a* μ*-negligible set in* Ω *we have:*

$$\rho\big((f_k)_{k \in N}\big)(\omega) = \big(\rho(f_k)\big)_{k \in N}(\omega).$$

Proof: Let $f = (f_k)_{k \in N}$ and let $T = \prod_{k \in N} T_k$. For each $n \in N$ let $L_n \supset T_n$ be a compactification of T_n and let $L = \prod_{n \in N} L_n$; L is a compactification of T. Then

$$i_{T,L} \circ f = (i_{T_n,L_n} \circ f_n)_{n \in N}: \Omega \to L$$

is Baire measurable, since $i_{T_n,L_n} \circ f_n : \Omega \to L_n$ is Baire measurable for each $n \in N$ (note that since the L_n's are compact, the Baire σ-field in L is just the product of the Baire σ-fields).

Let now $pr_n: L \to L_n$ be the coordinate projection for each $n \in N$. The set:

$$\mathcal{H} = \{u \circ pr_n \mid u \in C(L_n) \text{ and } n \in N\}$$

separates the points of L. Let A be the countable union of the μ-negligible sets corresponding to the functions $\rho(f_n)$, $n \in N$ (see the Note following Proposition 4.1). By the Remark following Definition 1 it suffices to check that for $\omega \notin A$ and for each $u \circ pr_n \in \mathcal{H}$ we have:

$$\rho(u \circ pr_n \circ (i_{T,L} \circ f))(\omega) = \rho(u \circ (i_{T_n,L_n} \circ f_n))(\omega)$$

$$= u \circ \rho(f_n)(\omega) = (u \circ pr_n) \circ \big(\rho(f_k)\big)_{k \in N}(\omega);$$

but this is obvious and hence the proof is complete.

Proposition 4.6. - *Let T be a completely regular space whose topology is finer than a separable metric topology $\mathcal{T}$. Let $f: \Omega \to T$ be lifting compact in T relative to ρ. Then, with the exception of a μ-negligible set in Ω, we have*

$$\rho(f)(\omega) = f(\omega);$$

in particular, $f: \Omega \to T$ is Borel measurable.

Proof: It suffices to note that there is a *countable* set $\mathcal{H}$ consisting of bounded, $\mathcal{T}$-continuous functions, *separating* the points of T (if d is the metric defining the topology $\mathcal{T}$, and (x_n) is a sequence dense in $\mathcal{T}$, set $h_n(x) = \dfrac{d(x,x_n)}{1 + d(x,x_n)}$, for $n \in N$). The rest of the argument is the same as in the proof of Proposition 4.3.

§5. Examples

Throughout these examples $(\Omega,\mathcal{F},\mu)$ is the Lebesgue space of $\Omega = [0,1]$.

I) Let $Y \subset [0,1]$ be a Lebesgue non-measurable set, i.e., $Y \notin \mathcal{F}$ with $\mu_*(Y) = 0$ and $\mu^*(Y) = 1$. Let $(Y, \mathcal{F} \cap Y, \mu_Y)$ be the induced measure space; this is a complete probability space. Let $T = Y$ and let $f: Y \to Y$ be the identity mapping. Then $f(\mu_Y) = \mu_Y$ is not a tight measure (for every compact set $C \subset Y$, $\mu_Y(C) = 0$), but for any lifting ρ, f is «lifting compact in Y relative to ρ» (see Proposition 4.3). Thus the implication 2) $\Rightarrow$ 1) in Theorem 4.2 is false.

This example is due to Edgar and Talagrand [3].

II) Here and below, if f is a real-valued function defined on R or an interval of R, we write

$$f(z+) = \lim_{\substack{y\to z\\ y>z}} f(y), \quad f(z-) = \lim_{\substack{y\to z\\ y<z}} f(y),$$

the limits being taken in the ordinary topology of R. Denote by $C^b([0,1],+)$ the algebra of all bounded real-valued functions defined on $[0,1]$, which are continuous on the right in the ordinary topology of $[0,1]$, i.e. such that $f(z) = f(z+)$ for each $z \in [0,1)$; denote by $C^b([0,1],-)$ the algebra of all bounded real-valued functions defined on $[0,1]$, which are continuous on the left in the ordinary topology of $[0,1]$. The notation $C^b(R,+)$ (respectively, $C^b(R,-)$) has a similar meaning.

Let ρ_1 (respectively ρ_2) be a lifting of $M^\infty_R(\Omega,\mathcal{F},\mu)$ such that

(1) $\rho_1(f)(\omega) = f(\omega)$ for each $f \in C^b([0,1],+)$ and $\omega \in [0,1)$ (respectively such that

(2) $\rho_2(f)(\omega) = f(\omega)$ for each $f \in C^b([0,1],-)$ and $\omega \in (0,1]$).

Such liftings exist. We shall give the argument for ρ_1 (the argument for ρ_2 is entirely similar). Let ρ be a lifting of the algebra of all bounded Lebesgue-measurable functions defined on R, satisfying

$$\rho(f) = f \quad \text{for every } f \in C^b(R,+)$$

(for the existence of such a lifting see [4], p. 123, Theorem 6). Let χ be a character of the algebra $L^\infty_R(\Omega,\mathcal{F},\mu)$. For $f \in M^\infty_R(\Omega,\mathcal{F},\mu)$ define $\hat{f}: R \to R$ by $\hat{f} = f$ on $[0,1)$ and $\hat{f} = 0$ outside the interval $[0,1)$, and set

$$\rho_1(f)(\omega) = \begin{cases} \rho(\hat{f})(\omega) & \text{for } \omega \in [0,1) \\ \chi(\tilde{f}) & \text{for } \omega = 1 \end{cases}.$$

It is easily verified that ρ_1 is a lifting of $M^\infty_R(\Omega,\mathcal{F},\mu)$ and that ρ_1 satisfies (1).

For each $x \in R$ and $a > 0$ let

$$U_a(x) = [x,x+a) \cup (-x-a,-x)$$

and let Σ be the topology on R for which $\{U_a(x) \mid a > 0\}$ is a fundamental system of neighborhoods of x, for each $x \in R$. We denote by K the interval $[-1,1]$ endowed with the topology induced by Σ. Then K is compact, non-metrizable and a function $f: K \to R$ belongs to $C(K)$ if and only if $f(z) = f(z+) = f((-z)-)$, for each $z \in K$, $z \neq 1$. Let now $T = [0,1]$ as a (topological) subspace of K; note that a bounded function $f: T \to R$ belongs to $C^b(T)$ if and only if $f(t) = f(t+)$ for each $t \in T$, $t \neq 1$.

Let $e: \Omega \to T$ be the identity mapping $e(\omega) = \omega$ for $\omega \in \Omega$. It is clear that $e: \Omega \to T$ is Baire measurable. For $v \in C^b(T)$ and $\omega \in [0,1)$ we have by (1) above:

$$\rho_1(v \circ e)(\omega) = \rho_1(v)(\omega) = v(\omega) = v \circ e(\omega).$$

Thus e is lifting compact in T relative to ρ_1 and in fact $\rho_1(e)(\omega) = e(\omega)$ for each $\omega \in [0,1)$.

Let $i: \Omega \to K$ be the identity mapping $i(\omega) = \omega$ for $\omega \in \Omega$ (we have $i = i_{T,K} \circ e$). We show that

(3) $$\rho_2(i)(a) = -a \quad \text{for each } 0 < a < 1.$$

Let $0 < a < 1$ and let $h = 1_{[-a,a)} \in C(K)$. Then $h \circ i$, as an element of $M_R^\infty(\Omega,\mathcal{F},\mu)$ coincides with $1_{[0,a)}$ (in Ω) and hence by (2) above we have

$$\rho_2(h \circ i)(a) = \rho_2(1_{[0,a)})(a) = \rho_2(1_{(0,a]})(a) = 1$$

and

$$h \circ i(a) = 0.$$

Since $\rho_2(h \circ i)(a) = h \circ \rho_2(i)(a) = 1$, we deduce that $b = \rho_2(i)(a) \neq i(a) = a$. We show that $b = -a$. In fact, otherwise there is $u: [-1,1] \to R$ continuous in the ordinary topology, such that $u(-z) = u(z)$ and $u(b) \neq u(a)$. Then $u \in C(K)$ and by (2) again we have:

$$\rho_2(u \circ i)(a) = \rho_2(u|[0,1])(a) = u(a)$$

$$\rho_2(u \circ i)(a) = u \circ \rho_2(i)(a) = u(b),$$

a contradiction. Thus (3) is proved. This shows that e is not lifting compact in T relative to ρ_2. Thus the implication 3) $\Rightarrow$ 2) in Theorem 4.2 is false.

III) Let T be $[0,1]$ equipped with a completely regular topology satisfying the following conditions:

a) The topology of T is finer than the ordinary Euclidean topology of $[0,1]$.

b) Every $h \in C^b(T)$ is $\mathcal{F}$-measurable.

c) There are open sets in T that do not belong to $\mathcal{F}$.

Let $e: \Omega \to T$ be the identity map $e(\omega) = \omega$; then $e: \Omega \to T$ is Baire measurable, but e is not Borel measurable and hence (see Proposition 4.6), for any lifting ρ of $M_R^\infty(\Omega,\mathcal{F},\mu)$, e cannot be lifting compact in T relative to ρ.

We now show how to construct spaces T satisfying a), b), c).

Let $\mathcal{C}$ be the usual Euclidean topology on $[0,1]$. When we refer to sets in $[0,1]$ or functions on $[0,1]$, it is understood that we speak of $[0,1]$ endowed with the topology $\mathcal{C}$.

Let $Y \subset [0,1]$ be a Lebesgue non-measurable set, i.e. $Y \notin \mathcal{F}$ with $\mu_*(Y) = 0$ (and of course $\mu^*(Y) > 0$). Let T be [0,1] endowed with the topology $\mathcal{G}'$ generated from $\mathcal{G}$ by adding each singleton $\{s\}$, $s \in Y$, as an open set; the open sets in $\mathcal{G}'$ are all sets of the form $U \cup A$ where U is $\mathcal{G}$-open and A is an arbitrary subset of Y. $\mathcal{G}'$ can also be described as the weakest topology on [0,1] making continuous every function in the set $\{1_{\{s\}} \mid s \in Y\} \cup \{e\}$. Hence T is completely regular. We now check that T satisfies a), b), c) above:

a) and c) are obvious from the definition.

b) Let $h \in C^b(T)$. The set D_h of all $x \in [0,1]$ where h is not $\mathcal{G}$-continuous is an F_σ in [0,1] (this has nothing to do with h being in $C^b(T)$ and is true for an arbitrary mapping $h: [0,1] \to R$). Since h is $\mathcal{G}'$-continuous we must have $D_h \subset Y$ and since $\mu_*(Y) = 0$ it follows that $\mu(D_h) = 0$. Hence $1_{D_h^c} h$ is a Borel function and in particular h is $\mathcal{F}$-measurable.

We now note that the Borel sets in T are all sets of the form

$$(B \cap Y^c) \cup A$$

where B is Borel in [0,1] and A an arbitrary subset of Y (clearly this is a σ-field and it contains all the sets that are open in T). We define the Borel measure on T, λ', as follows:

$$\lambda'((B \cap Y^c) \cup A) = \mu(B). \tag{4}$$

Note that on Y^c, $\mathcal{G}$ and $\mathcal{G}'$ coincide; the Borel measure μ_{Y^c} induced by μ on Y^c is then τ-additive (see statement 3°) in §1). We recall that $i_{Y^c,T}$ denotes the canonical embedding of Y^c into T; then it is easily checked that:

$$\lambda' = i_{Y^c,T}(\mu_{Y^c}) \tag{5}$$

and hence (see statement 4°) in §1) λ' is τ-additive.

From (4) it follows that if $u:[0,1] \to R$ is bounded and Borel on [0,1], then

$$\int u d\lambda' = \int u d\mu. \tag{6}$$

We now show that λ' is a Borel extension of the Baire image measure $\lambda = e(\mu)$ on T. It suffices to check that λ' and λ agree on every $h \in C^b(T)$. Note that $\lambda'(Y) = 0$ and with the notation used in the proof of b), $1_{D_h^c} h$ is Borel on [0,1]; by (6) we then have

$$\int h d\lambda' = \int 1_{D_h^c} h d\lambda' = \int 1_{D_h^c} h d\mu = \int h d\mu = \int h \circ e \, d\mu = \int h d\lambda.$$

This proves our assertion.

The above example shows that the implication 4) $\Rightarrow$ 3) in Theorem 4.2 is in general false.

Remark. - The above example is similar to, but more general than, the example given in [3]. In fact, if the set Y is in addition a Bernstein set, i.e. Y has the property that both Y and Y^c meet every uncountable closed set in [0,1] (for the construction of a Bernstein set see [10], pp. 23-24), then the space T has the further property:

d) The topology of T is Lindelöf.

In this case every Baire measure on T admits a Borel extension which is τ-additive (see Remark 1) following Proposition 3.4).

If the set Y contains an uncountable closed (in [0,1]) set F (Y is not a Bernstein set), then the space T constructed above is not Lindelöf; the collection $\{\{s\} \mid s \in F\} \cup \{F^c\}$ is an open cover of T containing no countable subcover.

§6. Lifting compact spaces

The considerations of [3] led us to set the following definition (see [1]):

Definition 2. - A completely regular space T is said to be <<lifting compact>> if: For any complete probability space $(\Omega,\mathcal{F},\mu)$ and for any lifting ρ of $M_R^\infty(\Omega,\mathcal{F},\mu)$, every Baire measurable mapping $f: \Omega \to T$ is lifting compact in T relative to ρ.

We recall (see [9]) that a completely regular space T is called <<strongly measure-compact>> if every Baire measure on T is "tight" (equivalently if every Baire measure on T admits a Borel extension which is a Radon measure: see Proposition 3.3 and the Note following it). We recall also (see [9]) that a completely regular space T is called <<measure-compact>> if every Baire measure on T is "τ-smooth" (equivalently if every Baire measure on T admits a Borel extension which is τ-additive; see Proposition 3.4 and the Note following it).

The following result of [3] is then a direct consequence of Theorem 4.2 (if λ is a Baire measure on T let $(\Omega,\mathcal{F},\mu)$ be the completion of the measure space (T, σ-field of Baire sets in T, λ) and let $f: \Omega \to T$ be the identity mapping):

Corollary 6.1. - Consider the following assertions concerning a completely regular space T:

i) T is strongly measure-compact.

ii) T is lifting compact.

iii) T is measure-compact.

Then i) $\Rightarrow$ ii) $\Rightarrow$ iii). If T is universally measurable then i) $\equiv$ ii) $\equiv$ iii).

We recall that a Baire measure λ on a completely regular space T is called <<completion regular>> if for every Borel set B in T there are Baire sets A_1 and A_2 such that $A_1 \subset B \subset A_2$ and $\lambda(A_2 - A_1) = 0$.

From Proposition 4.6 we immediately derive the following result (if λ is a Baire measure on T, let $(\Omega,\mathcal{F},\mu)$ be the completion of the measure space (T, σ-field of Baire sets in T, λ) and let $f: \Omega \to T$ be the identity mapping).

Corollary 6.2. - Let T be a completely regular space whose topology is finer than a separable metric topology. If T is lifting compact, then every Baire measure on T is <<completion regular>>.

The notion of <<lifting compact>> space has good stability properties. From Proposition 4.3 we immediately obtain:

Theorem 6.3. - An arbitrary subspace of a compact metrizable space is lifting compact.

We also have:

Theorem 6.4. - I) Every Baire set in a lifting compact space is lifting compact.

II) Let S, T be completely regular spaces, $u: S \to T$ a continuous bijection such that $u^{-1}: T \to S$ is Baire measurable. If S is lifting compact then T is lifting compact.

<u>Proof</u>: I) Let T be a completely regular, lifting compact space, and let $X \subset T$ be a Baire set. Let $K \supset T$ be a compactification of T. Let $f: \Omega \to X$ be Baire measurable. Then $i_{X,T} \circ f: \Omega \to T$ is Baire measurable and thus is lifting compact in T relative to ρ. Let $\lambda = (i_{X,T} \circ f)(\mu)$ be the Baire image measure on T and $\nu = \rho(i_{X,T} \circ f)(\mu)$ be the Borel image measure on T (see Remark 3) following Theorem 4.2); then ν and λ agree on Baire sets. In particular

$$\nu(X) = \lambda(X) = \mu(f^{-1}(X)) = \mu(\Omega) = 1,$$

that is, $\mu(\rho(i_{X,T} \circ f)^{-1}(X)) = 1$. Let $\Omega_1 = \rho(i_{X,T} \circ f)^{-1}(X)$. For $\omega \in \Omega_1$ we have:

$$\rho_K(i_{X,K} \circ f)(\omega) = \rho_K(i_{T,K} \circ (i_{X,T} \circ f))(\omega) \quad = \quad \rho(i_{X,T} \circ f)(\omega) \in X.$$

This shows that f is lifting compact in X relative to ρ.

II) Let $f: \Omega \to T$ be Baire measurable. Then $g = u^{-1} \circ f: \Omega \to S$ is Baire measurable and hence is lifting compact in S relative to ρ. Since $u: S \to T$ is continuous, it follows by Proposition 4.4 that $u \circ g = u \circ (u^{-1} \circ f) = f$ is lifting compact in T relative to ρ. This completes the proof of Theorem 6.4.

<u>Theorem 6.5</u>. - <u>Let</u> $(T_k)_{k \in N}$ <u>be a sequence of lifting compact spaces. Then</u>

$$T = \prod_{k \in N} T_k$$

<u>is a lifting compact space</u>.

<u>Proof</u>: Let $f: \Omega \to T$ be Baire measurable. Then $f = (f_k)_{k \in N}$ where for each $k \in N$, $f_k: \Omega \to T_k$ is Baire measurable and hence is lifting compact in T_k relative to ρ. From Proposition 4.5 it follows then that the product map f is lifting compact in T relative to ρ. This proves Theorem 6.5.

<u>Remarks</u>. - 1) Theorems 6.4 and 6.5 are the analogs of the results proved by Moran for <<measure-compact>> and <<strongly measure-compact>> spaces (see [9]).

2) The space T of Example II in §5 is not lifting compact as the discussion in Example II shows. This can also be seen as follows: If T were lifting compact, then $S = T \times T$ would also be lifting compact and hence measure-compact. But it is well-known that T is measure-compact (T is Lindelöf), while $S = T \times T$ is not (see [8],[14]).

As mentioned in the introduction, applications to Banach spaces and locally convex spaces will be given in a forthcoming paper. Here we content ourselves with the following remark.

Let E be <u>a Banach space</u>. Note that for $f: \Omega \to E$ saying that f is scalarly measurable is equivalent with saying that $f: \Omega \to (E, \sigma(E,E'))$ is Baire measurable (see [2]). Then the following assertions about E are equivalent: (i) The space $(E, \sigma(E,E'))$ is lifting compact. (ii) Every E-valued scalarly measurable function is scalarly equivalent to a Bochner measurable function (see [6],[15]).

Bibliography

[1] A. Bellow, "Mesures de Radon et espaces relèvement compacts," to appear in C.R. Acad. Sci. Paris.

[2] G. A. Edgar, "Measurability in a Banach space," Indiana Univ. Math. J., (4) vol. 26, 1977, pp. 663-677.

[3] G. A. Edgar and M. Talagrand, "Liftings of functions with values in a completely regular space," to appear in Proc. A.M.S.

[4] A. Ionescu Tulcea and C. Ionescu Tulcea, Topics in the theory of lifting, Ergebnisse der Math. und ihrer Grenzgebiete, Band 48, Springer-Verlag, New York, 1969.

[5] C. Ionescu Tulcea, "Liftings of functions with values in a completely regular space," Math. Annalen, 187 (1970), pp. 200-206.

[6] A. Ionescu Tulcea, "On pointwise convergence, compactness and equicontinuity II," Advances in Math., (2) vol. 12, 1974, pp. 171-177.

[7] J. D. Knowles, "Measures on topological spaces," Proc. London Math. Soc. (3) 17 (1967), pp. 139-156.

[8] W. Moran, "The additivity of measures in completely regular spaces," J. London Math. Soc. (43) 1968, pp. 633-639.

[9] W. Moran, "Measures and mappings on topological spaces," Proc. London Math. Soc. (3) 19 (1969), pp. 493-508.

[10] J. C. Oxtoby, Measure and Category, Graduate Texts in Mathematics 2, Springer-Verlag, New York 1971.

[11] L. Schwartz, "Certaines propriétés des mesures sur les espaces de Banach," Exposé No. 23, Séminaire Maurey-Schwartz (1975-1976), École Polytechnique.

[12] Chr. Sunyach, "Une caractérisation des espaces universellement mesurables," C.R. Acad. Sci. Paris, 268 (1969), pp. 864-866.

[13] A. Tortrat, "Calcul des Probabilités," Masson et Cie, Paris 1971.

[14] A. Tortrat, "Sur les lois τ-régulières, leurs produits et leur convolution," Trans. Sixth Prague Conference on Information Theory, Academia Publishing House, Prague 1973.

[15] A. Tortrat, "τ-régularité des lois, séparation au sens de A. Tulcea et propriété de Radon-Nikodym," Ann. Inst. Henri Poincaré, XII, No. 2, 1976, pp. 131-150.

[16] H. von Weizsäcker, "Strong measurability, liftings and the Choquet-Edgar theorem," Proc. Dublin Conference 1977, Springer-Verlag Lecture Notes in Math. #645, vol. II, pp. 209-218.

ON " IDEMPOTENT " LIFTINGS

P. Georgiou
Mathematical Institute, University of Athens
57, Solonos Street, Athens-Greece

Dedicated to Professor D.A. Kappos on the occasion of his 75th birthday, September 27, 1979

1. Notations and lemmata.

Let T be a compact (Hausdorff) space, μ a positive Radom measure on T with support $\operatorname{Supp}\mu = T$, $\mathcal{L}^{\infty}_{\mathbb{R}}(T,\mu)$ the seminormed algebra of bounded measurable real valued functions on T, $L^{\infty}_{\mathbb{R}}(T,\mu)$ the Banach algebra $\mathcal{L}^{\infty}_{\mathbb{R}}(T,\mu)$ modulo μ-negligible functions and $f \mapsto \tilde{f}$ the canonical mapping of $\mathcal{L}^{\infty}_{\mathbb{R}}(T,\mu)$ onto $L^{\infty}_{\mathbb{R}}(T,\mu)$. For a lifting r of $\mathcal{L}^{\infty}_{\mathbb{R}}(T,\mu)$ we define $\mathcal{T}_r := \{rB : B \in \mathcal{T}\}$, where $\mathcal{T} := \{B \subseteq T : \phi_B \in \mathcal{L}^{\infty}_{\mathbb{R}}(T,\mu)\}$.

Furthermore let (S,ν) be the hyperstonean space associated with (T,μ). Let U_T be the isomorphism of $L^{\infty}_{\mathbb{R}}(T,\mu)$ onto $L^{\infty}_{\mathbb{R}}(S\ \nu)$ such that $\mu(f) = \nu(U_T\tilde{f})$ for every $f \in \mathcal{C}_{\mathbb{R}}(T)$. Then there exists a continuous surjection $\pi_T : S \to T$ such that $\mu = \pi_T(\nu)$. We denote by ρ the strong lifting of $\mathcal{L}^{\infty}_{\mathbb{R}}(S,\nu)$ and we put $\mathcal{S}_\rho := \{\rho C : C \in \mathcal{S}\}$, where $\mathcal{S} := \{C \subseteq S : \phi_C \in \mathcal{L}^{\infty}_{\mathbb{R}}(S,\nu)\}$. Note that $\mathcal{S}_\rho$ is the class of all simultaneously open and closed subset of S.

We also consider the following boolean homomorphisms

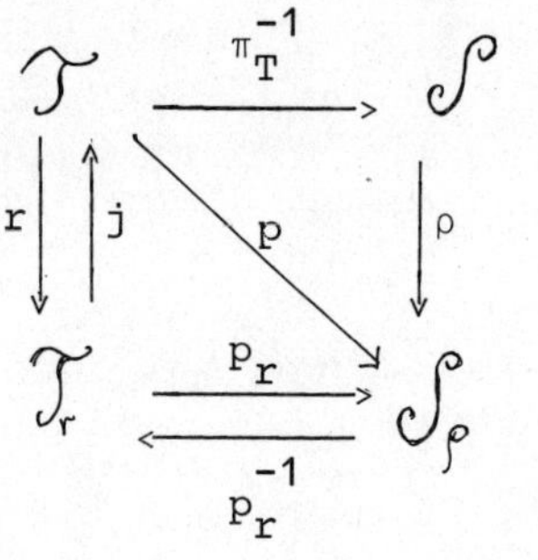

where j denotes the canonical injection $\mathcal{T}_r \overset{\subseteq}{\to} \mathcal{T}$, $p_r = \rho \circ \pi_T^{-1} \circ j$ and $p = \rho \circ \pi_T^{-1}$. One can show that π_T^{-1} is injective, p_r and p_r^{-1} are isometries and $\mu = \nu \circ p$.

From these statements we deduce (cf. [1]), that the "special" disintegration $\lambda_r : t \mapsto \lambda_{rt}$ of ν with repect to p, r and ρ has the property that λ_{rt} are Dirac measures on S. We put $\operatorname{Supp} \lambda_{rt} = \{s_{rt}\}$. θ_r denotes now the mapping

$$t \mapsto s_{rt} : T \to S.$$

Note that $\theta_r^{-1} = p_r^{-1}$.

The main properties of this disintegration are the following (cf. [1]):

(1) $\quad < \rho U_T \tilde{f}, \lambda_r > = rf$ for every $f \in \mathcal{L}^\infty_{\mathbb{R}}(T, \mu)$; i.e.

$$rf = \rho U \tilde{f} \circ \theta_r.$$

(2) $\quad rf = f \circ \pi_T \circ \theta_r$ for every $f \in \mathcal{C}_{\mathbb{R}}(T)$.

(3) $\quad \psi_r^{-1}(B) \underset{\mu}{\equiv} B$ for every $B \in \mathcal{T}$, where $\psi_r := \pi_T \circ \theta_r$.

(4) $\quad \mu^*(\psi_r(T)) = \mu(T)$.

(5) The mapping θ_r has the property
hh) a ν-measurable set $C \subseteq S$ is ν-negligible if and only if $\theta_r^{-1}(C)$ is μ-negligible.

Let r_1 and r_2 be two filtings of $\mathcal{L}^\infty_{\mathbb{R}}(T, \mu)$. Then (cf. [2]) the mapping.

$$r_1 * r_2 : f \in \mathcal{L}^\infty_{\mathbb{R}}(T, \mu) \mapsto < r_1 f \circ \pi_T, \lambda_{r_2} > \in \mathcal{L}^\infty_{\mathbb{R}}(T, \mu)$$

is a lifting and the set of all liftings of $\mathcal{L}^\infty_{\mathbb{R}}(T, \mu)$ $\Lambda(T, \mu)$ is a semigroup with respect to the operation $* : (r_1, r_2) \mapsto r_1 * r_2$. We note that (cf. [2]):

(i) r_o is a strong lifting of $\mathcal{L}^\infty_{\mathbb{R}}(T, \mu)$ if and only if r_o is a right unit element of the semigroup $(\Lambda(T, \mu), *)$;

(ii) $\psi_{r_1 * r_2} = \psi_{r_1} \circ \psi_{r_2}$.

We call a lifting r of $\mathcal{L}^\infty_{\mathbb{R}}(T, \mu)$ idempotent, if

$$r * r = r.$$

An idempotent lifing r has the properties (cf. [2]):
$rf \circ \psi_r = f \circ \psi_r$ for every $f \in \mathcal{C}_{\mathbb{R}}(T)$

or

if B is open, then $B \cap \psi_r(T) \subseteq rB \cap \psi_r(T)$

etc.

2. Measure spaces with the same hyperstonean space.

Let now X be a compact (Hausdorff) space, m a positive Radon measure on X with support $\operatorname{Supp} m = X$ and let us assume that (S, ν) is also the hyperstonean space associated with (X, m). $\mathcal{X}$, U_X, π_X etc. are defined as the $\mathcal{T}$, U_T, π_T etc.

We define for $f \in \mathcal{L}^{\infty}_{\mathbb{R}}(T, \mu)$ and $g \in \mathcal{L}^{\infty}_{\mathbb{R}}(X, m)$ the "relation" $\approx$ as following:

$f \approx g$ iff $U_T\tilde{f} = U_X\tilde{g}$ in $L^{\infty}_{\mathbb{R}}(S, \nu)$.

Then we have the

Proposition 1: *If* $f \in \mathcal{L}^{\infty}_{\mathbb{R}}(T, \mu)$, $g \in \mathcal{C}_{\mathbb{R}}(X)$ *and* $f \approx g$, *then*

$$rf = g \circ \pi_X \circ \theta_r .$$

Proof: Clear from the relation

$$g \circ \pi_X = \rho(f \circ \pi_T). \quad \square$$

From this proposition we have the following corollaries:

Corollary 1: *The mapping*

$$\pi_X \circ \theta_r : T \to X$$

is weakly measurable (cf. definition [3; p.51]).

Corollary 2: *Let* r *be also the induced lifting* (from the lifting r of $\mathcal{L}^{\infty}_{\mathbb{R}}(T, \mu)$) *on the set of weakly measurable mappings from* T *to* X (cf. [3; p.51]). *Then*

$$r(\pi_X \circ \theta_r) = \pi_X \circ \theta_r .$$

Corollary 3: *The mapping*

$$\pi_X \circ \theta_r \circ \pi_T : S \to X$$

is weakly measurable and

$$\rho(\pi_X \circ \theta_r \circ \pi_T) = \pi_X ,$$

where now ρ *denotes the induced lifting on the set of weakly measurable mappings from* S *to* X.

Proof: Let $g \in \mathcal{C}_{\mathbb{R}}(X)$ and $f \approx g$. Then

$$g \circ \pi_X = \rho(g \circ \pi_X) = \rho(f \circ \pi_T) = \rho(rf \circ \pi_T)$$

$$= \rho(g \circ \pi_X \circ \theta_r \circ \pi_T) = g \circ \rho(\pi_X \circ \theta_r \circ \pi_T). \square$$

Remark: It is clear that $\rho(\pi_T) = \pi_T$.

Corollary 4: *If* $f \in \mathcal{L}^{\infty}_{\mathbb{R}}(T, \mu)$, $g \in \mathcal{C}_{\mathbb{R}}(X)$ *and* $f \approx g$, *then*

$$m(g) = \int f \, d\mu.$$

Proof: From [1; 5 Proposition (i)] follows that for $h \in \mathcal{C}_{\mathbb{R}}(T)$ and $k \in \mathcal{C}_{\mathbb{R}}(T)$ is $\int h(k \circ \pi_T) d\nu = \int k < h, \lambda_r > d\mu$.

Now for $h = g \circ \pi_X$ is

$$\int k < g \circ \pi_X, \lambda_r > d\mu = \int k(g \circ \pi_X \circ \theta_r) d\mu$$

$$= \int (g \circ \pi_X) \cdot (k \circ \pi_T) d\nu$$

and for $k = 1$ (constant) is

$$\int g \circ \pi_X \circ \theta_r \, d\mu = \int (g \circ \pi_X) d\nu = \int g \, dm.$$

But

$$f \underset{\mu}{\equiv} rf = g \circ \pi_X \circ \theta_r$$

and hence

$$\int g \, dm = \int f \, d\mu. \square$$

We assume that the relation $\approx$ induces an injection from $\mathcal{C}_{\mathbb{R}}(T)$ into $\mathcal{C}_{\mathbb{R}}(X)$.

Clearly there exists then a continuous surjection π_{XT} from X onto T such that

$$\pi_{XT}(m) = \mu$$

and

$$\pi_T = \pi_{XT} \circ \pi_X.$$

This relation implies (cf. [1; 1 Corollary]) the

Proposition 2: *If* r *is a strong lifting of* $\mathcal{L}^{\infty}_{\mathbb{R}}(T, \mu)$, *then*

$I_T = \pi_T \circ \theta_r = \pi_{XT} \circ \pi_X \circ \theta_r$, *where* I_T *denotes the identity mapping* $T \to T$.

So we have the

Proposition 3: *If* r *is a strong lifting of* $\mathcal{L}^\infty_{\mathbb{R}}(T, \mu)$, *then*

$$g = g \circ \pi_X \circ \theta_r \circ \pi_{XT}$$

for every $g \in \mathcal{C}_{\mathbb{R}}(X)$ *for which there is* $f \in \mathcal{C}_{\mathbb{R}}(T)$ *such that* $g \approx f$.

Proof: If f is as in Proposition, then (cf. Proposition 1) we have

$$g = f \circ \pi_{XT} = rf \circ \pi_{XT} = g \circ \pi_X \circ \theta_r \circ \pi_{XT} . \ \square$$

This is now the main

Lemma 1: *Let* $f \in \mathcal{L}^\infty_{\mathbb{R}}(T, \mu)$, $g_f \in \mathcal{L}^\infty_{\mathbb{R}}(X, m)$ *and* $f \approx g_f$. *Then*

$$f \circ \pi_{XT} \underset{m}{\equiv} g_f .$$

Proof: If we assume that $f \circ \pi_{XT} \underset{m}{\not\equiv} g_f$, then there is a non negligible measurable set, on which $f \circ \pi_{XT}$ and g_f are different and also the functions

$$f \circ \pi_{XT} \circ \pi_X = f \circ \pi_T \quad \text{and} \quad g_f \circ \pi_X$$

are different on a set $A \in \mathcal{S}_\rho$ $(A \neq \emptyset)$, because the relation $f \approx g_f$ induce

$$f \circ \pi_T \underset{\nu}{\equiv} g_f \circ \pi_X . \ \square$$

From this Lemma we have also that: $rf \circ \pi_{XT} \underset{m}{\equiv} g_f$ for every $f \in \mathcal{L}^\infty_{\mathbb{R}}(T, \mu)$.

3. An idempotent lifting.

Definition: Let l be the mapping

$$l : g \in \mathcal{L}^\infty_{\mathbb{R}}(X, m) \mapsto lg := \rho(g \circ \pi_X) \circ \theta_r \circ \pi_{XT}$$
$$= rf \circ \pi_{XT} \in \mathcal{L}^\infty_{\mathbb{R}}(X, m) ,$$

where $f \approx g$

From Lemma 1 it is obvious that:

<u>Proposition 4</u>: *The mapping* l *is a lifting of* $\mathcal{L}^{\infty}_{\mathbb{R}}(X, m)$.

<u>Proof</u>: Because $rf = \rho U_T \tilde{\tilde{f}} \circ \theta_r$ (cf. [1] and [2]). □

<u>Lemma 2</u>: *The relation is hold*

$$\pi_X \circ \theta_l = \pi_X \circ \theta_r \circ \pi_{XT} .$$

<u>Proof</u>: For $h \in \mathcal{C}_{\mathbb{R}}(X)$ we have (cf. [1])

$$lh = h \circ \pi_X \circ \theta_l$$

and (from the definition of l)

$$lh = h \circ \pi_X \circ \theta_r \circ \pi_{XT}$$

and therefore

$$\pi_X \circ \theta_l = \pi_X \circ \theta_r \circ \pi_{XT}$$

Moreover (cf. Corollary 2) this mapping is equal to

$$l(\pi_X \circ \theta_l) = l(\pi_X \circ \theta_r \circ \pi_{XT}). \quad \square$$

And now we can prove the

<u>Theorem :</u> *If* r *is a strong lifting of* $\mathcal{L}^{\infty}_{\mathbb{R}}(T, \mu)$, *then* l *is an idempotent lifting of* $\mathcal{L}^{\infty}_{\mathbb{R}}(X, m)$.

<u>Proof</u>: Let

$$\mathcal{L}^{\infty}_{\mathbb{R}}(X, m) \ni g \approx f_g \in \mathcal{L}^{\infty}_{\mathbb{R}} (T, \mu).$$

Then from the definition of the operation $*$ we have

$$\begin{aligned} l * lg &= lg \circ \pi_X \circ \theta_l \\ &= lg \circ \pi_X \circ \theta_r \circ \pi_{XT} \\ &= rf_g \circ \pi_{XT} \circ \pi_X \circ \theta_r \circ \pi_{XT} \\ &= rf_g \circ I_T \circ \pi_{XT} \\ \text{(from Proposition 2)} \qquad &= rf_g \circ \pi_{XT} \\ &= lg . \quad \square \end{aligned}$$

This idempotent lifting l has the following property

<u>Proposition 5</u>: *If* r *is a strong lifting of* $\mathcal{L}^{\infty}_{\mathbb{R}}(T, \mu)$, $g \in \mathcal{C}_{\mathbb{R}}(X)$,

$f_g \in \mathcal{C}_{\mathbb{R}}(T)$ *and* $g \approx f_g$, *then*

$$lg = g.$$

Proof: Plainly, because (cf. lemma 1)

$$lg = rf_g \circ \pi_{XT} = f_g \circ \pi_{XT} = g. \quad \square$$

References

1. *Georgiou, P.*: Über eine spezielle Desintegration. Math. Ann. 197, 279-285(1972)

2. *Georgiou, P.*: A semigroup sturcture in the space of liftings. Math. Ann. 208, 195-202(1974).

3. *Ionescu Tulcea, A and C.* : Topics in the theory of lifting. Berlin-Heidelberg - New york: Springer 1969.

DIFFERENTIATION OF MEASURES ON UNIFORM SPACES

Pertti Mattila*

1. Introduction

Let X be a separable metric space and μ a Borel regular measure on X such that bounded sets have finite μ measure. Well-known results in the theory of differentiation of integrals tell that

$$\lim_{r\downarrow 0} \int_{B(x,r)} f d\mu/\mu B(x,r) = f(x) \qquad \text{for } \mu \text{ a.a. } x \in X$$

for any μ summable f, if either X is finite dimensional in a suitable sense (e.g., a euclidean n-space) or μ is finite dimensional in the sense that

$$\limsup_{r\downarrow 0} \mu B(x, 5r)/\mu B(x,r) < \infty \qquad \text{for } \mu \text{ a.a. } x \in X.$$

Here $B(x,r)$ is the closed ball with centre x and radius r. (See for example Sections 2.8 and 2.9 in [3].) The typically infinite-dimensional cases are not covered by these results. In this paper we consider the question what kind of differentiation theorems can be obtained if no assumptions referring to finite dimensionality are made, but instead μ is supposed to satisfy certain homogeneity conditions. The following is a special case of our results:

If there is a nondecreasing function $h : (0,\infty) \to (0,\infty)$ and $c > 0$ such that $\lim_{r\downarrow 0} h(r) = 0$ and

$$(1) \qquad c h(r) \le \mu B(x,r) \le h(r)$$

for μ a.a. $x \in X$ and for $r > 0$, then for any $f \in L^p(\mu)$, $1 \le p < \infty$,

$$\int_{B(x,r)} f d\mu/\mu B(x,r) \to f(x) \text{ in } L^p(\mu) \text{ as } r\downarrow 0.$$

Under the weaker homogeneity condition

$$(2) \qquad \liminf_{r\downarrow 0} \mu B(x,r)/h(r) > 0 \text{ and } \mu B(x,r) \le h(r)$$

for μ a.a. $x \in X$ and for $r > 0$, this is false (Example 5.3), but then the

*Supported in part by NSF Grant MCS77-18723(02).

convergence takes place in measure over sets with finite μ measure. In fact, we shall prove slightly stronger results, which correspond to the Lebesgue set theorems in the pointwise theory. Examples 5.1 and 5.2 show that in (1) and (2) the uniform upper bound $\mu B(x, r) \leq h(r)$ cannot be replaced by $\limsup_{r \downarrow 0} \mu B(x, r)/h(r) \leq 1$ and that the mean convergence cannot be replaced by pointwise convergence almost everywhere, even when f is a characteristic function of a Borel set. In the latter example μ satisfies the stronger homogeneity condition $\mu B(x, r) = h(r)$ for $x \in X$ and $r > 0$; moreover, X is a compact abelian group and μ a Haar measure on X.

We shall also show that if μ satisfies (2) and ν is finite, Borel regular and singular with respect to μ, then

$$\nu B(x, r)/\mu B(x, r) \to 0$$

in μ measure over sets with finite μ measure. These results combined with the Radon-Nikodym theorem give differentiation theorems for general measures ν.

The proofs of the above-mentioned results are based on Fubini's theorem. No covering theorems are used, but some Vitali-type covering theorems are consequences of our results. One of them is stated in 3.8.

We shall develop our theory in a more general context by considering measures μ on uniform spaces. Then, for example, Radon measures on locally compact uniform spaces and Haar measures on locally compact groups will be included in our theory. Balls $B(x, r)$ will be replaced by open symmetric neighborhoods $U[x]$ of x. We shall assume on μ that it has a uniform approximation property defined in Section 2 and that Fubini's theorem is applicable.

J. Bocle [1] has earlier proved similar global differentiation theorems for Haar measures on locally compact groups.

In the final Section 6 we shall show that similar results in the euclidean n-space are valid without any homogeneity conditions on μ at all.

2. Preliminaries

2.1. Notation. Throughout this paper X will be a uniform space with a fixed base $\mathcal{U}$, whose members are symmetric open subsets of $X \times X$. If $U \in \mathcal{U}$, the U neighborhood of $x \in X$ is

$$U[x] = \{y : (x, y) \in U\}.$$

The symmetricity of U means that $y \in U[x]$ if and only if $x \in U[y]$. For any $A \subset X$ and $U \in \mathcal{U}$, the U neighborhood of A,

$$A(U) = \bigcup \{U[x] : x \in A\},$$

is an open set. If $U \in \mathcal{U}$, there is, by definition, $V \in \mathcal{U}$ such that

$$V \circ V = \{(x, y) : (x, z) \in V \text{ and } (z, y) \in V \text{ for some } z \in X\} \subset U.$$

Then $(A(V))(V) \subset A(U)$ for any set $A \subset X$.

If $f : \mathcal{U} \to \overline{R} = \{t : -\infty \leq t \leq \infty\}$, then we denote by $\liminf_{U \in \mathcal{U}} f(U)$, $\limsup_{U \in \mathcal{U}} f(U)$ and $\lim_{U \in \mathcal{U}} f(U)$ the upper limit, lower limit and limit of f as U tends to the diagonal of $X \times X$ through $\mathcal{U}$. For example,

$$\limsup_{U \in \mathcal{U}} f(U) = \inf_{V \in \mathcal{U}} \sup\{f(U) : U \in \mathcal{U},\ U \subset V\}.$$

We denote by H the set of all functions $h : \mathcal{U} \to (0, \infty)$ such that $U \subset V$ implies $h(U) \leq h(V)$ and $\lim_{U \in \mathcal{U}} h(U) = 0$.

If $1 \leq p < \infty$, $A \subset X$ and μ is an (outer) measure over X, $L^p(\mu, A)$ will stand for the set of all μ measurable functions $f : A \to \overline{R}$ such that $\int_A |f|^p d\mu < \infty$, and $L^p(\mu) = L^p(\mu, X)$. If f_U, $f : X \to \overline{R}$ for $U \in \mathcal{U}$, μ is a measure over X, and A is a μ measurable set, then we denote

$$f_U \to f \text{ in } L^p(\mu, A)$$

if $\lim_{U \in \mathcal{U}} \int_A |f_U - f|^p d\mu = 0$, and

$$f_U \to f \text{ in } \operatorname{meas}(\mu, A)$$

if for every $\varepsilon > 0$ $\lim_{U \in \mathcal{U}} \mu\{x \in A : |f_U(x) - f(x)| \geq \varepsilon\} = 0$.

The characteristic function of a set A is denoted by χ_A.

2.2. Definition. By the family of Borel sets of a topological space Y we mean the smallest σ-algebra containing the open sets of Y. We shall say that a measure μ over X has the uniform approximation property if for any Borel set E with $\mu(E) < \infty$ and for any $\varepsilon > 0$, there are a Borel set $F \subset E$ and $U \in \mathcal{U}$ such that

$$\mu(E\setminus F) < \varepsilon \text{ and } \mu(F(U)\setminus F) < \varepsilon.$$

We give two examples:

2.3. Proposition. Suppose that either

(1) X is locally compact and μ is a Radon measure over X, or

(2) X is separable with a metric d,

$$\mathcal{U} = \{\{(x, y) : d(x, y) < r\} : 0 < r < \infty\}$$

is the standard base induced by d, open sets are μ measurable, and there are open sets $G_1, G_2, \ldots$ such that $X = \bigcup_{i=1} G_i$ and $\mu(G_i) < \infty$ for all i.

Then μ has the uniform approximation property.

Proof. Let E be a Borel set with $\mu(E) < \infty$ and $\varepsilon > 0$.

If (1) holds, there are a compact set F and an open set G such that $F \subset E \subset G$ and $\mu(G\setminus F) < \varepsilon$ (see [3, 2.2.5]). Since F is compact and X is locally compact, there is $U \in \mathcal{U}$ for which $F(U) \subset G$. Then F and U have the required properties.

Assume (2). By [3, 2.2.2(2)] there is an open set G such that $E \subset G$ and $\mu(G\setminus E) < \varepsilon/2$. Set

$$G_r = \{x : \operatorname{dist}(x, X\setminus G) > r\} \text{ for } r > 0.$$

Then $\lim_{r\downarrow 0} \mu(G\setminus G_r) = 0$, and we may take r so small that $\mu(G\setminus G_r) < \varepsilon/4$. By [3, 2.2.2(1)] there is a closed set $F \subset E \cap G_r$ such that $\mu(E\cap G_r\setminus F) < \varepsilon/4$. Then

$$F(r) = \{x : \operatorname{dist}(x, F) < r\} \subset G,$$

$$\mu(E\setminus F) \leq \mu(G\setminus G_r) + \mu(E\cap G_r\setminus F) < \varepsilon/2.$$

and

$$\mu(F(r)\setminus F) \leq \mu(G\setminus E) + \mu(E\setminus F) < \varepsilon.$$

The proofs of the main results will be based on Fubini's theorem in the form of the following lemma.

2.4. Lemma. *Suppose that μ and ν are measures on X such that open sets are μ and ν measurable and that A and B are Borel sets of X such that A has σ-finite μ measure and B has σ-finite ν measure. If $U \in \mathcal{U}$ is such that the Borel subsets of $C(U) \times C(U)$, where $C = A \cup B$, are $\mu \times \nu$ measurable, then the function $x \mapsto \nu(U[x] \cap E)$ is μ measurable in A for any Borel set $E \subset X$ and*

$$\int_A \mu(U[x] \cap B) d\nu x = \int_B \nu(U[x] \cap A) d\mu x.$$

In particular, if $f : X \to \overline{R}$ is a nonnegative Borel function such that $\int_A f d\mu < \infty$, then

$$\int_A \mu(U[x] \cap B) f(x) d\mu x = \int_B \int_{U[x] \cap A} f d\mu d\mu x.$$

Proof. Since

$$U[x] \cap E = \{y : (x, y) \in U \cap (X \times E) \cap (C(U) \times C(U))\}$$

for $x \in A$ and $U \cap (X \times E) \cap (C(U) \times C(U))$ is $\mu \times \nu$ measurable, it follows from Fubini's theorem that $x \mapsto \nu(U[x] \cap E)$ is μ measurable in A. Since $U \cap (A \times B)$ is $\mu \times \nu$ measurable, Fubini's theorem gives

$$\int_A \mu(U[x] \cap B) d\nu x = \int_A \int_B \chi_{U \cap (A \times B)}(x, y) d\mu y d\nu x =$$

$$\int_B \int_A \chi_{U \cap (A \times B)}(x, y) d\nu x d\mu y = \int_B \nu(U[x] \cap A) d\mu x.$$

The last assertion follows now with $\nu(D) = \int_D f d\mu$.

The assumption that the Borel subsets of $C(U) \times C(U)$ are $\mu \times \nu$ measurable is satisfied if either X is locally compact, μ and ν are Radon measures and Clos $C(U)$ is compact, or if the open sets are μ and ν measurable and the subspace $C(U)$ is separable and metrizable.

3. Differentation of Integrals

Throughout this section we assume that μ is a measure on X such that open sets are μ measurable and μ has the uniform approximation property, A is a Borel set in X, $U_o \in \mathcal{U}$ and the Borel subsets of $A(U_o) \times A(U_o)$ are $\mu \times \mu$ measurable.

We first prove a density lemma:

3.1. Lemma. *Suppose that* B *and* E *are Borel sets in* X *and there is* $h \in H$ *such that*

$$\mu U[x] \leq h(U)$$

for μ *a.a.* $x \in B(U)$ *and for* $U \in \mathcal{U}$, $U \subset U_o$. *If* $U_1 \in \mathcal{U}$, $U_1 \circ U_1 \subset U_o$, $B \subset E \cap A(U_1)$ *and* $\mu(B) < \infty$, *then*

$$\lim_{U \in \mathcal{U}} h(U)^{-1} \int_B \mu(U[x] \setminus E) d\mu x = 0.$$

Proof. Let $\varepsilon > 0$. By the uniform approximation property there are a Borel set $F \subset B$ and $U_2 \in \mathcal{U}$, $U_2 \subset U_1$ such that

$$\mu(B \setminus F) < \varepsilon \text{ and } \mu(F(U_2) \setminus F) < \varepsilon.$$

Let $U \in \mathcal{U}$, $U \subset U_2$. Then $U[x] \subset F(U)$ for all $x \in F$, and we obtain from Lemma 2.4

$$h(U)^{-1} \int_B \mu(U[x] \setminus E) d\mu x \leq h(U)^{-1} \int_F \mu(U[x] \setminus F) d\mu x + \varepsilon =$$

$$h(U)^{-1} \int_F \mu(U[x] \cap F(U) \setminus F) d\mu x + \varepsilon = h(U)^{-1} \int_{F(U) \setminus F} \mu(U[x] \cap F) d\mu x + \varepsilon \leq \mu(F(U) \setminus F) + \varepsilon < 2\varepsilon.$$

This proves the lemma.

3.2. Theorem. *Suppose that there is* $h \in H$ *such that*

$$\mu U[x] \leq h(U)$$

for μ *a.a.* $x \in A(U)$ *and for* $U \in \mathcal{U}$, $U \subset U_o$. *If* $f : X \to \overline{R}$ *is a Borel function such that* $\int |f|^p d\mu < \infty$ *for some* $1 \leq p < \infty$, *and if* A *has* σ*-finite* μ *measure, then*

$$\lim_{U \in \mathcal{U}} h(U)^{-1} \int_A \int_{U[x]} |f(y) - f(x)|^p d\mu y \, d\mu x = 0.$$

Proof. We first assume that $f = \chi_E$ where E is a Borel set and $\mu(E) < \infty$. There is $U_1 \in \mathcal{U}$ such that $U_1 \circ U_1 \subset U_o$. Let $U \in \mathcal{U}$, $U \subset U_1$. Then

$$h(U)^{-1} \int_A \int_{U[x]} |\chi_E(y) - \chi_E(x)|^p d\mu y \, d\mu x =$$

$$h(U)^{-1} \int_{A \cap E} \mu(U[x] \setminus E) d\mu x + h(U)^{-1} \int_{A \setminus E} \mu(U[x] \cap A(U) \cap E) d\mu x.$$

We may apply 3.1 with $B = A(U_1) \cap E$ to obtain

$$\lim_{U \in \mathcal{U}} h(U)^{-1} \int_{A(U_1) \cap E} \mu(U[x] \setminus E) d\mu x = 0.$$

Lemma 2.4 gives

$$\int_{A\setminus E}\mu(U[x]\cap A(U)\cap E)d\mu x = \int_{A(U)\cap E}\mu(U[x]\cap A\setminus E)d\mu x \le \int_{A(U_1)\cap E}\mu(U[x]\setminus E)d\mu x.$$

It follows that

$$\lim_{U\in\mathcal{U}} h(U)^{-1}\int_A\int_{U[x]}|\chi_E(y)-\chi_E(x)|^p d\mu y d\mu x = 0.$$

Next we assume that f is a nonnegative simple function,

$$f = \sum_{i=1}^{n} a_i\chi_{E_i},$$

where the sets E_i are Borel sets with $\mu(E_i) < \infty$. Applying twice Minkowski's inequality, we obtain

$$h(U)^{-1}\int_A\int_{U[x]}|f(y)-f(x)|^p d\mu y d\mu x =$$

$$h(U)^{-1}\int_A\int_{U[x]}\Big|\sum_{i=1}^{n} a_i(\chi_{E_i}(y)-\chi_{E_i}(x))\Big|^p d\mu y d\mu x \le$$

$$h(U)^{-1}\int_A\Big(\sum_{i=1}^{n}\Big(\int_{U[x]}|a_i(\chi_{E_i}(y)-\chi_{E_i}(x))|^p d\mu y\Big)^{1/p}\Big)^p d\mu x \le$$

$$h(U)^{-1}\Big(\sum_{i=1}^{n}\Big(\int_A\int_{U[x]}|a_i(\chi_{E_i}(y)-\chi_{E_i}(x))|^p d\mu y d\mu x\Big)^{1/p}\Big)^p =$$

$$\Big(\sum_{i=1}^{n} a_i\Big(h(U)^{-1}\int_A\int_{U[x]}|\chi_{E_i}(y)-\chi_{E_i}(x)|^p d\mu y d\mu x\Big)^{1/p}\Big)^p.$$

This tends to zero by the first part of the proof.

If f is a nonnegative Borel function in $L^p(\mu)$, there is an increasing sequence (f_i) of simple Borel functions such that

$$\lim_{i\to\infty}\int|f-f_i|^p d\mu = 0.$$

Let $\varepsilon > 0$. Choose i so large that

$$\Big(\int|f-f_i|^p d\mu\Big)^{1/p} < \varepsilon.$$

By the second part of the proof there is $U_2 \in \mathcal{U}$, $U_2 \subset U_o$, such that

$$(h(U)^{-1}\int_A\int_{U[x]}|f_i(y)-f_i(x)|^p d\mu y d\mu x)^{1/p} < \varepsilon$$

for $U \in \mathcal{U}$, $U \subset U_2$. Then we can use Minkowski's inequality and Lemma 2.4 to obtain

$$(h(U)^{-1}\int_A\int_{U[x]}|f(y)-f(x)|^p d\mu y d\mu x)^{1/p} \leq$$

$$h(U)^{-1/p}(\int_A((\int_{U[x]}|f(y)-f_i(y)|^p d\mu y)^{1/p} +$$

$$(\int_{U[x]}|f_i(y)-f_i(x)|^p d\mu y)^{1/p} + (\int_{U[x]}|f_i(x)-f(x)|^p d\mu y)^{1/p})^p d\mu x)^{1/p} \leq$$

$$h(U)^{-1/p}((\int_A\int_{U[x]}|f(y)-f_i(y)|^p d\mu y d\mu x)^{1/p} +$$

$$(\int_A\int_{U[x]}|f_i(y)-f_i(x)|^p d\mu y d\mu x)^{1/p} + (\int_A\int_{U[x]}|f_i(x)-f(x)|^p d\mu y d\mu x)^{1/p}) =$$

$$h(U)^{-1/p}((\int \mu(U[x]\cap A)|f(x)-f_i(x)|^p d\mu x)^{1/p} +$$

$$(\int_A\int_{U[x]}|f_i(y)-f_i(x)|^p d\mu y d\mu x)^{1/p} + (\int_A \mu U[x]|f_i(x)-f(x)|^p d\mu x)^{1/p}) \leq$$

$$2(\int|f-f_i|^p d\mu)^{1/p} + (h(U)^{-1}\int_A\int_{U[x]}|f_i(y)-f_i(x)|^p d\mu y d\mu x)^{1/p} < 3\varepsilon.$$

In the general case the result follows by the representation $f = f^+ - f^-$ and Minkowski's inequality.

3.3. Corollary. *If in addition to the assumptions of Theorem 3.2 there is* $c > 0$ *such that*

$$\mu U[x] \geq ch(U) \text{ for } \mu \text{ a.a. } x \in A,$$

then

$$(\mu U[x])^{-1}\int_{U[x]} f d\mu \to f(x) \text{ in } L^p(\mu, A).$$

Proof. By Hölder's inequality

$$\int_A |(\mu U[x])^{-1}\int_{U[x]} f d\mu - f(x)|^p d\mu x =$$

$$\int_A (\mu U[x])^{-p} | \int_{U[x]} (f(y)-f(x)) d\mu y |^p d\mu x \le$$

$$\int_A (\mu U[x])^{-1} \int_{U[x]} |f(y)-f(x)|^p d\mu y d\mu x \le$$

$$c^{-1} h(U)^{-1} \int_A \int_{U[x]} |f(y)-f(x)|^p d\mu y d\mu x,$$

from which the assertion follows.

3.4. Corollary. Suppose that X has a countable base for its uniformity. If in addition to the assumptions of Theorem 3.2 $\mu(A) < \infty$ and

$$\liminf_{U \in \mathcal{U}} h(U)^{-1} \mu U[x] > 0 \quad \text{for } \mu \text{ a.a. } x \in A,$$

then

$$(\mu U[x])^{-1} \int_{U[x]} |f(y)-f(x)|^p d\mu y \to 0 \quad \text{in meas}(\mu, A),$$

and

$$(\mu U[x])^{-1} \int_{U[x]} f d\mu \to f(x) \quad \text{in meas}(\mu, A).$$

Proof. Let $\{U_i : i = 1, 2, \dots\}$ be a base for the uniformity of X. If the first assertion is false, then there is $\varepsilon > 0$ and for each i, $V_i \in \mathcal{U}$ such that $V_i \subset U_i \cap U_o$ and

$$\mu\{x \in A : (\mu V_i[x])^{-1} \int_{V_i[x]} |f(y)-f(x)|^p d\mu y \ge \varepsilon\} \ge \varepsilon. \tag{1}$$

Let

$$A_i = \{x \in A : \mu V_j[x] \ge h(V_j)/i \text{ for } j \ge i\}, \ i=1,2,\dots .$$

Then the sets A_i are Borel sets and $\lim_{i\to\infty} \mu(A \setminus A_i) = 0$. Choose i so large that $\mu(A\setminus A_i) < \varepsilon/2$. It follows from Theorem 3.2 that

$$\lim_{j\to\infty} \int_{A_i} (\mu V_j[x])^{-1} \int_{V_j[x]} |f(y)-f(x)|^p d\mu y d\mu x = 0.$$

Hence there is $j \ge i$ for which

$$\mu\{x \in A_i : (\mu V_j[x])^{-1} \int_{V_j[x]} |f(y)-f(x)|^p d\mu y \ge \varepsilon\} < \varepsilon/2.$$

Since $\mu(A\setminus A_i) < \varepsilon/2$, we obtain a contradiction with (1). This proves the first assertion. The second follows from the first.

Next we give a condition on μ which makes it possible to remove the symmetricity assumption on $\mathcal{U}$. For $B \subset X$ we let

$$B^{-1} = \{(x, y) : (y, x) \in B\} \text{ and } B_s = B \cup B^{-1}.$$

Then B_s is symmetric.

3.5. Corollary. Let $\mathcal{V}$ be a base for the uniformity of X consisting of open sets. If there is $C < \infty$ such that

$$\mu V^{-1}[x] \leq C \mu V[x]$$

for μ a.a. $x \in A(U_o)$ and for $V \in \mathcal{V}$, then the results 3.1-4 hold with $\mathcal{U}$ replaced by $\mathcal{V}$.

Proof. The family $\mathcal{V}_s = \{V_s : V \in \mathcal{V}\}$ is a base consisting of open symmetric sets, $V \subset V_s$ and $\mu(V_s) \leq (C+1)\mu(V)$. Then 3.1-4 hold for $\mathcal{V}_s$, and it follows that they also hold for $\mathcal{V}$.

3.6. Remark. Let X be a locally compact group and μ a left or right Haar measure on X. Then X has a natural uniform structure such that $\mu U[x]$ is independent of $x \in X$ [5, p. 210]. Thus μ satisfies the above homogeneity conditions. The condition of 3.5 is also satisfied if $A(U_o)$ has compact closure. This follows from the fact that the modular function of X is continuous and thus bounded on compact sets.

3.7. Remark. The following covering theorem can be deduced from 3.4 with the methods of [4, III. 2.5-7]:

Suppose that μ and A are as in 3.4, ν is a measure on X such that Borel sets are ν measurable, $\nu(X) < \infty$ and ν is absolutely continuous with respect to μ. If $U_1 \in \mathcal{U}$ and

$$\mathcal{V} = \{U[x] : x \in A,\ U \in \mathcal{U},\ U \subset U_1\},$$

then for any $\varepsilon > 0$ there is a countable subfamily $\{U_i[x_i] : i=1,2,\ldots\} \subset \mathcal{V}$ such that

$$\nu(A \setminus \bigcup_{i=1}^{\infty} U_i[x_i]) = 0$$

and

$$\sum_{i=1}^{\infty} \nu U_i[x_i] \leq \nu(\bigcup_{i=1}^{\infty} U_i[x_i]) + \varepsilon.$$

An example of Davies [2] shows that it is not always possible to choose the sets $U_i[x_i]$ disjoint, and the Example 5.2 combined with [4, II.2.7] shows that it is not sufficient to assume that $\mathcal{V}$ contains arbitrarily small neighborhoods $U[x]$ for each $x \in A$. In both examples X is a compact metric space, and $\mu B(x, r) = h(r)$ for $x \in X$ and $r > 0$.

4. Differentiation of Singular and General Measures

In this section we assume that μ and ν are measures on X, $\nu(X) < \infty$, A is a Borel set in X, $\mu(A) < \infty$, and either (1) X is locally compact, μ and ν are Radon measures and $\operatorname{Clos} A$ is compact, or (2) X is a separable metric space, μ and ν are Borel regular and bounded sets have finite μ measure.

It is clear from the proofs that the results remain valid in a somewhat more general setting.

4.1. Theorem. *If ν is singular with respect to μ and if there are $h \in H$, $c > 0$ and $U_o \in \mathcal{U}$ such that for $U \in \mathcal{U}$, $U \subset U_o$,*

$$\mu U[x] \leq h(U) \quad \text{for } \mu \text{ a.a. } x \in A(U)$$

and

$$\mu U[x] \geq ch(U) \quad \text{for } \mu \text{ a.a. } x \in A,$$

then

$$\nu U[x]/\mu U[x] \to 0 \quad \text{in meas}(\mu, A).$$

Proof. Since ν is singular, there is a Borel set $B \subset X$ such that $\mu(B) = 0 = \nu(X \setminus B)$. Let $\varepsilon > 0$, $\delta > 0$. We shall show that there are open sets V, W and $U_1 \in \mathcal{U}$, $U_1 \subset U_o$, such that

$$B \subset V, \ \mu(V) < \delta, \ W(U_1) \subset V, \ \nu(V \setminus W) < \delta.$$

In the case (1), we find V such that $B \subset V$ and $\mu(V) < \delta$, because μ is a Radon measure. Then there are a compact set $C \subset V$ and $U_1' \in \mathcal{U}$, $U_1' \subset U_o$, for which $\nu(V \setminus C) < \delta$ and $C(U_1') \subset V$. We can now take $W = C(U_1)$ for any $U_1 \in \mathcal{U}$ such that $U_1 \circ U_1 \subset U_1'$. Moreover, we may choose U_1 so that $\operatorname{Clos} A(U_1)$ is compact.

In the case (2), we use [3, 2.2.2(2)] to find V such that $B \subset V$ and $\mu(V) < \delta$. Setting

$$V_U = \{x \in V : U[x] \subset V\} \quad \text{for } U \in \mathcal{U},$$

we have $\bigcup_{U \in \mathcal{U}} V_U = V$ and $\lim_{U \in \mathcal{U}} \nu(V \setminus V_U) = 0$, since $\nu(V) < \infty$, and the uniformity of X has a countable base. Thus there is $U_1 \in \mathcal{U}$, $U_1 \subset U_o$, such that $W(U_1) \subset V$ and $\mu(V \setminus W) < \delta$ with $W = V_{U_1}$.

By Lemma 2.4 we get for $U \in \mathcal{U}$, $U \subset U_1$,

$$\int_{A \setminus V} \nu U[x]/\mu U[x] d\mu x \leq (ch(U))^{-1} \int_{A \setminus V} \nu(U[x] \cap B \setminus W) d\mu x =$$

$$(ch(U))^{-1} \int_{B \setminus W} \mu(U[x] \cap A \setminus V) d\nu x \leq c^{-1} \nu(B \setminus W) < c^{-1} \delta.$$

Since $\mu(V) < \delta$, we obtain

$$\mu\{x \in A : \nu U[x]/\mu U[x] \geq \varepsilon\} < \delta(c^{-1}\varepsilon^{-1}+1).$$

This proves the theorem.

We now drop the assumption on the singularity of ν. By Lebesgue decomposition theorem there are then a nonnegative Borel function f on X and a singular measure ν_1, which satisfies the same conditions as ν, such that

$$\nu(B) = \int_B f d\mu + \nu_1(B).$$

Combining 3.3 and 4.1 we obtain

4.2. Theorem. *If μ satisfies the conditions of Theorem 4.1 and ν is as above, then*

$$\nu U[x]/\mu U[x] \to f(x) \quad \text{in meas}(\mu, A).$$

The following corollaries can be proved as 3.4 and 3.6.

4.3. Corollary. *Suppose that X has a countable base for its uniformity. If there are $h \in H$ and $U_o \in \mathcal{U}$ such that*

$$\mu U[x] \leq h(U)$$

for μ a.a. $x \in A(U)$ and for $U \in \mathcal{U}$, $U \subset U_o$, and

$$\liminf_{U \in \mathcal{U}} h(U)^{-1} \mu U[x] > 0$$

for μ a.a. $x \in A$, then

$$\nu U[x]/\mu U[x] \to f(x) \quad \text{in meas}(\mu, A).$$

4.4. Corollary. *Let $\mathcal{V}$ be a base for the uniformity of X consisting of open sets.*

If there is $C < \infty$ *such that*

$$\mu V^{-1}[x] \leq C\mu V[x]$$

for μ *a.a.* $x \in A(U_o)$ *and for* $V \in \mathcal{V}$, *then the results* 4.1-3 *hold with* $\mathcal{U}$ *replaced by* $\mathcal{V}$.

4.5. *Remarks.* In the above results convergence in measure cannot be replaced by mean convergence. This follows from the simple example where X is the real line with its usual uniformity, μ is the Lebesgue measure, and ν is the Dirac measure at the origin.

All the above results immediately extend to signed measures by the Jordan decomposition theorem.

5. *Examples*

Here we construct three examples to illustrate the sharpness of the preceding differentiation theorems.

5.1. *Example.* *We construct a compact metric space* X *and a Radon measure* μ *on* X *such that there are* $h \in H$ *and Borel sets* $E, F \subset X$ *for which*

$$\mu(E) < 1 = \mu(X) = \mu(F),$$

$$\lim_{r \downarrow 0} \frac{\mu B(x,r)}{h(r)} = 1 \text{ for } x \in F$$

and

$$\lim_{r \downarrow 0} \frac{\mu(E \cap B(x,r))}{\mu B(x,r)} = 1 \text{ for } x \in F.$$

It follows that the ratios $\mu(E \cap B(x,r))/\mu B(x,r)$ do not converge in the mean (nor in measure) to the characteristic function of E.

For $n = 1, 2, \ldots$ set $X_n = \{1, 2, \ldots, 6^n\}$, $Y_n = \{1, 2, \ldots, 2^n\}$ and define a metric d_n in X_n such that for $x \neq y$, $d_n(x,y) = 1/2$ if $x \in Y_n$ or $y \in Y_n$, and $d_n(x,y) = 1$ if $x, y \in X_n \setminus Y_n$. Let the measure μ_n on X_n be defined by $\mu_n\{x\} = 6^{-n}$ for $x \in X_n$. Set

$$X = \prod_{n=1}^{\infty} X_n$$

and

$$d(x,y) = \sup\{2^{-n} d_n(x_n, y_n) : n = 1, 2, \ldots\}$$

for $x = (x_1, x_2, \ldots)$, $y = (y_1, y_2, \ldots) \in X$. Then (X, d) is a compact metric space. There exists a unique Radon measure

$$\mu = \prod_{n=1}^{\infty} \mu_n,$$

the product of the measures μ_n, for which

$$\mu\left(\prod_{n=1}^{\infty} E_n\right) = \prod_{n=1}^{\infty} \mu_n(E_n) \quad \text{for } E_n \subset X_n.$$

Then $\mu(X) = 1$. Let

$$E_n = \prod_{i=1}^{n-1} X_i \times Y_n \times \prod_{i=n+1}^{\infty} X_i, \quad E = \bigcup_{n=1}^{\infty} E_n,$$

$$F_n = \prod_{i=1}^{n-1} X_i \times (X_n \setminus Y_n) \times \prod_{i=n+1}^{\infty} X_i, \quad F = \bigcup_{k=1}^{\infty} \bigcap_{n=k}^{\infty} F_n.$$

Then

$$\mu(E_n) = 3^{-n}, \quad \mu(E) \leq \sum_{n=1}^{\infty} 3^{-n} < 1,$$

$$\mu(F_n) = 1 - 3^{-n},$$

$$\mu(F) = \lim_{k \to \infty} \mu\left(\bigcap_{n=k}^{\infty} F_n\right) = \lim_{k \to \infty} \prod_{n=k}^{\infty} (1 - 3^{-n}) = 1.$$

If $2^{-n-1} \leq r < 2^{-n}$, then for $x = (x_1, x_2, \ldots) \in F_n$

$$E \cap B(x, r) \supset E_n \cap B(x, r) = \prod_{i=1}^{n-1} B(x_i, 2^i r) \times Y_n \times \prod_{i=n+1}^{\infty} B(x_i, 2^i r),$$

$$B(x, r) = \prod_{i=1}^{n-1} B(x_i, 2^i r) \times (Y_n \cup \{x_n\}) \times \prod_{i=n+1}^{\infty} B(x_i, 2^i r),$$

whence

$$\frac{\mu(E \cap B(x, r))}{\mu B(x, r)} \geq \frac{\mu_n(Y_n)}{\mu_n(Y_n \cup \{x_n\})} = \frac{3^n}{3^n + 1}.$$

Therefore, for any $x \in F$

$$\lim_{r \downarrow 0} \frac{\mu(E \cap B(x, r))}{\mu B(x, r)} = 1.$$

Take any $x_o \in F$ and define $h(r) = \mu B(x_o, r)$. Then for all $x \in F$, $\mu B(x, r) = h(r)$

for small r.

5.2. <u>Example.</u> <u>We construct a compact metric space</u> X <u>and a Radon measure</u> μ <u>on</u> X <u>such that there are</u> $h \in H$ <u>and a Borel set</u> $E \subset X$ <u>for which</u>

$$\mu(E) < \mu(X) = 1,$$

$$\mu B(x, r) = h(r) \text{ for } x \in X,\ r > 0,$$

<u>and</u>

$$\limsup_{r \downarrow 0} \frac{\mu(E \cap B(x, r))}{\mu B(x, r)} = 1 \text{ \underline{for} } \mu \text{ \underline{a.a.} } x \in X.$$

Let (n_k) be a sequence of positive integers such that

$$s_k = (n_k/(n_k+1))^k \to 1 \text{ as } k \to \infty,$$

and set

$$A_k = \{0, 2, \ldots, 2n_k-2\},\quad B_k = \{1, 3, \ldots, 2n_k-1\},\quad X_k = A_k \cup B_k.$$

Define a metric d_k on X_k such that for $x \neq y$, $d_k(x, y) = 1$ if $(x, y) \in (A_k \times A_k) \cup (B_k \times B_k)$, and $d_k(x, y) = 1/2$ otherwise. Let the measure μ_k on X_k be defined by $\mu_k\{x\} = 1/(2n_k)$ for $x \in X_k$. Then $\mu_k(A_k) = \mu_k(B_k) = 1/2$ and $\mu_k B(x, r)$ is independent of $x \in X_k$ for all $r > 0$. Put $h_k(r) = \mu_k B(x, r)$. Then

(1) $$\mu_k(A_k \cap B(x, r)) = h_k(r) \text{ for } x \in A_k,\ 0 < r < 1/2,$$

(2) $$\mu_k(A_k \cap B(x, r)) = (n_k/(n_k+1))h_k(r) \text{ for } x \in B_k,\ 1/2 \leq r < 1.$$

By A^k and μ^k we mean the k-fold product of a set A and a measure μ, respectively. Let (t_i) be a strictly increasing sequence with $1/2 < t_i < 1$. Set

$$Y_k = X_k^k,\quad C_k = A_k^k,\quad \nu_k = \mu_k^k,$$

and define a metric ρ_k on Y_k by

$$\rho_k(x, y) = \max\{d_k(x_i, y_i)/t_i : 1 \leq i \leq k\}$$

for $x = (x_1, \ldots, x_k)$, $y = (y_1, \ldots, y_k) \in Y_k$. Then

$$\nu_k(C_k) = 2^{-k},\quad \nu_k B(x, r) = \prod_{i=1}^{k} h_k(rt_i)$$

and if $x = (x_1, \ldots, x_k) \in A_k^j \times B_k^{k-j}$, $1 \le j \le k$, $1/(2t_{j+1}) \le r < 1/(2t_j)$,

$$\nu_k(C_k \cap B(x, r)) = \prod_{i=1}^{k} \mu_i(A_k \cap B(x_i, rt_i)) = (n_k/(n_k+1))^{k-j} \prod_{i=1}^{k} h_k(rt_i) \ge s_k \nu_k B(x, r)$$

by (1) and (2). Therefore, for every x in

$$D_k = \bigcup_{j=1}^{k} A_k^j \times B_k^{k-j}$$

there exists r, $0 < r < 1$, such that

$$\nu_k(C_k \cap B(x, r)) \ge s_k \nu_k B(x, r). \tag{3}$$

Clearly $\nu_k(D_k) = k2^{-k}$. We can choose strictly increasing sequences (k_i) and (m_i) of positive integers such that with $p_i = m_{i+1} - m_i$

$$\sum_{i=1}^{\infty} p_i 2^{-k_i} < 1 \text{ and } \sum_{i=1}^{\infty} p_i k_i 2^{-k_i} = \infty.$$

Setting $m_o = 0$, we define for $m_i < n \le m_{i+1}$, $i = 0, 1, \ldots$

$$Z_n = Y_{k_i}, \quad \lambda_n = \nu_{k_i}, \quad E_n = C_{k_i}, \quad F_n = D_{k_i}, \quad \sigma_n = \rho_{k_i},$$

$$X = \prod_{n=1}^{\infty} Z_n, \quad \mu = \prod_{n=1}^{\infty} \lambda_n,$$

$$E'_n = \prod_{i=1}^{n-1} Z_i \times E_n \times \prod_{i=n+1}^{\infty} Z_i, \quad E = \bigcup_{n=1}^{\infty} E'_n,$$

$$F'_n = \prod_{i=1}^{n-1} Z_i \times F_n \times \prod_{i=n+1}^{\infty} Z_i, \quad F = \bigcap_{k=1}^{\infty} \bigcup_{n=k}^{\infty} F'_n$$

and

$$d(x, y) = \sup\{2^{-n} \sigma_n(x_n, y_n) : n = 1, 2, \ldots\}$$

for $x = (x_1, x_2, \ldots)$, $y = (y_1, y_2, \ldots) \in X$. Then (X, d) is a compact metric space, $\mu(X) = 1$, $\mu B(x, r)$ is independent of x for all $r > 0$,

$$\mu(E) \leq \sum_{n=1}^{\infty} \lambda_n(E_n) = \sum_{i=1}^{\infty} p_i \nu_{k_i}(C_{k_i}) = \sum_{i=1}^{\infty} p_i 2^{-k_i} < 1,$$

$$\sum_{n=1}^{\infty} \mu(F'_n) = \sum_{i=1}^{\infty} p_i \nu_{k_i}(D_{k_i}) = \sum_{i=1}^{\infty} p_i k_i 2^{-k_i} = \infty,$$

whence by the Borel-Cantelli lemma

$$\mu(F) = 1.$$

If $x \in F$, then for each k there is $j_k \geq k$ such that $x_{j_k} \in F_{j_k}$. If i_k is the integer for which $m_{i_k} < j_k \leq m_{i_k+1}$, then $E_{j_k} = C_{i_k}$, $F_{j_k} = D_{i_k}$, $\lambda_{j_k} = \nu_{i_k}$, and thus by (3) there exists r_k, $0 < r_k < 1$, such that

$$\lambda_{j_k}(E_{j_k} \cap B(x_{j_k}, r_k)) = \nu_{i_k}(C_{i_k} \cap B(x_{j_k}, r_k)) \geq s_{i_k} \nu_{i_k} B(x_{j_k}, r_k) = s_{i_k} \lambda_{j_k} B(x_{j_k}, r_k).$$

Setting $r'_k = 2^{-j_k} r_k$ we have $\lim_{k\to\infty} r'_k = 0$ and

$$\mu(E \cap B(x, r'_k)) \geq \mu(E'_{j_k} \cap B(x, r'_k)) = \lambda_{j_k}(E_{j_k} \cap B(x_{j_k}, r_k)) \prod_{i \neq j_k} \lambda_i B(x_i, 2^i r'_k) \geq s_{i_k} \mu B(x, r'_k).$$

Since $\lim_{k\to\infty} s_{i_k} = 1$, we obtain

$$\limsup_{r \downarrow 0} \frac{\mu(E \cap B(x, r))}{\mu B(x, r)} = 1 \text{ for } x \in F,$$

as required.

If $x, y \in X$, then $x = (x_n)$, $y = (y_n)$ where for each n, x_n and y_n belong to some X_k. We define

$$x+y = (z_n)$$

where $z_n = x_n + y_n \pmod{2n_k}$ if $x_n, y_n \in X_k$. Then X is an abelian group such that the translations are isometries and μ is a Haar measure on X.

5.3. Example. We construct a complete, separable, locally compact metric space X, a Radon measure μ on X and $f \in L^1(\mu)$ such that

$$(1)\qquad \lim_{r\downarrow 0} \frac{\mu B(x,r)}{2r} = 1 \text{ and } \mu B(x,r) \leq 2r$$

for μ a.a. $x \in X$ and for $r > 0$, and that

$$(2)\qquad \lim_{r\downarrow 0} \int \frac{1}{\mu B(x,r)} \int_{B(x,r)} f d\mu d\mu x = \infty .$$

For each $k = 1, 2, \ldots$ we choose disjoint compact subintervals $I_{k,j}$, $j = 1, \ldots, k^3$, of the real line with length k^{-2}. We denote

$$A_k = \bigcup_{j=0}^{k^3} I_{k,j}, \quad B_k = \bigcup_{j=1}^{k^3} I_{k,j}$$

and choose the intervals so that the sets A_k are also disjoint. We define a metric d on

$$X = \bigcup_{k=1}^{\infty} A_k$$

by requiring that for $x \in I_{k,j}$

$$\begin{aligned} d(x,y) &= k^{-3}|x-y|, \text{ if } y \in I_{k,j}, \\ &= k^{-2}, \text{ if } j = 0 \text{ and } y \in I_{k,i} \text{ for some } i \geq 1, \\ &= 2k^{-2}, \text{ if } j \neq 0 \text{ and } y \in I_{k,i} \text{ for some } i \geq 1,\ i \neq j, \\ &= 4, \text{ if } y \in X \setminus A_k. \end{aligned}$$

Letting $\mathcal{L}^1$ be the one-dimensional Lebesgue measure, we define a Borel regular measure μ on X by

$$\mu(A) = \sum_{k=1}^{\infty} k^{-3} \mathcal{L}^1(A \cap A_k) \text{ for } A \subset X,$$

and a Borel function $f \in L^1(\mu)$ by

$$f = \sum_{k=1}^{\infty} k^3 \chi_{I_{k,0}} .$$

Then one easily verifies (1). If $k^{-2} \leq r < 2k^{-2}$ and $x \in I_{k,j}$ for some $j \geq 1$, then $B(x,r) = I_{k,0} \cup I_{k,j}$ and

$$\frac{1}{\mu B(x, r)} \int_{B(x, r)} f d\mu = k^3/2.$$

Thus

$$\int \frac{1}{\mu B(x, r)} \int_{B(x, r)} f d\mu d\mu x \geq (k^3/2)\mu(B_k) = k/2,$$

and (2) follows.

6. Differentiation in Euclidean Spaces

In this section we shall show that the results which in Sections 3 and 4 were proved assuming that μ satisfies certain homogeneity conditions hold in the euclidean n-space R^n metrized with some norm without any such conditions. Since the results concerning convergence in measure follow readily from the well-known pointwise theory (see [3, Section 2.9 combined with 2.8.18]), we only consider the differentiation of integrals. The proof is based on a covering principle and a corresponding pointwise differentiation theorem.

6.1. Lemma. There is a constant $C < \infty$ such that whenever μ and ν are Borel regular measures on R^n, $A \subset R^n$ and $0 < r < \infty$, then

$$\int \frac{\nu(A \cap B(x, r))}{\mu B(x, r)} d\mu x \leq C\nu(A).$$

Proof. We may assume that $0 < \nu(A) < \infty$ and, by Borel regularity and the monotone convergence theorem, that A is a bounded Borel set. For $m = 1, 2, \ldots$, set

$$B_m = \{x : A \cap B(x, r) \neq \phi,\ 1/m \leq \mu B(x, r) < \infty\}.$$

Since $\mu\{x : \mu B(x, r) = 0\} = 0$, we have

$$\lim_{m \to \infty} \int_{B_m} \frac{\nu(A \cap B(x, r))}{\mu B(x, r)} d\mu x = \int \frac{\nu(A \cap B(x, r))}{\mu B(x, r)} d\mu x.$$

Fix m and suppose $B_m \neq \phi$. Then for $x \in B_m$

$$\frac{\nu(A \cap B(x, r))}{\mu B(x, r)} \leq m\nu(A) < \infty;$$

hence we may define inductively $x_1, \ldots, x_k \in B_m$ such that

$$\frac{\nu(A \cap B(x_1, r))}{\mu B(x_1, r)} > \tfrac{1}{2}\sup\{\frac{\nu(A \cap B(x, r))}{\mu B(x, r)} : x \in B_m\},$$

and for $i \geq 2$,

$$x_i \in B_m \setminus \bigcup_{j=1}^{i-1} B(x_j, r),$$

$$\frac{\nu(A \cap B(x_i, r))}{\mu B(x_i, r)} > \tfrac{1}{2}\sup\{\frac{\nu(A \cap B(x, r))}{\mu B(x, r)} : x \in B_m \setminus \bigcup_{j=1}^{i-1} B(x_j, r)\}.$$

Then the balls $B(x_j, r/2)$ are disjoint and lie in a bounded set; therefore, the process has to stop. Moreover,

$$\nu(A \cap B(x, r)) = 0 \quad \text{for} \quad x \in B_m \setminus \bigcup_{i=1}^{k} B(x_i, r),$$

for otherwise we could define x_{k+1}. Set

$$A_1 = B_m \cap B(x_1, r), \quad A_i = B_m \cap B(x_i, r) \setminus \bigcup_{j=1}^{i-1} B(x_j, r), \quad i = 2, \ldots, k.$$

Then $B_m \cap \bigcup_{i=1}^{k} B(x_i, r) = \bigcup_{i=1}^{k} A_i$, the sets A_i are disjoint Borel sets, $A_i \subset B(x_i, r)$ and

$$\frac{\nu(A \cap B(x, r))}{\mu B(x, r)} \leq 2\frac{\nu(A \cap B(x_i, r))}{\mu B(x_i, r)} \quad \text{for} \quad x \in A_i.$$

Since the centers x_i are outside every ball $B(x_j, r)$ for $j \neq i$, there is a positive integer N depending only on the norm of R^n such that at most N of the balls $B(x_i, r)$ may have a point in common. Thus we get

$$\int_{B_m} \frac{\nu(A \cap B(x, r))}{\mu B(x, r)} d\mu x = \int_{\bigcup_i A_i} \frac{\nu(A \cap B(x, r))}{\mu B(x, r)} d\mu x = \sum_{i=1}^{k} \int_{A_i} \frac{\nu(A \cap B(x, r))}{\mu B(x, r)} d\mu x \leq$$

$$\sum_{i=1}^{k} 2\frac{\nu(A \cap B(x_i, r))}{\mu B(x_i, r)} \mu(A_i) \leq 2\sum_{i=1}^{k} \nu(A \cap B(x_i, r)) \leq 2N\nu(A).$$

This proves the lemma.

6.2. Theorem. Let μ be a Borel regular measure on R^n such that bounded sets have finite μ measure. If $f \in L^p(\mu)$ for some $1 \leq p < \infty$, then

$$\lim_{r \downarrow 0} \int (\mu B(x, r))^{-1} \int_{B(x, r)} |f(y)-f(x)|^p d\mu y d\mu x = 0.$$

Proof. Let $\varepsilon > 0$. For $0 < M < \infty$ define

$$\begin{aligned} f_M(x) &= f(x) \text{ if } x \in B(0, M) \text{ and } |f(x)| \leq M, \\ &= M \text{ if } x \in B(0, M) \text{ and } |f(x)| > M, \\ &= 0 \text{ if } x \in R^n \setminus B(0, M). \end{aligned}$$

We may choose M so large that

$$\int |g|^p d\mu < \varepsilon \text{ where } g = f - f_M.$$

Applying twice Minkowski's inequality, we obtain

$$\begin{aligned} &\left(\int (\mu B(x, r))^{-1} \int_{B(x, r)} |f(y)-f(x)|^p d\mu y d\mu x\right)^{1/p} \leq \\ (1) \quad &\left(\int (\mu B(x, r))^{-1} \int_{B(x, r)} |f_M(y)-f_M(x)|^p d\mu y d\mu x\right)^{1/p} + \\ &\left(\int (\mu B(x, r))^{-1} \int_{B(x, r)} |g(y)-g(x)|^p d\mu y d\mu x\right)^{1/p}. \end{aligned}$$

It follows from [3, 2.8.18 and 2.9.9] and Hölder's inequality that

$$\lim_{r \downarrow 0} (\mu B(x, r))^{-1} \int_{B(x, r)} |f_M(y)-f_M(x)|^p d\mu y = 0$$

for μ a.a. $x \in R^n$. By Minkowski's inequality

$$(\mu B(x, r))^{-1} \int_{B(x, r)} |f_M(y)-f_M(x)|^p d\mu y \leq 2^p M^p.$$

Since $\mu B(0, M+1) < \infty$ and since $f_M(y) = 0$, if $y \in B(x, r)$, $x \in R^n \setminus B(0, M+1)$ and $0 < r < 1$, Lebesgue's bounded convergence theorem yields

$$\begin{aligned} &\lim_{r \downarrow 0} \int (\mu B(x, r))^{-1} \int_{B(x, r)} |f_M(y)-f_M(x)|^p d\mu y d\mu x = \\ &\lim_{r \downarrow 0} \int_{B(0, M+1)} (\mu B(x, r))^{-1} \int_{B(x, r)} |f_M(y)-f_M(x)|^p d\mu y d\mu x = 0. \end{aligned}$$

To estimate the corresponding integral for g, we use Minkowski's inequality and Lemma 6.1 to obtain

$$\int (\mu B(x, r))^{-1} \int_{B(x, r)} |g(y)-g(x)|^p d\mu y d\mu x \leq$$

$$\int (\mu B(x, r))^{-1} \Big(\Big(\int_{B(x, r)} |g(y)|^p d\mu y\Big)^{1/p} + \Big(\int_{B(x, r)} |g(x)|^p d\mu y\Big)^{1/p}\Big)^p d\mu x =$$

$$\int \Big(\Big((\mu B(x, r))^{-1} \int_{B(x, r)} |g(y)|^p d\mu y\Big)^{1/p} + |g(x)|\Big)^p d\mu x \leq$$

$$\Big(\Big(\int (\mu B(x, r))^{-1} \int_{B(x, r)} |g(y)|^p d\mu y d\mu x\Big)^{1/p} + \Big(\int |g|^p d\mu\Big)^{1/p}\Big)^p \leq$$

$$\Big(\Big(C\int |g|^p d\mu\Big)^{1/p} + \Big(\int |g|^p d\mu\Big)^{1/p}\Big)^p \leq (C^{1/p}+1)^p \varepsilon .$$

It follows that the left-hand side of (1) tends to zero, which proves the theorem.

6.3. Corollary. *With the assumptions of* 6.2

$$(\mu B(x, r))^{-1} \int_{B(x, r)} f d\mu \to f(x) \ \text{ in } \ L^p(\mu) \ \text{ as } \ r \downarrow 0.$$

This follows from Hölder's inequality as in 3.3.

References

[1] Boclé, J., Sur la théorie ergodique, Ann. Inst. Fourier (Grenoble), 10 (1960), 1-45.

[2] Davies, R. O., Measures not approximable or not specifiable by means of balls, Mathematika, 18 (1971), 157-160.

[3] Federer, H., Geometric Measure Theory, Springer-Verlag, 1969.

[4] Hayes, C. A., and C. Y. Pauc, Derivation and Martingales, Springer-Verlag, 1970.

[5] Kelley, J. L., General Topology, Van Nostrand, 1955.

The Institute for Advanced Study
Princeton, New Jersey 08540

and

University of Helsinki
Helsinki, Finland

DIFFERENTIATION OF DANIELL INTEGRALS

A. Volčič
Istituto di Matematica Applicata
Piazzale Europa 1, Trieste-Italia

Measure and integration theory for extended real valued functions is frequently presented as a chapter of functional analysis defining the integral as a linear form, continuous on monotone sequences (sometimes on nets) defined on a suitable linear lattice of functions. This presentation is usefull from the didactic point of view, but is also important because of the existence "in nature" of such linear forms, which are not deducible from a measure. The conical measures introduced by G. Choquêt (2) provide a significant example.

The aim of this paper is to contribute to the differentiation theory (in the sense of Radon-Nikodym derivatives) of Daniell integrals, without any use of measure theoretic methods. Some results can be applied (see (14)) in the resolution of a problem posed by D. Kölzow (5) .

Let R be a linear lattice (a Riesz space) of extended real valued functions with domain X. A Daniell integral on R is a linear functional I , which is continuous on monotone sequences. As every Daniell integral has a Jordan decomposition (7), we suppose (if not otherwise stated) that all the integrals are positive. We can suppose also that the completion (as in (9)) is already performed and that I is defined on L_1 , which is a Riesz-Lebesgue space (in the sense of (1)). It is well known that the Stone-measurable functions M_1 are defined as follows:

$f \varepsilon M_1$ iff $\mathrm{mid}(f,g',g'') = \sup(\inf(f,g'),\inf(f,g''))\varepsilon L_1$, $\forall\, g',g''\varepsilon L_1$

and that M_1 is a Riesz-Baire space (1).

We want now to define a collection of functions which can represent the dual of L_1. Of course if the Stone's condition

$\inf(\chi_X,f)\varepsilon L_1$,$\forall\, f\varepsilon L_1$

holds, then the problem can be reduced to a well known measure-theoretic situation ((6),(8),(10),(11),(16)). But we do not suppose that this condition holds. Let us consider then the following collection L_∞ of extended real valued functions:

$k\varepsilon L_\infty$ iff $fk\varepsilon L_1$,$\forall\, f\varepsilon L_1$.

Proposition 1

L_∞ is a linear space, a lattice and an algebra.

The lattice operations are the only one to check, but it is easy to prove that for each $k \varepsilon L_\infty$, k^+ and k^- (and therefore $|k|$) belong to L_∞, observing that it is enough to verify the definition on the positive cone of L_1.

Let us consider the following collection A_∞ of sets:

$A \varepsilon A_\infty$ iff $\chi_A \varepsilon L_\infty$,

having T as a subcollection of (locally null) sets:

$T \varepsilon T$ iff $I(\chi_T f)=0$, $f \varepsilon L_1$.

Proposition 2

(a) if $g \varepsilon M_1$, then $A=\{x: g(x)=0\} \varepsilon A_\infty$

(b) if $k \varepsilon L_\infty$, then $B=\{x: k(x)<\alpha\} \varepsilon A_\infty$, $\alpha \varepsilon \mathbb{R}$.

(a): g can be taken positive. By the Lebesgue's theorem from $\text{mid}(ng,f,0) \uparrow \chi_{CA} f$ we draw the desired conclusion.

(b): as $k-\alpha \; \varepsilon L_\infty$, B can be taken as a set, on which a function of L_∞ has non positive values. Again by the Lebesgue's theorem and from $\inf((nk^-),\chi_X) f \uparrow \chi_B f$ we establish the result.

It is easy to prove that T is a hereditary σ-ideal.

Proposition 3

T containes the family N of null sets.

Let be N such that $\chi_N \varepsilon L_1$ and $I(\chi_N)=0$. If $\phi_N = \lim n\chi_N$, $\phi_N \varepsilon L_1$ and $I(\phi_N)=0$ by Beppo Levi's theorem. If $f \varepsilon L_1$,then $|f\chi_N| \leqslant \phi_N$, therefore $f\chi_N \varepsilon L_1$ and $I(f\chi_N)=0$, so N is locally null.

The converse is not true (14), but:

a) if $A=\{x: f(x)>0\}$ and $f \varepsilon L_1^+$, then A is null iff it is locally null. If A is locally null, then $0 \leqslant \chi_A \leqslant \sum_1^\infty n f_n$ with $f_n = \chi_A f$ for each $n \varepsilon \mathbb{N}$. Again by Beppo Levi, as $f_n \varepsilon L_1$ and $I(f_n)=0, \sum_1^\infty n f_n \varepsilon L_1$ and $I(\sum_1^\infty n f_n)=0$, so $\chi_A \varepsilon L_1$ and $I(\chi_A)=0$.

b) if $f_1, f_2 \varepsilon L_1$ and $A=\{x: f_1(x)>f_2(x)\}$, then A is null iff it is locally null. It follows from a), as $A=\{x: f_1(x) - \inf(f_1(x),f_2(x))>0\}$.

Definition

A function $g: X \to \bar{R}$ is said to be essentially bounded, if for some $\beta \in R$, the set $\{x: |g(x)| > \beta\}$ is locally null.

Let us define for each essentially bounded function g,

$|g|_\infty = \inf\{ \beta: \{x: |g(x)| > \beta\} \in T\}$.

It is easy to see that this is a seminorm (and a norm on L_∞ / T).

Proposition 4

If $k \in L_\infty$, then k is essentially bounded.

Suppose $k > 0$ and that k is not essentially bounded. Then the sets $A_n = \{x: k(x) > n\}$ are in A_∞, but are not locally null. Take for each n a function $f_n \in L$ such that $I(f_n \chi_{A_n}) > 0$ and multiply it by a suitable positive real number c_n in order that $g_n = c_n f_n$ is such that $I(g_n \chi_{A_n}) = 1/n^2$

By Beppo Levi's theorem $f = \Sigma_1^\infty g_n \chi_{A_n} \in L_1$, but

$I(kf) = I(k \Sigma_1^\infty g_n \chi_{A_n}) \geq \Sigma_1^\infty n I(g_n \chi_{A_n}) = \Sigma_1^\infty 1/n$

and so we have a contradiction.

Note that it can be proved by standard arguments that

$|k|_\infty = \sup \{I(|k| f)/I(f) \; ; \; I(f) \neq 0\}$.

Of course the sup can be taken over the f's such that $I(f) = 1$.

Proposition 5

If the sequence $\{k_n\}$ of functions of L_∞ is essentially uniformly bounded (i.e. for some β $\{x: |k_n(x)| > \beta\}$ is locally null for each n) and converges pointwise to k, then $k \in L_\infty$.

Proof: apply the Lebesgue dominated convergence theorem to the sequence $\{k_n f\}$. Up to a null set, it is $|k_n f| \leq \beta |f|$ and $k_n f \to kf$.

Corollary

A is a σ-algebra.

The σ-completness is the only non evident thing to prove: but we can apply the proposition 5, as the sequence of characteristic functions of $\bigcup_1^n A_i$ is uniformly bounded and monotone.

Let us denote by M_∞ the A_∞-measurable functions.

Proposition 6

$A \varepsilon A_\infty$ iff A is locally measurable with respect to the σ-ring of sets $\{x: g(x)>0\}$, $g \varepsilon M_1$.

For the proof, see (13).

Proposition 7

$k \varepsilon L_\infty$ iff $k \varepsilon M_\infty$ and is essentially bounded.

The "only if" part has been aleady proved.

The converse can be proved by an approximation argument: divide $[-|k|_\infty, |k|_\infty]$ into 2^n intervals $I^{n,m} = [a_{n,m-1}, a_{n,m}[$, where $a_{n,m} = -|k|_\infty + m|k|_\infty / 2^{n-1}$; the interval is closed if $m = 2^n$. Define

$$k_n(x) = \Sigma_m \chi_{I^{n,m}}$$

Then $|k(x)-k_n(x)| \leq 1/2^n$ and $k_n \varepsilon L_\infty$. By proposition 5, $k \varepsilon L_\infty$.

Proposition 8

$(L_\infty / T, |\cdot|_\infty)$ is a Banach space.

For each $r \varepsilon \mathbb{N}$ there is a natural number $n(r)$ such that for $n,m>n(r)$ $|k_n - k_m|_\infty < 1/r$. Let be $A_{n,m,r} = \{x: |k_n(x)-k_m(x)| \geq 1/r\}$ and consider the set A, union of the $A_{n,m,r}$'s with $n,m>n(r)$, which is locally null. On CA the sequence $\{k_n\}$ is pointwise convergent and the limit k (0 on A) belongs (as the sequence $\{|k_n|_\infty\}$is Cauchy) by proposition 5 to L_∞. Now $I(|k_n - k_m| f) < \varepsilon$ for n,m sufficiently large and for every $f \varepsilon L_1^+$ such that $I(f)=1$. Taking the limit as $m \to \infty$ we have that $|k_n - k|_\infty < \varepsilon$ for n sufficiently large.

Proposition 9

$k \varepsilon M_\infty$ iff $kg \varepsilon M_1$ for each $g \varepsilon M_1$.

Let be $k \varepsilon M_\infty$ and $k \geq 0$. Let us prove that for each $g \geq 0$ in M_1 $kg \varepsilon M_1$ i.e. that $\inf(kg,f) \varepsilon L_1$,for each $f \varepsilon L_1$. But

$$\inf(kg,f) = \begin{cases} k \cdot \inf(g,(f/k)) & \text{if } k>0 \\ 0 & \text{otherwise} \end{cases}$$

and $f/k = \lim f/k_n$, where $k_n = \inf(\sup(k,1/n),n)$; k_n is A_∞-measurable and essentially bounded, therefore $k_n\ L_\infty$. From this we deduce that

$$h_n = \begin{cases} k_n \cdot \inf(g,(f/k_n) & \text{if } k>0 \\ 0 & \text{otherwise} \end{cases}$$

belongs to L_1 and $|h_n| \leq f$ for each n. As $h_n \to \inf(kg,f)$, $kg \varepsilon M_1$.

Let us prove the converse: let be $kg\varepsilon M_1$ for each $g\varepsilon M_1$ (suppose $k\geqslant 0$ and $g\geqslant 0$) and $f\varepsilon L_1^+$. Then for each $\alpha\varepsilon R$ $h_n = \inf((k/\alpha)^n f, f)\varepsilon L_1$, as the collection of functions k such that $kg\varepsilon M_1$ for each $g\varepsilon M_1$ is obviously an algebra. But $\lim h_n = f\chi_A$, where $A = \{x: k(x)\geqslant\alpha\}$. As $|h_n|\leqslant f$ for each n, $f\chi_A\varepsilon L_1$ and therefore $A\varepsilon A_\infty$, so $k\varepsilon M_\infty$.

Thecrem 1

The following properties are equivalent:

(a) χ_X is Stone-measurable

(b) $M_1 = M_\infty$

(c) $L_1 \subset M_\infty$

(d) $L_\infty \subset M_1$

(e) $M_1 \subset M_\infty$

(f) $M_\infty \subset M_1$

The implications (a)→(b)→(c), (f)→(b)→(c) and (b)→(e)→(c) are known or easy to prove. It remains to prove that (d)→(f) and (c)→(a).

(d)→(f): if $f\varepsilon M_\infty$, then $f_n = \inf(f^+, n) - \inf(f^-, n)$ belongs to L_∞ and therefore $f_n\varepsilon M_1$. But $f = \lim f_n$ and $f\varepsilon M_1$, so $M_\infty \subset M_1$.

(c)→(a): if $f\varepsilon L_1^+$, then $f\varepsilon M_\infty$ and therefore $\inf(f,n)\varepsilon L_\infty$ and so $\inf(f,n)f\varepsilon L_1$ for any positive integer n and from this fact we deduce that $f^2\varepsilon M_1$. Moreover, if $f\varepsilon L_1^+$ and $g\varepsilon M_1^+$, $\mathrm{mid}(f,g,0) = \inf(f,g)\varepsilon L_1$. So we have the hypotheses of a Zaanen's theorem (15) which guarenties that the constants are measurable.

In M_1 and M_∞ two partial orders can be introduced. If $g_1, g_2\varepsilon M_1$, we put $g_1 > g_2$, if $\{x: g_1(x) < g_2(x)\}$ is a null set. If $h_1, h_2\varepsilon M_\infty$, we put $h_1 > h_2$, if $\{x: h_1(x) < h_2(x)\}$ is locally null. If $g_1 > g_2$ ($h_1 > h_2$) and $g_2 > g_1$ ($h_2 > h_1$), g_1 and g_2 (h_1 and h_2) are said to be equivalent: $g_1 \sim g_2$ ($h_1 \sim h_2$).

Definition

A Daniell integral I is said to be localizable, if the lattice $M_\infty/\sim$ is complete (compare (16)).

Theorem 2

I is localizable iff the lattice $A_\infty/\sim$ is complete.

The proof is the same as in (16).

We are now going to prove the main results of the paper: theorem 3 (a Hahn-type decomposition), theorem 4 (a Kelley-type characterization of localizability; for the measure case see (4)) and theorem 5 (Radon-Nikodym theorem).

From now on we shall have to deal with two Daniell integrals I and J on the same space. If we want to compare them (add them, evaluate the absolute continuity and so on) they must have a sufficient common domain. We shall suppose that $L_1(I) \cap L_1(J)$ is I- and J- dense in $L_1(I)$ and $L_1(J)$ respectively.

For a signed integral K a set $A \varepsilon A_\infty(K)$ is said to be positive (negative), if $K(f\chi_A) \geqslant 0$ ($\leqslant 0$) for each $f \varepsilon L_1^+(K)$. If H^+, H^- are two disjoint sets whose union is X, such that H^+ is positive and H^- is negative, the sets H^+ and H^- are said to form a Hahn decomposition of X with respect to K.

__Lemma 1__

If $J \ll I$, then $M_\infty(I) \subset M_\infty(J)$ and $A_\infty(I) \subset A_\infty(J)$.

The proof can be found in (14).

__Lemma 2__

If $f' \varepsilon L_1^+$ and $H \varepsilon A_\infty$ are such that $I(f\chi_H) \geqslant 0$ (or $\leqslant 0$) for each $f \varepsilon L_1$ such that $0 \leqslant f \leqslant f'$, then $E = H \cap \{x : f'(x) \neq 0\}$ is a positive (negative) set.

If $g \varepsilon L_1^+$ were such that $I(g\chi_E) < 0$, then from

$n \inf(f', (g/n))\chi_E = \inf((nf'), g)\chi_E \uparrow g\chi_E$

and $0 \leqslant \inf(f', (g/n)) \leqslant f'$, we could deduce a contradiction, as

$I(g\chi_E) = \lim I((\inf(nf', g)\chi_E)$.

__Lemma 3__

If K is a signed integral such that $K \ll I$ and $A \varepsilon A_\infty(I)$ then for each $f' \varepsilon L_1^+(K)$ such that $K(f'\chi_A) < 0$ (>0) there is an $f \varepsilon L_1^+(K) \cap L_1^+(I) = L$ such that $K(f\chi_A) < 0$ (>0).

We have supposed that L is K-dense in $L_1(K)$. It is easy to see that L^+ is K-dense in $L_1^+(K)$. Now, if $\{f_n\}$ is a sequence of functions of L^+ K-convergent to f', $K(f_n\chi_A)$ converges to $K(f'\chi_A)$, so for n large enough, $K(f_n\chi_A)$ is positive (negative).

Theorem 3

If I is localizable, every K which is absolutely continuous with respect to I has a Hahn decomposition.

Let be $\mathcal{H}^+$ the collection of sets belonging to the complete lattice $\mathcal{A}_\infty(I)$ and positive for K. Let $H^+=\sup\mathcal{H}^+$. We shall prove that H^+ is positive and $H^- = H^+$ is negative.

Let us suppose that H^+ is not positive. Then there exists $f'\varepsilon\mathcal{L}^+(K)$ such that $K(f'\chi_{H^+})<0$. By lemma 3, there exists $f\varepsilon\mathcal{L}^+$ such that $K(f\chi_{H^+})<0$.

The set $E= H^+\cap\{x: f(x)\neq 0\}$ cannot be negative because it is not locally null and therefore H^+-E would be an upper bound for $\mathcal{H}^+$, strictly smaller than H^+. On the other hand; if A is any positive s , $A\cap E\sim\emptyset$ and from $A<H^+$ we deduce that $A<H^+-E$ and contradict the definition of H^+.

Now a standard argument can be used to prove that there is a function $f'\varepsilon\mathcal{L}^+$ such that $H^+\cap\{x: f'(x)\neq 0\}$ is negative and not locally null, contradicting so the definition of H^+ , as $H^+-\{x: f'(x)\neq 0\}$ is a strictly smaller upper bound for $\mathcal{H}^+$. The function f' is defined as $f'=f-g$, where $g=\sum_1^\infty{}_i g_i$, the function g_i being selected in such a way that

$0\leqslant g_i\leqslant f-\Sigma_j g_j \quad (j<i-1)$

$K(g_i\chi_{H^+})\geqslant 1/n_i$

and n_i is the smallest positive integer verifying the last inequality (see for example (3), par. 29, theorem A).

In the same way it can be proved that H^- is negative.

The next theorem proves the opposite implication.

Theorem 4

If each K which is absolutely continuous with respect to I has a Hahn decomposition, then I is localizable.

In a σ-algebra a collection of sets $\mathcal{G}$ has the least upper bound iff the generated σ-ideal has it (8). Let be $\mathcal{G}\subset\mathcal{A}_\infty(I)$ a σ-ideal and let us define, for each $f\varepsilon\mathcal{L}_1^+(I)$:

$J(f)= \sup\{I(f\chi_G);\ G\varepsilon\mathcal{G}\}$

Define J on $\mathcal{L}_1(I)$ by linearity. Obviously J is a Daniell integral, absolutely continuous with respect to I.

Let be G^- the negative set of the Hahn decomposition with respect to $K=I-2J$ and let us prove that $G^-=\sup\mathcal{G}$.

If $G\varepsilon\mathcal{G}$, $K(f\chi_G)=-I(f\chi_G)$ for each $f\varepsilon L_1^+$, so G is negative for K and therefore $G<G^-$. On the other hand, if G' is strictly smaller than G^-, the set $A=G^- -G'$ is negative and not locally null for K, so there exists $f\varepsilon L_1^+$ such that $0<I(f\chi_A)<2J(f\chi_A)$. From the definition of J it follows that there is a set G such that

$$I(f\chi_A\chi_G)>J(f\chi_A)/2>0,$$

so $(G^- -G')\cap G$ is not locally null and therefore G^- is not an upper bound for $\mathcal{G}$.

Lemma 4

Let be I and J two positive Daniell integrals and I be localizable. Let be $\{a_n\}$ a sequence of real numbers such that $a_n>1$ and $a_n\downarrow 1$. If A_n belongs to $\mathcal{A}_\infty(I)$ and is the positive part of a Hahn decomposition with respect to $K_n=I-a_nJ$, then $A=\cup A_n$ $(n\varepsilon\mathbb{N})$ is the positive part of the Hahn decomposition with respect to $K=I-J$.

The set A is obviously positive, as for each $A\subset E$ it follows that $(E_n=A_n\cap E)$:

$$I(f\chi_{E_n})\geq a_nJ(f\chi_{E_n})\geq J(f\chi_{E_n}) \quad \text{for each } f\varepsilon L_1^+ .$$

Being $E=\cup E_n$, we have $I(f\chi_E)\geq J(f\chi_E)$ for each $f\varepsilon L_1^+$.

Now let be $E\cap A=\emptyset$ and let us suppose that there exists $f\varepsilon L_1^+$ such that $I(f\chi_E)>J(f\chi_E)$. In this case there is an n such that

$$I(f\chi_E)>a_nJ(f\chi_E).$$

We have obtained a contradiction, as $A_n\subset A$ and therefore $A_n\cap E=\emptyset$.

Lemma 5

Let I and J be positive Daniell integrals, let be I localizable and $J<<I$. If $\mathcal{G}=\{g:\ g\varepsilon\mathcal{M}_\infty(I),\ I(fg)<J(f),\ \forall f\varepsilon L_1^+\}$ and $g'=\sup\mathcal{G}$, then $g'\varepsilon\mathcal{G}$.

Suppose the contrary and let be $f'\varepsilon L_1^+$ such that $I(f'g')>J(f')$. By proposition 2, $B=\{x:\ f'(x)g'(x)>0\}\varepsilon\mathcal{A}_\infty$. Let us consider the signed integral $K_n(f)=I(fg'\chi_B)-(1+1/n)J(f\chi_B)$, which is absolutely continuous with respect to I. Apply theorem 3 and let be A_n the positive part of the Hahn decomposition with respect to K_n. By lemma 5 there is an integer n such that A_n is not locally null. The function

$$g''(x)=\begin{cases} g'(x) & \text{for } x\notin A_n\cap B \\ g'(x)/(1+1/n) & \text{for } x\varepsilon A_n\cap B \end{cases}$$

belongs to $M_\infty(I)$ and is strictly smaller than g'. On the other hand, as it is easy to verify, $g''>g$ for each $g\varepsilon G$ and this contradicts the definition of g'.

Theorem 5

If I is localizable and $J<<I$, there exists $g'\varepsilon M_\infty(I)$ such that $I(fg')=J(f)$, $\forall f\varepsilon L_1$.

We need only to prove the desired identity for non negative functions. The function g' defined in lemma 5 satisfies the inequality. Let us prove that this is in fact an equality. Suppose the contrary and let be $f'\varepsilon L_1^+$ such that $I(f'g')<J(f')$.

Let be n such that $(1+1/n)I(f'g')<J(f')$ and $B=\{x:g'(x)f'(x)>0\}$. If A_n is the positive part of the Hahn decomposition with respect to

$K_n(f)=J(f\chi_B)-(1+1/n)I(g'f\chi_B)$,

the function $g=(1+1/n)g'\chi_B\chi_{A_n}$ belongs to G , but is not bounded by g'. This is a contradiction.

The localizability is also necessary for the validity of the above Radon-Nikodym type theorem. We can limit ourselves to the integrals which are defined on the whole of $L_1(I)$, which are bounded, as the followig proposition shows (compare the results of (12)).

Proposition 10

If in the hypotheses of theorem 5 $L_1(J)$ $L_1(I)$, then $g'\varepsilon L_\infty(I)$.

Suppose the contrary: let be

$\sup\{J(f)=I(fg');\ I(f)=1,\ f\varepsilon L_1^+\}=\infty$.

For each n, let be $f_n\varepsilon L_1^+$ such that $I(f_n)=1$ and $J(f_n)>n^2$; put $g_n=f_n/n^2$ and $g=\sum_1^\infty{}_n g_n$. Then

$\sum_1^\infty{}_n I(g_n)=\sum_1^\infty{}_n 1/n^2<\infty$

and by Beppo Levi's theorem $g\varepsilon L_1(I)$. On the other hand, $J(g_n)>1$ and therefore $\sum_1^\infty{}_n J(g_n)=\infty$.

Theorem 6

If for every $J<<I$ such that J is defined on $L_1(I)$ there exists

$g'\varepsilon L_\infty(I)$ such that $I(fg')=J(f)$ for each $f\varepsilon L_1$, then I is localizable.

Let be G a σ-ideal in $A_\infty(I)$; for each $f\varepsilon L_1^+(I)$ define

$J(f)=\sup\{I(f\chi_G);\ G\varepsilon G\}$.

Extending by linearity J on $L_1(I)$ we obtaine an I-absolutely continuous Daniell integral. By hypothesis, there exists $g'\varepsilon L_\infty(I)$ such that $J(f)= =I(fg')$ for each $f\varepsilon L_1(I)$. Simple calculations show that the set $G'= = \{x:g'(x)>0\}$ is the least upper bound of G in $A_\infty(I)$, so I is localizable.

REFERENCES

(1) G. Aquaro- Alcuni aspetti della teoria dell'integrale di Daniell-Stone; Conf. Sem. di Mat. Univ. Bari (1965)

(2) G. Choquét- Étude des mesures coniques. Cônes convexes saillants faiblement complets sans génératrices extrémales, Comptes rendus 445-447 (1962)

(3) P.R. Halmos- Measure theory;Van Nostrand (1950)

(4) J.L. Kelley- Decomposition and representation theorems in measure theory; Math. Ann., 163 89-94 (1966).

(5) D. Kölzow- Differentiation von Massen ; Lecture Notes in Mathematics 65 (1968)

(6) O. Nikodym- Sur une généralisation des intégrales de M.J. Radon, Fund. Math. 131-179 (1930)

(7) F. Papangelou- Note on the functional analogue of the Jordan decomposition of measures; Arch. der Math. vol. XVII,fasc.3 253-255 (1966)

(8) I.E. Segal- Equivalences of measure spaces; Am. Jour. of Math. 73, 275-313 (1951)

(9) M.H. Stone- Notes on integration; Proc. Nat. Acad. Sci. U.S.A.vol. XXXIV, 336-342, 447-455, 483-490 (1948), vol. XXXV, 50-58 (1949)

(0) A. Volčič- Teoremi di decomposizione per misure localizzabili;Rend. di Matem. (2) vol. 6, serie VI, 307-336 (1973)

(11) A. Volčič- Sulla differenziazione delle misure; Rend. Ist. Matem. Univ. Trieste vol. VI fasc. II 156-177 (1974)

(12) A. Volčič- Sull'insieme di definizione delle misure; Boll. U.M.I. (5) 16-A, 1, 186-189 (1979)

(13) A. Volčič- Un confronto tra l'integrale di Daniell-Stone e quello

di Lebesgue; to appear in Rend. Circolo Mat. Palermo (1979)

(14) A. Volčič- Sulla differenziazione degli integrali di Daniell-Stone; to appear in Rend. Sem. Mat. Padova (1979)

(15) A.C. Zaanen- A note on the Daniell-Stone integral; Colloque sur l'Analyse Fonctionelle 63-69 (1961)

(16) A.C. Zaanen- The Radon-Nikodym theorem I,II; Indag. Math. XXIII 157-187 (1961)

AN OUTLINE OF THE THEORY OF STATIONARY MEASURES OVER $\mathbb{R}^q$

P. Masani*

University of Pittsburgh, Pittsburgh, Pa. 15260

1. Introduction

In [1] and [13] we developed the theory of helices in Hilbert and Banach spaces in terms of an "average vector" and a finitely additive measure defined in terms of its propagator. In effect, these papers indirectly presented the theory of vector-valued finitely additive stationary measures over $\mathbb{R}$ 1).
The present paper focusses directly on the finitely additive (f.a.) stationary vector measures over $\mathbb{R}^q$, its principal objective being to establish a full-fledged analogue for $q>1$ of the main theorem given in [13, 3.12] for $q=1$ (§2). Another objective is to discuss the countable additivity (c.a.) of such stationary measures. It will become apparent (§3) that only in rather special cases will a f.a., stationary (and inner regular) measure be c.a. This fact is established for Hilbert spaces by exploiting a result emanating essentially from an example due to R.S. Phillips and P. Sarnak. We also provide some sufficient conditions for countable additivity.

---

*Alexander von Humboldt Senior Visiting Scientist at the University of Cologne, Fall 1979.

1) In this paper $\mathbb{R}$ and $\mathbb{C}$ will stand for the real and complex number fields, respectively, and $\mathbb{F}$ for any one of them. $\mathbb{Z}$ will denote the set of integers. $\mathbb{R}_+$, $\mathbb{Z}_+$ and $\mathbb{R}_{o+}$, $\mathbb{Z}_{o+}$ will refer to the subsets of positive elements, and subsets of non-negative elements of $\mathbb{R}$ and $\mathbb{Z}$.

But these are far from being necessary and several pertinent questions remain unresolved. We also point out that the (obviously translation-invariant) semi-variation of a stationary measure can be infinite even when the measure itself is bounded.

In this paper only the main results of the theory will be stated and their proofs will either be omitted or only very briefly indicated. A more complete presentation will be given in another paper.

Although our presentation will be confined to stationary measures over $\mathbb{R}^q$, it will be convenient to state initially the definitions for the more general case in which an arbitrary locally compact group replaces $\mathbb{R}^q$. For this we shall need the following notation:

(1.1)
- (a) $\mathfrak{X}$ is a Banach-space over the field $\mathbb{F}$;
- (b) Γ is an additive (not necessarily abelian) locally compact group;
- (c) $\mathcal{P}$ is a pre-ring of subsets of Γ;
- (d) FA $(\mathcal{P}, \mathfrak{X})$ is the set of all f.a. measures on $\mathcal{P}$ with values in $\mathfrak{X}$.

The concept of stationarity hinges on the following concept of a unitary representation:

1.2 Def. A function $U(\cdot)$ is called a strongly continuous unitary representation of Γ over $\mathfrak{X}$, iff its domain is Γ and $\forall\, s,t \in \Gamma$, $U(t)$ is an isometry on $\mathfrak{X}$ onto $\mathfrak{X}$ such that

$$U(s)U(t) = U(s+t), \qquad U(0) = I, \qquad \operatorname*{slim}_{t\to s} U(t) = U(s).^{2)}$$

(It follows, of course, that $U(-t) = U(t)^{-1}$.)

The concept of stationary measure is then defined as follows:

1.3 Def. Let $\rho \in FA(\mathcal{P}, \mathfrak{X})$. Then[3]

(a) $$\mathcal{S}_\rho \underset{d}{=} \mathfrak{S}\{\text{Range } \rho(\cdot)\} \subseteq \mathfrak{X}$$

---

2) The symbol "slim" (read "strong limit") will refer to the limit in the strong operator topology.

3) For $A \subseteq \mathfrak{X}$, $\mathfrak{S}(A) \underset{d}{=}$ the (least closed linear) subspace of spanned by A.

is called <u>the subspace due to $\rho(\cdot)$;</u>

(b) we say that $\rho(\cdot)$ is <u>right-stationary,</u> if $\mathcal{P}$ is right-translation invariant, i.e.

$$\forall P \in \mathcal{P} \ \& \ \forall t \in \Gamma, \quad P+t \in \mathcal{P},$$

and there exists a strongly continuous unitary representation $U(\cdot)$ of Γ over $\mathcal{S}_\rho$ such that

$$\forall P \in \mathcal{P} \ \& \ \forall t \in \Gamma, \quad \rho(P+t) = U(t)\{\rho(P)\}.$$

The (trivially unique) representation $U(\cdot)$ in (b) is called <u>the propagator</u> of $\rho(\cdot)$.

There is an obvious corresponding concept of <u>left-stationarity</u>. For abelian groups Γ (such as $\mathbb{R}^q$), right- and left-stationarities are of course equivalent.

Stationary measures are most often encountered in situations in which $\mathcal{X}$ is a Hilbert space. For Hilbert spaces $\mathcal{X}$, the stationarity of $\rho(\cdot)$ is customarily defined by the condition 1.4 (β) below. But this definition is equivalent to the one just given for any Hilbert space:

1.4 <u>Prop.</u> Let (i) Γ, $\mathcal{P}$ be as in (1.1), (ii) $\mathcal{X}$ be a Hilbert space over $\mathbb{F}$. Then the following conditions are equivalent:

(α) $\rho(\cdot)$ is right-stationary;

(β) $\mathcal{P}$ is right-translation invariant,

$$\forall P,Q \in \mathcal{P} \ \& \ \forall t \in \Gamma, \quad (\rho(P+t),\rho(Q+t)) = (\rho(P),\rho(Q)),$$

$$\forall P \in \mathcal{P}, \quad \lim_{t\to 0} \rho(P+t) = \rho(P).$$

This proposition is a direct consequence of general theorems on the existence of propagators of Hilbertian varieties, cf. [14, 3.5, 3.6].

Before proceeding with the general theory, we shall briefly cite some important instances of stationary measures.

1.5 <u>Examples.</u> (a) The Wiener theory of the homogeneous chaos [18] is essentially based on a f.a. measure ξ on the δ-ring $\mathcal{P}$ of Borel subsets of $\mathbb{R}^q$ of finite Lebesgue measure taking values in $L_2(\Omega,\mathcal{A},P;\mathbb{F})$, where $(\Omega,\mathcal{A},P)$ is a probability space. One of its crucial properties, to wit

$$\forall B \in \mathcal{P} \ \& \ \forall t \in \mathbb{R}^q, \quad P\cdot\xi(B+t)^{-1} = P\circ\xi(B)^{-1},$$

ensures that ξ is stationary.

(b) As indicated in [13, pp. 2-4], every stochastic process $x(\cdot)$ gives rise to a f.a. measure ξ defined on the pre-ring $\mathcal{P}$ of intervals $(a,b] \subseteq \mathbb{R}$ by

$$\xi(a,b] = x(b)-x(a), \quad a<b, \quad a,b \in \mathbb{R}.$$

If all the $x(t)$ have a finite p^{th} moment ($1\le p<\infty$), then $\xi(\cdot) \in FA(\mathcal{P}, \mathcal{X})$, where $\mathcal{X} = L_p(\Omega, \mathcal{A}, P; \mathbb{F})$. If further the process $x(\cdot)$ has strictly stationary increments, i.e. $\forall n \in \mathbb{Z}_+$, $\forall h, a_1, b_1, \ldots, a_n, b_n \in \mathbb{R}$, the probability distribution of the n-variate

$$x(b_1+h) - x(a_1+h), \ \ldots, \ x(b_n+h) - x(a_n+h)$$

is independent of h, then by adapting the arguments in Doob [4, pp. 456, 509] we can show that the measure $\xi(\cdot)$ is stationary.

(c) The $L_2(\mathbb{R})$-valued measures $\xi_1(\cdot)$, $\xi_2(\cdot)$ defined by

$$\xi_1(a,b](t) \underset{d}{=} \frac{1}{\pi} \log \left|\frac{b-t}{a-t}\right|, \quad t \in \mathbb{R}$$

$$\xi_2(a,b](\lambda) \underset{d}{=} \frac{1}{\sqrt{2\pi}} \frac{e^{ib\lambda}-e^{ia\lambda}}{i\lambda}, \quad \lambda \in \mathbb{R},$$

which yield the explicit theory of the Hilbert and Fourier-Plancherel transforms over $L_2(\mathbb{R})$ are stationary, cf. [10, §§4,6] and also [9][4].

(d) The theory of $L_p(\mathbb{R})$-valued stationary measures over $\mathbb{R}$ (or rather of $L_p(\mathbb{R})$-valued helices) has been instrumental in simplifying the theory of the spaces $V_p(\mathbb{R})$ and $U_p(\mathbb{R})$ of functions of so-called bounded p-variation, which for p=2 are latent in Wiener's generalized harmonic analysis [17]. This is clear from the recent papers of Lau & Lee [7] and Nelson [16]. Moreover, the work of Wiener [17] can itself be appreciably simplified and cast in the setting of modern functional analysis by bringing helical ideas to bear on it, as the writer has shown recently in [15].

These varied examples are suggestive of the range of applicability of the theory to be outlined in the sequel. However, as we shall explain in §4 , for $p\ge 2$ the theory of such measures does not give adequate

---

4) The stationary measures ξ_1, ξ_2 of Ex. (c) as well as the ξ of Ex.(a) are Hilbert-space valued and moreover countably additive and orthogonally scattered in the sense of [8, Def. 1.2.].

insight into the geometrical properties of the associated helical hypersurfaces. From this geometric standpoint this paper is limited, and should be regarded as merely the first step towards a general theory of helical hypersurfaces over $\mathbb{R}^q$.

2. The main theorem for stationary measures over R^q and its consequences

To be able to enunciate the basic results for stationary vector-valued measures over $\mathbb{R}^q$, we must begin with some notation and some preliminary concepts.

2.1 Notation (a) $\forall m,n \in \mathbb{Z}$ such that $m<n$,

$$[m,n] \underset{d}{=} \{m,m+1,\ldots,n\},$$

$$[m,n) \underset{d}{=} \{m,m+1,\ldots,n-1\}.$$

(b) For $a = (a_1,\ldots,a_q)$ & $b = (b_1,\ldots,b_q) \in \mathbb{R}^q$, we shall write $a<b$, iff $\forall i \in [1,q]$, $a_i < b_i$, and we shall similarly interpret $a \le b$;

$$\text{for } a<b, \quad (a,b] \underset{d}{=} (a_1,b_1]\times \ldots \times(a_q,b_q].$$

(c) $$\mathcal{P} \underset{d}{=} \{(a,b]: a,b \in \mathbb{R} \ \&\ a\le b\}.$$

Letting $\mathcal{P}^q \underset{d}{=} \mathcal{P}\times \ldots \times \mathcal{P}$ (q times), it is obvious that

(2.2) $\mathcal{P}^q = \{(a,b]: a,b \in \mathbb{R}^q \ \&\ a\le b\}.$

Closer to our concern with stationary measures will be the following notation:

(2.3)
- (a) $\mathcal{X}$ is a Banach space over $\mathbb{F}$
- (b) $\rho \in FA(\mathcal{P}^q, \mathcal{X})$ is stationary, (cf. (1.1)-1.3)
- (c) $\mathcal{S}_\rho$ is the subspace of $\rho(\cdot)$, (cf. 1.3 (a))
- (d) $U(\cdot)$ is the propagator of $\rho(\cdot)$, (cf. 1.3 (b))
- (e) $\mathcal{N}_\rho(\cdot)$ is the semi-variation of ρ, cf. [5, p. 320].

Our definition of the average vector of $\rho(\cdot)$ hinges on the following lemma on the growth of $\rho(\cdot)$:

2.4 <u>Lemma</u> (Growth of $\rho(\cdot)$). $\forall (0,t] \in \mathcal{P}^q$,

$$|\rho(0,t]| \le \max_{s \in (0,1]^q} |\rho(0,s]| \prod_1^q (1+t_j).$$

<u>Gist of proof.</u> By [13,2.11] the result holds for $q=1$. Now suppose that it holds for a fixed p, where $1 \le p < q$. Then by introducing the sectional measures $\rho_A(\cdot)$, $\rho^B(\cdot)$ of $\rho(\cdot)$ defined by

$$\rho_A(B) \underset{d}{=} \rho(A \times B), \underset{d}{=} \rho^B(A),$$

where $A \in \mathcal{P}^p$ and $B \in \mathcal{P}^{q-p}$, and observing that they are stationary, we can deduce the result for p+1. Hence by induction the result holds for all $q \in \mathbb{Z}_+$. □

An immediate consequence of this lemma is the existence of the Bochner integral

$$(2.5) \qquad J_\rho \underset{d}{=} \int_{\mathbb{R}^q_{o+}} e^{-(t_1+\ldots+t_q)} \rho(0,t] m_q(dt) \in \mathcal{S}_\rho$$

where $m_q(\cdot)$ is the Lebesgue measure over $\mathbb{R}^q$. This allows us to give the following definition in the spirit of our earlier work, cf. [13,3.6]:

2.6 <u>Def.</u> The vector $\alpha_\rho \underset{d}{=} (-\sqrt{2})^q J_\rho$ is called <u>the average vector of the measure $\rho(\cdot)$.</u>

The proof of our main theorem will be by induction, to perform which we have to introduce the one-parameter sectional groups of the multi-parameter group $U(\cdot)$ of (2.3)(d):

2.7 <u>Def.</u> $\forall j \in [1,q]$ & $\forall t \in \mathbb{R}$, $U_j(t) \underset{d}{=} U(t \cdot \varepsilon_j)$, where $\varepsilon_1 = (1,0,\ldots,0),\ldots, \varepsilon_q = (0,0,\ldots,1) \in \mathbb{R}^q$.

Elementary considerations of multi-parameter semi-groups, cf. [6, p. 334, §10.10] or [1, p. 14], lead us to the results:

$$(2.8) \quad \begin{cases} U_j(\cdot) = \text{a strongly continuous unitary representation of } \mathbb{R} \\ \qquad\qquad \text{over } \mathcal{S}_\rho,\ j \in [1,q] \\ \forall t \in \mathbb{R}^q,\ U(t) = U_1(t_1) \cdot \ldots \cdot U_q(t_q). \end{cases}$$

Now as in the case q=1, cf. [13, p. 11], we associate with the propagator $U(\cdot)$ of $\rho(\cdot)$ an operator-valued measure $T_U(\cdot)$ as follows:

2.9 <u>Def.</u> (a) $\forall j \in [1,q]$ & $\forall (a,b] \in \mathcal{P}$,

$$T_j(a,b] = \frac{1}{\sqrt{2}} \{U_j(b) - U_j(a) - \int_a^b U_j(t)dt\};$$

(b) $\forall (a,b] \in \mathcal{P}^q$, $T_U(a,b] = \prod_{j=1}^q T_j(a_j,b_j]$,

i.e. $T_U = T_1 \times \ldots \times T_q$, the tensor product of $T_1, \ldots, T_q$.

It is easily seen that $T_U(\cdot)$ is a finitely additive measure on $\mathcal{P}^q$ with values in the space $CL(\mathcal{S}_\rho, \mathcal{S}_\rho)$ of continuous linear operators on $\mathcal{S}_\rho$ into $\mathcal{S}_\rho$. Our main theorem for stationary measures over $\mathbb{R}^q$ now reads as follows:

2.10 <u>Main Thm.</u> $\forall P \in \mathcal{P}^q$, $\rho(P) = T_U(P)(\alpha_\rho)$.

<u>Gist of proof.</u> First note that by [13,3.12] the result holds for q=1. Now let $p \in [1,q)$ and $B \in \mathcal{P}^{q-p}$. Then, as indicated in the proof of 2.4, the sectional measure $\rho^B(\cdot)$ on $\mathcal{P}^p$ is stationary. Let $\alpha^p(B)$ be its average vector, cf. 2.6. We can show that $\alpha^p(\cdot)$ is itself a stationary measure on $\mathcal{P}^{q-p}$, and that moreover its own average vector is precisely equal to the average vector α_ρ of $\rho(\cdot)$ itself. These results together with the earlier observed validity of the theorem for q=1, allow us to complete an argument by induction. □

This theorem reveals the structure of all finitely additive stationary measures on the pre-ring of half-open half-closed intervals over $\mathbb{R}^q$: their values in $\mathcal{H}$ are determined by the action of an easily defined, finitely additive operator-valued measure $T_U(\cdot)$ on a single "average vector" in $\mathcal{H}$. The propagator $U(\cdot)$ is retrievable from the measure $T_U(\cdot)$ by the inversion formula:

$$(2.11) \quad \forall a \in \mathbb{R}^q, \quad U(a) = (-\sqrt{2})^q \int_{\mathbb{R}^q_{o+}} e^{-(t_1+\ldots+t_q)} T_U(a,a+t] m_q(dt).$$

It follows that $U(\cdot)$ and $T_U(\cdot)$ invariably span the same subspace of $\mathcal{H}$; more precisely,

$$(2.12) \quad \forall X \subseteq \mathcal{H}, \quad \mathfrak{S}\{U(t)(X): t \in \mathbb{R}^q\} = \mathfrak{S}\{T_U(P)(X): P \in \mathcal{P}^q\}.$$

On combining (2.12) with the Main Thm. 2.10 we get the following useful corollary:

2.13 <u>Cor.</u> $\mathcal{S}_\rho = \mathfrak{S}\{U(t)(\alpha_\rho):t \in \mathbb{R}^q\}$; thus $\mathcal{S}_\rho$ is a cyclic subspace of the propagator group $U(\cdot)$ of $\rho(\cdot)$.

Another useful consequence of the Main Thm. 2.10 is the inner regularity, in the sense of Dinculeanu [3, p. 303, Def. 2], of all finitely additive stationary measures on $\mathcal{P}^q$:

2.14 <u>Cor.</u> $\rho(\cdot)$ is inner regular on $\mathcal{P}^q$ in the sense of Dinculeanu, i.e. $\forall P \in \mathcal{P}^q$ & $\forall \varepsilon > 0$, $\exists$ a compact $K_\varepsilon \subseteq P$ such that

$$Q \in \mathcal{P}^q \ \&\ K_\varepsilon \subseteq Q \subseteq P \implies |\rho(Q) - \rho(P)| < \varepsilon.$$

<u>3. When are stationary measures countably additive?</u>

In studying the issue of the countable additivity of stationary measures over $\mathbb{R}^q$ we shall concentrate on the case $q=1$. Let $\mathcal{P}$ be as in 2.1 (c) and let $\rho \in FA(\mathcal{P}, \mathcal{X})$ be a stationary measure with propagator $U(\cdot)$ and average vector α. It will be convenient to associate with $\rho(\cdot)$ the measure $\zeta_\alpha(\cdot)$ defined by

$$(3.1) \quad \forall (a,b] \in \mathcal{P}, \qquad \zeta_\alpha(a,b] \underset{d}{=} \{U(b) - U(a)\}(\alpha).$$

Obviously $\zeta_\alpha \in FA(\mathcal{P}, \mathcal{X})$, and it is easy to check from our Main Thm. 2.10 (with $q=1$) that

$$(3.2) \qquad \rho(\cdot) \in CA(\mathcal{P}, \mathcal{X}) \ \text{ iff } \ \zeta_\alpha(\cdot) \in CA(\mathcal{P}, \mathcal{X}).$$

In the light of this equivalence, we may deal with the measure ζ_α instead of $\rho(\cdot)$ in the study of countable additivity.

Now it is a fundamental fact, cf. Hille & Phillips [6, §10.3], that if A is the infinitesimal generator of $U(\cdot)$; then $\forall a,b \in \mathbb{R}$,

$$(3.3) \qquad \text{Range} \int_a^b U(t)dt \subseteq \mathcal{D}_A \ \ \& \ \ A\cdot\int_a^b U(t)dt = U(b) - U(a).$$

It follows from this that

$$(3.4)\begin{cases} \forall P \in \mathcal{P} \ \& \ \forall \alpha \in \mathcal{X}, \ \int_P U(t)(\alpha)dt \in \mathcal{D}_A, \\ \qquad \& \quad \zeta_\alpha(P) = A\{\int_P U(t)(\alpha)dt\}. \end{cases}$$

Now consider the families of sets:

$$(3.5)\quad \begin{cases} \hat{\mathcal{P}} \underset{d}{=} \{\bigcup_1^\infty P_k : \quad P_k \in \mathcal{P} \;\&\; P_k \text{ are } \| \}, \\ \hat{\mathcal{P}}_o \underset{d}{=} \{S : \quad S \in \hat{\mathcal{P}} \;\&\; \exists P \in \mathcal{P} \ni S \subseteq P \;\&\; P\backslash S \in \hat{\mathcal{P}} \}. \end{cases}$$

It is easy to see that $\hat{\mathcal{P}}_o \subsetneq \hat{\mathcal{P}}$. For instance, if S is the union of half-open, half-closed "middle third" subintervals of (0,1], then (0,1]\S will resemble the Cantor set and not belong to $\hat{\mathcal{P}}$; thus $S \in \hat{\mathcal{P}} \backslash \hat{\mathcal{P}}_o$.

Now the issue of the countable additivity of ζ_α, and by (3.2) of $\rho(\cdot)$ itself, hinges on the question as to whether

$$(*) \qquad \forall\, S \in \hat{\mathcal{P}}_o, \quad \text{Range} \int_S U(t)dt \subseteq \mathcal{D}_A.$$

This is clear from the following theorem:

3.6 <u>Thm.</u> Let (i) $U(\cdot)$ be a strongly continuous unitary representation of $\mathbb{R}$ over $\mathcal{X}$ and A be its infinitesimal generator, (ii)

$$\forall\, (a,b] \in \mathcal{P} \;\&\; \forall\, \alpha \in \mathcal{X}, \qquad \zeta_\alpha(a,b] \underset{d}{=} \{U(b) - U(a)\}(\alpha).$$

Then $$\zeta_\alpha \in CA(\mathcal{P},\mathcal{X}) \implies \forall\, S \in \hat{\mathcal{P}}_o, \; \{\int_S U(t)dt\}(\alpha) \in \mathcal{D}_A.$$

The proof rests on the result that all the subseries of an unconditionally convergent series in $\mathcal{X}$ are unconditionally convergent in $\mathcal{X}$, cf. Day [2, p. 78]. There are indications that the converse of the last implication prevails, viz.

$$\forall\, S \in \hat{\mathcal{P}}_o, \; \{\int_S U(t)dt\}(\alpha) \in \mathcal{D}_A \implies \zeta_\alpha \in CA(\mathcal{P},\mathcal{X}).$$

But we are unable to prove this.

Theorem 3.6 confronts us with the question of the nature of the Banach spaces $\mathcal{X}$ for which the condition (*) fails. It is tempting to conjecture that it fails for all infinite dimensional $\mathcal{X}$:

3.7 <u>Conjecture.</u> Let $\mathcal{X}$ be any infinite-dimensional Banach space over $\mathbb{F}$, and $\hat{\mathcal{P}}_o$ be as in (3.5). Then there exists a strongly continuous unitary representation $U(\cdot)$ of $\mathbb{R}$ over $\mathcal{X}$ with (discontinuous) infinitesimal generator A and there exists a set $S \in \hat{\mathcal{P}}_o$ for which (unlike (3.4)) Range $\int_S U(t)dt \not\subseteq \mathcal{D}_A$.

We are unable to prove or disprove this conjecture in its full generality. But we can show its validity when $\mathcal{H}$ is taken to be an infinite-dimensional Hilbert space. Our proof depends on the following result on the Fourier-Plancherel measure for $L_2(\mathbb{R})$ defined by us in [9, §4] and also in [10, §6], which in turn is based essentially on an example communicated to us by R.S. Phillips and P. Sarnak:

3.8 Prop. Let $\xi(\cdot)$ be the Fourier-Plancherel measure for $L_2(\mathbb{R})$, i.e. let for all Borel subsets B of $\mathbb{R}$ of finite Lebesgue measure,

$$\xi(B)(\lambda) \underset{d}{=} \frac{1}{\sqrt{2\pi}} \int_B e^{i\lambda t} dt, \qquad \lambda \in \mathbb{R}.$$

Then $\exists S \in \hat{\mathcal{P}}_o \ni \sup_{\lambda \in \mathbb{R}} |\lambda \xi(S)(\lambda)| = \infty.$

In fact, with

$$S_n \underset{d}{=} \bigcup_o^n (1/3^k, 2/3^k], \qquad S \underset{d}{=} \bigcup_o^\infty (1/3^k, 2/3^k],$$

and $\lambda_j \underset{d}{=} 3^j(\pi/2)$, we have $S \in \hat{\mathcal{P}}_o$, and

$$\forall j \in \mathbb{Z}_+ \ \& \ \forall n > j, \qquad |\lambda_j \xi(S_n)(\lambda j)| > j/\sqrt{2\pi}$$

$$\& \qquad \forall j \in \mathbb{Z}_+, \qquad |\lambda_j \xi(S)(\lambda_j)| > j/\sqrt{2\pi}.$$

With the aid of this result we can assert:

3.9 Thm. The Conjecture 3.7 is valid for all infinite-dimensional Hilbert spaces $\mathcal{H}$.

Gist of proof. Let $\mathcal{H}$ be a Hilbert space. Then the unitary representation $U(\cdot)$ of $\mathbb{R}$ over $\mathcal{H}$ has a spectral measure $E(\cdot)$, so that $U(t) = \int_{\mathbb{R}} e^{it\lambda} E(d\lambda)$, $t \in \mathbb{R}$. Its infinitesimal generator A is then given by $A = \int_{\mathbb{R}} i\lambda E(d\lambda)$, and a simple Fubini-type theorem in the spectral calculus shows that

$$(1) \qquad \int_S U(t)dt = \sqrt{2\pi} \int_{\mathbb{R}} \xi(S)(\lambda) E(d\lambda),$$

for all bounded Borel subsets S of $\mathbb{R}$. Now let $(x_n)_1^\infty$ be an ortho-normal basis for $\mathcal{H}$, supposed to be separable, and take $E(\cdot)$ to be the atomic spectral measure with atoms at $\lambda_j \underset{d}{=} 3^j(\pi/2)$, $j \in \mathbb{Z}_+$, and $E\{\lambda_j\} \underset{d}{=} P_{x_j}$, the orthogonal projection on the 1-dimensional subspace $\mathfrak{S}\{x_j\}$. Then letting U(·) be the associated unitary representation,

$\alpha \underset{d}{=} \sum_1^\infty (1/j)x_j$ and S be as in 3.8, we can show, using (1), that

$$\int_S U(t)(\alpha)dt \notin \mathfrak{D}_A.$$

Thus, Range $\int_S U(t)dt \not\subseteq \mathfrak{D}_A$. □

By combining these results we get the following corollary, which sums up our knowledge in regard to the countable additivity of stationary measures over $\mathbb{R}$:

3.10 <u>Cor.</u> Let $\mathfrak{X}$ be an infinite-dimensional Banach space over $\mathbb{F}$ for which the conjectured result in 3.7 is valid; in paritcular let $\mathfrak{X}$ be any infinite-dimensional Hilbert space[5)]. Then there exists a stationary measure on $\mathfrak{P}$ to $\mathfrak{X}$ which is f.a. but not c.a. on $\mathfrak{P}$. More fully, let (i) U(·) be a strongly continuous unitary representation of $\mathbb{R}$ over with infinitesimal generator A, $S \in \hat{\mathfrak{P}}_o$ (cf. (3.5)) and $\alpha \in \mathfrak{X}$ be such that

$$\int_S U(t)(\alpha)dt \notin \mathfrak{D}_A.$$

(ii) $\forall P = (a,b] \in \mathfrak{P}$,

$$\zeta_\alpha(P) \underset{d}{=} \{U(b) - U(a)\}(\alpha), \qquad \rho_\alpha(P) \underset{d}{=} T_U(P)(\alpha).$$

Then none of the stationary measure $\zeta_\alpha(\cdot)$, $\rho_\alpha(\cdot)$ is c.a. on $\mathfrak{P}$.

We are now faced with the question as to the necessary and sufficient conditions under which a f.a. stationary measure will be c.a. This question may be posed for measures over $\mathbb{R}^q$, for $q \in \mathbb{Z}_+$. We are able to address this question only to the extent of providing some sufficient conditions. In the next proposition the condition given is suggested by our study of the differentiability of helices [13, Thm. 4.4]:

3.11 <u>Prop.</u> Let (i) $\rho \in FA(\mathfrak{P}^q,\mathfrak{X})$ be stationary with propagator U(·) and average vector α, (ii) $\forall j \in [1,q]$, A_j be the infinitesimal

---

5) After this paper was written we have been able to establish the validity of the conjecture for $\mathfrak{X}$ = any ℓ_p space, $1 \le p < \infty$.

generator of the associated representations U_j of $\mathbb{R}$ over $\mathcal{S}_\rho$, cf. 2.7 & (2.8), (iii) $\alpha \in \text{Dom.} \prod_1^q (A_j - I)$. Then

(a) $$\rho(\cdot) \in CA(\mathcal{P}^q, \mathcal{X}) \quad \& \quad \rho(\cdot) \ll m_q(\cdot),$$

where $m_q(\cdot)$ is the Lebesgue measure over $\mathbb{R}^q$;

(b) The total variation measure $|\rho|(\cdot)$ of $\rho(\cdot)$ satisfies

$$|\rho|(\cdot) = |\prod_1^q (A_j - I)| \; m_q(\cdot) \quad \in \quad CA(\mathcal{P}^q, \mathbb{R}_{o+}).$$

In particular (a) and (b) hold when $\alpha \in \text{Dom} \prod_1^q A_j$

The c.a. stationary measures $\rho(\cdot)$ demarcated in the last proposition have finite total variations on $\mathcal{P}^q$, and hence in all Radon-Nikodym Banach spaces they will possess locally integrable Radon-Nikodym derivatives $d\rho/d|\rho|$. Such $\rho(\cdot)$ will lack the linearly independent or basic character essential for their fulfilling a role as representing measures, cf. [10, p. 430 (3)] and [8, §4]. This raises the supplementary question as to the necessary and sufficient conditions for an $\mathcal{X}$-valued stationary measure to be <u>both</u> countably additive and robust enough to serve as a representing measure. We are able to answer this question only when $\mathcal{X}$ is a Hilbert space and when furthermore we stringently interpret "robustness" to mean "orthogonal scatteredness" in the sense of [8, Def. 1.2]. We must first recall the concept of a "wandering vector" of a Hilbert space operator-valued measure:

3.12 <u>Def.</u> Let (i) $\mathcal{X}$ and $\mathcal{H}$ be a Banach space and a Hilbert space over $\mathbb{F}$, (ii) $\mathcal{P}$ be a pre-ring over a set Ω, and (iii) $T(\cdot) \in FA(\mathcal{P}, CL(\mathcal{X}, \mathcal{H}))$, where $CL(\mathcal{X}, \mathcal{H})$ is the space of continuous linear operators on $\mathcal{X}$ to $\mathcal{H}$. We say that $x \in \mathcal{X}$ is a <u>wandering vector for $T(\cdot)$</u>, iff

$$A, B \in \mathcal{P} \quad \& \quad A \parallel B \implies T(A)(x) \perp T(B)(x).$$

We shall denote by $\mathcal{W}_T$ the set of all wandering vectors of $T(\cdot)$.

Our result then reads as follows:

3.13 <u>Prop.</u> Let (i) $\mathcal{H}$ be a Hilbert space over $\mathbb{F}$, (ii) $\rho \in FA(\mathcal{P}^q, \mathcal{H})$ be stationary with propagator $U(\cdot)$ and average vector α, and (iii) $\alpha \in \mathcal{W}_{T_U}$. Then[6)]

---

6) $\text{Rstr}_S F$ means the restriction of the function F to the domain S.

$$\rho(\cdot) \in \mathrm{CAOS}(\mathcal{P}^q, \mathcal{H}) \ \& \ |\rho(\cdot)|^2 = |\alpha|^2 \cdot \mathrm{Rstr}_{\mathcal{P}^q} m_q,$$

where $m_q(\cdot)$ is the Lebesgue measure over $\mathbb{R}^q$.

We turn finally to the question of the semi-variation of a stationary measure, cf. [5, p. 320]. It is easily checked that if $\rho(\cdot)$ is right-stationary in the sense of Def. 1.3(b), then its semi-variation $\mathcal{S}_\rho(\cdot)$ is right-translation invariant:

$$\forall P \in \mathcal{P} \ \& \ \forall t \in \Gamma, \qquad \mathcal{S}_\rho(P+t) = \mathcal{S}_\rho(P).$$

But it is possible that for some $P \in \mathcal{P}$, $\mathcal{S}_\rho(P) = \infty$. For measures over $\mathbb{R}$ this is a consequence of Prop. 3.8 and the following obvious result:

3.14 <u>Triv.</u> Let (i) $\zeta \in \mathrm{FA}(\mathcal{P}, \mathcal{X})$ and (ii) $P \in \mathcal{P}$ be such that $\exists$ $\|$ sets $P_k \in \mathcal{P} \cap 2^P$, $k \in \mathbb{Z}_+$, for which

$$\sup_{n \geq 1} \left| \sum_{k=1}^{n} \zeta(P_k) \right| = \infty.$$

Then $\mathcal{S}_\zeta(P) = \infty$, cf. 2.3(e).

On combining this triviality with Prop. 3.8, and readapting the ideas used in the proof of Thm. 3.9 we get:

3.15 <u>Thm.</u> Let $\mathcal{X}$ be an infinite-dimensional Hilbert space. Then there exists a bounded stationary measure $\zeta \in \mathrm{FA}(\mathcal{P}, \mathcal{X})$ for which the semi-variation $\mathcal{S}_\xi(\cdot)$ is not finite on $\mathcal{P}$.

As with the issue of countable additivity, it is likely that this theorem holds for all infinite-dimensional Banach spaces. But we have not proved this. However, it should be noted that once we have settled the question of the finiteness of the semi-variation of $\zeta_\alpha(\cdot)$, we have, in effect, settled this question for the measure $\rho_\alpha(\cdot)$, for it is a triviality that at any given $P \in \mathcal{P}$ the semi-variations of the two measures are either both finite or both infinite.

4. Prologue

To provide complete proofs of the results given above, several lemmas not stated in this paper are needed. These in turn depend on a large body of ancillary operator theory. Thus, although the basic ideas of the theory can be readily outlined, as on the preceeding pages, the technical completion of the paper is a much more arduous and lengthy undertaking. There is also a spectral theory for Hilbert space valued finitely additive stationary measures over $\mathbb{R}^q$, following the pattern in [11, §3], which we have not even touched upon. We should also point out that the theory of stationary measures over $\mathbb{R}^q$ for $q \geq 2$ does not adequately reveal the geometry of helical surfaces and hypersurfaces. For instance, for $q=2$ we have a helical surface $x(\cdot,\cdot)$ in a Banach space $\mathfrak{X}$, parametrised on $\mathbb{R}^2$. Our theory gives us an insight into the associated measure, i.e. in effect into differences of the form

$$x(b_1,b_2) - x(b_1,a_1) - x(a_1,b_2) + x(a_1, a_2),$$

where $a = (a_1,a_2)$ and $b = (b_1,b_2)$ are in $\mathbb{R}^2$. It sheds little light on the chordal differences

$$x(b_1,b_2) - x(a_1,a_2).$$

Insight into the latter is needed for an appreciation of the geometry of the surface $x(\cdot, \cdot)$ in $\mathfrak{X}$. Such geometrical understanding is necessary for a probe into the deeper properties of the nexus between such surfaces in a Hilbert space $\mathcal{H}$ and the infinitely decomposable probability measures over $\mathbb{R}^2$, cf. [12]. Thus the theory developed above should be viewed as merely a prelude to a deeper geometrical theory of helical hypersurfaces.

REFERENCES

1. P.L. Butzer and H. Behrens, Semi-groups of operators and approximation, Springer Verlag, Berlin, 1967.

2. M.M. Day, Normed linear spaces, Springer Verlag, Berlin, 1973.

3. N. Dinculeanu, Vector measures, Pergamon Press, Oxford, 1967.

4. J.L. Doob, Stochastic processes, Wiley, New York, 1953.

5. N. Dunford & J. Schwartz, Linear operators, I, Interscience, New York, 1958; 1963.

6. E. Hille and R.S. Phillips, Functional analysis and semi-groups, Amer.Math.Soc., Providence, R.I., 1957.

7. K.S. Lau & J.K. Lee, On generalized harmonic analysis, Trans.Amer.Math.Soc. (to appear).

8. P. Masani, Orthogonally scattered measures, Adv. in Math. 2 (1968), 61-117.

9. P. Masani, Explicit form for the Fourier-Plancherel transform over locally compact abelian groups, 162-182, in "Abstract Spaces and Approximation" edited by P.L. Butzer and B.Sz. Nagy, Birkhäuser, Basel, 1969.

10. P. Masani, Quasi-isometric measures and their applications, Bull.Amer.Math.Soc. 76 (1970), 427-528.

11. P. Masani, On helices in Hilbert space I, Theory of Probability & Appl. (USSR) 17 (1972), 3-20. (English Ed. SIAM 17 (1972),1-19).

12. P. Masani, On infinitely decomposable probability distributions and helical varieties in Hilbert space, 209-223 in "Multivariate Analysis", edited by P.R. Krishnaiah, Acad. Press, New York, 1973.

13. P. Masani, On helices in Banach spaces, Sankhya 38 (1976), 1-27.

14. P. Masani, Dilations as propagators of Hilbertian varieties, SIAM J. Analysis 9 (1978), 414-456.

15. P. Masani, Commentary on the memoire [30a] on generalized harmonic analysis in Norbert Wiener Collected Works, II (ed. by P. Masani) MIT Press, Cambridge, Mass., 1979.

16. R. Nelson, The spaces of functions of finite upper p-variation, Trans.Amer.Math.Soc. 253 (1979), 171-190.

17. N. Wiener, Generalized harmonic analysis, Acta Math. 55 (1930), 117-258.

18. N. Wiener, The homogeneous chaos, Amer.J.Math. 60 (1938), 897-936.

AN ELEMENTARY INTEGRAL

Paul McGill
Department of Mathematics
The New University of Ulster
Coleraine, N. Ireland.

1. Introduction

The purpose of this note is to clarify the connection between elementary integration procedures and the modern theory of Lebesgue-type integrals. The standard method for constructing the latter is to proceed as follows.

(a) First define a space $\mathcal{E}$ of simple functions and define the integration mapping $I : \mathcal{E} \to X$ where X is a suitable topological space.

(b) Imbed $\mathcal{E}$ in a suitable complete space Y in such a way as to make I uniformly continuous in the induced uniformity. Then I will extend to a mapping $\overline{I}$ defined on $\overline{\mathcal{E}}$ and taking values in the completion of X. $\overline{\mathcal{E}}$ is called the space of integrable functions.

The difficulty now is to study completeness properties of the space $\overline{\mathcal{E}}$ and to perhaps show that we can characterise its elements in some way other than as members of Y. The article [6] provides a good example of this type of approach.

For elementary integrals, such as the Henstock-Kurzweil construction of the Denjoy integral and McShane's elementary((but powerful [4])definition of the stochastic integral, the problems are slightly different. These integrals are defined more directly and require only completeness of the space of values. Consequently they tend to lack structure. This paper presents a study of some of the structure associated with one elementary integral and shows that, under suitable assumptions, it possesses fairly good properties. In particular we indicate dominated convergence and completeness results in a general setting.

2. Construction

Let T be a set, $\mathcal{K}$ a lattice of subsets of T (i.e. a collection of subsets containing the empty set which is closed under finite unions and finite intersections). We shall always assume that T is given the topology generated by the $\mathcal{K}$ open sets

$$\mathcal{G} = \mathcal{G}(\mathcal{K}) = \{G \subseteq T : K \backslash G \in \mathcal{K} \text{ for all } K\}$$

$\mathcal{K}_\sigma$ ($\mathcal{G}^\sigma$) will denote the closure of $\mathcal{K}$ (resp. $\mathcal{G}$) under countable intersections (resp. unions). Let $\mathcal{A}$ be the collection of all mappings

$$A : T \to \mathcal{G}$$

with countable range such that $x \in A(x)$ for all x. We write $A \prec A'$ (A <u>refines</u> A') if $A(x) \subseteq A'(x)$ for each x. This notation is also used, in the obvious way, for open sets. The letters A, K, and G shall always be used to represent members of the families $\mathcal{A}$, $\mathcal{K}$, and $\mathcal{G}$ respectively.

Suppose that H_i (i = 0, 1, 2, 3) are commutative complete Hausdorff topological groups and that there exists a continuous bi-additive mapping $\bullet : H_1 \times H_2 \to H_3$. We will let $\mathcal{N}_i$ (i = 0, 1, 2, 3) be a base for the neighbourhoods of zero in H_i. Then consider functions

$$f : T \to H_1 \qquad \mu : \mathcal{K} \to H_2 \qquad \eta : \mathcal{K} \to H_0$$

The function η is used as a control and will be assumed to satisfy the following conditions.

(1) η is σ-smooth ($K_n \downarrow \phi$ implies that $\lim_n \eta(K_n) = 0$).

(2) η is exhaustive ($\{K_n\}$ disjoint implies that $\lim_n \eta(K_n) = 0$).

(3) η is tight ($K_2 \subseteq K_1$ implies that $\eta(K_1) - \eta(K_2) = \lim \{\eta(K) : K \subseteq K_1 \backslash K_2\}$).

Under these conditions it is well known (e.g. [3]) that η has a unique $\mathcal{K}_\sigma$ inner-regular extension to some σ-algebra containing $\mathcal{G}(\mathcal{K})$. We shall write this extension as η also.

A collection of pairs (K_i, x_i) $1 \le i \le n$ is said to be <u>compatible</u> with $(A,V) \in \mathcal{A} \times \mathcal{N}_0$ if the following conditions are satisfied.

(a) $K_i \subseteq A(x_i)$ $1 \le i \le n$.

(b) $K_i \cap K_j = \phi$ $i \neq j$.

(c) $K \cap K_i = \phi$ $1 \le i \le n$ implies that $\eta(K) \in V$.

We shall denote compatibility by writing $(K_i, x_i) \sim (A,V)$. If conditions (a) and (b) only are satisfied, then we write $(K_i, x_i) \sim A$.

<u>Definition</u> Say that $u = (\eta)\int f\, d\mu$ if for each $U \in \mathcal{N}_3$ there exists (A,V) in $\mathcal{A} \times \mathcal{N}_0$ such that

$$\sum f(x_i) \bullet \mu(K_i) - u \in U$$

whenever $(K_i, x_i) \sim (A,V)$.

We will now prove that the above condition is non-trivial. First introduce the convention that if $U \in \mathcal{N}_i$ (i = 0, 1, 2, 3) then $U(\frac{1}{n})$ is a fixed element of $\mathcal{N}_i$ such that $nU(\frac{1}{n}) \subseteq U$.

<u>Lemma 1</u> For every pair $(A,V) \in \mathcal{A} \times \mathcal{N}_0$ there exists $(K_i, x_i) \sim (A,V)$.

<u>Proof</u>: If not choose some (A,V) such that this fails. Hence we can find some K such that $\eta(K) \notin V$. But by tightness and σ-smoothness, since $\{A(x) : x \in K\}$ forms a cover of K, there exists $(K_i^1, x_i^1) \sim A$ with $K_i^1 \subseteq K$ and

$$\eta(\bigcup K_i^1) = \sum \eta(K_i^1) \notin U$$

But again by the assumption there exists K_1, $K_1 \cap K_i^1 = \phi$ for all i, and $\eta(K_1) \notin U$. Now repeat the above argument with K replaced by K_1 and find a collection (K_i^2, x_i^2) compatible with A and disjoint from (K_i^1, x_i^1). By proceeding inductively we generate a sequence of disjoint collections each satisfying the above condition. This contradicts the exhaustivity of η.

3. Properties

One of the difficulties with an elementary integral is ensuring that there are enough integrable functions. Our principal existence result is the following.

Theorem 1 If $(\eta)\int f\,d\mu$ exists then $\lambda(E) = (\eta)\int_E f\,d\mu$ exists for every η-measurable set E. λ is tight, exhaustive and σ-smooth with $\lambda << \eta$.

We first prove the following two preliminary lemmas. The second one states the absolute continuity requirement which is implicit in the definition.

Lemma 2 Suppose that $(\eta)\int f\,d\mu$ exists. Then for each $O \in \mathcal{G}^\sigma$ $(\eta)\int_O f\,d\mu$ exists.

Proof: Choose $U \in \mathcal{N}_3$ and find (A,V) such that

$$\sum f(x_i)\mu(K_i) - (\eta)\int f\,d\mu \in U(\tfrac{1}{2})$$

if $(K_i, x_i) \sim (A,V)$ (note that from now on we omit the symbol • from our formulae). By exhaustivity there exists $K \subseteq O$ such that

$$\eta(K') \in V(\tfrac{1}{2})$$

if $K' \cap K = \phi$. Letting A' refine A, C(K) and O we fix $(K_i, x_i) \sim (A', V(\frac{1}{4}))$. This collection splits as

$$(K_i^1, x_i^1) \quad \text{and} \quad (K_i^2, x_i^2)$$

where $x_i^1 \in O$, $x_i^2 \notin O$. Now let A" refine A' and $C(\bigcup K_i^1)$. If $(K_i'', x_i'') \sim (A'', V(\frac{1}{4}))$ we can split this as

$$(K_i^3, x_i^3) \quad \text{and} \quad (K_i^4, x_i^4)$$

with $x_i^3 \in O$, $x_i^4 \notin O$. Then if K_o is disjoint from $\bigcup K_i^1$ and $\bigcup K_i^4$ it is easy to see that $\eta(K_o) \in V$, so that $(K_i^1, x_i^1) \cup (K_i^4, x_i^4) \sim (A,V)$. Therefore we get

$$\sum f(x_i^1)\mu(K_i^1) - \sum f(x_i^3)\mu(K_i^3) \in U$$

This leads to the required Cauchy condition and the proof is complete.

Lemma 3 Let $(\eta)\int f\,d\mu$ exist. Given $U \in \mathcal{N}_3$ there exists V such that if $\eta(K) \in V$ when $K \subseteq O \in \mathcal{G}^\sigma$ then for some A

$$\sum \chi_E(x_i) f(x_i)\mu(K_i) \in U$$

if $E \subseteq O$ and $(K_i, x_i) \sim A$.

<u>Proof</u>: Choose V' such that for some A' $(K_i,x_i) \sim (A',V')$ gives

$$\sum f(x_i)\mu(K_i) \quad - \quad (\eta)\int f\,d\mu \quad \in \quad U$$

Choose A refining A' and O, and notice that if $K' \subseteq \bigcup\{K_i : x_i \in E\}$ then $\eta(K') \in V'$. The result is now clear if we take $2V \subseteq V'$.

<u>Proof of Theorem</u>: Let E be η-measurable. Given $V \in \mathcal{N}_0$ there exists [3] $O \in \mathcal{G}^\sigma$ and $F \in \mathcal{K}_\sigma$, $O \supseteq E \supseteq F$ such that $\eta(K') \in V$ if $K' \subseteq O\backslash F$. By Lemma 2 $(\eta)\int_O f\,d\mu$ exists so by Lemma 3 we deduce that $(\eta)\int_E f\,d\mu$ exists since the approximating sums are Cauchy. Lemma 3 now proves that $\lambda << \eta$. Hence the rest of the statement follows from the properties of η.

It will not be true in general that $(\eta)\int_K d\mu = \mu(K)$ for all K so we require the following result.

<u>Definition</u> Write $\mu << \eta$ (loc.) if for each K there exists $O \in \mathcal{G}^\sigma$ containing K such that $\mu' << \eta$ where μ' denotes the restriction of μ to O.

<u>Lemma 4</u> Suppose that $(\eta)\int d\mu$ exists and that $\mu << \eta$ (loc.). Then we have $(\eta)\int_K d\mu = \mu(K)$ for every K.

<u>Proof</u>: Choose $O \in \mathcal{G}^\sigma$ as above and also such that $K'\cap K = \phi$ together with $K' \subseteq O$ gives $\eta(K') \in V(\frac{1}{2})$. Let A refine O and C(K). If $(K_i,x_i) \sim (A,V(\frac{1}{2}))$ it is clear that

$$\eta\Big(\bigcup_{x_i\in K} K_i\Big) \quad - \quad \eta(K) \quad \in \quad V$$

The result now follows by absolute continuity.

To extend Lemma 4 to arbitrary measurable sets is very easy. However the extension to integrable functions requires a dominated convergence result.

<u>Definition</u> μ is <u>convex bounded</u> on $\mathcal{K}$ if for each $U \in \mathcal{N}_3$ there exists $U' \in \mathcal{N}_1$ such that whenever $\alpha_i \in U'$ then

$$\sum \alpha_i\mu(K_i) \in U$$

for every finite disjoint collection $\{K_i\}$.

A detailed discussion of spaces in which every bounded measure is convex bounded can be found in [6]. The relationship to the dominated convergence theorem is also given there.

<u>Theorem 2</u> Let $\{f_n\}$ be a sequence of η-measurable functions for which $(\eta)\int f_n d\mu$ exists for all $n \geq 1$. Suppose that the following three conditions are satisfied.

(1) $f_n \to f \quad \eta$ a.e.

(2) μ is convex bounded.

(3) The conclusion of Lemma 3 holds uniformly in n when f is replaced by f_n.

Then f is integrable and $\lim (\eta)\int f_n\,d\mu \;=\; (\eta)\int f\,d\mu$.

<u>Proof</u>: By the Egoroff Theorem (see [6]) $\{f_n\}$ converges to f η-almost

uniformly. By (3) we are able to reduce to the situation where $\{f_n\}$ converges uniformly to f on T. The usual Riemann-type argument will now work.

Since many spaces of integrable functions lack completeness we will examine this problem in our setting. Let us suppose that μ and η are fixed and define

$$\mathcal{L} = \{f : f \text{ is } \eta\text{-measurable and } (\eta)\int f\, d\mu \text{ exists}\}$$

The set $\mathcal{L}$ is a group which we can topologise by requiring that $f_\alpha \xrightarrow{\tau} f$ if and only if

$$\lim\, (\eta)\int_E f_\alpha d\mu = (\eta)\int_E f\, d\mu$$

for each η-measurable set E.

Theorem 3 Let $\{f_n\}$ be τ-Cauchy with $\lim f_n = f$ η a.e. Then if μ is convex bounded and H_o is metrisable we have $f_n \xrightarrow{\tau} f$.

Proof: Assume first that f is integrable. By restricting ourselves to considering only one element of $\mathcal{N}_3$ we can reduce to the situation where f takes values in a metrisable group. Now apply the Egoroff Theorem to get an increasing sequence of measurable sets E_m such that the complement of their union is a η-null set and $f_n \longrightarrow f$ uniformly on E_m. Clearly

$$\lim\, (\eta)\int_{E_m} f_n d\mu = (\eta)\int_{E_m} f\, d\mu$$

By the Vitali-Hahn-Saks Theorem [2] since the measures $\lambda_n(E) = (\eta)\int_E f_n d\mu$ are uniformly η-continuous we get

$$\lim\, (\eta)\int f_n d\mu = (\eta)\int f\, d\mu$$

Notice that this is precisely what we require. It remains to prove that f is integrable. To do this we check that f satisfies the conclusion of Lemma 3 so that the result follows by Theorem 2 applied to $\chi_{E_m} f$. By the Vitali-Hahn-Saks Theorem choose V such that if $\eta(K) \in V$ for every $K \subseteq O \in \mathcal{G}^\sigma$ then

$$(\eta)\int_K f_n d\mu \in U \in \mathcal{N}_3 \qquad (n \geq 1)$$

An argument as in [1] shows that we can choose A_n such that if $(K_i, x_i) \sim A_n$ then for every $E \subseteq O$

$$\sum \chi_E(x_i) f_n(x_i)\mu(K_i) \in 4U$$

Assuming, as we may, that $f_n \longrightarrow f$ everywhere we choose $U_1 \in \mathcal{N}_1$ such that

$$\sum \alpha_i \mu(K_i) \in U$$

if $\alpha_i \in U_1$ and $\{K_i\}$ is a finite disjoint collection. There exists n(x) such that

$$f_n(x) - f(x) \in U_1 \qquad (n \geq n(x))$$

Define

$$A(x) = A_{n(x)}(x)$$

It is now clear that f satisfies the conclusion of Lemma 2 with U replaced by 6U. This completes the proof.

We are now able to consider the problems concerning the structure of $\mathcal{L}$. For example if H_3 is metrisable then the following result shows that τ is generated by a pseudo-metric.

Lemma 5 Let H be a commutative Hausdorff topological group. If $\{\lambda_n\}$ is a sequence of exhaustive H-valued measures defined on the σ-ring $\mathcal{R}$ and and converging to zero pointwise on $\mathcal{R}$ then $\{\lambda_n\}$ converges to zero uniformly on $\mathcal{R}$.

Proof: Suppose not. Hence there exists a neighbourhood U of zero in H for which we can perform the following construction. Find

$$\lambda_{n_1}(E_1) \notin 2U$$

Choose $n_2' > n_2 > n_1$ such that

$$\lambda_n(E_1) \in U(\tfrac{1}{2}) \quad (n \geq n_2) \quad ; \qquad \lambda_{n_2'}(E_2) \notin 2U$$

Proceeding inductively we obtain intertwined increasing sequences, $\{n_r\}$ and $\{n_r'\}$, such that

$$\lambda_n(E_{r-1}) \in U(2^{-r}) \quad (n \geq n_r) \quad ; \qquad \lambda_{n_r'}(E_r) \notin 2U$$

If $F_r = E_r \setminus (\bigcup_{j=1}^{r-1} E_j)$ then

$$\lambda_{n_r'}(F_r) \notin U \qquad (r \geq 1)$$

if we assume that U is closed. This however contradicts the Brooks-Jewett Theorem [2].

Consider now the problems involved in proving that $\mathcal{L}$ is complete. First we would like to show that L, the quotient of $\mathcal{L}$ w.r.t. the usual equivalence of functions equal η a.e., is a Hausdorff topological group. And secondly we wish to deduce that every τ-Cauchy sequence has a subsequence which converges η a.e. The latter is known to fail [5] for the space of Pettis integrable functions. However Turpin has pointed out to us that if $H_1 = R$, $H_3 = H_2 = X$ a topological vector space, such results can be obtained provided X has the bounded multiplier property. Details are to be found in [6].

References

1. R. B. Darst and E. J. McShane, The Deterministic Ito-Belated integral is equivalent to the Lebesgue integral, Proc. Amer. Math. Soc. Vol. 72 No. 2 (1978) 271-275.

2. L. Drewnowski, Equivalence of Brooks-Jewett, Vitali-Hahn-Saks and Nikodym Theorems, Bull. Acad. Pol. des Sci. Vol. XX No. 9 (1972) 725-731.

3. P. McGill, Measure Extensions and Decompositions, (Preprint).

4. P. Protter, A Comparison of Stochastic Integrals, Ann. Prob. Vol. 7 No. 2 (1979)

5. E. Thomas, Totally Summable functions with values in locally convex spaces, Lecture Notes in Mathematics 541, 117-131.

6. P. Turpin, Integration par rapport a une mesure a valeurs dans un espace vectoriel topologique non suppose localement convexe, Expose No. 8 Colloque : Integration Vectorielle et Multivoque (Caen, 1975).

REGULARITY AND EXTENSION OF SEMIGROUP-VALUED BAIRE MEASURES

Pedro Morales
Université de Sherbrooke
Sherbrooke, Québec, Canada

1. Introduction

It is well-known that any $[0,\infty]$-valued Baire measure is regular, and it can be extended uniquely to a regular Borel measure [7]. An analogue of the first result for a locally convex value Baire measure was established by Dinculeanu and Lewis [3], which, in turn, was generalized to a topological group valued Baire measure by Sundaresan and Day [10]. An analogue of the second result for a locally convex valued Baire measure - limited to relatively compact sets - was obtained by Dinculeanu and Kluvanek [2], which, in turn, was generalized to a topological group valued Baire measure by Khurana [8].

Our purpose is to generalize these results for a uniform semigroup valued Baire measure. We have found that local s-boundedness on relatively compact sets - a condition automatically verified for a topological group valued measure - has the essential effect of local finiteness in classical theory. The classical definition of regularity coincides, for relatively compact sets only, with the current definition of regularity for topological group valued Baire and Borel measures. We have set down a definition of regularity for uniform semigroup valued Baire and Borel measures, which coincides with each, in their respective contexts.

2. p-Semivariations

Let G be a commutative semigroup with neutral element 0. If there exists a uniformity on G such that the map $(x,y) \to x+y$ is uniformly continuous, G is said to be a uniform semigroup. Apart from the trivial example of a commutative topological group, there is $[0,\infty]$ of classical measure and integration theory, also the interesting example appearing in Sion's study of integral representations of topological group valued measures [9, p.3]. The following key result is due essentially to Weber [11, Hilfsatz (1.1), p.414] (see also [5]):

2.1 Theorem. The uniformity of a uniform semigroup G may be generated by a family P of semi-invariant continuous $[0,1]$-valued pseudo-metrics on G.

(A pseudo-metric p on G is semi-invariant if $p(x+z,y+z) \le p(x,y)$ for all x,y,z in G.) For each p in P, write $|x|_p = p(x,0)$. Then $|0|_p = 0$, $p(x+y,y) \le |x|_p$ and $|x+y|_p \le |x|_p+|y|_p$. In the rest of the paper, the uniform semigroup G is assumed to be Hausdorff.

Let μ be a G-valued additive set function on a ring R of subsets of a set X. For each p in P we define the p-semivariation $\tilde{\mu}_p$ of μ on 2^X by the formula

$\tilde{\mu}_p(M) = \sup\{|\mu(E)|_p : E \in R, E \subseteq M\}$. The following properties are easily verified: (1) $\tilde{\mu}_p$ is an increasing [0,1]-valued set function on 2^X vanishing at ϕ; (2) $\tilde{\mu}_p|R$ is subadditive and dominates $|\mu(.)|_p$; (3) If μ is σ-additive, then $\tilde{\mu}_p|R$ is σ-subadditive and $\tilde{\mu}_p(\bigcup_{n=1}^{\infty} E_n) = \sup_n \tilde{\mu}_p(\bigcup_{i=1}^{n} E_i)$ for every sequence (E_n) in R. The set function μ is said to be *locally s-bounded* if, for every $E \in R$ and every disjoint sequence (E_n) such that $E_n \subseteq E$, $(\mu(E_n))$ converges to 0. It is clear that if μ is locally s-bounded, then, for every p in P, $\tilde{\mu}_p|R$ is also locally s-bounded.

2.2 Theorem. If $\mu: R \to G$ is σ-additive and locally s-bounded, then, for every p in P, the restriction $\tilde{\mu}_p|R$ is continuous at ϕ.

Proof. Let $p \in P$. Let $E_n \downarrow \phi$, $E_n \in R$. Taking into account the second property of the semivariation mentioned in (3), it suffices to prove that $\lim_{m,n} \tilde{\mu}_p(E_m \Delta E_n) = 0$. Suppose the contrary. Then, for some $\delta > 0$, we can construct two sequences (m_k), (n_k) of positive integers such that $m_k < n_k < m_{k+1} < n_{k+1}$ and $\tilde{\mu}_p(E_{m_k} - E_{n_k}) \geq \delta$. This contradicts the local s-boundedness of $\tilde{\mu}_p|R$.

2.3 Corollary. Let $\mu: R \to G$ be σ-additive and locally s-bounded. If $E_n \downarrow \phi$, $E_n \in R$, then, for every neighbourhood V of 0 in G, there exists a positive integer N such that $E \subseteq E_N$, $E \in R$ implies $\mu(E) \in V$.

If A is a class of subsets of X, the symbols $\sigma(A)$, $\delta(A)$ will denote, respectively, the σ-ring, δ-ring generated by A.

3. Regularity

In the remainder of the paper, X is a locally compact Hausdorff space, and K, K_o denote, respectively, the class of compact, compact G_δ subsets of X. Thus, $\sigma(K)$, $\sigma(K_o)$ are, respectively, the class of Borel sets, class of Baire sets of X. Our definition of Borel measure, Baire measure will be motivated by the following

3.1 Lemma. Let $\mu: \sigma(K) \to [0,\infty]$ be σ-additive. Then μ is finite on every $K \in K$ if and only if μ if σ-finite and $\mu|\delta(K)$ is locally s-bounded.

Of the two conditions of the preceeding lemma, only the second has meaning for a G-valued set function. Accordingly, we extend the classical definitions as follows: A *Baire measure* (*Borel measure*) on X is a G-valued σ-additive set function μ on $\sigma(K_o)$ $(\sigma(K))$ whose restriction $\mu|\delta(K_o)$ $(\mu|\delta(K))$ is locally s-bounded. It is clear that every topological group valued σ-additive set function on $\sigma(K_o)$ $(\sigma(K))$ is a Baire measure (Borel measure) on X.

We can extend, for a G-valued function, the usual definition of regularity for topological group valued set functions. The inadequacy of this definition (called (RG)-regularity) is illustrated for the following trivial fact: Let $\mu: \sigma(K) \to [0,\infty]$ be σ-additive and finite on every $K \in K$. If $\mu(E) = \infty$, then μ is not (RG)-regular at E. However, for G-valued set functions on $\delta(K_o)$, (RG)-regularity has the following nice property (see [8, Corollary 4, p.895] and [11, Folgerung (6.2),p.422]):

3.2 Lemma. Let $\nu:\delta(K_o) \to G$ be additive. Then ν is σ-additive and locally s-bounded if and only if ν is (RG)-regular.

Proof. *Necessity*. Let $K \in K_o$. To prove that ν is (RG)-regular at K, let V be a neighbourhood of 0 in G. There exists a sequence (U_n) of open sets in $\delta(K_o)$ such that $U_n \downarrow K$. Let N be a positive integer with the property given by Corollary 2.3. Then the pair (K,U_N) works for (K,V). Let $\Sigma = \{E \in \delta(K_o): \nu$ is (RG)-regular at E$\}$. By the same arguments used in [3,p.93] and [10,p.611] we show that Σ is a δ-ring. So $\Sigma = \delta(K_o)$. *Sufficiency*. Let $E_n \downarrow \phi$, $E_n \in \delta(K_o)$. Let V be a neighbourhood of 0 in G. Choose a sequence (W_j) of neighbourhoods of 0 in G such that $\Sigma_{j=1}^n W_j \subseteq V$ for all n=1,2,3,... Let (K_j,U_j) work for (E_j,W_j). Then $\cap_{j=1}^{\infty} K_j = \phi$, and therefore there exists a positive integer N such that $\cap_{j=1}^{N} K_j = \phi$. It follows that $\nu(E_n) \in V$ for all $n \geq N$. So ν is σ-additive. To prove that ν is locally s-bounded, let $E \in \delta(K_o)$ and let (E_n) be a disjoint sequence in $\delta(K_o)$ such that $E_n \subseteq E$. Then $F_n = \cup_{j=n}^{\infty} E_j$ is a decreasing sequence in $\delta(K_o)$ such that $F_n \downarrow \phi$. So $\lim_n \nu(F_n) = 0$. Let $p \in P$. Since $|\nu(E_n)|_p \leq |\nu(F_n)|_p + |\nu(F_{n+1})|_p$, it follows that $(\nu(E_n))$ converges to 0.

The larger context $\sigma(K_o)$ imposes a weakening of (RG)-regularity. Accordingly, we propose the following definition: A G-valued Baire measure μ on X is *regular at* $E \in \sigma(K_o)$ if, for every neighbourhood V of 0 in G, there exists a set $H \in (K_o)_\sigma$ and an open set $U \in \sigma(K_o)$ such that $H \subseteq E \subseteq U$ and $\mu(F) \in V$ whenever $F \subseteq U-H$ and $F \in \sigma(K_o)$. We say that the pair (H,U) works for (E,V). The set function μ is *regular* if it is regular at every set of $\sigma(K_o)$. Regularity of a G-valued Borel measure on X is defined in the same way with K in the role of K_o. The following lemma shows the adequacy of this definition:

3.3 Lemma. Let μ be a G-valued Baire measure on X. Then: (1) If $G = [0,\infty]$ and $\mu(K) < \infty$ for all K in K_o, then μ is regular if and only if it is regular in the classical sense; (2) If G is a group, μ is regular if and only if it is (RG)-regular; and (3) Regularity of μ coincides with (RG)-regularity on $\delta(K_o)$.

The same lemma holds for a Borel measure on X. We are in position to establish the main result of the section:

3.4 Theorem. Every G-valued Baire measure on X is regular.

Proof! Let μ be a G-valued Baire measure on X, and let $\nu = \mu|\delta(K_o)$. Let $E \in \sigma(K_o)$. Let V be a closed neighbourhood of 0 in G. Choose a sequence (W_j) of neighbourhoods of 0 in G such that $\Sigma_{j=1}^n W_j \subseteq V$ for all n=1,2,3,... Write $E = \cup_{j=1}^{\infty} E_j$, $E_j \in \delta(K_o)$. Lemma 3.2 implies that there exists a set $K_j \in K_o$ and an open set $U_j \in \delta(K_o)$ such that $K_j \subseteq E_j \subseteq U_j$ and $\nu(F) \in W_j$ whenever $F \subseteq U_j - K_j$ and $F \in \delta(K_o)$, j=1,2,3,... Let $H = \cup_{j=1}^{\infty} K_j$ and $U = \cup_{j=1}^{\infty} U_j$. Then $H \in (K_o)_\sigma$, U is an open set in $\sigma(K_o)$ and $H \subseteq E \subseteq U$. It is easy to verify that (H,U) works for (E,V).

4. Extension

Let $\nu:\delta(K_o) \to G$ be σ-additive and locally s-bounded, and let $p \in P$. Taking into account Theorem 2.2, it follows that the restriction $\tilde{\nu}_p|\delta(K_o)$ is a subadditive submeasure in the sense of Dobrakov [4]. By Corollary 2 of [4,p.34], $\tilde{\nu}_p|\delta(K_o)$ extends uniquely to a subadditive submeasure $\bar{\nu}_p$ on $\delta(K)$ with the additional property: $\inf\{\bar{\nu}_p(U-K):K \in K,$ U is an open set in $\delta(K)$ and $K \subseteq E \subseteq U\} = 0$ for all $E \in \delta(K)$. Define $\rho_p(A,B) = \bar{\nu}_p(A\Delta B)$, A,B in $\delta(K)$. Then $(\rho_p)_{p\in P}$ is a family of pseudo-metrics on $\delta(K)$. Let V be the uniformity on $\delta(K)$ generated by this family.

4.1 Lemma. $\nu:(\delta(K),V) \to G$ is uniformly continuous.

4.2 Lemma. $\delta(K_o)$ is dense in $\delta(K)$.

In the remainder of the paper, G is assumed to be complete. The following result generalizes the second part of Corollary 4 of [8,p.895]:

4.3 Theorem. Let $\nu:\delta(K_o) \to G$ be σ-additive and locally s-bounded. Then ν extends uniquely to a (RG)-regular σ-additive locally s-bounded set function on $\delta(K)$.

Proof. By Lemmas 4.1 and 4.2, ν extends to a uniformly continuous set function $\bar{\nu}$ on $\delta(K)$. Let $p \in P$. Since $|\bar{\nu}(E)|_p \leq \bar{\nu}_p(E)$ for all $E \in \delta(K_o)$, it follows, by continuity, that this inequality holds when E varies in $\delta(K)$. To prove the (RG)-regularity of $\bar{\nu}$, let $E \in \delta(K)$ and let V be a neighbourhood of 0 in G. Then there exists p_i in P, $\varepsilon_i > 0$, $1 \leq i \leq n$, such that $\cap_{i=1}^n\{x\in G:|x|_{p_i} < \varepsilon_i\} \subseteq V$. Then there exist a set $K_i \in K$ and an open set $U_i \in \delta(K)$ such that $K_i \subseteq E \subseteq U_i$ and $\bar{\nu}_{p_i}(U_i-K_i) < \varepsilon_i$, $i=1,2,\ldots,n$. Write $K = \cup_{i=1}^n K_i$ and $U = \cap_{i=1}^n U_i$. Then, (K,U) works for (E,V). Since $\bar{\nu}(E\cup F)+\bar{\nu}(E\cap F) = \bar{\nu}(E)+\bar{\nu}(F)$ for all E,F in $\delta(K_o)$ and the finite Boolean operations are continuous, the equation holds when $E,F \in \delta(K)$. This proves that $\bar{\nu}$ is additive. From Lemma 3.2 it follows that $\bar{\nu}$ is σ-additive and locally s-bounded. It remains to prove the uniqueness. Let $\bar{\nu}_1,\bar{\nu}_2$ be two (RG)-regular σ-additive locally s-bounded set functions on $\delta(K)$ extending ν. Let $E \in \delta(K)$. Let $p \in P$, $\varepsilon > 0$. There exists a set $K_i \in K$ and an open set $U_i \in \delta(K)$ such that $K_i \subseteq E \subseteq U_i$ and $(\tilde{\bar{\nu}}_i)_p(U_i-K_i) < \varepsilon$, $i=1,2,$. Write $K = K_1\cup K_2$ and $U = U_1\cap U_2$, and choose a set K_o in K_o such that $K \subseteq K_o \subseteq U$. Then $\bar{\nu}_1(K_o) = \bar{\nu}_2(K_o)$. It follows that $p(\bar{\nu}_1(E),\bar{\nu}_2(E)) < 4\varepsilon$. So $\bar{\nu}_1(E) = \bar{\nu}_2(E)$.

Now let μ be a G-valued Baire measure on X. Consider the restriction $\nu = \mu|\delta(K_o)$. From Theorem 4.3 it follows that there exists a unique (RG)-regular σ-additive locally s-bounded set function $\bar{\nu}$ on $\delta(K)$ extending ν.

4.4 Lemma. If $K_n\uparrow$, $K_n \in K$, then $(\bar{\nu}(K_n))$ converges.

Proof. Let $p \in P$, $\varepsilon > 0$. Since $\bar{\nu}$ is (RG)-regular, for every $n=1,2,\ldots$ there exists an open set $U_n' \in \delta(K)$ containing K_n such that $(\tilde{\bar{\nu}})_p(U_n'-K_n) < \varepsilon/(2^{n+1})$. Write $U_n = \cup_{i=1}^n U_n'$. Then U_n is open, $U_n\uparrow$, $U_n \in \delta(K)$ and $U_n \supseteq K_n$. Since $U_n-K_n \subseteq \cup_{i=1}^n(U_i'-K_i)$, $(\tilde{\bar{\nu}})_p(U_n-K_n) < \varepsilon/2$. Choose $K_n' \in K_o$ such that $U_n \supseteq K_n' \supseteq K_n$.

Write $K_n'' = \cup_{i=1}^n K_i'$. Then $K_n'' \in K_o$, $K_n''\uparrow$, $U_n \supseteq K_n'' \supseteq K_n$ and $(\tilde{\tilde{\nu}})_p(K_n''-K_n) < \varepsilon/2$. Since $K_n''\uparrow \cup_{n=1}^\infty K_n'' \in \sigma(K_o)$ and μ is σ-additive, $\mu(K_n'') \to \mu(\cup_{n=1}^\infty K_n'')$. But $\mu(K_n'') = \nu(K_n'') = \bar{\nu}(K_n'')$, so $(\bar{\nu}(K_n''))$ is Cauchy. We have $p(\bar{\nu}(K_n),\bar{\nu}(K_m)) \le p(\bar{\nu}(K_n),\bar{\nu}(K_n'')) + p(\bar{\nu}(K_n''), \bar{\nu}(K_m'')) + p(\bar{\nu}(K_m''), \bar{\nu}(K_m)) \le |\bar{\nu}(K_n''-K_n)|_p + p(\bar{\nu}(K_n''),\bar{\nu}(K_m'')) + |\bar{\nu}(K_m''-K_m)|_p < \varepsilon+p(\bar{\nu}(K_n''),\bar{\nu}(K_m''))$, showing that $(\bar{\nu}(K_n))$ is Cauchy.

<u>4,5 Lemma.</u> If $E_n\uparrow$, $E_n \in \delta(K)$, then $(\bar{\nu}(E_n))$ converges.

<u>Proof.</u> Let $p \in P$, $\varepsilon > 0$. Since $\bar{\nu}$ is (RG)-regular, for $n=1,2,\ldots$ we can choose $K_n \in K$ such that $K_n \subseteq E_n$ and $(\tilde{\tilde{\nu}})_p(E_n-K_n) < \varepsilon/2$. Let $K_n' = \cup_{i=1}^n K_i$. Then $K_n' \in K$, $K_n'\uparrow$, $K_n' \subseteq E_n$ and $(\tilde{\tilde{\nu}})_p(E_n-K_n') < \varepsilon/2$. We have $p(\bar{\nu}(E_n), \bar{\nu}(E_m)) \le p(\bar{\nu}(E_n),\bar{\nu}(K_n')) + p(\bar{\nu}(K_n'),\bar{\nu}(K_m'))+p(\bar{\nu}(K_m'),\bar{\nu}(E_m)) \le \varepsilon+p(\bar{\nu}(K_n'),\bar{\nu}(K_m'))$. This, with Lemma 4.4, shows that $(\bar{\nu}(E_n))$ is Cauchy.

<u>4.6 Lemma.</u> If $E_n\uparrow$, $F_n\uparrow$ with E_n, $F_n \in \delta(K)$ then the iterated limits $\lim_n \lim_m \bar{\nu}(E_n \cap F_m)$, $\lim_m \lim_n \bar{\nu}(E_n\cap F_m)$ exist and are equal.

<u>Proof.</u> For fixed n, $(E_n \cap F_m)_{m=1,\ldots}\uparrow$ so by Lemma 4.5, $\lim_m \bar{\nu}(E_n\cap F_m)$ exists. Since $F = \lim_m F_m \in \sigma(K)$, $\lim_m (E_n\cap F_m) = E_n \cap F \in \sigma(K)$, so $\bar{\nu}(E_n\cap F)$ is defined. Since $\bar{\nu}$ is σ-additive, $\lim_m \bar{\nu}(E_n\cap F_m) = \bar{\nu}(E_n\cap F)$. Since $(E_n\cap F)\uparrow$, $\lim_n \bar{\nu}(E_n\cap F)$ exists. So we have $\lim_n \lim_m \bar{\nu}(E_n\cap F_m) = \lim_n \bar{\nu}(E_n\cap F)$ exists. Similarly the other iterated limit exists. We divide the proof of the equality of the iterated limits into several cases:

(i) $E_n,F_n \in \sigma(K_o)$. Since $E = \lim_n E_n$, $F = \lim_m F_m$ belong to $\sigma(K_o)$ and μ is σ-additive, $\lim_n \lim_m \bar{\nu}(E_n\cap F_m) = \lim_n \lim_m \mu(E_n\cap F_m) = \lim_n \mu(E_n\cap F) = \mu(E\cap F) = \lim_m \mu(E\cap F_m) = \lim_m \lim_n \mu(E_n\cap F_m) = \lim_m \lim_n \bar{\nu}(E_n\cap F_m)$.

(ii) $E_n \in \sigma(K_o)$, $F_m \in K$. Let $p \in P$, $\varepsilon > 0$. By the argument in the proof of Lemma 4.4, we can construct an increasing sequence (K_m) in K_o such that $K_m \supseteq F_m$ and $(\tilde{\tilde{\nu}})_p(K_m-F_m) < \varepsilon/2$ for all $m=1,2,\ldots$ In particular, $(\tilde{\tilde{\nu}})_p(E_n\cap(K_m-F_m)) < \varepsilon/2$ for all $m,n=1,2,\ldots$ Since $E_n\cap K_m = (E_n\cap F_m)\cup(E_n\cap(K_m-F_m))$, we have $p(\bar{\nu}(E_n\cap F_m), \bar{\nu}(E_n\cap K_m)) \le |\bar{\nu}(E_n\cap(K_m-F_m))|_p < \varepsilon/2$. Then, since p is continuous, $p(\lim_n \lim_m \bar{\nu}(E_n\cap F_m), \lim_n \lim_m \bar{\nu}(E_n\cap K_m))$, $p(\lim_m \lim_n \bar{\nu}(E_n\cap F_m), \lim_m \lim_n \bar{\nu}(E_n\cap K_m)) \le \varepsilon/2$. It follows that $p(\lim_n \lim_m \bar{\nu}(E_n\cap F_m), \lim_m \lim_n \bar{\nu}(E_n\cap F_m)) \le \varepsilon$. This with $p \in P$, $\varepsilon > 0$ arbitrary, establishes the equality in this case.

(iii) E_n, $F_n \in K$. We use the argument of (ii) with the roles of E_n,F_n interchanged and then apply the result of (ii).

(iv) $E_n \in K$, $F_n \in \sigma(K)$. Let $p \in P$, $\varepsilon > 0$. Since $\bar{\nu}$ is (RG)-regular, for $m=1,2,\ldots$ there exists $K_m \in K$ such that $K_m \subseteq F_m$ and $(\tilde{\tilde{\nu}})_p(F_m-K_m) < \varepsilon/2$. Write $K_m' = \cup_{i=1}^m K_i$. Then $K_m' \in K$, $K_m'\uparrow$, $K_m' \subseteq F_m$ and $(\tilde{\tilde{\nu}})_p(F_m-K_m') < \varepsilon/2$ for all $m=1,2,\ldots$ In particular, $(\tilde{\tilde{\nu}})_p(E_n\cap(F_m-K_m')) < \varepsilon/2$ for all $n,m=1,2,\ldots$ Since $E_n\cap F_m = (E_n\cap K_m')\cup(E_n\cap(F_m-K_m'))$ we have $p(\bar{\nu}(E_n\cap F_m), \bar{\nu}(E_n\cap K_m')) < \varepsilon/2$. Applying the continuity of p and case (iii), we deduce the inequality $p(\lim_n \lim_m \bar{\nu}(E_n\cap F_m), \lim_m \lim_n \bar{\nu}(E_n\cap F_m) < \varepsilon$. This, for $p \in P$,

$\varepsilon > 0$ arbitrary, establishes the equality in this case.

(v) $E_n, F_n \in \delta(K)$. We use the argument of (iv) with the roles of E_n, F_n interchanged and apply the result of (iv).

4.7 Corollary. If $E_n \uparrow E$, $F_n \uparrow E$ with $E_n, F_n \in \delta(K)$ then $\lim_n \bar{\nu}(E_n) = \lim_n \bar{\nu}(F_n)$.

We are in a position to establish the main result of the section:

4.8 Theorem. Every G-valued Baire measure μ on X extends uniquely to a regular Borel measure.

Proof. Let $E \in \sigma(K)$. Let $E_n \uparrow E$, $E_n \in \delta(K)$. By Lemma 4.5, $(\bar{\mu}(E_n))$ converges and, by Corollary 4.7, this limit is independent of the particular increasing sequence in $\delta(K)$ converging to E. The required extension is unambiguously defined by the formula

$$\bar{\mu}(E) = \lim_n \bar{\nu}(E_n), \; E_n \uparrow E, \; E_n \in \delta(K).$$

It is clear that $\bar{\mu}$ is an additive extension of μ such that $\bar{\mu}|\delta(K)$ is locally s-bounded. The rest of the proof will be divided into three steps:

(i) $\bar{\mu}$ is σ-additive. Let $E_n \uparrow E$, $E_n \in \sigma(K)$. For each $n=1,2,\ldots$ choose an increasing sequence $(E_{n,m})$ in $\delta(K)$ such that $E_{n,m} \uparrow E_n (m \to \infty)$. Write $F_n = \cup_{i=1}^{n} E_{i,n}$. Then $F_n \in \delta(K)$, $F_n \subseteq E_n$ and $F_n \uparrow E$. Let $p \in P$. Since $(F_1 \cap E_{1,m}) \uparrow E_1, \bar{\nu}(F_1 \cup E_{1,m}) \to \bar{\nu}(E_1)$. Hence there exists a positive integer k such that $p(\bar{\nu}(F_1 \cup E_{1,k}), \bar{\mu}(E_1)) < 1$. Write $D_1 = F_1 \cup E_{1,k}$. Then $F_1 \subseteq D_1 \subseteq E_1$, $D_1 \in \delta(K)$ and $p(\bar{\mu}(D_1), \bar{\mu}(E_1)) < 1$. Let n be a positive integer. As induction hypothesis, assume the existence of a finite increasing sequence $(D_i)_{1 \le i \le n}$ in $\delta(K)$ such that $F_i \subseteq D_i \subseteq E_i$ and $p(\bar{\mu}(D_i), \bar{\mu}(E_i)) < 1/i$, $1 \le i \le n$. Since $(D_n \cup F_{n+1} \cup E_{n+1,m}) \uparrow E_{n+1} (m \to \infty)$, $\bar{\mu}(D_n \cup F_{n+1} \cup E_{n+1,m}) \to \bar{\mu}(E_{n+1})$. Choose a positive integer k such that $p(\bar{\mu}(D_n \cup F_{n+1} \cup E_{n+1,k}), \bar{\mu}(E_{n+1})) < 1/(n+1)$. Let $D_{n+1} = D_n \cup F_{n+1} \cup E_{n+1,k}$. Then $F_{n+1} \subseteq D_{n+1} \subseteq E_{n+1}$, $D_n \subseteq D_{n+1}$, $D_{n+1} \in \delta(K)$ and $p(\bar{\mu}(D_{n+1}), \bar{\mu}(E_{n+1})) < 1/(n+1)$. We have constructed inductively an increasing sequence (D_n) in $\delta(K)$ such that $D_n \uparrow E$ and $p(\bar{\mu}(D_n), \bar{\mu}(E_n)) < 1/n$. Then $\bar{\mu}(D_n) \to \bar{\mu}(E)$. Since $p(\bar{\mu}(E_n), \bar{\mu}(E)) < 1/n + p(\bar{\mu}(D_n), \bar{\mu}(E)) \to 0$ and $p \in P$ is arbitrary $\bar{\mu}(E_n) \to \bar{\mu}(E)$.

(ii) $\bar{\mu}$ is regular. Let $E \in \delta(K)$. Let V be a closed neighbourhood of 0 in G. Choose a sequence (W_j) of neighbourhoods of 0 in G such that $\cup_{j=1}^{n} W_j \subseteq V$ for all $n=1,2,\ldots$ Let (E_j) be a sequence in $\delta(K)$ such that $E = \cup_{j=1}^{\infty} E_j$. Since $\bar{\nu}$ is (RG)-regular, for $j=1,2,\ldots$ there exists $K_j \in K$ and an open set $U_j \in \delta(K)$ such that $K_j \subseteq E_j \subseteq U_j$ and $\bar{\mu}(F) \in W_j$ whenever $F \in \delta(K)$ and $F \subseteq U_j - K_j$. Write $H = \cup_{j=1}^{\infty} K_j$ and $U = \cup_{j=1}^{\infty} U_j$. Then $H \in K_\sigma$, U is an open set in $\delta(K)$ and $H \subseteq E \subseteq U$. Since $U - H \subseteq \cup_{j=1}^{\infty} (U_j - K_j)$, it follows that (H,U) works for (E,V).

(iii) $\bar{\mu}$ is the only regular Borel measure extending μ. Let $\bar{\mu}_1$, $\bar{\mu}_2$ be two regular Borel measures extending μ. Then $\bar{\mu}_1|\delta(K)$, $\bar{\mu}_2|\delta(K)$ are (RG)-regular σ-additive locally s-bounded extensions of $\mu|\delta(K_o)$. By theorem 4.3, $\bar{\mu}_1|\delta(K) = \bar{\mu}_2|\delta(K)$ and it follows that $\bar{\mu}_1 = \bar{\mu}_2$.

References

1. N.Dinculeanu, Vector Measures, Pergamon Press, New York (1967).

2. N.Dinculeanu and I.Kluvanek, On vector measures, Proc.London Math.Soc., III. Ser.17, 505-512 (1967).

3. N.Dinculeanu and P.W.Lewis, Regularity of Baire measures, Proc.Amer.math.Soc. 26, 92-94 (1970).

4. I.Dobrakov, On submeasures I, Dissertationes math., Warszawa 112, 35 p. (1974).

5. G.Fox and P.Morales, Extension of a compact semigroup-valued set function, submitted to Proc.Royal Soc. Edinburgh, sect.A.

6. G.G.Gould, Integration over vector-valued measures, Proc.London math.Soc. sect.A, III. Ser.15, 193-225 (1965).

7. P.Halmos, Measure Theory, D.Van Nostrand Company Inc., New York (1950).

8. S.S.Khurana, Extension and regularity of group-valued Baire measures, Bull.Acad. Polon.Sci., Sér.Sci.math.astron.phys. 22, 891-895 (1974).

9. M.Sion, A Theory of Semi-group-valued Measures. Lecrure Notes in Mathematics 355. Berlin-Heidelberg-New York: Springer-Verlag V, 140 p. (1973).

10. K.Sundaresan and P.W.Day, Regularity of group-valued Baire and Borel measures, Proc.Amer.math.Soc. 36, 609-612 (1972).

11. H.Weber, Fortsetzung von Massen mit Werten in uniformen Halbgruppen, Arch. der Math. 27, 412-423 (1976).

MARTINGALES OF PETTIS INTEGRABLE FUNCTIONS

Kazimierz Musiał
Wrocław University and Polish Academy of Sciences

1. Introduction

In this paper we shall deal with the properties of Banach space valued martingales of Pettis integrable functions.

Chatterji proved [1] that a Banach space X has the Radon-Nikodym property if and only if each uniformly integrable (or uniformly bounded) martingale of X-valued Bochner integrable functions in L_1-convergent.

We prove that replacing Bochner integrable functions by Pettis integrable ones we get a characterization of Banach spaces possessing the weak Radon-Nikodym property.

We give also a martingale characterization of Banach spaces in which all measures of σ-finite variation have norm relatively compact ranges. Moreover we give a necessary and sufficient condition for the X-valued simple functions to be dense in the space of all X-valued Pettis integrable functions.

2. Definitions and notations.

Throughout X stands for a Banach space (real or complex), B(X) for its unit ball and X^* for the conjugate space. S denotes a non-empty set, Σ is a σ-algebra of subsets of S and μ is a probability measure defined on Σ.

If Ξ is a sub-algebra of Σ and $\nu: \Xi \to X$ is an additive set function, then $\nu(\Xi)$ will denote the set of values of ν and $|\nu|$ will denote its variation.

If Σ_o is a sub-σ-algebra of Σ, then a function $f: S \to X$ is weakly Σ_o-measurable iff the function x^*f is Σ_o-measurable for every $x^* \in X^*$.

A weakly Σ-measurable function is called weakly measurable.

A weakly measurable function $f: S \to X$ is called Pettis integrable on Σ_o iff there exists a set function $\nu: \Sigma_o \to X$ such that

$$x^*\nu(E) = \int_E x^* f\, d\mu$$

for all $x^* \in X^*$ and $E \in \Sigma_o$. In that case we write

$$\nu(E) = \int_E f\, d\mu$$

ν is called the indefinite integral of f on Σ_o. f is called Pettis integrable iff it is Pettis integrable on Σ.

The space of all X-valued Pettis integrable functions on (S,Σ,μ) is denoted by $\mathcal{P}(S,\Sigma,\mu; X)$. $P(S,\Sigma,\mu; X)$, or $P(\mu;X)$ for short, will denote the space obtained by identifying functions which are weakly equivalent, endowed with the following norm:

$$|f| = \sup\{ \int_S |x^*f|\,d\mu : x^* \in B(X^*)\}$$

It is well known that

$$\sup\{ \|\int_E f d\mu\| : E \in \Sigma\}$$

defines an equivalent norm in $P(\mu;X)$.

The following extension property of Pettis integrable functions will be used offen in this paper:

PROPOSITION 1. *Let Σ_o be a sub-σ-algebra of Σ and let $f: S \to X$ be a weakly Σ_o-measurable function which is also Pettis integrable on Σ_o. Then f is Pettis integrable on Σ.*

<u>Proof</u>.[*] Let $\nu : \Sigma_o \to X$ be a measure given by

$$\nu(E) = \int_E f\,d\mu$$

Put for every $F \in \Sigma$

$$\tilde{\nu}(F) = \int_S E(\chi_F | \Sigma_o) d\nu$$

Then for each $x^* \in X^*$ and $F \in \Sigma$

$$x^*\tilde{\nu}(F) = \int_S E(\chi_F | \Sigma_o) dx^*\nu =$$

$$= \int_S E(\chi_F | \Sigma_o) x^* f d\mu = \int_S E(\chi_F \cdot x^* f | \Sigma_o) d\mu = \int_F x^* f d\mu .$$

This proves that $\tilde{\nu}(F) = \int_F f d\mu$.

If Σ_o is a sub-σ-algebra of Σ, f is Pettis integrable and g is Pettis integrable on Σ_o then g is called the <u>conditional expectation</u> of f with respect to Σ_o iff

(a) g is weakly Σ_o-measurable and

(b) $\int_E f d\mu = \int_E g d\mu$ for every $E \in \Sigma_o$. In that case we write $g = E(f | \Sigma_o)$.

Given a directed set $(\Pi, <)$ and a family of σ-algebras $\Sigma_\pi \subset \Sigma$, $\pi \in \Pi$, the system $\{f_\pi, \Sigma_\pi ; \pi \in \Pi\}$ forms an <u>X-valued martingale</u> on (S, Σ, μ) iff the following conditions are satisfied:

(a) If $\pi < \rho$, then $\Sigma_\pi \subset \Sigma_\rho$;

(b) f_π is Pettis integrable on Σ_π ;

(c) If $\pi < \rho$ then $E(f_\rho | \Sigma_\pi) = f_\pi$.

A martingale $\{f_\pi, \Sigma_\pi ; \pi \in \Pi\}$ is <u>convergent</u> in $P(S, \Sigma, \mu ; X)$ if there exists an $f \in P(S, \Sigma, \mu ; X)$ such that

$$\lim_\pi \left| f_\pi - f \right| = 0$$

$\nu_\pi : \Sigma_\pi \to X$ is a measure given by

$$\nu_\pi(E) = \int_E f_\pi d\mu$$

---

[*] The proof presented here is due to Fremlin

for $E \in \Sigma_\pi$, and $\tilde{\nu}_\pi : \Sigma \to X$ is its extension to the whole Σ given by

$$\tilde{\nu}_\pi(E) = \int_E f_\pi d\mu$$

(In virtue of Proposition 1 such an extension does exist).

A set function $\nu : \bigcup_{\pi\in\Pi} \Sigma_\pi \to X$ given by

$$\nu(E) = \lim_\pi \nu_\pi(E)$$

is called <u>the limit set function</u> of $\{f_\pi, \Sigma_\pi\}$.

It is easily seen that the equalities

$$|\nu|(E) = \lim_\pi |\nu_\pi|(E) = \sup_\pi |\nu_\pi|(E)$$

hold for every $E \in \bigcup_{\pi\in\Pi} \Sigma_\pi$.

We say that an X-valued martingale $\{f_\pi, \Sigma_\pi; \pi\in\Pi\}$ is:

<u>Uniformly bounded</u> if there is M such that for every $x^* \in X^*$ and $\pi \in \Pi$ the inequality $|x^* f_\pi| \le M\|x^*\|$ holds μ-a.e.

<u>Variationally bounded</u> if $\sup_\pi |\nu_\pi|(S) < \infty$.

<u>Terminally uniformly continuous</u> if for every $\varepsilon > 0$ there exists $\delta > 0$ and $\pi_0 \in \Pi$ such that if $\pi \ge \pi_0$ and $E \in \Sigma$ is such that $\mu(E) < \delta$, then

$$\left\|\int_E f_\pi d\mu\right\| < \varepsilon$$

<u>Uniformly continuous</u> if

$$\lim_{\mu(E)\to 0} \int_E f_\pi d\mu = 0$$

uniformly with respect to $\pi \in \Pi$

(<u>Terminally</u>) <u>uniformly integrable</u> if it is variationally bounded and (terminally) uniformly continuous.

If $\Pi = N = \{1,2,\dots\}$ then the uniform and the terminal uniform integrability are equivalent.

For a given algebra Σ we shall denote by Π_Σ the set of all collections π, consisting of a finite number of disjoint elements of Σ with strictly positive measure. Π_Σ is a directed set when $\pi' \ge \pi$ is defined to mean that every element of π is, except for a null set, a union of elements of π'.

For each $\pi \in \Pi_\Sigma$ by f_π we shall mean an X-valued simple function given by

$$f_\pi = \sum_{E\in\Pi} \frac{\nu(E)}{\mu(E)} \cdot \chi_E$$

and Σ_π will denote the σ-algebra generated by the sets of the partition π. $\{f_\pi, \Sigma_\pi;\ \pi\in\Pi_\Sigma\}$ is an X-valued martingale.

3. The range of the limit set function.

The main result of this section is the following partial generalization of a result of Uhl ([13], Theorem 2).

PROPOSITION 2. *Let $\{f_\pi, \Sigma_\pi;\ \pi\in\Pi\}$ be a martingale of X-valued functions defined on (S,Σ,μ). Then, the limit set function ν of $\{f_\pi, \Sigma_\pi;\ \pi\in\Pi\}$ is μ-continuous and has norm relatively compact range if and only if the martingale is Cauchy in $P(S,\Sigma,\mu;X)$ and all measures ν_π, $\pi\in\Pi$ have norm relatively compact ranges.*

Proof. Necessity. Without loss of generality we may assume that Σ is the completion of $\sigma(\bigcup_{\pi\in\Pi}\Sigma_\pi)$ with respect to the restriction of μ to $\sigma(\bigcup_{\pi\in\Pi}\Sigma_\pi)$.

Let $\nu: \bigcup_{\pi\in\Pi}\Sigma_\pi \to X$ be the limit set function of the martingale.

Since the compactness of $\nu_\pi(\Sigma_\pi)$, $\pi\in\Pi$, is obvious, it is sufficient to show that the martingale satisfies the Cauchy condition.

An appeal to a result of Hoffmann-Jørgensen ([7], Theorem 9) shows that for each positive ε there exists a function $h_\varepsilon: S\to X$ of the form

$$h_\varepsilon = \sum_{i=1}^{k} x_i \chi_{E_i}$$

where $E_i \in \bigcup_{\pi\in\Pi}\Sigma_\pi$ and $x_i\in X$, such that

$$\sup\{\|\nu(E) - \int_E h_\varepsilon d\mu\| : E\in \bigcup_{\pi\in\Pi}\Sigma_\pi\} < \varepsilon$$

We have to show that

$$\lim_{\pi,\rho} \sup\{\int_S |x^*f_\pi - x^*f_\rho|d\mu:\ x^*\in B(X^*)\} = 0$$

Take $\varepsilon > 0$ and $h = h_{\varepsilon/4}$. By the tringle inequality, it suffices to show that there exists $\pi_o\in\Pi$ such that for all $\pi \geq \pi_o$, we have

$$\sup\{ \int_S |x^*f_\pi - x^*h| d\mu : x^* \in B(X^*)\} \le \varepsilon$$

To do it take $\pi_o \in \Pi$ such that h is Σ_{π_o}-measurable. Then for each $\pi \ge \pi_o$ and $x^* \in B(X^*)$, we have

$$\int_S |x^*f_\pi - x^*h| d\mu \le 4 \sup_{E\in\Sigma_\pi} | \int_E x^*f_\pi d\mu - \int_E x^*h d\mu| \le$$

$$\le 4 \sup_{E\in\Sigma_\pi} \| \int_E f_\pi d\mu - \int_E h d\mu\| \le 4 \sup_{E\in\Sigma_\pi} \| \nu(E) - \int_E h d\mu\| < \varepsilon$$

by the definitions of ν and h.

This completes the proof.

Sufficiency. If $\pi, \rho \in \Pi$, then we have for each $E \in \Sigma$

$$|f_\pi - f_\rho| \ge \|\tilde{\nu}_\pi(E) - \tilde{\nu}_\rho(E)\|$$

Hence, there exists a set function $\nu: \Sigma \to X$ such that

$$\lim_\pi \tilde{\nu}_\pi(E) = \tilde{\nu}(E)$$

uniformly on Σ.

It is clear that $\tilde{\nu}$ is additive and μ-continuous. It follows that ν is a measure. Since ν is the restriction of $\tilde{\nu}$ to $\bigcup_{\pi\in\Pi} \Sigma_\pi$, it is μ-continuous as well.

If $\nu_\pi(\Sigma_\pi)$ is a relatively norm compact set, then it follows from Lemma IV.6.1 of [8], that the range of $\tilde{\nu}_\pi$ is relatively norm compact too.

Together with the uniform convergence of $\tilde{\nu}_\pi$ to $\tilde{\nu}$ this yields the relative norm compact of $\tilde{\nu}(\Sigma)$. But $\nu(\Sigma) \subset \tilde{\nu}(\Sigma)$ and so the range of ν is a norm relatively compact set.

REMARK 1. It follows from the proof of the sufficiency part of Proposition 2 that the following result is true)

If $\{f_\pi\}_{\pi\in\Pi}$ is a directed net of Pettis integrable functions on (S,Σ,μ) which is Cauchy in $P(S,\Sigma,\mu;X)$ and the indefinite integrals of all f_π, $\pi \in \Pi$, have norm relatively compact ranges, then $\nu: \Sigma \to X$ given by

$$\nu(E) = \lim_\pi \int_E f_\pi d\mu$$

<u>is a</u> μ-<u>continuous measure possessing a norm relatively compact range</u>.

COROLLARY 1. <u>Let</u> $f\colon S \to X$ <u>be a Pettis integrable function on</u> (S,Σ,μ) <u>and let</u> $\nu\colon \Sigma \to X$ <u>be its indefinite integral</u>. <u>Moreover, let</u> $\{\Sigma_\pi;\ \pi \in \Pi\}$ <u>be a directed family of sub</u>-σ-<u>algebras of</u> Σ (<u>i.e.</u> $\pi < \rho$ <u>implies</u> $\Sigma_\pi \subset \Sigma_\rho$) <u>generating</u> Σ. <u>If the range of</u> ν <u>is a relatively norm compact set and all</u> $E(f|\Sigma_\pi)$, $\pi \in \Pi$, <u>exist then</u>

$$\lim_{\pi} \left| E(f|\Sigma_\pi) - f \right| = 0$$

<u>Proof</u>. In virtue of Proposition 2 the martingale $\{E(f|\Sigma_\pi),\Sigma_\pi;\pi\in\Pi\}$ is Cauchy in $P(S,\Sigma,\mu;X)$. Applying the Doob-Helms theorem for the scalar valued case we get the required convergence.

4. <u>Simple functions in</u> $P(\mu,X)$.

As a particular case of Corollary 1 we get the following

PROPOSITION 3. <u>Let</u> $f\colon S \to X$ <u>be a Pettis integrable function on</u> (S,Σ,μ) <u>and let</u> $\nu\colon \Sigma \to X$ <u>be its indefinite integral</u>. <u>Then the range of</u> ν <u>is a norm relatively compact set if and only if</u>

$$\lim_{\pi\in\Pi_\Sigma} \left| f_\pi - f \right| = 0.$$

To formulate the next result we introduce the following property of a Banach space X:

<u>DEFINION</u> 1. X has the <u>Pettis Compactness Property with respect to</u> (S,Σ,μ) (μ-<u>PCP</u>) if each X-valued measure which is Pettis differentiable with respect to μ has a norm relatively compact range (it is known, cf. [10], that a Pettis defferentiable measure is always of σ-finite variation). X has the <u>PCP</u> iff X has the μ-PCP with respect to all (S,Σ,μ).

THEOREM 1. X <u>has the Pettis Compactness Property with respect to</u> (S,Σ,μ) <u>if and only if</u> X-<u>valued simple functions are dense in</u> $P(S,\Sigma,\mu;X)$.

<u>Proof</u>. The assertion is a direct consequence of Proposition 2 and Remark 1.

The example of a Pettis integrable function constructed by Fremlin and Talagrand [4] shows that l_∞ does not have the PCP.

On the other hand, Stegall [4] has proved that every X has the PCP with respect to perfect measure spaces. Hence, we get

COROLLARY 2. _If_ (S,Σ,μ) _is perfect then simple functions are dense in_ $P(S,\Sigma,\mu;X)$.

The following spaces have the PCP:

(1) Separable Banach spaces;

(2) AL-spaces (Grothendieck [5], p.308, Ex.13). In particular $C^*[0,1]$ has the PCP.

(3) Separably complementable spaces (X is separably complementable if for every separable spaces $Y \subset X$ there exists a separable Z such that $Y \subset Z \subset X$ and there is a bounded projection of X onto Z). In particular WCG spaces and their subspaces have the PCP.

(4) Banach spaces possessing the WRNP (Proposition 4).

As the next consequence of Remark 1 and Proposition 2 we get the following.

THEOREM 2. _If_ X _has the_ μ_-PCP, then the completion of_ $P(S,\Sigma,\mu;X)$ _is isomorphically isometric to the space of all_ μ_-continuous_ X_-valued measures possessing norm relatively compact ranges endlowed with the semivariation norm_

$$\|\nu\| = \sup\{|x^*\nu|(S): x^* \in B(X^*)\}$$

5. _Banach spaces in which every measure of finite variation has norm relatively compact range._

DEFINITION 2. X has the _Compact Range Property with respect to_ (S,Σ,μ) (X has μ-_CRP_) iff every μ-continuous X-valued measure of finite (or σ-finite) variation has norm relatively compact range. X has the _Compact Range Property_ (_CRP_) iff X has the μ-CRP for all μ.

As a corollary from Proposition 2 we get the following characterization of Banach spaces possessing the μ-CRP.

THEOREM 3. *For a Banach space* X *and a probability space* (S,Σ,μ) *the following conditions are equivalent:*

(i) X *has the* μ-*CRP*;

(ii) *Given any directed set* Π *and a terminally uniformly integrable martingale* $\{f_\pi,\Sigma_\pi;\ \pi\in\Pi\}$ *of* X-*valued functions on* (S,Σ,μ), *then* $\{f_\pi,\Sigma_\pi;\ \pi\in\Pi\}$ *is Cauchy in* $P(S,\Sigma,\mu;X)$;

(iii) *Each uniformly integrable martingale* $\{f_n,\Sigma_n;\ n\in N\}$ *of* X-*valued simple functions on* (S,Σ,μ) *is Cauchy in* $P(S,\Sigma,\mu;X)$;

(iv) *Given any directed set* Π *and a uniformly bounded martingale* $\{f_\pi,\Sigma_\pi;\ \pi\in\Pi\}$ *of* X-*valued functions on* (S,Σ,μ), *then* $\{f_\pi,\Sigma_\pi;\ \pi\in\Pi\}$ *is Cauchy in* $P(S,\Sigma,\mu;X)$;

(v) *Each uniformly bounded martingale* $\{f_n,\Sigma_n;\ n\in N\}$ *of* X-*valued simple functions on* (S,Σ,μ) *is Cauchy in* $P(S,\Sigma,\mu;X)$.

Proof. (i) => (ii) Assume (i) and take a terminally uniformly integrable martingale $\{f_\pi,\Sigma_\pi;\ \pi\in\Pi\}$. Without loss of generality we may assume that $\Sigma=\sigma(\bigcup_{\pi\in\Pi}\Sigma_\pi)$.

Let $\nu\colon \bigcup_{\pi\in\Pi}\Sigma_\pi\to X$ be the limit set function of the martingale. It follows from the terminal uniform integrability of $\{f_\pi,\Sigma_\pi;\ \pi\in\Pi\}$ that ν is a μ-continuous set function of finite variation. Thus, there exists a μ-continuous measure $\tilde{\nu}\colon\Sigma\to X$ of finite variation which is the unique extension of ν to the whole Σ. By assumption $\tilde{\nu}(\Sigma)$ is a conditionally compact set, and so, in virtue of Proposition 2, the martingale $\{f_\pi,\Sigma_\pi;\ \pi\in\Pi\}$ is Cauchy in $P(\mu;X)$.

(ii) => (iv) Let $\{f_\pi,\Sigma_\pi;\ \pi\in\Pi\}$ be a uniformly bounded martingale of X-valued functions on (S,Σ,μ) and let M be such that given $x^*\in X^*$ and $\pi\in\Pi$ the inequality

$$|x^*f_\pi|\le M\,\|x^*\|$$

holds μ-a.e.
Then, we have for every $E\in\Sigma$

$$\|\tilde{\nu}_\pi(E)\|\le\sup\{\int_E|x^*f_\pi|\,d\mu:\ x^*\in B(X^*)\}\le M\mu(E)$$

and so the martingale is uniformly integrable. Applying (ii) we get the desired property of $\{f_\pi,\Sigma_\pi;\ \pi\in\Pi\}$.

The implication (iii) => (v) can be proved in a similar way. Since (v) is a particular case of (iv), to complete the proof it is sufficient to show that (i) is a consequence of (v).

(v) => (i) Let $\nu: \Sigma \to X$ be a μ-continuous measure of finite variation. In virtue of a result of Phillips ([11], Lemma 5.4) we may assume that

$$\|\nu(E)\| \leq M\mu(E)$$

for all $E \in \Sigma$ and a positive M.

In that case $\{f_\pi, \Sigma_\pi;\ \pi \in \Pi_\Sigma\}$ is a uniformly bounded martingale of simple functions.

If $\pi_1 < \pi_2 < \ldots$ then, by assumption, $\{f_{\pi_n}, \Sigma_{\pi_n};\ n \in N\}$ is Cauchy in $P(\mu;X)$, and hence $\{f_\pi, \Sigma_\pi;\ \pi \in \Pi_\Sigma\}$ is Cauchy.

In virtue of Proposition 2 the limit set function of $\{f_\pi, \Sigma_\pi;\ \pi \in \Pi_\Sigma\}$ (i.e. ν) has a norm relatively compact range.

This completes the proof of the Theorem.

REMARK 2. The equivalence of the conditions (i), (v) and (iv) for strongly measurable functions f_π has been proved by Egghe [3].

Applying a theorem of Halmos and von Neumann [6, p.173] we get the following result

THEOREM 4. *If X has the μ-CRP for a non-atomic μ, then X has the CRP.*

In particular we have

COROLLARY *) 3. *If X has the μ-CRP for a separable non-atomic μ, then the completions of $P(S,\Sigma,\mu;X)$ and $P([0,1],\mathcal{L},\lambda;X)$ are isomorphically isometric.*

6. A martingale characterization of the Weak Radon-Nikodym Property.

DEFINITION 3. [10]. X has the Weak Radon-Nikodym Property (WRNP) iff for every complete (S,Σ,μ) and every μ-continuous measure $\nu: \Sigma \to X$ finite variation there exists $f \in P(S,\Sigma,\mu;X)$ such that

---

*) λ denotes the Lebesgue measure on [0,1] and $\mathcal{L}$ is the σ-algebra of λ measurable sets.

$$\nu(E) = \int_E f d\mu$$

for every $E \in \Sigma$.

The following theorem gives a partial solition of Problem 6 posed in [10].

PROPOSITION 4. _If_ X _has the Weak Radon-Nikodym Property then it has also the Compact Range Property_.

Proof. Let (S,Σ,μ) be an arbitrary probability space and let $\nu: \Sigma \to X$ be a μ-continuous measure of finite variation. Without loss of generality we may assume that μ is complete. Moreover, there exists a complete perfect probability space $(\tilde{S},\tilde{\Sigma},\tilde{\mu})$ such that

(a) $S \subset \tilde{S}$ and $\Sigma = \tilde{\Sigma} \cap S$

(b) $\mu(E \cap S) = \tilde{\mu}(E)$ for every $E \in \tilde{\Sigma}$

Define a measure $\tilde{\nu}: \tilde{\Sigma} \to X$ by setting

$$\tilde{\nu}(E) = \nu(E \cap S)$$

for every $E \in \tilde{\Sigma}$.

It is obvious that $\tilde{\nu}$ is $\tilde{\mu}$-continuous and $\tilde{\nu}(\tilde{\Sigma}) = \nu(\Sigma)$.

Since X has the WRNP there exists $f \in P(\tilde{S},\tilde{\Sigma},\tilde{\mu};X)$ such that

$$\tilde{\nu}(E) = \int_E f d\tilde{\mu}$$

for every $E \in \tilde{\Sigma}$.

In virtue of a result of Stegall [4] the range of $\tilde{\nu}$ is a norm relatively compact subset of X.

Thus, the range of ν is norm relatively compact as well.

Now we are in a position to prove the main result of this paper.

THEOREM 5. _For a Banach space_ X _the following conditions are equivalent when holding for all complete probability spaces_ (S,Σ,μ):

(i) X _has the weak Radon-Nikodym property_;

(ii) _Given any directed set_ Π _and a terminally uniformly integrable martingale_ $\{f_\pi,\Sigma_\pi;\ \pi\in\Pi\}$ _of_ X-_valued functions on_ (S,Σ,μ), _then_ $\{f_\pi,\Sigma_\pi;\ \pi\in\Pi\}$ _is convergent in_ $P(S,\Sigma,\mu;X)$;

(iii) _Each uniformly integrable martingale_ $\{f_n,\Sigma_n;\ n\in N\}$ _of simple_ X-_valued functions on_ (S,Σ,μ) _is convergent in_ $P(S,\Sigma,\mu;X)$;

(iv) <u>Given any directed set</u> Π <u>and a uniformly bounded martingale</u> $\{f_\pi, \Sigma_\pi;\ \pi\in\Pi\}$ <u>of</u> X-<u>valued functions on</u> (S,Σ,μ), <u>then</u> $\{f_\pi,\Sigma_\pi,\ \pi\in\Pi\}$ <u>is convergent in</u> $P(S,\Sigma,\mu;X)$;

(v) <u>Each uniformly bounded martingale</u> $\{f_n,\Sigma_n;\ n\in N\}$ <u>of simple</u> X-<u>valued functions on</u> (S,Σ,μ) <u>is convergent in</u> $P(S,\Sigma,\mu;X)$.

<u>Proof</u>. (i) => (ii) Let $\{f_\pi,\Sigma_\pi;\ \pi\in\Pi\}$ be a terminally uniformly integrable martingale of X-valued functions defined on (S,Σ,μ). We have to show the existence of a function $f\in P(S,\Sigma,\mu;X)$ which satisfies the relation

$$\lim_\pi |f_\pi - f| = 0$$

In virtue of Proposition 1 we may assume, without loss of generality, that Σ is the completion of $\sigma(\bigcup_{\pi\in\Pi}\Sigma_\pi)$ with respect to the restriction of μ to $\sigma(\bigcup_{\pi\in\Pi}\Sigma_\pi)$.

Let $\nu: \bigcup_\pi \Sigma_\pi \to X$ be the limit set function of $\{f_\pi,\Sigma_\pi\}$. The terminal uniform integrability yields the μ-continuity of ν and the boundedness of its variation.

Hence, there exists a measure $\tilde{\nu}: \Sigma\to X$ being the unique extension of ν to the whole of Σ. $\tilde{\nu}$ is μ-continuous and of finite variation.

By assumption (i) there is an $f: S\to X$ such that

$$\tilde{\nu}(E) = \int_E f d\mu$$

whenever $E\in\Sigma$.

In virtue of Proposition 4 $\tilde{\nu}(\Sigma)$ is a norm relatively compact set and so it follows from Proposition 2 that $\{f_\pi,\Sigma_\pi;\ \pi\in\Pi\}$ is Cauchy in $P(S,\Sigma,\mu;X)$.

Applying the theorem of Doob and Helms to the scalar valued martingales $\{x^*f_\pi,\Sigma_\pi;\ \pi\in\Pi\}$, $x^*\in X^*$, we get the convergence

$$\lim_\pi |f_\pi - f| = 0$$

The implications (ii) => (iv) and (iii) => (v) are the consequences of the uniform integrability of every uniformly bounded martingale. (iii) is a particular case of (ii) and (v) is a particular case of (iv). Thus, to complete the whole proof it is sufficient to show yet that (v) implies (i).

(v) => (i) Let $\nu: \Sigma\to X$ be a μ-continuous measure of finite variation. Without loss of generality we may assume that

$$\|\nu(E)\| \le M\mu(E)$$

for all $E \in \Sigma$.

In that case $\{f_\pi, \Sigma_\pi;\ \pi \in \Pi_\Sigma\}$ is a uniformly bounded martingale of X-valued simple functions. Moreover, it follows from the assumption that $\{f_\pi, \Sigma_\pi;\ \pi \in \Pi_\Sigma\}$ is Cauchy in $P(\mu;X)$.

It follows from the weak* Radon-Nikodym theorem (cf. [12]) that there exists a weak* measurable function $f\colon S \to X^{**}$ such that

$$x^*\nu(E) = \int_E x^* f\,d\mu$$

for every $E \in \Sigma$ and $x^* \in X^*$.

It is easily seen that the Cauchy condition for the martingale and the Doob and Helms theorem applied to $\{x^*f_\pi, \Sigma_\pi;\ \pi \in \Pi_\Sigma\}$ implies the relation

$$\lim_\pi \sup\{\int_S |x^*(f - f_\pi)|\,d\mu :\ x^* \in B(X^*)\} = 0$$

It follows that there exists a sequence $\pi_1 < \pi_2 < \ldots$ such that

$$\lim_n \int_E x^* f_{\pi_n} d\mu = \int_E x^* f\,d\mu = x^*\nu(E)$$

for every $E \in \Sigma$ and $x^* \in X^*$.

Since by assumption there is $g \in P(S,\Sigma,\mu;X)$ such that

$$\lim_n |f_{\pi_n} - g| = 0$$

we get the equality

$$\nu(E) = \int_E g\,d\mu$$

for every $E \in \Sigma$.

Tjis compleres the proof.

<u>REMARK</u> 3. Observe that in the proof of the implication (v) => (i) we need only the following weaker form of (v):

(v') For every uniformly bounded martingale $\{f_n, \Sigma_n;\ n \in N\}$ of simple X-valued functions on (S,Σ,μ) there exists $f \in P(S,\Sigma,\mu;X)$ such that

$$(*) \qquad \lim_{n\to\infty} \int_E x^* f_n\,d\mu = \int_E x^* f\,d\mu$$

for all $E \in \Sigma$ and $x^{*} \in X^{*}$

In the similar way the other conditions of Theorem 5 may be weakened.

Replacing (*) by the convergence in $L_1(S,\Sigma,\mu)$ we get a characterization of the WRNP obtained by Egghe ([2], Proposition 1). In this case the proof is completely trivial and it is based only on the Doob-Helms theorem.

REMARK 4. It has been observed by Lindenstrauss and Stegall [9] that the space B (constructed in [9]) does not have the RNP. Since it is separable it does not have the WRNP as well. On the other hand B^{**} has the WRNP (Musiał [10]) and so all B-valued measures of finite variation have norm relatively compact ranges, but not all of them are Pettis differentiable.

B is also an example of a space possessing a property (A) (see below) and not possessing the WRNP.

REMARK 5. We can prove that the WRNP depends only on the Lebesgue measure, i.e. the space possessing the WRNP with respect to $([0,1],\mathcal{L},\lambda)$ has the WRNP. The space $([0,1],\mathcal{L},\lambda)$ may be replaced by any perfect non-atomic measure space. We do not know whether it may be replaced by an arbitrary non-atomic measure space.

Theorem 5 can be used to give a short proof of the following fact, posed by the author as Problem 3 in [10] (and solved already by Jeurnink and Janicka):

THEOREM 6. *If X has the WRNP, then X does not contain any isomorphic copy of* c_o.

Proof. We shall prove more, namely that no Banach space possessing the CRP can contain c_o.

Indeed, if r_n, n=1,... are the Rademacher functions on the unit interval

$$f_n(s) = (r_1(s),r_2(s),\dots,r_n(s),0,\dots)$$

and Σ_n is the σ-algebra generated by the intervals of the form $(k2^{-n},(k+1)2^{-n})$, $0 \le k \le 2^n-1$, then $\{f_n,\Sigma_n\}_{n\in N}$ is a c_o-valued martingale. Since

$$|f_n(s)| = 1$$

the martingale is uniformly bounded.
But

$$\|f_n - f_{n+1}\| = \sup\{\int_0^1 |x^*(0,\ldots,0,r_{n+1}(s),0,\ldots)|ds:\ x^* \in B(l_1)\} =$$

$$= \sup\{|\alpha| \int_0^1 |r_{n+1}(s)ds:\ |\alpha| \le 1\} = 1$$

This shows that $\{f_n,\Sigma_n\}_{n=1}^{\infty}$ does not satisfy the Cauchy condition in $P([0,1],\mathcal{L},\lambda;c_o)$.

If X contains an isomorphic copy of c_o and $U: c_o \to X$ is an isomorphic embedding, then $\{Uf_n,\Sigma_n\}_{n=1}^{\infty}$ is an X-valued martingale which does not satisfy the Cauchy condition in $P([0,1],\mathcal{L},\lambda;X)$.

In virtue of Theorem 3. X does not have the CRP (and hence also the WRNP).

7. <u>Generalizations</u>.

6.1. The following two properties (considered by Janicka) generalize the notion of the WRNP:

(A) Given a complete (S,Σ,μ) and a μ-continuous measure $\nu: \Sigma \to X$ of finite variation there exists a function $f: S \to X^{**}$ such that

$$z\nu(E) = \int_E zfd\mu$$

for every $z \in X^{***}$ and $E \in \Sigma$;

(B) There exists a Banach space $Y \supset X$ such that for every complete (S,Σ,μ) and every μ-continuous measure $\nu: \Sigma \to X$ of finite varaiation there exists a function $f: S \to Y$ such that

$$y^*\nu(E) = \int_E y^*fd\mu$$

for every $y^* \in Y^*$ and $E \in \Sigma$.

It has been proved by Drewnowski that permutations of the quantifiers in (B) give nothing new.

It is not known whether (A) and (B) define the same property of X.

Replacing in Theorem 5 the convergence of X-valued martingales in $\mathbf{P}(S,\Sigma,P;X)$ by the convergence in $\mathbf{P}(S,\Sigma,P;X^{**})$ or in $\mathbf{P}(S,\Sigma,P;Y)$ respectively, we get martingale characterizations of the properties (A) and (B).

6.2. Theorem 3 and Theorem 5 hold also for amarts (only the proof of Proposition 2 needs an essential change, cf. Uhl [13])

References

[1] S.D. Chatterji, Martingale convergence and the Radon-Nikodym theorem in Banach spaces, Math. Scand. 22(1968), 21-41.

[2] L. Egghe, On (WRNP) in Banach spaces and Radon-Nikodym properties in locally convex spaces, (preprint).

[3] L. Egghe, On Banach spaces in which every uniformly bounded amart is Pettis Cauchy, (preprint).

[4] D.H. Fremlin and M. Talagrand, A decomposition theorem for additive set functions with applications to Pettis integrals and ergodic means (preprint).

[5] A. Grothendieck, Espaces vectoriels topologiques, Sao Paulo (1964).

[6] P. Halmos, Measure theory. New York (1950).

[7] J. Hoffmann-Jørgensen, Vector measures, Math. Scand. 28(1971), 5-32.

[8] I. Kluvanek, G. Knowles, Vector Measures and Control Systems, North-Holland Mathematics Studies, vol.20(1976).

[9] J. Lindenstrauss and C. Stegall, Examples of separable spaces which do not contain l_1 and whose duals are non-separable. Studia Math. LIV(1975), 81-105.

[10] K. Musiał, The Weak Radon-Nikodym Property in Banach spaces, Studia Math. 64(1978), 151-174.

[11] R.S. Phillips, On weakly compact subsets of a Banach space, Amer. J. Math. 65(1943), 108-136.

[12] V.I. Rybakov, On vector measures (in Russian), Izv. Vyss. Ucebn. Zav. Matiematika 79(1968), 92-101.

[13] J.J. Uhl, Jr., Pettis Mean Convergence of Vector-valued Asymptotic Martingales, Z. Wahr. verw. Geb. 37(1977), 291-295.

INTEGRATION OF FUNCTIONS WITH VALUES IN COMPLETE SEMI-VECTOR SPACE

by

Endre Pap

Institute of Mathematics,Novi Sad,Yugoslavia

Introduction

When we start to do some theory of measure or integrals (or more general from functional analysis) for semigroup valued functions, we can very simply sketch the problem which often arise in translations of theorems and its proofs: u - v. But we work in a semigroup and we have not always an inverse element.But in most of the situations we have $|u - v|$ instead of $u - v$, or sometimes $|u| - |v|$ (where $|.|$ is a norm or a quasi-norm). Then we can take $|u - v|$ as a metric $d(u,v)$ or $|u|$ as a function $f(u)$ with some special properties - E.Pap |2| (M.Sion |6| has worked with the uniformity).We make a hypothesis,that the Classical Analysis with set,function,limit, continuity,series,integral,etc. and on other hand the Functional Analysis with metric,functional,measure,distribution,operator,etc. can be incorporated in a semigroup enriched with some additional structures.

In this paper we are going to use such metric approach and some ideas of J.Mikusiński |1| obtained for the Bochner integral.First,we are going to give an axiomatic treatmant of integral of semigroup valued functions.Then we are going to construct a model for such axiomatic theory in a semi-vector space - E.Pap |3| . On the end of the paper we are going to give a previous general result on functions with the values in a semigroup - E.Pap |5| .

1. The ZED-integral

Let X be a commutative semigroup with neutral element 0 and with a metric d, such that

$$(d_+) \qquad d(x+y,x'+y') \leq d(x,x') + d(y,y')$$

holds for every $x,x',y,y' \in X$.We assume that (X,d) is a complete metric space.Let K be an arbitrary set and let U be a family of functions from K to X. We assume that a function $\int$ (called *integral*) with values in X (or R) is defined on U.We assume that U, $\int$ satisfy the following axioms:

(Z) $0 \in U$ and $\int 0 = 0$,

(D) If $f,g \in U$, then $d(f,g) \in U$ and

$$d\left(\int f, \int g\right) \leqslant \int d(f,g) ,$$

(E) If $f_n \in U$ $n=1,2,\ldots$, $\sum_n \int d(f_n,0) < \infty$, and the equality $f(x) = f_1(x) + f_2(x) + \ldots$ holds at every point $x \in X$ at which $\sum_n d(f_n(x),0) < \infty$,then $f \in U$ and $\int f = \int f_1 + \int f_2 + \ldots$,we write $f \simeq f_1+f_2+\ldots$,

where $d(f,g)(t) = d(f(t),g(t))$.

The integral $\int$ which satisfies axioms Z (zero property), D (distance property) and E (expansion property) we call ZED-integral. We consider the theory in two interpretation simultaneosly for X-valued and for real valued functions.In the real case,d is the usuly distance.We can extend the integral $\int$ on $-f$ for group valued function (so also for real valued function) in the following way

(I) If $f \in U$, then $-f \in U$ and $\int(-f) = -\int f$.

In the following the integrals will be always ZED-integrals for X--valued functions and IED-integrals for real valued functions.

<u>Remark 1.1.</u> In special cases the ZED-integral reduces on J.Mikusinski´s HEM-integral |1| where

(H) If $f \in U$ and $\lambda \in R$,then $\lambda f \in U$ and $\int \lambda f = \lambda \int f$,

and the Bochner integral and the Lebesgue integral.

2. Some properties of the ZED-integral

We obtain from E in special case

(A) If $f \in U$ and $g \in U$,then $f+g \in U$, and $\int(f+g) = \int f + \int g$.

So in the case of complete semigroup with cancelation, if $0 \in U$, then the axiom Z follows from A. By D it follows that $\int d(f,g) = 0$ implies $\int f = \int g$. If $f \geqslant 0$,then $d(f,0) = f$ and by D it follows that $f \geq 0$ implies $\int f \geq 0$.

A function $h \in U$ is called a <u>null function</u>, if $\int d(h,0) = 0$. Two arbitrary functions f and g from U are called <u>equivalent,</u> if

$$\int d(f,g) = 0.$$

We than write $f \sim g$. By Z and D if $f \sim g$, then $\int f = \int g$. If $f_1 \sim f_2$ and $g_1 \sim g_2$, then $f_1+g_1 \sim f_2+g_2$. $\sim$ is a relation of equivalence. We denote by $\tilde{f}$ the class of all functions equivalent to f and by $\tilde{U}$ the set of all classes $\tilde{f}$ such that $f \in U$. Let us introduce the metric in $\tilde{U}$

$$d(\tilde{f},\tilde{g}) = \int d(f,g) \quad ,$$

which satisfies also (d_+). We say that the sequence of functions f_n converges in metric to f, $f_n \to f$ i.m., if

$$\int d(f_n,f) \to 0 \quad \text{as} \quad n \to \infty \quad .$$

By (d_+) we obtain that if $f_n \to f$ i.m. and $g_n \to g$ i.m. then $f_n+g_n \to f+g$ i.m. . By D, if $f_n \to f$ i.m. , then $\int f_n \to \int f$. We introduce on the usuly way the notions of null set and convergence almost everywhere - |1| . Theorems 7.1 , 7.2 and 7.4 with proofs hold also in our case as in |1| (we take only $d(.,0)$ instead of $|.|$).

<u>Theorem 2.1.</u> If $\int d(f_1,0) + \int d(f_2,0) + \ldots < \infty$ $(f_n \in U)$, then

$d(f_1(x),0) + d(f_2(x),0) + \ldots$ converges almost everywhere.

<u>Proof.</u> Let Z be the set of the points at which $d(f_1(x),0) + d(f_2(x),0) + \ldots$ does not converges. Let K_z be the characteristice function of Z. Then we have

$$K_z \simeq d(f_1,0) - d(f_1,0) + d(f_2,0) - d(f_2,0) + \ldots \quad .$$

Hence $K_z = 0$.

<u>Theorem 2.2</u>. If $f \simeq f_1+f_2+\ldots$ $(f_n \in U)$, then $f = f_1+f_2+\ldots$ i.m. and $f = f_1+f_2+\ldots$ a.e. .

<u>Proof</u>. We have

$$d(f_{n+1}+f_{n+2}+\ldots,0) \geq d(f,f_1+\ldots+f_n)$$

at any point at which $d(f_1(x),0) + d(f_2(x),0) + \ldots < \infty$.

Since $f \simeq f_1+f_2+\ldots$, for a given $\varepsilon > 0$, we can choose an index n_o such that

$$\int d(f_{n+1},0) + \int d(f_{n+2},0) + \ldots < \varepsilon \quad \text{for } n > n_o.$$

Hence and by analog theorem 7.4 from |1|

$$\int d(f,f_1+\ldots+f_n) \leq \int d(f_{n+1}+f_{n+2}+\ldots ,0) \leq$$

$\leqslant \int d(f_{n+1},0) + \int d(f_{n+2},0) + \ldots < \varepsilon$ for $n > n_o$. The second part follows by theorem 2.1.

Theorem 2.3. If $f \simeq f_1+f_2+\ldots$ ($f_n \in U$) and $g \simeq f_1+f_2+\ldots$,then $f \sim g$.

Proof. By theorem 2.2 $f = f_1+f_2+\ldots$ i.m. and $g=f_1+f_2+\ldots$ i.m. Hence by the inequality

$$\int d(f,g) \leqslant \int d(f,f_1+f_2+\ldots) + \int d(g,f_1+f_2+\ldots)$$

we obtain the assertion $f \sim g$.

Theorem 2.4. The space $\tilde{U}$ is complete.

The proof is analog to the proof of theorem 9.1 from |1| -p.33. We use only the theorem 2.2 instead of theorem 8.6 from |1| .

Riesz theorem (theorem 9.3, |1| ,p.33) can be also adapted for our case.The Lebesgue dominated convergence theorem (also the generalized theorem 10,|1| ,p.36) hold for ZED-integral which are HED-integral for real valued functions.

We have obtained in the paper |4| Fubini theorem for a double integral and with an additional property on metric

$$(d_-) \qquad |\, d(x,x') - d(y,y')| \leqslant d(x+y,x'+y')$$

we have obtained the double ZED-integral.

3. A model for the HED-integral

Now we shall construct a nontrivial model for HED-integral with the values in a semi-vector space and $K = R^q$.

Let $(X,+)$ be a commutative semigroup with a neutral element 0. X is a semi-vector space ,if for every $x \in X$ and every $\lambda \in R_+$ the product λx of λ with x is defined as an element of X,and for all $x,y \in X$, $\lambda, \eta \in R_+$ we have: 1. either $\lambda(x+y) = \lambda x + \lambda y$ or $\lambda(x+y) = \lambda x+y$ on the whole X, 2. $\lambda(\eta x) = (\lambda\eta)x$, 3. $1x = x$, 4. $0x = 0$.

Remark 3.1. We had defined in the paper |3| the notion of semi-vector space less general as in this paper.

We take a metric d on X with the properties (d_+), (d_-) and

$$(d_h) \qquad d(\lambda x,\lambda y) = \lambda d(x,y) \quad .$$

In the following we assume that (X,d) is a complete metric space.It is obvious that a metric which satisfies (d_+) and (d_-) is also translation invariant.Hence in X must be hold the cancelation low.

Now we start to construct our model of HED-integral. By a <u>semi--vector valued step function</u> f we mean a function which can be represented in the form

(3.1) $$f = \lambda_1 f_1 + \ldots + \lambda_n f_n \quad ,$$

where $f_1,\ldots,f_n$ are brick functions (the characteristic function of a brick $a \leqslant x < b$) and $\lambda_i \in X$ $(i=1,\ldots,n)$. It is easily seen that the set of all step functions is a semi-vector space. We always can choose the brick functions $f_1,\ldots,f_n$ in the representation (3.1) so that their carriers are disjoint.By the integral $\int$ of the step function (3.1) we mean

(3.2) $$\int f = \lambda_1 \int f_1 + \ldots + \lambda_n \int f_n \quad ,$$

where the integral $\int f_i$ of a brick function f_i isthe volume(in general sense) of the brick.

<u>Theorem 3.1</u>. The integral, defined with the formula (3.2), is a HAD--integral.The integral is indepedent of the representation (3.1).

<u>Proof</u>. We prove only the property D. Let $f = \lambda_1 f_1+\ldots+\lambda_n f_n$ and $g = k_1 g_1+\ldots+k_m g_m$ be step functions. We can choose a finite system of brick functions f_{ij}, f_i', g_j' such that

$$d(f,g) = \sum_{i,j} d(\lambda_i,k_j) f_{ij} + \sum_i d(\lambda_i,0) f_i' + \sum_j d(0,k_j) g_j' \quad .$$

So $d(f,g)$ is also a step function. Hence and by (d_+) and (d_h) we obtain the property D. Using D we obtain that the integral of step functions is independent of the representation (3.1).

A function f from R^q to a given complete semi-vector space X is <u>integrable</u> if

1. $d(\lambda_1,0)\int f_1 + d(\lambda_2,0)\int f_2 + \ldots < \infty$

and

2. $f(x) = \lambda_1 f_1(x) + \lambda_2 f_2(x) + \ldots$ at those points x

at which $\sum_n d(\lambda_n f_n(x),0) < \infty$,

where f_i are brick functions and $\lambda_i \in X$.Then we shall write

(3.3) $$f \simeq \lambda_1 f_1 + \lambda_2 f_2 + \ldots \quad .$$

The set of all integrable functions is a semi-vector space.We define for integrable function f the integral $\int f$ as

(3.4) $$\int f = \lambda_1 \int f_1 + \lambda_2 \int f_2 + \ldots \quad .$$

<u>Theorem 3.2</u>. The distance $d(f,g)$ of two integrable function is a Lebesgue integrable function and satisfies D. The integral $\int$, defined by (3.4),is independent of representation (3.3).

Proof. First, we shall prove that the distance $d(f,g)$ is Lebesgue integrable for $f \simeq \lambda_1 f_1 + \lambda_2 f_2 + \dots$ and $g \simeq k_1 g_1 + k_2 g_2 + \dots$. We denote with Z the set of all points where the equalities $f = \lambda_1 f_1 + \lambda_2 f_2 + \dots$ and $g = k_1 g_1 + k_2 g_2 + \dots$ hold (points at which $\Sigma\ d(f_n(x),0) < \infty$ and $\Sigma\ d(g_n(x),0) < \infty$). We have $d(f,g) = \lim_{n\to\infty} d(s_n,u_n)$ for $x \in Z$ where $s_n = \lambda_1 f_1 + \dots + \lambda_n f_n$ and $u_n = k_1 g_1 + \dots + k_n g_n$. We can write

$$(3.5) \qquad d(f,g) = d(s_1,u_1) + (\, d(s_2,u_2) - d(s_1,u_1)\,) + \dots$$

on Z. By the properties of step functions, we can write

$$(3.6) \qquad \begin{aligned} & d(s_1,u_1) = t_1 h_1 \\ & d(s_{n+1},u_{n+1}) - d(s_n,u_n) = t_{p_n+1} h_{p_n+1} + \dots + t_{p_{n+1}} h_{p_{n+1}} \end{aligned}$$

for $n = 1,2,\dots$, where the brick functions h_i are with disjoint carriers and $t_i \in R$ $(i=1,2,\dots)$ and $p_1 = 1$. By (d_+) and (d_-) we obtain the following inequalities

$$|d(s_{n+1},u_{n+1}) - d(s_n,u_n)| \leqslant d(s_n + u_n + \lambda_{n+1} f_{n+1}, u_n + s_n + k_{n+1} g_{n+1}) \leqslant$$

$$\leqslant d(\lambda_{n+1} f_{n+1}, k_{n+1} g_{n+1}).$$

Hence and by (3.6) we obtain

$$d(t_{p_n+1},0) h_{p_n+1} + \dots + d(t_{p_{n+1}},0) h_{p_{n+1}} \leqslant d(\lambda_{n+1},0) f_{n+1} + d(k_{n+1},0) g_{n+1}$$

($n=1,2,\dots$). Adding all these inequalities and also $d(t_1,0) h_1 \leqslant d(\lambda_1,0) f_1 + d(k_1,0) g_1$ we get

$$(3.7) \qquad d(t_1,0) h_1 + d(t_2,0) h_2 + \dots \leqslant (\, d(\lambda_1,0) f_1 + \dots\,) + (\, d(k_1,0) g_1 + \dots\,) .$$

If we first integrate all these inequalities, and then sum them up we get

$$(3.8) \qquad d(\lambda_1,0) \int h_1 + \dots \leqslant d(\lambda_1,0) \int f_1 + \dots + d(k_1,0) \int g_1 + \dots .$$

Hence there exists a Lebesgue integrable function h such that

$$h \simeq t_1 h_1 + t_2 h_2 + \dots .$$

The series on the right side, by (3.7) converges absolutely on Z. By (3.5) and (3.6) its sum on Z is $d(f,g)$. So we have $d(f,g) = h$ on Z. We can write

$$d(f,g) \simeq t_1 h_1 + d(\lambda_1 f_1, k_1 g_1) - d(\lambda_1 f_1, k_1 g_1) + t_2 h_2 + \dots .$$

Thus d(f,g) is a Lebesgue integrable function (|1|). So we have

$$(3.9)\qquad \int d(f,g) = t_1 \int h_1 + t_2 \int h_2 + \dots \quad .$$

Hence by (3.8)

$$\int d(f,g) \leqslant \quad (d(\lambda_1,0)\int f_1 + \dots \quad) + (d(k_1,0)\int g_1 + \dots \quad) \quad .$$

We add (3.6) for n=1,...,m-1

$$d(s_m,u_m) = t_1 h_1 + \dots + t_{p_m} h_{p_m} \quad .$$

By theorem 3.1 we have

$$d(\lambda_1 \int f_1 + \dots + \lambda_m \int f_m, k_1 \int g_1 + \dots + k_m \int g_m) \leqslant$$

$$\leqslant t_1 \int h_1 + \dots + t_{p_m} \int h_{p_m} \quad .$$

Letting $m \to \infty$ we obtain by (3.9) D. By D and Z we obtain the indepedence of the integral from the function representation.

Theorem 3.3. The integral defined by (3.4) satisfies E.

The proof is same as the proof of the theorem 5.1 from |1|. We use only the lemma 3.4 and composition theorem for series from |4|.

Lemma 3.4. Given any integrable function f and any number $\varepsilon > 0$, there exists an expansion $f \simeq \lambda_1 f_1 + \lambda_2 f_2 + \dots$ such that

$$d(\lambda_1,0)\int f_1 + d(\lambda_2,0)\int f_2 + \dots < \int d(f,0) + \varepsilon \quad .$$

The proof of this lemma is analog to the proof of theorem 3.2 from |1| with elimination of considering the function f - s instead of which we considere d(f,s) and we use the theorem3.2.

So we have constructed a model for HED-integral,which in special cases reduces on HEM-integral,Bochner and Lebesgue integrals.

4. An other investigation on the functions with values in a semigroup

We conclude this paper with a important theorem for semigroup valued functions,obtained in the paper |5| . Namely,this theorem have many consequences and applications in Measure Theory and Functional Analysis.

Let (X,+) be a additive semigroup with the neutral element 0 (we do not assume commutativity as well as continuity of the semigroup operation) and the convergence $H: N^X \to 2^X$ satisfies the following conditions

(S) If $x_n = x$ for each n=1,2,... , then $x_n \to x$ (H) ,

(F) If $x_n \to x$ (H),then $x_{p_n} \to x$ (H),

(K) If $x_n \to 0$ (H), then there exists a subsequence y_n of x_n such that $\sum_{k=1}^{n} y_k \to y$ (H) for some $y \in X$.

Let $(S,\oplus)$ be a semigroup with a family F of functionals f which have the following property

$$(F_2) \qquad f(x \oplus y) \geqslant |f(x) - f(y)|$$

for each $x,y \in S$ - |2| . Every functional f induces the pseudometric $d_f(x,y) = |f(x) - f(y)|$. The function $g: X \to S$ is said to be <u>additive</u> iff for each $x,y \in X$ we have $g(x + y) = g(x) \oplus g(y)$. The function $g: X \to S$ is said to be <u>F-continuous</u> iff for each sequence $x_n \to x$ (H) we have

$$d_f(g(x_n),g(x)) \to 0 \quad \text{as } n \to \infty \quad \text{for each } f \in F.$$

<u>Theorem</u> (Main Theorem - |5|). Let $g_i: X \to S$ $(i \in N)$ be additive, F-continuous, $f(g_i(0)) = 0$ $(i \in N)$ and let

$$\lim_{i \to \infty} f(g_i(x)) = 0 \quad \text{for each fixed } x \in X \text{ and}$$

and every $f \in F$. If $x_j \to 0$ (H), then

$$\lim_{i,j \to \infty} f(g_i(x_j)) = 0 \qquad \text{for each } f \in F.$$

In special cases, we obtain many applications of this theorem, for example: the Bourbaki theorem on joint continuity, the Orlicz--Pettis theorem on convergence, the Kernel theorem for Köthe´s co--echelon spaces, the Banach-Steinhaus theorem, Nikodym theorem on pointwise convergent sequences of measures, the Nikodym theorem on the uniform boundedness of measures.

References

|1| J.Mikusiński, The Bochner Integral, Birkhäuser Verlag, 1978.

|2| E.Pap, A generalization of the diagonal theorem on a block--matrix, Mat.ves., 11(26) (1974), 66-71.

|3| E.Pap, General Spaces of Sequences and Boundedness Theorem, Math.Balk. 5 (1975), 216-221.

|4| E.Pap, On the ZED-integral, Zbornik PMF u N.Sadu (to appear)

|5| E.Pap, On the Continuous Functions with Values in a Semigroup Topol.Coll. 1978 - Budapest (to appear)

|6| M.Sion, A Theory of Semigroup Valued Measures, Springer-Verlag, 1973.

THE STOCHASTIC INTEGRAL AS A VECTOR MEASURE.

By Klaus Bichteler

To consider the stochastic integral $\int XdZ$ as a vector measure is, of course, an old hat. ITO took this point of view in 1944 [I1] and started a success story. However, ITO's integrators Z were Wiener processes, and during the search for generalizations -- which produced the technical notion of a semimartingale 'in the old sense' [M3] -- that point of view faded a bit. But from Professor DELLACHERIE's new definition of a semimartingale, the one he gave in the previous talk, we see that the vector measure point of view is in full focus again. (It should perhaps be noted that several authors had concentrated all the while on this aspect, most notably METIVIER and PELLAUMAIL [M2,MP,P1]; see also KUSSMAUL [K1]).

The purpose of this talk is to explain how some of the techniques of 'abstract' vector measure theory can yet improve the theory of stochastic integration -- and not only by simplifying this rather technical subject. Daniell's underrated [H1] way of going about measure theory, in particular, furnishes valuable new a.s. and mean convergence results for stochastic differential equations; they are due to the numerical control of the size of both integrand and integrator that his method provides.

A vector measure is a continuous linear map $\mu: \mathcal{R} \to L$ from a space $\mathcal{R}$ of the type $C_{00}(K)$, K locally compact, to some topological vector space L. In the situation at hand, $\mathcal{R}$ is the vector lattice of step functions over the ring $\mathcal{A}$ generated by stochastic intervals of the form (S,T], where S and T are bounded stopping times taking only finitely many values. The processes in $\mathcal{R}$ are termed 'elementary integrands'. For any cadlag adapted process Z one defines $dZ((S,T]) = Z_T - Z_S$ and extends this by linearity. The result is a linear map, or vector measure,

$$dZ: \mathcal{R} \to L^0(\Omega,\mathcal{F},P) .$$

Following the spirit of [MP,P1,K1] and of Professor DELLACHERIE's talk, we set forth the

Definition. Let $0 \leq p < \infty$. The cadlag process Z is an L^p-integrator if dZ is continuous as a map into L^p, on each of the sup-normed spaces

$$\mathcal{R}_t = \{X \in \mathcal{R};\ X = 0 \text{ after } t\}\ .$$

An L^0-integrator is thus a semimartingale in Professor DELLACHERIE's new definition. I shall sketch below a different proof of the result of DELLACHERIE, MOKOBODSKI and LETTA [M4] that L^0-integrators are semimartingales 'in the old sense'. From the view taken in this talk this is a peripheral point, though, inasmuch as the integration theory of a semimartingale in the old sense is rather more technical and laborious than that of an L^p-integrator, $0 \leq p < \infty$. Here is how the latter works.

With ρ_p the usual metric on L^p, $0 \leq p < \infty$, set

$$G_p(H) = \sup\{\rho_p(dZ(X)):\ X \in \mathcal{R},\ |X| \leq H\}$$

when H is the supremum of a sequence of elementary integrands, and

$$G_p(F) = G_p^Z(F) = \inf\{G_p(H):\ F \leq H \text{ as above}\}$$

for arbitrary processes F. $\mathcal{L}^1(dZ;p)$ is the closure of $\mathcal{R}$ in G_p^Z-mean, and the integral

$$\int \cdot\, dZ:\ \mathcal{L}^1(dZ;p) \to L^p$$

is the extension of dZ by continuity. The DCT holds since L^p is Σ-complete [T1], which in turn is a consequence of Khintchine's inequality [Z1,Y1,B2]. Function metrics that entail the DCT have been introduced under the name 'upper gauges' in a previous Oberwolfach Conference on Vector Measures ([B5]; also [B3]). Other upper gauges dominating dZ have been studied by MEYER [M5], YOR [Y1-3], EMERY [E1], PELLAUMAIL [P1], METIVIER [M2], and others, but to my knowledge only in the range $1 \leq p < \infty$.

From the DCT the usual results of Stochastic Integration follow rather easily, including ITO's formula, local time, and the existence of solutions of stochastic differential equations. In particular, an L^0-integrator Z need not be split locally as (finite-variation process) + (square-integrable martingale) to obtain most of these results, and this saves labor.

I want to explain now how careful management of the upper gauges G_p^Z permits to show that certain stochastic integrals and solutions to stochastic differential equations can be evaluated path by path.

To start with, we assume that Z is an L^2-integrator and that X is a cadlag process with left-continuous version X_-, whose maximal function

$$X_t^* = \sup\{ |X_s| : s \leq t\}$$

is in $L^2(P)$ at the instant t Then $X_-\cdot 1_{[0,t]}$ is in $\mathcal{L}^1(dZ;1)$, and

$$Y_t = \int_0^t X_- dZ$$

exists. For any n define a sequence $T_0 = 0$, $T_{k+1} = t \wedge \inf\{s > T_k: |X_s - X_{T_k}| > 2^{-n}\}$ of stopping times and cadlag processes $X^n = \Sigma X_{T_k} \cdot 1_{[T_k, T_{k+1})}$,

$$Y_t^n = \int_0^t X_-^n dZ = \Sigma X_{T^k}(Z_{T_{k+1}\wedge t} - Z_{T_k \wedge t}) .$$

Note that $|X_- - X_-^n| \leq 2^{-n}$ on $[0,t]$, and consequently $G_2^Z[(X_- - X_-^n)\cdot 1_{[0,t]}) \leq 2^{-n}\gamma_t^2[Z]$. Here

$$\gamma_t^p[Z] = (G_p^Z([0,t]))^{1 \vee 1/p} \quad , \quad 0 < p < \infty$$

is a measure for the size of Z as an L^p-integrator. Using a maximal inequality

(*) $$\|Z_t^*\|_{L^p} \leq C_p \gamma_t^p[Z] \qquad 0 < p < \infty$$

we arrive at

$$\| |Y - Y^n|_t^* \|_{L^1} \leq 2^{-n} C_1 \gamma_t^1[Z]$$

and $Y^n \to Y$ a.s. uniformly on $[0,t]$. This argument actually provides an algorithm for the pathwise computation of an approximation $Y_.^n(\omega)$ of $Y.(\omega)$ from the signals $X.(\omega)$ and $Z.(\omega)$, together with an error estimate:

$$P[|Y-Y^n|_t^* > \varepsilon] \leq 3\cdot 2^{-n}\gamma_t^1[Z]/\varepsilon \ .$$

Even in the classical case that Z is standard Brownian motion and $\gamma_t^1[Z] \leq t$ this seems to have escaped notice so far.

Even if Z is merely an L^0-integrator one can establish the

<u>Theorem 0.</u> $Y_t^n \to Y_t$ a.s. and

$$\Sigma|Y^{n+1}-Y^n|_t^* \to 0 \ \text{ a.s. for all } \ t > 0 \ . \tag{0}$$

To see this observe that the hypotheses and conclusion do not change if P is replaced by an equivalent measure P' and apply the following lemma with $p = 0$ and $q = 2$.

<u>Lemma</u>. Let $0 \leq p < q \leq 2$, and let Z be an L^p-integrator. For every $t > 0$ there exists a probability P' equivalent with P so that the stopped process Z^t is an $L^q(P')$-integrator. If $0 < p < q \leq 2$ one has, moreover, control over the size of Z^t as an $L^q(P')$-integrator: With universal constant C_{pq}

$$\gamma_t^q[Z;P'] \leq C_{pq}\gamma_t^p[Z;P]$$

$$\|dP/dP'\|_{L^{p/(q-p)}(P)} \leq C_{pq}$$

$$\|dP'/dP\|_{L^\infty} \leq C_{pq} \ .$$

Proof. MAUREY [M1] and ROSENTHAL [R1] have a generalization of GROTHENDIECK'S Fundamental Theorem, to this effect: Let $\zeta\colon C \to L^p(P)$ be a continuous linear map from a space C having the finite-dimensional subspace structure of a space $L^\infty(\nu)$, and let $0 \leq p < q \leq 2$. Then ζ factors as $\zeta(f) = \Phi\cdot V(f)$, $f \in C$, where $V\colon C \to L^q(P)$ is continuous and $0 < \Phi \in L^{pq/(q-p)}$. We apply this with $\zeta = dZ$ and $C = \mathcal{R}_t$ and set $P' = c\Phi^{-q}P$ where c is chosen so that $P'(\Omega) = 1$. The control comes from the fact that for $0 < p < q \leq 2$

$$\|V\|_{L(C,L^q(P))} \cdot \|\Phi\|_{L^{pq/(q-p)}} \le m_{pq} \|\zeta\|_{L(C,L^p(P))} ,$$

with universal constants m_{pq}.

Let Z be an $L^0(P)$-integrator. Professor DELLACHERIE has pointed out to me during this conference that his arguments in [D2] together with the lemma above show that there exists a measure $P' \sim P$ that works for all times t: Z is an $L^2(P')$-integrator. As such it is a P'-semimartingale in the old sense [MP] and thus a P-semimartingale in the old sense [folklore]. The converse, that a semimartingale in the old sense is an L^0-integrator, is quite elementary.

The controls of the lemma permit one to extend several classical inequalities to the range $0 < p < 1$. For instance, $\|Z_t^*\|_{L^p} \le C_p \gamma_t^p[Z]$. One can use them to reduce L^p-estimates of the growth of the solution of a stochastic differential equation

$$X = C + F(X_-) * Z \tag{1}$$

to the L^2-case without losing valuable information.

Equation (1) is usually attacked with Picard's method [DD, DM, P3-6, E1, MP]. Keeping track of the G_p^Z's allows one to establish that the iterates $X^{n+1} = C + F(X_-^n) * Z$ converge pathwise a.s. to the solution. In fact, an argument quite similar to that leading to theorem 0 yields the first part of

Theorem 1 Suppose F is Lipschitz with Lipschitz constant K and the jumps of Z are a.s. uniformly bounded: $L = \| |X-X_-|_\infty^* \|_{L^\infty} < \infty$. Then (1) has a unique solution X, which is a.s. uniformly on bounded intervals the limit of the iterates; in fact,

$$\Sigma |X^{n+1} - X^n|_T^* < \infty$$

for any a.s. finite time T. The solution X satisfies the following *a priori* growth condition: Let $0 < p,q,r < \infty$. For any a.s. finite time U and every $\varepsilon > 0$ there is a stopping time $T \le U$ with $P[T < U] < \varepsilon$ such that

$$(2) \qquad \|X_T^*\|_{L^q} \leq C + a_{pq}(C + F(0)/K)\exp[C_{pq}K\gamma_U^p[Z]\varepsilon^{-2/p}]) \quad \text{and}$$

$$(3) \qquad \| |X-\bar{X}|_T^* \|_{L^r} \leq K^{-1} \| |FX-\bar{F}X|_{-T}^* \|_{L^r} \cdot (b_{pr} + d_{pr}\exp[c_{pr}K\gamma_U^p[Z]\varepsilon^{-2/p}) ,$$

whenever $\bar{X}$ is the solution of a second differential equation $Y = C + \int \bar{F}(Y_-)dZ$ with the same (constant) initial condition C and driving term Z, and with (different) K-Lipschitz correspondence $\bar{F}$. The constants a-d are all of the form $\alpha + \beta KL$, with universal constants α, β depending only on the indices indicated.

Remarks. The theorem persists if (1) is replaced by a system of differential equations and C by a non-constant process. The correspondence F need not be of functional type; it suffices that it be Lipschitz and that $(FX_-)^T$ depends only on X_-^T for all times T [E1]. This important generalization is due to MEYER [DM,E1].

Since X^{n+1} can be computed pathwise from X^n, and X pathwise from (X^n), the correspondence $Z.(\omega) \to X.(\omega)$ is pathwise a.s. In fact, it is easily seen to be measurable. However, since two consecutive limits are involved, it seems impossible to extract from the arguments numerical estimates on the step size 2^{-n} of (0) and on N that will assure us that the 2^{-n}-approximate (0) of X^N is close to the solution X, with given probability and tolerance. For practical computations in, say, linear filtering problems such estimates would be good to have, in particular if they were good enough to replace costly Monte Carlo simulations.

One is led to Peano's method of solving differential equations, inasmuch as it involves only one limit. EMERY [E1] has shown that it does converge, in probability. Let us show that it converges uniformly a.s., and that one can obtain *a priori* error estimates. To this end let $U < \infty$ be the optional time up to which we wish to solve (1), let $\delta > 0$ be the step size, and set $S_0 = 0$, $S_{n+1} = U \wedge \inf\{t > S_n : |Z_t - Z_{S_n}| \geq \delta\}$. We obtain an optional partition $\sigma: 0 = S_0 \leq S_1 \leq S_2 \cdots \leq U$, which we use to define progressively

$$\bar{X}_0 = C \quad \text{and} \quad \bar{X}_t = \bar{X}_{S_n} + F(\bar{X}_{S_n})(Z_t - Z_{S_n})$$

for $S_n < t \leq S_{n+1}$. The process $\bar{X}$ is nothing but the Peano approximate for the step size δ. It is also easily seen to be the solution of a second differential equation

$$(\bar{1}) \qquad Y = C + \int \bar{F}(Y_-)dZ$$

[E1], with K-Lipschitz function $\bar{F}$ defined as follows. For any process Y set

$$Y^\sigma = \sum_n Y_{S_n} \cdot 1_{[[S_n, S_{n+1}))} \quad \text{and} \quad \bar{F}(Y) = F(Y^\sigma) .$$

Clearly $\bar{X}$ satisfies $(\bar{1})$. (True, $Y \to \bar{F}Y$ is not of functional type, but theorem 1 covers this case as well.) As

$$|FX - \bar{F}X| \leq K|X - X^\sigma| ,$$

theorem 1 yields the following estimate for the deviation of the Peano δ-approximate $\bar{X}$ from the solution X:

$$(4) \qquad \| |X-\bar{X}|^*_T \|_{L^r} \leq \| |X-X^\sigma|^*_{-T} \|_{L^r} \cdot (b_{pr} + d_{pr} \exp[c_{pr} K \gamma^p_\gamma [Z] \varepsilon^{-2/p}])$$

for T as specified there.

Now, if there were a pathwise mean value theorem -- of which there is, of course no hope -- we would proceed thus: since Z varies on $[S_n, S_{n+1})$ by less than δ, $X = C + \int F(X_-)dZ$ would vary there by less than $F(X)^*_{S_{n+1}} \cdot \delta \leq K \cdot X^*_{S_{n+1}} \cdot \delta$. The growth condition (2) would show that $|X-X^\sigma|^*_T < \delta \cdot A$ on a set of measure $\geq 1 - \varepsilon$ for some suitable constant A, and the desired estimate would follow.

It turns out that in dimension 1 and if F is of functional type there does exist a replacement for the mean value theorem, which permits us to argue along these lines. To simplify, we make an additional assumption on the function F (and leave it to the reader to generalize by localization).

Theorem 3. Suppose F is of class C^1 and never vanishes. Let $\varepsilon > 0$, $\tau > 0$, $0 < p < \infty$, and let U be an optional time such that $\gamma_U^p[Z] < \infty$ and $L = \| |\Delta Z|_U^* \|_{L^\infty} < \infty$. There exists a number

$$A = A[C,F(0),K,L,\gamma_U^p[Z],p,\varepsilon] < \infty ,$$

computable from the indicated arguments, such that with probability $\geq 1-\varepsilon$ the Peano approximate $\overline{X}$ for the step size δ differs from the solution X by less τ, uniformly until time U, provided that

$$\delta \leq \tau\varepsilon^{1/r}/A .$$

Proof. Given $\delta > 0$ we define σ and Z^σ as above, and investigate how small δ has to be chosen to produce the desired inequalities. We start by preparing an estimate of the variation of the square function of Z on the intervals of σ: with C_p the universal constant of (*)

$$\| |[Z,Z]-[Z,Z]^\sigma|_U^* \|_{L^p} \leq 12\delta\, C_p\, \gamma_U^p[Z] . \tag{5}$$

Proof of (5). The bracket $[Z,Z]$ is defined by $[Z,Z]_t = Z_t^2 - 2\int_0^t Z_- dZ$. Clearly $|[Z,Z]-[Z^\sigma,Z^\sigma]|_U^* \leq |Z^2-Z^{\sigma 2}|_U^* + 2|\int(Z_- -Z_-^\sigma)dZ|_U^* + 2|\int Z_-^\sigma\, dZ - Z_-^\sigma dZ^\sigma|_U^* =$ I + II + III. Using $|Z_- - Z_-^\sigma| \leq \delta$ and (*), we get

$$\|I\|_{L^p} \leq \| |(Z-Z^\sigma)|_U^* |Z+Z^\sigma|_U^* \|_{L^p} \leq 2\delta \|Z_U^*\|_{L^p} \leq 2\delta\, C_p\, \gamma_U^p[Z] .$$

By the very definition of $\gamma_U^p[Z]$,

$$\|II\|_{L^p} \leq 2C_p \gamma_U^p[\int(Z_- -Z_-^\sigma)dZ] \leq 2\delta\, C_p\, \gamma_U^p[Z] .$$

It is easily seen that, a.s. on $[S_n \leq t < S_{n+1})$, and thus a.s. everywhere, $|\int_0^t Z_-^\sigma(dZ-dZ^\sigma)| \leq \delta Z_t^*$ and consequently $\|III\|_{L^p} \leq 2\delta\, C_p\, \gamma_U^p[Z]$ once again. Hence

$$\| |[Z,Z]-[Z^\sigma,Z^\sigma]|_U^* \|_{L^p} \leq 6\delta\, C_p\, \gamma_U^p[Z] .$$

Now both $[Z,Z]^\sigma$ and $[Z^\sigma,Z^\sigma]$ are constant on each of the intervals $[S_n, S_{n+1})$, the former coinciding with $[Z,Z]$ in the left end point. Consequently $|[Z,Z]-[Z,Z]^\sigma| \leq 2|[ZZ]-[Z^\sigma,Z^\sigma]|$, and (5) follows.

We turn to the proof proper of the theorem: Let Φ be a function such that $\Phi'\cdot F = 1$. Such exists as $F \neq 0$, and it has $\Phi''\cdot F^2 = -F'$.

We apply ITO's theorem, with the notation of [B2], to Φ and the solution X of (1) and obtain for $S_n \leq t < S_{n+1}$

$$
\begin{aligned}
(*)\quad |\Phi(X_t)-\Phi(X_{S_n})| &= |\Phi'(\xi)||X_t-X_{S_n}| = |X_t-X_{S_n}|/F(\xi) \\
&= \Big|\int_{S_n}^t \Phi'(X_-)dX + \frac{1}{2}\int_{S_n}^t \Phi''(X_-)d\{X,X\} + \sum_{S_n<S\leq t} \Phi(X_S)-\Phi(X_{-S})-\Phi'(X_S)\Delta_S X \\
&\leq \Big|\int_{S_n}^t \Phi'(X_-)F(X_-)dZ\Big| + \frac{1}{2}\int_{S_n}^t |\Phi''(X_-)(F^2(X_-)|d\{Z,Z\} + \sum_{\dots} (|\Phi'(\xi_S)F(X_{-S})-1||\Delta_S Z|) \\
&\leq |Z_t-Z_{S_n}| + \frac{1}{2}\int_{S_n}^t |F'(X_-)|d\{Z,Z\} + \sum_{\dots} .
\end{aligned}
$$

Here ξ is a point between X_t and X_{S_n}, and ξ_S a point between X_S and X_{-S}. These points exist by the ordinary mean value theorem. Now

$$
\begin{aligned}
|\Phi'(\xi_S)F(X_{-S})-1| &= \Big|\frac{F(X_{-S})-F(\xi_S)}{F(\xi_S)}\Big| \\
&\leq K\,\frac{|X_{-S}-X_S|}{F(\xi_S)} \leq K|F(X_{-S})||\Delta_S Z|/F(\xi_S) \leq K|\Delta_S Z|\Big|\frac{F(X_{-S})-F(\xi_S)}{F(\xi_S)}\Big| + K|\Delta_S Z| .
\end{aligned}
$$

Since for $S_n < S < S_{n+1}$, $|\Delta_S Z| < 2\delta$, we have thus

$$
|\Phi'(\xi_S)F(X_{-S})-1| \leq \frac{K|\Delta_S Z|}{1-K|\Delta_S Z|} \leq 2K|\Delta_S Z|
$$

provided that

$$
\delta \leq (4K)^{-1} .
$$

Multiplying (*) by $|F(\xi)|$ yields a.s. on $[S_n \leq t \leq S_{n+1}]$

$$|X_t - X_t^\sigma| = |X_t - X_{S_n}| \le |FX|_t^*(\delta + \frac{K}{2}(\{Z;Z\}_t - \{Z;Z\}_{S_n}) + 2K \sum_{S_n<s<t} (\Delta_s Z)^2)$$

$$\le |FX|_t^*(\delta + 2K([Z;Z]_t - [Z,Z]_{S_n})) \le |FX|_t^*(\delta + 2K([Z;Z]_t - [Z;Z]_t^\sigma)) .$$

Thus

$$\text{(6)} \qquad |X - X^\sigma|^* \le (KX^* + |F(0)|)(\delta + 2K|[Z;Z] - [Z,Z]^\sigma|^*)$$

except on an evanescent set: it holds on each of the countable numbers of sets

$$([S_n \le t < S_{n+1}] \times [0,t]) \cap [S_n, t] , \qquad t \in \mathbb{Q},\ n \in \mathbb{N} ,$$

which covers the stochastic interval $[0,U]$ a.s.

It is time to put together (2), (4), (5) and (6). Given $p,q > 0$, we choose r so that $r^{-1} = p^{-1} + q^{-1}$ and obtain

$$\text{(7)} \qquad \| |X - \overline{X}|_T^* \|_{L^r} \le \delta(b_{pr} + d_{pr} \exp[c_{pr} K\gamma_U^p[Z]\varepsilon^{-2/p})$$
$$\times (|F0| + (KC + a_{pq}(KC + |F0|)\exp[c_{pq} K\gamma_U^p[Z]\varepsilon^{-2/p})$$
$$\times (1 + 24KC_p^{(7..4)}\gamma_U^p[Z]) ,$$

where $T \le U$ is such that $P[T < U] < \varepsilon$. If $p < 1$ or $q < 1$, so that the norms are not subadditive, additional constants 2^p or 2^q will appear, but we won't bother to spell it out.

Inequality (7) is somewhat stronger than the statement of the theorem. Let $A_q = A_q[C, F(0), K, L, \gamma_U^p[Z], p, \varepsilon]$ denote the factor of δ in (7), and let $\Omega_0 = [T = U] \cap [|X - \overline{X}|_U^* \le \tau]$. The Peano δ-approximate $\overline{X}$ clearly does not ever before time U deviate from the solution X by more than τ. And it is elementary to check that if δ is chosen smaller than

$$\tau\varepsilon^{1/r}/A_q$$

then $P(\Omega_0) > 1 - 2\varepsilon$. Replacing ε by $\varepsilon/2$ and choosing $A = \inf A_q$ proves the theorem.

Corollary. Suppose again that F never vanishes, but let Z be any semimartingale. The Peano approximates X^n for (1) of step size 2^{-n} converge a.s. uniformly on compact intervals to the solution X of (1). In fact,

$$\Sigma |X^n - X^{n-1}|_S^* < \infty \quad \text{a.s.}$$

for any a.s. finite time S.

Proof. We may assume that $\gamma_S^2[Z] < \infty$, by changing the measure. Given $\varepsilon > 0$, there is a stopping time $U \leq S$ with $P[U < S] < \varepsilon$ such that $|\Delta Z|^*$ is bounded on $[0,U)$. Then $L_U[Z^{-U}] < \infty$ and $\gamma_U^2[Z^{-U}] < \infty$ for the process $Z^{-U} = Z \cdot 1_{[0,U)}$. Choose $q = 2$ and observe that (7) yields

$$\Sigma E |X^{-} - X^{n-}|_T^* < \infty$$

for arbitrarily large times $T \leq U$, where X^{-} and X^{n-} denote the solution of (1) and the 2^{-n}-approximate driven by Z^{-U}, respectively. These coincide a.s. with X, X^n, respectively, on $[0,U)$, on account of the local nature of the integral. That is to say,

$$\Sigma |X - X^n|_{-T}^* < \infty \quad \text{a.s.}$$

for arbitrarily large times T, and the claim follows.

Remark. The argument goes through for some systems of differential equations; namely, if the matrix $F = (F_b^a)$ satisfies $\sum_r \partial F_b^a/\partial x^r F_c^r = \sum_r \partial F_c^a/\partial x^r F_b^r$, because then there exists a vector field (Φ^a) with $\Sigma \partial \Phi^a/\partial r F_b^r = \delta_b^a$. I conjecture that the argument presented is unnecessarily clumsy, and that someone will find a shorter one providing a better result at the same time.

References

[B1] BICHTELER, K. "Stochastic Integrators", Bull. A.M.S. 1/2, September 1979.

[B2] BICHTELER, K. "Stochastic integration and L^p-theory of semimartingales", (to appear).

[B3] BICHTELER, K. "Function norms and function metrics satisfying the dominated convergence theorem, and their applications", (to appear).

[B4] BICHTELER, K. "Integration theory", Lecture Notes Math #315, Springer, Berlin, 1973.

[B5] BICHTELER, K. "Measures with values in non-locally convex spaces", In Proc. Oberwolfach Conf. on Vector Measures, Lecture Notes Math. #541, Springer, 1976.

[D1] DELLACHERIE, C. Capacités et Processus Stochastiques , Springer, Berlin 1972.

[D2] DELLACHERIE, C. "Quelques applications du lemme de Borel-Cantelli à la théorie des semimartingales", Sém. Prob. XII, 742-745, Lecture Notes #649, Springer, Berlin 1978.

[D3] DOLEANS-DADE, C. "On the existence and unicity of solutions of a stochastic differential equation", Z. Wahrsch., 34(1976), 93-101.

[DM] DOLEANS-DADE, C. and MEYER, P.A. "Équations différentielles stochastiques", Séminaire de Probabilités XI, Lecture Notes in Math, #581, 376-382, Springer, Berlin 1977.

[E1] EMERY, M. "Stabilité des solution des équations différentielles stochastiques; application aux intégrales multiplicatives stochastiques", Z. Wahrsch., 41(1978), 241-262.

[E2] EMERY, M. "Équations différentielles stochastiques: La méthode de Métivier et Pellaumail", Séminaire de Probabilités de Strasbourg XIV, (to appear).

[H1] HALMOS, P.R. Measure Theory, Van Nostrand, New York, 1950.

[I1] ITÔ, K. "Stochastic integral", Proc. Imperial Acad. Tokyo, 20(1944), 519-524.

[J1] JACOD, J. "Calcul stochastique et problèmes de martingales", Lect. Notes Math. #714, Springer, 1979.

[K1] KUSSMAUL, . "Stochastic integration and generalized martingales", Research Notes, Pittman, London, 1977.

[M1] MAUREY, B. "Theorèmes de factorization pour les operateurs linéaires a valeurs dans les espaces L^p", Astérisque, 11(1974), 1-163.

[M2] METIVIER, M. "Reelle und vektorwertige Quasimartingale und die Theorie der stochastischen Integration", Lect. Notes Math. #607, Springer, Berlin 1977.

[MP] METIVIER, M. and PELLAUMAIL, J. "Mesure stochastique a valeur dans les espaces L^0", Z. Wahrsch., 40/2 (1977).

[MP'] METIVIER, M. and PELLAUMAIL, J. "Une formule de majoration pour martingales; et sur une équation stochastique assez générale", C.R. Acad. Sci. Paris, 285 and 921(197), 685.

[M3] MEYER, P.-A. "Un cours sur les intégrales stochastiques", Lect. Notes Math. #511, Springer 1976.

[M4] MEYER, P.-A. "Charactérisation des semimartingales d'après Dellacherie", Séminaire de Probabilités XIII, 620-623, Lect. Notes Math. #721, Springer, Berlin 1979.

[M5] MEYER, P.-A. "Inégalités de normes pour les intégrales stochastiques", Séminaire de Probabilités XII, 757-762, Lect. Notes Math. #XII, Springer, Berlin 1978.

[P1] PELLAUMAIL, J. "Sur l'intégrale stochastique et la décomposition de Doob-Meyre", Asterisque 9(1973), 1-124.

[P2] PELLAUMAIL, J. "Une example d'intégrale d'une fonction réelle par rapport à une mesure vectorielle: l'intégrale stochastique", C.R. Acad. Sci. Paris, 274(1972), 1369-1372.

[P3] PROTTER, P. "On the existence, uniqueness, convergence and explosions of solutions of systems of stochastic differential equations", Ann. Prob., 5(1977), 243-261.

[P4] PROTTER, P. "Right-continuous solutions of systems of stochastic integral equations", J. Multivariate Analysis, 7(1977), 204-214.

[P5] PROTTER, P. "Markov solutions of stochastic differential equations", Z. Wahrsch., 41(1977), 39-58.

[P6] PROTTER, P. "H^p-stability of solutions of stochastic differential equations", Z. Wahrsch., 44(1978), 337-352.

[R1] ROSENTHAL, H. "On subspaces of L^p", Ann. Math., 97(1973), #2, 344-373.

[T1] THOMAS, E. "L'intégration par rapport à une mesure de Radon vectorielle", Ann. Inst. Fourier, 20(1970), 55-191.

[Y1] YOR, M. "Inégalites entre processus minces et applications", Comptes Rendus Acad. Sci., 286(1978), #(A), 799-802.

[Y2] YOR, M. "Quelques interactions entre mesures vectorielles et intégrales stochastiques", Lect. Notes Math. #713, Springer 1979.

[Y3] YOR, M. "Remarques sur les normes H^p de (semi) martingales", C.R. Acad. Sci. Paris, 287(1978), 461-465.

[Z1] ZYGMUND, A. <u>Trigonometric Series</u>, Cambridge 1959.

Klaus Bichteler
Department of Mathematics
The University of Texas at Austin
Austin, Texas 78712 U.S.A.

SOME COMMENTS ON THE MAXIMAL INEQUALITY IN MARTINGALE THEORY

S.D. CHATTERJI

§1. Introduction.

Let (Ω,Σ, P) be a probability space, I a finite or denumerable directed set and Σ_α, $\alpha \in I$, an increasing family of sub-σ-algebras of Σ. We say that the maximal inequality holds for the family $\{\Sigma_\alpha\}_{\alpha\in I}$ if for any $f \in L^1$, we have

$$P(A_\lambda) \leq \frac{1}{\lambda} \int_{A_\lambda} |f|\, dP \qquad \text{.... (1)}$$

where $A_\lambda = \{\omega | \ f^*(\omega) \geq \lambda\}$, $\lambda > 0$,

$$f^*(\omega) = \sup_\alpha |f_\alpha(\omega)| \ , \quad f_\alpha = E\{f|\Sigma_\alpha\}.$$

The following facts are by now well-known:

(a) the validity of (1) implies that $\lim_\alpha f_\alpha(\omega)$ exists a.e.;

(b) (1) holds if I is totally ordered but not generally;

(c) (1) is not necessary for the a.e. convergence of $f_\alpha(\omega)$.

The purpose of the present note is to illustrate (b) and (c) by a simple example in §2. As regards (a) an explicit reference is [1] p.22. A detailed discussion of related facts is contained in the work of Sucheston and his collaborators (cf. references in this volume). In §3 we point out briefly (as we have already done in [1]p.27) that (1) contains some non-trivial classical inequalities of analysis.

§2. A counter-example.

Our discussion will be based on the fact that (1) implies that

$$||f^*||_p \leq p' \cdot ||f||_p \qquad \text{.... (2)}$$

where $1 < p < \infty$, $p' = p / (p-1)$.

For our counter-example, we take a finite set $\Omega = \{1,2,\dots,n\}$, Σ = all subsets of Ω, P = the uniform probability on Ω, I = set of all partitions of Ω and Σ_α = σ-algebra generated by the partition α. For any $f : \Omega \to \mathbb{R}$, $f_\alpha = E\{f|\Sigma_\alpha\}$ is clearly defined by

$f_\alpha(\omega) = \sum_{j \in A} f(j) / |A|$ where A is the unique element of the partition α containing ω and $|A|$ = cardinality of A. If the maximal inequality were true, we should have (2) for any f and in particular for f = the indicator function of the one point set $\{1\}$. However, in this case,

$$f^*(\omega) = \sup_\alpha f_\alpha(\omega) = \begin{cases} \frac{1}{2} & \text{if } \omega \neq 1 \\ 1 & \text{if } \omega = 1 \end{cases}$$

so that (2) would yield :

$$\{1 + (n-1)(\tfrac{1}{2})^p\}^{1/p} \leq p'$$

for all $p > 1$; but for any $n \geq 2$, this is impossible since the left hand side tends to 1 and the right ànd side to 0 as $p \to \infty$. Thus the maximal inequality must be false (for any $n \geq 2$).

Notice however that any martingale $\{f_\alpha, \Sigma_\alpha\}$ on a finite space Ω must converge since the Σ_α's become constant from some point on.

§3. Some inequalities contained in (2).

Note first that (2) holds for any measure space (Ω,Σ,P) , $P(\Omega) \leq \infty$, provided that we are dealing with a monotone sequence of σ-algebras Σ_n (increasing or decreasing) and $f \in L^p \cap L^1$ with f_n defined so that

$$\int_A f\,dP = \int_A f_n\,dP \quad , \qquad A \in \Sigma_n \quad .$$

For any set $B \in \Sigma$, $0 < P(B) < \infty$, let Σ_B be the σ-algebra of sets $A \in \Sigma$ such that either $A \supset B$ or $A \subset \complement B$. For any $f \in L^1$ define f_B as follows:

$$f_B(\omega) = \begin{cases} f(\omega) & \text{if } \omega \in \complement B \\ \frac{1}{P(B)} \cdot \int_B f\,dP & \text{if } \omega \in B \end{cases}$$

If $\{B_n\}_{n\geq 1}$ is a monotone sequence of sets in Σ with $0 < P(B_n) < \infty$ then $\{f_n, \Sigma_n\}_{n\geq 1}$ clearly satisfies the martingale property where $\Sigma_n = \Sigma_{B_n}$ and $f_n = f_{B_n}$. If $f \in L^p \cap L^1$, we shall have (with $B_0 = \emptyset$) :

$$f^*(\omega) \geq \frac{1}{P(B_n)} \left|\int_{B_n} f\,dP\right|$$

for $\omega \in B_n \setminus B_{n-1}$ in case $B_1 \subset B_2 \subset \dots$ and for $\omega \in B_n \setminus B_{n+1}$ in case $B_1 \supset B_2 \supset \dots$. Thus, setting $\alpha_n = P(B_n)$, we shall have :

$$\sum_{n=1}^{\infty} (\alpha_n - \alpha_{n-1}) \frac{1}{\alpha_n^p} \left|\int_{B_n} f\,dP\right|^p \leq (p')^p \cdot ||f||_p^p \qquad \dots (3)$$

in case $B_1 \subset B_2 \subset \dots$ and in case $B_1 \supset B_2 \supset \dots$,

$$\sum_{n=1}^{\infty} (\alpha_n - \alpha_{n+1}) \frac{1}{\alpha_n^p} \left|\int_{B_n} f\,dP\right|^p \leq (p')^p \cdot ||f||_p^p \qquad \dots (4)$$

The inequalities (3) and (4) contain as special cases some inequalities of Hardy ([2] p.239-248). Take, for example, $\Omega = [0, \infty[$, P = Lebesgue measure and $B_n = [0, x_n]$, $x_1 < x_2 < \dots$ in (3) ; then

$$\sum_{n=1}^{\infty} (x_n - x_{n-1}) \cdot \frac{1}{x_n^p} \left|\int_0^{x_n} f\right|^p \leq (p')^p \int_0^{\infty} |f|^p$$

whence a suitable limit argument gives

$$\int_0^{\infty} \left|\frac{1}{x} \int_0^x f\right|^p dx \leq (p')^p \int_0^{\infty} |f|^p .$$

Other inequalities as well as the possibility of using the analogue of (2) for continuous parameter martingales remain to be explored.

References.

[1] Chatterji, S.D.
Les martingales et leurs applications analytiques. Lecture Notes In Maths. NO.3o7,p.27-164, Springer-Verlag, Berlin (1973).

[2] Hardy, G.H., Littlewood, J.E. and Polya, G.
Inequalities.Cambridge Univer. Press 1934.

S.D. Chatterji
Professeur, Dépt. de Math.
Ecole Polytechnique Fédérale
de Lausanne
61, av. de Cour
CH-1007 LAUSANNE
Suisse

UN SURVOL DE LA THEORIE DE L'INTEGRALE
STOCHASTIQUE

par C. Dellacherie

Disposant ici d'un nombre confortable de pages, je voudrais développer la conférence de même titre écrite pour le Congrès International d'Helsinki, conférence qui avait pour but de présenter, de manière compréhensible pour un lecteur n'ayant que quelques notions de théorie des processus (stochastiques), les principaux résultats d'une grande part du calcul différentiel stochastique (plus précisément, de la part où la probabilité de base ne joue que par la classe des ensembles négligeables qu'elle définit). Même en laissant ainsi de côté des parts importantes de la théorie (par exemple, tout ce qui touche à la notion, pourtant fondamentale, de martingale locale), il ne me serait pas possible de donner in extenso les démonstrations de tous les résultats cités (tout le volume me serait nécessaire !) ; j'ai cependant tâché d'expliquer, avec clarté et précision, le "pourquoi" des théorèmes fondamentaux, tout en évitant d'introduire trop de concepts nouveaux pour la plupart des lecteurs. Le lecteur désireux d'approfondir le sujet pourra se reporter aux ouvrages cités dans la bibliographie.

PRELIMINAIRES

On se donne au départ un <u>espace filtré</u> $(\Omega,\mathbb{F},P)$ <u>vérifiant les conditions habituelles</u> : la <u>filtration</u> $\mathbb{F}=(\underline{\underline{F}}_t)_{t\in\mathbb{R}_+}$ est une famille croissante de tribus sur Ω (intuitivement, $\mathbb{R}_+$ est l'échelle des temps et $\underline{\underline{F}}_t$ la tribu des événements connus à l'instant t), P est une mesure de probabilité sur $\underline{\underline{F}}_\infty = \bigvee_t \underline{\underline{F}}_t$; les conditions habituelles expriment que la filtration est complète (i.e. $\underline{\underline{F}}_\infty$ est complète et $\underline{\underline{F}}_0$ contient tous les ensembles négligeables) et continue à droite (i.e. on a $\underline{\underline{F}}_t = \bigcap_{s>t} \underline{\underline{F}}_s$ pour tout t ; autrement dit, $\underline{\underline{F}}_t$ contient aussi le futur infinitésimal après t). En fait, les conditions habituelles ne sont pas réellement indispensables pour ce que nous allons faire ; mais, d'une part, elles sont fréquemment vérifiées (c'est le cas pour les filtrations associées aux bons processus de Markov), et, d'autre part, elles simplifient considérablement et la technique et la présentation des résultats.

Un processus X (à valeurs réelles ; nous ne considérerons que ce cas) est intuitivement une fonction du temps t dépendant de l'aléa ω, et donc, finalement, une application $(t,\omega)\to X(t,\omega)$ de $\mathbb{R}_+ \times \Omega$ dans $\mathbb{R}$; on note le plus souvent $X_t(\omega)$ la valeur en (t,ω), d'où la notation courante $X=(X_t)_{t\in\mathbb{R}_+}$. On appelle trajectoire associée à $\omega\in\Omega$ la fonction $t\to X_t(\omega)$. On dira que le processus X est càdlàg (resp càglàd) si toutes ses trajectoires sont des fonctions Continues A Droite et pourvues d'une Limite A Gauche (dans $\mathbb{R}$) pour $t>0$ (resp Continues A Gauche pour $t>0$ et pourvues d'une Limite A Droite pour $t\geq 0$). Deux processus X et Y sont dits indistinguables si $\{\omega\in\Omega : \exists t\ X_t(\omega)\neq Y_t(\omega)\}$ est un ensemble négligeable ; dans toute la suite, nous travaillons "à l'indistinguabilité près" (c'est en particulier ainsi qu'il faudra entendre les résultats d'unicité). Il est facile de voir que X et Y, tous deux càdlàg (ou càglàd), sont indistinguables dès que Y est une modification de X (i.e. on a $Y_t = X_t$ p.s. pour chaque t) ; en particulier, un processus admet au plus une modification càdlàg. Un processus V sera dit croissant (resp à variation finie) si ses trajectoires sont croissantes (resp ont une variation finie sur tout intervalle fini) et si, de plus, V est càdlàg (ceci pour pouvoir écrire $V_t(\omega) = V_0(\omega) + \int_0^t dV_s(\omega) = V_0(\omega) + \int_{]0,t]} dV_s(\omega)$, où l'intégrale est, trajectoire par trajectoire, une intégrale de Stieljes).

Etant donnée notre filtration $\mathbb{F}$, il est naturel de se restreindre à étudier les processus X tels que X_t soit une v.a. $\underline{\underline{F}}_t$-mesurable pour tout t et que X lui-même soit mesurable lorsque $\mathbb{R}$ est muni de sa tribu borélienne $\underline{\underline{B}}(\mathbb{R})$ et $\mathbb{R}_+ \times \Omega$ de la tribu produit $\underline{\underline{B}}(\mathbb{R}_+)\times \underline{\underline{F}}_\infty$; nous dirons, pour abréger, qu'un tel X est un $\mathbb{F}$-processus (la terminologie officielle étant "X est mesurable et adapté à la filtration"). Il n'est pas difficile de voir qu'un processus càdlàg ou càglàd $X=(X_t)$ est un $\mathbb{F}$-processus dès que X_t est $\underline{\underline{F}}_t$-mesurable pour chaque t.

Enfin, un processus M est une $\mathbb{F}$-martingale (resp $\mathbb{F}$-surmartingale, $\mathbb{F}$-sousmartingale ; on omet "$\mathbb{F}$" s'il n'y a pas ambiguité) si

a) M_t est une v.a. $\underline{\underline{F}}_t$-mesurable et intégrable pour chaque t ;

b) on a, pour tout couple (s,t) tel que $s\leq t$,

$$M_s = E[M_t | \underline{\underline{F}}_s] \text{ p.s. (resp } \geq, \leq),$$

ce qui équivaut à dire, étant donné a), que l'on a

$$\forall A\in\underline{\underline{F}}_s \quad \int_A M_s\, dP = \int_A M_t\, dP \quad \text{(resp } \geq, \leq).$$

Notez que M est une (sur)martingale ssi -M est une (sous)martingale. Il est classique que, sous les conditions habituelles, toute martingale admet une (unique) modification càdlàg ainsi que, plus généralement, toute surmartingale ou sousmartingale X telle que la fonction

$t \to E[X_t]$ soit continue à droite. Toutes les (sous,sur)martingales considérées dans la suite seront càdlàg, et nous omettrons de le mentionner. L'exemple fondamental de martingale continue (i.e. à trajectoires continues) est le mouvement brownien unidimensionnel issu de 0 ; d'après un théorème célèbre de P. Lévy, il est caractérisé en loi par le fait que c'est une martingale continue B telle que $B_0 = 0$ et que le processus $(B_t^2 - t)$ soit une martingale. Bien entendu, il n'est pas vrai que toute filtration $\mathbb{F}$ "contient" un mouvement brownien : lorsque $\mathbb{F}$ est déterministe (i.e. $\underline{\underline{F}}_0 = \underline{\underline{F}}_\infty$), toutes les martingales sont à trajectoires constantes et donc triviales. Par ailleurs, "le" mouvement brownien n'est unique qu'en loi : une filtration $\mathbb{F}$ peut être assez riche pour "contenir" différents mouvements browniens indépendants. Rappelons enfin que (presque) toutes les trajectoires d'un mouvement brownien ont une variation infinie sur tout intervalle non dégénéré.

INTRODUCTION

Le problème de la définition de l'intégrale stochastique est, en gros, le suivant : savoir, pour une large classe $\mathbb{X}$ de $\mathbb{F}$-processus càdlàg(1) "intégrants" X et pour une large classe $\mathbb{Y}$ de $\mathbb{F}$-processus "intégrés" Y, définir un $\mathbb{F}$-processus càdlàg Z par

$$Z_t = Z_0 + \int_0^t Y_s \, dX_s$$

où Z_0 est une v.a. $\underline{\underline{F}}_0$-mesurable et où l'opération symbolisée par $\int$ soit "digne" d'être appelée une intégrale. Pour fixer les idées, disons que cette opération doit non seulement être linéaire en Y mais aussi quelque peu continue et que, par ailleurs, même si on ne peut exiger de pouvoir en général intégrer trajectoire par trajectoire (penser au brownien !), elle doit ressembler quelque peu à une intégrale trajectorielle - par exemple vérifier $Z_v - Z_u = X_v - X_u$ p.s. sur $A \in \underline{\underline{F}}_\infty$ si Y vaut 1 sur le rectangle $]u,v] \times A$.

Nous verrons d'abord que ce problème a une solution en quelque sorte optimale, la classe $\mathbb{X}$ contenant en particulier tous les $\mathbb{F}$-processus à variation finie et toutes les (sur,sous)martingales, la classe $\mathbb{Y}$ contenant en particulier tous les $\mathbb{F}$-processus càglàd (mais pas tous les $\mathbb{F}$-processus càdlàg !). De plus, il se trouvera que $Z = \int Y\,dX$, pour $X \in \mathbb{X}$ et $Y \in \mathbb{Y}$, sera lui-même un élément de $\mathbb{X}$, ce qui nous permettra de donner un sens à l'écriture infinitésimale $dZ = YdX$. On peut alors développer tout un calcul différentiel stochastique (contenant le calcul différentiel classique, mais bien plus riche - le cas classique revenant à prendre Ω réduit à un point et à considérer $X_t = t$) dans lequel on sait faire des changements de variables (formule d'Ito) et

---

(1) Nous reviendrons plus loin sur cette restriction assez naturelle.

intégrer des équations différentielles.

Encore quelques mots avant de se mettre au travail. D'abord, la démarche que nous adoptons ici pour présenter l'intégrale stochastique n'est pas celle qui a été adoptée, si je puis dire, par l'Histoire : la classe $\mathbb{X}$ des processus intégrants - appelés semimartingales - a été découverte petit à petit en partant du seul mouvement brownien et en passant par les martingales de carré intégrable alors que nous allons prendre pour définition une caractérisation assez nouvelle de cette classe (je donnerai quelques détails d'ordre historique à l'occasion de la bibliographie). Par ailleurs, comme on peut trouver maintenant dans de nombreux ouvrages le calcul intégral stochastique exposé dans toute sa généralité (intégrale du type "Lebesgue-Stieljes", où $\mathbb{Y}$ contient les $\mathbb{F}$-processus càglàd et est stable par convergence simple dominée), j'ai eu la coquetterie ici de développer une petite partie de ce calcul (intégrale du type, disons, "Riemann-Dieudonné", où $\mathbb{Y}$, stable par convergence uniforme, est réduit en fait à la classe des $\mathbb{F}$-processus càglàd), suffisante pour traiter le calcul différentiel.

SEMIMARTINGALES

Si x est une fonction càdlàg de $\mathbb{R}_+$ dans $\mathbb{R}$ et y une fonction càglàd de $\mathbb{R}_+$ dans $\mathbb{R}$, en escalier (i.e. y est une combinaison linéaire - finie - d'indicatrices d'intervalles), il est naturel de définir, pour $t\in\mathbb{R}_+$, l'intégrale élémentaire

$$\int 1_{]0,t]}(s)\ y(s)\ dx(s) = \int_0^t y(s)\ dx(s)$$

en posant $\int 1_{]u,v]}(s)\ dx(s) = x(v) - x(u)$ et en étendant par linéarité, le prolongement étant bien défini (exercice !). Maintenant, si X est un $\mathbb{F}$-processus càdlàg et Y un $\mathbb{F}$-processus càglàd (à trajectoires) en escalier, il est clair que l'on définit un nouveau $\mathbb{F}$-processus càdlàg Z, nul en 0, en posant

$$Z_t = \int_0^t Y_s\ dX_s \quad \text{pour tout } t,$$

l'intégrale étant l'intégrale élémentaire prise trajectoire par trajectoire. Mais une intégrale digne de ce nom doit vérifier quelque continuité par rapport à la chose intégrée, et cela va nous mener à notre définition des semimartingales.

Nous dirons qu'un $\mathbb{F}$-processus càglàd, en escalier, est en fait un processus prévisible élémentaire (terminologie officielle, justifiée plus loin) s'il est une combinaison linéaire d'indicatrices de rectangles de $\mathbb{R}_+\times\Omega$ de la forme $]u,+\infty[\times A$ avec $A\in\underline{\underline{F}}_u$ (noter qu'une telle indicatrice est un $\mathbb{F}$-processus càglàd), et nous désignerons par $\underline{\underline{E}}$ l'ensemble de ces processus prévisibles élémentaires. Regardant, pour t et X fixés, l'intégrale élémentaire comme opérateur linéaire de $\underline{\underline{E}}$ dans L^0, espace des (classes de) v.a. finies sur Ω, nous allons demander que, pour chaque t , cet opérateur soit continu aussi peu que possible, compte tenu de ce que l'on veut finalement obtenir comme intégrale stochastique (voir les commentaires suivant la définition).

1 DEFINITION. Un $\mathbb{F}$-processus càdlàg X est appelé une semimartingale (une $\mathbb{F}$-semimartingale s'il y a ambiguité sur la filtration) si, pour chaque $t \in \mathbb{R}_+$, l'intégrale élémentaire $Y \to \int_0^t Y_s \, dX_s$ définit un opérateur linéaire continu de $\underline{\underline{E}}$, muni de la topologie de la convergence uniforme (en (t,ω)), dans L^0, muni de la topologie de la convergence en probabilité.

REMARQUE. Si on suppose seulement que X est un processus continu à droite en probabilité tel que X_t soit $\underline{\underline{F}}_t$-mesurable pour chaque t , l'intégrale élémentaire étant définie comme précédemment, on peut montrer que X admet une modification qui est càdlàg et donc une semimartingale si l'intégrale élémentaire est continue pour les topologies précitées.

Il est clair que la topologie choisie sur $\underline{\underline{E}}$ est la topologie "raisonnable" la plus forte possible si on veut avoir une bonne intégrale ; par ailleurs, la topologie choisie sur L^0 est la topologie "raisonnable" la plus faible possible qui ait quelque rapport avec la convergence simple. En fait, si Ω est réduit à un point, on retrouve exactement les fonctions continues à droite à variation finie sur tout intervalle fini (exercice !) ; plus généralement, si $\mathbb{F}$ est déterministe, les semimartingales sont les processus mesurables à variation finie (rappelons que, par convention, cela inclut la "càdlàgité"), comme nous le verrons un peu plus loin.

Avant de donner des exemples (rassurons tout de suite le lecteur, cependant : toute martingale est une semimartingale, même si ce n'est pas évident avec notre définition), nous commençons par dégager quelques propriétés simples (avec notre définition) mais importantes des semimartingales.

2 PROPRIETES. a)(Changement de loi) Supposons que l'on remplace la probabilité P par une probabilité équivalente Q . Comme cela ne change pas la topologie de L^0 (ni le fait que $\mathbb{F}$ vérifie les conditions habituelles), une semimartingale X pour P reste une semimartingale pour Q . Noter, en contraste, que la notion de martingale dépend fortement de la probabilité de base, à cause de la condition d'intégrabilité et surtout de la présence d'espérances conditionnelles dans la définition. Plus généralement, on a un résultat analogue lorsque l'on suppose seulement que Q est absolument continue par rapport à P (L^0 a alors une topologie plus faible), sauf qu'il faut alors compléter $\mathbb{F}$ par rapport à Q si l'on veut retrouver les conditions habituelles.

b)(<u>Changement de filtration</u>) Supposons que l'on remplace $\mathbb{F}$ par une filtration $\mathbb{G}$ plus petite (i.e. $\underline{\underline{G}}_t$ est une sous-tribu de $\underline{\underline{F}}_t$ pour tout t) vérifiant les conditions habituelles. Alors une $\mathbb{F}$-semimartingale X est encore une $\mathbb{G}$-semimartingale dès que X est un $\mathbb{G}$-processus ; en effet, on ne fait alors que diminuer les espaces $\underline{\underline{E}}$ et L^0 considérés.

c)(<u>Localisation</u>) Soit X un $\mathbb{F}$-processus et supposons qu'il existe une suite croissante (T_n) de v.a. ≥ 0, tendant vers $+\infty$, vérifiant la condition suivante : pour tout n, il existe une semimartingale X^n (attention, n n'est pas une puissance, mais un indice !) telle que l'on ait $X = X^n$ sur l'intervalle stochastique $[\![0,T_n[\![$, i.e.

$$X_t(\omega) = X^n_t(\omega) \text{ pour tout } \omega \text{ et tout } t\varepsilon[0,T_n(\omega)[\ .$$

Alors X est une semimartingale. En effet, il est clair que X est alors càdlàg, et la continuité de l'intégrale élémentaire résulte aisément du fait que, pour t fixé, on a $\int_0^t Y_s\, dX_s = \int_0^t Y_s\, dX^n_s$ sur $\{T_n > t\}$ pour tout $Y\varepsilon\underline{\underline{E}}$.

d)Enfin, il est clair que l'ensemble $\underline{\underline{S}}$ des semimartingales est un espace vectoriel. On peut en fait montrer que c'est une <u>algèbre</u> (pour le produit ordinaire) et une <u>lattice</u> (pour l'ordre ordinaire), et même mieux que cela. Nous reviendrons sur ce point quand nous verrons la formule de changement de variables.

3 EXEMPLES. a) Il est clair que <u>tout $\mathbb{F}$-processus à variation finie est une semimartingale</u>, l'intégrale élémentaire coincidant alors avec l'intégrale de Stieljes. Réciproquement, supposons que la $\mathbb{F}$-semimartingale X soit encore une $\mathbb{F}^\varepsilon$-semimartingale pour la filtration plus grosse $\mathbb{F}^\varepsilon$ définie comme suit : on a $\varepsilon>0$ fixé et $\underline{\underline{F}}^\varepsilon_t = \underline{\underline{F}}_{t+\varepsilon}$ pour tout t. Alors, X est à variation finie. En effet on a alors, parmi les $\mathbb{F}^\varepsilon$-processus prévisibles élémentaires, des processus Y tels que

$$Y_s(\omega) = \text{signe de } (X_{t_{i+1}}(\omega) - X_{t_i}(\omega)) \quad \text{pour } t_i < s \leq t_{i+1}$$

lorsque $(t_0 = 0, t_1, \ldots, t_i, t_{i+1}, \ldots, t_n = t)$ est une subdivision finie de l'intervalle $[0,t]$, de pas $< \varepsilon$. Comme l'image de la boule unité de $\underline{\underline{E}}$ par l'intégrale élémentaire doit être, pour chaque t, un borné de L^0, on en conclut sans peine que X doit être à variation finie.

b) <u>Toute martingale est une semimartingale</u>. Soit d'abord M une martingale telle que M^2_t (ici 2 est un exposant) soit intégrable pour tout t. Alors, pour $u\leq v\leq s\leq t$, on a $E[1_A(M_v-M_u)1_B(M_t-M_s)] = 0$ pour tout $A\varepsilon\underline{\underline{F}}_u$ et tout $B\varepsilon\underline{\underline{F}}_s$, car la v.a. entre crochets est intégrable et la v.a. intégrable $1_A(M_v-M_u)1_B$ est $\underline{\underline{F}}_s$-mesurable (cf le b) de la définition d'une martingale). Par ailleurs, quitte à remplacer M_t par M_t-M_0 pour tout t, on peut supposer $M_0 = 0$, et on a alors $E[(M_t-M_s)^2] = E[M^2_t-M^2_s]$; un calcul élémentaire montre alors que, pour $Y\varepsilon\underline{\underline{E}}$, on a

$$E[(\int_0^t Y_s\, dX_s)^2] \leq \|Y\|^2_\infty E[M^2_t - M^2_0]$$

si bien que l'intégrale élémentaire est, pour chaque t , continue dans L^2 et donc a fortiori dans L^0 . Lorsque la martingale M n'est pas de carré intégrable, on montre, à l'aide de résultats de la théorie des martingales dont nous parlerons plus loin, la chose suivante : il existe une suite croissante (T_n) de v.a. ≥ 0 tendant vers $+\infty$ et, pour chaque n , une martingale de carré intégrable M^n et un $\mathbb{F}$-processus à variation finie V^n (les n sont des indices) tels que $M = M^n + V^n$ sur l'intervalle stochastique $[\![0,T_n[\![$, ce qui permet de conclure par localisation. Notez qu'en général les trajectoires d'une martingale, même bornée, ne sont pas à variation finie, soit qu'elles comportent trop de petits sauts, soit que, continues, elles oscillent trop fortement (penser au mouvement brownien ; en fait, (presque) toute trajectoire d'une martingale continue a une variation infinie sur tout intervalle sur lequel elle n'est pas constante).

c) Plus généralement, <u>toute</u> (<u>sous</u>,<u>sur</u>)<u>martingale est une semimartingale</u>. Soit, par exemple, X une surmartingale. D'abord, par localisation, on se ramène à montrer que la surmartingale X^n telle que l'on ait $X^n_t = X_t$ pour $t \leq n$ et $X^n_t = X_n$ pour $t > n$, est une semimartingale, si bien que l'on peut supposer X à trajectoires constantes pour $t \geq n$. Ensuite, quitte à considérer les surmartingales $X \wedge n$, $n \in \mathbb{N}$ et à utiliser encore un argument de localisation (toute trajectoire d'un processus càdlàg, ou càglàd, étant bornée sur tout compact), on peut aussi supposer X borné supérieurement. Les hypothèses sur X sont alors suffisantes pour pouvoir appliquer un théorème fondamental de la théorie des martingales (décomposition de Doob-Meyer) : on a $X = M - A$ où M est une martingale et A un $\mathbb{F}$-processus croissant.

d) <u>Tout processus càdlàg</u>, <u>homogène et à accroissements indépendants</u>, <u>est une semimartingale pour sa filtration naturelle complétée</u>. En effet, il résulte aisément de la décomposition de Lévy du processus que ce dernier est somme d'une martingale de carré intégrable et d'un processus à variation finie.

Tous nos exemples de $\mathbb{F}$-semimartingales sont du type "somme locale" d'une $\mathbb{F}$-martingale (même de carré intégrable) et d'un $\mathbb{F}$-processus à variation finie. En fait, il n'y en a point d'autres (et l'on retrouve alors une définition plus commune des semimartingales) : c'est ce que nous allons essayer d'expliquer maintenant.

4 THEOREME. <u>Un</u> $\mathbb{F}$-<u>processus càdlàg</u> X <u>est une semimartingale si et seulement s'il vérifie la condition</u> ($\underline{L}$) <u>suivante</u> : <u>il existe une suite croissante</u> (T_n) <u>de v.a.</u> ≥ 0 <u>tendant vers</u> $+\infty$ <u>et</u>, <u>pour chaque</u> n , <u>une</u> $\mathbb{F}$-<u>martingale</u> M^n <u>et un</u> $\mathbb{F}$-<u>processus à variation finie</u> V^n, <u>tels que l'on ait</u> $X = M^n + V^n$ <u>sur</u> $[\![0,T_n[\![$ <u>pour chaque</u> n . <u>De plus</u>, <u>la condition</u> ($\underline{L}$)

peut être renforcée en la condition ($\underline{\underline{L}}_b$) obtenue en demandant que, pour chaque n, T_n soit un $\mathbb{F}$-temps d'arrêt[1], M^n une $\mathbb{F}$-martingale bornée[2] et V^n un $\mathbb{F}$-processus à variation bornée[2].

DEMONSTRATION. Il résulte des n°2 et 3 que la condition est suffisante. La démonstration "ex nihilo" de la condition nécessaire est fort longue, mais comporte trois grandes étapes distinctes que nous allons décrire maintenant, sans entrer dans les détails techniques demandant trop de connaissances en théorie des martingales.

Etape A : (Théorie des martingales) D'abord, on montre que les propriétés ($\underline{\underline{L}}$) et ($\underline{\underline{L}}_b$) sont des propriétés "locales", ce qui est à peu près évident (pour l'initié, il n'est pas difficile de passer, dans ces situations, de "T v.a. ≥ 0" à "T $\mathbb{F}$-temps d'arrêt" car, si Z^1 et Z^2 sont deux $\mathbb{F}$-processus càdlàg, la v.a. T définie par

$$T(\omega) = \inf \{t \geq 0 : Z^1_t(\omega) \neq Z^2_t(\omega)\}$$

est un $\mathbb{F}$-temps d'arrêt). Ensuite, on montre que les propriétés ($\underline{\underline{L}}$) et ($\underline{\underline{L}}_b$) sont en fait équivalentes. Désignant par Z un $\mathbb{F}$-processus vérifiant ($\underline{\underline{L}}$), on se ramène aisément à considérer séparément le cas où Z est à variation finie, qui est très facile à traiter, et le cas où Z est une martingale, qui est plus coriace à cause des "grands" sauts possibles : il faut toute une technique (appelée compensation des sauts) et pas mal d'astuce pour y arriver - pendant longtemps d'ailleurs, on s'était contenté d'un résultat plus faible que ($\underline{\underline{L}}_b$), où l'on avait "martingale de carré intégrable" à la place de "martingale bornée". Ceci fait, on montre alors que si un $\mathbb{F}$-processus Z vérifie ($\underline{\underline{L}}_b$) relativement à notre probabilité P, il vérifie encore ($\underline{\underline{L}}_b$) relativement à toute probabilité Q équivalente à P. Ici encore, on se ramène aisément à considérer séparément le cas où Z est à variation bornée, qui est trivial, et le cas où Z est une martingale bornée, qui est plus délicat : on est amené à faire joujou avec la martingale M telle que M_t soit, pour chaque t, une densité de Q par rapport à P, Q et P étant restreintes à la tribu $\underline{\underline{F}}_t$, petit jeu au cours duquel on est amené à utiliser le théorème de décomposition de Doob-Meyer. Revenant finalement à notre semimartingale X, on peut alors se permettre deux choses pour se simplifier la vie : d'une part, puisque

---

(1) Cela signifie que l'indicatrice de $[\![0,T_n[\![$ est un $\mathbb{F}$-processus. Nous avons cependant évité (non sans difficultés !) d'introduire cette notion, en fait éminemment importante, car elle demande pas mal de connaissances pour être manipulée efficacement. Le lecteur peut donc ignorer cette précision sans inconvénient majeur.

(2) Cela doit être entendu au sens le plus fort : la borne ne dépend pas de la trajectoire, et chaque trajectoire "admet" cette borne sur tout $\mathbb{R}_+$.

la propriété ($\underline{\underline{L}}$) est "locale", on peut supposer que X est à trajectoires constantes pour $t \geq n$, ce qui permet de définir $\int_0^\infty Y_s\, dX_s$ pour tout $Y \in \underline{\underline{E}}$. D'autre part, puisque seule la classe d'équivalence de la probabilité P entre en jeu, on peut supposer de plus que, pour tout $Y \in \underline{\underline{E}}$, $\int_0^\infty Y_s\, dX_s$ est une v.a. intégrable : en effet, chacune de ces v.a. est intégrable si chaque X_t est intégrable, et chaque X_t est intégrable si $X_\infty^\# = \sup_t |X_t|$ l'est (noter que $X_\infty^\#$ est une v.a., car on peut prendre le sup sur les rationnels) ; or, comme X est càdlàg et à trajectoires constantes après n , $X_\infty^\#$ est une v.a. finie et peut donc être supposée intégrable, quitte à remplacer P par une loi équivalente.

Etape B : (Analyse fonctionnelle) Grâce à l'étape A , nous pouvons supposer que $Y \rightarrow \int_0^\infty Y_s\, dX_s$ est un opérateur continu de $\underline{\underline{E}}$ dans L_0 , à valeurs dans L^1. Dans ce cas, l'image de la boule unité B de $\underline{\underline{E}}$ par cet opérateur est un convexe symétrique de L^1, borné dans L^0. Une application judicieuse du théorème de Hahn-Banach, que nous détaillerons après avoir vu la dernière étape, montre alors qu'il existe une probabilité Q équivalente à notre probabilité P telle que l'on ait

$$\sup_{Y \in B} |E_Q[\int_0^\infty Y_s\, dX_s]| < +\infty$$

Autrement dit, nous pouvons maintenant supposer en plus que la forme linéaire $Y \rightarrow E[\int_0^\infty Y_s\, dX_s]$ sur $\underline{\underline{E}}$ est continue. La dernière étape va montrer que, dans ces conditions, X est la différence de deux surmartingales et donc un processus vérifiant ($\underline{\underline{L}}$) d'après le n°3 (il est clair que les processus vérifiant ($\underline{\underline{L}}$) forment un espace vectoriel).

Etape C : (Théorie des martingales) En fait, nous allons seulement utiliser la continuité de la forme $E[\int_0^\infty \ldots]$, laquelle suffit donc à entrainer la continuité de l'opérateur $\int_0^\infty \ldots$ dans L^0 (mais pas dans L^1). Avouons que cela est encore mal compris, encore qu'il soit tout à fait naturel que la théorie des martingales s'introduise à ce stade : pour un $\mathbb{F}$-processus càdlàg H tel que H_t soit intégrable pour tout t , dire que H est une martingale équivaut à dire (c'est très facile à voir) que, pour chaque t , la forme $E[\int_0^t Y_s\, dH_s]$ sur $\underline{\underline{E}}$ est nulle (et donc continue !). Un $\mathbb{F}$-processus càdlàg H tel que, pour chaque t , on ait

$$\mathrm{Var}_t(H) = \sup_{Y \in B} |E[\int_0^t Y_s\, dH_s]| < +\infty ,$$

B étant la boule unité de $\underline{\underline{E}}$, et H_t étant supposé intégrable pour tout t, est appelé une quasimartingale[(1)]. La notation "$\mathrm{Var}_t(H)$" - qui rappelle "variation" - se comprend mieux lorqu'on écrit cette quantité sous une autre forme. Appelons $\mathbb{S}(t)$ l'ensemble des subdivisions finies $\sigma =$

---

(1) La terminologie est fluctuante. Pour certains auteurs, une quasimartingale H doit vérifier $\sup_t \mathrm{Var}_t(H) < +\infty$; d'autres, encore plus exigeants, demandent $\sup_t [\mathrm{Var}_t(H) + E[|H_t|]] < +\infty$.

($t_0 = 0, t_1, \dots, t_i, t_{i+1}, \dots, t_n = t$) de l'intervalle $[0,t]$; un calcul élémentaire sur les espérances conditionnelles montre alors que, pour tout t , l'on a

$$\mathrm{Var}_t(H) = \sup_{\sigma\in S(t)} E[\sum_{\sigma} |E[H_{t_{i+1}} - H_{t_i} | \underline{\underline{F}}_{t_i}]|]$$

Sur cette expression, il est facile de voir que toute (sous,sur)martingale est une quasimartingale, $\mathrm{Var}_t(H)$ valant alors $|E[H_t - H_0]|$. Et le théorème de décomposition de Rao (analogue au théorème de décomposition d'une fonction à variation finie en différence de deux fonctions monotones) assure que toute quasimartingale est la différence de deux surmartingales (ou, si l'on veut, de deux sousmartingales). C'est fini.

Nous complétons maintenant la démonstration de l'étape B . Nous avons un espace probabilisé $(\Omega,\underline{\underline{F}},P)$ et un convexe symétrique C de $L^1(P)$, borné dans L^0. Il s'agit de trouver une probabilité Q équivalente à P telle que $\sup_{f\in C} |E_Q[f]| < +\infty$. Dire que C est borné dans L^0 revient à dire que, pour tout $\varepsilon>0$, il existe $c_\varepsilon>0$ telle que

$$\sup_{f\in C} P\{|f| \geq c_\varepsilon\} \leq \varepsilon$$

Soit alors $A\in\underline{\underline{F}}$ tel que $P(A) > 0$, prenons $\varepsilon = P(A)/2$ et posons $c = 2c_\varepsilon$; on a alors en particulier, en posant $h^+ = \sup(h,0)$ comme d'habitude,

$$\inf_{f\in C} E[(c1_A - f)^+] > 0 ,$$

situation que nous exprimerons en disant que $c1_A$ <u>est loin d'être majoré par un élément de</u> C . Comme notre convexe est symétrique, l'existence de notre probabilité Q est alors assurée par la belle application suivante(1) du théorème de Hahn-Banach

5 THEOREME. <u>Soit</u> $(\Omega,\underline{\underline{F}},P)$ <u>un espace probabilisé et soit</u> C <u>un convexe de</u> $L^1(P)$ <u>tel que</u> $0\in C$. <u>Les conditions suivantes sont équivalentes</u>

a) <u>Pour tout</u> $A\in\underline{\underline{F}}$ <u>tel que</u> $P(A) > 0$, <u>il existe une constante</u> $c>0$ <u>telle que</u> $c1_A$ <u>soit loin d'être majoré par un élément de</u> C .

b) <u>Il existe</u> $g\in L^\infty$ <u>tel que l'on ait</u> $g > 0$ p.s. <u>et</u> $\sup_{f\in C} E[fg] < +\infty$.

DEMONSTRATION. Montrons d'abord que non a) $\Rightarrow$ non b) (et même un peu mieux). Supposons qu'il existe $h\in L^1_+$ (cône positif de L^1) tel que $E[h] > 0$ et $\inf_{f\in C}E[(ch - f)^+] = 0$ pour tout $c>0$; on peut alors écrire $nh = f_n - u_n - v_n$ avec $f_n\in C$, $u_n\in L^1_+$ et $E[|v_n|] < 1/n$ pour tout n si bien que l'on a $E[f_n g] \geq nE[hg] - 1/n$ pour tout $g\in L^\infty_+$ d'où l'impossibilité

---

(1) Il s'agit d'un résultat tout récent de YAN (Jia-An), que nous avons préféré donner ici plutôt qu'un argument plus ancien (et tout aussi beau) de MOKOBODZKI, ce dernier étant exposé maintenant en de nombreux endroits. Signalons cependant que la démonstration de MOKOBODZKI, qui repose, comme un résultat encore plus ancien de NIKICHINE, sur un théorème de minimax, permet de montrer que, si (C_n) est une suite de convexes de $L^1(P)$, chacun d'eux étant borné dans L^0, alors il existe une probabilité Q équivalente à P , à densité bornée, telle que l'on ait $\sup_{f\in C_n}|E_Q[f]| < +\infty$ pour tout n .

d'avoir b) satisfaite. Voyons maintenant a) $\Rightarrow$ b). Soit $\underline{\underline{G}}$ l'ensemble des $g \varepsilon L^{\infty}_{+}$ tels que $\sup_{f \varepsilon C} E[fg] < +\infty$ (on a $0 \varepsilon \underline{\underline{G}}$), et soit (g_n) une suite d'éléments de $\underline{\underline{G}}$; choisissons des $c_n > 0$ tels que les séries $\sum c_n \|g_n\|_{\infty}$ et $\sum c_n k_n$ convergent, où $k_n = \sup(0, \sup_{f \varepsilon C} E[fg_n])$: alors, $g = \sum c_n g_n$ appartient encore à $\underline{\underline{G}}$ et $\{g > 0\}$ est la réunion des $\{g_n > 0\}$. On en déduit l'existence, à un ensemble négligeable près, d'un plus grand ensemble $G \varepsilon \underline{\underline{F}}$ tel qu'il existe $g \varepsilon \underline{\underline{G}}$ vérifiant $G = \{g > 0\}$. Nous allons montrer, en raisonnant par l'absurde, que $P(G) = 1$, ce qui établira le théorème. Supposons donc que $P(G^c)$ est > 0. Alors, posant $G^c = A$, on sait qu'il existe $c > 0$ tel que $c1_A$ soit loin d'être majoré par C ; autrement dit, $c1_A$ n'appartient pas à l'adhérence dans L^1 du convexe $C - L^1_+ = \{f - h,\ f \varepsilon C,\ h \varepsilon L^1_+\}$. D'après le théorème de Hahn-Banach, il existe donc $g_A \varepsilon L^{\infty}$ tel que l'on ait

$$\sup_{f \varepsilon C, h \varepsilon L^1_+} E[(f-h)g_A] < c\,E[1_A g_A]$$

Remplaçant h par $n1_{\Omega}$, on voit que g_A est ≥ 0 ; prenant $h = 0$, on voit que l'on a $\sup_{f \varepsilon C} E[f g_A] < +\infty$ et, aussi $E[1_A g_A] > 0$ en prenant de plus $f = 0$ (il se peut que a) $\not\Rightarrow$ b) si on ne suppose pas $0 \varepsilon C$ - exercice !). Mais alors, si $g \varepsilon \underline{\underline{G}}$ vérifie $\{g > 0\} = G$, on a $g + g_A \varepsilon \underline{\underline{G}}$ et $P\{g + g_A > 0\} > P(G)$, ce qui contredit la maximalité de G.

Revenons maintenant à notre semimartingale X. Nous avons vu que, pour toute probabilité Q équivalente à P, X se décompose "localement" en la somme d'une martingale et d'un processus à variation finie, l'écriture de X sous une telle forme dépendant de Q (et n'étant pas unique par ailleurs). Nous verrons plus loin que, connaissant une telle écriture locale de X relativement à P, on peut trouver explicitement une écriture locale de X relativement à Q (formule de Girsanov). On peut aussi se demander si, en choisissant judicieusement Q, on peut arriver à obtenir une décomposition "globale" de X relativement à Q. Cela est effectivement possible, dans un sens très fort, comme l'affirme le théorème suivant (que nous ne tenterons pas de démontrer ici)

6 THEOREME. Soit X une $\mathbb{F}$-semimartingale. Il existe une probabilité Q équivalente à P telle que X s'écrive $X = M + V$ où M est une $\mathbb{F}$-martingale relativement à Q et V un $\mathbb{F}$-processus à variation finie tels que, pour tout t et tout $p \varepsilon [1, \infty[$, la v.a. M_t appartienne à $L^p(Q)$ ainsi que la variation $\int_0^t |dV_s|$ de V sur $[0,t]$.

REMARQUE. Pour $p \varepsilon [1, \infty[$ fixé, on peut montrer qu'une semimartingale X peut s'écrire $X = M + V$ avec, pour tout t, $\int_0^t |dV_s| \varepsilon L^p$ et $\sup_{s \leq t} |M_s| \varepsilon L^p$ (qui équivaut à $M_t \varepsilon L^p$ pour $p > 1$), si et seulement si, pour tout t, l'intégrale élémentaire est un opérateur continu de $\underline{\underline{E}}$ dans L^p. Revenant alors à notre théorème (dont la démonstration utilise le n°4 et donc le n°5), on voit qu'il implique l'existence de Q équivalente à P telle que, pour tout $p\varepsilon[1,\infty[$, on ait $\sup_{Y\varepsilon B} E_Q[|\int_0^t Y_s\, dX_s|^p] < +\infty$ pour tout t, B étant la boule unité de $\underline{\underline{E}}$. Cela est beaucoup plus fort, en se contentant même de $p=1$, que ce que nous avions obtenu au n°5, soit $\sup_{Y\varepsilon B}|E_Q[\int_0^t Y_s\, dX_s]| < +\infty$. Signalons à ce sujet que l'on déduit aisément d'un résultat de NIKICHINE ([1],th.16) l'existence d'un espace probabilisé $(\Omega,\underline{\underline{F}},P)$ et d'un convexe symétrique C de $L^1(P)$, borné dans L^0, tel que C ne puisse être borné dans aucun $L^1(Q)$ avec Q équivalente à P.

INTEGRALE STOCHASTIQUE A LA RIEMANN

L'idée de départ est très simple : puisque, pour chaque t, l'intégrale élémentaire définit un opérateur linéaire continu de $\underline{\underline{E}}$, muni de la convergence uniforme, dans L^0, muni de la convergence en probabilité, on peut, L^0 étant complet pour cette structure, prolonger la définition de l'intégrale au complété de $\underline{\underline{E}}$ par uniforme continuité. C'est bien ainsi que nous allons procéder, mais pas tout de suite, car, d'une part la convergence uniforme sur $\underline{\underline{E}}$ est trop forte (il serait plus naturel d'avoir une structure uniforme qui soit quelque-chose comme la convergence uniforme en t et la convergence en probabilité en ω) et, d'autre part, $\underline{\underline{E}}$ lui-même est trop pauvre (tout $Y\varepsilon\underline{\underline{E}}$ saute trop régulièrement en ce sens que, si T est une v.a. ≥ 0 à loi diffuse, il n'y a p.s. pas de saut de Y à l'instant aléatoire T), pour que le complété de $\underline{\underline{E}}$ contienne tous les $\mathbb{F}$-processus càglàd. Aussi allons nous d'abord définir une bonne structure uniforme sur les processus, puis étudier la continuité de l'intégrale élémentaire sur un domaine plus gros que $\underline{\underline{E}}$. Par ailleurs, comme nous n'allons plus toucher à notre filtration $\mathbb{F}$ et que nous n'allons considérer que des $\mathbb{F}$-processus, nous convenons de dire "processus" tout court au lieu de "$\mathbb{F}$-processus" et cela, jusqu'à la fin de notre exposé.

A/ Choix d'une bonne structure uniforme

A tout processus Z à trajectoires localement bornées on associe le processus(1) Z^+ défini par $Z^+_t = \sup_{s\leq t}|Z_s|$; il est clair que Z^+ est un processus càglàd (resp càdlàg) si Z l'est. Nous définissons alors sur l'espace des processus la structure de la convergence uniforme sur tout compact en probabilité - en abrégé (ouf !), la convergence ucp -

----

(1) Il n'est pas évident, qu'en général, Z^+ soit un $\mathbb{F}$-processus (Z en étant un), mais c'est vrai. Dans le cas où Z est càdlàg ou càglàd, le sup peut être pris sur les rationnels, et c'est alors trivial. Par ailleurs, la notation officielle comporte une étoile, dont le signe "+" est la meilleure approximation sur mon clavier...

par la distance suivante, invariante par translation,

$$d(Z^1,Z^2) = \sum_n 2^{-n}\, E[(Z^1 - Z^2)^{*}_n \wedge 1] \,.$$

Il est facile de voir qu'une suite de processus (Z^k) – à trajectoires localement bornées, mais c'est en fait sans importance – converge vers 0 pour la topologie ucp si et seulement si, pour toute v.a. $T \geq 0$, la suite en k des v.a.[(1)] Z^{k*}_T tend vers 0 en probabilité, et qu'il suffit de vérifier que, pour une suite croissante (T_n) de v.a. ≥ 0 tendant vers $+\infty$, la suite en k des $Z^{k*}_{T_n}$ tend vers 0 en probabilité pour tout n. Cette structure est donc "locale" au sens où nous[(2)] entendons ce mot depuis le début de notre exposé, ce qui la rend très maniable. Par ailleurs, il est clair que l'espace $\underline{\underline{D}}$ des processus càdlàg et l'espace $\underline{\underline{G}}$ des processus càglàd sont complets, et, avec deux doigts de métier, on vérifie sans peine que l'espace $\underline{\underline{G}}^e$ des processus càglàd, bornés, en escalier, est dense dans $\underline{\underline{G}}$.

B/ Continuité de l'intégrale élémentaire

Nous nous donnons une semimartingale X et nous considérons maintenant l'intégrale élémentaire par rapport à X comme <u>définie sur</u> $\underline{\underline{G}}^e$, muni de la structure ucp, et <u>à valeur dans</u> $\underline{\underline{D}}$, muni aussi de la structure ucp (autrement dit, nous ne regardons plus $Z_t = \int_0^t Y_s\, dX_s$ pour chaque t, mais tout le processus $Z = (Z_t)$). Et nous allons démontrer que l'intégrale élémentaire est alors un opérateur linéaire <u>continu</u> de $\underline{\underline{G}}^e$ dans $\underline{\underline{D}}$, ce qui est nettement mieux que ce que nous donne a priori notre définition d'une semimartingale.

A cette fin, nous introduisons d'abord un procédé d'approximation qui nous permettra d'approcher convenablement les éléments de $\underline{\underline{G}}^e$ par des éléments de $\underline{\underline{E}}$. Nous associons, pour tout $k \in \mathbb{N}$, un processus ${}^kZ \in \underline{\underline{G}}^e$ à tout processus Z en posant ${}^kZ_0 = Z_0$ et

$${}^kZ_t = Z_{(n-1)2^{-k}} \qquad \text{pour } (n-1)2^{-k} < t \leq n2^{-k} \leq k$$

$${}^kZ_t = Z_k \qquad \text{pour } t > k\,.$$

Notez que, pour k fixé, $Z \to {}^kZ$ est un opérateur linéaire, de norme ≤ 1 pour la convergence uniforme et que, pour Z càglàd, Z est la limite simple des ${}^kZ$ (si Z est càdlàg, la limite simple des ${}^kZ$ est égal au processus Z_- tel que $Z_{-0} = Z_0$ et $Z_{-t} = Z_{t-}$, limite à gauche de Z en t, pour $t > 0$; nous utiliserons ce résultat plus tard). Par ailleurs, il est très facile de voir que, si Z est borné et $Z_0 = 0$, alors ${}^kZ$ appartient, pour tout k, au complété de $\underline{\underline{E}}$ pour la convergence uniforme. Le lecteur aura compris que notre opérateur $Z \to {}^kZ$ va nous servir à

---

(1) Si Z est un processus et T une v.a. ≥ 0, on note Z_T la v.a. définie par $Z_T(\omega) = Z_{T(\omega)}(\omega)$

(2) Le mot "local" a officiellement un sens voisin, mais distinct.

introduire des sommes riemaniennes, le fait que nous ayons choisi un procédé dyadique d'approximation ne jouant qu'un rôle mineur d'ordre typographique.

7 THEOREME. L'intégrale élémentaire par rapport à une semimartingale X définit un opérateur linéaire continu $Y \to \int Y\,dX$ de $\underline{\underline{G}}^e$, muni de la topologie ucp, dans $\underline{\underline{D}}$, muni de la topologie ucp.

DEMONSTRATION. Nous découpons la démonstration en trois petites étapes.

1) Nous démontrons d'abord que, pour t fixé, $Y \to \int_0^t Y_s\,dX_s$ est continu de $\underline{\underline{G}}^e$ dans L^0 quand $\underline{\underline{G}}^e$ est muni de la convergence uniforme. Soit (Y^n) une suite dans $\underline{\underline{G}}^e$ convergeant uniformément vers 0 et écrivons

$$\int_0^t Y_s^n\,dX_s = \int_0^t (Y_s^n - {}^kY_s^n)\,dX_s + \int_0^t {}^kY_s^n\,dX_s$$

Comme on a $\|{}^kY^n\|_\infty \leq \|Y^n\|_\infty$ et que les ${}^kY^n$ appartiennent au complété de $\underline{\underline{E}}$ pour la convergence uniforme, on peut choisir un n de sorte que la dernière intégrale soit petite dans L^0, uniformément en k. Par ailleurs, pour n fixé, et pour tout ω, on a $\lim_k \int_0^t {}^kY_s^n(\omega)\,dX_s(\omega)$ égal à $\int_0^t Y_s^n(\omega)\,dX_s(\omega)$ car X est continu à droite tandis que les trajectoires de Y^n sont en escalier et continues à gauche. C'est fini.

2) Nous supposons toujours $\underline{\underline{G}}^e$ muni de la convergence uniforme et montrons maintenant que $Y \to \int Y\,dX$ est continu de $\underline{\underline{G}}^e$ dans $\underline{\underline{D}}$, muni quant à lui de la convergence ucp. L'espace $\underline{\underline{D}}$ étant un espace vectoriel topologique, il nous suffit de montrer que l'image de la boule unité B de $\underline{\underline{G}}^e$ par notre intégrale est un borné de $\underline{\underline{D}}$, soit encore que, pour tout t, l'ensemble des v.a. $(\int Y\,dX)_t^*$ avec $Y \in B$ est un borné de L^0. Or nous savons, grâce à 1), que, pour tout $\varepsilon > 0$ et tout t, il existe une constante $c_\varepsilon > 0$ telle que

$$\sup_{Y\in B} P\{|\int_0^t Y_s\,dX_s| \geq c_\varepsilon\} \leq \varepsilon$$

Fixons t et $Y \in B$, et regardons l'ensemble $A = \{Z_t^* > c_\varepsilon\}$ avec $Z = \int Y\,dX$. Posons, pour tout $\omega \in A$, $T(\omega) = \inf\,\{u \leq t : |Z_u| > c_\varepsilon\}$ et, pour tout $\omega \notin A^c$, $T(\omega) = t$; on vérifie sans peine que T ainsi définie est une v.a. (noter que l'inf peut être pris sur $(\mathbb{Q} \cap [0,t]) \cup \{t\}$) et que l'indicatrice de l'intervalle stochastique $]\!]0,T]\!]$ est un élément de B ; d'autre part, comme Z est càdlàg, on a $|Z_T| \geq c_\varepsilon$ sur A. Maintenant, on a aussi

$$Z_T = \int_0^t 1_{]\!]0,T]\!]}(s)\,Y_s\,dX_s \;;$$

comme $1_{]\!]0,T]\!]}Y$ est un élément de B, il est alors clair que l'on a

$$\sup_{Y\in B} P\{(\int Y\,dX)_t^* > c_\varepsilon\} \leq \varepsilon \;.$$

3) Nous munissons finalement $\underline{\underline{G}}^e$ de la topologie ucp et démontrons le théorème. Soit (Y^n) une suite dans $\underline{\underline{G}}^e$ convergeant ucp vers 0. Posons $Z^n = \int Y_n\,dX_n$ et raisonnons par l'absurde. Quitte à extraire à deux reprises une sous-suite, on peut alors supposer qu'il existe un t et un $\varepsilon > 0$ tels que (Y_t^{n*}) converge p.s. vers 0 tandis que l'on a

$P\{Z_t^{n\#} > \varepsilon\} > \varepsilon$ pour tout n. Soit alors $A \in \underline{\underline{F}}_\infty$, de probabilité $> 1-\frac{\varepsilon}{2}$, tel que les $Y_t^{n\#}$ convergent uniformément vers 0 sur A. Posons, pour tout n, $\varepsilon_n = \|1_A Y_t^{n\#}\|_\infty$ et définissons un élément U^n de $\underline{\underline{G}}^e$ en prenant, pour tout s et tout ω, $U_s^n(\omega) = Y_s^n(\omega)$ si on a $-\varepsilon_n \leq Y_s^n(\omega) \leq \varepsilon_n$ et $U_s^n(\omega) = 0$ sinon. D'après 2), on a lim $\int U^n dX = 0$ dans $\underline{\underline{D}}$ et, comme l'intégrale élémentaire est prise trajectoire par trajectoire, on a pour tout n $\int_0^t U_s^n dX_s = \int_0^t Y_s^n dX_s$ sur A. On en déduit que les v.a. $1_A Z_t^{n\#}$ tendent vers 0 en probabilité, d'où une contradiction.

C/ Intégrale stochastique

Les espaces $\underline{\underline{G}}^e$, $\underline{\underline{G}}$ et $\underline{\underline{D}}$ seront désormais munis tacitement de la structure ucp. Rappelons que $\underline{\underline{G}}^e$ est dense dans $\underline{\underline{G}}$ et que $\underline{\underline{G}}$ et $\underline{\underline{D}}$ sont complets.

7 DEFINITION. Soit X une semimartingale. On appelle intégrale stochastique (de Riemann) par rapport à X l'opérateur linéaire continu $Y \to Z = \int Y\,dX$ de $\underline{\underline{G}}$ dans $\underline{\underline{D}}$ obtenu en prolongeant l'intégrale élémentaire de $\underline{\underline{G}}^e$ dans $\underline{\underline{D}}$ par continuité.

Il est clair que $\int Y\,dX$ est aussi linéaire en X. Nous passons maintenant en revue un certain nombre de propriétés fondamentales de l'intégrale stochastique (en abrégé, i.s.).

8 PROPRIETES. a) (Invariance) Il est clair que l'on ne change pas l'i.s. si on remplace P par une probabilité équivalente, ou si l'on remplace la filtration $\mathbb{F}$ par une filtration $\mathbb{G}$ plus petite telle que X reste un $\mathbb{G}$-processus.

b) (Associativité) Pour $Y \in \underline{\underline{G}}$ fixé, le processus $Z = \int Y\,dX$ est une semimartingale et, pour tout $U \in \underline{\underline{G}}$, on a $\int U\,dZ = \int UY\,dX$, ce qui donne un sens à l'écriture différentielle $dZ = Y\,dX$. On démontre cela en supposant d'abord $Y \in \underline{\underline{G}}^e$ et U parcourant $\underline{\underline{G}}^e$ (c'est alors très facile), puis en passant à la limite (point n'est besoin d'invoquer Banach-Steinhaus ; il suffit d'utiliser les n°1, 6 et 7 en remarquant que $\underline{\underline{G}}$ est une algèbre topologique).

c) (Sauts de l'i.s.) Si T est une v.a. > 0, on a $\Delta Z_T = Y_T\,\Delta X_T$ p.s., où ΔZ est le processus $Z - Z_-$ des sauts de Z (et de même pour X). On démontre cela encore une fois en supposant d'abord $Y \in \underline{\underline{G}}^e$, puis en passant à la limite.

d) (Rapport avec l'intégrale de Stieljes) Soit $A \in \underline{\underline{F}}_\infty$ tel que, pour (presque) tout $\omega \in A$, la trajectoire $t \to X_t(\omega)$ soit à variation finie. Alors, pour presque tout $\omega \in A$, la trajectoire $t \to Z_t(\omega)$ est égale à ce que donne l'intégrale de Stieljes de $t \to Y_t(\omega)$ par rapport à $dX_t(\omega)$. Cela se démontre comme les points précédents.

e) (Caractère local) Soit X' une autre semimartingale et supposons que l'on ait $X = X'$ sur un intervalle stochastique $]S,T]$, où S et T sont deux v.a. ≥ 0 telles que $S \leq T$; alors, si Y et Y' sont deux éléments de $\underline{\underline{G}}$ tels que $Y = Y'$ sur $]S,T]$, on a $Z_T - Z_S = Z'_T - Z'_S$ p.s. (où l'on a posé $Z = \int Y\,dX$ et $Z' = \int Y'\,dX'$). Cela se démontre en employant le même argument que précédemment, mais il faut un peu de métier pour traiter le cas où l'indicatrice de $]S,T]$ n'appartient pas à $\underline{\underline{G}}$ (i.e. le cas où l'un des S,T n'est pas un temps d'arrêt).

f) (Approximation riemannienne) Si Y est un processus càglàd (resp càdlàg), alors $\int {}^kY\,dX$ converge dans $\underline{\underline{D}}$ vers $\int Y\,dX$ (resp $\int Y_-\,dX$), d'où une approximation pour chaque t de l'intégrale susdite par des sommes riemanniennes en développant $\int_0^t {}^kY_s\,dX_s$. Pour voir cela, en nous contentant du cas où Y est càdlàg, on considère une suite (Y^n) dans $\underline{\underline{G}}^e$ convergeant vers Y_- dans $\underline{\underline{G}}$ et on écrit

$$\int (Y_- - {}^kY)\,dX = \int (Y_- - Y^n)\,dX + \int (Y^n - {}^kY^n)\,dX + \int ({}^kY^n - {}^kY)\,dX$$

Dans le membre de droite, le premier terme est petit dans $\underline{\underline{D}}$ quand n est grand, ainsi que le troisième, uniformément en k ; quant au second, pour n et ω fixés, il converge vers 0 uniformément sur tout compact de $\mathbb{R}_+$ car X est continu à droite et Y^n continu à gauche, en escalier.

Nous allons introduire maintenant (à l'aide de notre i.s., mais on peut procéder autrement, quoique d'une manière moins élégante) un concept très important - le crochet droit $[X,X]$ d'une semimartingale X - qui se trouve au coeur de tout le développement "moderne" de la théorie des martingales (inégalités du type Burkholder, espaces H^1 et BMO, etc) ; c'est d'ailleurs grâce à cette notion que l'on peut par exemple démontrer le théorème 6.

9 DEFINITION. On appelle crochet droit de deux semimartingales X, Y le processus càdlàg $[X,Y]$ défini par

$$[X,Y] = XY - \int Y_-\,dX - \int X_-\,dY ,$$

expression appelée "formule d'intégration par parties".

Quand X et Y sont des processus à variation finie, les intégrales sont des intégrales de Stieljes, et on retrouve alors la formule classique d'intégration par parties : $[X,Y]_t = X_0Y_0 + \sum_{0<s\leq t} \Delta X_s \Delta Y_s$, la famille des produits des sauts de X et Y étant sommable pour tout ω. Mais, on peut avoir $[X,Y] \neq 0$ pour des semimartingales continues X,Y telles que $X_0 = Y_0 = 0$: par exemple, si X est le mouvement brownien (unidimensionnel, issu de 0), on a $[X,X]_t = t$ pour tout t.

Avant de donner quelques propriétés du crochet, nous en établissons une approximation discrète qui en fera mieux comprendre la structure.

10 THEOREME. Soient X et Y deux semimartingales. Pour chaque t, la v.a. $[X,Y]_t$ est limite en probabilité, quand k tend vers $+\infty$, de

sommes de la forme

$$X_0Y_0 + \sum_{0\leq i\leq n} (X_{t_{i+1}} - X_{t_i})(Y_{t_{i+1}} - Y_{t_i})$$

avec n = partie entière de $2^k t$, $t_i = i2^{-k}$ pour $0\leq i\leq n$ et $t_{n+1} = t$.

DEMONSTRATION. On sait que $[X,Y]_t$ est limite en probabilité, pour k tendant vers $+\infty$, de $X_tY_t - \int_0^t {}^kY_s\, dX_s - \int_0^t {}^kX_s\, dY_s$. On écrit alors les deux intégrales sous forme de sommes finies, on développe X_tY_t sous la forme $X_0Y_0 + \sum_i (X_{t_{i+1}}Y_{t_{i+1}} - X_{t_i}Y_{t_i})$, et on trouve l'approximation voulue pour $k\geq t$.

REMARQUE. Plus généralement, on peut montrer que $[X,Y]_t$ est limite en probabilité de sommes de la forme considérée où (t_i) parcourt toutes les subdivisions finies de $[0,t]$, quand le pas des subdivisions tend vers 0. Prenant $X = Y$, on voit que, si une semimartingale n'a pas en général une variation finie, elle a toujours, en un certain sens, une 2-variation finie. Le "en un certain sens" provient du fait que les subdivisions considérées ne dépendent pas des trajectoires : prise trajectoire par trajectoire, la "vraie" 2-variation du mouvement brownien est p.s. infinie sur tout intervalle non dégénéré !

11 PROPRIETES. a) $[X,X]$ est un processus croissant (cela se voit immédiatement sur l'approximation discrète), et $[X,Y]$ est un processus à variation finie (en effet, on a $[X,Y] = \frac{1}{4}([X+Y,X+Y] - [X-Y,X-Y])$. Revenant à la formule de définition de $[X,Y]$, on en déduit que le produit de deux semimartingales est une semimartingale.

b) $[X,X]$ se comporte comme une forme quadratique positive, mais à valeur dans L^0. En particulier, posant $[X,Y]_s^t = [X,Y]_t - [X,Y]_s$, on a l'inégalité de Kunita-Watanabé

$$|[X,Y]_s^t| \leq ([X,X]_s^t)^{1/2} ([Y,Y]_s^t)^{1/2} .$$

c) Pour toute v.a. $T > 0$, on a $\Delta[X,Y]_T = \Delta X_T \Delta Y_T$ p.s. . Cela résulte aisément de l'identification des sauts d'une i.s..

d) On peut montrer que, si Y est à variation finie, on a

$$[X,Y]_t = X_0Y_0 + \sum_{0<s\leq t} \Delta X_s \Delta Y_s ,$$

la famille des produits des sauts étant sommable pour (presque) tout ω, même si X n'est pas à variation finie.

e) On peut aussi montrer que l'on a $[X,X] = 0$ si et seulement si X est un processus continu à variation finie tel que $X_0 = 0$ et que, plus généralement, la partie "continue" du processus croissant $[X,X]$ provient de la partie "martingale continue" de X lorsque ce dernier est décomposé localement en somme d'une martingale et d'un processus à variation finie.

f) Si Y est de la forme $\int H\, dU$, où H appartient à $\underline{\underline{G}}$ et U est une semimartingale, alors on a $[X,Y] = \int H\, d[X,U]$ et donc, en particulier,

$[\int H\,dU, \int H\,dU] = \int H^2\,d[U,U]$. Il suffit de traiter le cas $X = U$, le cas général s'y ramenant par polarisation. On établit d'abord la formule dans le cas simple où H est l'indicatrice d'un rectangle $]u,\infty[\times A$ avec $A\in\underline{\underline{F}}_u$, puis pour $H\in\underline{\underline{E}}$ par linéarité ; on obtient ensuite le cas où H appartient au complété de $\underline{\underline{E}}$ pour la convergence uniforme, puis, finalement, le cas où H appartient à $\underline{\underline{G}}$, par passage à la limite. Nous détaillons le dernier pas (le précédent étant analogue). Soit donc $H\in\underline{\underline{G}}$ que l'on approche par les ${}^kH$; posant $Y^k = \int {}^kH\,dX$, on sait que les Y^k convergent dans $\underline{\underline{D}}$ vers $Y = \int H\,dX$. Ecrivons

$$\int {}^kH\,d[X,X] = [X,Y^k] = XY^k - \int Y^k_-\,dX - \int X_-\,dY^k$$

Le seul terme posant un petit problème dans le passage à la limite est le dernier $\int X_-\,dY^k$. Mais, grâce à l'associativité, nous avons

$$\int X_-\,dY^k = \int X_-\,({}^kH\,dX) = \int {}^kH\,(X_-\,dX)$$

où $X_-\,dX$ est la différentielle d'une semimartingale ; il n'y a alors plus de difficulté pour passer à la limite.

Avant d'aborder le calcul différentiel stochastique, qui n'utilise que notre "petite" intégrale stochastique, nous allons dire maintenant quelques mots de la "grande" .

INTEGRALE STOCHASTIQUE A LA LEBESGUE

Nous partons toujours de l'intégrale élémentaire $\int Y\,dX$ par rapport à une semimartingale X, Y appartenant à $\underline{\underline{E}}$, et nous voulons étendre la définition de l'intégrale à une vaste algèbre $\underline{\underline{Y}}$ de processus, de sorte à avoir un "théorème de convergence simple dominée" , ce qui est beaucoup mieux que ce que nous avons pour l'instant.

Nous considérons maintenant la tribu $\underline{\underline{P}}$ sur $\mathbb{R}_+\times\Omega$ engendrée par les processus prévisibles élémentaires (rappelons qu'un processus est une application de $\mathbb{R}_+\times\Omega$ dans $\mathbb{R}$; $\mathbb{R}$ est muni de sa tribu borélienne) ou, plus précisément, par les processus Z de la forme $Z = Z_0 + Y$, Z_0 v.a. $\underline{\underline{F}}_0$-mesurable et $Y\in\underline{\underline{E}}$ (ceci, parce que nos éléments de $\underline{\underline{E}}$ sont tous nuls à l'instant 0). La tribu $\underline{\underline{P}}$ est appelée <u>tribu des ensembles prévisibles</u> et les processus mesurables par rapport à $\underline{\underline{P}}$ sont appelés <u>processus prévisibles</u> ; le nom provient en particulier du fait que les générateurs de $\underline{\underline{P}}$ sont des processus càglàd, et donc des processus bien connus à l'instant t si on les connait pour tout $s < t$. Il est facile de voir que <u>tout processus càglàd est prévisible</u> (Z càglàd est approché simplement par les ${}^kZ$), mais il est faux, en général, qu'un processus càdlàg soit prévisible.

Un processus prévisible Y est dit <u>localement borné</u> si $Y^\#_t = \sup_{s\leq t}|Y_s|$ est fini pour tout t ; on voit aisément que Y est localement borné si et seulement s'il existe une suite croissante (T_n) de v.a. ≥ 0 tendant vers $+\infty$ et, pour chaque n, une constante $c_n > 0$, telles que Y soit majoré en module par c_n sur l'intervalle stochastique $[\![0,T_n[\![$ pour tout n[1]. Nous noterons $\underline{\underline{P}}^1_b$ l'algèbre des processus prévisibles localement bornés, qui contient l'algèbre $\underline{\underline{G}}$ des processus càglàd.

La construction de l'intégrale stochastique (en abrégé, i.s.), dont nous indiquerons les grandes lignes, aboutit finalement au résultat suivant

12 THEOREME. Soit X une semimartingale. Il existe un unique opérateur linéaire $Y \to \int Y\,dX$ de $\underline{\underline{P}}^1_b$ dans $\underline{\underline{D}}$ vérifiant les conditions

1) Pour $Y \in \underline{\underline{E}}$, $\int Y\,dX$ est donné par l'intégrale élémentaire

2) Si (Y^n) est une suite dans $\underline{\underline{P}}^1_b$ majorée en module par un élément de $\underline{\underline{P}}^1_b$ et convergeant simplement vers Y ($\in \underline{\underline{P}}^1_b$), alors, pour chaque t, $(\int Y^n\,dX)_t$ converge en probabilité vers $(\int Y\,dX)_t$ (écrit aussi $\int_0^t Y_s\,dX_s$).

De plus, sous l'hypothèse de 2), $\int Y\,dX$ est une semimartingale et les semimartingales $\int Y^n\,dX$ convergent vers $\int Y\,dX$ dans $\underline{\underline{D}}$, muni de la topologie ucp. Enfin, l'i.s. $\int H\,dX$ vérifie les propriétés a),b),c), d) et e) du n°8 et la propriété f) du n°11 avec $H \in \underline{\underline{P}}^1_b$ au lieu de $H \in \underline{\underline{G}}$.

DEMONSTRATION. D'abord l'unicité résulte immédiatement du théorème des classes monotones, $\underline{\underline{E}}$ étant une algèbre engendrant la tribu $\underline{\underline{P}}$. Pour l'existence, on a dès le départ deux voies possibles : la voie "rennaise", où l'on considère "sérieusement" l'intégrale élémentaire comme une mesure vectorielle (à valeur ici dans L^0, pour t fixé, ou dans $\underline{\underline{D}}$, si on regarde tous les t à la fois) définie sur le petit domaine $\underline{\underline{E}}$ et qu'il s'agit d'étendre au grand domaine $\underline{\underline{P}}^1_b$, voie qui ne nécessite guère de théorie des martingales (sauf pour donner des exemples !), mais exige par contre quelques connaissances en analyse fonctionnelle ; et puis, la voie "strasbourgeoise", plus ancienne, où l'on considère "sérieusement" une semimartingale comme somme locale d'une martingale de carré intégrable et d'un processus à variation finie, voie qui nécessite donc quelques connaissances en théorie des martingales, mais peu de choses en dehors. Comme je ne renie pas mes origines, c'est cette seconde voie que je vais esquisser maintenant. D'abord, il n'y a aucune difficulté d'intégrer un $Y \in \underline{\underline{P}}^1_b$ par rapport à un processus V à variation finie : on prend trajectoire par trajectoire l'intégrale de Stieltjes, qui fournit bien une intégrale ayant les propriétés voulues. Supposons que l'on veuille maintenant intégrer

---

(1) Le processus Y étant prévisible, LENGLART a démontré récemment que l'on pouvait trouver des T_n, c_n de sorte que les T_n soient des temps d'arrêt et (toute la difficulté est dans ce "et") que Y soit majoré en module par c_n sur l'intervalle stochastique $[\![0,T_n]\!]$ (fermé à droite), ce qui, pour un spécialiste, est beaucoup mieux.

par rapport à une martingale M de carré intégrable, telle que, pour simplifier, on ait $M_0 = 0$ et $M_t = M_n$ pour $t \geq n$, pour un certain n (nous noterons $\underline{\underline{M}}^2_n$ l'espace vectoriel des martingales de ce type). Regardons alors plus précisément qu'au n°3-b) le processus $Z = \int Y\,dM$ pour $Y \in \underline{\underline{E}}$. Ecrivant chaque $\int_0^t Y_s\,dM_s$ comme une somme $\sum Y_i\,(M_{t_{i+1}} - M_{t_i})$, où (t_i) est une subdivion finie de $[0,t]$ et Y_i est, pour chaque i, une variable étagée $\underline{\underline{F}}_{t_i}$-mesurable, on voit aisément, grâce à un petit calcul d'espérances conditionnelles, que Z est une martingale de carré intégrable, et en fait un élément de $\underline{\underline{M}}^2_n$. D'autre part on peut montrer que, pour toute martingale de carré intégrable N, on a $E[N_t^2] = E[[N,N]_t]$ pour tout t (regarder l'approximation discrète du crochet droit ; on peut montrer qu'elle converge dans L^1 dans ce cas) et on peut donc écrire pour tout t

$$E[Z_t^2] = E[[Z,Z]_t] = E[\int_0^t Y_s^2\,d[M,M]_s]$$

si bien que, pour M fixée et Y parcourant $\underline{\underline{E}}$, l'intégrale $\int Y\,dM$ définit une isométrie de l'espace $\underline{\underline{E}}$ muni de la seminorme hilbertienne $\|Y\| = (E[\int_0^n Y_s^2\,d[M,M]_s])^{1/2}$ dans l'espace $\underline{\underline{M}}^2_n$ muni de la norme hilbertienne $\|N\| = (E[N_n^2])^{1/2}$, qui est complet pour cette norme. Soit $L^2(M)$ l'espace des processus prévisibles Y tels que $E[\int_0^n Y_s^2\,d[M,M]_s]$ soit fini, muni de la seminorme hilbertienne $(\ldots)^{1/2}$ associée (c'est en fait l'espace L^2 de la mesure sur $\underline{\underline{P}}$ définie par $Y \to E[\int_0^n Y_s\,d[M,M]_s]$) : $L^2(M)$ contient tous les processus prévisibles bornés, et $\underline{\underline{E}}$ est dense dans $L^2(M)$. On peut alors prolonger de manière unique l'isométrie définie par l'intégrale élémentaire en une isométrie de $L^2(M)$ dans $\underline{\underline{M}}^2_n$, ce qui donne une bonne i.s. $\int Y\,dM$ pour $Y \in L^2(M)$. Revenons finalement à X semimartingale et Y prévisible localement borné : on peut trouver une suite croissante (T_n) de v.a. ≥ 0 tendant vers $+\infty$ telle que, pour chaque n, il existe un processus prévisible borné Y^n, une martingale de carré intégrable M^n et un processus à variation finie V^n de sorte que l'on ait $Y = Y^n$ et $X = M^n + V^n$ sur $[\![0,T_n[\![$; on peut évidemment supposer de plus $T_n \leq n$, $M_0^n = 0$ et $M_t^n = M_n^n$ pour $t \geq n$. On définit alors $\int Y\,dX$ comme étant le processus égal à $\int Y^n\,dM^n + \int Y^n\,dV^n$ sur $[\![0,T_n[\![$ pour chaque n, l'unicité permettant de montrer que tout cela se recolle bien. Nous arrêtons là notre survol de la démonstration.

REMARQUE. On sait aussi prolonger l'i.s. par rapport à une semimartingale X à des processus non prévisibles, ou non localement bornés, mais nous n'en parlerons pas ici.

Nous terminons cette section en disant juste deux mots, en passant, sur la notion de martingale locale, que je ne me résouds pas à passer complètement sous silence. Lorsque Y est un processus prévisible borné et M une martingale de carré intégrable, on a vu que $\int Y\,dM$ est encore une martingale, de carré intégrable. Mais, si M est une martingale quelconque, ou encore si Y est seulement localement borné, alors $\int Y\,dM$ n'est plus en général une martingale, mais est ce qu'on appelle une <u>martingale locale</u>, i.e. un processus càdlàg L pour lequel existe une suite croissante (T_n) de temps d'arrêt tendant vers $+\infty$ telle que, pour chaque n, existe une martingale M^n de sorte que l'on ait $L = M^n$ sur $[\![0,T_n]\!]$ (fermé à droite : c'est très important !). Avec un peu de métier, on montre aisément qu'une semimartingale L est une martingale locale si et seulement s'il existe un processus càglàd Y ne s'annulant jamais tel que $\int Y\,dL$ soit une martingale (de carré intégrable, si on le désire).

La notion de martingale locale permet, en particulier, d'écrire des décompositions "globales" : on peut montrer, par exemple, qu'un processus càdlàg X est une semimartingale si et seulement si l'on peut écrire $X = L + V$ où L est une martingale locale et V un processus à variation finie - décomposition non unique, dépendant par ailleurs de la probabilité de base P. Et, si l'on a une telle décomposition $X = L + V$ relativement à P, on sait écrire une décomposition relativement à Q, probabilité équivalente à P, de la manière suivante. Appelons M la martingale relativement à P telle que M_t soit pour chaque t une densité de Q par rapport à P quand Q et P sont restreintes à $\underline{\underline{F}}_t$; on sait que M ne s'annule nulle part. On a alors ce qui est appelé la "formule de Girsanov"

$$X = (L - \int \frac{1}{M}\,d[L,M]) + (\int \frac{1}{M}\,d[L,M] + V)$$

où, dans la première parenthèse, on a une martingale locale relativement à Q et, dans la seconde, un processus à variation finie, les intégrales étant des intégrales de Stieljes (1/M n'est pas prévisible en général).

LA FORMULE DE CHANGEMENT DE VARIABLES

Si V est un processus continu à variation finie et si f est une fonction de classe $\underline{\underline{C}}^1$ sur $\mathbb{R}$, il est bien connu que f(V) est encore un processus continu à variation finie et que l'on a la formule $f(V) = f(V_0) + \int f'(V)\,dV$; c'est, trajectoire par trajectoire, la formule classique de "changement de variable", qui se visualise bien en écrivant, pour chaque t, $f(V_t) - f(V_0) = \sum_i (f(V_{t_{i+1}} - V_{t_i}))$ pour une subdivision finie (t_i) de [0,t] et en développant chaque parenthèse par la formule de Taylor à l'ordre 1. Si V, à variation finie, est seulement càdlàg, on obtient la formule, moins classique,

$$f(V_t) - f(V_0) = \int_0^t f'(V_{s-})\,dV_s + \sum_{0<s\leq t} (f(V_s) - f(V_{s-}) - f'(V_{s-})\,\Delta V_s)$$

où, dans la famille sommable qui tient compte des sauts, on reconnait un avatar du développement de Taylor à l'ordre 1 . Dans le cas d'une semimartingale X , on a une formule analogue pour f(X), avec f de classe $\underline{\underline{C}}^2$ cette fois, appelée souvent **formule d'Ito**, où il apparait cette fois un développement de Taylor à l'ordre 2 dû au fait, qu'en un certain sens (cf le n°10), X a une 2-variation finie.

13 THEOREME. *Soient* X *une semimartingale et* f *une fonction de classe* $\underline{\underline{C}}^2$ *sur* $\mathbb{R}$. *Alors* f(X) *est encore une semimartingale, et on a, pour tout* t,

$$f(X_t)-f(X_0) = \int_0^t f'(X_{s-})\,dX_s + \frac{1}{2}\int_0^t f''(X_{s-})\,d[X,X]_s + \sum_{0<s\leq t} U_s$$

où $U_s = \{f(X_s)-f(X_{s-})-f'(X_{s-})\Delta X_s - \frac{1}{2}f''(X_{s-})(\Delta X_s)^2\}$, *la première intégrale étant une i.s. et la famille* $(U_s)_{s\leq t}$ *p.s. sommable*.

DEMONSTRATION. Nous montrons d'abord que le processus

$$V = f(X) - f(X_0) - \int f'(X_-)\,dX$$

est à variation finie ; nous l'identifierons après. Pour montrer cela, il suffit de prouver qu'il existe une suite croissante (T_n) de v.a. ≥ 0 tendant vers $+\infty$ telle que, pour chaque n , V soit à variation finie sur $[\![0,T_n[\![$. Notons, au passage, que ceci est valable aussi pour l'identification de V . Nous prenons

$$T_n(\omega) = \inf\{t\geq 0 : t\leq n \text{ ou } |X_t(\omega)| > n\}$$

Dans ces conditions, l'indicatrice de $[\![0,T_n[\![$ est un processus si bien que $X1_{[\![0,T_n[\![}$ est encore une semimartingale, et nous nous sommes donc ramenés à démontrer le théorème quand X est une semimartingale à valeurs dans [-n,+n] . Soit alors C une constante de Lipschitz pour f' sur [-n,+n] ; on a, pour x,y dans l'intervalle,

$$(1) \qquad |f(y)-f(x)-(y-x)f'(x)| \leq C(y-x)^2$$

et donc, si (t_i) est une subdivision finie de l'intervalle [s,t] ,

$$|f(X_t)-f(X_s)-\sum_i f'(X_{t_i})(X_{t_{i+1}}-X_{t_i})| \leq C\sum_i (X_{t_{i+1}}-X_{t_i})^2$$

Prenant pour (t_i) la subdivision de [s,t] associée à une approximation ${}^kX$ de X et appliquant les n°8-f) et 10 , on obtient, en passant à la limite en probabilité,

$$|f(X_t)-f(X_s)-\int_s^t f'(X_{u-})\,dX_u| \leq C([X,X]_t-[X,X]_s)$$

d'où V est à variation finie. Pour l'identification de V , que nous ne détaillerons pas, on vérifie d'abord, grâce à la formule d'intégration par parties et les propriétés du crochet droit, la chose suivante : si la formule d'Ito est vraie pour une certaine fonction g de classe $\underline{\underline{C}}^2$, elle est encore vraie pour f(x) = xg(x). On en déduit que la formule est vraie lorsque f est une fonction polynome. Enfin, on approche f de classe $\underline{\underline{C}}^2$ uniformément sur [-n,+n] par des fonctions polynomes f_k de sorte que f' (resp f") soit aussi approchée uniformément par les f_k' (resp f_k'') , et on montre que l'on peut passer à la limite.

REMARQUES. a) On écrit le plus souvent la formule d'Ito en remplaçant $\int_0^t f''(X_{s-})\,d[X,X]_s$ par $\int_0^t f''(X_{s-})\,d[X,X]_s^c$ où $[X,X]^c$ est la partie "continue" du processus croissant $[X,X]$; dans ce cas, le terme tenant compte des sauts s'écrit plus simplement : U_s est remplacé par $\{f(X_s)-f(X_{s-})-f'(X_{s-})\Delta X_s\}$. Cette formule est en fait plus belle car on peut montrer que $[X,X]^c$ est le crochet droit de la partie "martingale continue" - notée X^c malgré l'ambiguité - de X, si bien que l'on trouve souvent $[X,X]^c$ écrit $[X^c,X^c]$ (ou même $\langle X^c,X^c\rangle$, mais nous ne parlerons pas ici du "crochet oblique"). Par ailleurs, on ne peut décomposer $\{f(X_s)-f(X_{s-})-f'(X_{s-})\Delta X_s\}$ en $\{f(X_s)-f(X_{s-})\}$ et $\{f'(X_{s-})\Delta X_s\}$ car, en général, les deux familles obtenues ne sont pas sommables séparément.

b) Supposons que f soit une fonction <u>convexe</u> au lieu de la supposer de classe $\underline{\underline{C}}^2$, et revenons à la première partie de la démonstration : au lieu de (1), nous écrivons maintenant

$$(2)\quad f(y)-f(x)-(y-x)f'(x) \geq 0 \quad \text{si } y \geq x$$

où f' est la dérivée à gauche de f. En procédant par approximation discrète comme précédemment, on voit que $V = f(X)-f(X_0)-\int f'(X_-)\,dX$ est cette fois un processus croissant. On en déduit, en particulier, que $f(X)$ est encore une semimartingale ; prenant $f(x)=x^2$ et $f(x)=x^+$, on peut alors établir aisément la fin du n°3-d). Par ailleurs, l'élucidation de la structure de V pour $f(x)=|x|$ mène à la théorie importante du <u>temps local</u> d'une semimartingale.

c) Dans de nombreuses applications du calcul différentiel stochastique (en géométrie différentielle, par exemple), on n'a affaire qu'à des semimartingales <u>continues</u>, ce qui fait disparaitre l'encombrante partie de la formule relative aux sauts. On utilise alors assez souvent une variante de l'i.s. appelée <u>intégrale de Stratonovitch</u> : si X et Y sont deux semimartingales continues, on pose par définition

$$(S)\int_0^t Y_s\,dX_s = \int_0^t Y_s\,dX_s + \frac{1}{2}[Y,X]_t$$

La formule de changement de variable, pour f de classe $\underline{\underline{C}}^3$, s'écrit alors comme en calcul différentiel classique

$$f(X_t)-f(X_0) = (S)\int_0^t f'(X_s)\,dX_s \;;$$

cela se déduit aisément de la formule d'Ito et des propriétés du crochet droit.

La formule de changement de variables qui, malgré son air un peu rébarbatif dans le cas où il y a des sauts, est d'une grande efficacité en calcul différentiel stochastique, admet une extension elle-même très utile aux fonctions de classe $\underline{\underline{C}}^2$ à plusieurs variables. Cette extension est un peu lourde à écrire sans introduction d'un formalisme convenable ; mais, comme nous n'allons pas l'utiliser par la suite, nous nous contenterons de noter $\overline{X} = (X^1,\dots,X^n)$ la donnée de n semimartingales $X^1,\dots,X^n$ et d'autre part de noter $D_i f$ (resp $D_i D_j f$) les dérivées partielles au premier ordre (resp second ordre) d'une fonction $f(\overline{x}) = f(x^1,\dots,x^n)$ de classe $\underline{\underline{C}}^2$ sur $\mathbb{R}^n$. On a alors

$$f(\overline{X}_t) = f(\overline{X}_0) + \sum_i \int_0^t D_i f(\overline{X}_{s-})\, dX^i_s + \frac{1}{2}\sum_{i,j} \int_0^t D_i D_j(\overline{X}_{s-})\, d[X^i,X^j]_s$$
$$+ \sum_{0<s\leq t} \{f(\overline{X}_s) - f(\overline{X}_{s-}) - \sum_i D_i f(\overline{X}_{s-})\,\Delta X^i_s - \tfrac{1}{2}\sum_{i,j} D_i D_j f(X_{s-})\,\Delta X^i_s\,\Delta X^j_s\} ,$$

la famille intervenant dans le dernier terme étant p.s. sommable. En particulier, si $X^1,\dots,X^n$ sont n semimartingales et si f est de classe $\underline{\underline{C}}^2$ sur $\mathbb{R}^n$, alors $f(X^1,\dots,X^n)$ est encore une semimartingale (on a un résultat analogue si on suppose f convexe au lieu de classe $\underline{\underline{C}}^2$). La démonstration est analogue en tout point à celle du n°13 ; lorsque l'on prend $f(x,y) = xy$, on retrouve la formule d'intégration par parties (que nous utilisons dans la démonstration, mais on peut aussi procéder autrement).

EQUATIONS DIFFERENTIELLES STOCHASTIQUES

Nous nous bornerons à énoncer un théorème d'existence et d'unicité "global" sous des hypothèses analytiques analogues à celles du théorème classique de Cauchy-Lipschitz.

14 THEOREME. Soient X une semimartingale et f une fonction définie sur $\mathbb{R}_+ \times \Omega \times \mathbb{R}$ et vérifiant

1) Pour x fixé, $(t,\omega) \to f(t,\omega,x)$ est un processus càglàd

2) Pour (t,ω) fixé, $x \to f(t,\omega,x)$ est une fonction lipschitzienne, avec une constante de Lipschitz k ne dépendant pas de t mais pouvant dépendre de ω.

Dans ces conditions, l'équation différentielle stochastique

$$dZ = f(.,.,Z_-)\, dX$$

admet une solution Z, semimartingale, unique si l'on se donne la condition initiale Z_0, v.a. $\underline{\underline{F}}_0$-mesurable.

On vérifie facilement que, si Z est un processus càdlàg, alors l'application $(t,\omega) \to f(t,\omega,Z_{t-}(\omega))$ est un processus càglàd, qui peut donc être intégré par rapport à la semimartingale X. Notre équation différentielle a donc bien un sens : elle signifie que le processus càdlàg Z doit être solution de l'équation intégrale

$$Z_t = Z_0 + \int_0^t f(s,.,Z_{s-})\, dX_s \quad ;$$

une solution Z de cette équation intégrale étant nécessairement une semimartingale, la propriété d'associativité de l'i.s. légitime alors l'écriture différentielle de l'équation.

Nous donnons maintenant quelques idées sur la démonstration du théorème. L'idée de départ est, comme dans le cas classique, de se ramener à une situation où l'on peut appliquer un théorème de point fixe à l'opérateur intégral associé à l'équation ; cependant elle est ici bien plus difficile à mettre en oeuvre. On connait pour cela deux voies, la "strasbourgeoise", qui procède par petits pas ramenant finalement l'énoncé à un cas particulier "maniable" grâce à des techniques de théories des martingales qu'il n'est pas question d'exposer ici, et la "rennaise", plus récente, qui a entr'autres l'avantage d'inclure tout le côté technique de théorie des martingales dans la belle inégalité de Métivier-Pellaumail que voici, sans démonstration,

15 THEOREME. Soit X une semimartingale. Il existe un processus croissant A contrôlant X au sens suivant : pour tout temps d'arrêt $T>0$ et tout processus càglàd Y, on a

$$E[(\int Y\,dX)^{*2}_{T-}] \leq E[A_{T-}(\int Y^2\,dA)_{T-}] \quad (\leq +\infty)$$

REMARQUE. L'inégalité est encore vraie pour Y prévisible localement borné. Par ailleurs, on peut montrer que la propriété de contrôle de l'énoncé caractérise les semimartingales : un processus càdlàg X est une semimartingale ssi il existe un processus croissant A tel que l'inégalité de l'énoncé soit vérifiée pour tout $Y\in\underline{\underline{E}}$ (l'intégrale $\int Y\,dX$ étant donnée par l'intégrale élémentaire).

DEMONSTRATION DU THEOREME 14. (Nous serons obligés ici de manipuler un peu les temps d'arrêt). Nous supposons d'abord que la constante de Lipschitz k de l'énoncé ne dépend pas de ω. Soit S un temps d'arrêt et supposons que l'on connaisse une solution U de l'équation sur l'intervalle $[\![0,S[\![$; cela signifie que U est une semimartingale qui coincide avec $Z_0+\int f(.,.,U_-)\,dX$ sur $[\![0,S[\![$ (par exemple, si S est constant et égal à 0, on peut prendre U constant égal à 0). L'identification des sauts d'une i.s. permet alors de trouver une solution V sur l'intervalle $[\![0,S]\!]$: il suffit de poser $V=U1_{[\![0,S[\![}+Z_S1_{[\![S,\infty[\![}$ où Z_S est la v.a. valant Z_0 sur $\{S=0\}$ et $f(S,.,U_{S-})\Delta X_S$ sur $\{0<S<\infty\}$. Soit maintenant T le temps d'arrêt défini par

$$T(\omega) = \inf\{t\geq S(\omega) : A^2_t(\omega)-A^2_S(\omega) > 1/2k^2\}$$

où k est la constante de Lipschitz et A un processus croissant contrôlant X au sens du n°15 ; comme A est càdlàg, on a $T>S$ sur $\{S<\infty\}$, et, A étant croissant, on a partout

$$A_{T-}\int_{]\!]S,T[\![} dA_s = A_{T-}(A_{T-}-A_S) \leq A^2_{T-}-A^2_S \leq 1/2k^2$$

avec $A_\infty=A_{\infty-}=\lim A_t$, $t\to\infty$. Soient alors U, V deux semimartingales solutions de l'équation sur l'intervalle $[\![0,S]\!]$, Z_0 étant donné ; pour simplifier les notations, nous poserons $F(U)=\int f(.,.,U_-)\,dX$, et de même pour V. On déduit alors de l'inégalité de Métivier-Pellaumail et des inégalités précédentes la chaîne d'inégalités suivante

$$E[(F(U)-F(V))^{*2}_{T-}] \leq E[A_{T-} \int_{]\!]S,T[\![} k^2(U_- - V_-)^2 \, dA_s] \leq \frac{1}{2} E[(U-V)^{*2}_{T-}]$$

où, dans la première inégalité, on a tenu compte du fait que l'on a $U = V$ sur $[\![0,S]\!]$ et $|f(.,.,U_-) - f(.,.,V_-)| \leq k|U_- - V_-|$. On voit alors poindre à l'horizon la possibilité de définir une contraction sur un bon espace de processus, ce qui permet finalement (nous ne donnons pas les détails) d'obtenir une solution de l'équation sur $[\![0,T[\![$, unique sur cet intervalle. Posant $S = T_1$, $T = T_2$, on recommence l'opération avec T_2 à la place de T_1 et, comme A est càdlàg, la suite des temps d'arrêt T_n ainsi construits tend en croissant vers $+\infty$, ce qui permet, par recollement, d'obtenir la solution unique de l'équation. Voyons rapidement, pour finir, comment traiter le cas où la constante de Lipschitz est une fonction $k(\omega)$ de ω. Quitte à remplacer $k(\omega)$ par $\sup |f(t,\omega,y) - f(t,\omega,x)||y-x|^{-1}$ quand t parcourt $\mathbb{R}_+$ et (x,y) le complémentaire de la diagonale de $\mathbb{R} \times \mathbb{R}$, on peut supposer que $k(.)$ est une v.a.. Soit alors c une constante telle que $P\{k \leq c\} > 0$; posons, pour tout entier n, $\Omega_n = \{k \leq c+n\}$ et désignons par P_n la probabilité $P(. \cap \Omega_n)/P(\Omega_n)$ sur Ω_n muni de la filtration trace $\mathbb{F}^n$ de $\mathbb{F}$. La restriction de X à $\mathbb{R}_+ \times \Omega_n$ est une $\mathbb{F}^n$-semimartingale d'après le n°1, et on peut considérer l'équation restreinte à Ω_n. Sur Ω_n, on a une solution unique Z^n d'après la première partie de la démonstration, et un procédé de recollement reposant sur l'invariance et le caractère local de l'i.s. (cf n°8) permet finalement d'avoir une solution unique Z sur Ω.

L'énoncé 14 ne fait intervenir qu'une seule semimartingale X. En fait, il est possible de remplacer l'unique intégrale dans l'équation intégrale associée par une somme finie d'intégrales du même type et, plus généralement, de considérer des systèmes d'équations différentielles stochastiques. D'autre part, on sait étudier la stabilité de la solution en fonction des données Z_0, f, X (nous donnerons des références dans la bibliographie). Enfin, on sait aussi traiter le cas où f est seulement localement lipschitzienne, avec étude des "explosions".

Une équation particulièrement importante est l'équation $dZ = Z_- dX$ avec $Z_0 = 1$, qui définit une semimartingale notée $\varepsilon(X)$ et appelée <u>l'exponentielle de</u> X <u>au sens des semimartingales</u>. On sait écrire explicitement $\varepsilon(X)$: si X est continu, $\varepsilon(X) = \exp(X - \frac{1}{2}[X,X])$, où "exp" est la fonction exponentielle ordinaire et, en général, on a

$$\varepsilon(X)_t = \exp(X_t - \frac{1}{2}[X,X]_t) \prod_{0<s\leq t} (1+\Delta X_s) \exp(-\Delta X_s + \frac{1}{2}(\Delta X_s)^2)$$

avec un produit infini p.s. absolument convergent. Par ailleurs, si X et Y sont deux semimartingales, on a $\varepsilon(X)\varepsilon(Y) = \varepsilon(X+Y+[X,Y])$.

BIBLIOGRAPHIE COMMENTEE

Je vais essayer ici de donner quelques idées sur le développement historique des notions d'intégrale stochastique et de semimartingale. En fait, je ne pourrai donner de tout cela qu'une vue bien partiale (je connais très mal d'autres approches de l'intégrale stochastique comme celles de MILLAR, ou McSHANE, ou SKOROKHOD), et aussi partielle (le peu que j'ai dit sur les martingales locales ne me permettant pas de leur rendre justice).

L'intégrale stochastique par rapport au mouvement brownien doit ses débuts à WIENER (intégration de processus "déterministes"), mais c'est à ITO que revient la création d'un véritable calcul différentiel stochastique attaché au mouvement brownien, dans une série d'articles allant de 1944 à 1961. Ce calcul, qui a des liens étroits avec la théorie des diffusions, n'a cessé depuis sa création de susciter de nombreux travaux, y compris en géométrie différentielle. Mais, parler de ces travaux nous entrainerait loin du coeur de notre exposé (et aussi de mon "domaine de compétence"...) ; aussi me bornerai-je à citer une référence classique (qui date déjà)

McKEAN (H.P.) : Stochastics Integrals, Academic Press, New York 1969

Après des travaux préliminaires de DOOB, MEYER et COURREGE, le premier travail fondamental sur l'extension du calcul d'Ito aux martingales de carré intégrable est

KUNITA (H.), WATANABE (S.) : On square integrable martingales (Nagoya Math. J. 30, 1967, p. 209-245)

qui n'aborde cependant pas les équations différentielles. L'extension aux martingales locales (introduites par ITO et WATANABE à l'occasion de leur travail sur le théorème de décomposition de Doob-Meyer) et aux semimartingales (définies par MEYER comme sommes d'une martingale locale et d'un processus à variation finie) débute dans

MEYER (P.A.) : Intégrales stochastiques (Sém. Prob. I , Lecture Notes in Math. 39 , p. 72-162, Springer, Berlin 1967)

où apparaissent l'utilisation des processus prévisibles (notion alors bien récente introduite par MEYER en théorie générale des processus sous le nom de processus très bien mesurables) et la définition et utilisation en temps continu du crochet droit (à la place du crochet oblique de MOTOO et WATANABE) ; elle trouve sa forme quasidéfinitive dans

DOLEANS-DADE (C.), MEYER (P.A.) : Intégrales stochastiques par rapport aux martingales locales (Sém. Proba. IV, LN. 124, p. 77-107, Springer, Berlin 1970)

Mais l'invariance de la notion de semimartingale (définie à la Meyer)

par changement de loi équivalente a été découverte beaucoup plus tard (mais moins tardivement qu'il n'y parait !) par

JACOD (J.), MEMIN (J.) : Caractéristiques locales et conditions de continuité absolue pour les semimartingales (Z. Wahrscheinlichkeitstheorie 35, 1976, p. 1-37)

précédé d'un travail de VAN SCHUPPEN et WONG sur la "formule de Girsanov" . On trouve aussi dans cet article la décomposition "locale" d'une semimartingale en une martingale bornée et un processus à variation bornée, qui a été aussi découverte indépendamment par DOLEANS-DADE et YEN (qui s'appelle YAN désormais). Enfin, après un travail préliminaire de PROTTER, l'étude générale des équations différentielles stochastiques sort de ses balbutiements avec

DOLEANS-DADE (C.) : On the existence and unicity of solutions of stochastic differential equations (Z. Wahr. 36, 1976, p. 93-101)

mais "l'astuce" pour traiter le cas où la constante de Lipschitz dépend de ω est due à LENGLART.

Par ailleurs, la première interprétation de l'intégrale stochastique comme intégrale vectorielle apparait dans

PELLAUMAIL (J.) : Sur l'intégrale stochastique et la décomposition de Doob-Meyer (Astérisque 9, Soc. Math. France 1973)

point de vue développé par METIVIER et PELLAUMAIL dans une série d'articles abordant aussi le calcul différentiel stochastique pour des processus à valeurs banachiques (et en particulier hilbertiennes). C'est la lecture de

METIVIER (M.), PELLAUMAIL (J.) : Mesures stochastiques à valeurs dans les espaces L^0 (Z. Wahr. 40, 1977, p. 101-114)

qui a provoqué la démonstration par MOKOBODZKI et moi-même du théorème 4 de notre exposé, i.e. de l'équivalence de la définition à la Meyer des semimartingales et de celle adoptée au n°1 de l'exposé. La démonstration originale, avec un énoncé amélioré par MEYER, est dans

MEYER (P.A.) : Caractérisation des semimartingales, d'après Dellacherie (Sém. Proba. XIII, LN 721, p. 620-623, Springer, Berlin 1979)

mais l'importance de la notion de quasimartingale comme "auxiliaire technique" avait été mise auparavant en évidence par

STRICKER (C.) : Quasimartingales, martingales locales, semimartingales et filtrations naturelles (Z. Wahr. 39, 1977, p. 55-64)

où se trouve aussi établi le premier résultat du type de notre théorème 6 (ce dernier résulte de conversations entre BICHTELER et moi-même à Oberwolfach ; une belle démonstration due à LENGLART paraitra dans Sém. Proba. XIV). L'équivalence des deux définitions de la notion de semimartingale a été aussi découverte, au moyen d'un théorème de factorisation de MAUREY et ROSENTHAL, par

BICHTELER (K.) : Stochastic integration, L^p-theory of semimartingales (à paraitre, sans doute dans Ann. Prob.)

qui, reprenant le point de vue "intégrale vectorielle", retrouve aussi d'importantes équivalences de normes "à la Burkholder" (évoquées pour les martingales au n°6) établies pour les semimartingales par

YOR (M.) : Inégalités entre processus minces et applications (C.R. Acad. Sc. Paris, t. 286, 1978, p. 799-801)

après que EMERY ait introduit des normes maniables sur les semimartingales.

On doit à ce dernier l'introduction systématique de topologies à caractère "local" (en particulier, la plus simple de toutes : notre topologie ucp) ; voir

EMERY (M.) : Une topologie sur l'espace des semimartingales (Sém. Proba. XIII, LN 721, p. 260-280, Springer, Berlin 1979)

L'idée de localiser sur des intervalles du type $[\![0,T[\![$ au lieu du type $[\![0,T]\!]$ (qui est celui rencontré en théorie des martingales locales) remonte en fait à

KAZAMAKI (N.) : Changes of time, stochastic integrals, and weak martingales (Z. Wahr. 22, 1972, p. 25-32)

mais ne s'était révélée efficace qu'après l'introduction d'une classe spéciale de semimartingales dans

YOEURP (C.) : Décompositions des martingales locales et formules exponentielles (Sém. Proba. X, LN 511, p. 432-480, Springer, Berlin, 1976)

étudiée de manière détaillée par MEYER. Ceci dit, l'introduction de bonnes topologies sur les surmartingales permet d'étudier la stabilité des solutions des équations différentielles stochastiques dans

PROTTER (P.) : $\underline{\underline{H}}^p$-stability of solutions of stochastic differential equations (Z. Wahr. 44, 1978, p. 337-352)

EMERY (M.) : Equations differentielles lipschitziennes. Etude de la stabilité (Sém. Proba. XIII, LN 721, p. 281-293, Springer, Berlin 1979)

Signalons encore, sur les équations différentielles stochastiques, les travaux de METIVIER et PELLAUMAIL (référence plus loin) et

DOSS (H.), LENGLART (E.) : Sur l'existence, l'unicité et le comportement asymptotique des solutions d'équations différentielles stochastiques (Ann. Inst. Henri Poincaré, vol. XIV, 1978, p. 189-214)

où l'on montre en particulier que la résolution de certaines équations différentielles stochastiques se ramène à celle d'équations différentielles ordinaires. Enfin, l'équation définissant "l'exponentielle" d'une semimartingale a été la première étudiée (DOLEANS-DADE 1970) et a suscité de nombreux travaux.

L'intégrale à la Riemann présentée dans notre exposé n'a, pour les spécialistes, qu'un intérêt essentiellement pédagogique. Je tiens cependant à remercier LENGLART pour le partage de ce souci pédagogique : il m'a bien aidé à surmonter quelques points épineux, insoupçonnés dans mes nombreux exposés oraux. Je signale au passage que PELLAUMAIL a été le premier à définir le crochet droit à l'aide de l'i.s. . Par ailleurs, BICHTELER (locus cité) a montré que l'on avait, pour Y càglàd, une approximation de $\int Y\,dX$ pour la convergence p.s. par des sommes riemanniennes plus sophistiquées que les nôtres. Enfin, Meyer m'a signalé que BRENNAN avait aussi défini récemment une intégrale de Riemann par rapport à une quasimartingale.

Avant de donner une liste de monographies récentes sur le sujet, j'ajouterai encore quelques mots sur des développements en cours de la théorie pour lesquels on trouvera des références dans le volume XIV du Séminaire de Probabilités de Strasbourg. D'abord, l'étude du comportement des semimartingales après grossissement de la filtration, mise en oeuvre depuis plusieurs années par BARLOW, JEULIN, MEYER et YOR (avec une modeste contribution de ma part) ; puis l'extension du domaine de validité de l'i.s. par YOR (au delà de la prévisibilité) et par JACOD (au delà de la bornitude locale), étudiée par MEMIN et YAN . Enfin, L. SCHWARTZ est en train de développer le calcul différentiel stochastique sur les variétés, ce qui a suscité de nouveaux travaux sur les semimartingales, définies seulement sur une partie de $\mathbb{R}_+ \times \Omega$ (SCHWARTZ, MEYER, STRICKER).

A seigneur tout honneur, nous commencerons notre liste de monographies par

MEYER (P.A.) : Un cours sur les intégrales stochastiques (in Sém. Proba. X , LN. 511, p. 246-400, Springer, Berlin 1976)

qui a redonné une impulsion à l'étude du sujet dans toutes ses dimensions. La palme pédagogique revient cependant à

METIVIER (M.) : Reelle und Vektorwertige Quasimartingale und die Theorie der Stochastischen Integration (Lecture Notes in Math. 607, Springer, Berlin 1977)

qui traite aussi des semimartingales à valeurs hilbertiennes. Un ouvrage tout récent, qui donne la vue la plus complète du calcul stochastique et de ses applications à l'heure actuelle,

JACOD (J.) : Calcul stochastique et Problèmes de martingales (Lecture Notes in Math. 714, Springer, Berlin 1979)

Enfin, seront parus quand paraitront ces lignes

DELLACHERIE (C.), MEYER (P.A.) : Probabilités et Potentiel, 2e volume (chapitres V à VIII), chez Hermann, Paris

qui n'aborde pas les équations différentielles, au contraire de

METIVIER (M.), PELLAUMAIL (J.) : Stochastic Integration, chez Academic Press, New York

et aussi, ouvrage dont je ne connais pas le contenu,

RAO (M.M.) : Stochastic Processes and Integration, chez Noordhoff, Groningen

Claude DELLACHERIE

Départment de Mathématique
Université de Rouen

B.P. n°67

76130 MONT-SAINT-AIGNAN

OPTIMAL CONTROL OF CONTINUOUS AND DISCONTINUOUS PROCESSES IN A RIEMANNIAN TANGENT BUNDLE*

T. E. Duncan
Division of Applied Sciences
Harvard University, Cambridge, Ma. 02138/USA
and
Department of Mathematics
University of Kansas, Lawrence, Ks. 66045/USA

1. Introduction

In this paper some formulations of stochastic systems with values in the tangent bundle of a Riemannian manifold will be given in terms of stochastic differential equations that contain jump processes and these formulations will serve to model stochastic control problems for which necessary and sufficient conditions for optimality will be given. In stochastic systems both discontinuous and continuous processes often appear and it frequently occurs that the system evolves in a smooth manifold that is not a linear space. With suitable conditions on the manifold there are continuous and discontinuous processes that respect the geometry and these properties can be lost by an "abstract" formulation. Traditionally, many problems in control have been modelled by differential equations in Euclidean spaces to show the dynamical property of the physical systems. The differential geometric formulation preserves the geometric interpretations of the differential equations while also providing a more mathematically reasonable formulation of the physical system.

The mathematical models that will be given in this paper will include jump processes or discontinuous martingales in the fibres of the tangent bundle. Mathematically, this formulation will allow the use of some of the techniques that are employed in Euclidean spaces where the base space and each tangent space are isomorphic.

Besides the mathematical reasons for this formulation there are physical problems which justify such a model. For example, consider a particle whose motion evolves on a smooth manifold and whose velocity is subjected to discontinuous changes by the collision with other particles or objects. The observations could be the motion of the particle and the times of the collisions and the particle could be controlled in the differential equation that represents the position of the particle. Examples such as this one could be imagined from other disciplines.

To have a more geometric view of the formulation of a jump process in the fibres of the tangent bundle consider an observer travelling along the motion in the base manifold. This observer sees the fibres of the tangent bundle along the curve as a family of tangent spaces, that is, Euclidean spaces. In fact to this observer it seems that the jump process evolves in a fixed Euclidean space. Many of the notions that are used to characterize jump processes that have values in a Euclidean space can be used when the fibres of the tangent bundle are connected along a curve in the base manifold. Basically, a global view of differential geometry has to be adopted rather than a local view that is obtained by charts because the complete paths of the processes are often used.

An example of the formulation in this paper occurs when the manifold is a compact, connected Riemann Lie group, for example, the linear Lie group SO(n). An important geometrical property of Lie groups is that these manifolds are parallelizable, that is, their tangent bundles are globally trivializable. For Lie groups, vectors in the various fibres can be invariantly transported to the tangent space at the identity, the Lie algebra, using the group structure. Consider a process in the Lie algebra that is obtained by solving a stochastic differential equation that contains a term which is a jump process or a discontinuous martingale. The control can be assumed to appear in the equation in the Lie algebra and the observations can be solved by the techniques in this paper.

Some recent previous work on the control of stochastic system containing discontinuous processes has been done by Boel-Varaiya [3] and Rishel [19]. Boel-Varaiya consider an abstract formulation of a stochastic system as a jump process that takes values in a Blackwell space. A family of stochastic systems is formed from a family of probability measures on the sample paths that are piecewise constant, right continuous and have only a finite number of jumps in any finite interval. This family of probability measures is indexed by a family of controls so that the problem is formulated as a controlled probability space. Optimality conditions are obtained for this abstract control problem. Rishel uses a sample path approach to obtain dynamic programming conditions that an optimal control must satisfy for a stochastic system whose solution is obtained from a family of vector fields that is indexed by a finite state Markov process.

The approach to be employed here will use techniques from both a family of probability measures and a sample path approach. The use of a family of probability measures allows for a general notion of a solution of a stochastic system while the sample path approach allows for the use of the geometrical structure in which the system evolves. The approach that is used here cannot be directly imbedded into only a family of probability measures approach because explicit use is made of the differential geometric setting of the problem.

The control will enter the stochastic differential equation that describes the process in the base manifold. The solution to this equation for each control will be defined by transforming an initial measure by a Radon-Nikodym derivative. In some other formulations this technique has been used in [1, 7, 8, 9]. While it could be assumed that the control appears in the stochastic differential equation that describes the process in the fibres of the tangent bundle, the model would require additional assumptions to ensure that the solution to the equation is defined (cf. e. g. [19]).

2. Preliminaries

Various machinery will be introduced to formulate a stochastic system in the tangent bundle of a Riemannian manifold.

Let M be a compact, connected, smooth Riemannian manifold of dimension m. The Riemannian connection will be used. TM will denote the tangent bundle of M, $T_a M$ will denote the fibre of TM over $a \in M$, $O(M)$ will denote the bundle of orthonormal frames over M and $O(m)$ will denote the Lie group of $m \times m$ orthogonal matrices.

In this paper the probability spaces, often denoted as $(\Omega, \mathfrak{F}, P)$, will be assumed to be complete and any increasing family of sub-σ-algebras of $\mathfrak{F}$, often denoted as $(\mathfrak{F}_t)$, will be assumed to be right continuous such that $\mathfrak{F}_0$ contains all the P-null sets.

To describe jump processes in the fibres of the tangent bundle it is necessary to use parallelism along stochastic processes in the base manifold. A convenient, geometric approach to parallelism is to form the horizontal lift of an M-valued process to a frame bundle [15]. The horizontal lift of an M-valued Brownian motion to the bundle of orthonormal frames over M which defines parallelism along the Brownian paths has been constructed in [9]. Let ℓ be a smooth vector field in TM. Using a generalization of the absolutely continuous transformation of measures technique of Cameron-Martin [4] and Girsanov [13] as given in [10] the measure for the stochastic differential equation written formally in $T_{Y_t} M$ as

$$dY_t = \ell_t \, dt + dB_t$$

is defined using a probability measure for an M-valued Brownian motion and the Radon-Nikodym derivative

$$\varphi_t = \exp\left[\int_0^t \langle \ell_s, dY_s \rangle_{Y_s} - \frac{1}{2}\int_0^t \langle \ell_s, \ell_s \rangle_{Y_s} ds\right]$$

where $\langle \cdot, \cdot \rangle_x$ is the Riemannian metric evaluated at $x \in M$, and $(dY_t, \mathfrak{F}_t, P)$ and $(dB_t, \mathfrak{F}_t, \tilde{P})$ are Brownian motions where $d\tilde{P} = \varphi dP$. More precisely we have the following result (Proposition 2 [10]).

PROPOSITION 1. Let $(Y_t, \mathfrak{F}_t, \Omega, \mathfrak{F}, P)_{t\in I}$ be a standard M-valued Brownian motion and let $(\ell_t)_{t\in I}$ be a TM-valued process that is $(\mathfrak{F}_t)$-predictable and uniformly bounded such that $\ell_t \in T_{Y_t}M$. Then the real-valued process (φ_t) given by

$$\varphi_t = \exp\left[\int_0^t \langle \ell_s, dY_s \rangle_{Y_s} - \frac{1}{2}\int_0^t \langle \ell_s, \ell_s \rangle_{Y_s} ds\right]$$

is a continuous martingale on $(\Omega, \mathfrak{F}, \bar{P})$ where $d\bar{P} = \varphi dP$, the formal vectors (dY_t) satisfy the stochastic differential equation in $T_{Y_t}M$

$$dY_t = \ell_t \, dt + dB_t$$

where $(dB_t, \mathfrak{F}_t, \bar{P})_{t\in I}$ are the formal vectors of a Brownian motion.

An M-valued process (K_t) is said to be a geometric (M-valued) process if it is continuous and if a (horizontal) lift to the bundle of orthonormal frames is defined for this process. An M-valued Brownian motion is an example of a geometric process [9].

If $J_t = (K_t, L_t)$ is a TM-valued process where $\pi(J_t) = K_t$ is a geometric M-valued process and $\pi : TM \to M$ is the projection then $L_t \in T_{K_t}M$ is said to be a jump process relative to (K_t) if the process is piecewise constant when it is represented in a family of orthonormal frames that are parallel along (K_t). The process (L_t) is said to be a discontinuous martingale relative to (K_t) if for each bounded $(\sigma(K_u, 0 \leq u \leq t))$-predictable family of vectors (ℓ_t) that are parallel along (K_t), the real-valued process $(\int_0^t \langle \ell, dL \rangle)$ exists and is a discontinuous martingale. These notions were introduced in [11].

Some mathematical models will be considered that describe a stochastic system that evolves in TM and that has the property that the fibre-valued part of the process is obtained from a discontinuous process.

The first model is a system whose evolution depends on the current state and the control which is allowed to depend on all the past observations. This system can be formally described in an orthonormal frame of $T_{Z_t}(TM)$ as

$$dX_t = a(t, X_{t-}, Y_t)dt + b(t, X_{t-}, Y_t)dM_t \tag{1}$$

$$dY_t = c(t, X_{t-}, Y_t, u_t)dt + dB_t \tag{2}$$

where $Z_t = (Y_t, X_t)$, $Z_0 = (Y_0, 0)$, $t \in [0,1]$, $\pi(Z_t) = Y_t$, (c, a) is a smooth vector field in $T(TM)$ that is uniformly bounded, b is a smooth field of linear transformations, (dB_t) are the formal vectors of a standard Brownian motion and (dM_t) is a discontinuous martingale relative to (Y_t).

Recall that the solution of the stochastic differential equation (2) is given by a measure which transforms an M-valued Brownian motion to the given stochastic differential equation using Proposition 1. Thus it is necessary to define a discontinuous martingale relative to a Brownian motion and to show that the computations required to define the solutions of (1 - 2) can be made. The discontinuous martingale will have the geometric property that its probability law is invariant under $O(m)$. From a probabilistic viewpoint it will be a fairly elementary discontinuous process obtained from a compound Poisson process. More general jump processes have been studied (e.g. [2,14]). Let (N_t) be a Poisson counting process with parameter or rate λ. Let $(X_n)_{n \geq 1}$ be a sequence of independent normal $\mathbb{R}^m$-valued random variables each with mean 0 and variance 1. Let (M_t) be the parallel transport along the M-valued Brownian motion of the process (L_t) where $L_t = \Sigma_{n=1}^{N(t)} X_n$ and $\Sigma_{n=1}^{0} X_n \equiv 0$. Since (N_t) and (X_n) are independent families of random variables it is easy to verify that (L_t) is a right continuous discontinuous martingale. Thus if (Y_t) is an M-valued Brownian motion on the complete probability space $(\Omega, \mathfrak{F}, P)$ which includes (N_t) and (X_n) then the process (Z_t) is defined because the fibre part of the process can be solved. Let $\mathfrak{F}_t$ be the P-completion of $\sigma(Y_u, M_u; 0 \leq u \leq t)$. For the drift term c in (2) the solution is defined by transforming the measure P by a Radon-Nikodym derivative of the form given in Proposition 1. This method of solution for (2) has been used in previous work on stochastic control ([1, 7, 9]).

It is important to note that the control enters only in the equation for the base process. If the control entered in the drift a( ) then there would be many additional technical conditions that would need to be imposed to ensure that the solution to this equation was well defined. However, the Poisson rate for the jump process can depend on the control as Boel-Varaiya [3] have done without making significant additional assumptions.

A second type of stochastic system that will be considered will be called a Markov system. It has the same form as the system described in (1 - 2) except that the control $u(t)$ is measurable with respect to the P-completion of $\sigma(Z_{t-})$ and b is assumed to be $GL(m)$-valued.

One other stochastic system formulation that can be solved by the subsequent techniques but will not be explicitly considered is parallelizable manifolds, that is, the tangent bundle is globally trivializable. Probably the most important class of such models besides Euclidean spaces are linear Lie groups. The group structure provides a natural way to identify a tangent space at some element of the group with the tangent space at the identity, the Lie algebra.

To complete the description of the stochastic control problem it is necessary to introduce some additional terminology. Let I be the unit interval $[0,1]$, $\mathcal{B}_I$ be the Borel σ-algebra on I, U be a complete metric space and $\mathcal{B}_U$ be the Borel σ-algebra on U. Recall that $(\Omega, \mathfrak{F}, P)$ is a complete probability space for the processes $(B_t, M_t)_{t\in I}$. Let $(\mathcal{G}_t)_{t\in I}$ be an increasing family of sub-σ-algebras of $\mathfrak{F}$ such that $\mathcal{G}_0$ contains all P-null sets and $\mathcal{G}_t \subset \mathfrak{F}_t$ for $t \in I$. $(\mathcal{G}_t)$ will often be called the observations of the stochastic system.

A family of controls for the stochastic system has to be introduced.

DEFINITION 1. The family of controls, $\mathcal{U}(s,t)$, is a family of maps from $(s,t] \times \Omega$ to U such that for each $u \in \mathcal{U}(s,t)$, $(u_r)_{r\in(s,t]}$ is a measurable process that is adapted to $(\mathcal{G}_r)$ and for $u, v \in \mathcal{U}(s,t)$ and $r \in (s,t]$ the control given by $w(q) = u(q)$ $s < q \leq r$ and $w(q) = v(q)$ $r < q \leq t$ is also in $\mathcal{U}(s,t)$.

Remark. If $s = 0$ in the definition above then by convention it is assumed that the controls are defined on $[0,t] \times \Omega$. $\mathcal{U}$ will denote $\mathcal{U}(0,1)$.

The cost, $J(u)$, associated with the control $u \in \mathcal{U}$ is given by

$$J(u) = E_u \left[\int_0^1 r_0^t \, C(t, u(t))\, dt + r_0^1 \, J_1 \right] \qquad (3)$$

where E_u is expectation when the control u is used in the stochastic system (1 - 2). The following conditions are assumed on the terms in the equation for J :

i) $C : I \times U \times \Omega \to \mathbb{R}_+$ is jointly measurable on $\mathcal{B}_I \times \mathcal{B}_U \times \mathcal{F}$ and is uniformly bounded. $C(t, \cdot, \omega) : U \to \mathbb{R}_+$ is continuous and $(C(t, u(t)))$ is adapted to $(\mathcal{F}_t)$.

ii) $r_s^t : \Omega \to \mathbb{R}_+$ is the discounting rate where $s, t \in I$ and $s \leq t$. For fixed $s \in I$, (r_s^t) is jointly $\mathcal{B}_I \times \mathcal{Y}_1$ measurable, is adapted to $(\mathcal{Y}_t)$ and is uniformly bounded.

$$r_s^{\boldsymbol{\cdot}} : [s, 1] \to \mathbb{R}_+ \quad \text{is continuous and} \quad r_{t_1}^{t_3} = r_{t_1}^{t_2} \, r_{t_2}^{t_3}$$

for $t_1 \leq t_2 \leq t_3$ and $r_t^t = 1$ for $t \in I$.

iii) $J_1 : \Omega \to \mathbb{R}_+$, the terminal cost, is a uniformly bounded random variable.

For $w \in \mathcal{U}$ the terms u, v in the triple $(t; u, v)$ where $t \in I$ denote the controls $u \in \mathcal{U}(0, t)$ and $v \in \mathcal{U}(t, 1)$ obtained by restricting w to $[0, t] \times \Omega$ and $(t, 1] \times \Omega$ respectively. Define ψ as

$$\psi(t; u, v) = E_{uv} \left[\int_t^1 r_t^s \, C(s, v(s))\, ds + r_t^1 \, J_1 \,\middle|\, \mathcal{Y}_t \right] \qquad (4)$$

where E_{uv} denotes expectation using the control $u \in \mathcal{U}(0, t)$ and $v \in \mathcal{U}(t, 1)$.

ψ can also be expressed as

$$\psi(t; u, v) = \frac{E[\rho(0, t; u)\, \rho(t, 1; v)\, (\int_t^1 r_t^{\boldsymbol{\cdot}} \, C(\cdot\, , v) + r_t^1 \, J_1) \,|\, \mathcal{Y}_t]}{E[\rho(0, t; u) \,|\, \mathcal{Y}_t]}$$

where

$$\rho(0, t; u) = \exp \left[\int_0^t \langle c(s, X_{s-}, Y_s), dY_s \rangle_{Y_s} - \frac{1}{2} \int_0^t \langle c(s, X_{s-}, Y_s), c(s, X_{s-}, Y_s) \rangle \, ds \right]$$

The verification that $L^1(\mu)$ is a complete lattice for a probability measure μ (p. 302 [12]) employs the Radon-Nikodym theorem and verifies the following equality (cf. Lemma 3 [10])

$$\bigwedge_{u \in \mathcal{U}(t,1)} \psi(t;u,v) = \frac{E\left[\bigwedge_{v \in \mathcal{U}(t,1)} \rho(0,t;u)\rho(t,1;v)\left(\int_t^1 r_t^{\cdot}\, C(\cdot\,,v) + r_t^1 J_1\right) \mid \mathcal{G}_t\right]}{E[\rho(0,t;u) \mid \mathcal{G}_t]} \tag{5}$$

$$= E_u\left[\bigwedge_{v \in \mathcal{U}(t,1)} \left(\int_t^1 r_t^{\cdot}\, C(\cdot\,,v) + r_t^1 J_1\right) \mid \mathcal{G}_t\right] \qquad \text{a.s. P.}$$

Let W be defined as

$$W(t,u) = \bigwedge_{v \in \mathcal{U}(t,1)} \psi(t;u,v)\ . \tag{6}$$

Since (5) is verified by the Radon-Nikodym theorem it is necessary to have a measure with respect to which all the measures $(P_u)_{u \in \mathcal{U}}$ are absolutely continuous. Without this property it is necessary to show that the family of controls $\mathcal{U}$ are complete with respect to W which is more difficult to verify (cf. [5,18]) and places an additional restriction for the Markov control problem (cf. p. 110 [3]).

3. Optimality criteria

Initially the stochastic systems described by (1-2) using the controls $\mathcal{U}$ will be considered. As we shall see two levels of procedure are evident to obtain necessary and sufficient conditions for optimality. The first level is an abstract level where it is shown that the value function for a suitable subfamily of the controls is a supermartingale and that the value function for the optimal control is a martingale. The second level is a local level where the specific description of the system via differential equations is used to represent the aforementioned supermartingales in terms of the martingales that describe the stochastic system.

As part of the abstract level to give necessary and sufficient conditions for optimality we have the following result.

LEMMA 1. Let $t_1, t_2 \in I$ and $t_1 \leq t_2$ and let $u \in \mathcal{U}(0,1)$. The following inequality is satisfied

$$W(t_1,u) \leq E_u\left[\int_{t_1}^{t_2} r_{t_1}^{\cdot}\, C(\cdot\,,u) \mid \mathcal{G}_{t_1}\right] + E_u[r_{t_1}^{t_2} W(t_2,u) \mid \mathcal{G}_{t_1}]\ . \tag{6}$$

The control $u \in \mathcal{U}$ is optimal on $[t_1, t_2]$ if and only if there is equality in (6).

Proof: Use (5). ■

For notational simplicity the restriction of $u \in \mathcal{U}$ to $(s,t] \times \Omega$ will often be denoted by the same symbol u. The correct domain for the control u should be clear from the context.

A value decreasing control $u \in \mathcal{U}$ is one for which the value function W has the property that for all $s, t \in I$ $s \leq t$ $W(t,u) \leq W(s,u)$. Since conditional expectation preserves order the value function for a value decreasing control is a supermartingale.

If $(A_t(u))$ is the increasing process associated with the supermartingale $(W(t,u))$ in the Doob-Meyer decomposition [16] then it easily follows that for $0 \leq s \leq t \leq 1$

$$0 \leq E_u[A_t(u) - A_s(u) | \mathscr{B}_s]$$

$$\leq E_u[\int_s^t r_0^{\cdot} C(\cdot, u) | \mathscr{B}_s] \qquad \text{a.s.} P_u .$$

By the Radon-Nikodym theorem it follows that

$$A_t(u) = \int_0^t \alpha(s,u) E[r_0^s C(s,u(s)) | \mathscr{B}_s] ds \tag{7}$$

where $0 \leq \alpha \leq 1$ a.e. $dt \times dP_u$. This observation has been made by Boel-Varaiya (p. 99 [3]).

For the local properties of optimality it is necessary to identify the martingales on the probability space that are adapted to the observations and to show that these martingales can be represented by some martingales that appear in the description of the stochastic system.

Let (dY_t, P) be the formal vectors of the M-valued Brownian motion that is used to define the solution of (2) and let (N_t) be the counting process that corresponds to the jump times (T_n) of the discontinuous process (M_t). Let $\mathcal{E}_t$ be the P-completion of the σ-algebra $\sigma(dY_s, N_s; 0 \leq s \leq t)$. This σ-algebra can be formed from real-valued stochastic integrals using smooth vector fields in TM and indicator functions for the jump times.

LEMMA 2. If $(W_t, \mathcal{E}_t, P)_{t \in I}$ is a real-valued square integrable martingale then there is a unique (equivalence class) representation of (W_t) as

$$W_t = W_0 + \int_0^t \langle k, dY\rangle + \int_0^t \ell d\tilde{N} \qquad (8)$$

where (k_t) is an $(\mathcal{E}_t)$-predictable TM-valued process such that $k(t) \in T_{Y_t}M$, (ℓ_t) is a real-valued $(\mathcal{E}_t)$-predictable process and $(\tilde{N}_t)$ is the martingale formed from (N_t) by compensation. Furthermore,

$$E\int(\langle k, k\rangle + \ell^2) < \infty .$$

Proof: Let F(resp. G) be a family of smooth real-valued functions on M(resp $\mathbb{R}$) that separate points and are uniformly bounded as well as their first two derivatives. Form an algebra $\mathcal{a}$ of functions from the constants and random variables $f(Y_t)$ and $g(N_s)$ where $f \in F$ and $g \in G$. By a result of Segal (Lemma 2.1 [20]) this algebra is dense in $L^2(\Omega, \mathcal{E}_1, P)$. Using the change of variables formula for semimartingales [17] an element in the algebra can be represented as a sum of a stochastic integral with respect to the M-valued Brownian motion (Y_t), a stochastic integral with respect to $(\tilde{N}_t)$, an integral that is of bounded variation and a constant. Note that the two stochastic integrals are orthogonal martingales [17]. Let (f_n) be a sequence from $\mathcal{a}$ that converges in $L^2(P)$ to $W_1 - W_0$. By conditional expectation we can require that the sequence (f_n) is a sequence of martingales. Each term in the sequence can be expressed as a sum of two stochastic integrals from the representation above for each element in $\mathcal{a}$. Since these pairs of stochastic integrals are orthogonal it easily follows that they converge to a pair of stochastic integrals that are square integrable, one a stochastic integral with respect to (Y_t) and the other a stochastic integral with respect to $(\tilde{N}_t)$. ■

Recall that the observations $(\mathcal{G}_t)$ are the P-completion of $\sigma(Y_s, T_n(s): 0 \leq s \leq t)$ where (T_n) are the jump times of (M_t).

For a real-valued square integrable martingale $(W_t, \mathcal{G}_t, P_u)$ we can use a so-called innovations result (Theorem 2 [10]) to show that the continuous martingale given in Lemma 2 can be represented $\int \langle k, dC\rangle$ where dC_t is represented in $T_{Y_t}M$ as $dC_t = dY_t - E[c_t | \mathcal{G}_t]dt$.

An expression will be given for the value function W for value decreasing controls using the description of the stochastic system.

THEOREM 1. For each $u \in \mathcal{U}$ that is value decreasing there are two real-valued $(\mathscr{Y}_t)$-predictable processes $(\alpha(t,u))$ and $(\gamma(t,u))$ and a TM-valued $(\mathscr{Y}_t)$-predictable process $(\beta(t,u))$ where $\beta(t,u) \in T_{Y_t}M$ such that

$$r_0^t W(t,u) = J^* - \int_0^t \alpha(\cdot,u) + \int_0^t \langle \beta(\cdot,u), dC\rangle + \int_0^t \gamma(\cdot,u)d\tilde{N} \tag{9}$$

where

i) $\int_0^t \langle \beta(\cdot,u), dC\rangle$ is a real-valued square integrable continuous martingale on $(\Omega, \mathfrak{F}, P_u)$ and $dC_t = dY_t - E[c_t | \mathscr{Y}_t]dt$.

ii) $\int_0^t \gamma(\cdot,u)d\tilde{N}$ is a real-valued square integrable discontinuous martingale on $(\Omega, \mathfrak{F}, P_u)$

iii) $$\int_{t_1}^{t_2} \{E[r_0^s \; C(s,u(s)) | \mathscr{Y}_s] - \alpha(s,u(s))\} ds \geq 0 \tag{10}$$

for $t_1, t_2 \in I$ and $t_1 \leq t_2$. There is equality in (10) for the optimal control $u^* \in \mathcal{U}$ and $J^* = J(u^*)$.

Proof: Let $u \in \mathcal{U}$ be a value decreasing control. The process $(r_0^t W(t,u), \mathscr{Y}_t, P_u)$ is a supermartingale. By the Doob-Meyer decomposition [16] it can be represented as the difference of a martingale and an increasing process. The increasing process $(A(t,u))$ has been described in (7). From the boundedness properties of the cost it follows that the martingale is square integrable. The remarks after Lemma 2 show that this $(\mathscr{Y}_t)$-martingale can be expressed as the sum of two stochastic integrals using (dC_t) and $(\tilde{N}_t)$.

The equality in (10) for an optimal control follows from optimality by Lemma 1. ■

For completely observable systems, that is, systems where the observations are the TM-valued process (Z_t), the value function W does not depend on the control and by the uniqueness of the decomposition of supermartingales the increasing process and the martingale in this decomposition do not depend on the controls.

THEOREM 2. Let the observations $\mathscr{Y}_t$ at time $t \in I$ be the P-completion of $\sigma(Z_u : 0 \leq u \leq t)$. A control $u^* \in \mathcal{U}$ is optimal for the system (1-2) and the cost (3) if and only if there are a constant J^*, an increasing process (A_t) and a martingale (K_t) such that

$$A_t = \int_0^t \alpha$$

$$K_t = \int_0^t \langle \beta, dC\rangle + \int_0^t \langle \gamma, dM\rangle$$

$$r_0^t W(t) = J^* - \int_0^t \alpha + K(t)$$

where $(\beta_t, \gamma_t)_{t\in I}$ is a $(\mathcal{G}_t)$-predictable $T(TM)$-valued process, $(\beta(t), \gamma(t)) \in T_{Z_t}(TM)$ and $\alpha(t) = r_0^t\, C(t, u^*(t))$.

The proof of this theorem follows directly from the proof of Theorem 1 noting that W does not depend on the control and from the fact that Lemma 2 can be extended by the same proof to represent a square integrable martingale with respect to these observations.

Finally, a Markovian optimization problem will be considered. Let $\mathcal{M}(s,t)$ be the Markovian controls on $(s,t] \times \Omega$ to U where we use the conventions that $\mathcal{M}(0,t)$ are the controls on $[0,t] \times \Omega$ and $\mathcal{M} = \mathcal{M}(0,1)$. A Markovian control u has the properties that u(t) is $\overline{\sigma(Z_{t-})}$ measurable where $\overline{\sigma(Z_{t-})}$ is the P-completion of the σ-algebra generated by the TM-valued random variable Z_{t-} and that u is jointly measurable. It will be assumed that the observations of the system at time $t \in I$ are $\overline{\sigma(Z_{t-})}$.

The cost for the Markovian optimization problem will be

$$J(u) = E_u \left[\int_0^t C(s, u(s), Z_{s-})ds + J_1(Z_1) \right] \tag{11}$$

so that C_s is $\overline{\sigma(Z_{s-})}$ measurable and J_1 is $\overline{\sigma(Z_1)}$ measurable. For notational convenience the discount rate has been eliminated but one could be inserted that has the same measurability property as C. The Markovian value function V is defined as

$$V(t, Z_t) = \bigwedge_{v\in\mathcal{M}(t,1)} \eta(t, v, Z_t) \tag{12}$$

where

$$\eta(t, v, Z_t) = E_v \left[\int_t^1 C(s, v(s), Z_{s-})ds + J_1(Z_1) \middle| Z_t \right] \tag{13}$$

Using the proof of Lemma 1 we have the principle of optimality for Markov controls.

LEMMA 3. Let $t_1, t_2 \in I$ and $t_1 \le t_2$. For $u \in \mathcal{M}$ the following inequality is satisfied

$$V(t_1, Z_{t_1}) \le E_u\left[\int_{t_1}^{t_2} C(s, u(s), Z_{s-})ds \,|\, Z_{t_1}\right] + E_u[V(t_2, Z_{t_2}) \,|\, Z_{t_1}] \tag{14}$$

A control $u^* \in \mathcal{M}$ is optimal if and only if there is equality in (14) for all $t_1, t_2 \in I$ such that $t_1 \le t_2$.

For the Markovian optimization problem we have the following result.

THEOREM 3. Let V be the Markovian value function. A control $u^* \in \mathcal{M}$ is optimal for V if and only if

$$V(t) = J_m - E\left[\int_0^t C(s, u^*(s), Z_s)ds \,|\, Z_t\right]$$

$$+ E\left[\int_0^t (\langle \beta, dC\rangle + \langle \gamma, dM\rangle) \,|\, Z_t\right] \tag{15}$$

where $(\beta(t, Z_{t-}), \gamma(t, Z_{t-}))$ is a $T(TM)$-valued process that is $(\mathcal{F}_t)$-predictable and $((\sigma(Z_{t-}))$ adapted such that $(\beta(t, Z_{t-}), \gamma(t, Z_{t-})) \in T_{Z_t}(TM)$ and

$$E\left[\int_0^1 (\langle \beta, \beta\rangle + \langle \gamma, \gamma\rangle)\right] < \infty .$$

Proof: Let $u^* \in \mathcal{M}$ be an optimal control. Using the Markov property and the bounded convergence theorem we have

$$E_{u^*}\left[\int_t^1 C(s, u^*(s), Z_{s-})ds + J_1(Z_1) \,|\, Z_t\right]$$

$$= \lim_{\delta_n \downarrow 0} E_{u^*}\left[\int_{t+\delta_n}^1 C(s, u^*(s), Z_{s-})ds + J_1(Z_1) \,|\, Z_t\right]$$

$$= \lim_{\delta_n \downarrow 0} E_{u^*}\left[\int_{t+\delta_n}^{1} C(s,u^*(s),Z_{s-})ds + J_1(Z_1)|\mathfrak{F}_t\right]$$

$$= E_{u^*}\left[\int_{t}^{1} C(s,u^*(s),Z_{s-})ds + J_1(Z_1)|\mathfrak{F}_t)\right] \quad \text{a.s.} \qquad (16)$$

where (δ_n) is a suitable sequence.

From (16) we can use Theorem 2 to represent V as

$$V(t) = J(u^*) - \int_0^t \alpha + \int_0^t \langle\beta, dC\rangle + \int_0^t \langle\gamma, dM\rangle$$

where $\alpha(t) = C(t,u^*(t),Z_{t-})$.

The simple functions which form the integrands of the stochastic integrals that generate all the square integrable stochastic integrals are of the form (p. 82 [6])

$$\theta_s = \sum_{i\in \mathbb{N}} H_i 1_{(t_i,t_{i+1}]}(s) \qquad (17)$$

where (t_i) is a dyadic partition of I and H_i is $\mathfrak{F}_{t_i}$ measurable. For this dyadic partition the process $(Y_t)_{t\in[t_i,t_{i+1}]}$ can be developed on $T_{Y_{t_i}}M$ and by parallelism the vectors in the fibres of TM along this part of the base process can be transported to $T_{Y_{t_i}}M$. Thus integrands of the form (17) generate the square integrable stochastic integrals for (dC_t) and (M_t). If (φ_t) is a $T(TM)$-valued process that is the integrand for (dC_t, dM_t) it follows by a suitable modification on a t-null set that φ_t is F_{t-} measurable for all $t \in I$ where $\mathfrak{F}_{t-} = \bigvee_{s<t} \mathfrak{F}_s$.

Fix $t \in I$ and let $h > 0$. The family of random variables $V(t) - V(t-h) - \int_{t-h}^{t} \langle\gamma, dM\rangle$ have been studied in [10]. In particular the quadratic oscillation of this process can be computed explicitly using a countable cover of M so that $T(TM)$ is locally trivialized. Let $h \downarrow 0$ and choosing an appropriate subsequence it follows that $\beta(t)$ is measurable with respect to $\bigcap_{h>0} \mathfrak{F}(t-h,t) = \overline{\sigma(Z_{t-})}$ where $\mathfrak{F}(t-h,t)$ is the P-completion of $\sigma(Z_s, t-h \leq s < t)$. Therefore there is a Borel measurable function such that $\beta(t,\omega) = \beta(t, Z_{t-})$.

Let (T_n) be the jump times of (M_t) such that $T_n < T_{n+1}$ for all $n \in \mathbb{N}$. Considering each of these jump times it follows by the measurability property for processes of the form (17) that $\gamma(T_n, \omega) = \gamma(T_n, Z_{T_{n-}})$.

Conversely if (15) is satisfied for $u^* \in \mathcal{M}$ then u^* is optimal by Lemma 1.■

References

1. V. Beneš, Existence of optimal stochastic control laws, SIAM J. Control 9 (1971) 446-475.

2. R. Boel, P. Varaiya and E. Wong, Martingales on jump processes I and II, SIAM J. Control 13 (1975) 999-1061.

3. R. Boel and P. Varaiya, Optimal control of jump processes, SIAM J. Control and Optimization, 15 (1977) 92-119.

4. R. H. Cameron and W. T. Martin, Transformations of Wiener integrals under a general class of linear transformations, Trans. Amer. Math. Soc. 38 (1945) 184-219.

5. M. H. A. Davis and P. Varaiya, Dynamic programming conditions for partially observable stochastic systems, SIAM J. Control 11 (1973) 226-261.

6. C. Doléans-Dade and P. A. Meyer, Intégrales stochastiques par rapport aux martingales locales, Séminaire de Probabilités IV, Lecture Notes in Math. 124 (1970) Springer-Verlag.

7. T. E. Duncan and P. Varaiya, On the solutions of a stochastic control system, SIAM J. Control 9 (1971) 354-371.

8. T. E. Duncan, Some stochastic systems on manifolds, Lecture Notes in Econ. and Math. Sys. 107 (1975) 262-270, Springer-Verlag.

9. T. E. Duncan, Stochastic systems in Riemannian manifolds, J. Optimization Theory and Applications, 27 (1979), 399-426.

10. T. E. Duncan, Dynamic programming optimality criteria for stochastic systems in Riemannian manifolds, Applied Math. and Optimization, 3 (1977), 191-208.

11. T. E. Duncan, Estimation for jump processes in the tangent bundle of a Riemann manifold, Applied Math. and Optimization 4 (1978) 265 -274.

12. N. Dunford and J. T. Schwartz, Linear Operators, Part I. Interscience, New York, 1958.

13. I. V. Girsanov, On transforming a certain class of stochastic processes by absolutely continuous substitution of measures, Theor. Probability Appl. 5 (1960) 285-301.

14. J. Jacod, Multivariate point processes: Predictable projection, Radon-Nikodym derivatives, representation of martingales, Z. Wahrscheinlichkeitstheorie 31 (1975) 235 - 253.

15. S. Kobayashi and K. Nomizu, Foundations of Differential Geometry V. I, Interscience, New York, 1963.

16. P. A. Meyer, Probability and Potentials, Blaisdell, Waltham, Mass. 1966.

17. P. A. Meyer, Intégrales stochastiques, Séminaire de Probabilités 1, Lecture Notes in Math. 39 Springer-Verlag, 1967.

18. R. Rishel, Necessary and sufficient dynamic programming conditions for continuous-time stochastic optimal control, SIAM J. Control 8 (1970) 559-571.

19. R. Rishel, Dynamic programming and minimum principles for systems with jump Markov disturbances, SIAM J. Control 13 (1975) 338-371.

20. I. E. Segal, Tensor algebras over Hilbert space I, Trans. Amer. Math. Soc. 81 (1956) 106 -134 .

* Research supported by AFOSR Grant 77-3177.

CONSTRUCTION OF STOPPING TIMES T SUCH THAT $\sigma(X_T) = \underline{\underline{F}}_T$ MOD P

by

Neil FALKNER

Laboratoire de Probabilités
Université de PARIS VI

1. Introduction.

Throughout this paper, $(\Omega,\underline{\underline{F}},P)$ will denote a probability space and $(\underline{\underline{F}}_t)_{0\leq t<\infty}$ will denote a filtration of $(\Omega,\underline{\underline{F}})$; that is, a non-decreasing family of sub-σ-fields of $\underline{\underline{F}}$. Intuitively, $\underline{\underline{F}}_t$ should be thought of as consisting of those events which depend only on what happens up to time t . For clarification of this and of other remarks about intuition made below, see [3];IV.94-103.

Recall that a stopping time (with respect to $(\underline{\underline{F}}_t)$) is a map $T : \Omega\to[0,\infty]$ such that for $0\leq t<\infty$ we have $\{T\leq t\}\in\underline{\underline{F}}_t$. To each stopping time T is associated the σ-field

$$\underline{\underline{F}}_T = \{F \in \sigma(\bigcup_t \underline{\underline{F}}_t) : F\cap\{T\leq t\}\in\underline{\underline{F}}_t \text{ for } 0\leq t<\infty\} .$$

Intuitively, a random time T is a stopping time iff it does not depend on the future, and in this case $\underline{\underline{F}}_T$ should be thought of as consisting of those events which depend only on what happens up to time T .

Things such as the optional sampling theorem and the strong Markov property lead one to think that stopping times behave like fixed times. However, in one obvious respect, stopping times are not like fixed times : they depend on the past. The object of this paper is to show that a stopping time can depend on <u>everything</u> in its past, in the sense that if (X_t) is a continuous $(\underline{\underline{F}}_t)$-adapted process which is reasonably non-trivial and if $(\underline{\underline{F}}_t)$ satisfies certain mild conditions then there exist stopping times T such that $P(0<T<\infty) = 1$ and

$$\underline{\underline{F}}_T = \sigma(X_T) \mod P .$$

It is also of interest, in constructing certain counterexamples, to know that such stopping times exist which have the additional property that they are not too big. For example, one might ask that the process $(X_{T\wedge t})$ be uniformly integrable, or that T have finite expectation, or perhaps that T be dominated by some stopping time ζ fixed in advance. We show that such additional constraints can often be satisfied too.

2. Some Terminology.

Given a sub-σ-field $\underline{\underline{B}}$ of $\underline{\underline{F}}$, let

$$\underline{\underline{B}}^P = \{F \in \underline{\underline{F}} : \exists B \in \underline{\underline{B}} \text{ such that } P(F\Delta B) = 0\} .$$

$\underline{\underline{B}}^P$ is a σ-field and is called the P-completion of $\underline{\underline{B}}$ in $\underline{\underline{F}}$. Let $\underline{\underline{A}}$ and $\underline{\underline{B}}$ be sub-σ-fields of $\underline{\underline{F}}$. We say $\underline{\underline{A}} \subseteq \underline{\underline{B}}$ mod P iff for every $A \in \underline{\underline{A}}$ there exists $B \in \underline{\underline{B}}$ such that $P(A\Delta B) = 0$. Evidently $\underline{\underline{A}} \subset \underline{\underline{B}}$ mod P iff $\underline{\underline{A}} \subseteq \underline{\underline{B}}^P$ iff $\underline{\underline{A}}^P \subseteq \underline{\underline{B}}^P$. If $\underline{\underline{A}}$ is countably generated then one can show that $\underline{\underline{A}} \subseteq \underline{\underline{B}}$ mod P iff there exists $\Lambda \in \underline{\underline{F}}$ with $P(\Lambda)=1$ such that $\underline{\underline{A}}|\Lambda \subseteq \underline{\underline{B}}|\Lambda$; evidently Λ can be taken to belong to $\underline{\underline{C}} \equiv \sigma(\underline{\underline{A}} \cup \underline{\underline{B}})$; If for every $C \in \underline{\underline{C}}$ there exists $B \in \underline{\underline{B}}$ with $B \subseteq C$ and $P(C\backslash B) = 0$ then Λ can be taken to belong to $\underline{\underline{B}}$. We say $\underline{\underline{A}}=\underline{\underline{B}}$ mod P iff $\underline{\underline{A}} \subseteq \underline{\underline{B}}$ mod P and $\underline{\underline{B}} \subseteq \underline{\underline{A}}$ mod P. Clearly $\underline{\underline{A}}=\underline{\underline{B}}$ mod P iff $\underline{\underline{A}}^P=\underline{\underline{B}}^P$. If $\underline{\underline{A}}$ and $\underline{\underline{B}}$ are countably generated then $\underline{\underline{A}}=\underline{\underline{B}}$ mod P iff there exists $\Lambda \in \underline{\underline{F}}$ with $P(\Lambda)=1$ such that $\underline{\underline{A}}|\Lambda = \underline{\underline{B}}|\Lambda$; of course Λ can be taken to belong to $\underline{\underline{C}}$; if for every $C \in \underline{\underline{C}}$ there exist $A \in \underline{\underline{A}}$, $B \in \underline{\underline{B}}$ with $A \subseteq C$, $B \subseteq C$, $P(C\backslash A) = 0$, $P(C\backslash B) = 0$ then choosing $A_n \in \underline{\underline{A}}$, $B_n \in \underline{\underline{B}}$ with

$$\Lambda \supseteq A_o \supseteq B_o \supseteq A_1 \supseteq B_1 \supseteq \dots$$

and

$$0 = P(\Lambda\backslash A_o) = P(A_o\backslash B_o) = \dots$$

and replacing Λ by $\bigcap_n A_n = \bigcap_n B_n$ we see that Λ can be taken to belong to to $\underline{\underline{A}} \cap \underline{\underline{B}}$. We say $\underline{\underline{A}}$ is countably generated mod P iff there exists a countable set $\underline{\underline{A}}_o \subseteq \underline{\underline{A}}$ such that $\underline{\underline{A}} = \sigma(\underline{\underline{A}}_o)$ mod P. By [4],III.7.1, $\underline{\underline{A}}$ is countably generated mod P iff it is separable as a pseudometric space when equipped with the pseudometric $(A_1,A_2) \to P(A_1\Delta A_2)$. It follows that if $\underline{\underline{F}}$ itself is countably generated mod P, so is every sub-σ-field of $\underline{\underline{F}}$.

For a recent treatment of the theory of filtrations, see [3] ; some older references are [2] and [6] .

Recall ([2],III,T51) that at least when $(\underline{\underline{F}}_t)$ satisfies "les conditions habituelles" (and we take $\underline{\underline{F}}_{o-} = \underline{\underline{F}}_o$ as we always shall) then $(\underline{\underline{F}}_t)$ is quasi-left-continuous iff whenever (T_k) is an increasing sequence of stopping times with limit T then

$$\underline{\underline{F}}_T = \sigma(\bigcup_k \underline{\underline{F}}_{T_k}) \quad .$$

Many of the filtrations one encounters in the theory of Markov process are quasi-left-continuous ; see [7],XIV,T36,[1],IV,4.2, and [5],6.7 and 13.2. Since we do not assume $(\underline{\underline{F}}_t)$ satisfies "les conditions habituelles", we shall work with a slight modification of the notion of quasi-left-continuity : we say $(\underline{\underline{F}}_t)$ is quasi-left-continuous mod P iff whenever (T_k) is an increasing sequence of stopping times with limit T we have

$$\underline{\underline{F}}_T = \sigma(\bigcup_k \underline{\underline{F}}_{T_k}) \mod P \ .$$

3. Theorem. Suppose $\underline{\underline{F}}$ is countably generated mod P and we are given a real-valued $(\underline{\underline{F}}_t)$-progressively measurable process (X_t) , an $\underline{\underline{F}}_o$-measurable function $\Gamma : \Omega \to (0,\infty)$, and an $(\underline{\underline{F}}_t)$-stopping time ζ such that for all $\omega \in \Omega, t \to X_t(\omega)$ is continuous on $[0,\zeta(\omega))$ and for P-a.a. $\omega \in \Omega$,

$$(3.1) \qquad \left[\sup_{0 \leq t < \zeta(\omega)} X_t(\omega)\right] - \left[\inf_{0 \leq t < \zeta(\omega)} X_t(\omega)\right] > \Gamma(\omega) \quad .$$

Suppose also that $(\underline{\underline{F}}_t)$ is quasi-left-continuous mod P . Then there exists an $(\underline{\underline{F}}_t)$-stopping time T such that :

$$(3.2) \qquad \underline{\underline{F}}_T = \sigma(X_T) \mod P$$

$$(3.3) \qquad 0<T<\zeta \quad \text{P-a.s.}$$

$$(3.4) \qquad \sup_{0 \leq t < \infty} |X_{T \wedge t} - X_o| < \Gamma/2 \qquad \text{P-a.s.}$$

Proof. Let $(G_n)_{n \in \mathbb{N}}$ be an open base for $\mathbb{R}$ consisting of non-empty sets. Let $A = \{-1,0,1\}^{\mathbb{N}}$ equipped with its usual product topology and define $h : A \to \mathbb{R}$ by

$$h(a) = \sum_k a_k/4^k$$

Then h is one-to-one and continuous so, as A is compact, h is a homeomorphism of A onto $C \equiv h[A]$. As A is totally disconnected, C has empty interior. Now for $n \in \mathbb{N}$, choose $\chi_n \in \mathbb{R}$ and $r_n \in (0,\infty)$ such that

$$\chi_n + r_n C \subseteq G_n \quad ;$$

$m \neq n$ implies $(\chi_m + r_m C) \cap (\chi_n + r_n C) = \phi$.

This can be done since if $\chi_o, r_o, \ldots, \chi_n, r_n$ have already been chosen then

$$C_n \equiv \bigcap_{m=o}^{n} \chi_m + r_m C$$

is closed with empty interior so $G_{n+1} \setminus C_n$ is open and non-empty.
Now $(n,a) \to \chi_n + r_n h(a)$ is a one-to-one continuous map of $\mathbb{N}\times A$ into $\mathbb{R}$ and its inverse, which we shall denote by g, is a Borel map of $\bigcup_{n\in\mathbb{N}} \chi_n + r_n C$ onto $\mathbb{N}\times A$.

Now for a momentary digression. If S_1 and S_2 are stopping times satisfying $S_1 \leq S_2$ and if $F \in \underline{\underline{F}}_{S_1}$ and if we let

$$S_3 = \begin{cases} S_2 & \text{on } F \\ S_1 & \text{on } \Omega\backslash F \end{cases}$$

then S_3 is also a stopping time. Next, if $B \subseteq [0,\infty)\times\Omega$ then D_B, the debut of B, is defined on Ω by

$$D_B(\omega) = \inf\{t \geq 0 : (t,\omega) \in B\} .$$

If $B = \bigcap_{n\in\mathbb{N}} U_n$ where each U_n is an $(\underline{\underline{F}}_t)$-progressively measurable subset of $[0,\zeta)\times\Omega$ such that $U_n(\omega)$ is open in $[0,\infty)$ and contains

$$\overline{U_{n+1}(\omega)} \cap [0,\zeta(\omega)) \quad \text{for all} \quad \omega \in \Omega$$

then it is easy to verify that D_B is a stopping time. Using these observations it will be easy to see that the functions T_k we are about to define are stopping times.
Define N_+, $N_- : \Omega \to \mathbb{N}$ by $N_+ = \min\{n : G_n \subseteq [X_o, X_o + \Gamma/2]\}$,

$$N_- = \min\{n : G_n \subseteq [X_o - \Gamma/2, X_o]\} .$$

Then N_+ and N_- are $\underline{\underline{F}}_o$-measurable. Let

$$T_o = \inf\{t \in [0,\zeta) : X_t = \chi_{N_+} \text{ or } \chi_{N_-}\} .$$

Then T_o is a stopping time. Now define $N : \Omega \to \mathbb{N}$ by

$$N(\omega) = \begin{cases} N_+(\omega) & \text{if } T_o(\omega) < \infty \text{ and } X_{T_o}(\omega) = \chi_{N_+(\omega)} ; \\ N_-(\omega) & \text{otherwise.} \end{cases}$$

Then N is $\underline{\underline{F}}_{T_o}$-measurable, and $T_o(\omega) < \infty$ implies $X_{T_o}(\omega) = \chi_{N(\omega)}$.

Now let $k \to (i(k), j(k))$ be a one-to-one map of $\mathbb{N}$ onto $\mathbb{N} \times \mathbb{N}$ such that $i(k) \leq k$ for all k (the usual "zigzag" bijection satisfies both $i(k) \leq k$ and $j(k) \leq k$) and define stopping times $T_1, T_2, \ldots$ and sets $F_{0,0}, F_{0,1}, \ldots, F_{1,0}, F_{1,1}, \ldots$ inductively as follows :

If $k \in \mathbb{N}$ and if T_k and $F_{k',\ell}$ $(k', \ell \in \mathbb{N}\ ,\ k' < k)$ have been defined, choose $F_{k,0}, F_{k,1}, \ldots \in \underline{\underline{F}}_{T_k}$ such that $\underline{\underline{F}}_{T_k} = \sigma(F_{k,\ell} : \ell \in \mathbb{N})$ mod P and let

$$T_{k+1} = \begin{cases} \inf\{t \in [T_k, \zeta) : |X_t - X_{T_k}| = r_N/4^k\} \\ \text{on } F_{i(k),j(k)} \ ; \\ \\ T_k \ \text{ on } \Omega \setminus F_{i(k),j(k)} \ . \end{cases}$$

(As usual, we take $\inf \phi = \infty$ and $[a,b) = \phi$ if $a \geq b$.) For the stopping time of the theorem, we take

$$T = \lim_{k \to \infty} T_k \ .$$

Let Λ be the set of $\omega \in \Omega$ for which (3.1) holds. Fix $\omega \in \Lambda$. Then there exist $t_1, t_2 \in [0, \zeta(\omega))$ such that

$$X_{t_1}(\omega) - X_{t_2}(\omega) > \Gamma(\omega) \ .$$

Taking t_3 equal to the appropriate one of t_1, t_2 we have

$$|X_{t_3}(\omega) - X_o(\omega)| > \Gamma(\omega)/2 \ .$$

Now it is a simple matter to check that

$$\chi_{N_-(\omega)} < X_t(\omega) < \chi_{N_+(\omega)} \quad \text{for } 0 \leq t < T_o(\omega)$$

$$T_o(\omega) < t_3$$

$$X_{T_o}(\omega) = \chi_{N(\omega)}$$

and then by induction on k that

$$|X_t(\omega) - \chi_{N(\omega)}| < \sum_{o \leq \ell \leq k} r_{N(\omega)}/4^\ell \ \text{ for } T_o(\omega) \leq t < T_{k+1}(\omega)$$

$$T_{k+1}(\omega) < t_3 \ .$$

It follows that $T(\omega) \leq t_3$ (actually, this inequality is strict) and

$$X_T(\omega) = \lim_{k \to \infty} X_{T_k}(\omega) \quad .$$

Now from the fact that $T_k(\omega)$ is finite for all k, it follows that there exists a (unique) sequence $a(\omega) \in A$ such that for all k,

$$X_{T_k}(\omega) = \chi_{N(\omega)} + r_{N(\omega)} \sum_{o \leq \ell < k} a_\ell(\omega)/4^\ell \quad .$$

Thus

$$X_T(\omega) = \chi_{N(\omega)} + r_{N(\omega)} \sum_{\ell \in \mathbb{N}} a_\ell/4^\ell$$

and

$$g(X_T(\omega)) = (N(\omega) \;,\; a_o(\omega) \;,\; a_1(\omega) \;, \ldots) \quad .$$

Now $a_\ell(\omega) = 0$ iff $\omega \notin F_{i(\ell),j(\ell)}$.

Thus

$$\sigma(F_{k,\ell} : k,\ell \in \mathbb{N})|\Lambda \subseteq \sigma(X_T)|\Lambda \quad .$$

As $\{F_{k,\ell} : \ell \in \mathbb{N}\}$ generates $\underline{\underline{F}}_{T_k}$ mod P and $P(\Omega\backslash\Lambda) = 0$, it follows that

$$\sigma(\bigcup_k \underline{\underline{F}}_{T_k}) \subseteq \sigma(X_T) \quad \text{mod} \quad P$$

But X_T is $\underline{\underline{F}}_T$-measurable as (X_t) is progressively measurable. Hence

$$\underline{\underline{F}}_T = \sigma(X_T) \quad \text{mod} \quad P$$

as $(\underline{\underline{F}}_t)$ is quasi-left-continuous mod P. This proves (3.2). As for (3.3) and (3.4), it is clear from the above considerations that the inequalities in question hold everywhere on Λ.

4. Remark. In the proof of the above theorem, one may clearly choose the sets $F_{k,\ell}$ so that for every $k,\ell \in \mathbb{N}$ there exist $\ell' \in \mathbb{N}$ such that

$$F_{k,\ell'} = \Omega \setminus F_{k,\ell} \quad .$$

If this is done then a moment's thought reveals that for all k,

$$T_k < T \quad \text{on} \quad \{T<\infty\} \quad .$$

Thus T is announceable (the sequence $(k \wedge T_k)$ announces T) and hence (by [3],IV.71) previsible. It also follows (see [3],IV.56) that

$$\sigma(\bigcup_k \underline{\underline{F}}_{T_k}) = \underline{\underline{F}}_{T-} \quad .$$

(Regarding the definition of $\underline{\underline{F}}_{T-}$, we take $\underline{\underline{F}}_{o-} = \underline{\underline{F}}_o$ whence if R and S are stopping times and R<S on $\{0<S<\infty\}$ then $\underline{\underline{F}}_R \subseteq \underline{\underline{F}}_{S-}$.)

Now evidently $X_T|\{T<\zeta\}$ is measurable with respect to the σ-field

$$\sigma(\bigcup_k \underline{\underline{F}}_{T_k})|\{T<\zeta\} \quad .$$

Thus we find that if we do not assume $(\underline{\underline{F}}_t)$ is quasi-left continuous mod P , we can nevertheless produce an announceable stopping time T satisfying (3.3) and (3.4) , such that

(4.1) $$\underline{\underline{F}}_{T-} = \sigma(X_T) \text{ mod } P \; .$$

5. Corollary. Suppose $\underline{\underline{F}}$ is countably generated mod P . Then there is an announceable stopping time T such that 0<T<1 everywhere on Ω and

$$\underline{\underline{F}}_{T-} = \sigma(T) \text{ mod } P \; .$$

(Remark. Any stopping time S is $\underline{\underline{F}}_{S-}$-measurable ; see [3],IV.56).

Proof. Take $X_t(\omega) = t$, $\Gamma(\omega) = 2$, and $\zeta(\omega) = \infty$, and observe that the set Λ in the proof of the theorem is all of Ω .

6. Corollary. Suppose $\underline{\underline{F}}$ is countably generated mod P and we are given an $(\underline{\underline{F}}_t)$-progressively measurable process (Y_t) taking values in a separable metric space (E,d) , a lower semicontinuous function $g : E\to(0,\infty)$, and an $(\underline{\underline{F}}_t)$-stopping time ζ such that for every $\omega\in\Omega$, $t\to Y_t(\omega)$ is continuous on $[0,\zeta(\omega))$ and for P-a.a. $\omega\in\Omega$,

(6.1) $$\sup_{0\leq t<\zeta(\omega)} d(Y_t(\omega) , Y_o(\omega)) > g(Y_o(\omega)) \quad .$$

Then there exists an announceable $(\underline{\underline{F}}_t)$-stopping time T such that :

(6.2) $$\underline{\underline{F}}_{T-} = \sigma(X_T) \text{ mod } P$$

(6.3) $$0<T<\zeta \text{ P-a.s.}$$

(6.4) $$\sup_{0\leq t<\infty} d(Y_{T\wedge t},Y_o) < g(Y_o) \text{ P-a.s.}$$

Proof. Let

$$\underline{\underline{U}} = \{U \text{ open} \subseteq E : P(Y_o\in\partial U) = 0 \text{ and } \forall\, y\in U, g(y)>\text{diam}(U)\} \quad .$$

Given $y \in E$, let $\gamma \in (0,g(y))$; then as g is lower semicontinuous, there exists $\varepsilon > 0$ such that $d(y,y') < \varepsilon$ implies $g(y') > \gamma$. Next, there exists $\varepsilon' > 0$ such that $\varepsilon' \leq \min\{\varepsilon, \gamma/2\}$ and $P(d(Y_o,y)=\varepsilon') = 0$. This is because $P(d(Y_o,y) = r)$ is strictly positive for at most countably many r . Then

$$\{y' \in E : d(y,y') < \varepsilon'\} \in \underline{\underline{U}} \quad .$$

Thus $\underline{\underline{U}}$ covers E . As E is second countable, there exist $U_o, U_1, \ldots \in \underline{\underline{U}}$ such that

$$E = \bigcup_n U_n \quad .$$

Let $V_o = U_o$ and for $n \geq 1$ let

$$V_n = U_n \setminus \bigcup_{m<n} \overline{U}_m \quad .$$

Then (V_n) is a pairwise disjoint sequence of open subsets of E and

$$P(Y_o \in E') = 1$$

where

$$E' = \bigcup_n V_n \quad .$$

Let

$$\zeta' = \zeta \wedge \inf\{t \in [0,\zeta) : Y_t \notin E'\} \quad .$$

By the observations on the construction of stopping times in the proof of the theorem, ζ' is a stopping time. Let

$$X_t = d(Y_t , E \setminus E')$$

$$\Gamma = \begin{cases} X_o/2 & \text{on } \{Y_o \in E'\} \\ 1 & \text{on } \{Y_o \notin E'\} \end{cases}$$

$$\Lambda = \{\omega \in \Omega : (6.1) \text{ holds and } Y_o(\omega) \in E'\} \quad .$$

As $d(\bullet , E\setminus E')$ is continuous, (X_t) is continuous on $[0,\zeta)$ and progressively measurable and Γ is $\underline{\underline{F}}_o$-measurable.
As we shall see in a moment , $E \setminus E' \neq \phi$ so (X_t) is real-valued and Γ takes values in $(0,\infty)$. Now $P(\Lambda) = 1$. Let $\omega \in \Lambda$. Then $Y_o(\omega) \in V_n$ for some n , and there exists $t_1 \in [0,\zeta(\omega))$ such that

$$d(Y_{t_1}(\omega) , Y_o(\omega)) > g(Y_o(\omega)) \quad .$$

Now $g(Y_o(\omega)) > \text{diam}(V_n)$ so $Y_{t_1}(\omega) \notin V_n$.

Thus if we let

$$t_o = \inf\{t : Y_t(\omega) \notin V_n\}$$

then $$t_o < \zeta(\omega) \quad ,$$

$$Y_t(\omega) \in V_n \quad \text{for} \quad 0 \le t < t_o \quad ,$$

and $$Y_{t_o}(\omega) \in \partial V_n \quad .$$

Now $V_m \cap \partial V_n = \phi$ for all m , so $Y_{t_o}(\omega) \in E \setminus E'$ (hence $E \setminus E' \neq \phi$, as promised) and $\zeta'(\omega) = t_o$. Thus

$$\inf_{0 \le t < \zeta'(\omega)} X_t(\omega) = 0$$

while $X_o(\omega) > 0$. For $0 \le t < \zeta'(\omega)$ we have

$$d(Y_t(\omega), Y_o(\omega)) \le \text{diam}(V_n) < g(Y_o(\omega)) \ .$$

Applying the theorem (and the remark which follows it) to (X_t) , Γ , and ζ' yields an announceable stopping time T such that $0 < T < \zeta'$ P-a.s. and $\sigma(X_T) = \underline{\underline{F}}_{T-}$ mod P .

Then $\underline{\underline{F}}_{T-} \subseteq \sigma(Y_T)$ mod P and, as T is announceable and $T < \zeta$ P-a.s. and (Y_t) is continuous on $[0, \zeta)$, $\sigma(Y_T) \subseteq \underline{\underline{F}}_{T-}$ mod P .

The bound (6.4) follows from the fact that $T < \zeta'$ P-a.s.

7. <u>Remark</u>. $\underline{\underline{F}}_{T-}$ is generated by

$$\underline{\underline{F}}_o \cup \{A \cap \{r < T\} : r \text{ is rational, } r > o \ , A \in \underline{\underline{F}}_{r-}\}$$

so if $\underline{\underline{F}}_{t-}$ is countably generated for all t then $\underline{\underline{F}}_{T-}$ is also countably generated. In this case (6.2) implies the existence of a set $\Omega_1 \in \underline{\underline{F}}$ such that $P(\Omega_1) = 1$ and

(7.1) $$\underline{\underline{F}}_{T-}|\Omega_1 = \sigma(Y_T|\Omega_1) \quad .$$

Evidently we may suppose $\Omega_1 \subseteq \{T<\zeta\}$. If this is done then for each $t \in [0,\infty)$, $Y_{t\wedge T}|\Omega_1$ is measurable with respect to $\underline{\underline{F}}_{T-}|\Omega_1$. Then, if we let $\underline{\underline{C}}$ be the space of continuous maps $f : [0,\infty) \to E$, endowed with the σ-field generated by the maps $f \to f(t)$ $(t \in [0,\infty))$, the map $\Pi_T : \Omega_1 \to \underline{\underline{C}} \times [0,\infty]$ defined by

$$\Pi_T(\omega) = ((Y_{t\wedge T}(\omega))_{0\leq t<\infty} , T(\omega))$$

is measurable with respect to $\underline{\underline{F}}_{T-}|\Omega_1$ and so with respect to $\sigma(Y_T|\Omega_1)$.

Now if in addition E is Polish then $\underline{\underline{C}}$ is a lusinien measurable space (and $\underline{\underline{C}} \times [0,\infty]$ is too) so by [3],I.18 in conjunction with III (appendice).80, there exists a Borel function $h : E \to \underline{\underline{C}} \times [0,\infty]$ such that $\Pi_T = h\circ Y_T$ on Ω_1 . Thus on a subset of Ω of measure one (namely Ω_1) , the time T and the process Y up to time T can be recovered (moreover , in a measurable fashion) from Y_T .

8. Corollary. Let $(\Omega,\underline{\underline{F}},(B_t),P)$ be a Brownian motion process in $\mathbb{R}^n$ (not necessarily starting from a point). Let D be an open subset of $\mathbb{R}^n$ such that $P(B_0 \in D)=1$ and let

$$\zeta = \inf\{t\geq 0 : B_t \notin D\} .$$

Let $\underline{\underline{F}}_t = \sigma(B_s : 0\leq s\leq t)$ for $0\leq t<\infty$.

Then there is an $(\underline{\underline{F}}_t)$-stopping time T such that

(8.1) $$\underline{\underline{F}}_T = \sigma(B_T) \mod P$$

(8.2) $$0<T<\zeta \quad P\text{-a.s.}$$

(8.3) $$\sup_{0\leq t<\infty} ||B_{T\wedge t} - B_0|| < 1 \quad P\text{-a.s.}$$

Proof. Let $\underline{\underline{F}}_\infty = \sigma(B_s : 0\leq s<\infty)$. By the method of [7],XIV.34 and 35, for each $g \in L^1(\Omega,\underline{\underline{F}}_\infty,P)$ there exists a process (M_t) continuous on $[0,\infty]$ and a set $\Lambda \in \underline{\underline{F}}_\infty$ with $P(\Lambda) = 1$ such that for every $t \in [0,\infty]$,

(8.4) $$\begin{cases} M_t|\Lambda \text{ is } \underline{\underline{F}}_t|\Lambda\text{-measurable and} \\ M_t = E(g|\underline{\underline{F}}_t) \quad P\text{-a.s.} \end{cases}$$

Then (8.4) also holds if t is replaced by any stopping time S. Hence if (R_k) is an increasing sequence of stopping times with limit R and g is $\underline{\underline{F}}_R$-measurable then

$$g = M_R = \lim_{k\to\infty} M_{R_k} \quad \text{P-a.s.}$$

so g is $\sigma(\bigcup_k \underline{\underline{F}}_{R_k})^P$-measurable. Thus $(\underline{\underline{F}}_t)$ is quasi-left-continuous mod P.

Now (B_s) is continuous and adapted so it is progressively measurable and ζ is a stopping time. Also, P-a.s.

$$\sup_{0\leq t<\zeta} \| B_t - B_o \| \geq \text{distance } (B_o, \mathbb{R}^n \backslash D) .$$

Thus by the previous corollary there exists an announceable stopping time T satisfying (8.2) and (8.3) such that

$$\underline{\underline{F}}_{T-} = \sigma(X_T) \quad \text{mod} \quad P .$$

As $(\underline{\underline{F}}_t)$ is quasi-left-continuous mod P and T is announceable, $\underline{\underline{F}}_{T-} = \underline{\underline{F}}_T$ mod P .

9. Remark. In the above corollary, if $(\Omega, \underline{\underline{F}})$ is a Blackwell space (as is the case if $(\Omega, \underline{\underline{F}}, (B_t), P)$ is the canonical realization of Brownian motion) then, by [3],IV.98(d), every $(\underline{\underline{F}}_t)$-stopping time is previsible. (In fact, every measurable $(\underline{\underline{F}}_t)$-adapted process is previsible !) Then by [3],IV.73, we have $\underline{\underline{F}}_{S-} = \underline{\underline{F}}_S$ for every $(\underline{\underline{F}}_t)$-stopping time S. In particular, $(\underline{\underline{F}}_t)$ is actually quasi-left-continuous, not just quasi-left-continuous mod P. More interestingly, it follows that for any $(\underline{\underline{F}}_t)$-stopping time S, $\underline{\underline{F}}_S$ is countably generated. Hence, for the stopping time T of the corollary, there exists $\Omega_1 \in \underline{\underline{F}}$ with $P(\Omega_1) = 1$ such that

$$\underline{\underline{F}}_T | \Omega_1 = \sigma(B_T | \Omega_1) .$$

REFERENCES

1. R.M. BLUMENTHAL and R.K. GETOOR : Markov Processes and Potential Theory, Academic Press, 1968.

2. C. DELLACHERIE : Capacités et processus stochastiques, Springer-Verlag 1972

3. C. DELLACHERIE and P.A. MEYER : Probabilités et potentiel, Hermann, 1975.

4. N. DUNFORD and J. SCHWARTZ : Linear Operators, part I, Interscience Publishers (a division of John Wiley and Sons),1963.

5. R.K. GETOOR : Markov Processes : Ray processes and right processes , Lecture Notes in Mathematics, vol. 440, Springer Verlag, 1975.

6. P.A. MEYER : Probabilités et potentiel , Hermann, 1966.

7. P.A. MEYER : Processus de Markov , Lecture Notes in Mathematics, vol. 26, Springer Verlag, 1967.

A NON-COMMUTATIVE STRASSEN DISINTEGRATION THEOREM

WOLFGANG HACKENBROCH
Dept. of Mathematics
University of Regensburg
84 Regensburg, Germany

Introduction. V. Strassen's well-known disintegration theorem ([10], theorem 1) has attracted attention in various fields of mathematics from measure theory to convex analysis and pure functional analysis (cf. [1], [5], [6], [7], [12], [13], [15]). All proofs of its abstract form seem to contain the same two ingredients: the Hahn-Banach theorem and some version of the Radon-Nikodym theorem (in [5] the latter is implicite on p. 505/506; in [13] p. 205 a Dunford-Pettis compactness argument is used instead). Wolff [15] gives an illuminating formulation of the pure "Hahn-Banach part" of the theorem. In [1] and in [6] p. 418 also vector valued versions are proved, and in [7] several of the standard applications are discussed under a unifying functional analytic point of view.

The aim of this note is to propose a "non commutative" version of the theorem applicable e.g. in the context of von Neumann algebras.

1. Preliminaries on ordered vector spaces.

We shall adopt the terminology and notation of Schaefer's book [9]. Throughout, $(E,\leq)$ denotes an ordered real vector space with generating positive cone E_+ and fixed *weak order unit* u (i.e. $u>0$, and $x=0$ for each $x\geq 0$ for which $x\wedge u=0$; here "$\wedge$" denotes the greatest lower bound in E whenever it exists. If even $x\leq\lambda u$ for some $\lambda>0$ (depending on x) for every $x\in E$, u is called an *order unit* for E). Moreover, E is supposed to be *monotone σ-complete* in the sense that the countable supremum $\bigvee_1^\infty x_n$ exists in E for every increasing majorized sequence (x_n) in E. - The interest in this class of ordered vector spaces is mainly due to the fact that, besides the obviuous lattice examples, also the selfadjoint parts of von Neumann algebras are included.

Let $P(E)$ denote the Boolean σ-algebra of all *split projections* of E (i.e. linear idempotent endomorphisms P with $0\leq P\leq 1$ pointwise on E_+; these are exactly the idempotent elements of the ideal center considered by Wils [14] p. 45. If E is the self-adjoint part of a W*-algebra A, $P(E)$ can be canonically identified with the set of all selfadjoint idempotents in the algebra center of A). We fix any measurable space (S,Σ) and a spectral measure (:= σ-homomorphism) $\pi: \Sigma\to P(E)$.

A natural example would be the *Loomis*-σ-homomorphism $\pi_o: \Sigma_o \to P(E)$ from the Baire σ-algebra Σ_o of the Stonean space S_o of $P(E)$ onto $P(E)$ (cf. [4] p. 102). We may obviuously think of $\pi: \Sigma \to P(E) \subset L(E)$ as well as of $\pi(\cdot)u: \Sigma \to E$ as *positive vector measures* (σ-additive with respect to order convergence) in the sense of J.D.M. Wright [16], p. 678, whose notion of an integral $\int f(t)\pi(dt)u$ ($\in E$) for real valued measurable f we shall use to construct the disintegrating mapping in the proof of our theorem.

Let us finally fix a σ-order continuous positive linear form μ on E. The crucial ordering "$\leq$ μ-a.e." on E with respect to which the disintegration will take place will then be defined through

$$x \leq y \ \mu\text{-a.e.} \ :\Leftrightarrow \mu(\pi(A)x) \leq \mu(\pi(A)y) \quad \text{for all } A \in \Sigma.$$

Let us illustrate this relation by two typical examples:

Example 1 ("Commutative" case): Assume E to be in addition a vector lattice and μ to be strictly positive (i.e. $\mu(x)>0$ for $x>0$). For π let us take the Loomis spectral measure π_o mentioned above. Then
$x \leq y$ μ-a.e. $\Leftrightarrow$ $x \leq y$.

For, otherwise there would exist $x, y \in E$ with $x \leq y$ μ-a.e. but with non-zero negative part $(y-x)_- =: z > 0$. This implies $px > py$ for the band projection $p \in P(E)$ of the band generated by z, and hence $\mu(px) > \mu(py)$, a contradiction.

Example 2 ("Non-commutative"case): Here we take E the self-adjoint part of an arbitrary W*-algebra A and for π the Loomis σ-homomorphism onto the (complete) Boolean algebra P of all selfadjoint projections in the center $Z(A)$ ($p \in P$ being viewed as the split projection $x \mapsto px$ of E). Furthermore fix any positive normal functional μ on A. Obviously we cannot expect that $x \leq y$ μ-a.e. implies $x \leq y$ unless A is commutative (mod μ).

On the other hand, the weaker relation $\leq\mu$-a.e. has the pleasant property that the mapping $x \mapsto |x| := \sqrt{x^*x}$ is sublinear with respect to $\leq$ μ-a.e. (which is known to be true for $\leq$ itself if and only if A is commutative, cf. [11] p.p. 19, 23). For the proof we must show that, for each fixed $p \in P$, $x \mapsto \mu(p|x|)$ is sublinear. Since $p|x| = |px|$, it is enough to consider the case $p=1$. But using polar decomposition in A, it is not hard to see that

$$\mu(|x|) = \sup\{\mu(yx):\ y\in A \text{ with } |y|\leq 1 \text{ and } yx\in E\}.$$

2. The disintegration theorem.

As in section 1, we consider

- a monotone σ-complete ordered real vector space $(E;\leq)$ with fixed weak unit u
- a spectral measure π: $\Sigma\to P(E)$ from some measure space (S,Σ) into the Boolean σ-algebra $P(E)$ of split projections of E
- a strictly positive σ-order continuous positive linear form μ on E.

Then, in addition to π itself, we have the measures

$$\pi(\cdot)u:\ \Sigma\to E_+ \quad \text{and} \quad \sigma:=\mu\circ\pi(\cdot)u:\ \Sigma\to \mathbb{R}_+$$
$$A\mapsto \pi(A)u \qquad\qquad A\mapsto \mu(\pi(A)u)$$

with associated spaces $L^1(\pi(\cdot)u)$ (cf. [16]) and $L^1(\sigma)$ of real valued integrable functions. Note that all these measures have the same null sets:

$$\pi(A) = 0 \Leftrightarrow \pi(A)u = 0 \Leftrightarrow \sigma(A) = 0$$

For, the only non-trivial implication is $\pi(A)u = 0 \Rightarrow \pi(A) = 0$, i.e. $\pi(A)x = 0$ (or equivalently: $(\pi(A)x)\wedge u = 0$) for all $x\in E_+$. But for any $y\leq\pi(A)x$ and $\leq u$ we obtain

$$y = \pi(A)y+\pi(CAy) \leq \pi(A)u+\pi(CA)\pi(A)x = 0.$$

In this situation we shall prove the following (in view of example 2 above "non-commutative") version of Strassen's theorem:

Theorem: Let F be any real vector space and p: $F\to E$ be sublinear with respect to $\leq$ μ-a.e. Consider a linear functional $\phi\in F^*$, dominated by $\mu\circ p$: $\phi(x)\leq\mu(p(x))$ for all $x\in F$. If the weak order unit u of E is not an order unit assume in addition $L^1(\sigma)\subset L^1(\pi(\cdot)u)$.

Then ϕ can be disintegrated below p, i.e. there exist linear φ: $F\to E$

$$\phi = \mu\circ\varphi \text{ and } \varphi(x) \leq p(x) \ \mu\text{-a.e. for each } x\in F.$$

Proof: Let us first follow Strassen's original argument and consider the space $\tilde{F}$ of F-valued Σ-measurable simple functions on S, $\tilde{F}$ containing F canonically as its set of constant functions. Then

$$\tilde{p}: \tilde{F} \to \mathbb{R}: \tilde{p}(\Sigma\chi_{A_i} x_i) = \Sigma\mu(\pi(A_i)p(x_i))$$

is welldefined if we take disjoint representations of the functions in $\tilde{F}$; and $\tilde{p}$ is a sublinear extension of $\mu \circ p$. For, subadditivity of p with respect to $\leq$ μ-a.e. implies for arbitrary functions $\Sigma\chi_{A_i} x_i$, $\Sigma\chi_{B_k} y_k \in \tilde{F}$ (with finite disjoint families (A_i) resp. (B_k) in Σ)

$$p(\sum_i \chi_{A_i} x_i + \sum_k \chi_{B_k} y_k) = \sum_{i,k} \mu(\pi(A_i \cap B_k) p(x_i + y_k)) \leq$$

$$\leq \sum_{i,k} \mu(\pi(A_i \cap B_k)(p(x_i) + p(y_k))) = \tilde{p}(\Sigma\chi_{A_i} x_i) + \tilde{p}(\Sigma\chi_{B_k} y_k).$$

Next, by Hahn-Banach's theorem take any linear functional $\tilde{\phi} \in \tilde{F}^*$ extending ϕ with $\tilde{\phi} \leq \tilde{p}$. In this way we obtain, for each fixed $x \in F$, an additive set function $\nu_x: \Sigma \to \mathbb{R}$ sending $A \in \Sigma$ to $\tilde{\phi}(\chi_A x)$. Taking any element $z \geq p(\pm x)$ in E (if u is an order unit for E, some positive multiple of u will do), we get the estimate

$$|\nu_x(A)| \leq \tilde{p}(\pm\chi_A x) = \mu(\pi(A)(\pm x)) \leq \mu(\pi(A)z), \quad A \in \Sigma,$$

showing in particular that ν_x is in fact σ-additive and absolutely continuous with respect to σ. Therefore the Radon-Nikodym will define a linear mapping

$$\psi: F \to \mathcal{L}^1(\sigma) \quad \text{with} \quad \nu_x(A) = \int_A (\psi(x))(t)\sigma(dt), \quad A \in \Sigma;\ x \in F.$$

If u is an order unit, the above estimate shows that even $\psi(x) \in L^\infty(\sigma)$ for all $x \in F$. In any case we have by hypothesis $\psi(x) \in \mathcal{L}^1(\pi(\cdot)u)$, so that we may define the linear map

$$\varphi: F \to E : \varphi(x) = \int (\psi(x))(t)\pi(dt)u.$$

Using the most elementary properties of this integral developed in [16], we see that φ has the required properties:

$$\mu(\pi(A)\varphi(x)) = \mu(\pi(A)\int(\psi(x)(t))\pi(dt)u) \overset{(*)}{=} \mu(\int\chi_A(t)(\psi(x))(t)\pi(dt)u) =$$

$$\overset{(+)}{=} \int\chi_A(t)(\psi(x))(t)\sigma(dt) = \nu_x(A) =$$

$$= \tilde{\phi}(\chi_A x) \leq \tilde{p}(\chi_A x) = \mu(\pi(A)p(x))$$

(in(*) σ-order continuity of split projections and in (+) the σ-order continuity of μ is used). This shows that $\varphi(x) \leq p(x)$ μ-a.e. and, for $A = S$, that $\mu \circ \varphi = \phi$.

□

Remark: The proof shows that the values $\varphi(x)$ of the disintegration map are all contained in the range of the integration map with respect to the vector measure $\pi(\cdot)u: \Sigma \to E_+$. If u is even an order unit for E, $\psi(x)$ is bounded and hence $\varphi(x) = \int(\psi(x))(t)\pi(dt)u$ is contained in the sublattice {Tu: T ∈ ideal center of E} of E studied in [14].

3. Some applications

First of all the classical "abstract" Strassen theorem is the special case of our theorem with $E = L^1(S,\Sigma;\mu;\mathbb{R})$ (for μ at least σ-finite so that there are obvious weak order units for E) and $\pi(A)$=multiplication by χ_A the canonical spectral measure; $\leq$ μ-a.e. then has the usual meaning if we take $\int d\mu$ as our strictly positive linear form. It is not hard to see how the Radon-Nikodym theorem itself can be recovered from this abstract Strassen theorem, taking into account the fact that for F the space of bounded measurable functions on S and $p: \begin{array}{l} F \to E \\ f \to |f| \end{array}$ the condition $\varphi(f) \leq p(f)$ immediately implies that φ is multiplication by (a representative of) the bounded measurable function $\varphi(\underline{1})$.

More generally, application of the theorem to the examples of section 1 leads to the following abstract Radon Nikodym theorems:

Corollary 1: Let E be a σ-complete vector lattice with order unit $u>0$ and μ a strictly positive σ-order continuous linear form on E. Then to every positive linear form $\lambda \leq \mu$ there exists some linear endomorphism T of E with $0 \leq T \leq \underline{1}$ such that $\lambda(x) = \mu(Tx)$ for all $x \in E$.

Proof: In the theorem take, as in example 1 of section 1, for π the Loomis spectral measure (so that $\leq$ μ-a.e. becomes equivalent to the ordering of E), and for $p: F = E \to E$ the sublinear lattice absolute value $p(x) = |x|$.

□

Remark: Our statement is a sharpening of what Freundenthal's spectral theorem (cf. [2]; [3] p. 61), applied to the order ideal E^*_μ generated by μ in the order dual of E, would easily give, namely some endomorphism S of E^*_μ with $0 \leq S \leq \underline{1}$ such that $\lambda = S\mu$.

Corollary 1 could also be obtained using the representation theory developed in [2].

Let us note that for a Dedekind complete vector lattice E the assumption that μ be *strictly* positive can be dispensed with by resorting to the support band $\{x\in E: \mu(|x|) = 0\}^d$ of μ. Our theorem in this case essentially generalizes Feyel's result ([1], p. 195).

Corollary 2: Let A be a W*-algebra with algebra center Z, and fix any normal, strictly positive linear functional μ on A. Then to every positive linear functional $\lambda\leq\mu$ on A there exists a positive linear mapping $T: A \to Z$ with $Tx < |x|$ μ-a.e. for each self-adjoint $x\in A$ such that $\lambda(x)=\mu(Tx)$ for all $x\in A$.

Proof: This is an immediate application of the discussion in example 2 of section 1 and the disintegration theorem applied to $F = E =$ self-adjoint part of A and π the Loomis spectral measure for the Boolean algebra of central projections; p is the functional $x \to |x|$ on E, sublinear with respect to $\leq$ μ-a.e.. The remark at the end of section 2 shows that the disintegrating mapping φ: $E \to E$ actually goes into Z, and its complex linear extension $T: A \to Z$ has the desired properties. □

Remark: This corollary should be compared with the Radon-Nikodym theorems in [8] p.p. 76, 77.

Note also that corollary 2 easily implies the existence of some $z\in Z$ with $0\leq z\leq 1$ such that $Tx = zx$ for all $x\in Z$.

If λ and μ are in addition both traces, we obtain the well known result that $\lambda(x) = \mu(zx)$ for all $x\in A$. Namely, by what has been already shown, this is true for $x\in Z$. That means that

$$\rho: x \to \lambda(x) - \mu(zx), \quad x\in A,$$

is a unitary invariant self adjoint functional in the predual of A, vanishing on Z. Let $\rho = \rho_1-\rho_2$ be its uniquelly determined *orthogonal decomposition* with $\rho_1\geq 0$, $\rho_2\geq 0$ and $|\rho| = |\rho_1|+|\rho_2|$ (cf. [8] p. 31). By the uniqueness also ρ_1 and ρ_2 must be unitary invariant and hence their supports p_1 resp. p_2 must belong to the center of A. Since p_1 and p_2 are mutually orthogonal, we obtain

$$0 = \rho(p_1) = \rho_1(p_1) - \rho_2(p_1) = \rho_1(p_1)$$

and thus $\rho_1=0$. In the same way $\rho_2=0$ and hence $\rho(x)=0$ for all $x\in A$.

References

[1] Feyel, D.: Deux applications d'une extension du théorème de Hahn-Banach, C.R. Acad. Sci. Paris 280 (1975) 193-196

[2] Hackenbroch, W.: Über den Freundenthalschen Spektralsatz,manuscripta math. 13 (1974) 83-99

[3] Hackenbroch, W.: Representation of vector lattices by spaces of real functions, in Functional Analysis: Surveys and Recent Results. Proceedings of the Paderborn Conference on Functional Analysis. K.-D. Bierstedt, B. Fuchssteiner (eds.) Amsterdam 1977

[4] Halmos, P.R.: Lectures on Boolean algebras. Van Nortrand Mathematical Studies 1, 1967

[5] König, H.: Sublineare Funktionale, Arch. Math.23 (1972) 500-508

[6] Neumann, M.: On the Strassen disintegration theorem, Arch. Math. 29 (1977) 413-420

[7] Neumann, M.: Exposê zum Desintegrationssatz von Strassen, unpublished manuscript, Saarbrücken 1977

[8] Sakai, S.: C*-Algebras and W*-algebras, Berlin-Heidelberg - New York 1971

[9] Schaefer, H.H.: Banach Lattices and Positive Operators, Berlin - Heidelberg - New York 1974

[10] Strassen, V.: The existence of probability measures with given marginals, Ann. Math. Statist. 36 (1965) 423-439

[11] Topping, D.M.: Vector lattices of self-adjoint operators, Trans. Amer. Math. Soc. 115 (1965) 14-30

[12] Valadier, M.: Sur le théorème de Strassen, C.R. Acad. Sc. Paris 278 (1974) 1021-1024

[13] Valadier, M.: On the Strassen theorem, Lect. Notes in Economics and Math. Systems 102 (1974) 203-215

[14] Wils, W.: The ideal center of partially ordered vector spaces, Acta math. 127 (1971) 41-77

[15] Wolff, M.: Eine Bemerkung zum Desintegrationssatz von Strassen, Arch. Math. 28 (1977), 98-101

[16] Wright, J.D.M.: Measures with values in partially ordered vector space, Proc. London Math. Soc. 25 (1972) 675-688

ON COVERING CONDITIONS AND CONVERGENCE[1]

Annie Millet and Louis Sucheston
The Ohio State University
Columbus, Ohio 43210

It was shown by J. Dieudonné [3] that Doob's martingale convergence theorem in general fails when the index set is not totally ordered. In 1956 K. Krickeberg introduced the "Vitali" condition V (also denoted V_0 and V_∞) on the σ-algebras, and proved that V was sufficient for essential convergence of L_1-bounded martingales ([7], or [15], p. 99). In the note [24], we showed that V was not necessary, replacing it by the condition SV. A still weaker condition C was introduced in [25], and shown sufficient for essential convergence of L_1-bounded martingales taking values in a Banach space with the Radon-Nikodym property. This result is here briefly proved again in Section 1, and applied to obtain a derivation theorem for finitely additive Banach space valued measures (Theorem 1.5).

In an independent work, K. Astbury introduced a condition A, sufficient for essential convergence of L_1-bounded martingales. In Section 2 we show that in the presence of a countable cofinal subset, A is equivalent with C; also other equivalent conditions are given, defined in terms of Orlicz norms.

Martingale theory in part traces its origins to the point-derivation theory. Thus Krickeberg's condition V was an adaptation of R. de Possel's abstraction of the classical Vitali property; similarly, SV is the stochastic version of the Besicovitch property (cf. [11] and [12]). We attempt to repay the debt, offering in Section 3 a point-derivation version of condition C, sufficient to obtain Lebesgue's theorem. In analogy to the stochastic case in Section 2, several equivalent formulations of C are discussed.

The outstanding remaining question is that of necessity of the condition C, in either stochastic or point-derivation setting.

The main results of this paper were presented by the second-named author at the Oberwolfach Conference on Measure Theory, on July 5, 1979.

---

[1]Research in part supported by the National Science Foundation (USA).

1. MAXIMAL INEQUALITIES AND CONVERGENCE OF MARTINGALES

Let J be a directed set filtering to the right, i.e., a set of indices partially ordered by $\leq$, such that for each pair t_1, t_2 of elements of J, there exists an element t_3 of J such that $t_1 \leq t_3$ and $t_2 \leq t_3$. Let $(\Omega, \mathfrak{F}, P)$ be a probability space. Sets and random variables are considered equal if they are equal almost surely. All considered sets and functions are measurable. Let $\underline{\underline{X}} = (X_t)$ be a family of random variables taking values in $\overline{\mathbb{R}}$. The <u>stochastic upper limit</u> of $\underline{\underline{X}}$, $\widetilde{X} = \text{s lim sup } X_t$, is the essential infimum of the set of random variables Y such that $\lim P(Y < X_t) = 0$. The <u>stochastic lower limit</u> of $\underline{\underline{X}}$ is $\text{s lim inf } X_t = -\text{s lim sup } (-X_t)$. The directed family is said to <u>converge stochastically</u> (or in probability) if $\text{s lim sup } X_t = \text{s lim inf } X_t$. The <u>essential upper limit</u> of $\underline{\underline{X}}$, $X^* = \lim\sup X_t$, is defined by $X^* = \underset{s}{\text{ess inf}}(\underset{t \geq s}{\text{ess sup}}\, X_t)$. The <u>essential lower limit</u> of $\underline{\underline{X}}$, $X_* = \lim\inf X_t$, is $-\lim\sup(-X_t)$. The directed family $\underline{\underline{X}}$ is said to <u>converge essentially</u> if $X^* = X_*$; this common value is called the <u>essential limit</u> of $\underline{\underline{X}}$, $\lim X_t$. If $\underline{\underline{A}} = (A_t)$ is a directed family of measurable sets, the <u>stochastic upper limit</u> of $\underline{\underline{A}}$, $\widetilde{A} = \text{s lim sup } A_t$, is the set defined by $1_{\widetilde{A}} = \text{s lim sup } 1_{A_t}$; the <u>essential upper limit</u> of $\underline{\underline{A}}$, $A^* = \lim\sup A_t$, is the set defined by $1_{A^*} = \lim\sup 1_{A_t}$. A <u>stochastic basis</u> $(\mathfrak{F}_t)$ is an increasing family of sub σ-algebras of $\mathfrak{F}$ (i.e., for every $s \leq t$, $\mathfrak{F}_s \subset \mathfrak{F}_t$). A <u>stochastic process</u> $\underline{\underline{X}}$ is a family of random variables $X_t: \Omega \to \mathbb{R}$ such that for every t, X_t is $\mathfrak{F}_t$ measurable. The process is called <u>integrable</u> (<u>positive</u>) if for every t, X_t is integrable (positive). A family of sets $\underline{\underline{A}}$ is <u>adapted</u> if for every $t \in J$, $A_t \in \mathfrak{F}_t$.

Denote by $\mathcal{J}$ the set of finite subsets of J. An (<u>incomplete</u>) <u>multivalued stopping time</u> is a map τ from Ω (from a subset of Ω called $D(\tau)$) to $\mathcal{J}$ such that $R(\tau) = \cup\tau(\omega)$ is finite, and such that for every $t \in J$, $\{\tau = t\} \overset{\text{Def}}{=} \{\omega \in \Omega: t \in \tau(\omega)\} \in \mathfrak{F}_t$; cf. [11], [12]. Denote by $M(IM)$ the set of (incomplete) multivalued stopping times. A <u>simple stopping time</u> is an element τ of M such that for every ω, $\tau(\omega)$ is a singleton; the set of simple stopping times is denoted T. An <u>ordered stopping time</u> is a simple stopping time τ such that the elements $t_1, t_2, \ldots, t_n$ in the range of τ are linearly ordered. Denote by T' the set of ordered stopping times. The <u>excess function</u> of $\tau \in IM$ is $e_\tau = \Sigma 1_{\{\tau = t\}} - 1_{D(\tau)}$. Let σ and τ be in IM; we say that $\sigma \leq \tau$ if $\forall s$,

$\forall t$, $\{\sigma = s\} \cap \{\tau = t\} \neq \emptyset$ implies $s \leq t$. For the order $\leq$, M is a directed set filtering to the right. Let $\tau \in \mathrm{IM}$; if $\underline{\underline{X}}$ is a positive stochastic process, we set $X(\tau) = \sup(1_{\{\tau = t\}} X_t)$; if $\underline{\underline{A}}$ is an adapted family of sets, we set $A(\tau) = \bigcup_t (\{\tau = t\} \cap A_t)$. Clearly $1_{A(\tau)} = 1_A(\tau)$ for every $\tau \in \mathrm{IM}$. The stochastic basis $(\mathfrak{F}_t)$ satisfies the <u>covering condition</u> C, if for every $\epsilon > 0$ there exists a constant $M_\epsilon > 0$ such that for every adapted family of sets $\underline{\underline{A}}$, there exists $\tau \in \mathrm{IM}$ with $e_\tau \leq M_\epsilon$ and $P[A^* \backslash A(\tau)] \leq \epsilon$. The following proposition states the validity of a stochastic maximal inequality.

<u>Proposition 1.1</u>: Let $\underline{\underline{X}}$ be a positive stochastic process such that $(X(\tau), \tau \in T')$ is uniformly integrable. Then for every $\lambda > 0$, letting $\widetilde{A} = \{s \limsup X_t \geq \lambda\}$, one has

$$P(\widetilde{A}) \leq \frac{1}{\lambda} \limsup_{T'} \int_{\widetilde{A}} X(\tau)\, dP .$$

<u>Proof</u>: Fix $\lambda > 0$, and for a number α with $0 < \alpha < \lambda/2$, set $B_t = \{X_t > \lambda - \alpha\}$; then $\widetilde{A} \subset \widetilde{B} = s \limsup B_t$. Given $\epsilon > 0$, choose $s \in J$ and $\tau \in T'$ with $\tau \geq s$ and $P[\widetilde{B} \,\Delta\, B(\tau)] < \epsilon$ ([23]). Then

$$P(\widetilde{B}) \leq P[B(\tau)] + \epsilon \leq \frac{1}{\lambda - \alpha} E[1_{B(\tau)} X(\tau)] + \epsilon .$$

Given $\delta > 0$, choose $\epsilon < \delta$ such that $P(C) < 2\epsilon$ implies $\sup_{\tau \in T'} E[1_C X(\tau)] \leq \delta\lambda$, and choose α such that $P[\widetilde{B} \backslash \{\widetilde{X} \geq \lambda\}] < \epsilon$. One has

$$P(\widetilde{A}) \leq P(\widetilde{B}) \leq \frac{1}{\lambda - \alpha} E[1_{\{\widetilde{X} \geq \lambda\}} X(\tau)] + \frac{\delta\lambda}{\lambda - \alpha} + \epsilon$$

$$\leq \frac{1}{\lambda - \alpha} E[1_{\{\widetilde{X} \geq \lambda\}} X(\tau)] + 3\delta .$$

The proposition follows on letting $s \to \infty$, $\delta \to 0$ and $\alpha \to 0$. □

The following theorem [25] characterizes convergence of Banach-valued martingales in terms of maximal inequalities. Of particular interest is the equivalence of (i) and (iii).

<u>Theorem 1.2</u>: Let $(\mathfrak{F}_t)$ be a stochastic basis; the following properties are equivalent:

(i) For every Banach space $(\underline{E}, |\ |)$, for every Bochner integrable $\underline{E}$-valued random variable X, and for every $\lambda > 0$, letting $A = \{\limsup |E^{\mathfrak{F}_t} X| \geq \lambda\}$, one has

$$P(A) \leq \frac{1}{\lambda} \int_A |X| dP .$$

(ii) There exists a function $M: \mathbb{R}_+ \times \mathbb{R}_+ \to \mathbb{R}_+$ such that for every $\lambda > 0$

(1) $\lim_{\epsilon \to 0} M(\lambda, \epsilon) = 0$.

(2) For every positive integrable random variable X, $EX \leq \epsilon$ implies $P(\limsup E^{\mathfrak{F}_t} X \geq \lambda) \leq M(\lambda, \epsilon)$.

(iii) For every Banach space $\underline{E}$ and every Bochner integrable $\underline{E}$-valued random variable X, the martingale $E^{\mathfrak{F}_t} X$ converges essentially.

Proof: Obviously (i) ⇒ (ii).

(ii) ⇒ (iii): Let $X \in L_1^{\underline{E}}$; fix $\alpha > 0$, $\lambda > 0$, choose $\epsilon < \alpha\lambda/2$ such that $M(\lambda/2, \epsilon) < \alpha$, and choose $Y \in \bigcup L_1^{\underline{E}}(\mathfrak{F}_t)$ such that $E|X - Y| < \epsilon$. Then if $\mathfrak{F}_\infty = \sigma(\bigcup \mathfrak{F}_t)$, we have

$$\begin{aligned} P(\limsup |E^{\mathfrak{F}_t} X - E^{\mathfrak{F}_\infty} X| \geq \lambda) &\leq P(\limsup E^{\mathfrak{F}_t} |X - Y| \geq \lambda/2) \\ &\quad + P(E^{\mathfrak{F}_\infty} |X - Y| \geq \lambda/2) \\ &\leq \alpha + 2\epsilon/\lambda \leq 2\alpha , \end{aligned}$$

which proves the essential convergence of $E^{\mathfrak{F}_t} X$ to $E^{\mathfrak{F}_\infty} X$.

(iii) ⇒ (i): Let $X \in L_1^{\underline{E}}$; since $E^{\mathfrak{F}_t} X$ converges essentially, $s \limsup |E^{\mathfrak{F}_t} X| = \limsup |E^{\mathfrak{F}_t} X|$. Applying Proposition 1.1 to the uniformly integrable positive submartingale $Y_t = |E^{\mathfrak{F}_t} X|$, and letting $A = \{\limsup |E^{\mathfrak{F}_t} X| \geq \lambda\}$, we obtain for every $\lambda > 0$

$$\begin{aligned} P(\limsup |E^{\mathfrak{F}_t} X| \geq \lambda) &\leq \frac{1}{\lambda} \limsup_{\tau \in T'} \int_A Y_\tau dP \\ &\leq \frac{1}{\lambda} \limsup_{\tau \in T'} \int_A E^{\mathfrak{F}_\tau} |X| dP \\ &\leq \frac{1}{\lambda} \int_A |X| dP . \qquad \square \end{aligned}$$

The following proposition states that the condition C implies a maximal inequality for positive submartingales.

<u>Proposition 1.3</u>: Let $(\mathfrak{F}_t)$ be a stochastic basis satisfying C. For every $\epsilon > 0$, there exists a number $L_\epsilon > 0$ such that for every positive submartingale $\underline{\underline{X}}$, one has

$$P(\limsup X_t \geq \lambda) \leq \left(\frac{L_\epsilon}{\lambda} \lim EX_t\right) \vee \epsilon .$$

<u>Proof</u>: Fix $\lambda > 0$, $\epsilon > 0$, and let $A_t = \{X_t > \lambda/2\}$. If $P(X^* \geq \lambda) > \epsilon$, using condition C choose $\tau \in \mathbb{M}$ satisfying $e_\tau \leq M_{\epsilon/2}$, and

$$P(\limsup X_t \geq \lambda) \leq P(\limsup A_t) \leq P[A(\tau)] + \frac{\epsilon}{2}$$
$$\leq \frac{2}{\lambda} E[X(\tau)] + \frac{\epsilon}{2} .$$

Let s be an index larger than τ; then

$$E[X(\tau)] \leq E[\Sigma\, 1_{\{\tau = t\}} X_t] \leq E[\Sigma\, 1_{\{\tau = t\}} X_s]$$
$$\leq E[(e_\tau + 1)X_s] \leq (M_{\epsilon/2} + 1) \lim EX_t .$$

Therefore

$$P(X^* \geq \lambda) \leq 2[P(X^* \geq \lambda) - \tfrac{\epsilon}{2}] \leq 4(M_{\epsilon/2} + 1) \lim EX_t/\lambda . \qquad \square$$

We now show that the covering condition C is sufficient for essential convergence of Banach-valued martingales.

<u>Theorem 1.4</u>: Let $(\mathfrak{F}_t)$ be a stochastic basis satisfying the condition C, and let $\underline{E}$ be a Banach space with the Radon-Nikodým property. Then every $L_1^{\underline{E}}$-bounded martingale converges essentially.

<u>Proof</u>: Set $\mathcal{A} = \cup \mathfrak{F}_t$, $\mathfrak{F}_\infty = \sigma(\mathcal{A})$ and let $\underline{\underline{X}}$ be an $L_1^{\underline{E}}$-bounded martingale. Define a finitely additive $\underline{E}$-valued measure λ on $\mathcal{A}$ by $\lambda(A) = \lim E[1_A X_t]$. We write $\mu \ll P$ if $\forall \epsilon > 0$, $\exists \delta > 0$ such that $P(A) < \delta$ implies $(\mathrm{Var}\ \mu)(A) < \epsilon$. Since $\mathrm{Var}\ \lambda = \lim E|X_t| < \infty$, λ can be decomposed as $\lambda = \mu + \pi$, where μ is a σ-additive measure on $\mathcal{A}$, $\mu \ll P$, $\mathrm{Var}\ \mu < \infty$, and π is a finitely additive measure on $\mathcal{A}$ such that the positive charges $\mathrm{Var}\ \pi$ and P are nearly orthogonal (for this result due to Chatterji and Uhl, see e.g. [6], p. 30-31).

Since $\underline{E}$ has the Radon-Nikodým property, there exists $X \in L_1^{\underline{E}}$ such that for each $A \in \mathcal{A}$, $\mu(A) = E[1_A X]$. The martingale $E^{\mathfrak{F}_t}X$ converges by Proposition 1.3 and the implication (ii) implies (iii) of Theorem 1.2. Set $Z_t = X_t - E^{\mathfrak{F}_t}X$, fix $\delta > 0$, and choose a sequence (A_k) of sets in $\mathcal{A}$ such that $\Sigma\, P(A_k) < \delta$ and $\Sigma\, k\, L_{k^{-2}} \mathrm{Var}\, \pi(A_k^c) < \infty$; here each set A_k is measurable with respect to some $\mathfrak{F}_t$, say $\mathfrak{F}_{t_k}$. For every k, the process $(|Z_t|\, 1_{A_k^c})_{t \geq t_k}$ is a positive L_1-bounded submartingale such that $\lim_t E[1_{A_k^c}|Z_t|] = \mathrm{Var}\, \pi(A_k^c)$. This process can clearly be extended to a positive submartingale $\underline{\underline{S}}$ (defined for all the values of the index t). Applying Proposition 1.3 to the extended process, we deduce that $P[\limsup_t 1_{A_k^c}|Z_t| > k^{-1}] \leq [L_{k^{-2}} k\, \mathrm{Var}\, \pi(A_k^c)] \vee k^{-2}$. Set $A = \cup A_k$; then $P(A) < \delta$, and by the Borel-Cantelli lemma, $\limsup |Z_t| = 0$ a.e. on A^c. Since δ is arbitrary, Z_t converges essentially to zero. □

As an application of Theorem 1.4, we show the existence of "derivatives" of Banach-valued finitely additive measures with respect to countable partitions.

<u>Theorem 1.5</u>: Let J be a family of countable partitions t of Ω ordered by refinement (i.e., if $s \leq t$, then every atom of s is a union of atoms of t), and filtering to the right. Let $(\mathfrak{F}_t)$ be the stochastic basis of σ-algebras generated by the partitions t satisfying C. Let $\underline{E}$ be a Banach space with the Radon-Nikodým property, and let λ be an $\underline{E}$-valued finitely additive measure of bounded variation, defined on the algebra $\mathcal{A} = \cup \mathfrak{F}_t$. If

$$X_t = \sum_{A \text{ atom of } t} \frac{\lambda(A)}{P(A)} 1_A$$

with the convention $\lambda(A)/P(A) = 0$ if $P(A) = 0$, then X_t converges essentially.

<u>Proof</u>: Denote by $\lambda = \mu + \pi$ the decomposition on $\mathcal{A}$ of λ into a σ-additive measure μ, $\mu \ll P$, and a finitely additive measure π such that the charges $\mathrm{Var}\, \pi$ and P are nearly orthogonal (cf. [6]). Let the extension of μ to $\sigma(\mathcal{A})$ still be denoted by μ. Since $\underline{E}$ has the Radon-Nikodým property, there exists $X \in L_1^{\underline{E}}$ such that $\mu(A) = \int_A X\, dP$, $\forall A \in \sigma(\mathcal{A})$. By Theorem 1.4, the uniformly integrable martingale $\underline{\underline{Y}}$ defined by

$$Y_t = \sum_{A \text{ atom of } \mathcal{F}_t} \left(\frac{1}{P(A)} \int_A X\,dP\right) 1_A$$

converges essentially. Hence it suffices to consider $X_t - Y_t$, or, equivalently X_t in the case where the charge $\operatorname{Var} \lambda$ is nearly orthogonal to P on $\mathcal{A}$; in the rest of the proof we make this assumption. Fix $\delta, \varepsilon > 0$ and let $A \in \mathcal{A}$ be such that $P(A^c) < \varepsilon$ and $(\operatorname{Var} \lambda)(A) < \delta$; A is in some σ-algebra $\mathfrak{F}_s$. Fix $M > 0$ and let $\sigma \geq s$ be a multivalued stopping time such that $e_\sigma \leq M$. We have

$$\begin{aligned}
E[1_A |X|(\sigma)] &\leq \sum_t E[1_{A \cap \{\sigma = t\}} |X_t|] \\
&\leq \sum_t \Big[\sum_{\substack{B \in t \\ B \subset \{\sigma = t\} \cap A}} |\lambda(B)| \Big] \\
&\leq \sum_t \Big[\sum_{\substack{B \in T \\ B \subset \{\sigma = t\} \cap A}} (\operatorname{Var} \lambda)(B) \Big] \\
&\leq \sum_t (\operatorname{Var} \lambda)(\{\sigma = t\} \cap A),
\end{aligned}$$

the last inequality holding because every charge is automatically countably superadditive. For every $k = 1, \ldots, M+1$, denote by E_k the set of points ω of $\bigcup (\{\sigma = t\} \cap A)$ that belong to k sets $\{\sigma = t\} \cap A$; then

$$\begin{aligned}
E[1_A |X|(\sigma)] &\leq \sum_t \sum_{k=1}^{M+1} (\operatorname{Var} \lambda)(\{\sigma = t\} \cap A \cap E_k) \\
&\leq (M + 1)(\operatorname{Var} \lambda)(A) \leq (M + 1)\delta .
\end{aligned}$$

Therefore

$$\sup_{\sigma \geq s,\, e_\sigma \leq M} E[1_A |X|(\sigma)] \leq (M + 1)\delta .$$

Let M_ε be given by the condition C, fix $\lambda > 0$, and set $B_t = A \cap \{|X_t| > \lambda/2\}$ if $t \geq s$, $B_t = \emptyset$ otherwise. There exists $\tau \in \mathbb{M}$ with $\tau \geq s$, $e_\tau \leq M_\varepsilon$ and

$$\begin{aligned}
P(\limsup |X_t| \geq \lambda) &\leq P(A^c) + P[\limsup(1_A |X_t|) > \lambda/2] \\
&\leq \varepsilon + P[B(\tau)] + \varepsilon \\
&\leq 2\varepsilon + \frac{2}{\lambda} E[1_A |X|(\tau)] \\
&\leq 2\varepsilon + \frac{2}{\lambda} (M_\varepsilon + 1)\delta .
\end{aligned}$$

We conclude that X_t converges essentially to zero. $\square$

2. COMPARISON OF SEVERAL COVERING CONDITIONS IN PRESENCE OF A COUNTABLE COFINAL SUBSET

In [2], K. Astbury introduced a new condition which we will call A, and proved by an upcrossing argument that under A real-valued L_1-bounded martingales converge essentially. It is easy to see that C implies A. We show here that conversely, if J has a countable cofinal subset, then A implies C. Also further covering conditions are introduced, namely a local version of C, and some conditions weaker than A, defined in terms of an Orlicz function Φ. In the presence of a countable cofinal subset, all these conditions are equivalent, and they can be also characterized by essential convergence of a class of random variables called C-potentials. Another characterization is by the equivalence of the essential and stochastic convergence of X_τ $(X_\tau = \Sigma\, 1_{\{\tau=t\}} X_t)$.

$(\mathfrak{F}_t)$ satisfies the <u>condition</u> <u>LC</u> (local C) if for every adapted family of sets $\underline{\underline{A}}$, $A^* = \lim_n \uparrow [\text{s} \limsup_{e_\tau \leq n} A(\tau)]$.

Let $\varphi\colon \mathbb{R}^+ \to \mathbb{R}^+$ be a continuous, strictly increasing function with $\varphi(0) = 0$ and $\lim_{t\to\infty} \varphi(t) = \infty$; let ψ be the inverse function of φ. Let Φ be the indefinite integral of φ, i.e., $\Phi(x) = \int_0^x \varphi(u)du$. Similarly, $\Psi(x) = \int_0^x \psi(u)du$. Φ and Ψ are <u>conjugate</u> <u>Orlicz</u> <u>functions</u>. The <u>Orlicz</u> <u>class</u> O_Φ is the set of random variables X such that $E[\Phi(|X|)] < \infty$. The <u>Orlicz</u> <u>space</u> L_Φ is the set of random variables X such that for every $Y \in O_\Psi$, $E(XY) < \infty$. The <u>Orlicz</u> <u>norm</u> $\|\ \|_\Phi$ on L_Φ is defined by $\|X\|_\Phi = \sup\{|E(XY)|: E[\Psi(|Y|)] \leq 1\}$. There exist positive constants a_1 and a_2 such that $a_1\|\ \|_1 \leq \|\ \|_\Phi \leq a_2\|\ \|_\infty$. The Orlicz function Φ has the <u>property</u> $\underline{\Delta_2}$ is $\limsup_{x\to\infty} \Phi(2x)/\Phi(x) < \infty$. The Orlicz function Φ has the <u>property</u> $\underline{\Delta}$ if for every $\epsilon > 0$, there exists a constant $K_\epsilon > 1$ such that

$$\liminf_{x\to\infty} \varphi[(1+\epsilon)x]/\varphi(x) > K_\epsilon .$$

If Φ has both properties Δ and Δ_2, then Ψ has the property Δ_2, and the functions Φ and $\Psi\circ\varphi$ are <u>equivalent</u>, i.e., there exist positive constants $a > 0$ and $A > 0$, and $x_0 > 0$ such that $\Phi(x) \leq A\Psi\circ\varphi(ax)$ and $\Psi\circ\varphi(x) \leq A\Phi(ax)$, $\forall x \geq x_0$ (cf. [26], [28] and e.g. [14]). The functions $\Phi(x) = x^p/p$,

$1 < p < \infty$, corresponding to L_p-spaces, as well as the Orlicz functions $\Phi(x) = p^{-1}x^p(\mathrm{Log}\, x)^{\alpha_1}[\mathrm{Log}(\mathrm{Log}\, x)]^{\alpha_2} \ldots [\mathrm{Log}(\ldots(\mathrm{Log}\, x))]^{\alpha_n}$, $p > 1$, $\alpha_i \geq 1$, $1 \leq i \leq n$, satisfy both conditions Δ and Δ_2.

Given $\tau \in \mathrm{IM}$ and a stochastic process $\underline{\underline{X}}$, set $X_\tau = \Sigma\, 1_{\{\tau = t\}} X_t$ and $n_\tau = 1_\tau = e_\tau + 1_{D(\tau)}$. Let $\|\ \|$ and $!\ !$ be norms defined on subspaces of equivalence classes of measurable functions, containing L_∞. We consider conditions expressed in terms of ratios of norms. We say that $(\mathfrak{F}_t)$ satisfies the condition $R(!\ !,\|\ \|)$ (resp. $B(!\ !,\|\ \|)$) if for every adapted family of sets $\underline{\underline{A}}$ with $PA^* > 0$, there exists a constant $M > 0$ such that for every index s, there exists $\tau \in \mathrm{IM}$ with $\tau \geq s$, $\{\tau = t\} \subset A_t$ for every t, and $!n_\tau! < M\|n_\tau\|$ (resp. $!n_\tau! < M\|1_{A(\tau)}\|$). We say that $(\mathfrak{F}_t)$ satisfies the condition $C(!\ !,\|\ \|)$ if for every $\epsilon > 0$, there exists a constant $M_\epsilon > 0$ such that for every adapted family of sets $\underline{\underline{A}}$ with $PA^* \geq \epsilon$, there exists $\tau \in \mathrm{IM}$ with $!n_\tau! < M_\epsilon\|1_{A(\tau)}\|$. To lighten the notation, we write $R(\Phi,\Phi')$ for $R(\|\ \|_\Phi,\|\ \|_{\Phi'})$, etc. K. Astbury introduced the condition $R(p,1)$ in the case where $1 < p \leq \infty$; in particular, Astbury's condition A is $R(\infty,1)$. $(\mathfrak{F}_t)$ satisfies the Vitali condition $V_{\|\ \|}$ if for every adapted family of sets $\underline{\underline{A}}$, and for every $\epsilon > 0$, there exists $\tau \in \mathrm{IM}$ with $P(A^*\setminus A(\tau)) < \epsilon$ and $\|e_\tau\| < \epsilon$. In the case $\|\ \|$ is $\|\ \|_p$, $1 \leq p \leq \infty$, (resp. $\|\ \|_\Phi$ for some Orlicz function Φ), we denote $V_{\|\ \|}$ by V_p (resp. V_Φ). It was proved by K. Krickeberg [16] that L_∞-bounded martingales converge essentially if and only if $(\mathfrak{F}_t)$ satisfies the condition V_1. Let Φ be an Orlicz function with the property Δ_2, and let Ψ be the conjugate Orlicz function. L_Ψ-bounded martingales converge essentially if and only if $(\mathfrak{F}_t)$ satisfies V_Φ ([18]; for the case of V_p, $1 < p < \infty$, see also [3]). The proof given in [18] also shows that V_Φ is equivalent with the essential convergence of martingales of the form $E^{\mathfrak{F}_t}X$, where X belongs to the Orlicz class O_Ψ.

<u>Remark 1</u>: The first-named author takes this opportunity to correct a misprint in [18]. In the statement of Lemma 3.1, the condition (5) should be $E[\varphi(n_{\mathfrak{H}})]/\varphi(1) + P(B\setminus A) \leq \epsilon 2^{-1}[1 + \varphi(1)]^{-1}P(B)$.

<u>Lemma 2.1</u>: Let Φ be an Orlicz function with the properties Δ and Δ_2. The condition $R(\infty,\Phi)$ implies the condition $R(\Phi,1)$.

<u>Proof</u>: Assume that $(\mathfrak{F}_t)$ is a stochastic basis satisfying $R(\infty,\Phi)$ but not $R(\Phi,1)$. There exists an adapted family of sets $\underline{\underline{A}}$ with $PA^* > 0$, and a constant $M > 0$, such that for every constant $K > 0$ there exists a stopping time τ satisfying $\{\tau = t\} \subset A_t$ for every t, $\|n_\tau\|_\infty < M\|n_\tau\|_\Phi$ and $\|n_\tau\|_\Phi \geq K\|n_\tau\|_1$. Since Φ has the property Δ_2, the solution $a > 0$ of the equation

$\int \Psi \circ \varphi(an_\tau)dP = 1$ satisfies $\|n_\tau\|_\Phi = \int n_\tau \varphi(an_\tau)dP = a^{-1}[1 + \int \Phi(an_\tau)dP]$ (see e.g. [14], p. 88). By the chocie of τ,

$$\begin{aligned}\|n_\tau\|_\Phi &\leq \|n_\tau\|_1 \|\varphi(an_\tau)\|_\infty \\ &\leq K^{-1}\|n_\tau\|_\Phi \varphi(a\|n_\tau\|_\infty) \\ &\leq K^{-1}\|n_\tau\|_\Phi \varphi(aM\|n_\tau\|_\Phi) .\end{aligned}$$

Since Φ and $\Psi \circ \varphi$ are equivalent and since $\int \Psi \circ \varphi(an_\tau)dP = 1$, $\int \Phi(an_\tau)dP$ remains bounded as well as $a\|n_\tau\|_\Phi$. The relation $K \leq \varphi(Ma\|n_\tau\|_\Phi)$ contradicts the assumption that K can be chosen arbitrary large. □

Proposition 2.2: Let J be a directed set with a countable cofinal subset, and let $(\mathfrak{F}_t)$ be a stochastic basis indexed by J. Let $\| \ \|$ and $! \ !$ be two norms defined on subspaces of (equivalence classes of) measurable functions, containing L_∞. Assume that the norm $! \ !$ is such that $0 \leq X \leq Y$ implies $!X! \leq !Y!$.

(i) Let $(\mathfrak{F}_t)$ satisfy $R(\infty, \| \ \|)$ and $V_{\| \ \|}$; then $(\mathfrak{F}_t)$ satisfies $B(\infty, \| \ \|)$.

(ii) Let $(\mathfrak{F}_t)$ satisfy $B(! \ !, \| \ \|)$; then $(\mathfrak{F}_t)$ satisfies $C(! \ !, \| \ \|)$.

Proof: We only study the case when J does not have a greatest element, since otherwise the proposition is trivial.

(i) Assume that $B(\infty, \| \ \|)$ fails; let $\underline{\underline{A}}$ be an adapted family of sets with $A^* \neq \emptyset$, and let (s_k) be an increasing sequence of indices such that for each $k > 0$, if $\tau \in IM$, $\tau \geq s_k$ and $\{\tau = t\} \subset A_t$ for every t, then $\|n_\tau\|_\infty \geq k^2 \|1_{A(\tau)}\|$. Since J has a countable cofinal subset, we may and do assume that the sequence (s_k) is cofinal. Since $(\mathfrak{F}_t)$ satisfies $V_{\| \ \|}$, for every k we may choose a stopping time $\sigma_k \in IM$ such that $\sigma_k \geq s_k$, $\|e_{\sigma_k}\| < 2^{-k}$, and $P[A(\sigma_k) \Delta A^*] \leq 2^{-k}$. The sequence of stopping times (σ_k) can be chosen in such a way that for every integer k there exists an index t_k with $\sigma_k < t_k < \sigma_{k+1}$. If $t \in R(\sigma_k)$, set $B_t = A_t \cap \{\sigma_k = t\}$; if $t \notin \cup R(\sigma_k)$, set $B_t = \emptyset$. Clearly $B(\sigma_k) = A(\sigma_k)$ for every k, and $B^* = A^*$. Let $M > 0$ be the constant given by the condition $R(\infty, \| \ \|)$ applied to the adapted family $\underline{\underline{B}}$. Fix i, and let $\tau \in IM$ satisfy $\tau \geq s_i$, $\{\tau = t\} \subset B_t$ for every t, and $\|n_\tau\|_\infty < M\|n_\tau\|$. For every $j \geq i$ define $\tau_j \in IM$ by $\{\tau_j = t\} = \{\tau = t\}$ if $t \in R(\tau) \cap R(\sigma_j)$. Then $n_\tau = \sum_{i \leq j \leq k} n_{\tau_j}$ for some $k \geq i$. Since $\|n_\tau\|_\infty \geq 1$, one has for large i

$$\|n_\tau\| \le \sum_{i\le j\le k} \|n_{\tau_j}\| \le \sum_{i\le j\le k} (\|e_{\sigma_j}\| + \|1_{A(\tau_j)}\|)$$

$$\le 2^{1-i} + \sum_{j\ge i} j^{-2}\|n_{\tau_j}\|_\infty \le 2^{1-i} + \|n_\tau\|_\infty(\sum_{j\ge i} j^{-2})$$

$$\le 2(\sum_{j\ge i} j^{-2})\|n_\tau\|_\infty \le 2M(\sum_{j\ge i} j^{-2})\|n_\tau\|.$$

Dividing by $\|n_\tau\| > 0$, and letting $i \to \infty$, one obtains a contradiction.

(2) Assume that $C(!\ !,\|\ \|)$ fails, and let $\epsilon > 0$ be such that for every integer M, there exists an adapted family of sets $A_{M,t}$, and an index s_M such that $P(A_M^*) \ge \epsilon$, and for every stopping time $\tau \in IM$, $\tau \ge s_M$ implies $!n_\tau! \ge M^2\|1_{A_M(\tau)}\|$. Since J has a countable cofinal subset, we may and do assume that the sequence (s_M) is increasing and cofinal.

Let $t_1 = s_1$, and define an increasing cofinal sequence (t_n) by induction as follows: t_{i+1} is an index such that $t_{i+1} > t_i$, $t_{i+1} \ge s_{i+1}$ and $P(\operatorname{ess\,sup}_{t_i\le t<t_{i+1}} A_{i,t}) \ge PA_i^* - \epsilon 2^{-i}$. Let $\underline{\underline{B}}$ be the adapted family of sets such that $B_t = A_{i,t}$ if $t_i \le t < t_{i+1}$, and $B_t = \emptyset$ if $t \notin \bigcup\{s: t_i \le s < t_{i+1}\}$. Then for every i, $P(\operatorname{ess\,sup}_{t\ge t_i} B_t) \ge P(\operatorname{ess\,sup}_{t_i\le t<t_{i+1}} A_{i,t}) \ge \epsilon - \epsilon 2^{-i}$, which implies $PB^* \ge \epsilon > 0$. Let $M > 0$ be the constant given by the definition of the condition $B(!\ !,\|\ \|)$ applied to the adapted family $\underline{\underline{B}}$, and fix $i > 0$. Let $\tau \in IM$ satisfy $\tau \ge t_i$, $\{\tau = t\} \subset A_t$ for every t, and $!n_\tau! < M\|1_{B(\tau)}\|$. For every $j \ge i$, define $\tau_j \in IM$ by $\{\tau_j = t\} = \{\tau = t\}$ if $t_j \le t < t_{j+1}$; then for some integer k,

$$0 < !n_\tau! < M\,\|\sum_{i\le j\le k} 1_{B(\tau_j)}\| \le M \sum_{i\le j\le k} \|1_{B(\tau_j)}\|$$

$$= M \sum_{i\le j\le k} \|1_{A_j(\tau_j)}\| \le M \sum_{j\ge i} j^{-2}!n_{\tau_j}!$$

$$\le M\,[\sum_{j\ge i} j^{-2}]!n_\tau!\,.$$

Since the inequality $M\sum_{j\ge i} j^{-2} \ge 1$ holds for every i, we get a contradiction. □

A stochastic process $\underline{\underline{X}}$ is a <u>C-semiamart</u> if $\limsup_{\tau \in M} (|EX_\tau| / \|n_\tau\|_\infty) < \infty$. A stochastic process $\underline{\underline{X}}$ is a <u>C-potential</u> if $\lim_{\tau \in M} (|EX_\tau| / \|n_\tau\|_\infty) = 0$.

<u>Proposition 2.3</u>: (Lattice property of C-semiamarts and C-potentials): Let (X_t) be an L_1-bounded C-semiamart (resp. C-potential); then $\underline{\underline{X}}^+$, $\underline{\underline{X}}^-$ and $|\underline{\underline{X}}|$ are C-semiamarts (resp. C-potentials).

<u>Proof</u>: Let $\underline{\underline{X}}$ be a C-semiamart. Clearly $\limsup_{\|e_\tau\|_\infty = 0} |EX_\tau| < \infty$, i.e., $\underline{\underline{X}}$ is a semi-amart. Hence $\underline{\underline{X}}^+$, $\underline{\underline{X}}^-$ and $|\underline{\underline{X}}|$ are semiamarts [9]. Fix $s \in J$ and $\sigma \in M$, $\sigma \geq s$; choose $t > \sigma$ and define $\tau \in M$ by $\{\tau = u\} = \{\sigma = u\} \cap \{X_u \geq 0\}$ if $u \in R(\sigma)$, and $\{\tau = t\} = \Omega$. Then $\|n_\tau\|_\infty \leq \|n_\sigma\|_\infty + 1 \leq 2\|n_\sigma\|_\infty$, and $\Sigma 1_{\{\sigma = u\}} X_u^+ = X_\tau - X_t$; this clearly implies that $\underline{\underline{X}}^+$ is a C-semiamart. Similarly $\underline{\underline{X}}^-$ is a C-semiamart, so that $|\underline{\underline{X}}|$ also is a C-semiamart.

Now let $\underline{\underline{X}}$ be a C-potential, then $\underline{\underline{X}}$ is also an amart, i.e., $\lim_T EX_\tau$ exists, where T is the subset of M composed of stopping times τ with $n_\tau = 1$. Since $\lim_T EX_\tau = 0$, the amart Riesz decomposition shows that $\lim_T E|X_\tau| = 0$, hence $\lim_J E|X_t| = 0$. The first part of the proof now shows that $\underline{\underline{X}}^+$, $\underline{\underline{X}}^-$, and $|\underline{\underline{X}}|$ are C-potentials. □

The following theorem gives several equivalent formulations of C.

<u>Theorem 2.4</u>: Let J be a directed set with a countable cofinal subset, and let $(\mathfrak{F}_t)$ be a stochastic basis indexed by J. The following conditions are equivalent:

(1) $(\mathfrak{F}_t)$ satisfies the condition C.

(2) $(\mathfrak{F}_t)$ satisfies the condition LC.

(3) $(\mathfrak{F}_t)$ satisfies the condition $C(\infty,1)$.

(4) There exists an Orlicz function Φ with the properties Δ and Δ_2 such that $(\mathfrak{F}_t)$ satisfies the condition $C(\infty,\Phi)$.

(5) $(\mathfrak{F}_t)$ satisfies the condition $B(\infty,1)$.

(6) There exists an Orlicz function Φ with the properties Δ and Δ_2 such that $(\mathfrak{F}_t)$ satisfies the condition $B(\infty,\Phi)$.

(7) $(\mathfrak{F}_t)$ satisfies Astbury's condition A.

(8) There exists an Orlicz function Φ with the properties Δ and Δ_2 such that $(\mathfrak{F}_t)$ satisfies the condition $R(\infty,\Phi)$.

(9) For every L_1-bounded C-semiamart $\underline{X}$, $|X|^* < \infty$ a.e.

(10) For every positive process $\underline{X}$, $s \limsup_{e_\tau \leq M} X(\tau) = 0$ for every $M > 0$ implies $\limsup_{e_\tau \leq M} X_\tau = 0$ for every $M > 0$.

(11) Every C-potential converges essentially (to zero).

<u>Proof</u>: Clearly it suffices to prove the equivalence when J has no greatest element. We show (1) $\Rightarrow$ (2) $\Rightarrow$ (5) $\Rightarrow$ (7) $\Rightarrow$ (8) $\Rightarrow$ (6) $\Rightarrow$ (4) $\Rightarrow$ (1), (1) $\Rightarrow$ (3) $\Rightarrow$ (5), (1) $\Rightarrow$ (9) $\Rightarrow$ (7), (2) $\Rightarrow$ (10) $\Rightarrow$ (11) $\Rightarrow$ (7). The implications (3) $\Rightarrow$ (5), (5) $\Rightarrow$ (7) and (7) $\Rightarrow$ (8) are obvious. The implications (1) $\Rightarrow$ (2) and (1) $\Rightarrow$ (3) are immediate consequences of the equivalent formulations (6) and (5) of C given in Theorem 1.1 [25]. The implication (6) $\Rightarrow$ (4) is a consequence of Proposition 2.2(ii).

<u>(2) $\Rightarrow$ (5)</u>: Assume that $B(\infty,1)$ fails, and let $\underline{A}$ be an adapted family of sets with $A^* \neq \emptyset$, such that for every $K > 0$ there exists an index s_K such that for every stopping time τ, $\tau \geq s_K$ implies $\|e_\tau\|_\infty \geq KPA(\tau)$. By (2), choose an integer N such that $P\tilde{A}_N > 2^{-1}PA^*$, where $\tilde{A}_N = s \limsup_{e_\tau \leq N} A(\tau)$. For every index s, applying Lemma 1.2 in [23] we obtain a stopping time τ satisfying $\tau \geq s$, $e_\tau \leq N$, and $P[A(\tau)] > 2^{-1}P\tilde{A}_N$. For every K, there exists a stopping time $\tau_K \geq s_K$ such that $PA^* \leq 2P\tilde{A}_N \leq 4PA(\tau_K) \leq 4\|e_{\tau_K}\|_\infty/K \leq 4N/K$, which contradicts the assumption $A^* \neq \emptyset$.

<u>(8) $\Rightarrow$ (6)</u>: Let Φ be an Orlicz function having the properties Δ and Δ_2 such that $(\mathfrak{F}_t)$ satisfies $R(\infty,\Phi)$. We at first show that the Vitali condition V_Φ holds. The Orlicz function Φ has the property Δ_2. By [18], it suffices to show that the martingale $X_t = E^{\mathfrak{F}_t}X$ converges essentially for every function $X \in L_\Psi^+$. The following argument generalizes one given in [2]. One may assume that for some function $X \in L_\Psi$ and some positive reals $a < b$, $A = \{X \leq a < b \leq \limsup X_t\} \neq \emptyset$. It is easy to see that there exists an adapted family of sets $\underline{A}$ such that $\lim 1_{A_t} = 1_A$. For every index t, set $B_t = A_t \cap \{X_t \geq b\}$, and $D_t = A^c \cap (e \sup_{u \geq t} B_u)$. Then $D_t \in \mathfrak{F}_\infty = \sigma(\cup \mathfrak{F}_u)$, and since $\limsup B_t = A$, one has $\limsup D_t = \emptyset$. Fix $\epsilon > 0$, and choose $D = D_s$ such that $\|1_D X\|_\Psi < \epsilon$. Then for every $t \geq s$ and every $\mathfrak{F}_t$-measurable subset $C_t \subset B_t \subset e \sup_{u \geq s} B_s$, one has

$$E[1_{C_t \cap D}X] = E[1_{C_t \cap A^c}X]$$

$$= E[1_{C_t}X] - E[1_{C_t \cap A}X]$$

$$\geq E[1_{C_t}X_t] - aP(C_t \cap A)$$

$$\geq bP(C_t) - aP(C_t \cap A)$$

$$\geq (b - a)P(C_t),$$

so that $E^{\mathfrak{F}_t}(1_D X) \geq (b - a)1_{B_t}$. Since $PB^* > 0$, there exists $M > 0$ and $\tau \in \mathrm{IM}$ such that $\tau \geq s$, $\{\tau = t\} \subset B_t$ for every t, and $\|n_\tau\|_\infty < M\|n_\tau\|_\Phi$. Let $\alpha > 0$ be the solution of the equation $\int \Psi \circ \varphi(\alpha n_\tau)\, dP = 1$. Then $\alpha\|n_\tau\|_\Phi$ remains bounded by a constant K which does not depend on τ. One has

$$(b - a)\|n_\tau\|_\Phi = (b - a)\int n_\tau \varphi(\alpha n_\tau)\, dP$$

$$\leq (b - a)\|n_\tau\|_1 \|\varphi(\alpha n_\tau)\|_\infty$$

$$\leq \sum_t E[1_{\{\tau = t\}}(b - a)] \quad \varphi(\alpha\|n_\tau\|_\infty)$$

$$\leq \sum_t E[1_{\{\tau = t\}}E^{\mathfrak{F}_t}(1_D X)] \; \varphi(\alpha M \|n_\tau\|_\Phi)$$

$$\leq \sum_t E[1_{\{\tau = t\}}1_D X] \; \varphi(KM)$$

$$\leq E(n_\tau 1_D X) \; \varphi(KM)$$

$$\leq \|n_\tau\|_\Phi \|1_D X\|_\Psi \; \varphi(KM)$$

$$\leq \epsilon \; \varphi(KM)\|n_\tau\|_\Phi .$$

Since $\|n_\tau\|_\Phi > 0$, one obtains a contradiction by choosing ϵ small. Therefore $(\mathfrak{F}_t)$ satisfies the condition V_Φ. By Proposition 2.2(i) applied with $\|\;\| = \|\;\|_\Phi$, one deduces that $(\mathfrak{F}_t)$ satisfies the condition $B(\infty, \Phi)$.

<u>(4) ⇒ (1)</u>: Let $X \in L_\Phi$; the Luxemburg norm of X is defined by $\|X\|_{(\Phi)} = \inf\{a > 0: \int \Phi(|X|/a)dP \leq 1\}$ [14]. Fix $\epsilon > 0$, let $\underline{A}$ be an adapted family of sets with $PA^* \geq \epsilon$, and let $\tau \in \mathrm{IM}$ satisfy $\|n_\tau\|_\infty < M_\epsilon \|1_{A(\tau)}\|_\Phi \leq 2M_\epsilon \|1_{A(\tau)}\|_{(\Phi)}$ (see e.g. [14] for the comparison between the two norms on L_Φ). Let a constant K be such that $\|\;\|_{(\Phi)} \leq K\|\;\|_\infty$ (see e.g. [14]). The stopping time τ satisfies

the inequalities

$$\|n_\tau\|_\infty \le 2KM_\epsilon$$

and

$$1 = \int \Phi(1_{A(\tau)}/\|1_{A(\tau)}\|_{(\Phi)})dP$$

$$\le \int \Phi(2M_\epsilon 1_{A(\tau)})dP = \Phi(2M_\epsilon)P[A(\tau)] .$$

By a standard argument similar to the one given in the proof of implication (5) ⇒ (2), Theorem 1.1 [25], we show that $(\mathfrak{F}_t)$ satisfies C.

(1) ⇒ (9): Let $\underline{\underline{X}}$ be a positive C-semiamart, and set $L = \limsup_{\tau \in M}(EX_\tau/\|n_\tau\|_\infty)$. Let s be an index such that for every $\tau \in IM$, $\tau \ge s$ implies $EX_\tau \le 2L\|n_\tau\|_\infty$. Fix $\lambda > 0$, and set $A_t = \{X_t > \lambda\}$; for every $\epsilon > 0$, let $\tau \in IM$ satisfy $\tau \ge s$, $e_\tau \le M_\epsilon$ and

$$P(X^* > \lambda) \le P(A^*) \le P[A(\tau)] + \epsilon$$

$$\le \frac{1}{\lambda} EX_\tau + \epsilon$$

$$\le \frac{2M_\epsilon}{\lambda} L + \epsilon .$$

By letting $\lambda \to \infty$, we deduce that for every $\epsilon > 0$, $P(X^* = +\infty) \le \epsilon$, so that $X^* < \infty$ a.e.

(9) ⇒ (7) and (11) ⇒ (7): Assume that (A) fails, and let $\underline{\underline{A}}$ be an adapted family of sets with $A^* \ne \emptyset$, such that for every integer K, there exists an index s_K such that every $\tau \in IM$ with $\tau \ge s_K$ and $\{\tau = t\} \subset A_t$ for every t, satisfies the inequality $\|n_\tau\|_\infty \ge K^3\|n_\tau\|_1$. Since J has a countable cofinal subset, one may assume that the sequence (s_k) is increasing, cofinal, and satisfies $P(A^* \setminus \operatorname{ess\,sup}_{s_K \le t < s_{K+1}} A_t) \le 2^{-K}$ for every K. For every t with $s_K \le t < s_{K+1}$, set $X_t = K1_{A_t}$; if $t \notin \cup\{u: s_K \le u < s_{K+1}\}$, set $X_t = 0$. Let τ be an incomplete stopping time, $\tau \ge s_i$; for every $j \ge i$, define

$\tau_j \in \mathrm{IM}$ by $\{\tau_j = t\} = \{\tau = t\}$ if $s_j \leq t < s_{j+1}$. Then

$$EX_\tau = \sum_{j \geq i} EX_{\tau_j} \leq \sum_{j \geq i} j \|n_{\tau_j}\|_1$$

$$\leq (\sum_{j \geq i} j^{-2}) \|n_{\tau_j}\|_\infty \leq (\sum_{j \geq i} j^{-2}) \|n_\tau\|_\infty ,$$

which proves that $\underline{\underline{X}}$ is a C-semiamart, even a C-potential. However, $A^* \subset \{X^* = \infty\}$.

(2) ⇒ (10): Let $\underline{\underline{X}}$ be a positive process such that $s \limsup_{e_\tau \leq M} X(\tau) = 0$ for every M. Fix $\lambda > 0$, and set $A_t = \{X_t > \lambda\}$. For every fixed $\epsilon > 0$, let n be such that $PA^* \leq P\widetilde{A}_n + \epsilon$, where $\widetilde{A}_n = s \limsup_{e_\tau \leq n} A(\tau)$. Choose s such that $P(X(\tau) \geq \lambda) \leq \epsilon$ if $\tau \geq s$ and $e_\tau \leq n$. Applying Lemma 1.2 in [23], we deduce that there exists $\tau \in \mathrm{IM}$ such that $\tau \geq s$, $e_\tau \leq n$ and $P[\widetilde{A}_n \backslash A(\tau)] \leq \epsilon$. Then $PA^* \leq P\widetilde{A}_n + \epsilon \leq P[A(\tau)] + 2\epsilon \leq P[X(\tau) \geq \lambda] + 2\epsilon \leq 3\epsilon$.

Since ϵ and λ are arbitrary, we have $X^* = 0$. Since for every M, $\limsup_{e_\tau \leq M} X_\tau \leq (M+1)X^*$, we deduce (10).

(10) ⇒ (11): Let $\underline{\underline{X}}$ be a positive C-potential, and fix $k > 0$. Then, since $\lim_{\tau \in M} (EX_\tau / \|n_\tau\|_\infty) = 0$, we have $\lim_{e_\tau \leq k} EX_\tau = 0$, which implies $\lim_{e_\tau \leq k} X_\tau = 0$. Therefore, X_t converges essentially to zero. □

Remark 2: The existence of a countable cofinal subset of J was used only in the proofs of the implications (8) ⇒ (6), (6) ⇒ (4), (9) ⇒ (7) and (11) ⇒ (7).

Remark 3: Also the following condition can be shown to be equivalent with LC: For every adapted family of sets $\underline{\underline{A}}$ such that $\lim_{\tau \in M} (P[A(\tau)] / \|n_\tau\|_\infty) = 0$, one has $A^* = \emptyset$.

3. LEBESGUE DERIVATION THEOREM UNDER CONDITION C

In the present section we obtain a derivation theorem for point bases under an analogue of the condition C. The notation and terminology are that of the elegant monograph of M. de Guzman [11].

Let X be a metric space, $\mathfrak{F}$ be the σ-algebra generated by the open sets, and let μ be a σ-finite complete measure on $\mathfrak{F}$. Assume that the class of continuous integrable functions is dense in L^1; such is the case if μ is the Lebesgue measure on $\mathbb{R}^n$. We say that a sequence of sets (A_k) contracts to a point x, if for each k $x \in A_k$, and given any $\epsilon > 0$ there exists k_0 such that $k \geq k_0$ implies $A_k \subset B(x,\epsilon)$, where $B(x,\epsilon)$ is the open ball centered at x and of radius ϵ. For every point $x \in X$, let $\mathcal{B}(x)$ be a collection of sequences of open sets (B_k) contracting to x, such that every subsequence of an element of $\mathcal{B}(x)$ also belongs to $\mathcal{B}(x)$. The family $\mathcal{B}$ of all $\mathcal{B}(x)$, $x \in X$, is called a derivation basis. Given a set A, we write $A \in \mathcal{B}(x)$ if A is an element of a sequence belonging to $\mathcal{B}(x)$, and we write $A \in \mathcal{B}$ if $A \in \mathcal{B}(x)$ for some x. A basis $\mathcal{B}$ is called a Buseman-Feller basis, if for every $A \in \mathcal{B}$ and every $x \in A$, $A \in \mathcal{B}(x)$. Let $\mathcal{B}$ be a Buseman-Feller basis; a collection of sets $\mathcal{V} \subset \mathcal{B}$ is a Vitali cover of a measurable set A if for every $x \in A$ there exists a sequence $(B_k(x)) \subset \mathcal{V}$ such that $\lim_k \operatorname{diam} B_k(x) = 0$ and $B_k(x) \in \mathcal{B}(x)$ for every k. For every function $f \in L_1$ and for every $x \in X$, set

$$D^*f(x) = \sup_{(B_k) \in \mathcal{B}(x)} \left[\limsup_k \frac{1}{\mu(B_k)} \int_{B_k} f\,d\mu\right],$$

$$D_*f(x) = \inf_{(B_k) \in \mathcal{B}(x)} \left[\liminf_k \frac{1}{\mu(B_k)} \int_{B_k} f\,d\mu\right].$$

The derivation basis derives f if $D^*f(x) = D_*f(x) = f(x)$ for μ almost all x. If $\mathcal{B}$ is a Buseman-Feller basis and if $f \in L_1$, then both functions D^*f and D_*f are measurable (Guzman [11], p. 66). Since the general σ-finite case can be easily reduced to the case $\mu(X) = 1$, we henceforth assume $\mu(X) = 1$.

We say that a basis satisfies the condition C if: For every $\epsilon > 0$ there exists $M_\epsilon > 0$ such that for every measurable set A with $\mu(A) > \epsilon$ and every Vitali cover $\mathcal{V}$ of A, there exists a finite sequence $(B_k) \subset \mathcal{V}$ such that $\|\Sigma 1_{B_k}\|_\infty < M_\epsilon \mu(\cup B_k)$.

Theorem 3.1: Let $\mathcal{B}$ be a Buseman-Feller basis satisfying the condition C. Then $\mathcal{B}$ derives every integrable function.

Proof: It is easy to see that any Buseman-Feller basis derives every continuous and integrable function. Fix $f \in L_1$, $a > 0$; we show that $\mu(\{x \in X : |D^*f(x) - f(x)| > a\}) = 0$. Let $\epsilon > 0$ be fixed. Let $\delta > 0$ be a constant to be determined later, and let g be a continuous integrable function such that $\|f - g\|_1 \leq \delta$; then

$$\{x \in X : |D^*f(x) - f(x)| > a\} \subset \{x \in X : |f - g|(x) > \frac{a}{2}\}$$
$$\cup \{x \in X : D^*|f - g|(x) > \frac{a}{2}\}.$$

Set $A = \{x \in X : D^*|f - g|(x) > \frac{a}{2}\}$; applying the definition of D^*, for every $x \in A$, we obtain a sequence $(B_k(x))$ in $\mathcal{B}(x)$ such that $\int_{B_k(x)} |f - g| d\mu >$ $\frac{a}{2} \mu(B_k(x))$. This defines a Vitali cover of A.

Suppose $\mu(A) > \epsilon$ and let $(B_k) \subset \mathcal{V}$ be a finite sequence of sets appearing in condition C. The strict inequality in C implies that at least one of the sets has positive measure. Hence

$$\begin{aligned} 1 &< M_\epsilon \mu(\cup B_k) \leq M_\epsilon \Sigma \mu(B_k) \\ &\leq M_\epsilon \frac{2}{a} \Sigma \int_{B_k} |f - g| d\mu \leq \frac{2M_\epsilon}{a} \|\Sigma 1_{B_k}\|_\infty \|f - g\|_1 \\ &\leq \frac{2M_\epsilon^2}{a} \delta . \end{aligned}$$

Choose $\delta < \inf(\frac{a}{2M_\epsilon^2}, \frac{\epsilon a}{2})$; one obtains a contradiction which implies that $\mu(A) \leq \epsilon$, and hence by Chebyshev's inequality

$$\mu(\{x \in X : |D^*f(x) - f(x)| > a\}) \leq \frac{2}{a} \delta + \epsilon \leq 2\epsilon .$$

Since this holds for every ϵ and a, $D^*f = f$ a.e. Considering $-f$ instead of f, one also obtains $D_*f = f$ a.e. □

We also extend Theorem 3.1 to the case when the function f takes values in a Banach space; the other assumptions remain the same. We say that <u>the derivation basis derives</u> f if

$$\sup_{(B_k)\in\mathcal{B}(x)} [\limsup | \frac{1}{\mu(B_k)} \int_{B_k} f d\mu - f(x) |] = 0$$

for μ almost all x. A slight modification of the above argument proves the following:

<u>Theorem 3.2</u>: Let $\mathcal{B}$ be a Buseman-Feller basis satisfying the condition C. Then $\mathcal{B}$ derives every strongly measurable Bochner integrable function.

We now introduce several conditions analogous to C, stated in terms of L^1-norms or Orlicz space norms. Theorem 3.3 below shows that under assumptions usually satisfied all these conditions are equivalent.

Let μ be a probability measure on $\mathfrak{F}$. Consider the following conditions on the Buseman-Feller basis $\mathcal{B}$:

(1) For every $\epsilon > 0$, there exists $M_\epsilon > 0$ such that for every measurable set A and every Vitali cover $\mathcal{V}$ of A, there exists a finite sequence $(B_i) \subset \mathcal{V}$ such that $\mu(A \setminus \cup B_i) < \epsilon$ and $\|\Sigma 1_{B_i}\|_\infty \leq M_\epsilon$.

(2) $\mathcal{B}$ satisfies the condition C.

(3) For every measurable set A with $\mu(A) > 0$ and every Vitali cover $\mathcal{V}$ of A, there exists a constant $M > 0$ such that for every $\epsilon > 0$, there exists a finite sequence $(B_i) \subseteq \mathcal{V}$ with diam $B_i < \epsilon$ and $\|\Sigma 1_{B_i}\|_\infty < M\mu(\cup B_i)$.

(4) For every measurable set A with $\mu(A) > 0$ and every Vitali cover $\mathcal{V}$ of A, there exists a constant $M > 0$ such that for every $\epsilon > 0$, there exists a finite sequence $(B_i) \subset \mathcal{V}$ with diam $B_i < \epsilon$ and $\|\Sigma 1_{B_i}\|_\infty < M\|\Sigma 1_{B_i}\|_1$.

(5) There exists an Orlicz function Φ with the properties Δ and Δ_2 such that for every measurable set A with $\mu(A) > 0$ and every Vitali cover

$\mathcal{V}$ of A, there exists a constant $M > 0$ such that for every $\epsilon > 0$, there exists a finite sequence $(B_i) \subset \mathcal{V}$ with $\operatorname{diam} B_i < \epsilon$ and $\|\Sigma 1_{B_i}\|_\infty < M\|\Sigma 1_{B_i}\|_\Phi$.

Theorem 3.3: (i) The conditions (2) and (3) are equivalent. (ii) If the measure μ is regular, then the conditions (2), (3), (4) and (5) are equivalent. (iii) If for every open set $B \in \mathcal{B}$ one has $\mu(B) = \mu(\overline{B})$, then the conditions (1), (2) and (3) are equivalent. ($\overline{B}$ is the closure of B.)

Proof: The implications (1) $\Rightarrow$ (2) $\Rightarrow$ (3) $\Rightarrow$ (4) $\Rightarrow$ (5) are obvious. We show the converse implications (5) $\Rightarrow$ (3) $\Rightarrow$ (2) $\Rightarrow$ (1).

(5) $\Rightarrow$ (3): Let $\mathcal{B}$ satisfy the condition (5); we at first show that $\mathcal{B}$ satisfies the Vitali condition V_Φ, i.e., for every measurable set A, for every Vitali cover $\mathcal{V}$ of A and for every $\epsilon > 0$, there exists a finite sequence $(B_i) \subset \mathcal{V}$ such that $\mu(A \setminus \cup B_i) < \epsilon$ and $\|\Sigma 1_{B_i} \setminus 1_{\cup B_i}\|_\Phi < \epsilon$. Assume the contrary; since Φ has the property Δ_2, by Hayes' theorem [12], there exists a function $f \in L_\Psi^+$ such that $\mathcal{B}$ fails to derive f. One may assume that for some positive reals $a < b$ the set $A = \{f \leq a < b \leq D^* f\}$ is non-null. The measure μ being regular, there exists a compact non-null set $K \subset A$. For every $\epsilon > 0$, let $\mathcal{V}_\epsilon$ be the family of sets B with $\operatorname{diam} B < \epsilon$, $\int 1_B f\, d\mu \geq b\mu(B)$, and such that for some $x \in K$, $B \in \mathcal{B}(x)$. Let $\mathcal{V} = \cup \mathcal{V}_\epsilon$; then $\mathcal{V}$ is a Vitali cover of K, and letting $\Omega_\epsilon = \cup\{B \mid B \in \mathcal{V}_\epsilon\}$, one has $K \subset \bigcap_n \Omega_{1/n} \subset \overline{K} = K$. Fix $\delta > 0$, and choose n such that $\int 1_{\Omega_{1/n} \setminus K} f \, d\mu \leq \delta$. Let $M > 0$ be the constant given by the condition (5) applied to the set K with Vitali cover $\mathcal{V}$, and let $(B_i) \subset \mathcal{V}_{1/n}$ be a finite sequence such that $\|\Sigma 1_{B_i}\|_\infty < M\|\Sigma 1_{B_i}\|_\Phi$. Let $\alpha > 0$ denote the solution of the equation $\int \Psi \circ \varphi(\alpha \Sigma 1_{B_i}) d\mu = 1$; then $\alpha\|\Sigma 1_{B_i}\|_\Phi$ remains bounded by a constant k which does not depend on the sets (B_i), and

$$
\begin{aligned}
(b - a)\|\Sigma 1_{B_i}\|_\Phi &= (b - a)\int (\Sigma 1_{B_i})\varphi(\alpha \Sigma 1_{B_i}) d\mu \\
&\leq (b - a)\|\Sigma 1_{B_i}\|_1 \|\varphi(\alpha \Sigma 1_{B_i})\|_\infty \\
&\leq \varphi(\alpha\|\Sigma 1_{B_i}\|_\infty)[\Sigma b\, \mu(B_i) - \Sigma a\, \mu(B_i \cap K)] \\
&\leq \varphi(\alpha M \|\Sigma 1_{B_i}\|_\Phi)[\Sigma \int 1_{B_i} f d\mu - \Sigma \int 1_{B_i \cap K} f d\mu] \\
&\leq \varphi(kM) \int (\Sigma 1_{B_i})\, 1_{K^c}\, f\, d\mu \\
&\leq \varphi(kM)\|\Sigma 1_{B_i}\|_\Phi \|1_{\cup B_i \setminus K} f\|_\Psi \leq \delta\, \varphi(kM)\|\Sigma 1_{B_i}\|_\Phi .
\end{aligned}
$$

Dividing by $\|\Sigma 1_{B_i}\|_\Phi > 0$, one obtains $\delta\varphi(kM) > b - a$ which brings a contradiction for small values of δ. Hence $\mathcal{B}$ satisfies V_Φ.

Suppose that there exists a measurable set A with $\mu(A) > 0$, a Vitali cover $\mathcal{V}$ of A and a sequence of numbers $\epsilon_k \downarrow 0$ such that for every finite sequence $(B_i) \subset \mathcal{V}$ with diam $B_i < \epsilon_k$, one has $\|\Sigma 1_{B_i}\|_\infty \geq k^2 \|1_{\cup B_i}\|_\Phi$. Fix the index k. Applying the Vitali condition V_Φ to the set A with Vitali cover $\mathcal{V}_k = \{B \in \mathcal{V} \mid \text{diam } B < \epsilon_k\}$, one obtains finitely many sets $\{B_i^k\} \subset \mathcal{V}_k$ such that $\mu(A \setminus \bigcup_i B_i^k) < 2^{-k}$ and $\|\Sigma_i 1_{B_i^k} - 1_{\bigcup_i B_i^k}\|_\Phi < 2^{-k}$. The construction being done step by step, one may assume that $\epsilon_{k+1} < \inf_i \text{diam } B_i^k$. Denote by $\mathcal{B}^k$ the finite family $\{B_i^k\}_i$. The family of sets $\mathcal{W} = \bigcup \mathcal{B}^k$ is a Vitali cover of $\widetilde{A} = \limsup_k \bigcup_i B_i^k$, and $\mu(A \setminus \widetilde{A}) = 0$. Then $\mu(\widetilde{A}) > 0$, and for any j and any finite family of sets $\{E_i\} \subset \mathcal{W}$ with diam $E_i \leq \epsilon_j$, one has for large values of j,

$$\begin{aligned}
\|\Sigma 1_{E_i}\|_\Phi &\leq \sum_{k \geq j} \|\sum_{E_i \in \mathcal{B}^k} 1_{E_i}\|_\Phi \\
&\leq \sum_{k \geq j} [\|1_{\cup\{E_i \mid E_i \in \mathcal{B}^k\}}\|_\Phi + 2^{-k}] \\
&\leq \sum_{k \geq j} k^{-2} \|\sum_{E_i \in \mathcal{B}^k} 1_{E_i}\|_\infty + 2^{-k} \\
&\leq (\sum_{k \geq j} k^{-2}) \|\Sigma 1_{E_i}\|_\infty + 2^{1-j} \\
&\leq 2(\sum_{k \geq j} k^{-2}) \|\Sigma 1_{E_i}\|_\infty .
\end{aligned}$$

This contradicts the condition (5) for the Vitali cover $\mathcal{W}$ of $\widetilde{A}$.

Let us now show that $\mathcal{B}$ satisfies the condition (3). Let A be a measurable non-null set with Vitali cover $\mathcal{V}$. Let $M > 0$ be a constant such that for every $\epsilon > 0$ there exists a finite sequence $(B_i) \subset \mathcal{V}$, such that diam $B_i < \epsilon$ and $\|\Sigma 1_{B_i}\|_\infty < M \|1_{\cup B_i}\|_\Phi$. Let (B_i) be a finite sequence satisfying the above inequalities. Given a function $f \in L_\Phi$, the Luxemburg norm of f is denoted by $\|f\|_{(\Phi)} = \inf\{k > 0 \mid \int \Phi(|f|/k)d\mu \leq 1\}$. Let $c > 0$ denote a constant such that $\|f\|_\Phi \leq 2\|f\|_{(\Phi)} \leq c\|f\|_\infty$. Then $\|\Sigma 1_{B_i}\|_\infty \leq 2cM$, and since $\|1_{\cup B_i}\|_{(\Phi)} \geq$

$1/(2M)$, one deduces that $\Phi(2M)\mu(\cup B_i) \geq 1$. Hence $\mathcal{B}$ satisfies the condition (3).

(3) ⇒ (2): Let $\mathcal{B}$ satisfy the condition (3); assume C fails. Let $\epsilon > 0$ be such that for every n there exists a measurable set A_n with $\mu(A_n) > \epsilon$, a Vitali cover $\mathcal{V}_n$ of A_n such that for every finite sequence $(B_i) \subset \mathcal{V}_n$ one has $\|\Sigma 1_{B_i}\|_\infty \geq n^2\mu(\cup B_i)$. Let $(B_i^1)_i$ be a finite sequence of elements of $\mathcal{V}_1$ such that $\mu(A_1 \setminus \bigcup_i B_i^1) \leq 2^{-1}$, and set $\alpha_1 = \inf_i \operatorname{diam} B_i^1$. If $\alpha_{n-1} > 0$ has been defined, let $(B_i^n)_i$ be a finite sequence of elements of $\mathcal{V}_n$ such that $\operatorname{diam} B_i^n \leq \inf(\alpha_{n-1}, 2^{-n})$, and $\mu(A_n \setminus \bigcup_i B_i^n) \leq 2^{-n}$; set $\alpha_n = \inf_i \operatorname{diam} B_i^n$, and set $\mathcal{B}^n = \{B_i^n\}_i$. The family of sets $\mathcal{V} = \cup \mathcal{B}^n$ is a Vitali cover of $B = \limsup_n (\bigcup_i B_i^n)$. Clearly $\mu(B) \geq \mu(\limsup A_n) \geq \epsilon$. Fix n; for any finite sequence of sets $(E_i) \subset \mathcal{V}$ with $\operatorname{diam} E_i \leq \alpha_n$, one has

$$\mu(\cup E_i) \leq \sum_{p \geq n} \mu(\cup\{E_i \mid E_i \in \mathcal{B}^p\})$$

$$\leq \sum_{p \geq n} p^{-2} \| \sum_{E_i \in \mathcal{B}^p} 1_{E_i} \|_\infty \leq (\sum_{p \geq n} p^{-2}) \|\Sigma 1_{E_i}\|_\infty .$$

This contradicts the condition (3) for the set A and its Vitali cover $\mathcal{V}$.

(2) ⇒ (1): Let $\mathcal{B}$ satisfy (2). Fix a measurable set A_1 with $\mu(A_1) > \epsilon$, and a Vitali cover $\mathcal{V}$ of A_1. Choose finitely many sets $(B_i^1)_i \in \mathcal{V}$ such that $\|\sum_i 1_{B_i^1}\|_\infty \leq M_\epsilon$, and $\mu(\bigcup_i B_i^1) \geq 1/M_\epsilon$. Set $F_1 = \bigcup_i \overline{B_i^1}$, $A_2 = A_1 \setminus F_1$, and let $\mathcal{V}_2$ be the family of all sets $B \in \mathcal{V}$ such that $B \cap F_1 = \emptyset$. Then F_1 is a closed set and $\mathcal{V}_2$ is a Vitali cover of A_2. Suppose that the measurable set A_k satisfies $\mu(A_k) > \epsilon$, and that $\mathcal{V}_k$ is a Vitali cover of A_k. Let $(B_i^k)_i$ be a finite family of elements of $\mathcal{V}_k$ such that $\mu(\bigcup_i B_i^k) \geq 1/M_\epsilon$, and $\|\Sigma_i 1_{B_i^k}\|_\infty \leq M_\epsilon$; set $F_{k+1} = \bigcup_i \overline{B_i^k}$, $A_{k+1} = A_k \setminus F_{k+1}$, and let $\mathcal{V}_{k+1}$ be the Vitali cover of A_{k+1} formed by the sets $B \in \mathcal{V}_k$ such that $B \cap F_{k+1} = \emptyset$. The sets $C_n = \bigcup_i B_i^n$ being pairwise disjoint, there exists an integer N such that $\mu(A \setminus \bigcup_{n \leq N} C_n) = \mu(A \setminus \bigcup_{n \leq N} F_n) \leq \epsilon$. The sets $\{B_i^n\}_{i, n \leq N}$ form the cover of A promised in condition (1).

REFERENCES

1. K. Astbury, Amarts indexed by directed sets, Ann. Prob., 6 (1978), 267-278.

2. K. Astbury, The order convergence of martingales indexed by directed sets, (to appear).

3. K. Astbury, The Weak Vitali and Dominated Sums Properties, (to appear).

4. R. V. Chacon and L. Sucheston, On convergence of vector-valued asymptotic martingales, Z. Wahrscheinlichkeitstheorie verw. Geb., 33 (1975), 55-59.

5. S. D. Chatterji, Martingale convergence theorem and the Radon-Nikodým theorem in Banach spaces, Math. Scand., 22 (1968), 21-41.

6. J. Diestel and J. J. Uhl, Vector Measures, Mathematical Surveys 15, 1977.

7. J. Dieudonné, Sur un théorème de Jessen, Fund. Math., 37 (1950), 242-248.

8. J. L. Doob, Stochastic Processes, Viley, New York, 1953.

9. G. A. Edgar and L. Sucheston, Amarts: A class of asymptotic martingales, J. Multivariate Anal., 6 (1976), 193-221; 572-591.

10. A. Garsia, Topics in Almost Everywhere Convergence, Markham, Chicago, 1970.

11. M. de Guzman, Differentiation of Integrals in R^n, Springer Verlag Lecture Notes in Mathematics, vol. 481, 1975.

12. C. A. Hayes, Necessary and sufficient conditions for the derivation of integrals of L_ψ-functions, Trans. Amer. Math. Soc., 223 (1976), 385-395.

13. A. Ionescu Tulcea and C. Ionescu Tulcea, Abstract ergodic theorems, Trans. Amer. Math. Soc., 107 (1963), 107-124.

14. M. A. Krasnolsel'skii and Ya. Rutickii, Convex Functions and Orlicz spaces, Noordhoff-Groningen-The Nederlands, 1961.

15. K. Krickeberg, Convergence of martingales with a directed index set, Trans. Amer. Math. Soc., 83 (1956), 313-337.

16. K. Krickeberg, Notwendige Konvergenzbedingungen bei Martingalen und verwandten Prozessen, Transactions of the Second Prague conference on information theory, statistical decision functions, random processes [1959 Prague], Publishing House of the Czechoslovak Academy of Sciences, 1960, 279-305.

17. M. Métivier, Martingales à valeurs vectorielles; applications à la dérivation des mesures vectorielles, Ann. Inst. Fourier, 17 (1967), 175-298.

18. A. Millet, Sur la caractérisation des conditions de Vitali par la convergence essentielle des martingales, C. R. Acad. Sci. Paris, Série A, 287 (1978), 887-890.

19. A. Millet and L. Sucheston, Classes d'amarts filtrants et conditions de Vitali, C. R. Acad. Sci. Paris, Série A, 286 (1977), 835-837.

20. A. Millet and L. Sucheston, Convergence of classes of amarts indexed by directed sets - Characterization in terms of Vitali conditions, Canadian J. Math, 32 (1980).

21. A. Millet and L. Sucheston, Sur la caractérisation des conditions de Vitali par la convergence essentielle de classes d'amarts, C. R. Acad. Sci. Paris, Série A, 286 (1977), 1015-1017.

22. A. Millet and L. Sucheston, Characterization of Vitali conditions with overlap in terms of convergence of classes of amarts, Canadian J. Math., 31 (1979).

23. A. Millet and L. Sucheston, A Characterization of Vitali Conditions in Terms of Maximal Inequalities, Annals of Probability, to appear.

24. A. Millet and L. Sucheston, La convergence des martingales bornées dans L^1 n'implique pas la condition de Vitali V, C. R. Acad. Sci. Paris, Série A, 288 (1979), 595-598.

25. A. Millet and L. Sucheston, On convergence of L_1-bounded martingales indexed by directed sets, Probability and Mathematical Statistics, to appear.

26. H. W. Milnes, Convexity of Orlicz spaces, Pacific J. Math., 7 (1957), 1451-1483.

27. J. Neveu, Discrete Parameter Martingales, North Holland, Amsterdam, 1975.

28. M. M. Rao, Smoothness of Orlicz spaces, Indag. Math. 27 (1965), 671-690.

29. C. Stegall, A proof of the martingale convergence theorem in Banach spaces, Springer Verlag Lecture Notes in Mathematics, 604 (1976), 138-141.

30. L. Sucheston, On existence of finite invariant measures, Math. Zeitschrift, 86 (1964), 327-336.

31. K. Yosida and E. Hewitt, Finitely additive measures, Trans. Amer. Math. Soc., 72 (1952), 46-66.

TAIL PROBABILITIES OF SUMS OF RANDOM VECTORS IN BANACH SPACES, AND RELATED MIXED NORMS

Wojbor A. Woyczyński

Cleveland State University
Cleveland, OH 44115
and
University of South Carolina
Columbia, S.C. 29208

1. Introduction and preliminaries

In the present paper $(E,||.||)$ is a real, separable Banach space, and (X_n) is a sequence of independent random vectors with values in E (strongly measurable). Whenever $E||X|| < \infty$ then EX stands for the Bochner integral. (r_i) will stand for the Rademacher sequence i.e. a sequence of real, independent-random variables with $P(r_i = \pm 1) = 1/2$.

If the array (a_{nk}), $k = 1, \cdots, n$, $n = 1,2,\cdots$ of real numbers is specified then we shall denote

$$(1.1) \qquad S_n = \sum_{K=1}^{n} a_{nk}X_k ,$$

and the goal of this paper is to study the asymptotic behavior of $P(||S_n|| > \varepsilon)$ as $n \to \infty$, under various restrictions on (a_{nk}), (X_k) and the geometry of the space E.

The asymptotic behavior we are looking for is described by the convergence of the series

$$(1.2) \qquad \sum c_n P(||S_n|| > \varepsilon) < \infty$$

where, preferably, (c_n) does not decrease too fast (certainly $n^{-1} = O(c_n)$). The idea of such a study goes back to a Paley's theorem in harmonic analysis which states that if (Φ_n) is an orthogonal system of functions on $[0,1]$, $|\Phi_n(t)| \le 1$, $\forall t$, and $f \in L_p[0,1]$, $1 < p \le 2$, then $\exists$ universal constant M_p such that

$$(1.3) \qquad (\sum n^{p-2}|\hat{f}(n)|^p)^{1/p} \le M_p \, ||f||_{L_p} ,$$

where $\hat{f}(n) = \int \Phi_n(t)f(t)dt$.

The analogues of this result for probability tails of sums of independent, identically distributed real random variables have been studied, among others, by Hsu, Robbins, Erdös, Spitzer, Baum and Katz (cf. e.g. [1] for references) and the typical result asserted that if $X \in L_p$ then $\Sigma n^{p-2} P[|X_1 + \cdots + X_n| > n\varepsilon] < \infty$, $\forall \varepsilon > 0$. However, the inequality of (1.3) type could not be obtained as the Marcinkiewicz interpolation theorem used to prove (1.3) was not applicable here (cf. [2] for some results in this direction). In what follows, we establish analogues of inequality (1.3) using a Lorentz type mixed quasi-norms introduced below.

We also obtain the almost sure convergence results related to a given rate of convergence to zero of $P(||S_n|| > \varepsilon)$. The relationship stems from the fact ([11]) that if $S_n = (X_1 + \cdots + X_n)/a_n$, $0 \le a_n \uparrow$ and X_n's are independent and symmetric then for any Banach space E, $\exists C$ such that

$$(1.4) \qquad ||(S_n)||_{\Lambda_1^*(b_n)} \le C\, ||(S_n)||_{\Lambda_1(c_n)}$$

as long as $b_n, c_n \downarrow 0$ and

$$\sum_{i=1}^{j} 2^i b_{2^i} = 0(2^j c_{2^j}) .$$

Above, and elsewhere in the paper for $p > 0$

$$(1.5) \qquad \Lambda_p^*(b_n) = \{(X_n) : ||(X_n)||_{\Lambda_p^*(b_n)} \overset{df}{=} \sup_{\varepsilon > 0} \varepsilon \Big(\sum_{n=1}^{\infty} b_n P[\sup_{k \ge n} ||X_k|| > \varepsilon]\Big)^{1/p} < \infty\}$$

and

$$(1.6) \qquad \Lambda_p(c_n) = \{(X_n) : ||(X_n)||_{\Lambda_p(b_n)} \overset{df}{=} \sup_{\varepsilon > 0} \varepsilon \Big(\sum_{n=1}^{\infty} a_n P[||X_n|| > \varepsilon]\Big)^{1/p} < \infty\}$$

Both $||.||_{\Lambda_p^*(b_n)}$ and $||.||_{\Lambda_p(c_n)}$ are quasi-norms ($|||.|||$ is a quasi-norm if $|||dx||| = |\alpha|\, |||x|||$, and $|||x+y||| \le A(|||x||| + |||y|||)$ for a certain constant A) and both $\Lambda_p^*(b_n)$ and $\Lambda_p(c_n)$ are complete quasi-normed spaces. Moreover if $p > 1$ one can define Banach-space norms equivalent to $||.||_{\Lambda_p(c_n)}$ in a fashion similar to the Calderon's construction of norms in the classical Lorentz spaces $L^{p,q}$ (cf. [1] p. 182-184, or [7]).

The inequality (1.4), in particular, implies that $\forall q > 1\ \exists\ C > 0$ such that

$$\sum_n n^{-q} P(\sup_{k \ge n} ||S_k|| > \varepsilon) \le C \sum_n n^{-q} P(||S_n|| > \varepsilon) , \quad \forall \varepsilon > 0 ,$$

and also

$$\sum_n n^{-1}P(\sup_{k\geq n}||S_k||>\varepsilon) \leq C\sum_n n^{-1}\log n\; P(||S_n||>\varepsilon)\;,\;\forall\varepsilon>0\;.$$

Notice that here, as in the rest of the paper, the constants A,B,C, ··· need not be equal even when denoted by the same letter. We keep track of them only if absolutely necessary.

We also introduce other Banach spaces of sequences of random vectors

$$(1.7)\quad L_p(a_n) = \{(X_n) : (\sum_{n=1}^{\infty} a_n E||X_n||^p)^{1/p} \overset{df}{=} ||(X_n)||_{L_p(a_n)} < \infty\}\;.$$

We shall also have need of the following geometric properties of the Banach space E (cf. [10] for more details):

<u>Definition</u> 1.1. If F is a (rearrangement invariant) Banach space of real sequences (α_i) with norm $||(\alpha_i)||_F$ then we say that $E \in$ R-type F ($E \in$ R-type p, $0 < p \leq 2$, if $F = \ell_p$) if for each $(x_i) \subset E$ with $(||x_i||) \in F$ the series $\Sigma r_i x_i$ converges almost surely in norm.

<u>Definition</u> 1.2. We shall say that F is finitely representable in E ($F \in$ f.r.E) if $\forall\varepsilon > 0\; \forall n \in \mathbb{N}\; \exists x_1, \cdots, x_n \in E\; \forall\alpha_1, \cdots, \alpha_n \in \mathbb{R}$

$$||(\alpha_i)||_F \leq ||\sum_{i=1}^{n} \alpha_i x_i|| \leq (1+\varepsilon)||(\alpha_i)||_F\;.$$

As usual $||X||_{L_p} = (E\,||X||^p)^{1/p}$, and $||(\alpha_i)||_{\ell_p} = (\Sigma|\alpha_i|^p)^{1/p}$.

2. <u>Inequalities of Marcinkiewicz-Zygmund type</u>.

THEOREM 2.1. Let $q \geq 1$. The following properties of E are equivalent

(i) $E \in$ R-type F,

(ii) $\exists C$ such that for any finite sequence of independent, zero mean (X_i) in E we have

$$E||\sum X_i||^q \leq CE||(||X_i||)||_F^q\;.$$

<u>Proof</u>: (i) $\Rightarrow$ (ii). By Kahane's theorem and the closed graph theorem if $E \in$ R-type F then $\forall q > 0\; \exists C_q$ such that

$$(E||\sum r_i x_i||^q)^{1/q} \leq C^q||(||x_i||)||_F\;.$$

Therefore, if $\tilde{X}_i$ denotes the symmetrization of X_i's then

$$E\|\sum X_i\|^q \le E\|\sum \tilde{X}_i\|^q = E\|\sum_{i=1}^n r_i\tilde{X}_i\|^q$$

$$\le C_q E\|(\|\tilde{X}_i\|)\|_F^q \le CE\|(\|X_i\|)\|_F^q . \qquad \text{Q.E.D.}$$

(ii) $\Rightarrow$ (i). This implication can be obtained using the method of proof of the Prop. 2.1 of [11] or Cor. 3.2 (b).

Now by the straightforward application of the Hölder's inequality we get

COROLLARY 2.1. Let $E \in$ R-type p and $q \ge p$. If $S = \Sigma a_k X_k$, $(a_k) \subset \mathbb{R}$ then

$$\|S\|_{L_q} \le C\|(a_k)\|_{\ell_s}\|(X_k)\|_{L_q(1)}$$

where $s = qp(q-p)^{-1}$, and $C = C(p,q,E)$.

3. Tail probabilities of sums of not necessarily uniformly tight summands and related strong laws.

THEOREM 3.1. Let $E \in$ R-type p, $q \ge 1$. Let $(a_{nk}) \subset \mathbb{R}$, $S_n = \Sigma_{k=1}^n a_{nk}X_k$, $\max_{1\le k\le n} |a_{nk}| = O(n^{-\gamma})$, $\gamma > 0$. Then for any $\alpha > q(1-\gamma p)$

$$\|(S_n)\|_{\Lambda_1(n^{-\alpha})} \le C\|(X_n)\|_{L_{pq}(n^{q(1-\gamma p)-\alpha})}$$

for a constant $C = C(p,q,E,\gamma,d)$.

Proof. By Corollary 2.1 and Chebyshev's inequality

$$\sum_{n=1}^\infty n^{-\alpha}P(\|\sum_{k=1}^n a_{nk}X_k\| > \varepsilon)$$

$$\le \sum_{n=1}^\infty n^{-\alpha}\varepsilon^{-pq}E\|\sum_{k=1}^n a_{nk}X_k\|^{pq}$$

$$\le C\varepsilon^{-pq}\sum_{k=1}^\infty n^{-\alpha}(\sum_{k=1}^n |a_{nk}|^{\frac{qp}{q-1}})^{q-1}(\sum_{k=1}^n E\|X_k\|^{pq})$$

$$\leq C\varepsilon^{-pq} \sum_{n=1}^{\infty} n^{-\alpha-\gamma qp+q-1} \sum_{k=1}^{n} E||X_k||^{pq}$$

$$= C\varepsilon^{-pq} \sum_{k=1}^{\infty} E||X_k||^{pq} \sum_{n=k}^{\infty} n^{-\alpha-\gamma qp+q-1}$$

$$\leq C\varepsilon^{-pq} \sum_{k=1}^{\infty} k^{-\alpha+q(1-\gamma p)} E||X_k||^{pq} \qquad \text{Q.E.D.}$$

Remark 3.1. R-type p of E is also the necessary condition in the above theorem.

In particular one obtains

COROLLARY 3.1. If $\ell_p \notin$ f.r.E and $r > p$, $2 > p \geq 1$, then

$$||(\frac{X_1+\cdots+X_n}{n^{1/p}})||_{\Lambda_1(n^{-1})} \leq C||(X_k)||_{L_r(n^{-r})} .$$

Proof. By Maurey-Pisier theorem (cf.[6] or [10] p. 371) $\exists\, \delta > 0$ such that $E \in$ R-type $(p+\delta)$. Now apply Theorem 3.1 with $d = 1$, $\gamma = 1/p$, p-replaced by $p+\delta$ and let $r = (p+\delta)q$. Now the corollary is immediate if you notice that $(\delta q+p)/[p(p+\delta)q]) \leq 1$. Q.E.D.

Remark 3.2. The above corollary implies that if $\ell_p \notin$ f.r.E and $r > p$ then if $\sup_k E||X_k||^r < \infty$ then

$$\sum n^{-1}P(||X_1+\cdots+X_n|| > n^{1/p}\varepsilon) < \infty , \quad \forall \varepsilon > 0 .$$

However, in this case one can obtain much stronger rate of convergence of tail probabilities (cf. [12], and Section 5 of this paper).

Here is the accompanying result concerning the almost sure convergence to zero of $S_n = \Sigma_{k=1}^{n} a_{nk}X_k$ in the special case when

$$(3.1) \qquad a_{nk} = \frac{w_k}{W_n} , \quad W_n = \sum_{k=1}^{n} w_k , \quad w_k > 0 ,$$

under the additional restriction that

$$(3.2) \qquad \sum_{j=n}^{\infty} d_j^{pq+1} j^{q-1} = O(d_n{}^{pq} n^{q-1})$$

(which certainly is satisfied if $d_n \sim n^{\alpha}$, $\alpha < 1-q$) where

(3.2) $$d_n = \max_{1\le k\le n} w_k/W_k .$$

Notice also that the condition $d_n \to 0$, $n \to \infty$, is equivalent to $w_n/W_n \to 0$ and $\Sigma w_n = \infty$ and therefore is necessary for S_n to converge to 0 a.s.

Remark 3.3. It is also possible to obtain results analogous to Theorem 3.2 with the second half of the condition (3.1) replaced by condition $W_n = (\Sigma_{k=1}^n w_k^{\gamma})^{1/\gamma}$, $0 < \gamma < 2$.

THEOREM 3.2. Let $E \in$ R-type p, $1 \le p \le 2$, $q \ge 1$, and let S_n, (a_{nk}) be as above. Then for any $(X_n) \in L_{pq}(d_n^{pq} n^{q-1})$ we have $S_n \to 0$ a.s.

Proof. By the Renyi-Hajek-Chow's inequality applied to the real submartingale $||\Sigma_{k=1}^n w_k X_k||^{pq}$ we get

$$\varepsilon^{pq} P(\sup_{j\ge n} ||S_n|| > \varepsilon) = \varepsilon^{pq} \lim_{m\to\infty} P(\max_{n\le j\le m} ||S_n||^{pq} > \varepsilon^{pq})$$

$$\le E||S_n||^{pq} + \sum_{j=n+1}^{\infty} W_j^{-pq} E(||\sum_{i=1}^{j} w_i X_i||^{pq} - ||\sum_{i=1}^{j-1} w_i X_i||^{pq}) .$$

By Corollary 2.1

$$E||S_n||^{pq} \le C(\sum_{k=1}^{n} (\frac{w_k}{W_n})^{\frac{pq}{q-1}})^{q-1} \sum_{k=1}^{n} E||X_k||^{pq}$$

$$\le C\, d_n^{pq} n^{q-1} \sum_{k=1}^{n} E||X_k||^{pq} .$$

In view of Kronecker's Lemma and (3.1) and (3.2) we get that $E||S_n||^{pq} \to 0$ as $n \to \infty$. Also the infinite series above converges by Corollary 2.1 and summing by parts we get the following estimate on partial sums of it:

$$\sum_{j=1}^{n} (W_{j-1}^{-pq} - W_j^{-pq}) E||\sum_{i=1}^{j} w_i X_i||^{pq}$$

$$\le \sum_{j=1}^{n} (W_{j-1}^{-pq} - W_j^{-pq}) \max_{1\le k\le j} |w_k|^{pq} j^{q-1} \sum_{i=1}^{j} E||X_i||^{pq}$$

$$= \sum_{i=1}^{n} E||X_i||^{pq} \sum_{j=1}^{n} d_j^{pq} j^{q-1} \left(\left(1 - \frac{w_j}{W_{j-1}}\right)^{pq} - 1\right)$$

$$\leq C_1 \sum_{i=1}^{n} E||X_i||^{pq} \sum_{j=i}^{n} d_j^{pq+1} j^{q-1}$$

$$\leq C_2 \left(d_n^{pq} n^{q-1} \sum_{i=1}^{n} E||X_i||^{pq} + \sum_{i=1}^{n} d_i^{pq} i^{q-1} E||X_i||^{pq}\right)$$

again by (3.2). Since, in view of our assumptions, $\sum_i d_i^{pq} i^{q-1} \times E||X_i||^{pq} < \infty$, we get again by Kronecker's Lemma that $\forall \varepsilon > 0$, $P(\sup_{j \geq n} ||S_j|| > \varepsilon) \to 0$ as $n \to \infty$. Q.E.D.

COROLLARY 3.2 (cf. also [11]). (a) Let $1 \leq p \leq 2$, $E \in$ R-type p, and $q \geq 1$. If (X_n) are independent, zero mean random vectors in E such that

$$\sum_{n=1}^{\infty} \frac{E||X_n||^{pq}}{n^{pq+1-q}} < \infty$$

then $(X_1 + \cdots + X_n)/n \to 0$ a.s. in norm.

(b) Conversely, if $q \geq 1$, $1 \leq p \leq 2$, and for each $(x_i) \subset E$ such that $\Sigma ||x_i||^{pq}/n^{pq+1-q} < \infty$ the sequence $\Sigma_{i=1}^n r_i x_i/n \to 0$, $n \to \infty$, a.s. in norm then $E \in$ R-type p.

<u>Proof</u>. (a) is a special case of Theorem 3.2. We prove (b). Kahane's theorem (e.g. [10], p. 275) states that for any Banach space E and any $0 \leq p < \infty$, all the $L_p(E)$-norms are equivalent on the span of $(r_i x_i)$, $(x_i) \subset E$. Hence, in view of the Closed Graph Theorem $\exists C \, \forall (x_i) \subset E$

$$E|| \sum_{i=1}^{n} r_i x_i n^{-1} || \leq C\left(\sum_{i=1}^{n} ||x_i||^{pq}/i^{pq+1-q}\right)^{1/pq}$$

so that $\forall (x_i) \subset E$

$$E|| \sum_{i=1}^{n} n^{-1} i^{1-(1-q)/(pq)} r_i x_i || \leq C\left(\sum_{i=1}^{n} ||x_i||^{pq}\right)^{1/pq} .$$

Hence

$$E|| \sum_{i=1}^{n} r_i x_i || = E|| \sum_{i=n+1}^{2n} r_i x_{i-n} ||$$

$$\leq n^{-(1-q)/(pq)} E\|\sum_{i=1}^{n} \frac{i^{(1+(1-q)/(pq))}}{2n} r_i x_i + \sum_{i=n+1}^{2n} \frac{i^{(1+(1-q)/(pq))}}{2n} r_i x_i\|$$

$$\leq n^{-(1-q)/(pq)} \, c \, 2^{1/pq} \left(\sum_{i=1}^{n} \|x_i\|^{pq}\right)^{1/pq} ,$$

and by Hölder's inequality we get that

$$E\|\sum_{i=1}^{n} r_i x_i\| \leq C 2^{1/pq} n^{-(1-q)/pq} n^{1/pq-1/p} \left(\sum_{i=1}^{n} \|x_i\|^{p}\right)^{1/p}$$

$$\leq C 2^{1/p} \left(\sum_{i=1}^{n} \|x_i\|^{p}\right)^{1/p} . \qquad \text{Q.E.D.}$$

4. Marcinkiewicz-Zygmund's type strong laws for random vectors with uniformly bounded tails.

From now on $S_n = X_1 + \cdots + X_n$.

Definition 4.1. A sequence (X_i) of random vectors in E is said to have uniformly bounded tail probabilities by tail probabilities of a positive real random variable (in short $(X_i) \prec X_0$) if $\exists C > 0 \ \forall t > 0 \ \forall i \in \mathbb{N}$, $P(\|X_i\| > t) \leq CP(X_0 > t)$.

THEOREM 4.1. Let $1 \leq p < 2$. Then the following properties of a Banach space E are equivalent:

(i) $\ell_p \notin$ f.r.E ,

(ii) For any zero mean, independent, E valued

$$(X_i) \prec X_0 \in \begin{matrix} L_p & \text{if } 1 < p < \infty \\ L \log^+ L & \text{if } p = 1 , \end{matrix}$$

the series $\Sigma_{n=1}^{\infty} X_n/n^{1/p}$ converges a.s. in norm.

(iii) For any sequence (X_i) as in (ii), the sequence $S_n/n^{1/p} \to 0$ a.s.

We just sketch the proof of (i) $\Rightarrow$ (ii) the full version thereof will appear in [11] . (ii) $\Rightarrow$ (iii) follows directly by Kronecker's Lemma and (iii) $\Rightarrow$ (i) is essentially due to Maurey and Pisier [6] (cf. also [11] and [10], p. 389).

Step I. The first step in the proof is to show that if $1 \le p < 2$, $\ell_p \not\subset$ f.r.E and if (X_i) satisfies assumptions of Theorem 4.1(ii) then the series $\Sigma(X_n - EY_n)n^{-1/p}$, where $Y_n = X_n I(||X_n|| \le n^{1/p})$, converges a.s. Since $\Sigma P(X_n \ne Y_n) \le C_1 EX_0^p < \infty$, in view of the Borel-Cantelli Lemma it suffices to show that the series $\Sigma(Y_n - EY_n)/n^{1/p}$ converges a.s. This is accomplished as follows. Let $r > p$. Then

$$\sum_{n=1}^{\infty} E||Y_n - EY_n||^r \le 2^{r+1} \sum_{n=1}^{\infty} n^{-r/p} \int_0^{n^{1/p}} t^r dP(||X_n|| \le t)$$

$$\le C_1 \sum_{n=1}^{\infty} (1 - rn^{-r/p} \int_0^{n^{1/p}} t^{r-1}(1 - P(X_0 > t))dt)$$

$$= C_1 \sum_{n=1}^{\infty} \int_0^1 P(X_0 s^{-1/r} > n^{1/p}) ds$$

$$\le C_2 EX_0^p \int_0^1 s^{-p/r} ds = C_2 \frac{r}{r-p} EX_0^p < \infty .$$

Now, by Maurey-Pisier theorem ([6],[10], p. 371) and by the assumptions it follows that $\exists r > p$ such that $E \in$ R-type r. Therefore, the above estimate and Th. V. 7.5 of [10] gives the desired a.s. convergence of $\Sigma(Y_n - EY_n)n^{-1/p}$.

Step II. In view of Step I it suffices to show the absolute convergence of the series $\Sigma EY_n n^{-1/p}$ for $X_0 \in L_p$ if $1 < p < \infty$ and for $X_0 \in L \log^+ L$ if $p = 1$. If $p > 1$ then

$$\sum_{n=1}^{\infty} ||EY_n|| n^{-1/p} \le - \sum_{n=1}^{\infty} n^{-1/p} \int_{n^{1/p}}^{\infty} t dP(||X_n|| > t)$$

$$= \sum_{n=1}^{\infty} P(X_0 > n^{1/p}) + \int_1^{\infty} P(X_0 s^{-1} > n^{1/p}) ds \le C_1 EX_0^p .$$

If $p = 1$, then, integrating by parts, we get

$$\sum_{n=1}^{\infty} ||EY_n|| n^{-1} \le \sum_{n=1}^{\infty} [P(||X_n|| > n) + n^{-1} \int_n^{\infty} P(||X_n|| > t) dt]$$

$$\le C_1 [EX_0 + \sum_{k=1}^{\infty} \sum_{n=1}^{k} n^{-1} P(X_0 > k)]$$

$$= C_1 [EX_0 + \sum_{k=1}^{\infty} \log k \; P(X_0 > k)]$$

$$\leq C[EX_0 + E(X_0 \log^+ X_0)] < \infty . \qquad \text{Q.E.D.}$$

Remark 4.1. For other properties of $S_n/n^{1/p}$ in infinite dimensional spaces see [5] and [8].

5. Tail probabilities for sums of random vectors with uniformly bounded tails.

The Lemma 5.1 and Theorem 5.1 present, essentially, a slight extension of N. Jain's results obtained in [4] in the case of i.i.d. (X_n). Throughout this section (X_n) are zero-mean (if $E||X_n|| < \infty$), independent and

$$S_n = X_1 + \ldots + X_n , \quad N_k = \max_{1 \leq i \leq k} ||X_i|| , \quad n,k \geq 1 .$$

Now, let $\phi, \psi : \mathbb{R}^+ \to \mathbb{R}^+$ be strictly increasing and ϕ^{-1} be the inverse function of ϕ. Denote

$$\theta = \phi \circ \psi \tag{5.1}$$

and

$$\beta(j) = \theta(j+1) - \theta(j) \tag{5.2}$$

and assume that $\exists C_1, C_2$ such that

$$C_1 \leq C_2 \beta(j+1) \leq \beta(j) , \quad j \geq 1 , \tag{5.3}$$

and that $\phi \in \Delta_2$ i.e. $\phi(2u) \leq C_\phi \phi(u)$, $u \geq 0$. In connection with the Δ_2-condition one observation is in order here. Let $\alpha > 0$. Then for some integer n, $2^{n-1} \leq \alpha \leq 2^n$. Therefore if $\phi \in \Delta_2$ we have

$$\phi(\alpha u) \leq \phi(2^n u) \leq C_\phi{}^n \phi(u) \leq C_\phi{}^{\log_2 \alpha + 1} \phi(u) \tag{5.4}$$

$$= (2\alpha)^{\log_2 C_\phi} \phi(u) \qquad , u > 0 .$$

We shall call

$$\log_2 \sup_u \frac{\phi(2u)}{\phi(u)} = e(\phi) ,$$

the characteristic exponent of ϕ.

For the sake of this section we'll also rephrase the condition $(X_n) \prec X_0 \in L_\phi$, where the Orlicz space $L_\phi = \{X : E_\phi|X| < \infty\}$ under the Δ_2-condition. Define for any random vector X the "non-increasing rearrangement" of its norm:

$$(5.5) \qquad J_X(t) = \inf\{\delta : P(||X|| > \delta) \le t\} , \quad 0 \le t \le 1 .$$

The random variable $||X(.)||$ on (Ω, P) and the function $J_X(.)$ on [0,1] with Lebesgue measure μ, have the same distributions (cf. [7], [1]) so that the existence of X_0, s.t. $(X_n) \prec X_0 \in L_\phi$ is equivalent with the condition

$$(5.6) \qquad \int_0^1 \phi[X_0^*(t)]dt < \infty$$

where

$$X_0^*(t) = \sup_n J_{X_n}(t) , \quad 0 \le t \le 1 .$$

or, in other words, to the condition

$$(5.7) \qquad ||(X_n)||_{L_\phi^\infty} \overset{df}{=} \inf\{\alpha : \int_0^1 \phi[\alpha^{-1}X_0^*(t)]dt \le 1\} < \infty .$$

This gives the inner description of such (X_n) for which a dominating $X_0 \in L_\phi$ exists.

To assure the quasi-normability of L_ϕ^∞ (cf. [13], p. 70) which in turn guarantees the validity of the closed graph theorem in L_ϕ^∞ (cf. [9], II. 6) we shall assume that for some $\delta > 0$

$$(5.8) \qquad \liminf_{u \to \infty} \inf\{\alpha : \frac{\phi(\alpha u)}{\phi(u)} \ge \delta\} > 0 .$$

LEMMA 5.1. Under the above assumptions $\exists C$ such that

$$||(N_k/\psi(k))||_{\Lambda_{e(\phi)}(\beta(k)/k)} \le C||(X_k)||_{L_\phi^\infty}$$

In particular if $(X_n) \prec X_0 \in L^p$, $r \ge 1$, $p > 0$, then

$$\sup_{\varepsilon > 0} \varepsilon(\sum_{n=1}^{\infty} n^{r-2}P(N_n > \varepsilon n^{r/p}))^{1/p} \le C(EX_0{}^p)^{1/p} .$$

Proof.

$$\sup_{\varepsilon>0} \varepsilon^{e(\phi)} \sum_{k=1}^{\infty} \frac{\beta(k)}{k} P(N_k > \varepsilon\psi(k))$$

$$= \sup_{\varepsilon>0} \varepsilon^{e(\phi)} \sum_{k=1}^{\infty} \frac{\beta(k)}{k} [1 - \prod_{\ell=1}^{k} (1 - P(||X_\ell|| > \varepsilon\psi(k)))]$$

$$\le \sup_{\varepsilon>0} \varepsilon^{e(\phi)} \sum_{k=1}^{\infty} \frac{\beta(k)}{k} [1-(1-\mu(X_0^* > \varepsilon\psi(k)))^k]$$

(and, since $1-(1-a)^k \le ka$, $0 \le a \le 1$,)

$$\le \sup_{\varepsilon>0} \varepsilon^{e(\phi)} \sum_{k=1}^{\infty} \beta(k)\mu(X_0^* > \varepsilon\psi(k))$$

(by (5.2))

$$\le C \sup_{\varepsilon>0} \varepsilon^{e(\phi)} \sum_{k=1}^{\infty} \beta(k)\mu(\phi(X_0^*/\varepsilon) > \theta(k))$$

(by (5.3))

$$\le C \sup_{\varepsilon>0} \varepsilon^{e(\phi)} \sum_{k=1}^{\infty} \int_{\theta(k-1)}^{\theta(k)} \mu(\phi(X_0^*/\varepsilon) > t)dt$$

$$= C \sup_{\varepsilon>0} \varepsilon^{e(\phi)} \int_0^{\infty} \mu(\phi(X_0^*/\varepsilon) > t)dt$$

$$= C \sup_{\varepsilon>0} \varepsilon^{e(\phi)} \int_0^{\infty} \phi(X_0^*/\varepsilon)dt \le C2^{e(\phi)} \int_0^1 \phi(X_0^*)dt$$

by (5.4). Under our assumptions (cf. 5.8), $\Lambda_{e(\phi)}(\beta(k)/k)$ and L_ϕ^∞ are complete quasi-normed spaces in the sense of Section 1, so that the closed graph theorem and the above estimates yield Lemma 5.1.

THEOREM 5.1. If there exists $(\gamma_k) \subset \mathbb{R}^+$ such that $(||S_k||/\gamma_k)$ is bounded in probability and

$$\text{(5.9)} \qquad \frac{k \vee \phi(\gamma_k)}{\theta(k)} = O((\log k)^{-\delta} \wedge (\beta(k))^{-\delta})$$

for some $\delta > 0$ then there exists C such that

$$||(S_n/\psi(n))||_{\Lambda_{e(\phi)}(\beta(n)/n)} \le C||(X_n)||_{L_\phi^\infty}.$$

Proof. It suffices to consider the case of symmetric (X_i). Assume $||(X_n)||_{L_\phi^\infty} < \infty$. By Chebyshev's inequality and (5.4)

$$(5.10)\quad (\varepsilon/2)^{e(\phi)}P(||S_n|| > \varepsilon\psi(n)) = (\varepsilon/2)^{e(\phi)}P(\phi||S_n/\varepsilon|| > \theta(n))$$

$$\le \frac{E\phi||S_n||}{\theta(n)}$$

which, by Theorem 3.1 of [4],

$$\le \frac{n\int_0^1 \phi(X_0^*)\,dt}{\theta(n)} + 8AC\phi(\gamma_n) = 0(\frac{n \vee \phi(\gamma_n)}{\theta(n)})) .$$

By Hoffmann-Jørgensen inequality ([3]) $\forall j \;\; \forall t > 0$

$$P(||S_n|| > 3^j t) \le A_j P(N_n > t) + B_j (P(||S_n|| > t))^{2^j} ,$$

so that if we choose an integer r so that $\delta 2^r > 2$, we get, by (5.9) and (5.10)

$$\sup_{\varepsilon > 0} \varepsilon^{e(\phi)} \sum_{n=1}^\infty \frac{\beta(n)}{n} P(||S_n|| > 2^r \varepsilon\psi(n))$$

$$\le A_r \sup_{\varepsilon > 0} \varepsilon^{e(\phi)} \sum_{n=1}^\infty \frac{\beta(n)}{n} P(N_n > \varepsilon\psi(n))$$

$$+ B_r \sum_{n=1}^\infty \frac{\beta(n)}{n}\, 0\,((\log n \vee \beta(n))^{-\delta 2^r}) < \infty$$

because of Lemma 5.1 and the fact that the terms of the second series are of the order $(1/n \log^{1+\varepsilon} n)$ for an $\varepsilon > 0$. Now, again, the standard application of the closed graph theorem gives Theorem 5.1.

Q.E.D.

THEOREM 5.2. Assume $\ell_p \not< f.r.E$, $1 \le p < 2$, and let independent, zero-mean E-valued

$$(X_n) \prec X_0 \in \begin{cases} L_p \cap L_\phi & \text{if } 1 < p < 2 \\ L \log^+ L \cap L_\phi & \text{if } p = 1 , \end{cases}$$

and let

$$\frac{K \vee \phi(k^{1/p})}{\theta(k)} = O((\log k)^{-\delta} \wedge (\beta(k))^{-\delta}) .$$

Then there exist a constant C (depending only on ϕ, ψ, p and E) such that

$$||(S_n/\psi(n))||_{\Lambda_{e(\phi)}(\beta(n)/n)} \leq C||X_0||_{L_\phi} .$$

Proof. By Theorem 4.1 under the above assumptions $S_n/n^{1/p} \to 0$ a.s. so that the sequence $(S_n/n^{1/p})$ is of course bounded in probability. Thus Theorem 5.1 gives the above result. Q.E.D.

Specifying ϕ, ψ one obtains immediately the following corollaries in which the necessity of the condition $\ell_p \notin$ f.r.E follows from Theorem 4.3 in [11].

COROLLARY 5.1. (a) Let E be a Banach space, $1 < p < 2$, and let $\alpha \geq 1/p$. Then $\ell_p \notin$ f.r.E if and only if for each zero-mean, independent E-valued $(X_n) \prec X_0$ and $r \geq 1$

$$\sup_{\varepsilon > 0} \varepsilon\left(\sum_{n=1}^{\infty} n^{r-2} P(||S_n|| > n^{r/p}\varepsilon)\right)^{1/p} \leq C(EX_0^p)^{1/p} .$$

(b) Let E be a Banach space and $1 \leq p < 2$. Then $\ell_p \notin$ f.r.E if and only if for each independent, zero-mean $(X_i) \prec X_0 \in L^p \log^+ L$ we have

$$\sum_{n=1}^{\infty} n^{-1} \log n \; P(||S_n|| > n^{1/p}\varepsilon) < \infty , \quad \varepsilon > 0 .$$

An inequality in (b) similar to that in (a) can also be obtained.

REFERENCES

[1] P. L. Butzer and H. Berens, Semigroups of operators and approximation, Springer-Verlag, Berlin 1967.

[2] Y. S. Chow and T. L. Lai, Paley-type inequalities and convergence rates related to the law of large numbers and extended renewal theory, Z. Wahrscheitilichkeitstheorie verw. Gebiete 45(1978), 1-19.

[3] J. Hoffmann-Jørgensen, Sums of independent Banach-space valued random variables, Studia Math. 52(1974), 159-186.

[4] N. Jain, Tail probabilities for sums of independent Banach space valued random variables, Z. Wahr. verw. Geb. 33(1975), 155-166.

[5] M. B. Marcus and W. A. Woyczynski, Stable measures and central limit theorems in spaces of stable type, Trans. Amer. Math. Soc. 251(1979), 71-102.

[6] B. Maurey and G. Pisier, Series de variables aléatoires vectorielles, independantes et proprietes geometriques des espaces de Banach, Studia Math. 58(1976), 45-90.

[7] E. M. Stein and G. Weiss, Introduction to Fourier Analysis on Euclidean Spaces, Princeton 1971.

[8] K. Sundaresan and W. A. Woyczynski, Laws of large numbers and Beck convexity in metric linear spaces, J. Mult. Analysis (to appear).

[9] K. Yosida, Functional analysis, Berlin 1965.

[10] W. A. Woyczynski, Geometry and martingales in Banach spaces, Part II. Independent increments, Advances in Probability (Dekker) 4(1978), 267-518.

[11] ________________, On Marcinkiewicz-Zygmund laws of large numbers in Banach spaces and related rates of convergence, (to appear).

[12] R. L. Taylor, Convergence of weighted sums of random elements in type p spaces (to appear).

[13] S. Rolewicz, Metric Linear Spaces, Warsaw 1972.

STONE SPACE REPRESENTATION OF VECTOR FUNCTIONS AND OPERATORS ON L^1

Dennis Sentilles

In the course of work with W. H. Graves [2] this writer realized a simply proven Stonean version of the Radon-Nikodym Theorem. Throughout let A denote a σ-complete Boolean algebra carrying a strictly positive countably additive measure μ and let S denote the Stone space of A, with $a \in A$ denoting its clopen counterpart in S as well.

THEOREM 1. If ν is countably additive and bounded on A, then for <u>every</u> $s \in S$ one has that the $\lim\limits_{a \searrow s} \frac{\nu(a)}{\mu(a)} = D_\mu \nu(S)$ exists and is continuous at s, where the limit is taken through the neighborhood filter of clopen sets a containing s.

PROOF. Suppose that $\liminf\limits_{a \searrow s} \frac{\nu(a)}{\mu(a)} < \beta < \limsup\limits_{a \searrow s} \frac{\nu(a)}{\mu(a)}$. Find a_1 containing s such that $(\nu - \beta\mu)(a_1) > 0$. Then find $b_1 \subset a_1$ so that $(\nu - \beta\mu)_{b_1}(c) \equiv (\nu - \beta\mu)(b_1 c) \geq 0$ for all c. Then $s \notin b_1$, because were $s \in b_1$ one could not have $\liminf\limits_{a \searrow s} \frac{\nu(a)}{\mu(a)} < \beta$ since $\frac{\nu(a)}{\mu(a)} \geq \beta$ for all $a \subset b_1$, $s \in a$. Since $s \notin b_1$ pick a_2 so that $s \in a_2$ and $a_2 \cap b_1 = \square$. Find b_2 similarly. Suppose $\{b_\alpha\}_{\alpha<\xi<\omega_1}$ have been so defined with $b_\alpha \cap b_{\alpha'} = \square$ for $\alpha \neq \alpha'$. Then $s \notin \overline{\bigcup\limits_{\alpha<\xi} b_\sigma} = \bigvee\limits_{\alpha<\xi} b_\alpha = b_0$. For if $s \in b_0$, then for any $a \subset b_0$ and $s \in a$, one has $(\nu - \beta\mu)(a) = (\nu - \beta\mu)(b_0 a) = \sum\limits_{\alpha<\xi} (\nu - \beta\mu)(b_\alpha a) \geq 0$, since $\nu - \beta\mu$ is countably additive and ξ is a countable ordinal. Thus find $a_\xi \ni s$, $a_\xi \cap b_0 = \square$ so that $(\nu - \beta\mu)(a_\xi) > 0$ and from this find $b_\xi \subset a_\xi$ so that $(\nu - \beta\mu)_{b_\xi} \geq 0$. By induction then one has pairwise disjoint elements $b_\alpha \in A$ for all $\alpha < \omega_1$ (the first uncountable ordinal) a contradiction to the strict positivity of μ. The proof of the continuity of $D_\mu \nu$ is easier. Find b, c $((D_\mu \nu(s) + \epsilon)\mu - \nu)_b \geq 0$ and $((D_\mu \nu(s) - \epsilon)\mu - \nu)_c \leq 0$. Then $s \in b - c$ and the proof follows.

This result raises two obvious questions to which the remainder of this note is devoted: (1) Does the ability to differentiate ν by μ at <u>every</u> point of S have any significance? (2) Since the above limit exists relative to the entire neighborhood system at $s \in S$ (and not for example "centered" neighborhoods, as in R^2) how do the limit values attained relate to the function values of f where $\nu(E) = \int_E f d\mu$, where $E \in \Sigma$, a σ-algebra on which μ and ν are defined? We first consider (1).

Let (Ω, Σ, μ) be a measure space with μ bounded and positive c.a., and let $A = \Sigma/\mu^{-1}(0)$ with μ also denoting the corresponding strictly positive measure on A. Let X be a Banach space and $T: L^1(\mu) \to X$ a continuous operator. We regard $\chi_B \in L^1(\mu)$ as the measure μ_b on A where $b = B \Delta \mu^{-1}(0)$. For $x' \in X'$ define $DT: S \to X''$ by $\langle x', DT(s)\rangle = \lim_{a \searrow s} \frac{\langle x', T\mu a\rangle}{\mu(a)}$. It is easy to see that $\langle x', T\nu\rangle = \int_S \langle x', DT\rangle d\nu$ where $\nu(E) = \int_E f d\mu$, $E \in \Sigma$, $f \in L^1(\mu)$ and that $||DT(s)|| \leq ||T||$. Finally by Theorem 1, $DT: S \to X''$ is continuous in the X'-topology.

Let $C_0(S, X_\sigma) = \{g: S \to X'': g$ is X'-continuous and $g^{-1}(X''\backslash X)$ is nowhere dense$\}$. Let $C(S, X_\sigma) = \{g \in C_0(S, X_\sigma): g(S) \subset X\}$. Membership in C rather than C_0 can of course only be determined for a function defined on all of S.

THEOREM 2. (a) T is a weakly compact operator iff $DT \in C(S, X_\sigma)$
(b) T is a Bochner-representable operator iff $DT \in C_0(S, X_\sigma)$.

PROOF. (a) This is clear because any $g \in C(S, X_\sigma)$ has weakly compact range and the weak closure of the convex hull of a weakly compact set in X remains in X and is weakly compact.

(b) If $h \in L^\infty(X, \Omega, \Sigma, \mu)$, then h is uniformly approximately $\sum_{n=1}^{\infty} x_n \chi_{A_n}$, $x_n \in X$ and A_n p.w.d. and $\cup A_n = \Omega$. Define

$$g: \bigcup_{n=1}^{\infty} a_n \subset S \to X \text{ by}$$

$g(s) = \sum_{n=1}^{\infty} x_n \chi_{a_n}(s)$. Since $\cup a_n$ is dense and open in S and S is extremally disconnected, one can weakly extend g to a weakly continuous function $g: S \to X''$ with $g^{-1}(X''\backslash X) \subset S\backslash \cup a_n$. Hence $g \in C_0(S, X_\sigma)$. If $Tf = \int fh d\mu$, then $DT = g$, noting that $C_0(S, X_\sigma)$ is closed under uniform convergence. The converse, that any $g \in C_0(S, X_\sigma)$ arises from a Bochner function h is deeper and depends on the following Stonian version of the Dunford-Pettis-Phillips Theorem:

THEOREM 3. (a) If $f \in C_0(S, X_\sigma)$, then $f(S) \cap X$ is norm separable.
(b) There is a closed nowhere dense set $M \subset S$ such that $f|S\backslash M$ is norm continuous.

The proof of (a) follows from a consideration of the metric $d(x'f, y'f) = \int_S |x'f - y'f| d\mu$ on the set $\{x'f: ||x'|| \leq 1\}$.
We point out that following our work, W. Sachermeyer [3] was able to

prove (a) using dentability and ordinal induction, but assuming only the countable chain condition and not a strictly positive μ, thus achieving a strengthening of Dunford-Pettis-Phillips.

For (b), if $\overline{\{x_n\}} \supset f(S) \cap X$, the functions $||f(\cdot) - x_n|| = \sup\{|\langle x', f(\cdot) - x_n\rangle| : ||x'|| \le 1\}$ are lower semi-continuous, hence continuous on all but a nowhere dense set M_n. Norm continuity on $S \setminus \cup M_n$ then follows, and with μ strictly positive, $\cup M_n$ is also nowhere dense.

The proof of 2(b) can now be completed by observing that if $f \in C_0(S, X_\sigma)$ and $a \supset M$, then $f(S \setminus a) \cap X$ is norm compact so that $f\chi_{S\setminus a} \simeq \sum_{n=1}^{P} x_n \chi_{a_n}$.

One can also go on to show that $DT(s) = ||\ ||\text{-}\lim_{a \searrow s} \frac{T\mu_a}{\mu(a)}$ outside a nowhere dense set. Note of course that the distinction between 2(a) and 2(b) demands the ability to differentiate at <u>every</u> point. It would perhaps be useful to characterize Dunford-Pettis operators and Pettis integrable functions along the same lines, but I have not been able to achieve any results.

Let now μ denote Lebesgue measure on $[0,1]$, and Σ the Lebesgue σ-algebra with $A = \Sigma/\mu^{-1}(0)$. We turn to the second of the two questions above. For $\omega \in [0,1]$ let $d(\omega,b)$ denote the Lebesgue density of B at ω where $b = B \,\Delta\, \mu^{-1}(0)$, if it exists. Then set $I_\omega = \{a \in A: d(\omega,a) = 0\}$ and $C_\omega = \{s \in S: a \in I_\omega \Rightarrow s \notin a\}$. The Lebesgue Density Theorem allows one to prove

COROLLARY 4. (a) C_ω is closed and nowhere dense in S.
(b) $d(\omega,a) = 0$ iff $a \cap C_\omega = \square$
(c) $d(\omega,a) = 1$ iff $C_\omega \subset a$.

The essence of the matter is as follows: If $\nu(E) = \int_E f d\mu$ on $[0,1]$, then $\lim_{\delta \searrow 0} \frac{1}{2\delta} \int_{\omega-\delta}^{\omega+\delta} |f(t) - k| d\mu(t) = 0$ iff the corresponding function $\hat{f}$ on S is constant on the set C_ω, with constant value k. We state however, a vector version:

THEOREM 5. Let $g \in L^\infty(X, \Omega, \Sigma, \mu)$ and let $\nu(E) = \text{Bochner} - \int_E g d\mu$ and let $\hat{g} = D_\mu \nu \in C_0(S, X_\sigma)$ be defined by weak differentiation of ν by μ. Then (a) $\hat{g}$ is norm continuous on C_ω, a.e.ω.
(b) If (a) holds and $\hat{g}|C_\omega = x_\omega$, then

(*) $\lim_{\Delta \searrow \omega} \frac{1}{\mu(\Delta)} \int_{\Delta} ||g(t) - x_{\omega}||d\mu(t) = 0$, where Δ is an open interval containing ω.

(c) If (*) holds, then $\hat{g}|C_{\omega} = x_{\omega}$.

The proof follows from the inequality

$$\frac{1}{\mu(\Delta)} \int_{\Delta} ||g - x_{\omega}||d\mu \le [1 - \frac{\mu(a\cap\Delta)}{\mu(\Delta)}](||g||_{\infty} + ||x_{\omega}||)$$

$$+ \frac{1}{\mu(\Delta)} \int_{a\cap\Delta} ||\hat{g} - x_{\omega}||d\mu$$

which holds for all a and Δ, and 4(c) and the norm continuity of $\hat{g}$ on C_{ω}. (Here Δ also denotes its clopen counterpart in S.)

We close with some observations about lifting $L^{\infty}(X, \Omega, \Sigma, \mu)$. Let $\rho: \Omega = [0,1] \to \cup C_{\omega}$ such that $\rho(\omega) \in C_{\omega}$. It follows that $(\hat{g}\circ\rho)(\omega) = g(\omega)$ a.e. ω, so that $\hat{g}\circ\rho$ is a lifting of L^{∞} which is X valued except on a set of measure zero (corresponding to the nowhere dense set $\hat{g}^{-1}(X''\setminus X)$). For X the real line, $\hat{g}\circ\rho$ lifts $L^{\infty}(\mu)$. It is not known whether one can find a ρ so that $\hat{g}\circ\rho(\omega) \in X$ for all ω and all $g \in L^{\infty}(X, \mu)$.

References

1. J. Diestel and J. J. Uhl, Vector measures, A.M.S. Surveys, No. 15, Providence, 1977.
2. W, H, Graves and D. Sentilles, The extension and completion of the universal measure and the dual of the space of measures, J. Math. Anal. and Applic., 68, No. 1(1979), 228-64.
3. W. Schachermeyer, On a paper of D. Sentilles, preprint.
4. D. Sentilles, Differentiation of measures on Boolean algebras: Lifting and Radon-Nikodym, preprint.

AN ISOMORPHISM THEOREM AND RELATED QUESTIONS.

S. Tomášek

Introduction. The purpose of this paper is to prove a general isomorphism theorem (Theorem 1) which generalizes the structural assertion of the Fubini theorem and two Grothendiecks theorems of Fubini type; especially, the stated theorem yields a suitable tool for simplification (i.e., the theory of integration has been essentially eliminated) of proofs of those theorems which describe the structure of the complete tensor products $L^1(\mu)\hat{\otimes}E$.

1. The Isomorphism Theorem.

A class $\mathcal{C}$ of topological vector spaces will be called regular, if it holds:

(a) if $H \in \mathcal{C}$ and if G is a linear subspace of H with the induced topology, then $G \in \mathcal{C}$;

(b) the completion $\hat{G}$ of any G in $\mathcal{C}$ is also in $\mathcal{C}$;

(c) for any $G \in \mathcal{C}$ the associated and separated quotient space $G/(\bar{0})$ with the usual quotient topology is in $\mathcal{C}$.

Let E , F and G be arbitrary vector spaces in $\mathcal{C}$ and let u denote a bilinear mapping of $E \times F$ into G ; $\tilde{u}$ means the corresponding associated linear mapping $\tilde{u} : E \otimes F \to G$. By $T_{\mathcal{C}}$ we understand the projective tensor topology in $E \otimes F$ defined by the family of all linear mappings $\tilde{u} : E \otimes F \to G$, G arbitrary in $\mathcal{C}$.

The tensor product $E \otimes F$ topologized in such a manner will be denoted by $E \otimes_{\mathcal{C}} F$.

Examples. 1. If E and F are normed and if $\mathcal{C}$ is the class of all normed spaces, then the topology $T_{\mathcal{C}}$ is defined by the cross-norm: $z \longrightarrow \|z\|_{\gamma}$.

2. Let E and F be two locally convex (locally p-convex, $0<p\leq 1$) spaces. Then $T_{\mathcal{C}}$ is identical with the projective tensor locally convex (locally p-convex) topology.

3. If $\mathcal{E}$ is the class of all topological vector spaces, then $T_{\mathcal{E}}$ is the W-topology (cf. [9]).

Lemma 1. Let E' be a dense linear subspace of E, F' a dense linear subspace of F. If E and F are separated and complete, then $E' \otimes_{\mathcal{C}} F'$ is a topological subspace of $E \otimes_{\mathcal{C}} F$.

The proof of this statement is quite elementary and may be carried out with the aid of the theory explained in [2].

Lemma 2. Let $\mathcal{C}$ be a regular class of topological vector spaces. If E' and F', respectively, are two dense linear subspaces of E and F, respectively, then also $E' \otimes F'$ is dense in $E \otimes_{\mathcal{C}} F$.

Proof. Because $T_{\mathcal{C}} \leq T_{\mathcal{E}}$, where $\mathcal{E}$ denotes the class of all topological vector spaces, it suffices to prove the statement of the Lemma 2 for the topology $T_{\mathcal{E}}$ only. Let $\Omega = \Omega((U_n),(V_n))$ be an arbitrary neighborhood for the topology $T_{\mathcal{E}}$ and let $z = x_1 \otimes y_1 + \dots + x_n \otimes y_n$ be an arbitrary element in $E \otimes F$. Recall that Ω consists of all $\sum u_i \otimes v_i \in E \otimes F$ with $u_i \in U_i$, $v_i \in V_i$ / cf. [9]/. Let us take two sequences (U'_n) and (V'_n) of neighborhoods in E and F with $U'_i \subseteq U_i$, $V'_i \subseteq V_i$ for each $i \in N$ and satisfying

$$(1) \qquad 2\cdot U'_i \subseteq U'_{i-1}, \quad 2\cdot V'_i \subseteq V'_{i-1} \quad (i = 2, 3, \dots)$$

Let for the integer m, $m > 2\cdot n$, the following conditions are satisfied

(2) $$2^{-m}\cdot y_i \in V_i \ , \quad 2^{n-m}\cdot x_i \in U_{i+n} \qquad (1 \leq i \leq n)$$

If $x_i' \in E'$, $y_i' \in F'$ $(1 \leq i \leq n)$ satisfy

$$x_i' - x_i \in U'_{m+i} \ , \quad y_i' - y_i \in V'_{m+i} \ ,$$

then we obtain for some $u_i \in U'_{m+i}$, $v_i \in V'_{m+i}$ $(1 \leq i \leq n)$

$$x_i' = x_i + u_i \ , \quad y_i' = y_i + v_i \qquad (1 \leq i \leq n) \ ,$$

hence

$$z' = \sum x_i' \otimes y_i' = \sum x_i \otimes y_i + \sum u_i \otimes y_i + \sum x_i \otimes v_i + \sum u_i \otimes v_i \ ,$$

consequently

$$z' \in z + \sum_{i=1}^{m} U'_{m+i} \otimes y_i + \sum_{i=1}^{m} x_i \otimes V'_{m+i} + \sum_{i=1}^{m} U'_{m+i} \otimes V'_{m+i}$$

From (1) it follows $(1 \leq i \leq n)$:

$$2^{m}\cdot U'_{m+i} \subseteq U_i \ , \quad 2^{m-n} V'_{i+m} \subseteq V_{i+n}$$

With the aid of (2) we now obtain

$$z' \in z + \sum_{i=1}^{m} U_i \otimes V_i + \sum_{i=1}^{m} U_{i+n} \otimes V_{i+n} + \sum_{i=1}^{m} U_{m+i} \otimes V_{m+i} \ ,$$

hence $z' \in z + \Omega$. Since $z' \in E' \otimes F'$, the proof is complete.

Recall that by $E \hat{\otimes} F$ we understand the completion of a separated topological tensor product $E \otimes F$.

Theorem 1. (The Isomorphism Theorem).

Let $\mathcal{C}$ be a regular class of topological vector spaces. Let E and F be two separated topological vector spaces in the class $\mathcal{C}$. Then the topological tensor products $(\hat{E}) \hat{\otimes}_{\mathcal{C}} (\hat{F})$ and $E \hat{\otimes}_{\mathcal{C}} F$ are (canonically) topologically isomorphic provided that $\hat{E} \otimes_{\mathcal{C}} \hat{F}$ is separated and in $\mathcal{C}$. If $\mathcal{C}$ is the class of all normed spaces, then this topological isomorphism is a linear isometry.

Proof. According to the Lemma 1 and 2 the space $E \otimes_{\mathcal{C}} F$ is a dense topological subspace in $\hat{E} \otimes_{\mathcal{C}} \hat{F}$. The canonical bilinear and continuous transformation $\varphi : E \times F \longrightarrow E \hat{\otimes}_{\mathcal{C}} F$, $\varphi(x,y) = x \otimes y$, admits a bilinear and continuous extension on the Cartesian product $\hat{E} \times \hat{F}$. Hence, there exists a linear and continuous extension I of φ , I : $(\hat{E}) \hat{\otimes}_{\mathcal{C}} (\hat{F}) \longrightarrow E \hat{\otimes}_{\mathcal{C}} F$. Let J be the injective mapping $J : E \hat{\otimes} F \rightarrow$

$\longrightarrow (\hat{E}) \hat{\otimes}_{\mathcal{C}} (\hat{F})$. It is now evident that I and J are mutually inverse. The rest of the proof is now clear.

Remarks. (1) If $\mathcal{C}$ denotes the class of all normed (locally convex; locally p-convex, $0 < p \leq 1$; topological vector) spaces, then $\hat{E} \otimes_{\mathcal{C}} \hat{F}$ is evidently also in $\mathcal{C}$ whenever E and F satisfy the first conditions of the Theorem 1 .

(2) If $\mathcal{C}$ means the class of all locally convex spaces and if E and F are separated, then from the duality theory it follows that $E \otimes_{\mathcal{C}} F$ is separated. In general, the problem of the separatedness of the tensor product $\hat{E} \otimes_{\mathcal{C}} \hat{F}$ rests open (cf. [8] for the W-topology) .

2. Two Theorems of Fubini Type.

Let X and Y be two locally compact spaces, μ and ν positive measures on X and Y . Under $\mathcal{E}(X)$, $\mathcal{E}(Y)$ and $\mathcal{E}(X \times Y)$ we understand the vector spaces of all step functions with respect to the ring of all integrable subsets determined by μ , ν and $\mu \otimes \nu$. The corresponding Banach spaces of all integrable functions (exactly: of all classes of integrable functions) with the usual topologies will be denoted by

$$L^1(\mu) = L^1(X,\mu) \quad , \quad L^1(\nu) = L^1(Y,\nu) \quad , \quad L^1(\mu \otimes \nu) = L^1(X \times Y, \mu \otimes \nu)$$

It is a well-known rudiment of the elementary measure theory that it holds (cf. [1]):

$$\hat{\mathcal{E}}(X) = L^1(\mu) \ , \quad \hat{\mathcal{E}}(Y) = L^1(\nu) \ , \quad \hat{\mathcal{E}}(X \times Y) = L^1(\mu \otimes \nu)$$

Lemma 1. The tensor product

$$\mathcal{E}(X) \otimes \mathcal{E}(Y)$$

is dense in $\mathcal{E}(X \times Y)$ under the topology induced by $L^1(\mu \otimes \nu)$.

Proof. Let $\varphi_K \in \mathcal{E}(X \times Y)$ be a characteristic function of a compact set K in $X \times Y$. Evidently, K may be arbitrarily approximated by open and inegrable subsets U in $X \times Y$ with $K \subseteq U$. For any such U there is a finite cover

$$K \subseteq \bigcup_{i=1}^{n} (U_i \times V_i) \subseteq U$$

with relatively compact and open subsets U_i and V_i $(1 \leq i \leq n)$ in X and Y. Further, it may be, without loss of the generality, supposed that $U_i \times V_i$ $(1 \leq i \leq n)$ are mutually disjoint and integrable (but not necessarily open). Consequently, the characteristic function φ_K may be arbitrarily approximated by the step functions $\sum \varphi_{U_i} \otimes \varphi_{V_i}$ in $\mathcal{E}(X) \otimes \mathcal{E}(Y)$. In general, any step function $f \in \mathcal{E}(X \times Y)$ being a finite linear combination of characteristic functions with respect to integrable subsets in $X \times Y$ may be approximated by functions of $\mathcal{E}(X) \otimes \mathcal{E}(Y)$ in the topology induced by $L^1(\mu \otimes \nu)$, because to any integrable subset $A \subseteq X \times Y$ there are a compact K and an open U such that $K \subseteq A \subseteq U$ and $\mu \otimes \nu\,(U - K)$ is arbitrarily small.

<u>Remark</u>. If $L^1(\mu)$ and $L^1(\nu)$ are defined by measure spaces $(X, \mathcal{S}, \mu)$ and $(Y, \mathcal{S}', \nu)$ (cf. [5]), then the statement of Lemma 1 is in the main contained in the elementary measure theory (cf. [7], p. 328).

<u>Lemma 2</u>. The space $\mathcal{E}(X) \otimes \mathcal{E}(Y)$ with the locally convex projective tensor topology is a topological subspace in $\mathcal{E}(X \times Y)$ under the topology induced by $L^1(\mu \otimes \nu)$.

Proof. Denote by $\tilde{I}$ the linear mapping of $\mathcal{E}(X) \otimes_\gamma \mathcal{E}(Y)$ (with the cross-norm) into $\mathcal{E}(X \times Y)$ defined by $I(f \otimes g) = f \cdot g$. Let h be an arbitrary function in $\mathcal{E}(X) \otimes \mathcal{E}(Y)$,

$$h = \sum_{i=1}^{m} \lambda_i \varphi_{A_i} \otimes \varphi_{B_i}$$

where A_i and B_i are integrable; evidently it may be supposed that $A_i \times B_i$ are mutually disjoint. Then it holds

$$\|h\|_\gamma = \sum_{i=1}^{m} |\lambda_i| \mu(A_i) \cdot \nu(B_i) = \int |\tilde{I}(h)| \, d(\mu \otimes \nu) = \|\tilde{I}(h)\|$$

Consequently, $\tilde{I}$ is an isometry. This completes the proof.

Since $\mathcal{E}(X) \otimes_\gamma \mathcal{E}(Y)$ is a dense subspace in $L^1(\mu \otimes \nu)$, we have

$$\mathcal{E}(X) \hat{\otimes}_\gamma \mathcal{E}(Y) = L^1(\mu \otimes \nu)$$

From The Isomorphism Theorem we may now conclude

Theorem 2 (Grothendieck) Assume X and Y to be two locally compact spaces, μ and ν two positive measures on X and Y. Then the spaces $L^1(\mu) \hat{\otimes} L^1(\nu)$ and $L^1(\mu \otimes \nu)$ are isometric.

Let X be locally compact, μ a positive measure on X. The tensor product $\mathcal{E}(\Phi) \otimes F$ of the space $\mathcal{E}(\Phi)$ of all step functions introduced above with a Banach (or Fréchet) space F is a dense subset in the space L^1_F of all integrable functions on X with values in F. Let be $h = \sum_i \varphi_{A_i} \otimes a_i \in \mathcal{E}(\Phi) \otimes F$, A_i disjoint. It is immediate that for a Banach space F (for a Fréchet space F the proof is in the main the same) the cross-norm $\|h\|_\gamma$ in $\mathcal{E}(\Phi) \otimes F$ satisfies

$$\|h\|_\gamma = \sum \|a_i\| \cdot \mu(A_i)$$

Hence, $\mathcal{E}(\Phi) \otimes_\gamma F$ is a dense topological subspace in L^1_F. We have established

Theorem 3 (Grothendieck, cf. [3], [4]). Let X be a locally compact space, μ a positive measure on X, F a Fréchet space. Then $L^1 \hat{\otimes} F$ is topologically isomorphic with L^1_F.

Remark. Using the same denotation as in Theorem 2 and taking into account the statement of Theorem 3 we obtain

$$L^1(\mu \otimes \nu) = L^1_{L^1(\nu)}(\mu).$$

This identity represents the structural assertion of the Fubini theorem (cf. [1], [5] etc.).

3. Tensor Product with a Köthe Space.

Suppose X to be a locally compact space with a positive measure μ. Consider the space of all systems $x = (x_\alpha ; \alpha \in A)$, x_α scalars, A a set of arbitrary cardinality. Let B be a set of indices and $\Lambda = (\lambda_{\alpha\beta} ; \alpha \in A, \beta \in B)$ a collection of positive real numbers. Under $K(\Lambda)$ we mean the Köthe space of all $x = (x_\alpha ; \alpha \in A)$ with

$$p_\beta(x) = \sum_{\alpha \in A} |x_\alpha| \cdot \lambda_{\alpha\beta} < \infty \quad , \beta \in B$$

Recall that $K(\Lambda)$ under the topology defined by the seminorms p_β , $\beta \in B$, is separated and complete. Denote by e_α , $\alpha \in A$, this point of $K(\Lambda)$ whose coordinates equal to zero except the α- coordinate which equals to the unit. Under $K_0(\Lambda)$ we shall now mean the subspace of $K(\Lambda)$ algebraically generated by $(e_\alpha ; \alpha \in A)$.

Let $K'(\Lambda)$ be the linear subspace in $\prod_{\alpha \in A} L^1_\alpha(\mu)$, $L^1_\alpha(\mu) = L^1(\mu)$ for each $\alpha \in A$, of all $(f_\alpha ; \alpha \in A)$ with the properties:

(α) $f_\alpha = 0$ for any $\alpha \in A$ except of a countable number of indices;

(β) for each $\beta \in B$ it holds

$$p'_\beta(f_\alpha ; \alpha \in A) = \int \sum_\alpha |f_\alpha(t)| \lambda_{\alpha\beta} \, d\mu(t) < \infty$$

The space $K'(\Lambda)$ will be topologized by the seminorms $(p'_\beta ; \beta \in B)$. $K'_0(\Lambda)$ stands for the subspace of $K'(\Lambda)$ algebraically identical with the direct sum $\sum_{\alpha \in A} L^1_\alpha(\mu)$. It holds

<u>Lemma</u>. (a) The space $K'(\Lambda)$ is separated.

(b) The space $K'_0(\Lambda)$ is dense in $K'(\Lambda)$.

(c) If B is countable, then $K'(\Lambda)$ is a Fréchet space.

Proof. The statements (a) and (b) are evident. To prove (c), it suffices to prove that $K'(\Lambda)$ is complete whenever B is countable. Let us briefly outline the idea of this proof. Firstly, we embed

$K'(\Lambda)$ into $\mathcal{F}_F^1/\mathcal{N}_F$, where $F = K(\Lambda)$ (for the denotation we refer to [1]): if $(f_\alpha ; \alpha \in A) \in K'(\Lambda)$, then its image in $\mathcal{F}_F^1/\mathcal{N}_F$ is the class of the last space cantaining $(f'_\alpha ; \alpha \in A) \in \mathcal{F}_F^1$, where $f'_\alpha \in \mathcal{L}^1(\mu)$ is a function representing the class $f_\alpha \in L^1(\mu)$. It is easy to prove that this correspondence is a topological isomorphism. Since $\mathcal{F}_F^1/\mathcal{N}_F$ is complete, the proof will be carried out, if we establish that the image of $K'(\Lambda)$ is closed in $\mathcal{F}_F^1/\mathcal{N}_F$.

Theorem 4. It holds

(a) The space $L^1(\mu)\hat{\otimes} K_o(\Lambda)$ with the projective topology is topologically isomorphic with $\hat{K}'_o(\Lambda)$ and with $\hat{K}'(\Lambda)$.

(b) If B is countable, then the space $L^1(\mu)\hat{\otimes} K(\Lambda)$ is topologically isomorphic with $K'(\Lambda)$.

Proof. Let $\mathcal{E}(X)$ be the dense subspace in $L^1(\mu)$ of all step functions of integrable subsets. It suffices to prove that $\mathcal{E}(X) \otimes K_o(\Lambda)$ with the locally convex projective topology is a topological subspace of $K'(\Lambda)$. Take

$$h = \sum_{i=1}^{m} f_{\alpha_i} \otimes e_{\alpha_i} \in \mathcal{E}(X) \otimes K_o(\Lambda)$$

and denote by $I(h)$ the image of h in $K'(\Lambda)$. It may be supposed, without loss of the generality, that $f_{\alpha_i} = \sum_{j=1}^{m} \mu_{ij}\varphi_{A_j}$, A_j mutually disjoint. Then

$$h = \sum_{j=1}^{m} (\varphi_{A_j} \otimes \sum_{i=1}^{m} \mu_{ij} e_{\alpha_i}) = \sum_{j=1}^{m} (\varphi_{A_j} \otimes x_j),$$

where $x_j = \sum_{i=1}^{m} \mu_{ij} e_{\alpha_i}$. Thus we obtain

$$p'_\beta(I(h)) = \int \sum_{i=1}^{m} |f_{\alpha_i}(t)|\, \lambda_{\alpha_i\beta}\, d\mu(t) = \int \sum_{i=1}^{m} |\sum_{j=1}^{m} \mu_{ij}\varphi_{A_j}|\lambda_{\alpha_i\beta}\, d\mu(t) =$$

$$= \int \sum_{j=1}^{m} (\sum_{i=1}^{m} |\mu_{ij}|\, \lambda_{\alpha_i\beta} \cdot \varphi_{A_j}\, d\mu(t) = \sum_{j=1}^{m} \mu(A_j)\cdot p_\beta(x_j) = \| h \|_\gamma^\beta$$

where the norm of the last expression is the cross - norm defined by $\mathcal{E}(X)$ and by $K_o(\Lambda)$, $K_o(\Lambda)$ with the seminorm p_β. The proof is complete.

Let $\mathfrak{Z}$ be the Fréchet space of all holomorphic functions on the entire open complex plane under the topology of compact convergence. If $f \in \mathfrak{Z}$, $f(z) = \sum_{n=0}^{\infty} a_n \cdot z^n$, then the correspondence $f \longleftrightarrow (a_n ; n = 0, 1, 2, \dots)$ is an algebraical isomorphism into the linear space of all sequences. We define the "gestuften Raum" as a Köthe space (cf. [6]) by means of a countable system Λ_0 of sequences $\lambda_{mn} = m^n$ ($n = 0,1,2,\dots$; $m = 1,2,\dots$). The correspondence $f \longleftrightarrow (a_n ; n = 0,1,2,\dots)$ is a topological isomorphism between the spaces $K(\Lambda_0)$ and $\mathfrak{Z}$.

Let $\mathfrak{Z}_{L^1(\mu)}$ be the space of all series

$$\{f_n\} = \sum_{n=1}^{\infty} f_n(t) \cdot z^n \quad ,$$

where $f_n \in L^1(\mu)$ ($n = 0,1,2,\dots$) satisfy the inequalities ($m = 1,2,\dots$)

$$p_m(\{f_m\}) = \int \sum_{n=1}^{\infty} |f_n(t)| \cdot m^n \cdot d\mu(t) < \infty$$

topologozed by the seminorms p_m ($m = 1,2,\dots$).

Corollary. The completed tensor product $L^1(\mu) \hat{\otimes} \mathfrak{Z}$ with the projective topology is topologically isomorphic with the space $\mathfrak{Z}_{L^1(\mu)}$.

Remark. Making use of the Theorem 1 we could also directly describe the structure of the complete tensor products $L^1(\mu) \hat{\otimes}_\gamma l^p(I)$, $1 \leq p < \infty$, $L^1(\mu) \hat{\otimes} \mathfrak{H}$, where $\mathfrak{H}$ is a Hilbert space, or, of the space $L^1(\mu) \hat{\otimes} c_0(I)$, where $c_0(I)$ is the space of all continuous functions vanishing at the infinity on the discrete space I.

B i b l i o g r a p h y

[1] Bourbaki, N.: Éléments de mathématique, Livre VI, Intégration, Chap. I - IV, Hermann, Paris, 1952

[2] Bourbaki, N.: Élément de mathématique, Livre III, Topologie générale, Hermann, Paris, 1951

[3] Grothendieck, A.: Produits tensoriels topologiques et espaces nucléaires. Mem. Amer. Math. Soc. 16, 1955

[4] Grothendieck, A.: La théorie de Fredholm. Bull. Soc. Math. France 84 (1956) , 319 - 384

[5] Halmos, P. R.: Measure Theory, New York, 1950

[6] Köthe, G.: Die Stufenräume, eine einfache Klasse linearer vollkommener Räume. Math. Z. 51 (1948), 317 - 345

[7] Sikorski, R.: Funkcje rzeczywiste. I. (Real-valued functions) Warszawa, 1958

[8] Tomášek, S.: Projectively generated topologies on tensor products. CMUC 11 (1970), 745- 768

[9] Tomášek, S.: Some remarks on tensor products. CMUC 6 (1965) 85 - 96 .

LOCAL FUNCTIONALS

M. M. Rao*

University of California, Riverside

1. Introduction. A representation theory of certain nonlinear functionals, as an extension of the linear case, has its origins from at least the 1930's---the earliest one seems to be due to Pinsker [31] under the name "partially additive functionals" including the local functionals to be described below. However, neither this theory nor its applications appears to have been noticed by others for the next two-and-a-half decades. Then the study began independently by Tichonov and Arsenin [38], and Chacón and Friedman [7] on the one hand, and by Martin and Mizel [26] on the other. Actually a representation of local functionals was needed in the study of generalized random processes, and this under the present name "local functional" has been noted and called for its solution by Gel'fand and Vilenkin [16]. Apparently the earlier interesting work of Kantarovitch and Pinsker ([18],[19],[31],[32]), even though it was briefly mentioned in [20], has been overlooked by all people who started to attack these problems in the 1960's. A study of these local functionals under different names, motivated by different applications (additive functionals, orthogonally or disjointly additive functions, etc.), is being actively pursued in two different directions by Friedman and his associates, and Mizel and his associates, among others.

From another point of view, motivated by the study of a representation of modular functions, Orlicz and his colleagues have also contributed much to this theory in the 1960's, and clarified the ultimate relation as being of the Lebesgue-Nikodým type. All the studies noted above may be arranged into two classes, based broadly on their origins. One, these functional representations are needed in the structure theory of characteristic functionals of generalized random processes with independent values ([16],[39]) and two, in the study of nonlinear differential equations and in the work of continuum mechanics (cf. [28],[29]).

In the next two sections an outline of the completed work on the representation of these functionals on different spaces of scalar and

---

*Prepared with a partial support of the NSF Grant, MCS76-15544A01 and a UCR sabbatical.

vector valued functions will be given. The fourth section contains in more detail the work on these functionals for probability theory, and their role there. The final section indicates a discussion of new problems and related remarks.

2. Local functionals on scalar function spaces. Let $\mathfrak{F}$ be a topological vector space of scalar functions on a set S. Then a mapping $\ell:\mathfrak{F}\to\mathbb{R}$ or $\mathbb{C}$ is called a local functional if $\ell(f_1+f_2) = \ell(f_1)+\ell(f_2)$ for all $f_i\in\mathfrak{F}$ with $f_1\cdot f_2 = 0$, and ℓ satisfies a suitable continuity condition. If $\mathfrak{F}$ is an abstract vector lattice (or a semi-ordered space) with a topology, then the same concept is meaningful if $f_1\cdot f_2 = 0$ is replaced by $\inf(|f_1|,|f_2|) = 0$. With this interpretation such functionals were termed partially additive by Kantorovich and Pinsker ([18],[19],[31]), orthogonally additive by Orlicz and Drenowski [9], additive functionals by Chacón, Friedman and Katz ([7],[12]) and disjointly additive by Mizel and his associates ([25],[26]). When $\mathfrak{F}$ is the Schwartz space $\mathcal{K}$ of infinitely differentiable real functions on $\mathbb{R}$ vanishing off compact sets, the term "local functional" was introduced by Gel'fand and Vilenkin ([16], p. 275) and was adopted by the author in his early work [33] and thereafter. Most of these names seem to originate from specific applications. Since probabilistic considerations are the main motivation here (and thus far these are the most known consequences of the theory) and since "local functional" was used in that context by Gel'fand in the 1950's, it will be utilized in what follows.

If $\mathfrak{F}$ is an $L^p(\mu)$-space, or an Orlicz or a Sobolev space, then the line of attack of the problem uses the existence of the base measure μ and the hypothesis on the local functional ℓ can be weakened rather than when no such base measure is at hand. The latter case obtains if $\mathfrak{F}$ is $C[0,1]$, $\mathcal{K}$, or $C_{oo}(G)$ which is the space of real continuous functions on a locally compact space G vanishing off compact sets. In these spaces the hypothesis on ℓ should be strengthened to manufacture a useful measure, generalizing the ideas of the classical Riesz-Markov theorem. The former spaces, on the other hand, appear to be interesting in the context of nonlinear differential equations and mechanics. The work of Mizel, Orlicz and their associates is related to this area. The spaces $\mathfrak{F}$ without a base measure are natural mostly in probability theory and certain other parts of functional analysis. The work in this area, starting with Tichonov and Arsenin [38] and Chacón-Friedman [7], is continued by others (cf. [3],[4],[15],[36]). However the "semi-ordered spaces" in [19] intersect both the abovestated areas and in fact these authors have given the following comprehensive result as a consequence of their general theory. For comparison this will be stated as:

Theorem 1. [19]. Let $L^p[0,1]$ be the usual L^p-space on the Lebesgue unit interval for $1 \le p \le \infty$ and $p = 0$ where L^0 is the space of equivalence classes of measurable real functions with the topology of convergence in measure. Then the general form of a continuous partially additive (= local) functional on $L^p[0,1]$ is given by the formula:

$$\ell(f) = \int_0^1 \varphi(f(t),t)dt \ , \quad f \in L^p[0,1] \ , \tag{1}$$

where

(a) $\varphi(x,\cdot) \in L^1(0,1)$, $x \in \mathbb{R}$,

(b) $\varphi(\cdot,t)$ is continuous for each t ,

(c) $\varphi(0,t) = 0$ for a·a·(t)

and if $1 \le p < \infty$, then

(d) $|\varphi(x,t)| \le k|x|^p + \tilde{\varphi}(t)$, $k = \text{const}, \tilde{\varphi} \in L^1(0,1)$;

if $p = \infty$, then (a)-(c), and

(d′) $|\varphi(x,t)| \le \tilde{\varphi}(t)$, for x in compact intervals and $\tilde{\varphi} \in L^1(0,1)$;

and finally if $p = 0$, then (a)-(c) and

(d″) $|\varphi(x,t)| \le \tilde{\varphi}(t)$, $x \in \mathbb{R}, \tilde{\varphi} \in L^1(0,1)$.

Unfortunately virtually all the later authors have overlooked this important work from [18]-[20], even though its existence was briefly noted in [20]. It seems to have been brought out only in [10]. Independently of the work of [18], Martin and Mizel [25] have obtained the following result in 1964, starting the theory *ab initio* and continuing thereafter.

Theorem 2. [25]. Let (Ω,Σ,μ) be a nonatomic probability space and $\ell: L^\infty(\mu) \to \mathbb{R}$ be a functional such that (i) $\ell(f+g) = \ell(f) + \ell(g)$ for $f \cdot g = 0$, (ii) $f_n \to f$ boundedly and pointwise a.e., then $\ell(f_n) \to \ell(f)$, and (iii) f,g are equimeasurable in that $\mu(f > x) = \mu(g > x)$, $x \in \mathbb{R}$, then $\ell(f) = \ell(g)$. Then there is a continuous $\varphi: \mathbb{R} \to \mathbb{R}$, $\varphi(0) = 0$, such that

$$\ell(f) = \int_\Omega \varphi(f(t))d\mu(t) \ , \quad f \in L^\infty(\mu) \ . \tag{2}$$

It may be checked that condition (iii) of the above result is strong enough to imply that ℓ is uniformly continuous on the bounded sets of $L^\infty(\mu)$ and more. The result has therefore been generalized by assuming the latter (weaker) condition (with also a weaker conclusion), and then extending to the $L^p(\mu)$-spaces for $0 < p < \infty$ by several authors (cf. [15],

[28],[29],[40]). In this line of reasoning the following result, from [11], is perhaps the most general. Some additional notation is needed for its presentation.

Let $\mathfrak{X} \subset L^0(\mu) = L^0(\Omega,\Sigma,\mu)$, $\mu(\Omega)<\infty$, be a Fréchet type space so that there is a complete absolutely continuous F-norm $\|\cdot\|$ on $\mathfrak{X}$ in the sense that (i) $\chi_A \in \mathfrak{X}$ for all $A \in \Sigma$, (ii) $f \in \mathfrak{X}$, $g \in L^0(\mu)$, $|g| \le |f| \Rightarrow g \in \mathfrak{X}$ and $\|g\| \le \|f\|$, (iii) $\lim_{\mu(A)\to 0} \|f\chi_A\| = 0$, $f \in \mathfrak{X}$, and (iv) $(\mathfrak{X},\|\cdot\|)$ is a complete metric space. For simplicity, let the support of $\mathfrak{X}$ be Ω. A mapping $\varphi:\mathbb{R}\times\Omega \to \mathbb{R}$ is a <u>Carathéodory function</u> (cf. [21]) if $\varphi(\cdot,\omega)$ is continuous, $\varphi(0,\omega) = 0$, for a.a.(ω), and $\varphi(t,\cdot)$ is measurable for $t \in \mathbb{R}$. Thus the kernel φ in (1) is a Carathéodory function. The result of Orlicz and Drewnowski can now be stated as (compare with Theorem 1):

<u>Theorem</u> 3. [11]. <u>Let</u> $(\mathfrak{X},\|\cdot\|)$ <u>be a Fréchet space as above. Let</u> $\ell:\mathfrak{X}\to\mathbb{R}$ <u>be a functional such that (i)</u> $\ell(f+g) = \ell(f)+\ell(g)$ <u>for</u> $f\cdot g = 0$ <u>and (ii)</u> ℓ <u>is continuous in the topology of</u> $\mathfrak{X}$. <u>Then there is a Carathéodory function</u> $\varphi:\mathbb{R}\times\Omega\to\mathbb{R}$ <u>such that</u>

$$\ell(f) = \int_\Omega \varphi(f(t),t)\,d\mu(t), \qquad f \in \mathfrak{X}. \tag{3}$$

<u>and conversely a functional</u> ℓ <u>defined by (4) for a Carathéodory function</u> φ <u>is a local functional satisfying (i) and (ii) on</u> $\mathfrak{X}$.

Several results in this direction have been obtained by Mizel and his associates if $\mathfrak{X}$ is replaced by a Sobolev space and others, all of which are based on a measure space (cf. [22]-[25],[29]-[30]). Other extensions include the papers ([1],[2],[41]).

Let us now consider the second point of view, namely for spaces without a base measure. The earliest result in this line seems to be due to Tichonov and Arsenin [38], considered on the space $C[0,1]$ of continuous real functions. A functional $V:C[0,1]\to\mathbb{R}$ is monotone if $V(f_1) \le V(f)$ for $f_1 \le f_2$. First it is observed that a monotone functional, such as V, can be defined also on the larger set of functions of the form $f\chi_A$, $f \in C[0,1]$ with A a Borel set, preserving the monotonicity property. The authors consider a subclass of functionals V, termed <u>capacitory</u>, such that (i) $V(f) \ge 0$ for $f \ge 0$, and $= 0$ for $f = 0$, (ii) V is monotone and (iii) completely additive, i.e., $V(\sum_{i=1}^{\infty} f\chi_{A_i}) = \sum_{i=1}^{\infty} V(f\chi_{A_i})$ where A_i are disjoint Borel sets, whenever this is defined.

Then the following result holds:

Theorem 4. [38]. Let $V:C[0,1]\to\mathbb{R}$ be a continuous functional of capacitory type. Then there exist a Radon measure μ on $[0,1]$ and a nonnegative function $\psi(\cdot,\cdot)$ such that $\psi(x,\cdot)$ is measurable, for each $x\in\mathbb{R}$, $\psi(\cdot,t)$ is continuous and monotonically increasing for a.a.(t) in terms of which one has

$$V(f) = \int_0^1 \psi(f(t),t)d\mu(t) , \qquad f\in C[0,1] . \tag{4}$$

It is thus evident that functionals of capacitory type form a subclass of local functionals. The point of this result is that it appears to be the first of its kind to consider local functionals on spaces not based on measure spaces. It is also interesting to note that the authors of [38] ended the paper by remarking that $V(f) = \max\{f(t):t\in[0,1]\}$ defines a functional on $C[0,1]$ which is not of capacitory type so that the class of problems opened up by the analysis here is much larger. Just as the early work of [18],[19], this paper was also overlooked by the later writers. It may be noted that the monotonicity of the capacitory functional (the name ostensibly comes from the fact that V is defined originally on a class of "impulse" functions and the representation problem can be solved if its "frequency characteristic" is known) insures its extendability to a larger class of functions containing $C[0,1]$. This property will appear in the following work also under different conditions and forms; without such an additional condition the stated representations do not obtain.

Independently of the paper [38], Chacón and Friedman [7] have established the following more general result (with a somewhat different hypothesis) which formed a basis for a class of problems.

Theorem 5. [7]. Let ℓ be a real functional on $\mathfrak{X} = C[0,1]$. Suppose ℓ satisfies (i) (boundedness) ℓ is bounded on balls; i.e., given $K>0$, there is $M_K>0$ such that $\|f\|\le K \Rightarrow |\ell(f)|<M_K$ ($\|\cdot\|$ = uniform norm), (ii) (uniform continuity) given $\varepsilon>0, K>0$ there is δ $(=\delta_{\varepsilon,K}>0)$ such that $\|f_i\|<K$, $i=1,2$ and $\|f_1-f_2\|<\delta \Rightarrow |\ell(f_1)-\ell(f_2)| < \varepsilon$, and (iii) (strong additivity) for all $f\in\mathfrak{X}$, $\ell_f = \ell(\cdot+f)$, $f_i\in\mathfrak{X}$, $i=1,2$, with $f_1\cdot f_2 = 0 \Rightarrow \ell_f(f_1+f_2) = \ell_f(f_1)+\ell_f(f_2)$. Then there is a (signed) Radon measure μ on $[0,1]$ and a real function $\psi(\cdot,\cdot)$ such that $\psi(x,\cdot)$ is Borel measurable, $\psi(\cdot,t)$ is continuous for a.a.(t), and $|\psi(f(t),t)|\le M_K$ a.a.(t) if $\|f\|\le K$ ($M_K>0$ depending only on K) in terms of which one has

$$\ell(f) = \int_0^1 \psi(f(t),t)d\mu(t) , \qquad f\in\mathfrak{X} . \tag{5}$$

Conversely, the pair (ψ,μ) with the above properties defines ℓ by

(5) which satisfies (i)-(iii).

This and Theorem 4 have stronger hypotheses than the first three results since the base measure μ has to be produced now in contrast to the earlier ones. The strengthening essentially allows one to extend the functionals V and ℓ to a larger class than C[0,1] and then the representations (4) and (5) hold. In subsequent papers Friedman and Katz extended (5) if [0,1] is replaced by a compact set, and the range space is more general.

3. Local functionals on vector function spaces. If $\mathfrak{X},\mathfrak{Y}$ are Banach spaces, let $M(\mathfrak{X},\mathfrak{Y})$ be the vector space of all mappings $F:\mathfrak{X}\to\mathfrak{Y}$ such that $F(0) = 0$, F is bounded on each ball of $\mathfrak{X}$, and F is uniformly Lipschitz on balls, i.e., $\sup\{\|F(x)-F(y)\|_{\mathfrak{Y}}:\|x-y\|_{\mathfrak{X}} < \delta\} \to 0$ as $\delta\to 0$. If $C(S,\mathfrak{X})$ is the space of $\mathfrak{X}$-valued continuous functions on a compact set S, $\mathcal{B}$ is the Borel σ-algebra of S, let $U:\mathcal{B}\to M(\mathfrak{X},\mathfrak{Y})$ be additive. Then for each $f = \sum_{i=1}^{n} x_i\chi_{A_i}$, $A_i\in\mathcal{B}$ disjoint, define $\int_S f\,dU = \sum_{i=1}^{n} U(A_i)x_i$, and extend this as in the Dunford-Schwartz integral, i.e., if $\|g_n-g\|\to 0$ (uniform norm), then $\int_E g\,dU = \lim_n \int_E g_n\,dU$, $E\in\mathcal{B}$, when this exists in $\mathfrak{Y}$. This extension is essentially due to Batt [4]. It satisfies the relation (cf. [4], pp. 147-149 for details):

$$\int_S (g+g_1+g_2)dU = \int_S (g+g_1)dU + \int_S (g+g_2)dU - \int_S g\,dU ,$$

for $g_1\cdot g_2 = 0$. Thus Batt obtained in [3],[4] that if $T\in M(C(S,\mathfrak{X}),\mathbb{C})$, then there exists a mapping $\mu:\mathcal{B}\to M(\mathfrak{X},\mathbb{C})$, additive, such that

$$Tf = \int_S f\,d\mu , \qquad f\in C(S,\mathfrak{X}) , \tag{6}$$

where μ has finite semivariation, and its variation on balls of radius δ is finite and tends to zero as $\delta\to 0$, and μ is regular, where the integral is as defined above. The integral is also briefly indicated by Friedman and Tong [15] where they show that if the adjoint space $\mathfrak{X}^*$ of $\mathfrak{X}$ is separable in the weak*-topology then there is a kernel $k:\mathfrak{X}^*\times\mathbb{R}\times S\to\mathbb{R}$ such that if $T:C(S,\mathbb{R})\to\mathfrak{X}$ is additive and uniformly continuous on bounded sets, then

$$x^*(Tf) = \int_S k(x^*,f(s),s)d\mu(s) , \qquad x^*\in\mathfrak{X}^*, f\in C(S) , \tag{7}$$

where μ is a regular measure.

These results, particularly those of Batt [4], have been extended for different types of Banach spaces by de Korvin, Alo, and Cheney [1],

[2], still keeping the domain as $C(S)$. On the other hand, the representation of $\ell: L^p_{\mathfrak{X}}(\mu) \to \mathcal{Y}$, $\mathfrak{X}$ separable, has been obtained by Mizel and his collaborators for $1 \le p \le \infty$, extending their earlier results, [27]-[30]. Some of this was again generalized by de Korvin and his associates. These seem to be the available results in the vector valued case.

4. Local functionals in probability theory. One of the fundamental problems of this theory appeared in the work of Gel'fand and Vilenkin [16] in the late 1950's. The preceding results still do not cover these questions and so it is appropriate to discuss them here in some detail.

Let $\mathcal{K}$ be the Schwartz space of infinitely differentiable real functions on $\mathbb{R}$ with compact supports. Let $L^0(P)$ be the metric space of equivalence classes of real random variables on a probability space (Ω,Σ,P), as usual (cf. Theorem 1). Then a continuous linear mapping $F:\mathcal{K} \to L^0(P)$ is termed by Gel'fand a generalized random process (g.r.p.) on $\mathcal{K}$. A g.r.p. has independent values at every point if for f,g in $\mathcal{K}$, with $f \cdot g = 0$, $F(f)$ and $F(g)$ are (statistically) independent. If $L^0(P)$ is replaced by its subspace $L^2(P)$ as the range of F, then F was called a random Schwartz distribution by K. Itô who introduced in 1954 this concept independently of Gel'fand (1955). However, both concepts agree on $L^2(P)$. An elegant exposition of the theory of g.-r.p.'s may be found in [16].

To analyze the structure of these processes, consider the characteristic functional (ch.f.) defined as $L(f) = E(e^{iF(f)}) = \int_\Omega e^{iF(f)} dP$. If F has independent values, then for f,g in $\mathcal{K}$, $f \cdot g = 0 \Rightarrow$

$$\begin{aligned} L(f+g) &= E(e^{i(F(f)+F(g))}) \\ &= E(e^{iF(f)}) \cdot E(e^{iF(g)}) = L(f)L(g) . \end{aligned} \tag{8}$$

Suppose L never vanishes. (It is not hard to give a condition on L for this to hold.) Let $M = \log L$ where, as usual, the log function is taken as the principal branch such that the continuity properties of L are translated to M. Then (8) becomes

$$M(f+g) = M(f) + M(g),, \qquad \{f,g\} \subset \mathcal{K}, f \cdot g = 0 . \tag{9}$$

Such a functional M, satisfying (9), was called a local functional by Gel'fand and Vilenkin ([16], p. 275) who raised the problem of its characterization. This problem in its complete generality is still unsolved. The ch.f. L, and hence the local functional M, is said to be of order $p \ge 0$, if $L(f)$ (or $M(f)$) depends only on f through its pth derivative. It can be checked that M is even strongly additive in the sense of Theorem 5. Using the various properties of the ch.f.'s and the nuclearity of the space $\mathcal{K}$, a representation of M , of order p , was given in ([33],[34]). Since $\mathcal{K}$ is not a Banach space, an extension of the theory of [7] and certain other properties of $\mathcal{K}$ were used for this work. Analogous representation under weaker conditions on a local functional when $\mathcal{K}$ is replaced by a Sobolev space (hence with a base measure) was given in [25] and [22] with a detailed analysis.

To present a general result in this direction, consider $C_{oo}(G)$, the space of real continuous functions, on a locally compact space G , with compact supports. It is a locally convex space with the inductive limit topology ($C_{oo}(G) = \cup\{C(S) : S \subset G$, compact$\}$). If $F : C_{oo}(G) \to L^0(P)$ is a g.r.p. with independent values at each point, then its ch.f. L satisfies (8) and defines a local functional M satisfying (9). The representation of M now presents a new problem.

A comprehensive solution of the problem can be given as follows:

Theorem 6. Let $M : C_{oo}(G) \to \mathbb{R}$ be a mapping. Then conditions A), B), and C) below are equivalent:

A) (i) (Sequential continuity) If $\{f_n, n \geq 1\} \subset C_{oo}(G)$ is a bounded pointwise convergent sequence, then $\{M(f_n), n \geq 1\} \subset \mathbb{R}$ is Cauchy;

(ii) (Additivity) $M(f_1+f_2) = M(f_1) + M(f_2)$ for all f_1, f_2 such that $f_1 \cdot f_2 = 0$ and for which the equation is defined;

(iii) (Bounded uniform continuity) For each $\varepsilon > 0, K_0 > 0$, there is a δ $(= \delta_{\varepsilon, K_0} > 0)$ such that $\|f_i\| \leq K_0$, $f_i \in C_{oo}(G)$, i=1,2 , $\|f_1 - f_2\| < \delta \Rightarrow |M(f_1) - M(f_2)| < \varepsilon$ where $\|\cdot\|$ is the sup-norm.

B) There exist a Radon measure μ on G and a function $\psi : \mathbb{R} \times G \to \mathbb{R}$ such that

(a) $\psi(0,t) = 0$, $\psi(\cdot,t)$ is continuous for a.a. $t \in G$;

(b) $\psi(x,\cdot)$ is μ-measurable for all $x \in \mathbb{R}$;

(c) for each $f \in C_{oo}(G)$, $\psi(f(t),t)$ is bounded for a.a. (t) , and for any $\{f_n, n \geq 1\}$, as in A)(i), $\{\psi \cdot f_n, n \geq 1\}$ is Cauchy in $L^1(\mu)$, in terms of which M is representable as

$$M(f) = \int_G \psi(f(t),t) d\mu(t) , \qquad f \in C_{oo}(G) . \tag{10}$$

C) (i′) (Strong additivity) For any $f_i \in C_{oo}(G)$, i=1,2,3 , $f_2 \cdot f_3 = 0$, one has $M(f_1+f_2+f_3) = M(f_1+f_2) + M(f_1+f_3) - M(f_1)$;

(ii′) (Boundedness) If $B_b = \{f \in C_{oo}(G) : \|f\| \leq b\}$, then $\{M(f), f \in B_b\}$ is a bounded subset of $\mathbb{R}$ for each $b > 0$, and $M(0) = 0$, where $\|\cdot\|$ is the sup-norm;

(iii′) (Bounded uniform continuity) Same as in A)(iii) above.

This result is proved in the order: A) $\Leftrightarrow$ B) $\Leftrightarrow$ C) . If, moreover, G is an amenable group and M translation invariant (i.e., $M(\tau_s f) = M(f)$ for all $s \in G$ where $(\tau_s f)(t) = f(s^{-1}t)$), then the representation can be improved. Namely in A) and C), one adds the translation invariance condition on M and in B) one can take μ as a (left) Haar measure and ψ as a continuous function only of the first variable satisfying the other conditions. The proof is again long and only an outline can (and will) be given here.

Outline of proof. If $B_o(G)$ is the class of all bounded pointwise limits of sequences from $C_{oo}(G)$, then $B_o(G)$ is a linear space containing $C_{oo}(G)$, and by A)(i), M can be uniquely extended onto $B_o(G)$. The class $\mathcal{C} = \{A : \chi_A \in B_o(G)\}$ is a ring containing the ring $\mathcal{R}$ generated by the compact Baire sets of G . Then A)(i) also allows one to extend M to a class consisting of simple functions based on $\mathcal{R}$. Call it $\tilde{M}$. This is analogous to the extension used in Tichonov-Arsenin [38]. Now for each $A \in \mathcal{C}, h \in \mathbb{R}$, define $\mu_h(A) = \tilde{M}(h\chi_A)$. It follows that $\mu_h : \mathcal{R} \to \mathbb{R}$ is additive. To extend μ_h for more sets, consider the Jordan decomposition $\mu_h = \mu_h^+ - \mu_h^-$, and let $\hat{\mu}_h^+$ be the additive set function onto the power set $\mathcal{P}(G)$, with the Hahn-Banach theorem. Next define for each closed set $E \subset G$,

$$\lambda_1^h(E) = \inf\{\hat{\mu}_h^+(O) : E \subset O \text{ , open}\} , \quad h \in \mathbb{R} , \tag{11}$$

$$\lambda_2^h(A) = \sup\{\lambda_1^h(E) : E \subset A \text{ , } E \text{ closed}\} , \quad A \in \mathcal{P}(G) . \tag{12}$$

It can be verified that if $\tilde{\mathcal{R}}^*$ is the class of λ_2^h-measurable sets in the sense of Carathéodory, and $\mathcal{R}^*$ is the algebra generated by the closed sets of G , then $\mathcal{R}^* \subset \tilde{\mathcal{R}}^*$ and $\lambda_2^h|\mathcal{R}^*$ ($= \lambda^h$, say) is additive, nonnegative (monotone) and regular. Moreover, for each closed set $E \subset G$, one has

$$\lambda^h(E) = \lambda_1^h(E) = \lambda_2^h(E) . \tag{13}$$

Then using A)(iii) one notes that λ^h is also bounded for each $h \in \mathbb{R}$. Thus λ^h is a regular content and can be used to generate a Radon measure $\tilde{\lambda}_0^h$ on the σ-algebra $\sigma(\mathcal{R}^*)$. Let $\bar{\lambda}_0^h$ be the corresponding measure obtained for μ_h^- with a similar procedure. Then $\lambda_0^h = \tilde{\lambda}_0^h - \bar{\lambda}_0^h$ is a (signed) Radon measure on G , and $\mu_h(A) = \lambda_0^h(A)$ for each compact $A \subset G$. Now using A)(iii) more decisively, one shows that λ_0^h depends on h continuously, uniformly relative to $\sigma(\mathcal{R}^*)$.

To obtain the desired μ , let $h_1, h_2, \ldots$ be an enumeration of rationals in $\mathbb{R}$, and define

$$\mu(\cdot) = \sum_{n=1}^{\infty} \frac{1}{2^n} \frac{|\lambda_0^{h_n}|(\cdot)}{1+|\lambda_0^{h_n}|(G)} \tag{14}$$

where $|\lambda|$ is the variation measure of λ. It follows that λ_0^h is μ-continuous for all $h \in \mathbb{R}$, and by the Radon-Nikodým theorem,

$$\tilde{M}(h\chi_A) = \lambda_0^h(A) = \int_A \psi_h(t)d\mu(t) = \int_G \psi(h\chi_A(t),t)d\mu(t) , \tag{15}$$

where $\psi(h,t) = \psi_h(t)$ for $t \in A$, $h \in \mathbb{R}$, and $= 0$ if $t \notin A$. It can be checked that the pair (ψ,μ) satisfies the conditions of B), and if $M_1(f) = \int_G \psi(f(t),t)d\mu(t)$, then $M_1|_{C_{oo}(G)} = \tilde{M}|_{C_{oo}(G)} = M$ and A) $\Rightarrow$ B) follows. The converse is similar and slightly simpler.

It is again easy to see that B) $\Rightarrow$ C). To establish the reverse implication, one uses the work of Friedman and Katz [14], for the compact case, and then the result is extended. The point here is, since M is not a priori extendable to $B_o(G)$, a stronger additivity condition is required. Thus one obtains λ_S^h for all compact sets $S \subset G$, and shows that this can be done consistently in that $\lambda_{S_1}^h, \lambda_{S_2}^h$ agree with $\lambda_{S_1 \cap S_2}^h$ on $S_1 \cap S_2$. Thus one determines a λ^h on G and proceeding as in A), (ψ,μ) can be obtained which will satisfy B). This will establish the theorem.

An additional argument is necessary, with the amenability assumption, for the refinement in the group case (cf. [36]).

A natural application of the above result is to the g.r.p.'s on $C_{oo}(G)$ with independent values as described at the beginning of this section. Since $C_{oo}(G)$ is not a nuclear space in contrast to that of $\mathcal{K}$, an additional problem arises. The continuity of the ch.f. L on $C_{oo}(G)$ is not sufficient to conclude that all such positive definite continuous functionals are ch.f.'s of regular probability measures on the adjoint space $(C_{oo}(G))^*$ of $C_{oo}(G)$. Generally, such an $L(\cdot)$ determines only a "cylindrical probability" (cf. [5]), which gives a probability on the algebraic dual of $(C_{oo}(G))^{**}$, and $(C_{oo}(G))^*$ can receive zero measure. Thus the desired condition for the measure to concentrate on $(C_{oo}(G))^*$ itself turns out to be the continuity of the positive definite $L(\cdot)$ in the Sazanov topology $\mathfrak{S}$ (cf. [6], p. 91 for a definition of $\mathfrak{S}$). For a precise statement of this application, see [35]. The result here even uses the differentiation theory [17] for its final form.

Applications to such linear processes, following [16], were made by the author [33] and Woyczinski ([42]-[44]). In [42], the representation was given for the cylindrical measures and not the g.r.p.'s. They correspond to "weak distributions" in the sense of Segal and Gross, and the $\mathcal{S}$-topology did not play a role. If $C_{oo}(G)$ is replaced by a Hilbert space, then according to a result of Prokhorov and Sazanov (cf. [6], p. 92), the continuity of $L(\cdot)$ in the $\mathcal{S}$-topology is the best condition for the problem. See also Urbanik [39] on related results.

5. Further problems and remarks. It is of interest to find conditions on the positive definite functions L on $C_{oo}(G)$, other than their continuity in the $\mathcal{S}$-topology, which will be easier to verify in real life applications. This may make the statement of [35] more symmetrical. It is also useful to extend this study to function spaces based on a fixed measure space so that weaker hypotheses suffice. This can undoubtedly be made, but the precise details are not yet available. Such a study has been hinted at, in the papers of Mizel and his associates. Also as noted there, applications to nonlinear differential equations seem more natural on this type of spaces. The work of Krasnoselskii (cf. [21]) and his associates indicates this more clearly when nonlinear integral equations (e.g., Hammerstein, Urysohn types) are studied.

In spaces based on measures, which are not "solid," a study of local functionals becomes more involved. With appropriate hypotheses, it will be interesting to study classes of g.r.p.'s with a possible weakening than independent values (e.g., martingale increments). To extend this work for stationary processes or those with orthogonal increments, further insight into the structure of these nonlinear functionals seems necessary. The bibliography below may help the interested reader to find the problems and tools in such analyses.

REFERENCES

1. R. A. Alo, C. A. Cheney, and A. de Korvin, "Nonlinear operators on sets of measures," Ann. Mat. Pura Appl. 109(1976), 1-22.

2. R. A. Alo and A. de Korvin, "Representation of Hammerstein operators by Nemytskii measures," J. Math. Anal. Appl. 52(1975), 490-513.

3. J. Batt, "Strongly additive transformations and integral representations with measures of nonlinear operators," Bull. Amer. Math. Soc. 78(1972), 474-478.

4. _____, "Nonlinear integral operators on $C(S,E)$," Studia Math. 48 (1973), 145-177.

5. S. Bochner, Harmonic Analysis and the Theory of Probability, Univ. California Press, Berkeley, 1955.

6. N. Bourbaki, Intégration, Chapitre IX, Hermann, Paris, 1969.

7. R. V. Chacón and N. Friedman, "Additive functionals," Arch. Rational Mech. Anal. 18(1965), 230-240.

8. L. Drewnowski and W. Orlicz, "A note on modular spaces-X, XI," Bull. Acad. Pol. Sci., Ser. Math. 16(1968), 809-814; 877-882.

9. ______, "On orthogonally additive functionals," ibid 16(1968), 883-888.

10. ______, "On representation of orthogonally additive functionals," ibid 17(1969), 167-173.

11. ______, "Continuity and representation of orthogonally additive functionals," ibid 17(1969), 647-653.

12. N. Friedman and M. Katz, "A representation theorem for additive functionals," Arch. Rational Mech. Anal. 21(1966), 49-57.

13. ______, "Additive functionals on L^p-spaces," Canadian J. Math. 18 (1966), 1264-1271.

14. ______, "On additive functionals," Proc. Amer. Math. Soc. 21(1969), 557-561.

15. N. Friedman and A. E. Tong, "Additive operators," Canadian J. Math. 23(1971), 468-480.

16. I. M. Gel'fand and N. Ya. Vilenkin, Generalized Functions, Vol. 4, (Translation) Academic Press, New York, 1964.

17. C. Ionescu Tulcea, "On liftings and derivation bases," J. Math. Anal. Appl. 35(1971), 449-466.

18. L. Kantorovitch and A. Pinsker, "Sur les fonctionnelles partiellement additives dans les espaces semi-ordonnés," C. R. Acad. Sci. (Paris) 207(1938), 1376-1378.

19. ______, "Sur les formes générales des fonctionnelles partiellement additives dan certains espaces semi-ordonnés," C. R. Acad. Sci. (Paris) 208(1939), 72-74.

20. L. V. Kantorovitch, B. Z. Vulich and A. G. Pinsker, Functional Analysis in Semi-ordered Spaces," Moscow, 1950.

21. M. A. Krasnoslskii, Topological Methods in the Theory of Nonlinear Integral Equations, (Translation) Pergamon Press, New York, 1964.

22. M. Marcus and V. J. Mizel, "A characterization of nonlinear functionals on W_1^p possessing autonomous kernels-I," Pacific J. Math. 65(1976), 135-158.

23. ______, "A Radon-Nikodým type theorem for functionals," J. Functional Anal. 23(1976), 285-309.

24. ______, "Extension theorems of Hahn-Banach type for nonlinear disjointly additive functionals and operators in Lebesgue spaces," J. Functional Anal. 24(1977), 303-335.

25. ______, "Representation theorems for nonlinear disjointly additive

functionals and operators on Sobolov spaces," Trans. Amer. Math. Soc. 228(1977), 1-45.

26. A. D. Martin and V. J. Mizel, "A representation theorem for certain nonlinear functionals," Arch. Rational Mech. Anal. 15(1964), 353-367.

27. V. J. Mizel, "Characterization of nonlinear transformations possessing kernels," Canadian J. Math. 22(1970), 449-471.

28. V. J. Mizel and K. Sundaresan, "Representation of additive and biadditive functions," Arch. Rational Mech. Anal. 30(1968), 102-126.

29. ______, "Additive functionals on spaces with nonabsolutely continuous norms," Bull. Acad. Pol. Sci., Ser. Math. 18(1970), 385-389.

30. ______, "Representation of vector-valued nonlinear functions," Trans. Amer. Math. Soc. 159(1971), 111-127.

31. A. G. Pinsker, "La represention analytique de quelque fonctionnelles partiellement additives," DAN (USSR) 18(1938), 339-403.

32. ______, "Sur la fonctionnelle dans l'espace de Hilbert," DAN (USSR) 20(1938), 411-414.

33. M. M. Rao, "Local functionals and generalized random fields," Bull. Amer. Math. Soc. 74(1968), 288-293.

34. ______, "Local functionals and generalized random fields with independent values," Theor. Prob. Appl. 16(1971) 457-473.

35. ______, "Processus linèaires sur $C_{oo}(G)$," C. R. Acad. Sci. (Paris) 289(1979).

36. ______, "Local functionals on $C_{oo}(G)$ and probability," Publ. de l'IRMA, ULP, Strasbourg, 1979.

37. K. Sundaresan, "Additive functionals on Orlicz spaces," Studia Math. 32(1969), 269-276.

38. A. N. Tichonov and V. Ja. Arsenin, "On some nonlinear functionals," Mat. Sbornik 65(1964), 512-521. (Russian)

39. K. Urbanik, "Generalized stochastic processes with independent values," Proc. Fourth Berkeley Symp. Math. Statist. and Prob. 2 (1961), 369-380.

40. W. A. Woyczyński, "Additive functionals on Orlicz spaces," Colloq. Math. 19(1968), 319-326.

41. ______, "Additive operators," Bull. Acad. Pol. Sci., Ser. Math. 17 (1969), 447-451.

42. ______, "On characteristic functionals of linear processes with independent pieces," ibid 17(1969), 551-557.

43. ______, "Representation of additive functionals and invariant characteristic functionals of linear processes with independent pieces," ibid 19(1971), 221-230.

44. ______, "Ind-additive functionals on random vectors," Dissertationes Math., No. 72(1970), 42 pp.

A converse to Edgar's theorem

by

E.G.F. Thomas

University of Groningen

Summary We prove that for suitable convex subsets B of a locally convex space, B has the Radon Nikodym property if and only if $B^{\mathbb{N}}$ has the integral representation property (i.e. the generalization of Choquet's theorem is valid for all closed convex subsets of $B^{\mathbb{N}}$). Analogous results are obtained for conuclear cones.

Introduction G.A. Edgar [5] has shown that if B is a separable closed bounded convex subset of a Banach space, and if B has the Radon Nikodym property, then every point in B is the barycenter of a Radon probability measure on the set of extreme points of B. R.D. Bourgin and Edgar [6] have shown that the representing measures are uniquely determined if and only if B is simplex.

In this paper we first generalize these results to the case where B is a closed bounded convex Suslin subset of a locally convex space (corollary 2 of theorem 1). If B has the Radon Nikodym property, so does $B^{\mathbb{N}}$, and so does every closed convex subset of $B^{\mathbb{N}}$. We then show that conversely, if for every closed convex subset A of $B^{\mathbb{N}}$ every point of A is the resultant of a Radon probability measure on the set of extreme points of A, the B has the Radon Nikodym property (corollary of theorem 2). (Throughout we assume that the closed convex hull of every compact subset of B is compact, a condition which is certainly satisfied if B is complete).

Actually convex cones, rather than bounded convex sets, are the main subject of the paper. Theorem 1 generalises the results of Edgar and of Bourgin and Edgar mentioned above, as well as my own previous results on conuclear cones [10] [11].

Converse theorems have been considered before in the case of simplices or lattices [7], [11].

The main tools of this paper are the conical measures first introduced by G. Choquet [2], [3].

In §1 and §2 we state the basic properties of localizable conical measures to be used in the sequel. In §3 the theory of conical measures is used to prove the fact, well known in the case of Banach Spaces [1], that if a set has the Radon-Nikodym property relative to, for instance, Lebesque measure, it also has the Radon-Nikodym property relative to every finite measure space.

Contents

1. Localizable conical measures.

Let F be a locally convex Hausdorff space over $\mathbb{R}$, F' its dual space and let S be the set of functions which are supremum of a finite number of continuous linear forms $\phi = \sup_i \ell_i$.

Let h(F) be the vector lattice S-S and put $h^+(F) = \{\phi \in h(F) : \phi(x) \geq 0 \ \forall x \in F\}$. Recall that a conical measure on F is a linear form $\mu : h(F) \to \mathbb{R}$ such that $\mu(\phi) \geq 0$ for all $\phi \in h^+(F)$. The resultant $r(\mu)$ of a conical measure μ is the point a belonging to the weak completion of F such that $\ell(a) = \mu(\ell)$ for all $\ell \in F'$ (See [2] or [3]).

Definition A conical measure μ will be said to be localizable if there exists a Radon measure m on $F \setminus \{0\}$ such that $\int |\ell| dm < +\infty$ for every $\ell \in F'$ and such that

(1) $\mu(\phi) = \int \phi \, dm \qquad \forall \phi \in h(F)$

The Radon measure m is said to be a localization of μ. If m is concentrated on a set $A \subset F$, μ is localizable on A.

Proposition 1 Let Γ be an arbitrary cone (i.e. a set such that $x \in \Gamma$ $\lambda \geq 0$ implies $\lambda x \in \Gamma$) and let m_1 and m_2 be localizations of μ. Then m_1 is concentrated on Γ if and only if m_2 is concentrated on Γ.

In this case we say that μ is concentrated on Γ. We denote by $M^+(\Gamma)$ the set of localizable conical measures concentrated of Γ and such that $r(\mu) \in F$.

Proposition 2 Let μ be a localizable conical measure concentrated on Γ and let $f : \Gamma \to \overline{\mathbb{R}}$ be positively homogeneous ($f(\lambda x) = \lambda f(x) \ \forall \lambda > 0$). Then, if m_1 and m_2 are localizations of μ, f is m_1-measurable if and only if f is m_2-measurable. If moreover $f \geq 0$, $\int f \, dm_1 = \int f \, dm_2 \leq +\infty$.

A homogeneous function which is measurable with respect to some localization of μ is said to be μ-measurable. For $f \geq 0$ the common value of the integral is denoted by $\int f \, d\mu$.

Proposition 3 Let μ be e localizable conical measure concentrated on Γ and let S be a section of Γ, i.e. a subset of $\Gamma \setminus \{0\}$ such that each ray of Γ encounters S in precisely one point. Let p be the positively homogeneous function defined on Γ equal to 1 on S. Then if p is μ-measurable μ has a unique localization on S. In particular, if there exists $\ell \in F'$ with $\ell(x) > 0 \ \forall x \in \Gamma \setminus \{0\}$, μ has a unique localization on the base $\{x \in \Gamma : \ell(x) = 1\}$.

These facts have been stated before (see e.g. [11] and [10] for detailed proofs).

We shall also need the following:

Proposition 4 Let Γ be a cone which is a Suslin subset of F. Then a conical measure μ is localizable on Γ (i.e. localizable and concentrated on Γ) if and only if μ has the Daniel property relative to Γ:

$$\phi_n \in h^+(F)\ ,\ \phi_n \geq \phi_{n+1}\ ,\ \inf_n \phi_n(x) = 0 \quad \forall x \in \Gamma \Rightarrow \inf_n \mu(\phi_n) = 0$$

Proof The condition is obviously necessary. For the proof of the sufficiency first assume Γ has a base S. Then the application of Daniel's theorem to the set of restrictions $L = \{\phi/_S : \phi \in h(F)\}$ gives a (unique) bounded measure m on the σ-algebra $\mathcal{A}$ generated by L such that $\mu(\phi) = \int_S \phi\, dm$. However, S being a Suslin space, and L separating the points of S, $\mathcal{A}$ is equal to the Borel σ-algebra of S. Again S being Suslin m is a Radon measure on S (cf. [9] chapter II). The general case may be reduced to the previous case by using the fact that Γ is the union of a countable set of cones having a base.

2. Integral representations.

Let Γ be a closed convex proper cone. We write $x \leq y$ if $y - x \in \Gamma$, let ext(Γ) be the cone of extreme generators of Γ, i.e. the set of elements $x \in \Gamma$ such that $0 \leq y \leq x$ implies $y = \lambda x$ for some number λ.

Definition A point $a \in \Gamma$ will be said to have a (unique) integral representation by means of extreme generators of Γ if there exists a (unique) $\mu \in M^+(\text{ext}\,\Gamma)$ such that $r(\mu) = a$.

If Γ has a bounded base S, $S \cap \text{ext}(\Gamma) = E(S)$ is the set of extreme points of S, and so according to proposition 3, the definition means that every point of S is the resultant of a unique probability measure concentrated on $E(S)$ (See [10] and [11] for further justifications of the definition).

It will be convenient to agree that Γ has the integral representation property if every closed convex cone Γ_1 of Γ satisfies the following conditions:

A) Every point of Γ_1 has an integral representation by means of extreme generators (i.e. the map $r: M^+(\text{ext}\,\Gamma_1) \longrightarrow \Gamma_1$ is onto)

B) Every point in Γ_1 has a unique integral representation (i.e. $r: M^+(\text{ext}\,\Gamma_1) \longrightarrow \Gamma_1$ is bijective) if and only if Γ_1 is a lattice in its own order.

Thus the well known theorem of G. Choquet can be stated by saying that a convex cone having a compact metrizable base has the integral representation property. The theorems of G.A. Edgar [5] and of R.D. Bourgin and G.A. Edgar [6] can be summarized by saying that a closed convex cone with a separable bounded base in a Banach space, having the Radon Nikodym property, has the integral representation property.

We have shown that if F is a quasi-complete conuclear space and the order intervals $\{x : 0 \leq x \leq a\} = \Gamma \cap (a - \Gamma)$ are bounded in the space F for each $a \in \Gamma$, then Γ has the integral representation property. In particular each weakly complete proper convex cone in such a space has the integral representation property. In the space $\mathbb{R}^{\mathbb{N}}$, which is both conuclear and weakly complete, every proper closed convex cone has the integral representation property (See [10] and [11]).

In this paper we propose a common 'generalization' of these results (§5 theorem 1) and a converse (§6 theorem 2).

3. The Radon Nikodym property.

Let $B \subset F$ be a closed bounded convex Suslin subset. We assume that the space F is quasi-complete, or more generally, that the closed convex hull of every compact subset of B is compact.

Definition B has the Radon Nikodym property relative to a finite measure space $(X, \mathcal{A}, P)$ if for every σ-additive P-continuous vector measure $M: \mathcal{A} \to F$ with the property that $M(A)/P(A) \in B$ for all $A \in \mathcal{A}$ with $P(A) > 0$, there exists a Borel measurable function $\rho: X \to B$ such that $M(A) = \int_A \rho dP$ for all $A \in \mathcal{A}$ (i.e. $\ell M(A) = \int_A \ell o \rho \, dP \quad \forall \ell \in F'$; briefly $M = \rho.P$)

Proposition 5 Let I be a compact metric space, let B(I) bet the set of Borel subsets of I and let λ be an atomless probability measure on B(I). If B has the Radon Nikodym property relative to (I, B(I), λ) then B has the Radon Nikodym property relative to every finite measure space $(X, \mathcal{A}, P)$.

Definition In this case we say that B has the Radon Nikodym property.

Remark The image measure $\rho(P)$ is a finite Borel measure on B, hence a Radon measure. Thus the above definition agrees with the general definition given by G.A Edgar [7].

Proof of proposition 5 We may and shall assume $0 \in B$. Let $P_B(x) = \inf\{\lambda \geq 0 : x \in \lambda B\}$. Then M is P-continuous and $M(A)/_{P(A)} \in B$ for all $A \in \mathcal{A}$ with $P(A) > 0$, if and only if

(3) $\quad P_B(M(A)) \leq P(A) \quad \forall A \in \mathcal{A}$

With M we associate a conical measure $\mu = \Delta(M)$ defined for $\phi \in S$ by

(4) $\quad \mu(\phi) = \sup_{\Sigma A_i = X} \Sigma \phi(M(A_i))$

the supremum being taken over all finite partitions of X by sets in $\mathcal{A}$. (Since $\phi \leq c^{st} P_B$ this is finite by (3)).

For $\phi \in h(F)$, $\mu(\phi)$ may be defined as the limit of the sums on the right of (4) as the partitions get finer.

Lemma 1 There exists a Borel function $\rho: X \to B$ such that $M = \rho.P$ if and only if μ is localizable on Γ_B (i.e. localizable and concentrated on Γ_B).

We omit the proof which is similar to the proof of theorem 14.1 and 14.2 in [10]. Note only that if $M = \rho.P$, $\mu(\phi) = \int \phi o \rho \, dP$ and so μ is localized in $\rho(P)$.

Lemma 2 Let $\mathcal{A}_1 \subset \mathcal{A}$ be a sub σ-algebra of $\mathcal{A}$. Let M_1 be the restriction of M to $\mathcal{A}_1$. Then $\Delta(M_1)(\phi) \leq \Delta(M)(\phi)$ for all $\phi \in S$. Conversely given any sequence $(\phi_n)_{n \geq 1}$ in S there exists a countably generated σ-algebra $\mathcal{A}_1 \subset \mathcal{A}$ such that $\Delta(M_1)(\phi_n) = \Delta(M)(\phi_n)$ for all n.

This is an immediate consequence of the definition (4).

Lemma 3 If B has the R.N.P. relative to every measure space $(X, \mathcal{A}_1, P)$ where $\mathcal{A}_1$ is a countably generated sub σ-algebra of $\mathcal{A}$, B has the R.N.P. relative to $(X, \mathcal{A}, P)$.

This easily follows from lemma 1, lemma 2 and proposition 4.

Let $\mathcal{L}^\infty(X, \mathcal{A}, P; \Gamma_B)$ be the set of Borel measurable functions $\rho: X \to \Gamma_B$ such that $\rho(x) \in \lambda B$ a.e for some λ. Let $L^\infty(X, \mathcal{A}, P; \Gamma_B)$ be the set of equivalence klasses. Denote by $||\rho||_\infty$ the essential supremum of $P_B(\rho(x))$.
We denote by $\mathcal{A}(P)$ the measure algebra in which sets are identified when their symmetric difference has measure zero.

Lemma 4 Let $(X_i, \mathcal{A}_i, P_i)$ be two probability spaces such that the measure algebras $\mathcal{A}_1(P_1)$ and $\mathcal{A}_2(P_2)$ are isomorphic. Then any isomorphism $T: \mathcal{A}_1(P_1) \to \mathcal{A}_2(P_2)$ may be extended to a map, again noted T, form $L^\infty(X_1, \mathcal{A}_1, P_1, \Gamma_B)$ to $L^\infty(X_2, \mathcal{A}_2, P_2; \Gamma_B)$ such that

(5) $$\int_{T(A)} T(\rho)\, dP_2 = \int_A \rho\, dP_1$$

for all $A \in \mathcal{A}_1(P_1)$ and $||T(\rho)||_\infty = ||\rho||_\infty$.

Proof Assume first that B is compact metrizable. Then B has the Radon Nikodym property. The equation

$$M(A_2) = \int_{T^{-1}(A_2)} \rho\, dP_1$$

defines a P_2-continuous vector measure with average range in $||\rho||_\infty B$. Thus $M = T(\rho)P_2$ for some $T(\rho) \in L^\infty(P_2; \Gamma_B)$ with $||T(\rho)||_\infty \leq ||\rho||_\infty$, (and by symmetry equality holds). In the general case, let $\rho: X \to \lambda B$ be Borel measurable. Then since $\rho(P_1)$ is a Radon measure there is a countable partition $X_1 = N + \sum_{i=1}^{\infty} A_i$ with sets belonging to $\mathcal{A}_1$ and $P_1(N) = 0$, such that ρ restricted to A_i takes values in a compact subset $B_i \subset \lambda B$, which by the assumption on B we may assume to be convex and to contain 0. This set, being a compact Suslin set, is moreover metrizable. Now using the above argument on each set A_i we can easily construct the required element $T(\rho)$ in the general case. (This type of decomposition, together with the fact that every series for which $\sum_n P_B(x_n) < +\infty$, converges in F, can be used to prove that the functions $\rho \in \mathcal{L}^\infty(P; \Gamma_B)$ are Pettis integrable [12]).

The proof of proposition 5 can now be carried out with the help of the isomorphism theorem (P. Halmos [8] §41, and S.D. Chatterdji [1]).

Proposition 6

1. Every closed convex subset of a Suslin set having the Radon Nikodym property also has the Radon Nikodym property.
2. Every countable product of Suslin sets having the Radon Nikodym property also has the Radon Nikodym property.

   This is obvious, given the fact that the subset and the product are again Suslin sets.

4. Conuclear cones.

Let $\Gamma \subset F$ be a closed convex proper cone. We assume that the closed convex envelope of a compact subset of Γ is compact, or equivalently that every Radon measure with compact support in Γ has a resultant belonging to Γ.

Let $\mathfrak{S}$ be a set of closed bounded convex subsets of Γ containing 0. We assume that $A \in \mathfrak{S}$ implies $\lambda A \in \mathfrak{S}$ for all $\lambda \geq 0$.

Definition Γ is $\mathfrak{S}$-conuclear if $\bigcup_{A \in \mathfrak{S}} A = \Gamma$ and if $\mathfrak{S}$ has the following equivalent properties:

i) For every $A \in \mathfrak{S}$ there exists $B \in \mathfrak{S}$ such that $A \cap co(\Gamma \setminus B) = \emptyset$

ii) For every $A \in \mathfrak{S}$ there exists $B \in \mathfrak{S}$ such that for every finite family $(x_i)_{i \in I}$ of elements of Γ with $\sum_{i \in I} x_i \in A$, one has $\sum_{i \in I} P_B(x_i) \leq 1$.

(Here co(E) denotes the convex envelope of the set E).

It is easy to see that for a given set A the properties i) en ii) are equivalent for B. We shall use the abbreviation $A << B$ to indicate this relation between A and B.

Examples

1) If Γ has a bounded base

$$S = \{x \in \Gamma : \ell(x) = 1\}$$

where $\ell \in F'$ and $\ell(x) > 0 \ \ \forall x \in \Gamma \setminus \{0\}$, Γ is $\mathfrak{S}$-conuclear $\mathfrak{S}$ being the set of sets $A_\alpha = \{x : \ell(x) \leq \alpha\}$, $\alpha \geq 0$. More generally:

2) If Γ is the union of closed bounded hats (i.e. convex sets A such that $\Gamma \setminus A$ is convex), Γ is $\mathfrak{S}$-conuclear $\mathfrak{S}$ being the set of closed bounded hats (note that $A << A$ for every hat A).

Note that in an $\mathfrak{S}$-conuclear cone the order intervals are bounded. In fact the relations $0 \leq x \leq a \in A << B$ imply $x \in B$. (since $P_B(x) + P_B(a - x) \leq 1$). Conversely:

3) If F is a quasi-complete conuclear space (e.g. strong dual of a nuclear baralled space) and if the order intervals of are bounded, Γ is $\mathfrak{S}$-conuclear, $\mathfrak{S}$ being the set of all compact convex subsets of Γ containing 0. [10].

Remarks 1. It is not known whether in example 3) Γ is actually the union of its compact hats.

2. In example 3) the compact subsets of Γ are metrizable and so Γ has the Radon-Nikodym property (cf. definition below).

Proposition 7

1. Let Γ be $\mathfrak{S}$-conuclear and let $\Gamma_1 \subset \Gamma$ be a closed convex subcone. Then Γ_1 is $\mathfrak{S}_1$-conuclear, where $\mathfrak{S}_1 = \{\Gamma_1 \cap A : A \in \mathfrak{S}\}$.
2. For $n \in \mathbb{N}$ let Γ_n be $\mathfrak{S}_n$-conuclear. Then $\Gamma = \prod_{n \in \mathbb{N}} \Gamma_n$ is $\mathfrak{S}$-conuclear, $\mathfrak{S}$ being the set of produkts $\prod_{n=1}^{\infty} A_n$ with $A_n \in \mathfrak{S}_n$.
3. Let $T : \Gamma_1 \longrightarrow \Gamma_2$ be a linear and continuous map. Assume Γ_2 is $\mathfrak{S}_2$-conuclear. Let $\mathfrak{S}_1 = \{T^{-1}(A) : A \in \mathfrak{S}_2\}$. If the sets in $\mathfrak{S}_1$ are bounded, Γ_1 is $\mathfrak{S}_1$-conuclear.

Proof 1 Clearly $A << B$ implies $\Gamma_1 \cap A << \Gamma_1 \cap B$.
2. If $A_n << B_n$, $A = \prod_{n \geq 1} A_n$ and $B = \prod_{n \geq 1} 2^n B_n$ we have $A << B$.
3. $A << B$ implies $T^{-1}(A) << T^{-1}(B)$.

Proposition 8 Let Γ be $\mathfrak{S}$-conuclear, the sets in $\mathfrak{S}$ being Suslin sets. The following conditions are equivalent:

1. Every set $B \in \mathfrak{S}$ has the Radon Nikodym property.
2. For every compact space K and every linear map $u : C^+(K) \rightarrow \Gamma$ there exists a positive Radon measure ν on K and a bounded (lusin) ν-measurable map $\rho : K \rightarrow \Gamma$ such that

(6) $\qquad u(f) = \int f\rho \, d\nu \qquad \forall f \in C(K)$

Proof 2) ⇒ 1) Let I be a compact metric space and let $M : B(I) \rightarrow \Gamma$ be a Borel measure such that $P_B(M(A)) \leq P(A)$ for all $A \in B(I)$. With M we associate a map u by integration: $u(f) = \int f dM$. By hypothesis there exists a Radon measure ν and a (lusin) ν-measurable function $\rho : I \rightarrow \Gamma$ such that $\int f dM = \int f\rho d\nu$, and ν may be assumed to be concentrated on $\{t : \rho(t) \neq 0\}$. Then ν is P-continuous and so, by the scalar Radon Nikodym theorem, we may without loss of generality assume $\nu = P$. Now since ρ is assumed to be Lusin P-measurable the image measure $\rho(P)$ is Radon. Now this implies that $\rho(t) \in B$ a.e. ([12] p.69). An appropriate modification of ρ gives the required density. Proposition 5 now shows that B has the Radon-Nikodym property.

1) ⇒ 2). Let $u(1) \in A$ and $A << B$. Then for $0 \leq f \leq 1$, $f \in C(K)$ and any finite partition $\Sigma f_i = f$ of f with $f_i \in C^+(K)$, one has $\Sigma P_B(u f_i) \leq 1$. Thus we may put for any $f \in C^+(K)$,

(7) $\qquad \nu(f) = \sup_{\Sigma f_i = f} \Sigma P_B(u f_i)$

the supremum being taken with respect to all finite partitions of f. It is easy to see that ν is a positive linear form on $C^+(K)$, i.e. a Radon measure. By (7) we have

(8) $\qquad P_B(u f) \leq \nu(f)$

for all $f \in C^+(K)$. It is not hard to see that u extends continuously to a linear map from $\mathcal{L}^1_+(\nu)$ to Γ and that the inequality (8) remains valid for the extension, in

particular for f the indicator function of a Borel subset of K.
Thus we get an inequality like (3) and so there exists a Borel measurable, hence (Lusin) ν-measurable, density $\rho : K \to B$ such that $u(f) = \int f\rho \, d\nu$.

Remark The collection $\mathfrak{S}$ does not occur in the condition 2. This is useful since Γ may be $\mathfrak{S}_1$ and $\mathfrak{S}_2$-conuclear for different collections $\mathfrak{S}_1$ and $\mathfrak{S}_2$.

Definition If Γ is $\mathfrak{S}$-conuclear for some collection $\mathfrak{S}$ of Suslin subsets having the Radon-Nikodym property we shall say that Γ is a conuclear cone having the Radon-Nikodym property.

Examples of conuclear cones with the Radon Nikodym property:

1. Any closed convex cone with a bounded base which is a Suslin set having the Radon Nikodym property.
2. Any closed convex cone Γ, in a quasi-complete conuclear space, such that the order intervals $\Gamma \cap (a - \Gamma)$ are bounded.

Proposition 9

1. Let Γ be a conuclear cone with the Radon Nikodym property and let $\Gamma_1 \subset \Gamma$ be a closed convex subcone. Then Γ_1 is a conuclear cone with the Radon Nikodym property.
2. Let $(\Gamma_n)_{n \in \mathbb{N}}$ be a sequence of conuclear cones with the Radon Nikodym property.

   Then $\prod_{n \in \mathbb{N}} \Gamma_n$ is a conuclear cone with the Radon Nikodym property.

This follows immediatly from propositions 6 and 7.

5. The direct theorem.

Theorem 1 Let Γ be a conuclear cone having the Radon Nikodym property, and such that the closed convex hull of every compact subset of Γ is compact. Then

A) Every point $a \in \Gamma$ has an integral representation by means of extreme generators (i.e. the map $r : M^+(\text{ext}\,\Gamma) \to \Gamma$ is surjective)

B) Every point $a \in \Gamma$ has a unique integral representation by means of extreme generators (i.e. $r : M^+(\text{ext}\,\Gamma) \to \Gamma$ is bijective) if and only if Γ is a Lattice in its own order.

Corollary 1 $\Gamma = \overline{co}\,(\text{ext}\,\Gamma)$.

Corollary 2 Let S be a bounded closed convex Suslin subset of F having the Radon Nikodym property, and such that the closed convex hull of every compact subset of S is compact. Then every point in S is the barycenter of a Radon probability measure on the set of extreme points of S. This probability measure is uniquely determined for each point if and only if S is a simplex.

In fact, it is sufficient to apply the above theorem to the cone in $F \times \mathbb{R}$ with base $\{(x,1) : x \in S\}$. As usual we call S a simplex if this cone is a lattice.

Remark The statement B) in theorem 1 may be improved as follows:

B') The point $a \in \Gamma$ has a unique integral representation if and only if the face $\Gamma(a) = \{x \in \Gamma : \exists \lambda \geq 0 \quad x \leq \lambda a\}$ is a lattice in its own order.

In the proof we may and shall assume that Γ is $\mathfrak{S}$-conuclear where $\mathfrak{S}$ is a hereditary set of Suslin subsets having the Radon Nikodym property, that is, we assume that if $A \in \mathfrak{S}$ and A_1 is a closed convex subset of A containing 0, A_1 also belongs to $\mathfrak{S}$. The Radon Nikodym property will not be necessary in the proof of lemmas 1 to 4.

Lemma 1 Let $M^+_{\mathfrak{S}}$ be the set of conical measures localizable in a bounded Radon measure on some set $A \in \mathfrak{S}$. Then if Γ is $\mathfrak{S}$-conuclear $M^+(\Gamma) = M^+_{\mathfrak{S}}$.

Proof Let $\mu \in M^+(\Gamma)$, $r(\mu) \in A$ and $A << B$. Then $\int P_B \, d\mu = \sup \sum_i P_B(x_i) \leq 1$, the supremum being taken over all finite families $(x_i)_{i \in I}$ of points of Γ such that $\sum \varepsilon_{x_i} < \mu$ (cf. [10] §17). Thus if m is a localization of μ, $\int P_B \, dm \leq 1$. Now if $g(x) = {}^x/P_B(x)$ and $m_1 = g(P_B m)$, m_1 is another localization of μ and m_1 is concentrated on B. Moreover $\int_B dm_1 \leq 1$, a fact to be used hereafter. This proves the inclusin $M^+(\Gamma) \subset M^+_{\mathfrak{S}}$. For the opposite inclusion we need to prove that for every $\mu \in M^+_{\mathfrak{S}}$, $r(\mu) \in F$. This easily follows from lemma 2.

Lemma 2 Let $\hat{B} = \{\mu \in M^+_{\mathfrak{S}} : \int P_B \, d\mu \leq 1\}$. Then $\hat{B}$ is a hat in the cone $M^+_{\mathfrak{S}}$. If B is compact metrizable, so is $\hat{B}$ (for the topology $\alpha(M^+_{\mathfrak{S}}, h(F))$). Moreover $M^+_{\mathfrak{S}} = \cup \hat{K}$ where K runs through the compact (hence metrizable) sets in $\mathfrak{S}$.

Proof The first assertion is obvious. Let $K \in \mathfrak{S}$ be compact let m be a Radon measure on K and let $\{m\}$ be the conical measure localized in m. Then the map $m \to \{m\}$ is continuous from the set of Radon measures on K with total mass ≤ 1, equipped with the vaque topology, onto $\hat{K}$. This proves the second assertion. Let $\mu \in M^+_{\mathfrak{S}}$ be localized in a bounded Radon measure m on a set $A \in \mathfrak{S}$. There exists a partition $A = N + \sum_{n=1}^{\infty} K_n$ of A with $m(N) = 0$ and K_n compact. Let $(\alpha_n)_{n \geq 1}$ be a sequence of positive numbers increasing to $+\infty$ such that $\sum_{n=1}^{\infty} \alpha_n \, m(K_n) \leq 1$. Let K be the closed convex hull of the compact set $\{0\} \cup \bigcup_{n \geq 1} \{\alpha_n^{-1} K_n\}$. Let m_n be the image of $1_{K_n} m$ under the map $x \to \alpha_n^{-1} x$, and let $m' = \sum_{n=1}^{\infty} \alpha_n m_n$. Then m' is a localization of μ on K and $\int dm' \leq 1$. Thus $\mu \in \hat{K}$. Now K is contained in λA for some number $\lambda > 0$, hence $K \in \mathfrak{S}$.

Lemma 3 Let $\nu, \mu \in M^+(\Gamma)$ with $\nu < \mu$ (i.e. $\nu(\phi) \leq \mu(\phi) \quad \forall \phi \in S$).

Then there exists a map $\rho : x \to \rho_x$ from Γ to $M^+(\Gamma)$, such that for each $\phi \in h^+(F)$, $x \to \rho_x(\phi)$ is positive homogeneous, ν-measurable and such that $\mu(\phi) = \int \rho_x(\phi) \, d\nu(x)$ for

all $\phi \in h^+(F)$.
(Such a map ρ will be called a <u>ν-dilation</u>).

<u>Proof</u> By lemma 1 and 2 $M^+(\Gamma)$ is the union of metrizable caps (i.e. compact hats). The same is true therefore of $M^+(\Gamma) \times M^+(\Gamma)$ and of the closed subcone $\tilde{\Gamma}$ consisting of all pairs (ν,μ) with $\nu < \mu$. Now the extreme generators of $\tilde{\Gamma}$ are pairs of the form (ε_x, μ) where $x = r(\mu)$ (see [4] proposition 5)
We can therefore write

$$(\nu,\mu) = \int (\varepsilon,\rho)\, dm\, (\varepsilon,\rho)$$

where m is Radon measure on the extreme points of a metrizable cap H of $\tilde{\Gamma}$. (Recall that the extreme points of H are contained in ext (Γ)). Any mass of m at 0 may be ignored and so we can assume that m is concentrated on the set $H_1 = \{x : P_H(x) = 1\}$. Let n be the image of m under the map $(\varepsilon,\rho) \to r(\rho)$ and let S be the image of H_1 under this map. Then n is concentrated on S and for

$$\phi \in h(F) \quad \int \phi dn = \int \phi(r(\varepsilon))dm(\varepsilon,\rho) = \int \varepsilon(\phi)dm(\varepsilon,\rho) = \nu(\phi)\ .$$

Thus n is a localization of ν. Now m has a desintegration $m = \int_S m_x\, dn(x)$ where m_x is a probability measure on the set of pairs (ε,ρ) such that $r(\rho) = x$. Thus

$$\mu(\phi) = \int \rho(\phi)dm(\varepsilon,\rho) = \int_S dn(x) \int \rho(\phi)dm_x(\varepsilon,\rho) = \int_S \rho_x(\phi)dn(x), \text{ where}$$

$\rho_x(\phi) = \int \rho(\phi)dm_x(\varepsilon,\rho)$. The resultant of ρ_x is x, in fact $\rho_x(\ell) = \int \ell(x)dm_x = \ell(x)$.
The function $x \to \rho_x$ may be extended to $\bigcup_{\lambda \geq 0} \lambda S$ by putting $\rho_{\lambda x} = \lambda \rho_x$, and in an arbitrary homogeneous manner to the remainder of Γ. The so constructed function then is a ν-dilation such that $\mu(\phi) = \int \rho_x(\phi)d\nu(x)$.

<u>Lemma 4</u> (Existence of a strict dilation).
Let $A \in \mathfrak{S}$ and let $\Gamma_A = \bigcup_{\lambda > 0} \lambda A$. Then there exists a map $x \to \rho_x$ from Γ_A to $M^+(\Gamma)$ with the following properties: i) $x \to \rho_x(\phi)$ is homogeneous and ν-measurable for every $\nu \in M^+(\Gamma)$ concentrated on Γ_A. ii) $\varepsilon_x < \rho_x$ for all $x \in \Gamma_A$, with $\varepsilon_x = \rho_x$ if and only if x belongs to ext(Γ).

<u>Proof</u> Let $A^* = A \cap (\Gamma - A)$ be the set of points 'between' 0 and points of A. Then if $A << B$, $A^* \subset B$, hence $A^* = A \cap (B - A)$. This implies that A^* is a Suslin set.

For $x,y \in F$ let $x \wedge y$ be the alternating bilinear form on $F' \times F'$ defined by $x \wedge y\, (\ell,k) = \ell(x)\, k(y) - k(x)\ell(y)$. Then x and y are proportional if and only if $x \wedge y = 0$. Hence the set of pairs (x,y) such that x and y are proportional is a closed subset of $F \times F$.

Now let $E = \{(x,y) \in A \times A^* : 0 \leq y \leq x,\ x \wedge y \neq 0\}$. This set, the intersection of a closed and an open subset of $A \times A^*$ is a Suslin set, and so is, consequently, its first projection $A \setminus \text{ext}(\Gamma)$ (this, incidentally, proves, with the help of lemma 1, that ext(Γ) is universsally measurable). By the von Neumann selection theorem ([9] chapter II theorem 13) there exists a universally measurable map $x \to (x, g(x))$ from

$A \setminus ext(\Gamma)$ to E. We extend g to A by putting for $x \in A \cap ext(\Gamma)$, $g(x) = x$, let g_1 be the restriction of g to $A_1 = \{x : P_A(x) = 1\}$ and let $f_1 : \Gamma_A \to \Gamma$ be the homogeneous extension of g_1 to Γ_A. Then the map $x \to \rho_x = \varepsilon_{f_1(x)} + \varepsilon_{x - f_1(x)}$ has the required property (note that, by proposition 3, every $\nu \in M^+(\Gamma)$ concentrated on Γ_A has a localization on A_1).

<u>Lemma 5</u> Let $u : C^+(K) \to \Gamma$ be a linear map and let $\mu = \Delta(u)$ be the corresponding conical measure, defined for $\phi \in S$ by

$$\mu(\phi) = \sup_{\Sigma f_i = 1} \sum_i \phi(uf_i)$$

where the supremum is taken over all finite partitions of unity. Then μ belongs to $M^+(\Gamma)$. Conversely, every $\mu \in M^+(\Gamma)$ is equal to $\Delta(u)$ for some u.

<u>Proof</u> We know (proposition 8) that $u = \rho\nu$ for some ρ and ν. Thus μ is localized in $\rho(\nu)$ ([10] §14.) Conversely, if $\mu \in M^+(\Gamma)$, μ is localized on some compact set belonging to $\mathfrak{G}$ (lemma 2). Thus $K_\mu \subset \Gamma$ and the converse assertion follows from [10] (theorem 13.4 and 13.3). In this construction $C^+(K)$ is equal to $\bigcup_{\lambda \geq 0} \lambda[0,\mu]$, where $[0,\mu] = \{\nu \in M^+(\Gamma) : 0 \leq \nu \leq \mu\}$. We denote this space by $C^+(\mu)$. The map $u : C^+(\mu) \to \Gamma$ is just the map $\nu \to r(\nu)$. We shall call this the canonical map. We then have $\mu = \Delta(u)$.

<u>Lemma 6</u> Let K_1 and K_2 be compact spaces, let $L : C^+(K_1) \to C^+(K_2)$ be a linear map such that $L\,1 = 1$. Let $u_2 : C^+(K_2) \to \Gamma$ be linear and let $u_1 = u_2 \circ L$. Then $\Delta(u_1) < \Delta(u_2)$. Conversely if $\mu_1, \mu_2 \in M^+(\Gamma)$ are such that $\mu_1 < \mu_2$ there exists a linear map $L : C^+(\mu_1) \to C^+(\mu_2)$, preserving the unit, such that the canonical maps satisfy the relation $u_1 = u_2 \circ L$.

<u>Proof</u> The first assertion follows immediatly from the definition since L transforms partitions of unity into partitions of unity. Conversely, if $\mu_1 < \mu_2$ these exists a μ_1-dilation ρ such that $\mu_2 = \rho(\mu_1)$ (lemma 3). Then ρ is also a ν-dilation for any $\nu \in C^+(\mu_1)$. If we put $L(\nu) = \rho(\nu) = \int \rho_x d\nu(x)$, L has the required property.

<u>Lemma 7</u> Let $(\mu_n)_{n \geq 0}$ be a sequence in $M^+(\Gamma)$ such that $\mu_n < \mu_{n+1}$ for all n. Then for $\mu \in h(F)$ the limit $\mu(\phi) = \lim_{n \to \infty} \mu_n(\phi)$ exists, μ belongs to $M^+(\Gamma)$ and $\mu_n < \mu$ for all n.

<u>Proof</u> Let $a = r(\mu_n) \in A$ and $A << B$. Then $\phi \leq M P_B$ for some M and $\mu_n(\phi) \leq M \int P_B d\mu_n \leq M$ (cf. proof lemma 1). For $\phi \in S$ the sequence $(\mu_n(\phi))_{n \geq 0}$ increases, hence converges. Since $h(F) = S - S$ the sequence converges for all $\phi \in h(F)$, μ is a conical measure and obviously $\mu_n < \mu$ for all n. We have to prove that μ is localizable. By lemma 5 it will be sufficient to prove that $\mu = \Delta(u)$ for some u. Let $C(K_n) = C(\mu_n)$ and let $u_n : C^+(K_n) \to \Gamma$ be the canonical map. We then have $\mu_n = \Delta(u_n)$. By lemma 5 there

exists linear maps $L_n : C^+(K_{n-1}) \to C^+(K_n)$, preserving 1, such that $u_{n-1} = u_n L_n$ for $n \geq 1$.

Let $K = \prod_{i=0}^{\infty} K_i$. We identify the space $C(\prod_{i=0}^{n} K_i)$ with the subspace C_n of C(K) depending only on the first n+1 variables. The union $C_f = \bigcup_n C_n$ is a dense subspace of C(K) and C_f^+ is dense in $C^+(K)$. For $f \in C_n$ we let L_1 work on $f(-,t_1,\ldots,t_n)$. This gives by the uniform continuity of f, a continuous function $L_1 f$ on $\prod_{i=1}^{n} K_i$. Continuing in this way we obtain a function $L_n L_{n-1} \ldots L_1 f$ which is a continuous function on K_n. Now put $uf = u_n L_n L_{n-1} \ldots L_1 f$. Then the map $u : C_f^+ \to \Gamma$ is well defined, linear, and extends continuously to $C^+(K)$. If f belongs to $C(K_n)$, regarded as subspace of C(K), $uf = u_n f$ (strictly $u_n = u \circ j_n$ where $j_n : C(K_n) \hookrightarrow C(K)$ is the natural embedding obtained by composition with the projection onto K_n). Thus by lemma 6 $\Delta(u_n) < \Delta(u)$. Conversely, let $\phi \in S$ and let $\alpha < \Delta(u)(\phi)$. Then there exists a partition of unity $(f_i)_{i \in I}$ such that $\alpha < \sum_{i=I} \phi(uf_i)$, and a suitable approximation shows that the f_i may be chosen from one of the subspaces C_n. Then $uf_i = u_n g_i$ where $g_i = L_n L_{n-1} \ldots L_1 f_i$ Since the g_i form a partition of unity in $C(K_n)$ it follows that $\alpha < \Delta(u_n)(\phi)$. Thus for all $\phi \in S$, $\Delta(u)(\phi) = \sup_n \Delta(u_n)(\phi) = \sup_n \mu_n(\phi) = \mu(\phi)$, i.e. $\mu = \Delta(u)$. This proves that μ is localizable i.e. $\mu \in \mathcal{M}^+(\Gamma)$.

<u>Lemma 8</u> Let X be an ordered set and let $F : X \to \mathbb{R}$ be a bounded non decreasing function. We assume that for every sequence $(x_n)_{n \geq 1}$ in X such that $x_n \leq x_{n+1}$ for all n, there exists $x \in X$ such that $x_n \leq x$ for all n. Then for every $x_o \in X$ there exists $x \geq x_o$ such that $F(x) = \sup_{x \leq y} F(y)$.

The following simple proof, which replaces the earlier proof by transfinite induction, has been kindly indicated to me during the Tagung by H. von Weizsäcker (cf. [13] proof of theorem 2): Let $M(x) = \sup_{y \geq x} F(y)$. Define a sequence $(x_n)_{n \geq 0}$ inductively such that $x_{n-1} \leq x_n$ and $F(x_n) \geq M(x_{n-1}) - \frac{1}{n}$. Now let x majorise x_n for all n. Then $F(x) \geq M(x_{n-1}) - \frac{1}{n} \geq M(x) - \frac{1}{n}$ for all n, hence $F(x) \geq M(x)$, i.e. $F(x) = \sup_{y \geq x} F(y)$.

<u>Proof of theorem 1</u>

A) Let $a \in A \in \mathfrak{C}$ and $A << B$. Let $(\ell_n)_{n \geq 1}$ be a sequence of continuous linear forms separating the points of B* (cf. proof lemma 4) such that $|\ell_n(x)| \leq {}^1/_{n^2}$ for all $x \in B^*$. Let $f(x) = \sum_{n=1}^{\infty} |\ell_n(x)|$. Then f is continuous, strictly subadditive and bounded by a constantM on B*. Let $X = \{\mu \in \mathcal{M}^+(\Gamma) : r(\mu) = a\}$, ordered by the Choquet ordering $\nu < \mu$, and let $F(\mu) = \int f d\mu$. Since every $\mu \in X$ is localized on B in a measure of total mass ≤ 1 (cf. proof lemma 1), we have $F(\mu) \leq M$ for all $\mu \in X$. Also, if $\nu < \mu, F(\nu) \leq F(\mu)$. By lemma 7 we may apply lemma 8 to X and F. Thus for any $\mu_o \in X$, in particular for $\mu_o = \varepsilon_a$, there exists $\mu > \mu_o$ such that $\int f d\mu = \sup_{\mu < \nu} \int f d\nu$. Now let $x \to \rho_x$ be a strict dilation defined on Γ_B (cf. lemma 4). Then if $\nu = \int \rho_x d\mu(x), \nu > \mu$, and $\int f d\nu = \int d\mu(x) \int f d\rho_x \geq \int f(x) d\mu(x)$. Thus since $\int f d\nu = \int f d\mu$, $\int f d\rho_x = f(x)$ μ a.e. (i.e. m a.e. if m is a localization of μ on B). This shows that m, and therefor μ, is concentrated on ext(Γ) (and in particular ext(Γ) $\neq$ (0)).

B') Assume the cone $\Gamma(a) = \{x \in \Gamma : \exists\, \lambda \geq 0,\ x \leq \lambda a\}$ is a lattice in its own order, which is also the order induced by Γ. For $\phi \in S$ let

$$\mu_a(\phi) = \sup_{\nu \in D_a} \nu(\phi)$$

where D_a is the set of conical measures $\sum_{i \in I} \varepsilon_{x_i}$, with I finite, $x_i \in \Gamma$, and $\sum_{i \in I} x_i = a$. Then we also have $\mu_a(\phi) = \lim_{\nu \in D_a} \nu(\phi)$, D_a being ordered by the Choquet ordering. Hence the limit exists for all $\phi \in h(F)$ and defined a conical measure μ_a with resultant a. We shall prove that μ_a is localizable. Observe that $D_a = \bigcup_{n \geq 1} D_a^n$, where $D_a^n = \{\nu = \sum_{i=1}^{n} \varepsilon_{x_i},\ \nu \in D_a\}$. Also, D_a^n is the image of the set $[o,a]^n$, ($[o,a]$ being the order interval between o and a) under the continuous map $(x_1, \ldots, x_n) \to \sum_{i=1}^{n} \varepsilon_{x_i}$. This shows that D_a is a Suslin set, and in particular separable.

Let $(\nu_n)_{n \geq 1}$ be a dense sequence in D_a. Then for every $\phi \in S$, $\mu_a(\phi) = \sup_{n \geq 1} \nu_n(\phi)$. Using the Riesz decomposition property of $\Gamma(a)$ we can inductively define a sequence $(\mu_n)_{n \geq 1}$ in D_a such that $\nu_n < \mu_n$ and $\mu_{n-1} < \mu_n$. Then $\mu_a(\phi) = \sup_n \mu_n(\phi)$ for all $\phi \in S$ and μ_a is localizable by lemma 7. Now if μ is any other localizable conical measure with resultant a, we have, for every $\phi \in S$, $\mu(\phi) = \sup_{\substack{\nu \in D_a \\ \nu < \mu}} \nu(\phi) \leq \mu_a(\phi)$ (cf. [10] 17.4 and 17.5). Hence there exists, by lemma 3, a dilation ρ such that $\mu_a = \int \rho_x \, d\mu(x)$. Now if μ is concentrated on ext (Γ) this gives $\mu_a = \mu$, hence the uniqueness.

6. A converse theorem.

A consequence of proposition 7, part 1, is that under the hypothesis of theorem 1, Γ has the integral representation property (cf. § 2). Also, by proposition 7, part 2, $\Gamma^{\mathbb{N}}$ has the integral representation property. This proves one of the implications in the following theorem.

Theorem 2. Let Γ be an $\mathfrak{S}$-conuclear cone, the sets in $\mathfrak{S}$ being Suslin sets. Assume that the closed convex hull of a compact subset of Γ is compact. Then Γ has the Radon Nikodym property if and only if $\Gamma^{\mathbb{N}}$ has the integral representation property.

Corollary. Let $S \subset F$ be a closed bounded convex Suslin set, such that the closed convex hull of a compact subset of S is compact. Then S has the Radon Nikodym property if and only if $S^{\mathbb{N}}$ has the integral representation property (i.e. the theorem of Choquet-Edgar is valid for all closed convex subsets of $S^{\mathbb{N}}$).

Proof of theorem 2. By the remarks above it suffices to prove that if $\Gamma^{\mathbb{N}}$ has the integral representation property, Γ has the Radon Nikodym property. We have to show that a Radon measure (u: $C(K)^+ \to \Gamma$) can be written in the form $u = \rho \cdot \nu$. By proposition 5 (see also the proof of proposition 8) it suffices to prove this in the case where $K = \{0,1\}^{\mathbb{N}}$. In this case $K = \varprojlim_{n \geq 1} K_n$ is the projective limit if a sequence of finite sets. The spaces $C(K_n)$ can be viewed as subspaces of $C(K)$. If $u_n = u_{/C^+(K_n)}$, the map $u \to (u_n)_{n\geq 1}$ is a injection of $L(C^+(K);\Gamma)$ into the projective limit $\lim\limits_{\leftarrow n} L(C^+(K_n);\Gamma) = \tilde{\Gamma}$, defined by the obvious compatibility conditions. Since $C_f^+ = \bigcup\limits_{n\geq 1} C(K_n)$ is dense in $C^+(K)$ and since every linear map $u : C_f^+ \to \Gamma$ can be continuously extended to $C^+(K)$, this map is actually onto. Since K_n is finite $L(C^+(K_n);\Gamma)$ is isomorphic to Γ^{K_n} (the vector measure is determined by its "atoms"). Hence $\tilde{\Gamma}$ is a closed subcone of $\prod\limits_{n\geq 1} \Gamma^{K_n} \simeq \Gamma^{\mathbb{N}}$. The corresponding topology on $L(C^+(K);\Gamma)$ is the topology of pointwise convergence on $C_f^+(K)$. The extreme generators of $L(C^+(K);\Gamma)$ are obviously of the form $\delta_t \otimes e$ with $e \in \text{ext}(\Gamma)$ and $t \in K$. By proposition 7, 3) the cone $\tilde{\Gamma}$ with its projective limit topology, is $\tilde{\mathfrak{G}}$ conuclear, where $\tilde{\mathfrak{G}}$ is the set of subsets $\tilde{B}$ of the form

$$\tilde{B} = \{u : u(1) \in B\}$$

with $B \in \mathfrak{G}$. Moreover the map $(t,e) \to \delta_{(t)} \otimes e$ is a homeomorphism of $K \times [B \cap \text{ext}(\Gamma)]$ onto the set of extreme generators of $\tilde{\Gamma}$ belonging to $\tilde{B}$. Thus lemma 1 implies that every element $u \in L(C^+(K);\Gamma)$ has an integral representation

$$u = \iint_{K \times B} \delta_{(t)} \otimes e \, dm(t,e)$$

for some $B \in \mathfrak{G}$. This means

$$< u(f), \ell > = \iint_{K \times B} f(t) < \ell, e > dm(t,e)$$

for all $\ell \in F'$ and all $f \in C_f$, hence for all $f \in C(K)$.

If ν is the projection of m onto K we may desintegrate m as follows

$$m = \int m_t \, d\nu(t)$$

where m_t is a probability measure on B, and write

$$< u(f), \ell > = \int f(t) \, d\nu(t) \int < \ell, e > dm_t(e)$$

i.e.

$$u(f) = \int f(t) \, \rho(t) \, d\nu(t)$$

where $\rho(t) = \int e \, dm_t(e) = r(m_t) \in B$. The function $\rho : K \to B$ is scalarly measurable (and can be taken to be Scalarly Borel) hence measurable (or even Borel measurable), B being a Suslin space. Thus the theorem is proved.

Remark. For the converse we only use the existence of integral representations in closed convex subcones of $\Gamma^{\mathbb{N}}$.

Proof of the corollary: If S has the Radon-Nikodym property so does $S^{\mathbb{N}}$, hence $S^{\mathbb{N}}$ has the inte gral representation property by the corollary of theorem 1.

For the converse we assume that every closed convex subset A of $S^{\mathbb{N}}$ is such that every point of A is the resultant of a Radon measure concentrated on the set of extreme points of A. Let Γ be the cone with basis $S \times \{1\}$ in $F \times R$ and let $S_1 = \{(\lambda x, \lambda) : x \in S, 0 \leq \lambda \leq 1\}$. We may assume that S contains more than one point, and so contains a line segment isomorphic to the interval [0,1]. The set S_1 is isomorphic to a subset of $S \times [0,1]$ (it is a subset if $0 \in S$). Thus S_1^N is isomorphic to a closed convex subset of $S^N \times [0,1]^N$, itself isomorphic to a closed convex subset of $S^N \times S^N \approx S^N$. Thus, for every closed convex subset of S_1^N one has the existence of integral representations.The same is true for the sets $\prod_{n=1}^{\infty} \lambda_n S_1$, for any sequence of positive numbers $(\lambda_n)_{n \geq 1}$, since these products are isomorphic to S_1^N. Now these sets are contained in hats H of Γ^N,such that $\prod_{n=1}^{\infty} \lambda_n S_1 \subset H \subset \prod_{n=1}^{\infty} 2^n \lambda_n S_1$ and they cover Γ^N (cf.[14]p.95). Thus every closed convex subcone $\tilde{\Gamma}$ of Γ^N is the union of hats which are isomorphic to closed convex subsets of S^N.This implies that every point in $\tilde{\Gamma}$ has an integral representation by means of extreme generators.Also, Γ is $\mathfrak{S}$ -conuclear, $\mathfrak{S}$ being the set of homothetics of S_1, which are Suslin sets. Thus by theorem 2 (and the remark after the proof) we may conclude that Γ, and therefore S, has the Radon-Nikodym property.

REFERENCES

[1] S.D.Chatterji, Martingale convergence and the Radon-Nikodym theorem in Banach spaces.Math.Skand. 22 (1968) pp.21 -41.

[2] G.Choquet, Mesures Coniques, affines et cylindriques. Symposica Mathematica Vol.II pp.145-182 (Acad.Press 1969).

[3] G.Choquet,Lectures on Analysis (Benjamin 1969).

[4] R.Becker, Some consequences of a kind of Hahn-Banach theorem.Séminaire Choquet, 17e année 1977/78 no.2 .

[5] G.A.Edgar ,A noncompact Choquet theorem,Proc. Amer. Math. Soc. 49 (1975) pp.354-358.

[6] R.D.Bourgin and G.A.Edgar, Noncompact Simpexes in Banach spaces with the Radon-Nikodym property.Journ. Func.Anal.23 (1976) pp.162-176.

[7] G.A.Edgar, On the Radon-Nikodym-property and martingale convergence.Proceedings of the Conference on Vector Space Measures and Applications,Dublin 1977, Springer Lecture Notes 645.

[8] P. Halmos, Measure Theory (Van Nostrand).

[9] L.Schwartz, Radon Measures on Arbitrary Topological Spaces and Cylindrical Measures, (Oxford University Press 1973).

[10] E.G.F.Thomas ,Integral Representations in convex cones,Report no.ZW-7703, University of Groningen Mathematics Institute (1977).

[11] E.G.F. Thomas, Représentations intégrales dans les cones convexes conucléaires, et applications.Séminaire Choquet 17e année 1977/78 no.9 .

[12] E.G.F.Thomas, Integration of functions with values in locally convex Suslin spaces. Trans. Amer. Math. Soc. Vol. 212 (1975) pp. 61-81.

[13] H. von Weizsäcker and G. Winkler, Non-compact extremal integral representations: some probabilistic aspects. (To appear).

[14] R.Phelps, Lectures on Choquets theorem (Van Nostrand).

Mathematisch Instituut
Universiteit van Groningen
Postbus 800
9700 AV Groningen
Netherlands

GAUSSIAN SURFACE MEASURES AND THE RADON TRANSFORM ON SEPARABLE BANACH SPACES

Alexander Hertle

Introduction.

In this paper, we consider a real separable Banach space B together with a (fixed) non degenerate Gaussian measure μ on B. Our first aim is to show the existence of the surface measure induced by μ on the spheres and hyperplanes of B. Using a canonical desintegration, we derive an analog of Fubini's theorem and polar integration in Banach spaces. Next, we apply these results to generalize the Radon transform on $\mathbb{R}^n$ to separable Banach spaces. Finally, some applications of this Radon transform are indicated, e.g. uniquely determinedness of a function on B by its μ-surface integrals along all hyperplanes. Further, the determination of a Gaussian process from certain conditional expectations is shown.

These results are presented in detail for the case of a separable Hilbert space H. How they generalize to separable Banach spaces is sketched.

In Section 1, we show that for all spheres and hyperplanes in H the surface measures (induced by μ) exist, by proving the existence of the following limit for all $f \in C_b(H)$:

$$\int_M f \, d\mu_M = \lim_{h \to 0} \frac{1}{2h} \int_{d(x,M) \leq h} f(x) \, d\mu(x) ,$$

where d denotes the distance and M a sphere or hyperplane in H. We give an explicite expression for the μ-area of spheres and hyperplanes in H. Applying strict desintegration, we obtain a Fubini and polar decomposition of μ, which yields that all these surface measures have full support.

In Section 2, we use the results of Section 1 to generalize the Radon transform to separable Hilbert spaces. - Identifying the hyperplanes in $\mathbb{R}^n$ with $S^{n-1} \times \mathbb{R}$, the Radon transform of L^1-functions is defined as an operator R from $L^1(\mathbb{R}^n)$ to $L^1(S^{n-1} \times \mathbb{R})$ by

$$(1) \qquad (Rf)(x,p) = \int_{x \cdot y = p} f(y)\, dy \, .$$

This transform was studied for rapidly decreasing functions first by Radon [11] and John [7]. In [5], the author studied the Radon transform on L^1, and extended it from functions to measures via

$$(2) \qquad (Rf)(x,p) = \frac{\partial}{\partial p} \int_{x \cdot y < p} f(y)\, dy$$

as follows: The <u>Radon transform of measures on $\mathbb{R}^n$</u> is defined as an operator from $M(\mathbb{R}^n)$ to $M(S^{n-1} \times \mathbb{R})$ by (cf. [5])

$$(3) \qquad (Rm)(g) = - \int_{S^{n-1}} \int_{\mathbb{R}} \frac{\partial}{\partial p} g(x,p) \int_{x \cdot y < p} dm(y)\, dp\, dx \, .$$

Denoting by μ_S the Gaussian surface measure induced by μ on the unit sphere S of H (Section 1), we now call the operator R from M(H) to $M(S \times \mathbb{R})$, defined by

$$(3') \qquad (Rm)(g) = - \int_{S} \int_{\mathbb{R}} \frac{\partial}{\partial p} g(x,p) \int_{x \cdot y < p} dm(y)\, dp\, d\mu_S(x) \, ,$$

the <u>Radon transform of measures on (H,μ)</u>. If $H = \mathbb{R}^n$ and μ the normal distribution, this definition reduces to (3). The following connection between the Radon and the Fourier transform (the characteristic functional) holds:

$$(Rm)(g) = (2\pi)^{-1/2} \int_S \int_{\mathbb{R}} (F_p g)(x,p)(Fm)(px)\, dp\, d\mu_S(x) \ ,$$

and yields the injectivity of the Radon transform on (H,μ). Similarly, we define the Radon transform of functions on H: Denoting by $\mu_{x,p}$ the surface measure induced by μ on the hyperplane $x \cdot y = p$, we call the operator R from $C_b(H)$ to $L^1(S \times \mathbb{R})$, defined by

$$(1') \qquad (Rf)(x,p) = \int_{x \cdot y = p} f(y)\, d\mu_{x,p}(y) \ ,$$

the <u>Radon transform of functions on (H,μ)</u>. By an analog of formula (2), the definitions of the Radon transform of functions and measures are consistent. From the injectivity of R we obtain that a continuous function (resp. a measure) on H is uniquely determined by its surface integrals along all hyperplanes (resp. by its values on all half spaces).

Finally, as an application of the Radon transform on Hilbert spaces, we show by using desintegration that the reconstruction of a measure from its integrals along hyper surfaces (e.g. hyperplanes) gives (via Radon transform) a method to reconstruct a Gaussian process from conditional expectations.

In a forthcoming paper, the main theorems for the Radon transform on $\mathbb{R}^n$ (e.g. inversion formula, Helgason's support theorem, John's theorem) are generalized to the infinite dimensional case, and applications are treated more explicitely.

Throughout this paper we shall use the notations of Hörmander [6] and Ludwig [9]. In particular, we denote by M(X) the Banach space of finite signed Borel measures on a polish space X with the norm of total variation. The Fourier transform F is defined on $\mathbb{R}^n$ by

$$(Ff)(y) = (2\pi)^{-n/2} \int f(z) e^{-iy\cdot z}\, dz ,$$

and on a Banach space by

$$(Fm)(y) = \int e^{i(x,y)}\, dm(y) .$$

The basic ideas of this paper are part of the author's thesis [4], supervised by Professor D. Kölzow to whom the author is indebted for initiating the present work. The author thanks also Professor E. Thomas for the suggestion to apply the desintegration theorem.

1. Gaussian surface measures on Hilbert and Banach spaces.

Let H be a separable real Hilbert space with unit sphere S and μ a non degenerate Gaussian measure with mean 0 on H. We denote by A the positive (nuclear) covariance operator of μ and by λ_k, $\lambda_1 \geq \lambda_2 \geq \cdots$, the eigenvalues of A counted by their multiplicity. We remark that we can regard (H,μ) as an abstract Wiener space in the sense of Gross [3], μ being the abstract Wiener measure on H.

Let M be a hypersurface in H. One says that μ induces a Gaussian surface measure μ_M on M, if the following limit exists for all $f \in C_b(H)$:

$$\int_M f\, d\mu_M = \lim_{h\to 0} \frac{1}{2h} \int_{d(x,M)\leq h} f(x)\, d\mu(x) ,$$

d being the distance in H. If the limit exists, μ_M is a measure on H and $\mu_M(M)$ is called the <u>μ-area</u> of M (cf. Skorohod [14], §27).

We recall the desintegration theorem (cf. Schwartz [13]):

If X and Y are polish spaces, $m \in M(X)$ and $p: X \to Y$ measurable, then there exists a family $(m_y)_{y \in Y}$ of Borel probability measures on X such that

(1) $\quad supp(m_y) \subset p^{-1}(y)$

(2) $$\int_X f\, dm = \int_Y \int_{p^{-1}(y)} f(x)\, dm_y(x)\, d(pm)(y)\ , \quad f \in C_b(X).$$

In the notation of Skorohod [14], these m_y are exactly the conditional measures of m under p.

THEOREM 1.1. *Let* $S_r = \{(x,x)=r\}$ *be the sphere with radius* $r^{1/2}$ *in* H. *Then, the Gaussian surface measure* μ_r *(induced by* μ*) on* S_r *exists. This measure has full support, i.e.* $supp(\mu_r) = S_r$*, and the* μ*-area of* S_r *is given by*

$$\mu_r(S_r) = \frac{1}{2\pi i} \int_{c-i\infty}^{c+i\infty} \prod_{k=1}^{\infty} (1+2p\lambda_k)^{-1/2} e^{pr}\, dp\ , \quad c>0$$

$$= \mathop{\ast}_{k=1}^{\infty} (2\pi t\,\lambda_k)^{-1/2}\, e^{-t/2\lambda_k}\,(r)\ ,$$

where $\ast$ *denotes the convolution* $f \ast g(r) = \int_0^r f(t)g(r-t)\, dt$.

Proof: We first prove the existence of the μ-area on the spheres S_r. To do this, we consider the distribution function

$$F(r) = \mu(\{(x,x) \le r\})$$

of μ under the norm. We have to show that F is differentiable on $\mathbb{R}_+$. First, note that F is monotone, F(0)=0, and thus F is of bounded variation. Hence, we can compute the Laplace-Stieltjes transform of F (which is "formally" the Laplace transform L(F') of F') as follows:

$$L(F')(p) = \int_0^\infty e^{-pr}\, dF(r) = \int_H e^{-p(x,x)}\, d\mu(x)$$

$$= \prod_{k=1}^{\infty} (1 + 2p\lambda_k)^{-1/2} ,$$

since μ is a product of one dimensional Gaussian measures with variances λ_k on the system of eigenvectors of A. Of course, $L(F')$ is an analytic function in the half plane $p>0$. We now have to certify that we can apply the Laplace inversion to $L(F')$. Since we have for all $n \in \mathbb{N}$ and $t \in \mathbb{R}$:

$$\left| \prod_{k=1}^{\infty} (1 + 2it\,\lambda_k) \right| \geq 2^n \lambda_n^n t^n ,$$

we see that $L(F')$ is rapidly decreasing on the line $p=0$. This means that we can apply the Laplace inversion to $L(F')$ and obtain

$$\mu_r(S_r) = F'(r) = \frac{1}{2\pi i} \int_{c - i\infty}^{c + i\infty} \prod_{k=1}^{\infty} (1 + 2p\lambda_k)^{-1/2} e^{pr}\, dp ,$$

where c is an arbitrary positive constant. Observing that

$$L((2\pi t\,\lambda_k)^{-1/2} e^{-t/2\lambda_k}) = (1 + 2p\lambda_k)^{-1/2} ,$$

we obtain from the convolution property of the Laplace transform the assertion in the Theorem. Further, from the rapid decrease of $L(F')$ in the imaginary direction, we can conclude from the inversion formula that F' must be a continuous (even C^∞) function, in particular F' exists everywhere.

Next, we have to show the existence of the surface integrals on S_r for all $f \in C_b(H)$. Observe that this follows for almost all radii r from the desintegration theorem and from the existence of the μ-area on S_r (cf. Skorohod [14] §27, and Remark 1.1 in this paper). To show the existence for all r, first recall that on H the Borel algebras generated by the weak and norm topology coincide (cf. Schwartz [12]). Therefore, we restrict ourselves to weakly continuous functions f on H. Now the closed balls in H are weakly compact and metrizable. Thus, by the Stone-Weierstraß and Hahn-Banach theorem, the functions f of the form $f(x) = e^{(x,z)}$, $z \in H$, are uniformly dense in the weak continuous functions on balls. So we only prove the existence of the surface integrals for these special functions and to do this, let us again consider the corresponding monotone distribution function

$$F(r) = \int_{(x,x)\leq r} e^{(x,z)}\, d\mu(x) ,$$

and obtain now

$$L(F')(p) = \prod_{k=1}^{\infty} (1 + 2p\lambda_k)^{-1/2} \quad e^{(A(I + 2pA)^{-1}z,z)/2} .$$

Now, the following estimate holds:

$$\begin{aligned} e^{(A(I + 2pA)^{-1}z,z)} &\leq e^{\|A/(I + 2pA)\|(z,z)} \\ &= e^{\|A\|(z,z)/(1 + 2p\|A\|)} \\ &\leq e^{\|A\|(z,z)} . \end{aligned}$$

Thus, L(F') again is rapidly decreasing in the imaginary direction. The same reasoning as before gives that F' is a continuous function and hence the surface integral of $e^{(x,z)}$ over S_r exists for all r. In view of our above consideration and using the method of Skorohod [14], §21 Theorem 1, we obtain the existence for all $f \in C_b(H)$.

It remains to show that $\operatorname{supp}(\mu_r) = S_r$. From the desintegration we know that μ_r has support in S_r. In order to see that μ_r has full support in S_r, we consider without loss of generality only the unit sphere S. Therefore, let G be a non void open subset of S and assume

$\mu_1(G)=0$. We can find a vector $x \in G$ of the form

$$x = \sum_{k=1}^{n} x_k e_k \ .$$

Note that $x \in H_o = A^{1/2}(H)$ and H_o is dense in H. Now let

$$y = \sum_{k=1}^{m} y_k e_k$$

be another finite combination of the e_k, lying in S. Then there exists an orthogonal operator U_y on H, which rotates x onto y. Indeed, we can choose $U_y = U_o \oplus I_o$, where U_o is a (suitable) orthogonal operator on the span of $e_1, \dots, e_{max(n,m)}$ and I_o the identity on the orthogonal complement of that span. Note that $I - U_y$ has finite rank. For such U_y we now easily conclude from Skorohod [14], §25 Theorem 4, that $U_y(\mu)$ is equivalent to μ. Since H_o is dense in H, all vectors y of the above form are dense in S, i.e. $S = \bigcup_y U_y(G)$. But then the assumption $\mu_1(G) = \mu_1(U_y G) = 0$ would imply the contradiction $\mu_1(S)=0$. Thus, all the μ_r have full support in S_r and the Theorem is proved completely.

<u>COROLLARY 1.1</u> (Polar integration in Hilbert space). <u>Let</u> $f \in C_b(H)$, <u>then</u>

$$\int_H f \, d\mu = \int_{\mathbb{R}_+} \int_{S_r} f(y) \, d\mu_r(y) \, dr \ .$$

<u>Proof</u>: We apply the desintegration theorem mentioned above to the special case $X = H$, $Y = \mathbb{R}_+$ and $p(x)=(x,x)$. Then we have in this case $(p\mu)(r) = \mu_r(S_r)$. On the other hand, from Lebesgue's differentiation theorem on $\mathbb{R}_+$, we obtain that the desintegrated measures of μ are exactly the surface measures $\bar{\mu}_{S_r}$ in the following Remark 1.1. The

Corollary now follows from the connection between $\overline{\mu}_{S_r}$ and μ_r in Remark 1.1.

EXAMPLES. (1) Let $H = L^2[0,1]$ and μ the classical Wiener measure on L^2, which covariance operator A is given by

$$(Af)(t) = \int_0^1 \min(s,t)f(s)\,ds .$$

Then $\lambda_k = ((n - 1/2)\pi)^{-2}$, and we obtain

$$\prod_{k=1}^{\infty}(1 + 2p\lambda_k)^{-1} = 2/(e^{(2p)^{1/2}} + e^{-(2p)^{1/2}}) = \operatorname{sech}(2p)^{1/2} .$$

(2) If the eigenvalues of A are $\lambda_k = (2n^2\pi^2)^{-1}$, then

$$\prod_{k=1}^{\infty}(1 + 2p\lambda_k)^{-1} = 2p^{1/2}/(e^{p^{1/2}} - e^{-p^{1/2}}) .$$

THEOREM 1.2. *Let $x\cdot y=p$ be the hyperplane with normal $x \in S$ and distance $p \in \mathbb{R}$ from the origin in H. Then, the Gaussian surface measure $\mu_{x,p}$ (induced by μ) on $x\cdot y=p$ exists. This measure has full support, i.e. $\operatorname{supp}(\mu_{x,p}) = \{x\cdot y=p\}$, and the μ-area of $x\cdot y=p$ is given by*

$$\mu_{x,p}(x\cdot y=p) = (2\pi(Ax,x))^{-1/2}\, e^{-p^2/2(Ax,x)} .$$

Proof: Again, we first show the existence of the μ-area on all hyperplanes $x\cdot y=p$ in H. The rest of the Theorem is proved in the same way as in the proof of Theorem 1.1. Thus, let us consider the distribution function

$$F(p) = \int_{x\cdot y<p} d\mu(y) .$$

We want to show that F is differentiable (and even C^∞) on $\mathbb{R}$. We have

$$F'(p) = \frac{\partial}{\partial p}\int_{x\cdot y<p} d\mu(y) = \frac{\partial}{\partial p}\int_{-\infty}^{p} d(P_x\mu)(p) ,$$

where P_x denotes the orthogonal projection on the subspace of H, spanned by x. Since $(F\mu)(y) = e^{-(Ay,y)/2}$, we obtain

$$(FP_x\mu)(p) = e^{-p^2(Ax,x)/2} .$$

Hence, $P_x\mu$ is absolutely continuous with respect to μ and Fourier inversion gives that

$$(P_x\mu)(p) = (2\pi(Ax,x))^{-1/2} e^{-p^2/2(Ax,x)} .$$

Thus, F is C^∞ on $\mathbb{R}$ and is given as in the Theorem, which proves the existence of the μ-area on $x\cdot y=p$.

COROLLARY 1.2 ("Fubini's theorem" in Hilbert space). *Let* $f\in C_b(H)$ *and* $x\in S$; *then*

$$\int_H f\,d\mu = \int_{\mathbb{R}}\int_{x\cdot y=p} f(y)\,d\mu_{x,p}(y)\,dp .$$

Proof: The proof is the same as in Corollary 1.1.

REMARK 1.1. For the definition of a surface measure, we have used the direct generalization of the Euclidean surface measure on $\mathbb{R}^n$. Another definition is (cf. Skorohod [14], §27):

$$\int_M f\,d\overline{\mu}_M = \lim_{h\to 0}\frac{1}{\mu(d(x,M)\le h)}\int_{d(x,M)\le h} f(x)\,d\mu(x) .$$

Now, by Theorem 1.1 and 1.2 we have the following connection in the case of spheres or hyperplanes (cf. Skorohod [14]):

$$\int_M f \, d\mu_M = \mu_M(M) \int_M f \, d\bar{\mu}_M .$$

Note the following connection to the desintegration theorem: If $X = H$, $Y = \mathbb{R}$ and p a measurable function on H, then we obtain for the desintegrated measures μ_y of μ: $\mu_y = \bar{\mu}_{p^{-1}(y)}$ for almost all $y \in \mathbb{R}$.

We also remark that Goodman [2] has defined a surface measure, locally induced by a lower dimensional Wiener space.

REMARK 1.2. We now consider the case of a separable Banach space B. First note that the desintegration theorem can be applied here as well. Now let $x' \in B'$, $\|x'\| = 1$, and $x'(y)=p$ a hyperplane in B. The same reasoning as in Theorem 1.2 gives that

$$\mu_{x',p}(x'(y)=p) = (2\pi\sigma^2(x'))^{-1/2} e^{-p^2/2\sigma^2(x')} ,$$

where $\sigma^2(x') = \int_B x'(y) \, d\mu(y)$ denotes the variance of the Gaussian distribution of x'. Thus, we have the analog of Theorem 1.2 and Corollary 1.2 for B.

In the case of spheres, we cannot derive an explicite analog of Theorem 1.1. But if we consider the classical Wiener space $B = C_o[0,1]$ with the Wiener measure w, we have from Beekman [1] that

$$w(\|x\| \leq r) = \frac{4}{\pi} \sum_{k=0}^{\infty} \frac{(-1)^k}{2k+1} \exp\left(-\frac{(2k+1)^2\pi^2}{16r^2}\right) .$$

Differentiating this equation with respect to r gives a continuous function, which is exactly the w-area of the sphere $\{\|x\| = r\}$. Thus, for the classical Wiener space we also have an analog of Theorem 1.1 and Corollary 1.1.

2. The Radon transform on separable Hilbert and Banach spaces.

We identify the hyperplanes in $\mathbb{R}^n$ with $S^{n-1} \times \mathbb{R}$ and consider $S^{n-1} \times \mathbb{R}$ as a measure space, provided with the product of the Euclidean surface measure on S^{n-1} and the Lebesgue measure λ on $\mathbb{R}$. The Radon transform of L^1-functions is defined as an operator from $L^1(\mathbb{R}^n)$ to $L^1(S^{n-1} \times \mathbb{R})$ by

$$(2.1) \qquad (Rf)(x,p) = \int_{x\cdot y=p} f(y)\, dy .$$

Obviously, the question of reconstructing a function from its integrals along hyperplanes is the problem of inverting the Radon transform, which was studied for rapidly decreasing functions first by Radon [11] and John [7]. In [5], the author studied the Radon transform on L^1 and extended it from functions to measures via

$$(2.2) \qquad (Rf)(x,p) = \frac{\partial}{\partial p} \int_{x\cdot y<p} f(y)\, dy$$

as follows: The Radon transform of measures on $\mathbb{R}^n$ is defined as an operator from $M(\mathbb{R}^n)$ to $M(S^{n-1} \times \mathbb{R})$ by (cf. [5])

$$(2.3) \qquad (Rm)(g) = - \int_{S^{n-1}} \int_{\mathbb{R}} \frac{\partial}{\partial p} g(x,p) \int_{x\cdot y<p} dm(y)\, dp\, dx ,$$

where $g \in \mathcal{S}(S^{n-1} \times \mathbb{R})$. We now generalize, using the results of Section 1, the Radon transform to separable Hilbert spaces (H,μ). Denoting by μ_S the Gaussian surface measure induced by μ on the sphere S of H, we call the operator R from $M(H)$ to $M(S \times \mathbb{R})$, defined by

$$(2.3)' \qquad (Rm)(g) = - \int_{S} \int_{\mathbb{R}} \frac{\partial}{\partial p} g(x,p) \int_{x\cdot y<p} dm(y)\, dp\, d\mu_S(x) ,$$

the Radon transform of measures on (H,μ). Note that Rm is well defined: Rm is a densely defined functional on $C_b(S \times \mathbb{R})$ (the function g has to be C^1 in p), and we have the following equalities:

$$\int_S \int_{\mathbb{R}} \int_{x\cdot y<p} \frac{\partial}{\partial p} g(x,p)\ dm(y)\ dp\ d\mu_S(x) \quad =$$

$$\int_S \int_H \int_{x\cdot y}^{\infty} \frac{\partial}{\partial p} g(x,p)\ dp\ dm(y)\ d\mu_S(x) \quad =$$

$$- \int_S \int_H g(x,x\cdot y)\ dm(y)\ d\mu_S(x) \ .$$

Hence, from the estimate

$$|(Rm)(g)| \leq |m|(H)\ \mu_S(S)\ \|g\|_\infty < \infty$$

we obtain that Rm is a finite measure on $S \times \mathbb{R}$.

REMARK 2.1. If $H = \mathbb{R}^n$ and μ the normal distribution in $\mathbb{R}^n$, the definition of the Radon transform on (H,μ) reduces to (2.3).

COROLLARY 2.1. If $m \in M(H)$ and $g \in C_b(S \times \mathbb{R})$, then

$$(Rm)(g) \quad = \quad \int_S \int_H g(x,x\cdot y)\ dm(y)\ d\mu_S(x) \ .$$

PROPOSITION 2.1. The Radon transform R from M(H) to $M(S \times \mathbb{R})$ is weak-weak and norm-norm continuous, but weak-norm discontinuous.

Proof: The proof is, up to a few changes, analoguously as in the corresponding finite dimensional theorem (cf.[5]).

PROPOSITION 2.2. Let $m \in M(H)$, $g \in C_b(S \times \mathbb{R})$ and $g(x,\cdot) \in C_o^\infty(\mathbb{R})$; then

$$(Rm)(g) \quad = \quad (2\pi)^{-1/2} \int_S \int_{\mathbb{R}} (F_p g)(x,p)\ (Fm)(px)\ dp\ d\mu_S(x) \ ,$$

<u>where</u> F_p <u>denotes the Fourier transform with respect to</u> p <u>and</u> Fm <u>is the characteristic functional of</u> m.

<u>Proof</u>: Again, this proof is with some changes the same one as for the analoguous connection between F and R in $\mathbb{R}^n$ (cf. [5]).

COROLLARY 2.2. <u>The Radon transform</u> R <u>from</u> M(H) <u>to</u> M(S $\times$ $\mathbb{R}$) <u>is injective</u>.

<u>Proof</u>: The injectivity of R follows from the formula in Proposition 2.2, since F_p is injective on $C_o(\mathbb{R})$, the Fourier transform F is injective on M(H) (Prohorov [10]), and the measure μ_S has full support in S (Theorem 1.1).

REMARK 2.2. From the injectivity of the Radon transform we obtain that a finite measure on H is already determined by its values on all half spaces.

We now define, analoguously to (2.1), the Radon transform of functions on (H,μ). Denoting by $\mu_{x,p}$ the surface measure induced by μ on the hyperplane x·y=p in H, we call the operator R from $C_b(H)$ to $L^1_{\mu_S \otimes \lambda}(S \times \mathbb{R})$, defined by

$$(Rf)(x,p) = \int_{x\cdot y=p} f(y)\, d\mu_{x,p}(y) .$$

the <u>Radon transform of functions on (H,μ)</u>. Note that we have

$$(2.2)' \qquad (Rf)(x,p) = \frac{\partial}{\partial p} \int_{x\cdot y<p} f(y)\, d\mu(y) .$$

The following proposition implies the consistency of the Radon transform of functions and measures:

<u>PROPOSITION 2.3</u>. <u>Let</u> $m \in M(H)$ <u>be absolutely continuous with respect to</u> μ <u>and</u> $f = dm/d\mu$. <u>Then for all</u> $g \in C_b(S \times \mathbb{R})$

$$(Rm)(g) = \int_S \int_{\mathbb{R}} g \, Rf \; d(\mu_S \otimes \lambda) ,$$

<u>i.e.</u> $d(Rm)/d(\mu_S \otimes \lambda) = R(dm/d\mu)$.

<u>Proof</u>: Let $m = f\mu$ with $f \in L^1(\mu)$. We consider

$$\int_{x \cdot y < p} f(y) \, d\mu(y) = \int_{-\infty}^{p} d(P_x(f\mu))(p) ,$$

where P_x is as in the proof of Theorem 1.2. From this proof we obtain also that $P_x\mu$ is absolutely continuous with respect to μ and hence so is $P_x(f\mu)$. Therefore, we have for all $x \in S$ that

$$\int_{-\infty}^{p} d(P_x(f\mu))(p)$$

is a.e. differentiable with respect to p and is an element of $L^1(\mathbb{R})$. Thus, we obtain by definition of the Radon transform that $R(f\mu)$ is absolutely continuous with respect to $\mu_S \otimes \lambda$ and $R(f\mu)$ can be identified with the density (2.2)'.

<u>COROLLARY 2.3</u>. <u>A continuous function on</u> (H,μ) <u>is uniquely determined by its surface integrals along all hyperplanes</u>.

<u>PROPOSITION 2.4</u>. <u>Let</u> $f \in C_b(H)$, $x \in S$ <u>and</u> $p \in \mathbb{R}$; <u>then</u>

$$F(f\mu)(px) = \int_{\mathbb{R}} (Rf)(x,q) \, e^{ipq} \, dq .$$

<u>Proof</u>: We know from Proposition 2.2 and Proposition 2.3 that

$$(2\pi)^{-1/2}\int_S\int_{\mathbb{R}}(F_p g)(x,p)\ F(f\mu)(px)\ dp\ d\mu_S(x) \ = \int_S\int_{\mathbb{R}} g\ Rf\ \ d\lambda\ d\mu_S\ .$$

Since for all $x\in S$ we have $(Rf)(x,\cdot)\in L^1$, we can identify via Parseval's relation (with respect to p) the functions $(F\mu)(px)$ and $(2\pi)^{1/2}\ F_p(Rf)(x,p)$.

EXAMPLES. The Radon transform of the (basic) Gaussian measure μ. From Theorem 1.2 we obtain:

$$(R\mu)(g) \ = \ (2\pi)^{-1/2}\int_S\int_{\mathbb{R}} g(x,p)\ (Ax,x)^{-1/2}\ e^{-p^2/2(Ax,x)}\ dp\ d\mu_S(x)\ .$$

This means that $R\mu$ is absolutely continuous with respect to $\mu_S\otimes\lambda$ on $S\times\mathbb{R}$ and $R\mu$ can be identified with the function

$$(R\mu)(x,p) \ = \ (2\pi(Ax,x))^{-1/2}\ e^{-p^2/2(Ax,x)}\ .$$

The Radon transform of an (arbitrary) Gaussian measure on (H,μ). Let ν be another Gaussian measure (different from μ) on H and let C denote its covariance operator. Then, in spite of ν may be singular to μ, again $R\nu$ is absolutely continuous with respect to $\mu_S\otimes\lambda$ and $R\nu$ can be identified with the function

$$(R\nu)(x,p) \ = \ (2\pi(Cx,x))^{-1/2}\ e^{-p^2/2(Cx,x)}\ ,$$

which is exactly the ν-area of the hyperplane $x\cdot y=p$.

REMARK 2.3. We now consider the case of a separable Banach space B. In view of Remark 1.2, we can define the Radon transform on B in the same way as in (2.3)', S being now the sphere in B'. Thus, we have the analogs of the previous propositions and corollaries for B. In particular, a continuous function on B is determined by its surface integrals along all hyperplanes.

REMARK 2.4. Using polar integration (Corollary 1.1) and the connection to Fourier transform (Propositions 2.2 and 2.4), we can derive inversion formulas for the Radon transform on (H,μ) (see [4] and a forthcoming paper).

On the determination of a stochastic process. We now want to connect our results to the question of determining a process from its conditional expectations. As we have seen in Remark 1.1, the desintegrated (=conditional) measures of the (Gaussian) measure μ are exactly the corresponding surface measures of μ devided by the μ-area of the hypersurface. In the special case of hyperplanes, our point of view is as follows:

Once having the surface integrals of $f\mu$, $f \in C_b(H)$, we can determine via Radon transform the process $f\mu$ from the corresponding conditional expectation.

Especially, if we consider the classical Wiener space C_o, then a hyperplane in C_o is given by $m(f)=p$, m being a measure on $[0,1]$. Now, because the Dirac measures are weakly dense and by an analog of John's theorem for Banach spaces (cf. [4]), we obtain:

The Wiener process is determined by the conditional expectations that a particle is at a time t located at p, where t is from a small open subset of $[0,1]$ and $p \in \mathbb{R}$.

Finally, we remark that in a subsequent paper the Radon transform on (H,μ) will be used to transform a hyperbolic differential equation on $\mathbb{R} \times H$ into a two-dimensional problem. Further, the inversion of the Radon transform will give a "plane wave decomposition" for the solution (in the finite dimensional case this has been done by F. John, P. D. Lax, D. Ludwig and R. S. Phillips, see [8], [9], and references there).

References

1. BEEKMAN J. A., Two Stochastic Processes, Almqvist & Wiksell International, Stockholm 1974.

2. GOODMAN V., A divergence theorem for Hilbert space, Trans. AMS 164 (1972), 411-426.

3. GROSS L., Abstract Wiener spaces, Proc. Vth Berkeley Symp. Math. Stat. Prob. II(1) (1967), 31-42.

4. HERTLE A., Zur Radon-Transformation von Funktionen und Maßen, Thesis, Erlangen 1979.

5. HERTLE A., The Radon transform of measures on $\mathbb{R}^n$, Preprint 1979.

6. HÖRMANDER L., Linear Partial Differential Operators, Academic Press, New York, 1963.

7. JOHN F., Bestimmung einer Funktion aus ihren Integralen über gewisse Mannigfaltigkeiten, Math. Ann. 109 (1934), 488-520.

8. LAX P. D. and PHILLIPS R. S., Scattering theory, Bull. Amer. Math. Soc. 70 (1964), 130-142.

9. LUDWIG D., The Radon transform on Euclidean space, Comm. Pure Appl. Math. 19 (1966), 49-81.

10. PROHOROV YU. V., The method of characteristic functionals, Proc. IVth Berkeley Symp. Math. Stat. Prob. II (1960), 403-419.

11. RADON J., Über die Bestimmung von Funktionen durch ihre Integralwerte längs gewisser Mannigfaltigkeiten, Ber. Verh. Sächs. Akad. Wiss. Leipzig. Math. Natur. Kl. 69 (1917), 262-277.

12. SCHWARTZ L., Radon measures on arbitrary topological spaces and cylindrical measures, Oxford University Press, London 1973.

13. SCHWARTZ L., Surmartingales régulières à valeurs mesures et désintégration régulières d'une mesure, J. d'Anal. Math. 26 (1973), 1-168.

14. SKOROHOD A. V., Integration in Hilbert space, Springer, Berlin 1974.

Alexander Hertle
Fachbereich Mathematik
Universität Mainz
Saarstr. 21
D-6500 Mainz

SPACES OF MULTIPLIABLE FAMILIES IN HAUSDORFF TOPOLOGICAL GROUPS

Corneliu Constantinescu

1. Many theorems concerning the space of measures on a given σ-ring of sets, for instance Nikodym's convergence theorem, the Vitali-Hahn-Saks theorem, Nikodym's boundedness theorem, Orlicz-Pettis theorem and all of their generalizations, rely on two properties of summable sequences in Hausdorff additive groups, which are formulated below. In all of these theorems we denote by G a Hausdorff additive group, and we call a sequence in G *supersummable* if everyone of its subsequences is summable.

Theorem A. *If* $(x_n)_{n\in\mathbb{N}}$ *is a supersummable sequence in* G, *then the map*

$$\underline{\underline{P}}(\mathbb{N}) \to G\ ,\ A \mapsto \sum_{n\in A} x_n$$

is continuous. Here $\underline{\underline{P}}(\mathbb{N})$ *denotes the power set of* $\mathbb{N}$ *endowed with the compact topology obtained by identifying* $\underline{\underline{P}}(\mathbb{N})$ *with* $\{0,1\}^{\mathbb{N}}$ *via the map*

$$\{0,1\}^{\mathbb{N}} \to \underline{\underline{P}}(\mathbb{N})\ ,\ f \mapsto \{n\in\mathbb{N} \mid f(n) = 1\}$$

Theorem B. *If* $((x_{mn})_{n\in\mathbb{N}})_{m\in\mathbb{N}}$ *is a sequence of supersummable sequences in* G *such that*

$$(\sum_{n\in A} x_{mn})_{m\in\mathbb{N}}$$

converges for every $A \subset \mathbb{N}$, *then the convergence is uniform with respect to* A. *In particular, if we set*

$$x_n := \lim_{m\to\infty} x_{mn}$$

for every $n\in\mathbb{N}$, *then* $(x_n)_{n\in\mathbb{N}}$ *is a supersummable sequence in* G *and*

$$\sum_{n\in A} x_n = \lim_{m\to\infty} \sum_{n\in A} x_{mn}$$

for every $A \subset \mathbb{N}$.

N.J. Kalton and L. Drewnowski observed ([4], pages 516-517) that Theorem A still holds if we drop the commutativity of G . In this case the sums must be replaced by products, and the order in which the terms are multiplied is given by the order of $\mathbb{N}$. The aim of this paper is to show that the above theorems remain true if we replace $\mathbb{N}$ by an arbitrary linearly ordered set, and to give some connected results. Most of the results presented here were announced in [3] .

2. *Throughout this paper we shall denote by* G *a topological group and by* e *its identity element.* In fact we shall assume in the interesting cases that G is Hausdorff, but for the first part of this paper the results still hold without this assumption. G has three canonical uniformities, and all these uniformities are compatible with the topology of G . The left and the right uniformities are the most well known. The third one, called the *two-sided uniformity of* G is the least upper bound of the left and right uniformities of G in the set of all uniformities on G ([2], Ch.III, §3, Exercice 6). In specifying various properties involving one of the three canonical uniformities, for instance precompactness, we precede the name of the property in question by left, right or no modifier at all, according to whether we mean the left, the right or the two-sided uniformity.

For an arbitrary set X , we denote by $\underline{\underline{P}}(X)$ the power set of X, which we shall identify with $\{0,1\}^X$ via the map

$$\{0,1\}^X \longrightarrow \underline{\underline{P}}(X) , \quad f \longmapsto \{x\in X \mid f(x) = 1\} .$$

If we endow $\{0,1\}$ with the discrete topology, then by Tychonoff's theorem $\underline{\underline{P}}(X)$ will be a compact space. A subset U of $\underline{\underline{P}}(X)$ is open, iff for every $J\in U$ there exists a finite subset K of X such that

$$\{L\in\underline{\underline{P}}(X) \mid L\cap K = J\cap K\} \subset U \quad .$$

Throughout this paper we shall denote by I *a nonempty linearly ordered set.* An *interval* of I is a subset J of I such that

$$\iota',\iota''\in J,\ \iota\in I\ ,\quad \iota'\leq\iota\leq\iota'' \Longrightarrow \iota\in J\ .$$

An *initial (final) interval* of I is a subset J of I such that

$$\iota\in J,\ \lambda\in I,\quad \lambda<\iota \Longrightarrow \lambda\in J$$

$$(\iota\in J,\ \lambda\in I,\quad \lambda>\iota \Longrightarrow \lambda\in J)\ .$$

We denote by $\mathcal{H}$ the set of maps f of $\underline{\underline{P}}(I)$ into G such that

$$f(J\cup K) = f(J)f(K)$$

for all subsets J,K of I that satisfy

$$\iota\in J,\ \lambda\in K \qquad \iota<\lambda\ .$$

Such maps appear if one considers families $(x_\iota)_{\iota\in I}$ in G for which every subfamily is multipliable (G Hausdorff).

Proposition 1. Let $\mathcal{F}$ *be a subset of* $\mathcal{H}$ *. If* $\mathcal{F}$ *is left (right) equicontinuous at a point of* $\underline{\underline{P}}(I)$*, and if* I *has no greatest (smallest) element, then for any neighbourhood* U *of* e *there exists a nonempty final (initial) interval* L *of* I *such that* $f(J)\in U$ *for any* $J\subset L$ *and for any* $f\in\mathcal{F}$ *.*

Let us denote by J_o the point of $\underline{\underline{P}}(I)$ at which $\mathcal{F}$ is left (right) equicontinuous, and let V be a neighbourhood of e such that $V^{-1}V\subset U$ ($VV^{-1}\subset U$). There exists a finite subset K of I such that

$$f(J_o)^{-1}f(J)\in V \qquad (f(J)f(J_o)^{-1}\in V)$$

for any $J\in\underline{\underline{P}}(I)$ with $J\cap K = J_o\cap K$ and for any $f\in\mathcal{F}$. We set

$$L := \{\iota\in I \mid \forall\lambda\in K \Longrightarrow \lambda<\iota\}$$

$$(L := \{\iota\in I \mid \forall\lambda\in K \Longrightarrow \lambda<\iota\})\ .$$

L is a nonempty final (initial) interval of I. Let J be a subset of L. We have

$$f(J_o)^{-1} f(J_o \cap K) \in V \quad , \quad f(J_o)^{-1} f((J_o \cap K) \cup J) \in V$$

$$(f(J_o \cap K) f(J_o)^{-1} \in V \quad , \quad f(J \cup (J_o \cap K)) f(J_o)^{-1} \in V)$$

and therefore

$$f(J) = f(J_o \cap K)^{-1} f((J_o \cap K) \cup J) =$$

$$= (f(J_o \cap K)^{-1} f(J_o))(f(J_o)^{-1} f((J_o \cap K) \cup J)) \in V^{-1} V \subset U$$

$$(f(J) = f(J \cup (J_o \cap K)) f(J_o \cap K)^{-1} =$$

$$= (f(J \cup (J_o \cap K)) f(J_o)^{-1})(f(J_o) f(J_o \cap K)^{-1}) \in VV^{-1} \subset U)$$

for any $f \in F$. □

Proposition 2. Let M *be a left (right) precompact set of* G *, and let* U *be a neighbourhood of* e *. Then there exists a neighbourhood* V *of* e *such that*

$$x^{-1} V x \subset U \qquad (x V x^{-1} \subset U)$$

for any $x \in M$ *.*

Let W be a neighbourhood of e such that $W^{-1} W^2 \subset U$ $(W^2 W^{-1} \subset U)$. Since M is left (right) precompact, there exists a finite subset M_o of M such that

$$M \subset \bigcup_{y \in M_o} (yW) \qquad\qquad (M \subset \bigcup_{y \in M_o} (Wy)) .$$

Let V be a neighbourhood of e such that

$$\bigcup_{y \in M_o} (y^{-1} V y) \subset W \qquad\qquad \bigcup_{y \in M_o} (y V y^{-1}) \subset W .$$

Let $x \in M$. There exists $y \in M_o$ with

$$x \in yW \qquad (x \in Wy) .$$

We get

$$x^{-1}Vx \subset W^{-1}y^{-1}VyW \subset W^{-1}WW \subset U$$

$$(xVx^{-1} \subset WyVy^{-1}W^{-1} \subset WWW^{-1} \subset U) . \square$$

<u>*Proposition 3.*</u> *Let* L *be an interval of* I *and* F *a subset of* H *such that* $\{f(K) | f \in F\}$ *is a precompact set of* G *for any* $K \in \underline{\underline{P}}(I)$. *If* F *is equicontinuous at a point* $J \in \underline{\underline{P}}(I)$, *then the set of restrictions of the maps of* F *at* $\underline{\underline{P}}(L)$ *is equicontinuous at* $J \cap L$.

We set

$$I' := \{\iota \in I | \lambda \in L \Longrightarrow \iota < \lambda\} , \quad I'' := \{\iota \in I | \lambda \in L \Longrightarrow \iota > \lambda\}.$$

Let U be an arbitrary neighbourhood of e . By Proposition 2 there exists a neighbourhhod V of e such that

$$f(J \cap I')^{-1}Vf(J \cap I') \subset U, \quad f(J \cap I'')Vf(J \cap I'')^{-1} \subset U$$

for any $f \in F$. Since F is equicontinuous at J, there exists a finite subset K of I such that

$$f(J)^{-1}f(J') \in V , \quad f(J')f(J)^{-1} \in V$$

for any $J' \in \underline{\underline{P}}(I)$ with $J' \cap K = J \cap K$ and for any $f \in F$. Let $J' \in \underline{\underline{P}}(L)$ such that $J' \cap K = (J \cap L) \cap K$. Then

$$((J \cap I') \cup J' \cup (J \cap I'')) \cap K = J \cap K \quad ,$$

and therefore

$$f(J)^{-1}f((J \cap I') \cup J' \cup (J \cap I'')) \in V ,$$

$$f((J \cap I') \cup J' \cup (J \cap I''))f(J)^{-1} \in V$$

for any $f \in F$. We get

$$f(J \cap L)^{-1}f(J') = f(J \cap I'')f(J)^{-1}f((J \cap I') \cup J' \cup (J \cap I''))f(J \cap I'')^{-1} \in$$

$$\in f(J\cap I'')Vf(J\cap I'')^{-1}\subset U ,$$

$$f(J')f(J\cap L)^{-1} = f(J\cap I')^{-1}f((J\cap I')\cup J'\cup (J\cap I''))f(J)^{-1}f(J\cap I')\in$$

$$\in f(J\cap I')^{-1}Vf(J\cap I')\subset U$$

for any $f\in F$, which proves the proposition. □

Theorem 4. *Let* F *be a subset of* H *such that* $\{f(J)|f\in F\}$ *is a precompact set of* G *for any* $J\in \underline{\underline{P}}(I)$. *If* F *is equicontinuous at a point of* $\underline{\underline{P}}(I)$ *then it is uniformly equicontinuous.*

Let $J_o\in \underline{\underline{P}}(I)$ and let U be an arbitrary neighbourhood of e. Let U_o be another neighbourhood of e such that $U_oU_o\subset U$. We denote by $\underline{\underline{L}}$ the set of final intervals L of I possessing the following property: for any neighbourhood V of e there exists a finite subset K of L such that

$$f(J_o\cap L)^{-1}f(J)\in VU_o$$

for any $f\in F$ and for any $J\subset L$ with $J\cap K = J_o\cap L\cap K$. We set

$$L_o := \bigcup_{L\in \underline{\underline{L}}} L .$$

We want to show first that L_o is nonempty. If I possesses a greatest element, say ι , then it is obvious that $\{\iota\}\in\underline{\underline{L}}$, and therefore $L_o \neq \emptyset$. Assume now that I does not possess a greatest element. Let W be a neighbourhood of e such that $W^{-1}W\subset U_o$. By Proposition 1 there exists a nonempty final interval L of I such that $f(J)\in W$ for any $f\in F$ and for any $J\subset L$. We get

$$f(J_o\cap L)^{-1}f(J)\in W^{-1}W\subset U_o\subset VU_o$$

for any $f\in F$ and for any $J\subset L$. Hence $L\in \underline{\underline{L}}$ and therefore $L_o\neq \emptyset$.

Now we want to show that $L_o\in\underline{\underline{L}}$. Assume first L_o to possess a smallest element, say ι . Then there exists $L\in\underline{\underline{L}}$ with $\iota\in L$, and we deduce $L_o = L\in\underline{\underline{L}}$. Assume now that L_o does not possess a smallest element. Let V be an arbitrary neighbourhood of e, and let V_o be a neighbourhood of e, such that $V_oV_o^{-1}V_o \subset V$. By Proposition 2 there exists a neighbourhood W of e such that

$$f(J_o \cap L_o)^{-1} W f(J_o \cap L_o) \subset V_o$$

for any $f \in F$. By Proposition 3 the set of the restrictions of the functions of F to $\underline{\underline{P}}(L_o)$ is equicontinuous at a point of $\underline{\underline{P}}(L_o)$. By Proposition 1 there exists a nonempty initial interval L' of L_o such that $f(J) \in W$ for any $f \in F$ and for any $J \subset L'$. Let $\iota \in L'$. There exists $L \in \underline{\underline{L}}$ with $\iota \in L$. Let K be a finite subset of L such that

$$f(J_o \cap L)^{-1} f(J) \in V_o U_o$$

for any $f \in F$ and for any $J \subset L$ with $J \cap K = J_o \cap L \cap K$. Let J be a subset of L_o with $J \cap K = J_o \cap L_o \cap K$. Then

$$(J \cap L) \cap K = (J_o \cap L) \cap K \,, \quad J \setminus L \subset L' \quad J_o \cap L_o \setminus L \subset L'$$

and therefore

$$f(J_o \cap L_o)^{-1} f(J) = f(J_o \cap L_o)^{-1} f(J \setminus L) f(J_o \cap L_o) f(J_o \cap L_o)^{-1}$$

$$f(J_o \cap L_o \setminus L)^{-1} f(J_o \cap L_o \setminus L) f(J_o \cap L) f(J_o \cap L)^{-1} f(J \cap L) \in$$

$$\in f(J_o \cap L_o)^{-1} W f(J_o \cap L_o) f(J_o \cap L_o)^{-1} W^{-1} f(J_o \cap L_o) V_o U_o \subset$$

$$\subset V_o V_o^{-1} V_o U_o \subset V U_o$$

for any $f \in F$. Since L_o is a final interval of I it follows $L_o \in \underline{\underline{L}}$.

We want to show now that $I \setminus L_o$ does not possess a greatest element. Assume the contrary and let ι be the greatest element of $I \setminus L_o$. Let V be a neighbourhood of e and let K be a finite subset of L_o such that

$$f(J_o \cap L_o)^{-1} f(J) \in V U_o$$

for any $f \in F$ and for any $J \subset L_o$ with $J \cap K = J_o \cap L_o \cap K$.

Let J be a subset of $\{\iota\} \cup L_o$ such that

$$J \cap (\{\iota\} \cup K) = J_o \cap (\{\iota\} \cup L_o) \cap (\{\iota\} \cup K) .$$

We get

$$(J \cap L_o) \cap K = J_o \cap L_o \cap K$$

and therefore

$$f(J_o \cap (\{\iota\} \cup L_o))^{-1} f(J) = f(J_o \cap L_o)^{-1} f(J \cap L_o) \in VU_o$$

for any $f \in F$. Since $\{\iota\} \cup L_o$ is a final interval of I it follows $\{\iota\} \cup L_o \in \underline{\underline{L}}$ and therefore $\iota \in L_o$ which is a contradiction. Hence $I \setminus L_o$ does not possess a greatest element.

We want to prove now $L_o = I$. Assume the contrary. Let V be an arbitrary neighbourhood of e , and let V_o be a neighbourhood of e such that $V_o^{-1} V_o V_o \subset V$. By Proposition 2 there exists a neighbourhood W of e such that

$$f(J_o \cap L_o)^{-1} W f(J_o \cap L_o) \subset V_o$$

for any $f \in F$. By Proposition 3 there exists a point of $\underline{\underline{P}}(I \setminus L_o)$ at which the set of restrictions to $\underline{\underline{P}}(I \setminus L_o)$ of the functions of F is equicontinuous. By Proposition 1 there exists a nonempty final interval L of $I \setminus L_o$ such that $f(J) \in W$ for any $f \in F$ and for any $J \subset L$. Let K be a finite subset of L_o such that

$$f(J_o \cap L_o)^{-1} f(J) \in V_o U_o$$

for any $f \in F$ and for any $J_o \subset L_o$ with $J \cap K = J_o \cap L_o \cap K$. Let J be a subset of $L \cup L_o$ with

$$J \cap K = J_o \cap (L \cup L_o) \cap K .$$

We get

$$f(J_o \cap (L \cup L_o))^{-1} f(J) = f(J_o \cap L_o)^{-1} f(J_o \cap L)^{-1} f(J_o \cap L_o) f(J_o \cap L_o)^{-1}$$

$$f(J \cap L) f(J_o \cap L_o) f(J_o \cap L_o)^{-1} f(J \cap L_o) \in$$

$$\in f(J_o \cap L_o)^{-1} W^{-1} f(J_o \cap L_o) f(J_o \cap L_o)^{-1} W f(J_o \cap L_o) V_o U_o \subset$$

$$\subset V_o^{-1} V_o V_o U_o \subset V U_o$$

for any $f \in F$. Since $L \cup L_o$ is a final interval of I it follows $L \cup L_o \in \underline{\underline{L}}$. Hence $L = \emptyset$, and this is a contradiction. We have proved $L_o = I$.

By the definition of $\underline{\underline{L}}$ there exists a finite subset K of I such that

$$f(J_o)^{-1} f(J) \in U_o U_o \subset U$$

for any $f \in F$ and for any $J \subset I$ with $J \cap K = J_o \cap K$. Since U is arbitrary F is left equicontinuous at J_o . In a similar way it can be proved that F is right equicontinuous at J_o. Hence F is equicontinuous at J_o. Since J_o is arbitrary and $\underline{\underline{P}}(I)$ is compact, F is uniformly equicontinuous.□

3. *From now on we assume* G *to be Hausdorff and endow* G *with its two-sided uniformity*. A family $(x_\iota)_{\iota \in I}$ in G is called *supermultipliable* if for any $J \subset I$ the family $(x_\iota)_{\iota \in J}$ is multipliable. We denote by ℓ the set of supermultipliable families in G , indexed by I. *We call coarse uniformity of* ℓ the coarsest uniformity on ℓ for which the maps

$$\ell \rightarrow G , \quad (x_\iota)_{\iota \in I} \longmapsto \prod_{\iota \in J} x_\iota$$

are uniformly continuous for all countable subset J of I. We call *fine uniformity of* ℓ the finest uniformity on ℓ for which the set of all these maps, J running through $\underline{\underline{P}}(I)$, is uniformly equicontinuous. We denote by ℓ_f and ℓ_c the set ℓ endowed with the fine and the coarse uniformity respectively. The topologies of ℓ_f and ℓ_c will be called *fine* and *coarse topology* respectively. If G is $\mathbb{R}$, then ℓ is nothing else but the set of summable families of real numbers, indexed by I (denoted sometimes by $\ell^1(I)$), and the fine topology of ℓ is the usual norm topology on it. The coarse topology of ℓ is the weak topology of ℓ associated to the duality $\langle \ell, \ell^* \rangle$, where ℓ^* denotes the vector space of families of real numbers, indexed by I, taking a finite number of values only, and equal to 0 outside a countable set. The coarse topology is therefore in this special case coarser than the usual weak topology of ℓ'.

We denote, for any $x := (x_\iota)_{\iota\in I} \in \ell$ *, by* $\bar{x}$ *the map*

$$\underline{\underline{P}}(I) \longrightarrow G \;,\quad J \longmapsto \prod_{\iota\in J} x_\iota \;.$$

Theorem 5. $\bar{x}$ *belongs to* H *for any* $x\in\ell$ *. If we identify* ℓ *with a subset of* H *by way of the injection*

$$\ell \longrightarrow H \;,\quad x \longmapsto \bar{x} \;,$$

then ℓ *is the set of continuous maps of* H *. Moreover, the fine uniformity of* ℓ *is the uniformity induced on* ℓ *by the uniformity of uniform convergence of* H *, and the coarse uniformity of* ℓ *is the uniformity induced on* ℓ *by the uniformity of* H *of pointwise convergence on the countable subsets of* I *.*

For any $x\in\ell$ the map $\bar{x}$ is continuous at $I\in\underline{\underline{P}}(I)$. By Theorem 4 it is continuous at any point of $\underline{\underline{P}}(I)$. The orther assertions are obvious. □

Proposition 6. *If* G *is metrizable and* I *countable, then for every Cauchy sequence* $(x_n)_{n\in\mathbb{N}}$ *in* ℓ_c *there exists a point* $J\in\underline{\underline{P}}(I)$ *such that* $(\bar{x}_n)_{n\in\mathbb{N}}$ *is equicontinuous at* J *.*

Let d be a metric on G defining its uniformity. We set for any $m,p\in\mathbb{N}$

$$F_{m,p} := \{J\in\underline{\underline{P}}(I)\,|\,\forall n \geq m \Longrightarrow d(\bar{x}_m(J),\, \bar{x}_n(J)) \leq \frac{1}{p}\} \;.$$

Since the maps $\bar{x}_n (n\in\mathbb{N})$ are continuous, the sets $F_{m,p} (m,p\in\mathbb{N})$ are closed. $(\bar{x}_n)_{n\in\mathbb{N}}$ being a Cauchy sequence with respect to the uniformity of pointwise convergence, we have

$$\bigcup_{m\in\mathbb{N}} F_{m,p} = \underline{\underline{P}}(I)$$

for any $p \in \mathbb{N}$. Since $\underline{\underline{P}}(I)$ is a Baire space $\bigcup_{m \in \mathbb{N}} \mathring{F}_{m,p}$ is a dense open set of $\underline{\underline{P}}(I)$ for any $p \in \mathbb{N}$. Hence the set

$$A := \bigcap_{p \in \mathbb{N}} \bigcup_{m \in \mathbb{N}} \mathring{F}_{m,p}$$

is nonempty.

Let $J \in A$. We want to show $(\bar{x}_n)_{n \in \mathbb{N}}$ is equicontinuous at J. Let ε be a strictly positive real number. Then there exists $p \in \mathbb{N}$ such that $\frac{1}{p} < \frac{\varepsilon}{3}$. There exists further $m \in \mathbb{N}$ such that $J \in \mathring{F}_{m,p}$. Let B be a neighbourhood of J such that

$$d(\bar{x}_n(K) , \bar{x}_n(J)) < \frac{\varepsilon}{3}$$

for any $n \in \mathbb{N}$ with $n \leq m$ and for any $K \in B$. We get

$$d(\bar{x}_n(K), \bar{x}_n(J) \leqq d(\bar{x}_n(K), \bar{x}_m(K)) + d(\bar{x}_m(K), x_m(J)) + d(\bar{x}_m(J), \bar{x}_n(J)) < \varepsilon$$

for any $n \in \mathbb{N}$ with $n \geq m$ and for any $K \in B \cap \mathring{F}_{m,p}$. Hence

$$d(\bar{x}_n(K) , \bar{x}_n(J)) < \varepsilon$$

for any $n \in \mathbb{N}$ and for any $K \in B \cap \mathring{F}_{m,p}$. This shows that $(\bar{x}_n)_{n \in \mathbb{N}}$ is equicontinuous at J. □

Theorem 7. Every Cauchy sequence in ℓ_c is a Cauchy sequence in ℓ_f .

We shall prove here this proposition for countable I and metrizable G only. Let $(x_n)_{n \in \mathbb{N}}$ be a Cauchy sequence in ℓ_{c} . By the preceding proposition there exists $J \in \underline{\underline{P}}(I)$ such that $(\bar{x}_n)_{n \in \mathbb{N}}$ is equicontinuous at J . Since I is countable, $(\bar{x}_n(K))_{n \in \mathbb{N}}$ is a Cauchy sequence, and therefore

$$\{\bar{x}_n(K) \mid n \in \mathbb{N}\}$$

is a precompact set in G for any $K \in \underline{\underline{P}}(I)$. By Theorem 4 $(\bar{x}_n)_{n \in \mathbb{N}}$

is uniformly equicontinuous. It follows immediately via Theorem 5 that $(x_n)_{n\in\mathbb{N}}$ is a Cauchy sequence in ℓ_f. □

Corollary 8. *Let* $((x_{n\iota})_{\iota\in I})_{n\in\mathbb{N}}$ *be a sequence of supermultipliable families in* G *such that* $(\prod_{\iota\in J} x_{n\iota})_{n\in\mathbb{N}}$ *converges for any* $J\subset I$. *If we set*

$$x_\iota := \lim_{n\to\infty} x_{n\iota}$$

for any $\iota\in I$, *then* $(x_\iota)_{\iota\in I}$ *is supermultipliable and*

$$\prod_{\iota\in J} x_\iota = \lim_{n\to\infty} \prod_{\iota\in J} x_{n\iota}$$

for any $J\subset I$.

We set

$$f\colon \underline{P}(I) \longrightarrow G\,,\quad J \longmapsto \lim_{n\to\infty} \prod_{\iota\in J} x_{n\iota}\,.$$

By the theorem f is continuous. This implies $(x_\iota)_{\iota\in I}$ is supermultipliable and $\bar{x} = f$. □

Remark. This result was proved for G commutative and I countable by P. Antosik ([1], Corollary 2(b)).

Corollary 9. *If* G *is complete, or if* I *is countable and* G *sequentially complete, then* ℓ_c *is sequentially complete.*

The second assertion follows immediately from the preceding corollary. The first one can be deduced from the second one. □

Remark. There always exists a Hausdorff group G*, complete with respect to its two-sided uniformity, such that G is a dense subgroup of G*, and the two-sided uniformity of G is induced by the two-sided uniformity of G* .

4. A subset A of a topological space X is called *countably compact* (*relatively countably compact*) if any sequence in A possesses an adherent point in A (in X).

The proofs for the next theorems will appear elsewhere.

Theorem 10. *Let* F *be a relatively countably compact set of* ℓ_c, *and let* $\bar{F}$ *be its closure in* ℓ_c. *Then:*

a) $\bar{F}$ *is the closure of* F *in* ℓ_f ;

b) ℓ_f *and* ℓ_c *induce on* $\bar{F}$ *the same uniformity* $\underline{\underline{U}}$, *and the map*

$$\bar{F}\times\underline{\underline{P}}(I) \longrightarrow G\ , \quad (x,J) \longmapsto \prod_{\iota\in J} x_\iota$$

is uniformly continuous (here $\bar{F}$ *is endowed with* $\underline{\underline{U}}$) ;

c) *if* $\{e\}$ *is a* G_δ*-set of* G , *then* $\bar{F}$ *endowed with* $\underline{\underline{U}}$ *is metrizable and compact; if in addition* F *is countably compact then* $F = \bar{F}$.

A subset A of a uniform space X is called *semi-separable* if for any entourage U of X there exists a sequence $(A_n)_{n\in\mathbb{N}}$ of subsets of X such that

$$\bigcup_{n\in\mathbb{N}}(A_n\times A_n)\subset U\ , \quad A\subset\bigcup_{n\in\mathbb{N}}A_n\ .$$

Theorem 11. *Let* H *be a topological group, obtained by endowing the underlying group of* G *with a Hausdorff group topology such that there exists a fundamental system of neighbourhoods of* e *in* H *which are closed (sequentially closed if* I *is countable) in* G , *and let* $(x_\iota)_{\iota\in I}$ *be a supermultipliable family in* G . *Then the following assertions are equivalent:*

a) $(x_\iota)_{\iota\in I}$ *is supermultipliable in* H , *and the products in* G *and* H *of its subfamilies coincide;*

b) $\{\prod_{\iota\in J} x_\iota \mid J\subset I\}$ *is a compact set of* H ;

c) $\{\prod_{\iota \in J} x_\iota | J \subset I\}$ *is a semi-separable set of* H ;

d) *any countable subfamily of* $(x_\iota)_{\iota \in I}$ *is multipliable in* H , *and its products in* G *and* H *coincide.*

Remark. N.J. Kalton proved ([6], Theorem 7) that a) holds if H is separable and commutative, I countable, and the above neighbourhoods of e closed. L. Drewnowski improved it by replacing "closed" with "sequentially closed" and the separability of H with the separability of $\{\prod_{\iota \in J} x_\iota | J \subset I\}$ ([4], Theorem 1). Drewnowski and Kalton remarked too that the result still holds in the non-commutative case for $I=\mathbb{N}$ ([4], pages 516-517). In a later paper ([5], Theorem 1) Drewnowski replaced the hypothesis "separable" by c).

Theorem 12. Let T *be a topological space,* T_o *a dense set of* T , F *the group of continuous maps of* T *into* G , *and* F',F" *be* F *endowed with the topology of pointwise convergence in* T_o *and the topology of uniform convergence respectively. We assume one of the following conditions to be fulfilled:*

1) T *is sequentially compact (i.e. any sequence in* T *possesses a convergent subsequence); 2)* T *is countably compact and* $T \setminus T_o$ *countable; 3)* T *is locally compact and paracompact, and* $T_o = T$. *Then any supermultipliable family* $(f_\iota)_{\iota \in I}$ *in* F' *is multipliable in* F" .

Remark. This result was proved by E. Thomas ([9], Théorème II 4) in the case when G is a normed space or a normed space endowed with its weak topology, and for T compact and $T_o=T$, or for T metrizable and compact, and T_o dense. It was extended by J. Labuda to the case when G is a Hausdorff commutative group, T is sequentially compact, and $T_o=T$ ([7], Théorème 2.3), or when T is compact and metrizable, and T_o dense ([7], Théorème 2.4), or when T is compact, $T_o=T$, and I countable ([7], Théorème 2.5). J. Labuda proved this result too ([8], Corollary 2.2) for G locally convex, T locally compact and σ-compact, $T_o=T$, and I countable.

REFERENCES

[1] Antosik, P., Mappings from L-groups into topological groups I. Bull.Acad.Pol.Sci. Sér.Sci.Math.Astr.Phys. 21(1973), 145-152.

[2] Bourbaki, N., Topologie générale, Hermann, Paris (1971).

[3] Constantinescu, C., Familles multipliables dans les groupes topologiques séparés I,II, C.R.Acad.Sci. Paris, Sér. A, 282 (1976), 191-193; 271-274.

[4] Drewnowski, L., On Orlicz-Pettis type theorems of Kalton, Bull. Acad.Pol.Sci. Sér.Sci.Math.Astr.Phys. 21(1973), 515-518.

[5] Drewnowski, L., Another note on Kalton's theorem. Studia Math. 52(1975), 233-237.

[6] Kalton, N.J., Subseries convergence in topological groups and vector spaces. Israel J.Math. 10(1971), 402-412.

[7] Labuda, J., Sur quelques généralisations des théorèms de Nikodym et de Vitali-Hahn-Saks. Bull.Acad.Pol.Sci. Sér.Sci.Math.Astr.Phys. 20(1972), 447-456.

[8] Labuda, J., Sur quelques théorèmes du type d'Orlicz-Pettis III. Bull.Acad.Pol.Sci. Sér.Sci.Math.Astr.Phys. 21(1973), 599-605.

[9] Thomas, E., L'intégration par rapport à une mesure de Radon vectorielle. Ann.Inst.Fourier 20, 2(1970), 55-191.

Mathematisches Seminar
ETH-Zentrum
8092-Zürich
Switzerland

A CONVERGENCE PROPERTY FOR SOLUTIONS OF CERTAIN QUASI-LINEAR ELLIPTIC EQUATIONS

Fumi-Yuki MAEDA

1. Let L be a quasi-linear elliptic partial differential operator on an open set Ω in R^N and let $\mathcal{H}_L$ be the set of all "weak" solutions of $Lu = 0$. When we try to develop a potential theory with respect to such an operator L, it becomes important to know whether $\mathcal{H}_L$ is closed in the following sense (cf. B. Calvert [2] and F-Y. Maeda [4]): if $u_n \in \mathcal{H}_L$, $n = 1,2,\ldots$, $\{u_n\}$ is locally uniformly bounded and $u_n \to u$ (a.e.) on Ω, then $u \in \mathcal{H}_L$. We shall say that $\mathcal{H}_L$ is *BS-closed* if it is closed in the above sense. Our problem here is to find reasonably weak conditions on L under which $\mathcal{H}_L$ is BS-closed. Because of the nature of the problem, the use of measure theoretic arguments (in particular the use of Lebesgue's convergence theorem) enables us to obtain better results than purely functional-analytical method.

In the present note, we establish the following

THEOREM. *Let*

(1) $$Lu = -\operatorname{div} A(x, u, \nabla u) + B(x, u, \nabla u),$$

where $A: \Omega \times R \times R^N \to R^N$, $B = B_1 + B_2$ *with* $B_i: \Omega \times R \times R^N \to R$ $(i = 1,2)$ *and* A, B_1 *and* B_2 *satisfy the Carathéodory condition,* i.e., *they are measurable in* $x \in \Omega$ *for each fixed* $(t, \tau) \in R \times R^N$ *and continuous in* $(t, \tau) \in R \times R^N$ *for each fixed* $x \in \Omega$. *We assume:*

(I) L *is elliptic*, i.e.,

(2) $$\langle A(x, t, \tau_1) - A(x, t, \tau_2), \tau_1 - \tau_2 \rangle > 0$$

for any $x \in \Omega$, $t \in R$ *and* $\tau_1, \tau_2 \in R^N$ *with* $\tau_1 \neq \tau_2$, *where* $\langle .,. \rangle$ *denotes the ordinary inner product in* R^N.

(II) *For some* $p > 1$, *the following structural condition is satisfied: for each compact set* K *in* Ω *and each positive number* M, *there exist constants* $\alpha = \alpha(M,K) \geqq 0$, $\lambda = \lambda(M,K) > 0$, $q = q(M,K)$ *satisfying* $1 \leqq q < p$, $r = r(M,K) > 1$, $s = s(M,K) > p/(p-q)$ *and functions* $a = a(M,K) \in L^{p'}(K)$ $(1/p + 1/p' = 1)$, $b_o = b_o(M,K) \in L^s(K)$,

$b_1 = b_1(M,K) \in L^r(K)$, $b_2 = b_2(M,K) \in L^1(K)$, $c = c(M,K) \in L^1(K)$ *such that whenever* $x \in K$, $|t| \leqq M$ *and* $\tau \in R^N$,

(3) $$|A(x, t, \tau)| \leqq \alpha|\tau|^{p-1} + a(x)$$

(4) $$\langle A(x, t, \tau), \tau\rangle \geqq \lambda|\tau|^p - c(X)$$

(5) $$|B_1(x, t, \tau)| \leqq b_o(x)|\tau|^q + b_1(x)$$

(6) $$|B_2(x, t, \tau)| \leqq b_2(x) .$$

Let H_L *be the set of all* $u \in W^{1,p}_{loc}(\Omega)$ *satisfying*

(7) $$\int_\Omega \{\langle A(x, u, \nabla u), \nabla\phi\rangle + B(x, u, \nabla u)\phi\} \, dx = 0$$

for all $\phi \in C^1_o(\Omega)$. *Then* H_L *is BS-closed.*

Similar convergence properties were investigsted by B. Calvert [1] (also cf. [4; Appendix]). Note that in the above theorem, no growth conditions on t are required, so that, for instance, an operator of the form $Lu = \Delta u + f(u)$ satisfies our assumptions whenever $f \in C(R)$, while it satisfies conditions in [1] only when $|f(t)| \leqq \alpha|t| + \beta$ (α, β: constants).

2. For the proof of this theorem, the author was much inspired by the paper J. Leray-J.L. Lions [3]. First we remark

LEMMA 1. Under the assumption (II) in the theorem, if $u \in W^{1,p}_{loc}(\Omega)$ and $v \in L^\infty_{loc}(\Omega)$, then

$$|A(x, v(x), \nabla u(x))| \in L^{p'}_{loc}(\Omega)$$

and

$$B_1(x, v(x), \nabla u(x)) \in L^{\tilde q}(K)$$

for every compact set K in Ω, where $\tilde q = \tilde q(M,K) = \min(ps/(p+qs), r)$ with $M = \sup_K|v|$, $q = q(M,K)$, $r = r(M,K)$ and $s = s(M,K)$.

PROOF. The assertion for A is immediate from (3) and the assertion for B follows from (5) and Hölder's inequality.

COROLLARY. Under the assumptions of the theorem, if $u \in H_L$, then

$$\int_\Omega \{\langle A(x, u, \nabla u), \nabla v\rangle + B(x, u, \nabla u)v\}\, dx = 0$$

for any $v \in W^{1,p}_{loc}(\Omega) \cap L^\infty_{loc}(\Omega)$ having compact support in Ω.

LEMMA 2 (cf. [1; Lemma 2]). Let K be a compact set in Ω, U be a relatively compact open set such that $K \subset U \subset \overline{U} \subset \Omega$ and let $M > 0$. Then, under the assumptions of the theorem,

$$\{ \int_K |\nabla u|^p\, dx \mid u \in \mathcal{H}_L,\ \sup_U |u| \leqq M \}$$

is bounded.

We can prove this lemma by using the above corollary and modifying the proof of [1; Lemma 2], and so we omit the proof.

Here, we quote a lemma which is given in Leray-Lions [3]:

LEMMA 3 ([3; Lemma 3.2]). Let U be an open set in R^N and let $p > 1$. If $f_n \in L^p(U)$, $n = 1,2,\ldots$, $\{ \|f_n\|_{L^p(U)} \}$ is bounded and $f_n \to f$ a.e. on U, then $f_n \to f$ weakly in $L^p(U)$.

3. <u>Proof of the theorem</u>: Let $u_n \in \mathcal{H}_L$, $n = 1,2,\ldots$, $\{u_n\}$ be locally uniformly bounded and $u_n \to u$ a.e. on Ω. We are to show that $u \in \mathcal{H}_L$. We divide the proof into four steps.

<u>1st step</u>. $u \in W^{1,p}_{loc}(\Omega) \cap L^\infty_{loc}(\Omega)$ and

(8) $$\nabla u_n|U \to \nabla u|U \text{ weakly in } L^p(U)^N$$

for any relatively compact open set U such that $\overline{U} \subset \Omega$.

PROOF. Let U' be a relatively compact open set such that $\overline{U} \subset U' \subset \overline{U'} \subset \Omega$. Then, $\sup_n(\sup_{U'}|u_n|) < \infty$. Hence, by Lemma 2 $\{\nabla u_n|U\}$ is bounded in $L^p(U)^N$, so that it is weakly sequentially relatively compact. Since $u_n \to u$ a.e., it follows that $u|U \in W^{1,p}(U)$ and $\nabla u|U$ is the unique weak limit point of $\{\nabla u_n|U\}$ in $L^p(U)^N$, and hence $\nabla u_n|U \to \nabla u|U$ weakly in $L^p(U)^N$.

<u>2nd step</u>. Put

$$f_n(x) = \langle A(x, u_n(x), \nabla u_n(x)) - A(x, u_n(x), \nabla u(x)), \nabla u_n(x) - \nabla u(x)\rangle$$

for $x \in \Omega$. Then

$$\int_\Omega f_n \phi \, dx \to 0 \quad (n \to \infty)$$

for any $\phi \in C_o^1(\Omega)$.

PROOF. Let $K = \text{Supp}\,\phi$ and $M = \sup_n(\sup_K |u_n|)$. By (8) and Lemma 1,

$$\int_\Omega \langle A(x, u, \nabla u), \nabla u_n - \nabla u \rangle \phi \, dx \to 0 \quad (n \to \infty) \; . \tag{9}$$

By (3),

$$|A(x, u_n, \nabla u)|^{p'} \leqq \{\alpha |\nabla u|^{p-1} + a\}^{p'} \in L^1(K) \; ,$$

where $\alpha = \alpha(M,K)$, $a = a(M,K)$. By continuity of $A(x,t,\tau)$ in t,

$$A(x, u_n, \nabla u) \to A(x, u, \nabla u) \quad \text{a.e. on } K.$$

Hence, by Lebesgue's convergence theorem,

$$\int_K |A(x, u_n, \nabla u) - A(x, u, \nabla u)|^{p'} \, dx \to 0 \quad (n \to \infty) \; .$$

Hence

$$\left| \int_\Omega \langle A(x, u_n, \nabla u) - A(x, u, \nabla u), \; \nabla u_n - \nabla u \rangle \phi \, dx \right|$$

$$\leqq \left\{\int_K |A(x, u_n, \nabla u) - A(x, u, \nabla u)|^{p'} dx\right\}^{1/p'} \left[\left\{\int_K |\nabla u_n|^p \, dx\right\}^{1/p} + \left\{\int_K |\nabla u|^p \, dx\right\}^{1/p} \right]$$

$$\to 0 \quad (n \to \infty).$$

Thus, together with (9), we see that

$$\int_\Omega \langle A(x, u_n, \nabla u), \nabla u_n - \nabla u \rangle \phi \, dx \to 0 \quad (n \to \infty). \tag{10}$$

On the other hand, since $u_n \in H_L$,

$$\int_\Omega \langle A(x, u_n, \nabla u_n), \nabla[(u_n - u)\phi] \rangle \, dx + \int_\Omega B(x, u_n, \nabla u_n)(u_n - u)\phi \, dx = 0$$

by the corollary to Lemma 1. Hence

$$\begin{aligned} &\int_\Omega \langle A(x, u_n, \nabla u_n), \nabla u_n - \nabla u \rangle \phi \, dx \\ &= - \int_\Omega \langle A(x, u_n, \nabla u_n), \nabla \phi \rangle (u_n - u) \, dx - \int_\Omega B(x, u_n, \nabla u_n)(u_n - u)\phi \, dx \; . \end{aligned} \tag{11}$$

By Lebesgue's convergence theorem,

$$\int_K |u_n - u|^p \, dx \to 0 \quad (n \to \infty).$$

By (3) and Lemma 2, we see that $\{\int_K |\langle A(x, u_n, \nabla u_n), \nabla\phi\rangle|^{p'} dx\}$ is bounded. Hence, by Hölder's inequality, we have

$$(12) \qquad \int_\Omega \langle A(x, u_n, \nabla u_n), \nabla\phi\rangle (u_n - u) \, dx \to 0 \quad (n \to \infty) .$$

Similarly, for $\tilde{q} = \min\{ps/(p+qs), r\}$ with $q = q(M,K)$, $r = r(M,K)$ and $s = s(M,K)$, $\{\int_K |B_1(x, u_n, \nabla u_n)|^{\tilde{q}} \phi \, dx\}$ is bounded by (5) and Lemma 2, and $\int_K |u_n - u|^{\tilde{q}'} dx \to 0$ $(1/\tilde{q} + 1/\tilde{q}' = 1)$ by Lebesgue's convergence theorem. Hence

$$(13) \qquad \int_\Omega B_1(x, u_n, \nabla u_n)(u_n - u)\phi \, dx \to 0 \quad (n \to \infty) .$$

Finally, by (6), we can apply Lebesgue's convergence theorem again to obtain

$$(14) \qquad \int_\Omega B_2(x, u_n, \nabla u_n)(u_n - u)\phi \, dx \to 0 \quad (n \to \infty) .$$

By (11), (12), (13) and (14), we have

$$(15) \qquad \int_\Omega \langle A(x, u_n, \nabla u_n), \nabla u_n - \nabla u\rangle \, \phi \, dx \to 0 \quad (n \to \infty) .$$

Now (10) and (15) imply that $\int_\Omega f_n \phi \, dx \to 0$ $(n \to \infty)$.

<u>3rd step</u>. $\nabla u_{n_j} \to \nabla u$ a.e. on Ω for some subsequence $\{u_{n_j}\}$.

PROOF. Since $f_n \geqq 0$ on Ω by virtue of (2), the above result in the second step implies that there is a subsequence $\{u_{n_j}\}$ of $\{u_n\}$ such that $f_{n_j} \to 0$ a.e. on Ω. Let K be any compact set in Ω and let U be a relatively compact open set such that $K \subset U \subset \overline{U} \subset \Omega$. Put $M = \sup_n(\sup_U |u_n|)$ and

$$E_K = \left\{ x \in K \;\middle|\; \begin{array}{l} f_{n_j}(x) \to 0,\ |\nabla u(x)| < +\infty,\ |\nabla u_{n_j}(x)| < +\infty,\ j=1,2,\dots \\ a(x) < +\infty,\ b_i(x) < +\infty,\ i=0,1,2,\ c(x) < +\infty \end{array} \right\} .$$

Then $K \setminus E_K$ is of measure zero. If $x \in E_K$, then using (3), (4), (5) and (6), we see that $\{\nabla u_{n_j}(x)\}$ is bounded, and then that $\nabla u_{n_j}(x) \to \nabla u(x)$ $(j \to \infty)$ (cf. [3; the proof of Lemme 3.3]).

<u>The final step</u>. $u \in \mathcal{H}_L$.

PROOF. By the above result, we may assume that $\nabla u_n \to \nabla u$ a.e. on Ω. Then, by continuity, $A(x, u_n, \nabla u_n) \to A(x, u, \nabla u)$ a.e. and $B_i(x, u_n, \nabla u_n) \to B_i(x, u, \nabla u)$ a.e. $(i = 1,2)$. Let $\phi \in C_o^1(\Omega)$ and let U be a relatively compact open set such that $\mathrm{Supp}\,\phi \subset U \subset \overline{U} \subset \Omega$. Since $\{A(x, u_n, \nabla u_n)|U\}$ is bounded in $L^{p'}(U)^N$ by (3) and Lemma 2, Lemma 3 implies that $A(x, u_n, \nabla u_n)|U \to A(x, u, \nabla u)|U$ weakly in $L^{p'}(U)^N$. Hence

$$\int_\Omega \langle A(x, u_n, \nabla u_n), \nabla\phi\rangle \, dx \to \int_\Omega \langle A(x, u, \nabla u), \nabla\phi\rangle \, dx .$$

Similarly, $\{B_1(x, u_n, \nabla u_n)|U\}$ is bounded in $L^{\tilde{q}}(U)$ for some $\tilde{q} > 1$ on account of (5) and Lemma 2, which, together with Lemma 3, implies

$$\int_\Omega B_1(x, u_n, \nabla u_n)\phi \, dx \to \int_\Omega B_1(x, u, \nabla u)\phi \, dx .$$

Finally, by Lebesgue's convergence theorem, we have

$$\int_\Omega B_2(x, u_n, \nabla u_n)\phi \, dx \to \int_\Omega B_2(x, u, \nabla u)\phi \, dx .$$

Hence, $u_n \in \mathcal{H}_L$, $n = 1,2,\ldots$ imply $u \in \mathcal{H}_L$.

REFERENCES

[1] B. Calvert, Harnack's theorems on convergence for nonlinear operators, Atti. Acad. Naz. Lincei Rend. <u>52</u> (1972), 364-372.

[2] B. Calvert, Dirichlet problems without coercivity by the Perron-Ako-Constantinescu method, Math. Chronicle <u>6</u> (1977), 48-67.

[3] J. Leray and J.-L. Lions, Quelques résultats de Višik sur les problèmes elliptiques non linéaires par les méthodes de Minty-Browder, Bull. Soc. math. France <u>93</u> (1965), 97-107.

[4] F-Y. Maeda, Classification theory for nonlinear functional-harmonic spaces, Hiroshima Math. J. <u>8</u> (1978), 335-369.

Department of Mathematics
Faculty of Science
Hiroshima University
Hiroshima, Japan

This work has been partly done while the author was visiting Universität Erlangen-Nürnberg.

CONCERNING PRE-SUPPORTS OF LINEAR PROBABILITY MEASURES

W. Słowikowski
Mathematics Institute
Aarhus University
8000 Århus, Denmark

The analytical significance of factorization by null-sets in classical measure theory manifests on the level of function spaces. Statements proved on this level seldom translate into pure measure language.

The situation changes diametrically if the concepts of measure and linearity enter together on the primary level. In this case we introduce the notion of a linear probability measure which is a class of equivalence of measures over linear spaces relative to a certain specially designed equivalence relation. What then occurs is that there emerge certain special subsets of the spaces on which the measure sits, the so-called pre-supports. In the most interesting examples pre-supports are themselves of measure zero. Nevertheless they carry some essential information about the almost everywhere behaviour.

This fact is quite significant regarding interpretation of probability measures outside of mathematics. In the case of Brownian motion the actual probability measure sits on functions of no physical significance, while the minimal pre-support which consists of differentiable functions provides the link with experimental approach. Similarily, in quantum field theory the Gaussian measures concerned sit on Hilbert-Schmidt enlargements of the real part of the one-particle-space.

These Hilbert-Schmidt enlargements have no physical signigicance, while the minimal pre-support which in this case is the mentioned real part of the one particle space, has.

Here we shall provide a new definition of the concept of pre-support ([5], preprint) which is more along the lines of this intorduction and we certainly verify its equivalence with the original definition. We provide a short review of important properties of pre-supports and close with providing a method for linear almost everywhere extensions of linear bounded transformations between pre-supports.

Last but not least the proofs presented here rectify some flaws that occured in [4].

First we define some basic concepts. A pair (X,X') consisting of a real linear space X and a linear space X' of linear functionals over X vanishing simultaneously only in zero is said to be a <u>standard pair</u> if X is a countable union of its convex $\sigma(X,X')$ - compact subsets. Observe that if X is the union of an ascending sequence Q_n of its convex symmetric $\sigma(X,X')$-compact subsets, then any convex $\sigma(X,X')$-

compact subset of X is absorbed by at least one Q_n. A linear transformation from X shall be called <u>almost uniformly continuous</u> if it is continuous on every convex $\sigma(X,X')$-compact subset of X. Observe that adding to X', all almost uniformly continuous linear functionals do not change the almost uniform continuity. We shall say that (X,X') is <u>standard saturated</u> if X' contains all almost uniformly continuous linear functionals. A subset Y of X is said to be <u>standard</u> if it decomposes into countably many convex $\sigma(X,X')$-compact subsets. Assign to any standard Y its <u>adjoint</u> Y' consisting of all almost uniformly continuous linear functionals over Y. Observe that $\sigma(X,X')$ and $\sigma(Y,Y')$ topologies coincide on every convex $\sigma(Y,Y')$-compact subset of Y.

A triplet (X,X',ν) consisting of a standard saturated pair (X,X') and a regular probability measure ν defined over the field $\mathcal{B}(X,X')$ of all Borel subsets of $(X,\sigma(X,X'))$ is called a <u>representation of a linear probability measure</u> (ν). Given a representation (X,X',ν), a measure one standard linear subset of X shall be called a <u>standard support</u> of (ν).

The representations (X_1,X_1',ν_1) and (X_2,X_2',ν_2) are said to be <u>equivalent</u> if there exists a linear space S which simultaneously is a linear subset of X_1 and a linear subset of X_2, a standard support of ν_1 in X_1 and a standard support of ν_2 in X_2 such that the identical mapping from S onto S is almost uniformly continuous relative to pairs (X_1,X_1') and (X_2,X_2'), i.e. have the same convex compact subsets in $\sigma(X_1,X_1')$ and $\sigma(X_2,X_2')$ topologies, and such that the measures ν_1 and ν_2 coincide on S. The introduced relation is an equivalent relation, and the classes of equivalence of this relation shall be called <u>linear probability measures</u>. The class corresponding to a representation (X,X',ν) shall be denoted by (ν). Given equivalent representations (X_1,X_1',ν_1) and (X_2,X_2',ν_2), the first one is said to be <u>finer</u> than the second one, or, equivalently, the second one is said to be <u>coarser</u> than the first one if X_1 is a standard support of ν_2 in X_2. Clearly, ν_2 coincides with ν_1 on X_1. A representation (X,X',ν) is said to be <u>proper</u> if every $x' \in X'$, which vanishes almost everywhere, vanishes identically. Observe that given any representation (X,X',ν), the regularity of ν implies that the intersection Y of all zero sets of almost everywhere vanishing $x' \in X'$ constitutes a standard support of ν in X. Restricting ν to Y and considering (Y,Y',ν), we obtain a proper representation equivalent to (X,X',ν) and finer than (X,X',ν). Hence, once for all we shall assume that the <u>representations we consider are always proper.</u>

Consider the space $L^0(\nu)$ of all ν-measurable functions, where (ν) is a linear probability measure. We consider elements of $L^0(\nu)$ as classes of equivalence of almost everywhere identical functions so that the definition of $L^0(\nu)$ does not depend on the representation of (ν) we take.

Given a representation (X,X',ν) of (ν), we assign to each $x' \in X'$ the class $x^* \in L^0(\nu)$ corresponding to x' as a measurable function. Notice that the mapping

$$X' \ni x' \to x^* \in L^0(\nu)$$

constitutes an injection. The image of X' in $L^0(\nu)$ shall be denoted by X^*.

Given a linear probability measure (ν), we define the adjoint $(\nu)^*$ of (ν) as the subset of $L^0(\nu)$ which is the union of all X^*, where (X,X',ν) runs over all (proper) representations of (ν). It is easy to check that $(\nu)^*$ is a linear subset of $L^0(\nu)$.

Given a standard saturated pair (X,X'), a sequence $\{x'_n\} \subset X'$ is said to be <u>convergent almost uniformly</u> if it converges uniformly on every convex $\sigma(X,X')$-compact subset of X. Clearly, the pointwise limit of x'_n belongs to X'. Taking supremums on symmetric convex $\sigma(X,X')$-compact subsets of X as seminorms, we make out of X' a Fréchet space in which the convergence coincides with the almost uniform convergence.

It is easy to see that given a sequence $\{f_n\}$ of elements of (ν), there always exists a representation (X,X',ν) of (ν), such that $f_n = x^*_n$ for some $\{x'_n\} \subset X'$. Using the Jegorov theorem and observing that a sequence of elements of X' converging uniformly on a $\sigma(X,X')$-compact subset of X converges uniformly on its closed convex hull, we arrive to the following

<u>Theorem 1</u>. Given a sequence $\{f_n\} \in (\nu)^*$ converging almost everywhere, there esists a representation (X,X',ν) of (ν) and an almost uniformly convergent $\{x'_n\} \subset X'$ such that $f_n = x^*_n$.

In particular, this theorem ascertains that $(\nu)^*$ constisutes a closed subset of $L^0(\nu)$ provided with the stochastic convergence topology. Hence for the future we shall consider $(\nu)^*$ a complete metric space by providing it with the topology of stochastic convergence.

We have now come to the point where we can introduce a crucial concept connected with the notion of linear probability measure. A linear space U is said to be a <u>pre-support</u> of (ν) if there exists a representation (X,X',ν) of (ν) such that U is a standard linear subspace of X with the property that every u' from the adjoint U' of U admits the unique extension $u^* \in (\nu)^*$, i.e. u' extends to an almost uniformly continuous linear functional y' on a standard support Y of ν in X, Y containing U, and every two such extensions are equal almost everywhere. Since it is easy to verify that the mapping

$$U' \ni u' \rightarrow u^* \in (\nu)^*$$

is closed, we have the following

Theorem 2. If U is a pre-support of a linear probability measure, then the almost uniform convergence of a sequence $\{u'_n\} \subset U'$ on U implies the stochastic convergence of the sequence $\{u^*_n\}$.

More involving is to verify the converse,

Theorem 3. Take a representation (X,X',ν) of a linear probability measure (ν) and consider a standard linear subset U of X. Let Z' be a linear subset of X', where all $x' \in Z'$ vanish simultaneously only in the point zero. Provide Z' with the topology of almost uniform convergence on U, and let $(\nu)^*$ carry the topology of stochastic convergence. If the mapping $Z' \ni z' \rightarrow z^* \in (\nu)^*$ is continuous, then U constitutes a pre-support of (ν).

Proof. Take $u' \in U'$. Using Corollary I.1.5 of [1], we can approximate u' almost uniformly on U by $z'_n \in Z'$. Then $\{z^*_n\}$ converges stochastically, and by passing to a sub-sequence we can have $\{z^*_n\}$ converging almost everywhere. Then by Theorem 1 it converges almost uniformly on a standard support S of ν in X and thus also on the standard support $U+S$. Denote its limit by x'. Certainly, x' is an extension of u' to a standard support of ν in X. Were it not almost everywhere unique, there would exist an $x^* \neq 0$ with $x' \in Y'$ of a standard support Y of ν, in X, $Y \supset U$, vanishing on U. Again, by Corollary I.1.5. of [1], there is a sequence $\{x'_n\} \subset Z'$ approximating x' almost uniformly on Y. Hence $x^* = 0$ and the theorem follows.

Hence, we have the following trivial

Corollary 1. A standard U forms a pre-support of a given linear probability measure (ν) if and only if for a representation (X,X',ν) of (ν), ν is scalarly concentrated on U (cf. [2]).

Given a pre-support U of (ν), we write

$$U^* = \{u^* \in (\nu)^*: u' \in U'\}.$$

Since U' provided with the almost uniform convergence constitutes a Fréchet space, we shall provide U^* with Fréchet topology transferred from U' by the mapping $U' \ni u' \to u^* \in U^*$. From Theorem 2 it follows that the identical injection of U^* into $(\nu)^*$ is continuous. A pre-support U is said to be proper if the mapping $U' \ni u' \to u^* \in U^*$ is one-to-one.

Proposition 1. Every pre-support U contains a proper pre-support V such that $U^* = V^*$.

Proof. Let V be the intersection of all $u'^{-1}(0)$ for $u' \in U'$ with $u^* = 0$. Every functional form V' extends to a functional from U'. Moreover, U/V is standard, and u' with $u^* = 0$ separate while considered in U/V. Hence any $u \in U'$ vanishing on V can be approximated almost uniformly on U by $u_n' \in U'$ with $u_n^* = 0$ so that $u^* = 0$, and the proposition follows.

Proposition 1 shows that it is sufficient to consider only proper pre-supports. Hence, from now on we shall incorporate properness in the definition of pre-support, i.e. pre-support shall automatically be assumed proper.

Theorem 4. Consider a representation (X,X',ν) of (ν) and two standard subsets U and V of X which both are presupports of (ν). If $U^* \subset V^*$, then $U \supset V$. In particular, $U^* = V^*$ implies $U = V$.

Proof. Let $U^* \subset V^*$. Take $u' \in U'$ and approximate it almost uniformly on U with $x_n' \in X'$. Hence $\{x_n^*\}$ converges in U and consequently in V, and then $\{x_n'\}$ converges almost uniformly on V and thus on $U+V$ as well. Its limit extends u' to a functional from $(U+V)'$. The existense of such an extension means that U is closed in $(U+V,\sigma(U+V,(U+V)'))$. On the other hand, U and V being proper implies that $U+V$ is a proper pre-support so that extensions of functionals from U to $U+V$ are unique. But that means that U is dense in $(U+V,\sigma(U+V,(U+V)'))$ so that $U+V = U$, and the theorem follows.

Theorem 5. Let Θ be a linear subspace of $(\nu)^*$ which admits a Fréchet topology such that the identical inejction of Θ into $(\nu)^*$ is continuous. Due to the closed graph theorem, such topology is unique. Suppose that there exists a pre-support U of (ν) such that $U^* \subset \Theta$ and that U^* is dense in Θ. Then there exists exactly one pre-support V of (ν), $V \supset U$, such that $\Theta = V^*$.

Proof. Consider J fulfilling the commutative diagram $U' \overset{J}{\to} \Theta \to U^* \to U'$, where the second mapping is the identical injection, and the third mapping is the canonical correspondence $u' \to u^*$. The adjoint J' of J maps the adjoint Θ' of Θ into U identified with the adjoint of $(U',\sigma(U',U))$. Due to the density of JU' in Θ, the mapping J' is a monomorphism which is almost uniformly bicontinuous. Put $V = J'\Theta'$. For $f' \in \Theta'$ and $u' \in U'$, we have $(J'u')f' = f'u^*$, and if an $\{u_n'\} \subset U'$ tends to zero almost uniformly on V, then $J'u_n'$ tends to zero on Θ' which means that u_n tends to zero in Θ and thus in $(\nu)^*$ as well. Then Theorem 3 ascertains that V is a pre-support subject to separate verification of its properness. Since every $v' \in V'$ can be approximated almost uniformly on V by $u_n' \in U'$, to verify properness of V, it is sufficient to show that if $\{u_n^*\}$ converges stochastically to zero, then $v' = 0$. But almost uniform convergence on V implies the convergence of $\{u_n^*\}$ in Θ, and since it converges to zero in $(\nu)^*$, its limit in Θ must be zero. Passing to the limit on both sides of the equality $(J'u_n')f' = f'u_n^*$, we obtain $(J'v')f' = 0$ for every $f' \in \Theta'$, and then $v' = 0$. Hence V constitutes a pre-support with the required properties. Since its uniqueness follows from Theorem 4, the theorem follows.

A pre-support U of a linear probability measure is said to be the minimal pre-support if it is contained in all other pre-supports referring to the same representation. As an immediate consequence of Theorems 3 - 5, we observe that U constitutes the minimal pre-support if and only if $U^* = (\nu)^*$, and we obtain the following

Proposition 2. The following conditions are equivalent:

a) (ν) admits the minimal pre-support

b) $(\nu)^*$ is locally convex in the stochastic convergence topology.

Consequently, U constitutes the minimal pre-support of (ν) if and only if $(\nu)^* = U^*$.

We have the following easy

Proposition 3. If for some $p > r > 0$ we have $(\nu)^* \subset L^p(\nu) \cap L^r(\nu)$ and if both topologies, L^p and L^r, coincide on $(\nu)^*$, then (ν) admits the minimal pre-support.

Proof. This is an immediate consequence of the fact that if on a linear subset of both $L^p(\nu)$ and $L^r(\nu)$ the L^p and L^r topologies coincide, then all topologies L^q for $0 \leqq q \leqq p$ coincide on this subset.

It is easy to see that given $L^0(\mu)$ of an arbitrary measure space to every subspace of it can be cannonically assigned a linear probability measure (ν) such that the subspace coincides with $(\nu)^*$. If, then, the property of Proposition 3 is fulfilled, (ν) admits the minimal pre-support. This way the hypercontractive inequalities of Nelson [3] and the corresponding inequalities for Walsh functions point out many cases in the literature where linear probability measures admitting the minimal pre-supports appear indirectly.

It becomes obvious that the relation of equivalence between representations of linear probability measures should refer to the identity of points of a pre-support and not of a support. To do that, we must find a way of identifying points of supports containing a common pre-support.

If V is an arbitrary standard pre-support in some representation of a linear probability measure, one can always arrange to have V contained in a Hilbertian pre-support H_1 (might be in another representation) in which V is dense. Faced with two such pre-supports H_1 and H_2 both containing V, we can easily find a Hilbertian norm in V which completes within both H_1 and H_2. To do that, it is sufficient to consider everything in $(\nu)^*$. Define $H_3^* = H_1^* + H_2^*$ and provide it with the norm

$$|f|_3^* = (\inf\{|g|_1^{*2} + |h|_2^{*2} : g \in H_1^*,\ h \in H_2^*,\ f = g+h\})^{\frac{1}{2}},$$

where $|\cdot|_j^*$ are norms in H_j^*.

The obtained space coincides by the mapping

$$(H_1^* \times H_2^*)/K \ni (g,h)/K \leftrightarrow g + h \in H_3^*$$

with the product $H_1^* \times H_2^*$ factorized by $K = \{(g,h): f+g = 0\}$ and thus it is a Hilbert space. By Theorem 5, H_3^* corresponds to a Hilbertian pre-support H_3 and by virtue of Theorem 4, $H_1, H_2 \supset H_3 \supset V$. We identify points of H_1 and H_2 which coincide in the completion of $(V, |\cdot|_3)$ in H_1 and H_2 respectively. If both H_1 and H_2 were of measure one, we cannor know whether H_3 is of measure one. However, we know that every Hilbert space $F^{\sim}$ containing F, such that the identical injection of F into $F^{\sim}$ is Hilbert-Schmidt, must be of measure one whenever F is a pre-support (Corollary 1, [2], p.93). Call $F^{\sim}$ an H-S enlargement of F if the identical injection of F into $F^{\sim}$ is a Hilbert-Schmidt contraction. Given two enlargements F_1 and F_2 of F, we say that F_2 is finer than F_1 if the identity on F extends to a continuous injection of F_2 into F_1.

We shall be through once we have proved the following

Proposition 4. Given H-S enlargements F_i, $i=1,2$, of F, there exists an H-S enlargement of F which is finer than the two given enlargements.

To prove this proposition we shall need two lemmas.

Lemma 1. Given an H-S enlargement $(F_1, |\cdot|_1)$ of F, and a functional $a' \in F'$, there exists an H-S enlargement $(F_2, |\cdot|_2)$ of F finer than $(F_1, |\cdot|_1)$ and such that a' is continuous in $(F, |\cdot|_2)$.

Proof. Let $\{e_n'\}$ be a Gram-Schmidt orthonormalization in $(F', |\cdot|')$ of an orthogonal basis in $(F_1', |\cdot|_1')$ restricted to F. Hence $\{e_n'\}$ constitutes an orthonormal basis in $(F', |\cdot|')$, and there exists $\{t_n\}$, $\sum_n t_n^2 < \infty$, such that $a' = \sum_n t_n e_n'$ in $(F', |\cdot|')$. Fix $\{N_n\}$ such that $\sum N_n^{-2} < \infty$, $N_n > 0$, and choose increasing $\{m_n\}$ in such a way that

$$\sum_n N_n^2 \sum_{k=m_n}^{m_{n+1}-1} t_k^2 < \infty.$$

Put for $x \in F$

$$|x|_* = (\sum_n N_n^2 | \sum_{k=m_n}^{m_{n+1}-1} t_k e_k' x|^2)^{\frac{1}{2}}$$

and

$$|x|_2 = (|x|_*^2 + |x|_1^2)^{\frac{1}{2}}.$$

It is easy to see that

$$|a'x| \leqq (\sum_n N_n^{-2})^{\frac{1}{2}} |x|_* .$$

and

$$\sum_n |e_n|_*^2 < \infty.$$

If $|x_n|_1 \to 0$ and $|x_n - x_m|_2 \to 0$, then $|x_n|_2 \to 0$. Indeed, if $|x_n|_1 \to 0$, then $e_k' x_n \to 0$ for every k.

Writing $(F_2, |\cdot|_2)$ for the completion of $(F, |\cdot|_2)$, we obtain the desired result.

__Lemma 2__. If $\{(F_n, |\cdot|_n)\}$ is a sequence of H-S enlargements of F, each of them finer than an H-S enlargement $(F_0, |\cdot|_0)$, then there exists an H-S enlargement $(F^\sim, |\cdot|^\sim)$ of F finer than all the enlargements $(F_n, |\cdot|_n)$.

__Proof__. We just put $|x|^\sim = (\sum_n a_n |x|_n^2)^{\frac{1}{2}}$ for sufficiently fast decreasing $\{a_n\}$, $a_n > 0$.

__Proof of Proposition 4__. Let $\{e_{i,n}'\}$ be an orthonormal basis in $(F_i', |\cdot|_i')$ for $i = 1, -1$ respectively. By Lemmas 1 and 4, there exist enlargements $(F_i^\sim, |\cdot|_i^\sim)$ finer than $(F_i, |\cdot|_i)$ and such that all $e_{-i,n}'$ are continuous with respect to $|\cdot|_i^\sim$ for $i = 1, -1$ respectively. Setting

$$|x|_0 = (|x|_1^{\sim 2} + |x|_{-1}^{\sim 2})^{\frac{1}{2}},$$

we find that the completion of $(F, |\cdot|_0)$ fulfils the requirements of the proposition.

Now we can identify points of measure one Hilbert spaces containing the same Hilbertian pre-support F by taking for those Hilbert spaces only H-S enlargements of F and identifying x_1 of an enlargement F_1 with x_2 of an enlargement F_2 if there exists an enlargement F_3 finer than both F_1 and F_2 such that the extensions over F_3 of the identical injection of F into F_1 and F into F_2 have both x_1 and x_2 as images of the same point of F_3. Transitivity of this relation is a trivial consequence of the following observation. Consider enlargements F_j, $j = 1,2,3,4,5$ such that F_4 is finer than F_1, F_2 and F_5 is finer than F_2, F_3. Consider the sum $F_6 = F_4 + F_5$ contained in F_2. Providing F_6 with the quotient topology of $F_4 \times F_5/K$, where $K = \{(x_4, x_5): x_4 + x_5 = 0\}$, we find that F_6 is an H-S enlargement of F which will identify the points which were identified by either F_4 or F_5.

The introduced identification makes the simultaneous work with different H-S enlargement consistent.

We conclude by providing a method for constructing unique almost everywhere linear extensions of bounded linear transformations between pre-supports.

Lemma 3. Consider a Hilbert space $(H,|\cdot|)$, a trace-class selfadjoint monomorphism $A \geqq 0$, and continuous linear transformations $T_1,\ldots,T_{p-1}$. Then there exists a trace-class selfadjoint $B \geqq (2p)^{-1}A$ with $\|B\|_{tr} \leqq \|A\|_{tr}$ such that for $i = 1,\ldots,p-1$

$$T_i^* B T_i \leqq 2p(\max_j \|T_j\|)^2 B.$$

Proof. Preserving the generality, take $\|T_i\| \leqq 1$, $1 \leqq i \leqq p-1$. Write $\underline{n} = (n\ ,\ldots,n_k)$ for a k-tuple out of numbers $0,1,\ldots,p-1$ and denote $\#\underline{n} = k$. Then $\mathrm{Card}\{\underline{n}: \# n = k\} = p^k$. Define

$$B = \sum_{k=1}^{\infty} (2p)^{-1} \sum_{\#\underline{n}=k} T^*_{\underline{n}} A T_{\underline{n}},$$

where $T_{\underline{n}} = T_{n_2} \ldots T_{n_k}$ for $\underline{n} = (n_1,\ldots,n_k)$ and T_0 = the identity. Since

$$\|p^{-k} \sum_{\#\underline{n}=k} T^*_{\underline{n}} A\, T_{\underline{n}}\|_{tr} \leqq \|A\|_{tr},$$

we have $\|B\|_{tr} \leqq \|A\|_{tr}$, Defining $\underline{n}i = (i,n_1,\ldots,n_k)$ for $\underline{n} = (n_1,\ldots,n_k)$, we obtain

$$T_i^* B T_i = \sum_{k=1}^{\infty} (2p)^{-k} \sum_{\# n=k} T^*_{\underline{n}i} A\, T_{\underline{n}i}$$

$$= 2p \sum_{k=1}^{\infty} (2p)^{-(k+1)} \sum_{\#\underline{m}=k+1} T^*_{\underline{m}} A\, T_{\underline{m}} \leqq 2pB,$$

and this concludes the proof.

Lemma 4. Consider two Hilbert spaces $(H_i,|\cdot|_i)$, $i = 1,2$, a finite set $T_1,\ldots,T_{p-1}$ of linear continuous mappings of $(H_1,|\cdot|_1)$ into $(H_2,|\cdot|_2)$ and one-to-one positive trace-class selfadjoint operators $A_i: H_i \to H_i$, $i = 1,2$. Then there exist selfadjoint trace-class $B_i: H_i \to H_i$, $B_i \geqq A_i$, $i = 1,2$, such that

$$T_k B_1 T_k^* \leqq aB_2 \qquad \text{and} \qquad T_k B_2 T_k^* \leqq aB_1$$

for $k = 1,2,\ldots,p-1$, where a is a positive constant.

Proof. Consider the product $(H_1 \times H_2, (|\cdot|_1^2 + |\cdot|_2^2)^{\frac{1}{2}})$. The transformations $\underline{T}_k(h_1,h_2) =_{df} (T_k^* h_2, T_k h_1)$ are selfadjoint, and the transformation $\underline{A}(h_1,h_2) = (A_1h_1,A_2h_2)$ is one-to-one positive trace-class selfadjoint, so we can find $\underline{B}$ as in Lemma 4. Since $\underline{T_n}A\underline{T_n}$ leaves $0 \times H_2$ and $H_1 \times 0$ invariant, the same does $\underline{B}$, and we can find $B_j: H_i \to H_i$, $i = 1,2$, such that $\underline{B}(h_1,h_2) = (B_1h_1,B_2h_2)$. It is easy to see that the B_i fulfill the requirements of the lemma.

Theorem 6. Consider real Hilbert spaces F_1 and F_2 and a sequence $\{T_n\}$ of bounded linear operators from F_1 to F_2. Then to every H-S enlargement $\tilde{F_2}$ of F_2 there corresponds an H-S enlargement $\tilde{F_1}$ of F_1 such that each T_n extends to a bounded linear operator from $\tilde{F_1}$ to $\tilde{F_2}$. Moreover, if $\approx{F_1}$ constitutes another H-S enlargement of F_1 with the property that every T_n extends to a bounded linear operator from $F_1^{\approx}$ to $\tilde{F_2}$, then there exists an H-S enlargement of F_1 which is finer than the enlargements $F_1^{\approx}$ and $\tilde{F_1}$, on which all T_n coincide.

Proof. Let A_2 be a positive symmetric trace-class operator such that the norm in $\tilde{F_2}$ writes as $|A_2^{\frac{1}{2}}\cdot|_2$ on F_2, where $|\cdot|_2$ is the norm in F_2, and let $A_1: F_1 \to F_1$ be an arbitrary positive symmetric trace-class operator. Applying Lemma 4 to the first operator from the sequence, we obtain certain H-S enlargements of F_1 and F_2 respectively such that the latter is finer than $\tilde{F_2}$ and such that T_1 extends. Suppose we have produced a sequence of n subsequent H-S enlargements of F_1, each next finer than the preceding one, such that for each $i \leq n$, the operators $T_1,\ldots,T_i$ extend from the i-th enlargement of F_1 to the enlargement $\tilde{F_2}$. Let positive symmetric trace-class $A_1^{(i)}: F_1 \to F_2$ be such that $|(A^{(i)})^{\frac{1}{2}}\cdot|_1$ gives the norm of the i-th enlargement of F_1. Applying Lemma 4 to operators $T_1,\ldots,T_n,T_{n+1}$ relative to $A_1^{(n)}$ and A_2, we produce the (n+1)-th enlargement of F_1 finer than all previous enlargements such that $T_1,\ldots,T_{n+1}$ can be extended. Now, from Lemma 2 we obtain the existence of the enlargement good for all T_n from the sequence. Finally, Proposition 4 ascertains that given any two enlargements of F_1, admitting extensions of all T_n, there must still exist a finer enlargement on which those extensions coincide. This concludes the proof.

R E F E R E N C E S

1. E. Alfsen, Compact convex sets and boundary integrals, Springer 1971.

2. N. Bourbaki, Éléments de Mathématique, Chapter IX, Integration, Hermann, 1969

3. E. Nelson, The free Markov field, J.Funct.Anal. 12 (1973), 221-227.

4. W. Słowikowski, Pre-support of linear probability measures and linear Lusin measurable functionals, Dissertationes Mathematicae (Rozprawy Matematyczne) 93, Warszawa 1972.

5. W. Słowikowski, The second quantization, the stochastic integration and measures in linear spaces, Matematisk Institut, Aarhus Universitet, Preprint Series 1976/77 No. 6.

ON A SUITABLE NOTION OF CONVERGENCE FOR THE SPACE OF MATRIX SUMMATIONS

F. Terpe
Ernst-Moritz-Arnst-Universität
Sektion Mathematik
22 Greifswald, DDR

A general summation on X over T means a net $(\mu_t)_{t\in T}$ of bounded Radon measures on a locally compact but non compact Hausdorff space X. T also is a locally compact but non compact Hausdorff space filtered by the cocompact subsets. Fore more details of a theory of summations of such general kind look [2],[5]. There the behavior of summations has been described by weak convergence of the measures μ_t on some compactifications of X. This theory gives a framework, which contains as a special case the classical summations by Toeplitz matrices and has applications in stochastic processes [2]. $\mathcal{M}(X)$ denotes the space of bounded Radon measures on X.

A summation $S = (\mu_t)_{t\in T}$ on X over T is called convergence preserving iff $S\text{-lim } f := \lim_{t\to\infty} \mu_t(f)$ exists in R for every $f \in C_a(X)$, where $C_a(X)$ is the space of all continuous real functions on X having a limit at infinity. S is called permanent iff S is convergence preserving and $S\text{-lim } f = \lim_{x\to\infty} f(x)$ for all $f \in C_a(X)$. S is called convergence generating iff S-lim f exists in R for all $f \in C_b(X)$. Here $C_b(X)$ is the space of all bounded real continuous functions on X. S is called core-contracting iff

$$\liminf_{x\to\infty} f(x) \leq \liminf_{t\to\infty} \mu_t(f) \leq \limsup_{t\to\infty} \mu_t(f) \leq \limsup_{x\to\infty} f(x) \quad \text{for all } f\in C_b(X).$$

The set of all summations on X over T we denote by $\mathcal{T}(X,T)$. $\mathcal{T}(X,T)$ becomes a vector space by "coordinate-wise" addition and scalar multiplication.

A suitable convergence structure for nets of summations from $\mathcal{T}(X,T)$ must have the property that the set of convergence preserving resp. permanent resp. convergence generating resp. core-contracting summation becomes closed. In [3] there were introduced four suitable convergence structures.

The aim of this note is to analyse the third convergence structure of [3] in the special case of Toeplitz matrix summations using common sequences only.

In [3] a net $(S_\alpha)_{\alpha\in A}$ of summations $S_\alpha = (\mu_t^\alpha)_{t\in T}$ on X over T was called strongly terminal convergent to S iff the following conditions a and b are fullfilled:

a) $(S_\alpha)_{\alpha\in A}$ simple converges to S in the weak topology $\sigma(\mathcal{M}(X), C_b(X))$, i.e. $\mu_t^\alpha(f) \to \mu_t(f)$ for all $f \in C_b(X)$ and all $t \in T$.

b) $(S_\alpha)_{\alpha\in A}$ converges uniformly at infinity in the norm topology to S, i.e. for each $\varepsilon > 0$ there exists a compact set $K_\varepsilon \subseteqq T$ and a number $\alpha(\varepsilon)$, such that $\|\mu_t^\alpha - \mu_t\| < \varepsilon$ for all $t \notin K_\varepsilon$ and all $\alpha \geqq \alpha(\varepsilon)$.

We denote this convergence by $S_\alpha \underset{3}{\to} S$.

Now we take $X = T = N = \{1,2,\ldots\}$. The topology in N is the discrete topology.

Let be $(a_{ik})_{i,k\in N}$ a real Toeplitz matrix (i.e. $\sum_{k=1}^{\infty} a_{ik}$ is absolutely convergent for each i). $(a_{ik})_{i,k\in N}$ gives a summation $A = (\mu_i)_{i\in N}$ on N over N, where $\mu_i := \sum_{k=1}^{\infty} a_{ik}\cdot\delta_k$ (δ_k is the Dirac measure in the point $k\in N$).

Theorem 1

Let be $(A^j)_{j\in N}$, $A^j = (\mu_i^j)_{i\in N}$, $\mu_i^j = \sum_{k=1}^{\infty} a_{ik}^j\cdot\delta_k$, a sequence of real Toeplitz matrix summations, and $A = (\mu_i)_{i\in N}$, $\mu_i = \sum_{k=1}^{\infty} a_{ik}\cdot\delta_k$ a real Toeplitz matrix summation.

Then the following conditions are equivalent:

1) $A^j \underset{3}{\to} A$

2) For each $\varepsilon > 0$ there is a number $n_\varepsilon \in N$, such that for all $i \in N$ and all $j \geqq n_\varepsilon$

$$(*) \quad \sum_{k=1}^{\infty} |a_{ik}^j - a_{ik}| < \varepsilon .$$

Proof

At first we show that there are equivalent the conditions

a) $A^j \to A$ with respect to the topology $\sigma(\mathcal{M}(N), C_b(N))$

and

a') $\lim_{j\to\infty}(\sum_{k=1}^{\infty} |a_{ik}^j - a_{ik}|) = 0$ for each $i \in N$.

We regard an arbitrary i and the matrix

$$\begin{pmatrix} a_{i1}^1 & a_{i2}^1 \ldots \\ a_{i1}^2 & a_{i2}^2 \ldots \\ \vdots & \vdots \ldots \end{pmatrix}.$$

Clearly this matrix is a Toeplitz matrix. The matrix delivers the summation $S_i = (\mu_i^\rho)_{\rho\in N}$. Now we see that condition a is equivalent with the following condition

a") Every S_i is convergence generating.

Hence we get by the Theorem of Schur [7] and the Theorem of Kojima-Schur [7] the equivalence of condition a with the set of the following conditions:

A) $\sup_j (\sum_{k=1}^{\infty} |a_{ik}^j|) < +\infty$ for each $i \in N$.

B) $\lim_{j\to\infty} a_{ik}^j = a_{ik}$ for each $k \in N$ and each $i \in N$.

C) $\lim_{j\to\infty} (\sum_{k=1}^{\infty} a_{ik}^j) = : a_i$ exists for each $i \in N$.

a') from above.

But conditions A,B,C are consequences of condition a'. For we have

$|a_{ik}^j| \leqq |a_{ik}^j - a_{ik}| + |a_{ik}|$, $j\in N$. Hence

$\sum_{k=1}^{\infty} |a_{ik}^j| \leqq \sum_{k=1}^{\infty} |a_{ik}^j - a_{ik}| + \sum_{k=1}^{\infty} |a_{ik}|$. Therefore from a' we conclude

$\sup_j \sum_{k=1}^{\infty} |a_{ik}^j - a_{ik}| < +\infty$ and $\sup_j \sum_{k=1}^{\infty} |a_{ik}^j| \leqq \sup_j \sum_{k=1}^{\infty} |a_{ik}^j - a_{ik}| +$

$+ \sum_{k=1}^{\infty} |a_{ik}| < +\infty$. Therefore condition a' implies condition A. It is very easy to show that condition a' implies condition B.

By supposition $\sum_{k=1}^{\infty} a_{ik}^j$ converges for each $i \in N$ and each $j \in N$, and

$\sum_{k=1}^{\infty} a_{ik}$ is converging too. Therefore

$\sum_{k=1}^{\infty} a_{ik}^j - \sum_{k=1}^{\infty} a_{ik} \leqq \sum_{k=1}^{\infty} |a_{ik}^j - a_{ik}|$. Hence

$|\sum_{k=1}^{\infty} a_{ik}^j - \sum_{k=1}^{\infty} a_{ik}| \leqq \sum_{k=1}^{\infty} |a_{ik}^j - a_{ik}|$. Now we conclude from condition a' condition C.

Clearly condition a' is equivalent with condition

a"') For each $i \in N$ and each $\varepsilon > 0$ there exists a number $n(i,\varepsilon)$, such that $\sum_{k=1}^{\infty} |a_{ik}^j - a_{ik}| < \varepsilon$ for all $j \geqq n(i,\varepsilon)$.

Property b of the convergence $A^j \underset{3}{\rightarrow} A$ here is equivalent with

b") For each $\varepsilon > 0$ there exists a number $n_\varepsilon \in N$, such that for each $i \geqq n_\varepsilon$ and each $j \geqq n_\varepsilon$ $\sum_{k=1}^{\infty} |a_{ik}^j - a_{ik}| < \varepsilon$.

But the set of conditions a"' and b" is equivalent with condition 2. Clearly condition 2 implies a"' and b". Now let be $\varepsilon > 0$. We put $m_\varepsilon := \max(n(1,\varepsilon),\ldots,n(n_\varepsilon - 1,\varepsilon),n_\varepsilon)$. We get $\sum_{k=1}^{\infty} |a_{ik}^j - a_{ik}| < \varepsilon$ for all $i \leqq n_\varepsilon - 1$ and all $j \geqq m_\varepsilon$ by virtue of condition a"'. Condition b" implies $\sum_{k=1}^{\infty} |a_{ik}^j - a_{ik}| < \varepsilon$ for all $i \geqq n_\varepsilon$ and all $j \geqq m_\varepsilon$. Therefore we have $\sum_{k=1}^{\infty} |a_{ik}^j - a_{ik}| < \varepsilon$ for all $j \geqq m_\varepsilon$ and every $i \in N$.

Theorem 2

Let be $(A^j)_{j\in N}$, $A^j = (a_{ik}^j)_{i,k\in N}$, a sequence of Toeplitz matrix summations and $A = (a_{ik})_{i,k\in N}$ a Toeplitz matrix summation. If every A^j is convergence preserving resp. permanent resp. convergence generating resp. core-contracting and if condition 2 of Theorem 1 holds, then A also is convergence preserving resp. permanent resp. convergence generating resp. core-contracting.

Proof

It follows from Theorem 1 and from Theorem 1.3.[3].

References

[1] J. Flachsmeyer, Über lokalgleichmäßige Konvergenz in Funktionenräumen. Math. Nachr. 29 (1965), 201-204.

[2] J. Flachsmeyer, F. Terpe, On summation on locally compact spaces. Math. Nachr. 75 (1976), 255-270.

[3] J. Flachsmeyer, F. Terpe, On convergence in the space of summations. Proc. IV. Prague Topol. Symp. (1976). Part B. 119-124.

[4] J. Flachsmeyer, F. Terpe, Summations as linear maps and as curves. Proc. Conf. Topology and Measure I (1974, Zinnowitz) 1978. Part 1. 129-138.

[5] F. Terpe, J. Flachsmeyer, On an aspect of compactification theory and measure theory in questions of summability. (Russian). Dokl. Akad. Nauk SSSR 227 (1976) No.2. English translation: Soviet Math. Dokl. 17 (1967) No.2.

[6] F. Terpe, On convergence in the space of matrix summations. Proc. Conf. Topology and Measure II (1977, Warnemünde), in print.

[7] K. Zeller, Theorie der Limitierungsverfahren. Ergebnisse der Math. N.F. 15 (1958).

PROBLEM SECTION

D. MAHARAM STONE

Question 1

Let $\mathcal{A}$ be a Boolean algebra that is complete, satisfies the countable chain condition, and is countably generated; and let $\mathcal{B}$ be a complete subalgebra of $\mathcal{A}$ (that is, the $\mathcal{A}$-inf and $\mathcal{A}$-sup of every subset of $\mathcal{B}$ are in $\mathcal{B}$). Need $\mathcal{B}$ be countably generated?

[Remark D. Fremlin has since shown, by an (unpublished) counterexample, that the answer is "No", even under the additional assumption that $\mathcal{A}$ has a finitely additive, strictly positive, finite measure.]

Question 2

Let X be a compact Hausdorff space with a complete regular Borel probability measure μ such that each non-empty open set G has $\mu(G) > 0$. Say that a measure class h, in the measure algebra E of (X,μ), is "open" if h has an open set belonging to it, "closed" if it has a closed member, "ambiguous" if it is both open and closed. The following elementary facts have been noticed (independently) by S. Graf and myself (unpublished):

(1) each open $h \in E$ has a largest open member G_h, and dually each closed h has a smallest closed member F_h.

(2) If h is ambigous, $\overline{G_h} = F_h$ and $G_h = \text{Int}(F_h)$.

(3) The set A of all ambigous classes is a (finitely additive) subalgebra of E.

(4) Each open class is a sup of ambiguous classes (and dually).

(5) There is a continuous measure-preserving map $\emptyset$ of the representation space R of A onto X, defined by

$$\{\emptyset(\alpha)\} = \cap \{F_h : h \in \alpha,\ h \text{ closed}\}$$

for each $\alpha \in R$.

(6) The completion ν of $\mu\bullet\emptyset$ on R has $\emptyset^{-1}(A)$ as its "ambiguous algebra" (of ambiguous measure classes).

Question: Does (R,ν) always have a strong lifting? If not, does the existence of a strong lifting on either of the spaces (X,μ), (R,ν) imply the existence of a strong lifting on the other?

W. F. PFEFFER

Question. Suppose that the contiunuum $\mathfrak{c}$ is nonmeasurable and $\mathfrak{c} \geq \aleph_2$. Is then each first countable, compact, Hausdorff space Radon?

Remark 1. If $\mathfrak{c}$ is measurable, the answer is NO by the Alexandroff's duplicate of the unit interval. If $\mathfrak{c} = \aleph_1$ the negative answer was obtained by Juhasz, Kunen, and Rudin in Canad. J. Math. 5 (1976), pp. 998-1005.

Remark 2. In particular, with or without the previous set theoretical assumptions, is the Helly space Radon?

For the Definition of the Helly space see J. L. Kelley: General Topology, Chapter 5, Exercise M, p. 164.

R. WHEELER

Problem. Let X be a completely regular Hausdorff space, $\mu \in M_\sigma^+(X)$ [i.e., μ is a finite, non-negative, countably-additive Baire measure on X]. Let H be a subset of C(X), the continuous real-valued functions on X, which is uniformly bounded and compact for the topology of pointwise convergence on X (=P-compact).

Let $f \sim g$ if $\int_X |f - g| d\mu = 0$.

Under what conditions on H and μ is it possible to choose a subset $G \subset H$ which contains exactly one member of each μ-equivalence class and is still P-compact?

In particular, if H is also convex, can this be done for any μ?

D. KÖLZOW

Given a measure space $(X,\mathfrak{A},\mu)$, a measurable space $(Y,\mathfrak{B})$, and a map θ from Y into the non-empty subsets of X.

Find conditions under which exist a probability section $y \to D(y,\cdot)$ for θ and a measure ν on $\mathfrak{B}$ such that $\mu(A) = \int D(y,A)d\nu(y)$ holds for all $A \in \mathfrak{A}$.

G. MÄGERL

Let $(X,\mathfrak{A})$ be a measurable space, Y be a compact Hausdorff space and I be the unit interval.

Suppose Φ is a measurable set-valued map from X into the nonempty closed subsets of $Y \times I$ and $f_1 : X \to Y$ is Baire measurable such that $f_1(x) \in \pi\Phi(x)$ for all $x \in X$ (π denotes the projection from $Y \times I$ onto Y).

Give conditions (other than metrizability of Y) which guarantee that Φ has a Baire measurable selection f such that $\pi \circ f = f_1$.

D. SENTILLES

Let $(\Omega,\sum,\mu)$ a measure space, X a B-space, S = Stone space of $\sum/\mu^{-1}(0)$ and $f : \Omega \to X$ bounded and weakly measurable.
Define $\hat{f} : S \to X''$ by $< \hat{f}(s), x' > = (\widehat{x'f})(s)$ when $\widehat{x'f}$ is the Stone representation of the real measurable function $x'f$. Characterize Pettis integrability of f in terms of $\hat{f}$.

E. PAP

The Diagonal Theorem was first formulated by J. Mikusiński in 1970. We can distinguish Diagonal Theorems of two kinds (see for example, P. Antosik, On the Mikusiński Diagonal Theorem, Bull. Acad. Polon. Sci. Math. Astronom. Phys. 19 (1971), pp.305-310, and P. Antosik, A diagonal theorem for nonnegative matrices and equicontinuous sequences of mappings, ibid. 24 (1976), pp. 855-860). Diagonal Theorems have many important applications also in Measure Theory (see for example, E. Pap's paper in these Proceedings). All the proofs of the Diagonal Theorems are elementary and simple. Many of the proved theorems were first proved with Baire Category theorem.

What are the connections between Baire Category theorem and Diagonal Theorems?

Vol. 640: J. L. Dupont, Curvature and Characteristic Classes. X, 175 pages. 1978.

Vol. 641: Séminaire d'Algèbre Paul Dubreil, Proceedings Paris 1976–1977. Edité par M. P. Malliavin. IV, 367 pages. 1978.

Vol. 642: Theory and Applications of Graphs, Proceedings, Michigan 1976. Edited by Y. Alavi and D. R. Lick. XIV, 635 pages. 1978.

Vol. 643: M. Davis, Multiaxial Actions on Manifolds. VI, 141 pages. 1978.

Vol. 644: Vector Space Measures and Applications I, Proceedings 1977. Edited by R. M. Aron and S. Dineen. VIII, 451 pages. 1978.

Vol. 645: Vector Space Measures and Applications II, Proceedings 1977. Edited by R. M. Aron and S. Dineen. VIII, 218 pages. 1978.

Vol. 646: O. Tammi, Extremum Problems for Bounded Univalent Functions. VIII, 313 pages. 1978.

Vol. 647: L. J. Ratliff, Jr., Chain Conjectures in Ring Theory. VIII, 133 pages. 1978.

Vol. 648: Nonlinear Partial Differential Equations and Applications, Proceedings, Indiana 1976–1977. Edited by J. M. Chadam. VI, 206 pages. 1978.

Vol. 649: Séminaire de Probabilités XII, Proceedings, Strasbourg, 1976–1977. Edité par C. Dellacherie, P. A. Meyer et M. Weil. VIII, 805 pages. 1978.

Vol. 650: C*-Algebras and Applications to Physics. Proceedings 1977. Edited by H. Araki and R. V. Kadison. V, 192 pages. 1978.

Vol. 651: P. W. Michor, Functors and Categories of Banach Spaces. VI, 99 pages. 1978.

Vol. 652: Differential Topology, Foliations and Gelfand-Fuks-Cohomology, Proceedings 1976. Edited by P. A. Schweitzer. XIV, 252 pages. 1978.

Vol. 653: Locally Interacting Systems and Their Application in Biology. Proceedings, 1976. Edited by R. L. Dobrushin, V. I. Kryukov and A. L. Toom. XI, 202 pages. 1978.

Vol. 654: J. P. Buhler, Icosahedral Golois Representations. III, 143 pages. 1978.

Vol. 655: R. Baeza, Quadratic Forms Over Semilocal Rings. VI, 199 pages. 1978.

Vol. 656: Probability Theory on Vector Spaces. Proceedings, 1977. Edited by A. Weron. VIII, 274 pages. 1978.

Vol. 657: Geometric Applications of Homotopy Theory I, Proceedings 1977. Edited by M. G. Barratt and M. E. Mahowald. VIII, 459 pages. 1978.

Vol. 658: Geometric Applications of Homotopy Theory II, Proceedings 1977. Edited by M. G. Barratt and M. E. Mahowald. VIII, 487 pages. 1978.

Vol. 659: Bruckner, Differentiation of Real Functions. X, 247 pages. 1978.

Vol. 660: Equations aux Dérivée Partielles. Proceedings, 1977. Edité par Pham The Lai. VI, 216 pages. 1978.

Vol. 661: P. T. Johnstone, R. Paré, R. D. Rosebrugh, D. Schumacher, R. J. Wood, and G. C. Wraith, Indexed Categories and Their Applications. VII, 260 pages. 1978.

Vol. 662: Akin, The Metric Theory of Banach Manifolds. XIX, 306 pages. 1978.

Vol. 663: J. F. Berglund, H. D. Junghenn, P. Milnes, Compact Right Topological Semigroups and Generalizations of Almost Periodicity. X, 243 pages. 1978.

Vol. 664: Algebraic and Geometric Topology, Proceedings, 1977. Edited by K. C. Millett. XI, 240 pages. 1978.

Vol. 665: Journées d'Analyse Non Linéaire. Proceedings, 1977. Edité par P. Bénilan et J. Robert. VIII, 256 pages. 1978.

Vol. 666: B. Beauzamy, Espaces d'Interpolation Réels: Topologie et Géometrie. X, 104 pages. 1978.

Vol. 667: J. Gilewicz, Approximants de Padé. XIV, 511 pages. 1978.

Vol. 668: The Structure of Attractors in Dynamical Systems. Proceedings, 1977. Edited by J. C. Martin, N. G. Markley and W. Perrizo. VI, 264 pages. 1978.

Vol. 669: Higher Set Theory. Proceedings, 1977. Edited by G. H. Müller and D. S. Scott. XII, 476 pages. 1978.

Vol. 670: Fonctions de Plusieurs Variables Complexes III, Proceedings, 1977. Edité par F. Norguet. XII, 394 pages. 1978.

Vol. 671: R. T. Smythe and J. C. Wierman, First-Passage Perculation on the Square Lattice. VIII, 196 pages. 1978.

Vol. 672: R. L. Taylor, Stochastic Convergence of Weighted Sums of Random Elements in Linear Spaces. VII, 216 pages. 1978.

Vol. 673: Algebraic Topology, Proceedings 1977. Edited by P. Hoffman, R. Piccinini and D. Sjerve. VI, 278 pages. 1978.

Vol. 674: Z. Fiedorowicz and S. Priddy, Homology of Classical Groups Over Finite Fields and Their Associated Infinite Loop Spaces. VI, 434 pages. 1978.

Vol. 675: J. Galambos and S. Kotz, Characterizations of Probability Distributions. VIII, 169 pages. 1978.

Vol. 676: Differential Geometrical Methods in Mathematical Physics II, Proceedings, 1977. Edited by K. Bleuler, H. R. Petry and A. Reetz. VI, 626 pages. 1978.

Vol. 677: Séminaire Bourbaki, vol. 1976/77, Exposés 489–506. IV, 264 pages. 1978.

Vol. 678: D. Dacunha-Castelle, H. Heyer et B. Roynette. Ecole d'Eté de Probabilités de Saint-Flour. VII-1977. Edité par P. L. Hennequin. IX, 379 pages. 1978.

Vol. 679: Numerical Treatment of Differential Equations in Applications, Proceedings, 1977. Edited by R. Ansorge and W. Törnig. IX, 163 pages. 1978.

Vol. 680: Mathematical Control Theory, Proceedings, 1977. Edited by W. A. Coppel. IX, 257 pages. 1978.

Vol. 681: Séminaire de Théorie du Potentiel Paris, No. 3, Directeurs: M. Brelot, G. Choquet et J. Deny. Rédacteurs: F. Hirsch et G. Mokobodzki. VII, 294 pages. 1978.

Vol. 682: G. D. James, The Representation Theory of the Symmetric Groups. V, 156 pages. 1978.

Vol. 683: Variétés Analytiques Compactes, Proceedings, 1977. Edité par Y. Hervier et A. Hirschowitz. V, 248 pages. 1978.

Vol. 684: E. E. Rosinger, Distributions and Nonlinear Partial Differential Equations. XI, 146 pages. 1978.

Vol. 685: Knot Theory, Proceedings, 1977. Edited by J. C. Hausmann. VII, 311 pages. 1978.

Vol. 686: Combinatorial Mathematics, Proceedings, 1977. Edited by D. A. Holton and J. Seberry. IX, 353 pages. 1978.

Vol. 687: Algebraic Geometry, Proceedings, 1977. Edited by L. D. Olson. V, 244 pages. 1978.

Vol. 688: J. Dydak and J. Segal, Shape Theory. VI, 150 pages. 1978.

Vol. 689: Cabal Seminar 76–77, Proceedings, 1976–77. Edited by A.S. Kechris and Y. N. Moschovakis. V, 282 pages. 1978.

Vol. 690: W. J. J. Rey, Robust Statistical Methods. VI, 128 pages. 1978.

Vol. 691: G. Viennot, Algèbres de Lie Libres et Monoïdes Libres. III, 124 pages. 1978.

Vol. 692: T. Husain and S. M. Khaleelulla, Barrelledness in Topological and Ordered Vector Spaces. IX, 258 pages. 1978.

Vol. 693: Hilbert Space Operators, Proceedings, 1977. Edited by J. M. Bachar Jr. and D. W. Hadwin. VIII, 184 pages. 1978.

Vol. 694: Séminaire Pierre Lelong – Henri Skoda (Analyse) Année 1976/77. VII, 334 pages. 1978.

Vol. 695: Measure Theory Applications to Stochastic Analysis, Proceedings, 1977. Edited by G. Kallianpur and D. Kölzow. XII, 261 pages. 1978.

Vol. 696: P. J. Feinsilver, Special Functions, Probability Semigroups, and Hamiltonian Flows. VI, 112 pages. 1978.

Vol. 697: Topics in Algebra, Proceedings, 1978. Edited by M. F. Newman. XI, 229 pages. 1978.

Vol. 698: E. Grosswald, Bessel Polynomials. XIV, 182 pages. 1978.

Vol. 699: R. E. Greene and H.-H. Wu, Function Theory on Manifolds Which Possess a Pole. III, 215 pages. 1979.

Vol. 700: Module Theory, Proceedings, 1977. Edited by C. Faith and S. Wiegand. X, 239 pages. 1979.

Vol. 701: Functional Analysis Methods in Numerical Analysis, Proceedings, 1977. Edited by M. Zuhair Nashed. VII, 333 pages. 1979.

Vol. 702: Yuri N. Bibikov, Local Theory of Nonlinear Analytic Ordinary Differential Equations. IX, 147 pages. 1979.

Vol. 703: Equadiff IV, Proceedings, 1977. Edited by J. Fábera. XIX, 441 pages. 1979.

Vol. 704: Computing Methods in Applied Sciences and Engineering, 1977, I. Proceedings, 1977. Edited by R. Glowinski and J. L. Lions. VI, 391 pages. 1979.

Vol. 705: O. Forster und K. Knorr, Konstruktion verseller Familien kompakter komplexer Räume. VII, 141 Seiten. 1979.

Vol. 706: Probability Measures on Groups, Proceedings, 1978. Edited by H. Heyer. XIII, 348 pages. 1979.

Vol. 707: R. Zielke, Discontinuous Čebyšev Systems. VI, 111 pages. 1979.

Vol. 708: J. P. Jouanolou, Equations de Pfaff algébriques. V, 255 pages. 1979.

Vol. 709: Probability in Banach Spaces II. Proceedings, 1978. Edited by A. Beck. V, 205 pages. 1979.

Vol. 710: Séminaire Bourbaki vol. 1977/78, Exposés 507–524. IV, 328 pages. 1979.

Vol. 711: Asymptotic Analysis. Edited by F. Verhulst. V, 240 pages. 1979.

Vol. 712: Equations Différentielles et Systèmes de Pfaff dans le Champ Complexe. Edité par R. Gérard et J.-P. Ramis. V, 364 pages. 1979.

Vol. 713: Séminaire de Théorie du Potentiel, Paris No. 4. Edité par F. Hirsch et G. Mokobodzki. VII, 281 pages. 1979.

Vol. 714: J. Jacod, Calcul Stochastique et Problèmes de Martingales. X, 539 pages. 1979.

Vol. 715: Inder Bir S. Passi, Group Rings and Their Augmentation Ideals. VI, 137 pages. 1979.

Vol. 716: M. A. Scheunert, The Theory of Lie Superalgebras. X, 271 pages. 1979.

Vol. 717: Grosser, Bidualräume und Vervollständigungen von Banachmoduln. III, 209 pages. 1979.

Vol. 718: J. Ferrante and C. W. Rackoff, The Computational Complexity of Logical Theories. X, 243 pages. 1979.

Vol. 719: Categorial Topology, Proceedings, 1978. Edited by H. Herrlich and G. Preuß. XII, 420 pages. 1979.

Vol. 720: E. Dubinsky, The Structure of Nuclear Fréchet Spaces. V, 187 pages. 1979.

Vol. 721: Séminaire de Probabilités XIII. Proceedings, Strasbourg, 1977/78. Edité par C. Dellacherie, P. A. Meyer et M. Weil. VII, 647 pages. 1979.

Vol. 722: Topology of Low-Dimensional Manifolds. Proceedings, 1977. Edited by R. Fenn. VI, 154 pages. 1979.

Vol. 723: W. Brandal, Commutative Rings whose Finitely Generated Modules Decompose. II, 116 pages. 1979.

Vol. 724: D. Griffeath, Additive and Cancellative Interacting Particle Systems. V, 108 pages. 1979.

Vol. 725: Algèbres d'Opérateurs. Proceedings, 1978. Edité par P. de la Harpe. VII, 309 pages. 1979.

Vol. 726: Y.-C. Wong, Schwartz Spaces, Nuclear Spaces and Tensor Products. VI, 418 pages. 1979.

Vol. 727: Y. Saito, Spectral Representations for Schrödinger Operators With Long-Range Potentials. V, 149 pages. 1979.

Vol. 728: Non-Commutative Harmonic Analysis. Proceedings, 1978. Edited by J. Carmona and M. Vergne. V, 244 pages. 1979.

Vol. 729: Ergodic Theory. Proceedings, 1978. Edited by M. Denker and K. Jacobs. XII, 209 pages. 1979.

Vol. 730: Functional Differential Equations and Approximation of Fixed Points. Proceedings, 1978. Edited by H.-O. Peitgen and H.-O. Walther. XV, 503 pages. 1979.

Vol. 731: Y. Nakagami and M. Takesaki, Duality for Crossed Products of von Neumann Algebras. IX, 139 pages. 1979.

Vol. 732: Algebraic Geometry. Proceedings, 1978. Edited by K. Lønsted. IV, 658 pages. 1979.

Vol. 733: F. Bloom, Modern Differential Geometric Techniques in the Theory of Continuous Distributions of Dislocations. XII, 206 pages. 1979.

Vol. 734: Ring Theory, Waterloo, 1978. Proceedings, 1978. Edited by D. Handelman and J. Lawrence. XI, 352 pages. 1979.

Vol. 735: B. Aupetit, Propriétés Spectrales des Algèbres de Banach. XII, 192 pages. 1979.

Vol. 736: E. Behrends, M-Structure and the Banach-Stone Theorem. X, 217 pages. 1979.

Vol. 737: Volterra Equations. Proceedings 1978. Edited by S.-O. Londen and O. J. Staffans. VIII, 314 pages. 1979.

Vol. 738: P. E. Conner, Differentiable Periodic Maps. 2nd edition, IV, 181 pages. 1979.

Vol. 739: Analyse Harmonique sur les Groupes de Lie II. Proceedings, 1976–78. Edited by P. Eymard et al. VI, 646 pages. 1979.

Vol. 740: Séminaire d'Algèbre Paul Dubreil. Proceedings, 1977–78. Edited by M.-P. Malliavin. V, 456 pages. 1979.

Vol. 741: Algebraic Topology, Waterloo 1978. Proceedings. Edited by P. Hoffman and V. Snaith. XI, 655 pages. 1979.

Vol. 742: K. Clancey, Seminormal Operators. VII, 125 pages. 1979.

Vol. 743: Romanian-Finnish Seminar on Complex Analysis. Proceedings, 1976. Edited by C. Andreian Cazacu et al. XVI, 713 pages. 1979.

Vol. 744: I. Reiner and K. W. Roggenkamp, Integral Representations. VIII, 275 pages. 1979.

Vol. 745: D. K. Haley, Equational Compactness in Rings. III, 167 pages. 1979.

Vol. 746: P. Hoffman, τ-Rings and Wreath Product Representations. V, 148 pages. 1979.

Vol. 747: Complex Analysis, Joensuu 1978. Proceedings, 1978. Edited by I. Laine, O. Lehto and T. Sorvali. XV, 450 pages. 1979.

Vol. 748: Combinatorial Mathematics VI. Proceedings, 1978. Edited by A. F. Horadam and W. D. Wallis. IX, 206 pages. 1979.

Vol. 749: V. Girault and P.-A. Raviart, Finite Element Approximation of the Navier-Stokes Equations. VII, 200 pages. 1979.

Vol. 750: J. C. Jantzen, Moduln mit einem höchsten Gewicht. III, 195 Seiten. 1979.

Vol. 751: Number Theory, Carbondale 1979. Proceedings. Edited by M. B. Nathanson. V, 342 pages. 1979.

Vol. 752: M. Barr, *-Autonomous Categories. VI, 140 pages. 1979.

Vol. 753: Applications of Sheaves. Proceedings, 1977. Edited by M. Fourman, C. Mulvey and D. Scott. XIV, 779 pages. 1979.

Vol. 754: O. A. Laudal, Formal Moduli of Algebraic Structures. III, 161 pages. 1979.

Vol. 755: Global Analysis. Proceedings, 1978. Edited by M. Grmela and J. E. Marsden. VII, 377 pages. 1979.

Vol. 756: H. O. Cordes, Elliptic Pseudo-Differential Operators – An Abstract Theory. IX, 331 pages. 1979.

Vol. 757: Smoothing Techniques for Curve Estimation. Proceedings, 1979. Edited by Th. Gasser and M. Rosenblatt. V, 245 pages. 1979.

Vol. 758: C. Năstăsescu and F. Van Oystaeyen; Graded and Filtered Rings and Modules. X, 148 pages. 1979.